Perspectives from Europe and Asia on Engineering Design
and Manufacture

Perspectives from Europe and Asia on Engineering Design and Manufacture

A Comparison of Engineering Design and Manufacture in Europe and Asia

Edited by

XIU-TIAN YAN

University of Strathclyde,
Department of Design, Manufacture and Engineering Management,
Glasgow, Scotland, U.K.

CHENG-YU JIANG

Northwestern Polytechnical University,
China

and

NEAL P. JUSTER

University of Strathclyde,
Glasgow, Scotland, U.K.

KLUWER ACADEMIC PUBLISHERS

DORDRECHT / BOSTON / LONDON

A C.I.P. Catalogue record for this book is available from the Library of Congress.

ISBN 1-4020-2211-5 (HB)
ISBN 1-4020-2212-3 (e-book)

Published by Kluwer Academic Publishers,
P.O. Box 17, 3300 AA Dordrecht, The Netherlands.

Sold and distributed in North, Central and South America
by Kluwer Academic Publishers,
101 Philip Drive, Norwell, MA 02061, U.S.A.

In all other countries, sold and distributed
by Kluwer Academic Publishers,
P.O. Box 322, 3300 AH Dordrecht, The Netherlands.

Printed on acid-free paper

Printed in the Netherlands.

TABLE OF CONTENTS

PART 3

**MATERIALS DESIGN AND THEIR MANUFACTURE ,
MANUFACTURE AND MANUFACTURING AUTOMATION 347**

PART 4

PRODUCT DEVELOPMENT AND DESIGN PRACTICE 585

PREFACE

With collaborative product development in a geographically distributed environment and global outsourcing becoming normal for many companies, it is imperative to bring academics, researchers and industrialists together to share research ideas and best practice. The European-Asia Symposium on Engineering Design and Manufacture (EASED 2004) provides such a platform and aims to increase the exchange of ideas and best practice among practitioners and researchers from two major global regions – Europe and Asia. As the manufacturing activities, associated with the design activities in European, American and Japan, are being transferred to Asia, it is timely to organise this International Symposium. The Symposium brings together research experts and industrialists to focus on the issues related to these global changes. This geographical distribution of tasks involved in the whole engineering product realisation process brings great challenge as well as huge benefits. This Symposium provides a platform for academic researchers and industrial practitioners to exchange ideas used to address the challenges presented by this new global economic development.

This book presents 75 papers from 185 accepted refereed papers presented at EASED2004. This book is a reflection of the key papers presented in the four important areas of the Symposium:

1) Knowledge intensive engineering support in various stages of the product realisation process, including early conceptual design, virtual product modelling, virtual prototyping, rapid prototyping, concurrent modelling of design and manufacture and their processes, design information representation, product dimensional modelling, DFX, decision making support and intelligent diagnosis;

2) Global design education challenges and collaborative working, educating global designers, education practice and comparison, web-based education, new framework for higher engineering education, distributed and collaborative design and working, information sharing and support;

3) Material sciences focusing on functional ceramic material design and their manufacture and manufacturing operations, covering ceramic material design and manufacture, material modelling, advanced manufacturing techniques, distributed production control, optimisation of manufacturing processes, shop floor scheduling, system performance measurement, robotics and automation;

4) Product development, focusing on design practice, design case studies, design analysis, component and mechanism design, product life cycle design, recycling design approaches.

Based on the above main topic areas, the book is divided into four chapters, each containing the selected papers representing the main research advancements, industrial reality and new methodologies developed in each area presented at the Symposium:

Chapter 1 Knowledge intensive methodologies and tools supporting engineering design and manufacture;

Chapter 2 Design education and collaborative working engineering design and manufacture

Chapter 3 Materials functional design and manufacture, and robotics and manufacturing automation

Chapter 4 Product development and design practice and analysis.

This book is of interest to technical and economic policy developers in Europe and Asia, researchers, academics, engineers and postgraduate students who have special interests in engineering design, engineering higher education, material design, product development and analysis including component design and analysis.

The editors of the book:

Xiu-Tian Yan, Cheng-Yu Jiang and Neal P Juster

ACKNOWLEDGEMENTS

The editors of the book would like to thank the two EASED 2004 Symposium Secretariats, one based in the University of Strathclyde, Glasgow, UK for the European correspondence and Symposium organisation and the other based in Northwestern Polytechnical University, Xian, China for Asia correspondence and organisation. In particular the editors would like to thank Dr. Cato X. C. Tan, Professor Hong Tang, Professor Xiansheng Qin, Mr. Alasdair Downes, Mr. Fayyaz Rehman, Mr. Hongliang Chen. Ms Chunhong Zhao and Ms Zifang Wu for their hard work and support, without which, it would have been very difficult to compile this book. The editors of the book would also like to extend their gratitude to the members of Symposium Advisory Scientific Board for their support in steering the Symposium and reviewing the papers. The editors of the Book would also like to thank the sponsoring organisations for their support to the organisation of the Symposium.

The Organisers of the Symposium:

The University of Strathclyde
Northwestern Polytechnical University
The University of Technology Troyes
Huazhong University of Science and Technology

The Symposium Sponsors:

The Design Society – A Worldwide Community,
The Japan Society of Mechanical Engineers,
The Chinese Mechanical Engineering Society,
The Chinese Association of Automation,
The National Natural Science Foundation of China

Symposium Chairmen:

Professor Cheng-Yu Jiang, Northwestern Polytechnical University, China
Professor Alan Hendry, University of Strathclyde, UK
Professor Neal Juster, Strathclyde University, UK
Dr Xiu-Tian Yan, Strathclyde University of, UK

China
Professor Guoqing Li, University of Macau
Professor Hui Li, Chendu University of Technology
Professor Peigen Li, Huazhong University of Science and Technology, China
Professor Udo Lindemann, Muenchen University of Technology, Germany
Professor Tony Medland, Bath University, UK
Professor Xiansheng Qin, Northwestern Polytechnical University, China
Professor Dieter Roller, Stuttgart University, Germany
Professor Tetsuo Tomiyama, Delft University of Technology, The Netherlands
Mr William Thomson, Clyde Blowers Plc, UK
Professor Jinsong Wang, Tsinghua University
Professor Junbiao Wang, Northwestern Polytechnical University, China
Professor Runxiao Wang, Northwestern Polytechnical University, China
Professor Wei Wang, Northwestern Polytechnical University, China
Professor Zhiqian Weng, Northwestern Polytechnical University, China
Professor Richard Weston, Loughborough University, UK
Dr Chris HC Wong, Hong Kong Polytechnic University, Hong Kong, PRC
Professor Yongdong Xu, Northwestern Polytechnical University, China
Dr Xiu-Tian Yan, University of Strathclyde, UK
Professor Haichen Yang, Northwestern Polytechnical University, China
Professor Dinghua Zhang, Northwestern Polytechnical University, China
Professor Litong Zhang, Northwestern Polytechnical University, China
Professor Weihong Zhang, Northwestern Polytechnical University, China

The Symposium Event Organizing Committee

Chairman: **Professor Wei Wang**
Vice-Chairmen: **Professor Hong Tang**
Professor Junbiao Wang
Hongliang Chen and Chunhong Zhao

Symposium Secretary:
Professor Xiansheng Qin
Dr. Cato Xincai Tan

PART 1

KNOWLEDGE INTENSIVE METHODOLOGIES AND TOOLS SUPPORTING ENGINEERING DESIGN AND MANUFACTURE......1

2

KNOWLEDGE ENGINEERING IN PRODUCT AND PROCESS DESIGN

Keith Case

Mechanical and Manufacturing Engineering, Loughborough University, Loughborough, Leics, LE11 3TU, UK. K.Case@lboro.ac.uk

Abstract: A primary objective of knowledge engineering is to facilitate the application of knowledge arising from product design and the planning of the manufacturing process for new and modified products. Knowledge is an expensive and valuable commodity so the development of methods for its capture, representation and application are of considerable importance. The emphasis of the research described in this paper is on the re-use of knowledge for the creation of product variants. This implementation of 'predictive design' represents the very significant move from standard parts to standard knowledge constructs. Standard parts can be used in any application that requires a defined function where the shape and properties do not need to be altered. However, standard knowledge constructs can provide parts that can be used wherever the function is required. The Process Specification Language (PSL) provides one possible way of achieving this aim and is briefly outlined. A complementary method using 'knowledge fragments' for Failure Modes and Effects Analysis (FMEA) applied to electronic products is also described.

Key words: Knowledge Engineering, Predictive Design, Process Specification Language, FMEA.

1.   INTRODUCTION

There is currently a strong trend for industries to outsource many aspects of manufacture to their supply chains and in many cases this is generating concern over the potential loss of informal knowledge and experience that can have a serious impact on design capability. The capture and management of knowledge in a formalized way is thus of increasing importance. The development of genuinely new products is an important activity, but much 'new' development actually focuses on the generation of variants of existing products. This is a particularly fertile area for the capture of knowledge of a product so that it can be re-used at a future date in a variant of that product. The research described here addresses this area with a focus on a subset of predictive engineering, by capturing and representing design and manufacturing methods such that it can be re-used.

Extensive research effort on the use of Information Technology (IT) for design and manufacturing has resulted in advanced tools such as object-oriented programming and database/knowledge base management systems. Application of these advanced IT tools in design and manufacturing applications has been mainly devoted to the automation of the decision-making process and the integration of such systems through data transfer. Parameterised product models have associated design and manufacturing knowledge with geometric information, but problems are becoming apparent. Often, systems have been developed and used for specific applications without considering the potential re-use of their individual methods in new or different applications. Knowledge and information is represented in information structures not suited to distributed and dynamic environments. Knowledge gained in previous designs or projects is rarely used for new designs/variants due to a lack of knowledge standardisation. The research aims to make important contributions to the solutions of the above problems through the creation of new knowledge-based methodologies.

In the US, Grafton [1] describes a project aimed at design and process re-use. A set-based representation of early design descriptions permits approximate assessment using imprecise, partially complete design information [2]. Generic designer assistance tools combine graph-based function models with geometric models so that a design solution can be achieved by optimisation algorithms [3]. The US National Institute of Standards and Technologies (NIST) is also concerned with design and manufacturing integration [4]. Some US researchers are developing Internet/Web-based collaborative design/life-cycle support systems such as the enterprise-web portal [5] and the collaborative product conceptualisation tool using web technology [6]. However, these projects are not focused on knowledge representation and re-use but on design processes and integration through IT/Web technology.

In the knowledge management/engineering area, the research is mainly devoted to generic methodologies which are applicable to a wide range of business sectors such as finance, administration, medical and general engineering. A typical example is the knowledge management technology tool kit – CommonKADS [7]. Other well-known knowledge management systems are reported by Angele et al [8] and Brazier et al [9]. These systems are not specifically aimed at design and manufacturing applications, and therefore are not integrated into the design environment. Some work has also been done in Engineering Design Centres in the UK. For example, the Schemebuilder design environment developed by Lancaster University [10] provides a tool to assist multi-disciplinary system design from concept to embodiment stages with problem analysis and alternative evaluation functions, using Bond Graphs to represent a product scheme.

2. PREDICTIVE DESIGN RESEARCH

Predictive Design research is concerned with developing a design environment in which design knowledge can be standardised and represented. Predictive design particularly helps companies in well-established industries to maintain a competitive position. The reasons are several folds.

Because they are well established, they are more successful in setting up predictive design engines – they know a lot about their product and how to design it. Competitive advantage is particularly difficult to achieve in mature markets where low levels of innovation are possible. The use of predictive design frees time for the generation of innovative ideas. The use of formal methodologies will significantly reduce new product development time and enhance integration of business processes.

Specific objectives of the research are:

- To capture design knowledge and develop methodologies for the standardisation and representation of the captured knowledge so that it can be re-used for new products or variants;
- To develop a design environment for collaborative product development in a distributed and controlled manner across an extended enterprise;
- To capture manufacturing and costing knowledge and develop methodologies so that such knowledge can be used for design of new products or variants;
- To develop a predictive design engine consisting of a set of independent methods which can be adapted and re-used for similar products or services; and
- To test and evaluate the developed methodologies with collaborating companies.

2.1   Methodology

The research is acquiring and classifying knowledge needed by the design team, including:

- *Customer requirements and anticipation knowledge* based on previous products and/or services crucial to the correct interpretation of the customer requirements and thus maintaining high quality;
- *Project management knowledge,* such as best practice procedures, methodologies and business processes used in previous projects.
- *Product knowledge* about previous (similar) products including a product model together with knowledge about how the final product is determined and what assessment/evaluation methods are used.
- *Manufacturing methods knowledge* about previous products including how the methods are selected and evaluated.
- *Costing knowledge* including costs of parts and assembly of products or modules, and product development costs;
- *Test and failure mode knowledge* to predict the quality of produced products and to determine tolerances.

Established process planning knowledge capturing methodology [11,12] is being improved and used for the capture of the design knowledge. Industrial collaborators' products and design process are being used as case studies to evaluate the methodologies. A collaborating company designs and manufactures vacuum pumps, and although design and manufacturing knowledge is available in house, it is not formally captured and represented. A second company designs door latches for major car manufacturers, with the assembly of the final products from supplied components being the most important process. Although the company does not manufacture components, knowledge on manufacturability and cost is of central importance to the design process. The design and costing knowledge are being captured from our collaborator and the manufacturability knowledge from its suppliers. The captured knowledge will be analysed and as much as possible standardised based on the types and functions of products and problems to be solved. Standard knowledge constructs can then be developed. The knowledge constructs can be used to form the decision logic of a solution within the knowledge engine and integrated with the PDM/CAD design environment. The knowledge constructs are basic methods implemented and stored in a library and are developed as the result of the knowledge capture. The knowledge constructs and methods in the Knowledge Engine are being developed in a way that they can be re-used in their entirety or as part of new methods for future designs or projects. PDM control functions can be used to manage the information, the design process, and the procedures. It also manages the different variants of product structures and associated design knowledge and manufacturing cost information.

3.   KNOWLEDGE MODELLING

One of the most important aspects of the work is the establishment of a knowledge representation that is capable of representing the required design and manufacturing knowledge and which is amenable to knowledge re-use. In general, knowledge is often represented by logical statements describing the relationships between entities. For example a "tea-making" knowledge base may state that *"a kettle is needed before you can start boiling water"*. This describes (in natural language) the relationship between a kettle, water, and the activity of boiling. A further step is to formally describe such relationships, allowing artificially intelligent (as well as human) reasoning. Precise, executable definitions of entities and relationships are required for this purpose. In recent years, lexicons of such terms (and their semantics) have become known as "ontologies".

3.1   Knowledge Sharing

Knowledge based systems have traditionally been developed for specific domains (e.g. design, manufacturing, and costing). Further domains may exist for suppliers, customers, and legacy systems. Typically, domains are isolated and therefore replicate what they need to know about other domains (e.g. a designer's knowledge of manufacturing). This gives scope for inconsistencies that can lead to poor decision making.

Knowledge sharing is a possible solution (e.g. allowing a designer to re-use parts of a manufacturing knowledge base). This is not however straightforward. Domains that have been independently developed invariably use their own (usually bespoke) ontology. Translating between
Multiple ontologies can be a complex undertaking. Figure 1 shows two possible approaches. The meshed approach (left) requires a translator between each domain and every other domain in the system. This is analogous to an international business, where everyone understands everyone else's language. Most businesses follow the more efficient model of a shared language (usually American-English). Similarly within knowledge base systems, there is a drive towards shared/neutral ontologies (left).

Standards bodies are the most obvious source of neutral ontologies. Indeed, a neutral ontology would need the explicit support of one or more of the leading institutions to become established. To this end, the National

Institute for Standards and Technology (NIST) and the International Standards Organisation (ISO) have become actively involved in ontology research.

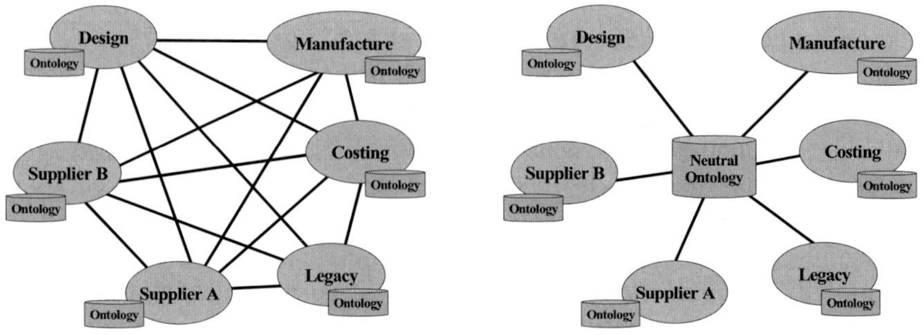

Figure 1. Meshed and Neutral Ontology Translation

3.2   The Process Specification Language

The Process Specification Language [13] is being proposed as a neutral ontology for process related knowledge (typical to the manufacturing, construction, and service industries). PSL provides formal definitions of entity kinds such as activities, objects, and states. The actual entities within a domain are represented by instantiated entity kinds. For our "tea-making" knowledge base, instantiated kinds may include:

- *Objects* such as: water, milk, sugar, teabag, spoon, cup, mug . . . .
- *States* such as: cold-water, hot-water, boiling-water, kettle-full, kettle-empty . . .
- *Activities* such as: filling, boiling, pouring, mashing, stirring . . .

PSL provides additional terms to describe the relationships between instantiated kinds, including: *prior*, *holds*, and *participates_in*. *Functions* are also defined for returning the start and end times of activities and the existence of objects (i.e. *beginof* and *endof*). The tea making example is extended in Figure 2, showing both natural language and PSL statements relevant to making tea.

A kettle and water are needed before you can start boiling:
forall (S) implies (occurrence_of((S, boiling)
        (AND   participates_in (kettle, boiling, beginof (S)
            participates_in (water, boiling, beginof (S)))

The activity of boiling results in boiling-water:
forall (S) implies (occurrence_of((S, boiling)
        Holds (boiling-water, S))

Filling a kettle is not possible when the kettle is full:
Forall (S) implies (poss(filling, S)
         prior(not kettle-full, S))

Figure 2. Example "Tea-Making" Knowledge Base

PSL was initially developed by NIST and has recently been adopted by the ISO. This adoption has established PSL as the focal point for research into ontologies for process related knowledge. PSL builds on several previous ontologies, including: A Language for Process Specification (ALPS), the Toronto Virtual Enterprise (TOVE), the Enterprise Ontology [14], Core Plan Representation (CPR), Shared Planning & Activity Representation (SPAR), The Process Interchange Format (PIF), and the WorkFlow Management Coalition (WfMC).

It should also be noted that PSL defines entities and relationships that are needed to represent processes. Other ontological efforts (e.g. the semantic web) concentrate on tools and formats for defining ontologies (e.g. OWL and RDF), leaving the entity and relationship definitions themselves to the end user. These tools/formats make it easier to develop ontologies for specific domains, but do not tackle the issue of knowledge sharing (which requires standardised entity/relationship definitions).

PSL promises a world where process related knowledge can be readily shared between business divisions, suppliers, customers, and legacy systems. Third party tools based on PSL can also be envisaged. PSL is however a relatively immature standard. Efforts have been made in the ISO documentation to include case material (based on a contrived engine example). Extensive, real life case histories are however needed if PSL to evolve into a widely accepted neutral ontology for process knowledge. The experimental platform described below sets out to establish the representational limits of PSL in a design environment, and provide at least some of the required case history.

3.3   Experimental Architecture

The experimental environment shown in Figure 3 is being used to establish the representational limits of PSL in a design environment, and provide a detailed case history of the use of PSL in a knowledge sharing environment.

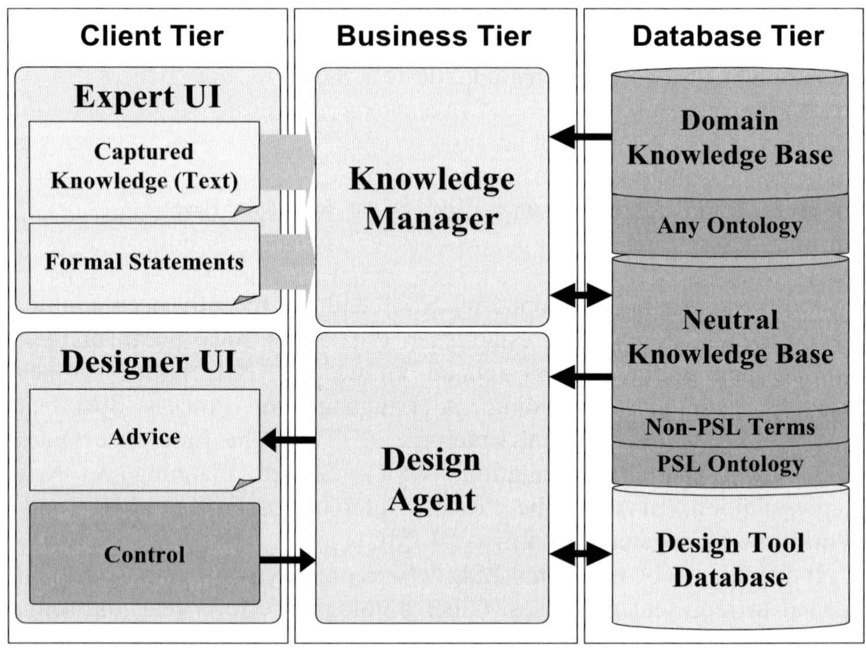

Figure 3. Proposed Experimental Architecture

The Neutral Knowledge Base (NKB) can be populated via two options. Firstly, an existing domain knowledge base (using any ontology) can be translated to the neutral format (based on PSL) by the Knowledge Manager. Secondly, captured knowledge (using any capture methodology), can be directly written into the NKB. Scope for adding non PSL terms will also be provided. To demonstrate the use of the shared knowledge in a design environment, an agent will interrogate the NKB, and provide design advice (based on the shared knowledge).

Initially, examples of manufacturing knowledge (e.g. process descriptions and rule bases) will be loaded into the NKB using one or both of the options described. The degree to which non PSL terms are required will help establish the representational limits of PSL. Further classes of knowledge (beyond manufacturing) will be identified and added to the NKB in due course.

4. FAILURE MODE AND EFFECT ANALYSIS (FMEA)

It is believed that PSL offers a way forward in representing knowledge but it is not the only feasible approach. A completed research study has used a knowledge fragment approach to the re-use of knowledge in the context of Failure Mode and Effect Analysis (FMEA). FMEA is a widely used concurrent engineering tool for quality improvement and risk assessment. The method originates from the US military and is now a recommended method for use in the ISO 9000 series. BS 5760 Part 5 [15] states that "FMEA is a method of reliability analysis intended to identify failures, which have consequences affecting the functioning of a system within the limits of a given application, thus enabling priorities for action to be set." However, implementation difficulties mean that in some instances FMEA is regarded as 'a tedious and time-consuming activity' [16] and may only be used due to contractual obligation [17]. The fundamental difficulties with FMEA are the lack of any formal guidelines as to its use and its use of natural language [18] resulting in unstructured knowledge and poor reasoning support [19]. Ways have therefore been sought to formalise the knowledge representation and to provide means for its effective re-use. In addition, although FMEA has conventionally been most closely associated with processes, this research has extended its application into the conceptual stages of product design to provide a greater extent of product life-cycle support.

4.1 FMEA Modelling and Reasoning

Functional and structural modelling are often associated with FMEA research [20,21]. The function expresses the design intent and the behaviour describes how this is achieved [22,23]. A structural model formally identifies the interactions amongst the components, entities, sub-processes or sub-systems of the product.

In conventional FMEA brainstorming techniques are used by teams of experts in attempts to identify all the possible failure modes and relate these to the causes and effects. Replacing this by intelligent computer reasoning is then the challenge if FMEA is to be effectively automated. Quantitative reasoning (based on analytic analysis) is a possibility, but qualitative (descriptive) reasoning has been selected as being comparable to human thought processes.

4.1.1   The FMEA Model

An object-oriented approach is used for FMEA modelling (Figure 4).

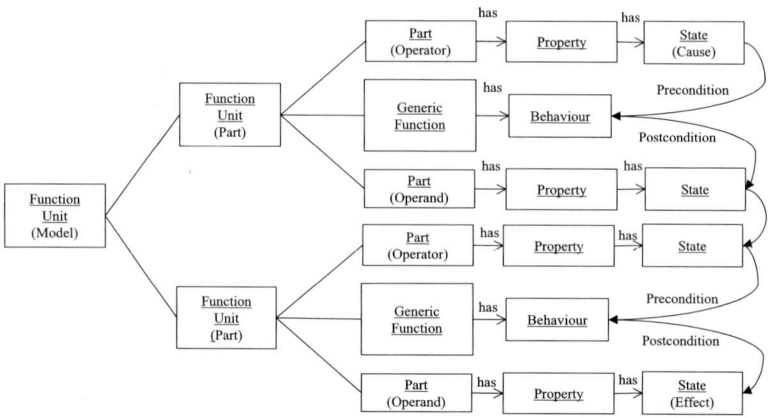

Figure 4. An Object-Oriented Model for FMEA

There are two particularly important relationships. In the precondition relationship the state of the operator determines the Behaviour of the Generic Function within a Function Unit. The Behaviour in turn decides the state of the operand within the Function Unit and this is the post-condition relationship. As operands are normally also an operator at the next level, state changes cause propagation throughout the system, providing the essential cause and effect chains. The information about various states of a part, and the behaviours corresponding to specific states are stored in the database. Information relating to the objects and the governing heuristic rules are maintained in a relational database. The operator, operand and generic function can be used as keys to search for matching states and behaviours when a new function unit is created.

4.1.2   Application in Automatic FMEA Generation

Figure 5 shows the design concept for a conveyor to handle printed circuit boards (PCBs) and Figure 6 shows a possible physical realisation of the functionality. The design concept is the starting point of the FMEA analysis

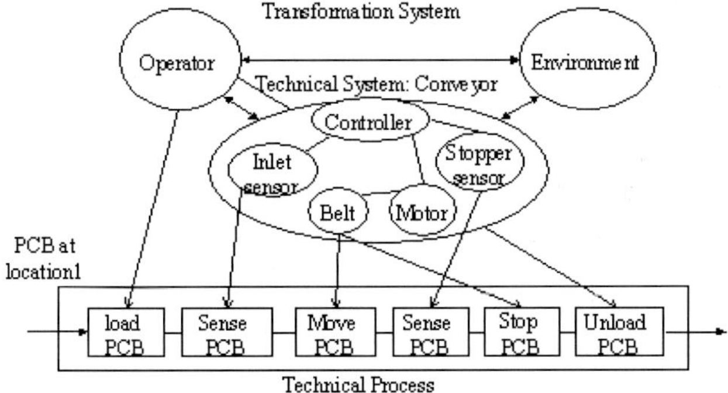

Figure 5. Transformation Model for Conveyor

Pre- and post-condition relationships between part states and behaviours are established either from prior knowledge or from failure reports which provide the source, cause and effect. A major objective is the re-use of knowledge and this is achieved in this context by the system 'learning' from a series of failure reports.

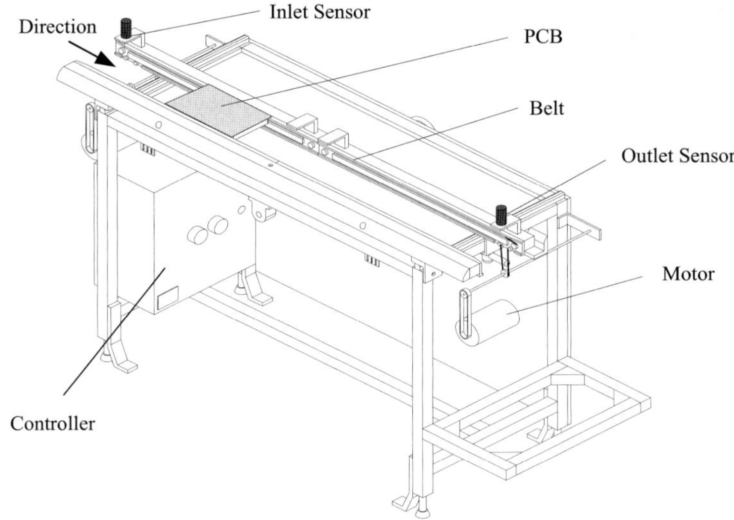

Figure 6. Physical Realisation of a Conveyor System

Traditionally the knowledge of an FMEA is contained within the case being recorded, but with this new FMEA model, the knowledge resides in the part. Hence, the system is able to respond to new cases not previously captured by the user. For example in PCB assembly, failure cases may have been recorded for 'motor moves belt' and 'belt moves PCB'. Hence, should the motor fail, then the belt will not move, and consequently the PCB fails to move.

Hence, the information captured in the database would be:

Function Unit 1:   motor moves belt

Operator: motor   Generic Function: move   Operand: belt

Precondition: State: motor failure   Behaviour: not moving

Postcondition: Behaviour: not moving   State: belt not moving

Function Unit 2:   belt moves PCB

Operator: belt   Generic Function: move   Operand: PCB

Precondition: State: belt not moving   Behaviour: not moving

Precondition: Behaviour: not moving   State: PCB not moving

If another user created a design with the function unit: "motor moves PCB", which has never been captured from a failure report, a search will be made for the operator "motor" with function "move" and the likely precondition retrieved (State: motor failure – Behaviour: not moving). The same process is carried out on the operand "PCB" and function "move". In this case, the likely postcondition (Behaviour: not moving - State: PCB not moving) is retrieved. The combination of this information will result in a new case: "motor fails, PCB not moving" – an effective re-use of knowledge.

Failure reports are used as the source for cause and effect chain building which eventually leads to the generation of an FMEA. The failure records of a conveyor and a chip-mounting machine have been used for a case study, courtesy of Motorola Technology Malaysia PLC. The main pieces of information identified by the failure reports identify are the 'item', 'problem', 'cause' and 'solution'.

All related parts for the conveyor are modelled and captured by the system. The full FMEA generation as created as shown in Figure 7.

In the case of the chip-mounting machine, not all parts are modelled, i.e. not all parts are provided with the properties, states and behaviours that are required for the FMEA generation. The chip-mounting machine has a conveyor which is similar to a conveyor used in a previous case. Hence, the data can be reused without creating additional parts for its conveyor. The result is as shown in Figure 8. Even if the machine is not fully modelled, it is capable of providing a generated result based on the historical data and user input from the limited failure reports. The user can then complete the FMEA manually.

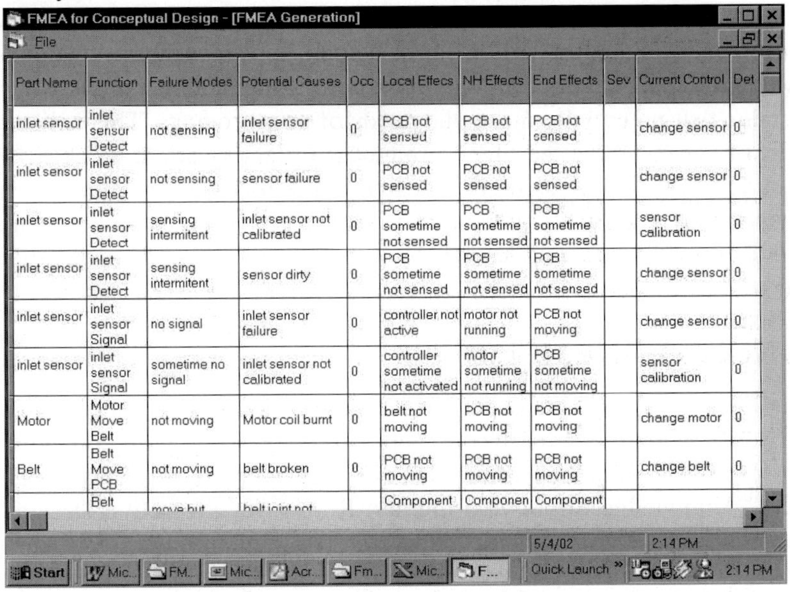

FMEA for Conceptual Design - [FMEA Generation]

File

Part Name	Function	Failure Modes	Potential Causes	Occ	Local Effecs	NH Effects	End Effects	Sev	Current Control	Det
inlet sensor	inlet sensor Detect	not sensing	inlet sensor failure	0	PCB not sensed	PCB not sensed	PCB not sensed		change sensor	0
inlet sensor	inlet sensor Detect	not sensing	sensor failure	0	PCB not sensed	PCB not sensed	PCB not sensed		change sensor	0
inlet sensor	inlet sensor Detect	sensing intermitent	inlet sensor not calibrated	0	PCB sometime not sensed	PCB sometime not sensed	PCB sometime not sensed		sensor calibration	0
inlet sensor	inlet sensor Detect	sensing intermitent	sensor dirty	0	PCB sometime not sensed	PCB sometime not sensed	PCB sometime not sensed		change sensor	0
inlet sensor	inlet sensor Signal	no signal	inlet sensor failure	0	controller not active	motor not running	PCB not moving		change sensor	0
inlet sensor	inlet sensor Signal	sometime no signal	inlet sensor not calibrated	0	controller sometime not activated	motor sometime not running	PCB sometime not moving		sensor calibration	0
Motor	Motor Move Belt	not moving	Motor coil burnt	0	belt not moving	PCB not moving	PCB not moving		change motor	0
Belt	Belt Move PCB	not moving	belt broken	0	PCB not moving	PCB not moving	PCB not moving		change belt	0
	Belt	move but	belt joint not		Component	Componen	Component			

5/4/02 2:14 PM

Start | W Mic.. | FM.. | Mic.. | Acr.. | Fm.. | Mic.. | F... | Quick Launch » | 2:14 PM

Figure 7. Generated FMEA for Conveyor

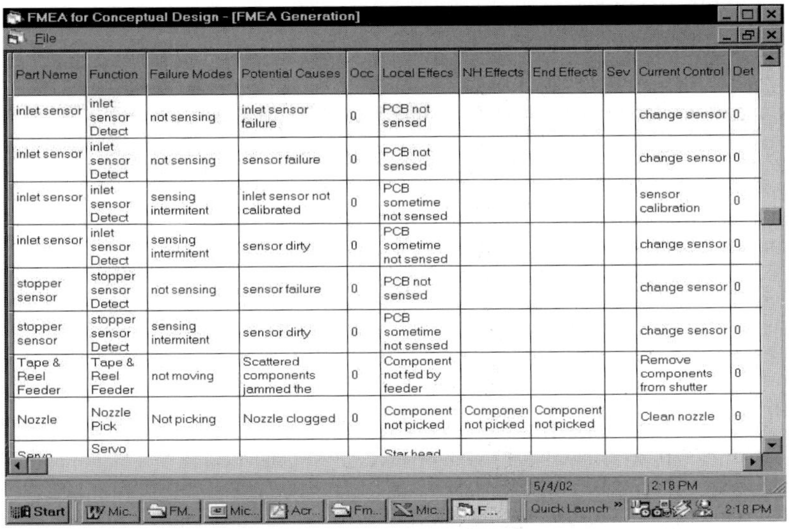

FMEA for Conceptual Design - [FMEA Generation]

File

Part Name	Function	Failure Modes	Potential Causes	Occ	Local Effecs	NH Effects	End Effects	Sev	Current Control	Det
inlet sensor	inlet sensor Detect	not sensing	inlet sensor failure	0	PCB not sensed				change sensor	0
inlet sensor	inlet sensor Detect	not sensing	sensor failure	0	PCB not sensed				change sensor	0
inlet sensor	inlet sensor Detect	sensing intermitent	inlet sensor not calibrated	0	PCB sometime not sensed				sensor calibration	0
inlet sensor	inlet sensor Detect	sensing intermitent	sensor dirty	0	PCB sometime not sensed				change sensor	0
stopper sensor	stopper sensor Detect	not sensing	sensor failure	0	PCB not sensed				change sensor	0
stopper sensor	stopper sensor Detect	sensing intermitent	sensor dirty	0	PCB sometime not sensed				change sensor	0
Tape & Reel Feeder	Tape & Reel Feeder	not moving	Scattered components jammed the	0	Component not fed by feeder				Remove components from shutter	0
Nozzle	Nozzle Pick	Not picking	Nozzle clogged	0	Component not picked	Componen not picked	Component not picked		Clean nozzle	0
Servo	Servo				Star head					

5/4/02 2:18 PM

Start | W Mic.. | FM.. | Mic.. | Acr.. | Fm.. | Mic.. | F... | Quick Launch » | 2:18 PM

Figure 8. FMEA for Chip-Mounting Machine

5.   CONCLUSIONS

The capture and re-use of design and manufacturing knowledge to effectively support product life-cycle activities through knowledge sharing and re-use remains a challenge. The paper has described a 'knowledge fragment' approach used in the context of Failure Modes and Effects Analysis which is capable of re-using existing knowledge to synthesise new knowledge that potentially meets the needs of new products. This is likely to be most effective where the 'new' product is in fact a variant on an existing product and this area of predictive design is the focus of on-going research which is in part investigating the use of the process specification language as a method for manufacturing knowledge representation and its application in product design.

ACKNOWLEDGEMENTS

The FMEA parts of this paper have been adapted from a paper presented to IMC'2002 and are primarily the work of Dr P.C.Teoh with collaboration from Motorola Technology Malaysia. The PSL work is part of an ongoing project - 'Knowledge Representation and Re-use for Predictive Design and Manufacturing Evaluation' funded by Loughborough University's Innovative Manufacturing Research Centre (IMRC) – itself funded by the Engineering and Physical Sciences Research Council (EPSRC). This is a joint research project with Cranfield University. The input from Dr J.Gao, D.Baxter, Dr R.I.Young, Dr J.A.Harding, S.D.Cochrane and S.Dani is gratefully acknowledged.

REFERENCES

1.   Grafton, R. B. (programme manager), Design Assessment and Re-use, NSF Award: #9624231, 1/7/1996 – 30/6/2001.
2.   Hazelrigg, G. A. (programme manager), Rapid Assessment of Early Design: RAED, NSF Award: #9813121, 1998 –2001.
3.   Hazelrigg, G. A. (programme manager): Generative Designer Assistant Tools, NSF Award: #9733434, 1/8/1998 – 31/7/2002.
4.   NIST Manufacturing Engineering Laboratory, Manufacturing Systems Integration division. Web site: www.nist.gov/public_affairs/guide/meltx04.htm
5.   Rezayat, M., "The Enterprise-Web protocol for life-cycle support", *Computer Aided Design*, **Vol 32**, pp.85-96, 2000.
6.   Roy, U. and Kodkani, S. S., "Collaborative product conceptualisation tool using web technology", *Computers in Industry*, **Vol. 41**, pp.195-209, 2000.
7.   Schreiber, G., Akkermans, H, Anjewierden, A., de Hoog, R.Shadbolt, N., de Velde, W. V. and Wielinga, R., Knowledge Engineering and Management – The CommonKADS Methodology, The MIT Press, SCH1KH 0-262-19300-0, 2000.
8.   Angele, J., Fensel, D., Landes, D. and Studer, R., "Developing Knowledge-based systems with MIKE", *Journal of Automated Software Engineering*, 1998.

9. Brazier, F., van Lange, P. H., Treur, J., Wijgaards, N. J. E. and Willems, M., "Modelling an elevator design task in DESIRE: the VT example", *International Journal of human-Computer Studies*, **44(3/4)**, pp.469-520, 1996.

10. Chaplin, R. V., Li, M., Oh, V. K., Sharpe, J. E. E. and Yan, X. T., "Integrated Computer Support for Interdisciplinary System Design", *Artificial Intelligence in Design'94*, ed. J S. Gero and F Sudweeks, Kluwer Academic Publishers, pp.591-608, 1994.

11. Gao, J.X., Tang, Y.S. and Sharma, R., "A feature model editor and process planning system for sheet metal products", *Journal of Materials Processing Technology*, 107/1-3, pp.88-95, 2000.

12. Tang, Y. S., The development of a user oriented process planning system for sheet metal fabrication, PhD thesis, Cranfield University, 2000.

13. Schlenoff C, Gruninger M, Tissot F, Valois J, Lubell J, Lee J. The Process Specification Language (PSL) Overview and Version 1.0 Specification. [Web Accessed 3rd October 2003]. http://ats.nist.gov/psl

14. Uschold M, King M, Moralee S, Zorgios Y, 1997. The Enterprise Ontology. Edinburgh University. Unpublished Document [Web Accessed 13th October 2003]. http://tmitwww.tm.tue.nl/staff/gwagner/EnterpriseOntology.pdf

15. BS 5760 Part 5, Reliability of systems, equipment and components. Guide to failure modes, effects and criticality analysis, 1991.

16. Price, C.J., Pugh, D.R., Wilson, M.S., and Snooke, N. "Flame system: automating electrical failure mode & effects analysis (FMEA)", *Proceedings of the Annual Reliability and Maintainability Symposium*, pp.90-95, 1995.

17. Dale, B.G., Shaw, P. "Failure Mode and Effects Analysis in the U.K. Motor Industry: A state-of-the-art Study", *Quality & Reliability International*, 1996.

18. Wirth, R., Berthold, B., Kramer, A. and Peter. G. "Knowledge-based Support of System Analysis for Analysis of Failure Modes and Effects", *Engineering Applied Artificial Intelligent*, **Vol. 9, No.9**, pp.219-229, 1996.

19. Lee, B.H. "Design FMEA for Mechatronic Systems Using Bayesian Network Causal Models", *Proceedings of the ASME Design Engineering Technical Conferences*, **Vol. 1**, pp.1235-1246, 1999.

20. Hunt, J.E., Pugh, D.R. and Price, C.J. "Failure mode effects analysis: a practical application of functional modelling", *Applied Artificial Intelligence*, **Vol 9, 1**, pp.33-44, Jan-Feb 1995.

21. Eubanks, C.F., Kmenta, S. and Ishii, K. "Advanced Failure Modes and Effects Analysis Using Behavior Modelling". *Proceeding of DETC'97, 1997 ASME Design Engineering Technical Conference and Design Theory and Methodology Conference*, Sacramento, California, 1997.

22. Gero, J.S., Tham, K.W. and Lee, H.S. "Behaviour: A Link Between Functional and Structure in Design". *Proceeding of the IFIP WG 5.2 Working Conference on Intelligent Computer Aided Design*, Elsevier Science Publisher, 1991.

23. Russomanno, D.J., Bonnell, R.D. and Bowles, J.B. "Functional reasoning in a failure modes and effects analysis (FMEA) expert-system". *Proceedings of the Annual Reliability and Maintainability Symposium*, pp. 339-347, 1993.

A PROPOSAL OF DESIGN VISION FOR ARTIFACT DESIGN AND PRODUCTION IN THE 21ST CENTURY

A Perspective from the Design Engineering Section of the National Committee for Artifact Design and Production, Science Council of Japan

Teruyuki Monnai

Professor of Waseda University, Dr. Eng., 3-4-1, Okubo, Shinjuku-ku, Tokyo 169-8555, Japan, email: monnai@waseda.jp,

Eiji Arai, Professor of Osaka University, Dr. Eng., Japan

Juhachi Oda, Professor of Kanazawa University, Dr. Eng., Japan

Tetsuo Tomiyama Professor of Delft University of Technology, Dr. Eng., The Netherlands

Akihiro Hotta, Professor of Chiba University , Dr. Eng., Japan

Abstract: The Design Engineering Section of the National Committee for Artifact Design and Production, the Science Council of Japan (SCJ), presented a proposal of design vision for artifact design and production in the 21st century to the Japanese government in 2003. This paper introduces the proposal , which consists of seven propositions on, respectively, the concept, process, system, collaboration, method, education, and science of design.

Key words: design vision, artifact, process, system, collaboration, tool, education, ethics

1. BACKGROUND AND OBJECTIVE

The Science Council of Japan (SCJ), born out of the firm conviction that science constitutes the foundation of culture, was established in January 1949, for the promotion and dissemination of science to all government agencies, industries and people's lives, as the organization to represent Japanese scientists. SCJ created the Design Engineering Section of National

Committee for Artifact Design and Production to deliberate important matters related to design sciences and to coordinate research programs.

We explored the qualitative transformation of design suitable to the post-industrialized society from 2001 to 2003. Based on this study, we presented a proposal of design vision for artifact design and production in the 21st century to the Japanese government in July 2003[1].

This paper introduces this proposal of design vision for artifact design and production in the 21st century, drafted by the Design Engineering Section of the National Committee for Artifact Design and Production, SCJ. The proposal consists of seven propositions on, respectively, the concept, process, system, collaboration, method, education, and science of design. We expect it to contribute to the transformation of design in Japan and various other post-industrialized societies.

2.   PRESENT STATE OF ARTIFACT DESIGN AND PRODUCTION

In the 20th century, science and technology underwent an amazing development, and artifacts such as cars, airplanes, electrical appliances, and computers were invented. As a result the lives of human beings have changed drastically. In many countries an industrialized, urbanized, and information oriented society has been realized, and the state of economy, society, and culture exhibits a different mode of existence from that of the last century. It must be emphasized, from the viewpoint of artifact design and production in advanced industrialized countries, that the nature-centered life has been transformed into the artifact-centered life. In such societies artifacts made by industrial production systems are used in everyday life, and the abundant society has been established in which many people enjoy the life based on physical artifacts. On the other hand, the abundance of artifacts supplied by the industrial production system stimulates the demand for them, and the mechanism of capitalism driven by those artifacts has produced a surplus above the requirement for them. The mass extraction of resources and mass production, as a consequence, has encouraged mass consumption and mass waste, and caused various problems, bringing serious damage to the human future of the human race, such as the destruction of the earth, the exhaustion of resources and energy, the pollution of the natural environment by chemical materials, and the loss of beautiful townscape and distinctive community culture.

Today the systems of politics, economy, and society cause many problems in Japan. Thus the restructuring of those systems and the regeneration of cities have been recently promoted, but the problems have not yet been solved. We think that these problems cannot be solved by the mere restructuring of institutions, and that we should understand they are, in fact, the consequence of the obsession of "industrialized society" with the pursuit of economical wealth. The industrialized society developed in the 20th century has brought about the destruction of the natural environment through mass extraction and mass waste, and the loss of community through mass production and mass consumption. We cannot maintain the viability of human life and global environment simply through the extension of the industrialized society.

In the 21st century, on the contrary, we should begin with the regeneration of the natural environment and community culture lost in the industrialized society, and strive to construct a sustainable society, in which symbiosis with nature and human interaction are promoted, in order to bring about an abundant life. The role of artifact design and production is to design and to produce artifacts which are in harmony with such a future society. It has an important role not only to produce materials and products but also to provide information and service which have heretofore been considered as secondary assets. From these viewpoints, the proposed new society could be called a "post-industrialized society" in contrast to "industrialized society".

It must be noted that the function of production has been given priority over the function of human life in the industrialized society; the relationship of those functions should be reversed in the post-industrialized society. When a comfortable environment with a beautiful landscape is realized, excellent people will inhabit it and new industries of information and service will be accumulated because of those human resources. That is, the function of human life is the key to the regeneration of environment and culture in the post-industrial society.

Thus, the day of the industrialized society based on mass production and mass consumption is over, and it has become an important task in the post-industrialized society to consider the environment and culture profoundly and to pursue new modes of artifact design and production in order to regenerate the field for human life.

3. BASIC VIEW OF ARTIFACT DESIGN AND PRODUCTION

A paradigm shift of artifact design and production, "from quantitative satisfaction to qualitative satisfaction", has been proposed based on the recognition of the situation discussed above. And it has been recognized that we must satisfy not only given overt requirements but also implicit conditions such as comfort, safety, and environment in today's design activities.

We focus on the qualitative change of design concepts in the post-industrialized society, and investigate new modes of designing which raise the quality of life not only by making individual artifacts but also by considering the relation among artifacts and moreover the relation between artifacts and man-environment systems. Today designs are expected to create the software such as information and services which has been previously regarded as of secondary value to hardware. The necessity for new design concepts is revealed clearly by the great interest in the life-cycle design and ecological design which consider the relation between artifacts and environmental problems, and universal designs which create artifacts which are useful to various users such as persons of advanced age and handicapped persons. It is necessary to develop the research and practice of the design which satisfies not only the quantitative needs but also the more wider qualitative needs considering the constraints from environment and society.

(The reason why we discuss the concept of artifact in this context is that we want to consider the following fact: in the contemporary society dominated by artifacts it becomes necessary to reconsider the artificial in relation to the natural, and to design not only various artifacts separately but also the interrelationship of those artifacts.)

In the 21st century, we need to transform the design concept of artifacts drastically against the background of the population explosion in developing countries, the aging of the population in advanced countries, the frequency of accidents, disasters, and terrorism, the advance of globalization, the intensification of global environmental problems, the change in human relations and societies caused by the revolution of information technology, and the rapid development of biotechnology. We must recognize that our living environment is full of artifacts and their waste products, and that our environment is finite. As the environment becomes more and more complex, uncertain, unstable, and fragile, the life of all animate beings is threatened everywhere.

The systems of politics, economy, culture, life, and technology, the components of modern society, cause various problems, and the advanced countries including Japan are moving from industrialized societies to post-industrialized societies, from an age of materialism to a post-materialism age. The Design Engineering Section of the National Committee for Artifact Design and Production held a symposium on "The Qualitative Change of Design" in 1999[2], which advocated the transformation from quantitative satisfaction to qualitative satisfaction. We decided to extend this idea and to propose "The Design Vision for the Artifact Design and Production in the 21st Century" for the design suitable to the post-industrialized society[3].

The following proposal of design vision designates a new mode of designing to create a total system of environment, economy, society, and culture by means of considering not only individual artifacts but also the relations among artifacts and the relations between artifacts and the natural environment. We think that it will contribute to changing the way of artifact design and production, and to raise the quality of life of human beings in the global context.

4. THE PROPOSAL OF DESIGN VISION FOR ARTIFACT DESIGN AND PRODUCTION IN THE 21ST CENTURY

The proposal of design vision consists of the following seven propositions. We show each proposition in italics, followed by a detailed explanation. The heading is the key word of the proposition.

4.1 Proposition 1: design concept

We should strive to transform the qualitative concept of design for the post-industrialized society. It becomes very important not only how we make things but also what things we make in the new context.

The capacity for artifact design and production has increased drastically with the progress of industrialization after the industrial revolution, and our surroundings are full of artifacts which satisfy various requirements. These artifacts satisfy most of the basic needs of our daily life. It is unnecessary to say that artifacts with new functions and performances are required from now on, but many people in advanced countries have become interested in the meaning and value of artifacts. Today qualitative satisfaction has become a more important aim than quantitative satisfaction. In such a context, designers are called upon to consider seriously what products they will make. This is a problem concerning given conditions and design programs, and means that designers, planners, clients, users, and legislators are called upon to take part in the design process.

In an age of global environment, it becomes an important role of design not only to create new artifacts but also to find the new ways of using and evaluating existing artifacts. Moreover included among the alternatives of designing will be the decision to make nothing and even to remove existing artifacts under certain circumstances. Thus in the post-industrialized society, we must extend the concept of design drastically, and work on design problems by means of inquiring both how we make things and what things we make.

4.2 Proposition 2: design process

Good artifacts are to be generated in a continuous process in which designing, production, and living are closely connected. It is necessary to expand the design process of artifacts in order to include not only the making process but also the living or breeding process.

It is necessary for us to understand the design process not only as a micro process which derives a design solution from given conditions, but also as a macro process which includes the process of inquiring into the given conditions themselves, the process of using designed objects, and the process of evaluating the consequences for the next design.

The history of design should be understood in the following way: at first the user was also the producer, then the producer was distinguished as the craftsman from the user, then the designer was distinguished from the producer in the context of the development of technology after the modern age, and finally the profession of designer playing a role of thinking was established. Here we have to notice the fact that there are many excellent designs by anonymous designers; these are craft products, for which we cannot specify the particular designer. These artifacts were designed through processes used by many people for a long time and evolved to adapt the surrounding environment. In contrast, in an industrialized society, it is usual that the artifacts are generated by design systems in which users, producers, and designers are separated. The artifacts generated by these design systems in a limited time often disregard the users and environments, with the result that frequent misfits arise.

In a post-industrialized society the provision of designed products and spaces to users should not be a one way process; rather it is necessary to restore the feedback circuit reflecting the experience of artifacts in the living environment. We must develop the design process by means of a dialogue among designers, producers, and users. It is indispensable to develop the design process based on such lively interaction among artifacts, users, and environments in order to generate artifacts which become more attractive with time, such as life cycle design, series design of products including grade up and version up, design of city, environment, and landscape, etc.

On the other hand it should not be forgotten that design is closely connected with production or product making, because the quality of design is seriously damaged by the loss of skilled craftsman and by the revolution of productive technology and industrial structure. In the case of car design, the production method to combine modules consisting of plural parts is now an accepted fact, and it seems to us that the process of design and maintenance of artifacts is changed drastically. It is necessary for designers and users to have an interest in the productive process in order to create and develop excellent artifacts.

4.3   Proposition 3: design system

Design in the 21ˢᵗ century should play an important role in enhancing the quality of life not only by making individual artifacts but also by improving the environmental and social system including the set of artifacts and natural objects. In this context, the object of design must be expanded from the hardware of artifacts to the software of services indispensable to the environmental and social system.

The design of individual artifacts is a fine thing, but a serious problem encountered by modern society is that the environmental and social system including the interrelationship among artifacts and the relations of artifacts to the nature and society has not been properly designed. In other words, the design of relations is more important matter than the design of elements. To be specific, the software of artifacts is mainly focused on the remarkable trends of contemporary design, such as the universal design concerning the physical and behavioural properties of various human beings, the ecological design concerning the sustainability of the environment, the design of landscape composed of an ensemble of various elements, the e-design using information networks, and the design for safety and security corresponding to diverse risks caused by the development of science and technology.

In fact the environmental and social system like townscape, having historical and total characters, cannot be constructed intentionally, and should be brought up as the plant or flower is grown in the garden. Thus the fundamental of the environmental and social design is not to create new artifacts, but it is essentially to cherish the things handed down from the preceding generation, to repair their damaged parts, and to hand over them to the next generation by adding the fruits in each time. Various design activities such as maintenance, conservation, restoration, regeneration, and creation are included in the above-mentioned process.

In fact environmental and social systems like townscapes, having historical and total characters, cannot be constructed from scratch, and must be brought up as plants or flowers are grown in the garden. Thus the fundamental of environmental and social design is not to create new artifacts, but essentially to cherish the things handed down from the preceding generation, to repair their damaged parts, and to hand them over to the next generation by adding the fruits of each time. Various design activities such as maintenance, conservation, restoration, regeneration, and creation are included in this process.

Moreover it is necessary in each individual design to deal with serious problems about life, environment, resource, and energy which human beings encounter. It becomes more and more important to study design engineering to build a foundation of extended design activities such as "service engineering" as the technology which realizes the transformation from quantitative satisfaction of artifacts to qualitative satisfaction at the social level and which makes not the owner but the society wealthy, "maintenance engineering" which makes use of existing artifacts according to the needs of an aging society with a decreasing population, and "inverse engineering" which aims to construct a sustainable recycled society.

4.4   Proposition 4: collaboration

The design problem is very complex, ambiguous, and unstable today. Therefore it is necessary to promote collaborative designs by various subjects in order to solve such nasty design problems.

Complex design problems cannot be solved only by the professionals acquiring the knowledge of a specific domain. There is a great need for the development of collaborative designs by various subjects overlaying their respective disciplinary knowledge, like user participation in design, emergent collaboration of different experts, and collaboration using the internet. Of especial significance for improving the design quality is the feedback response to design from the user's experience and environment; this is likely to be ignored in the industrialized society. When various subjects like user, designer, engineer, craftsman, artist, manager, and administrator collaborate in order to overcome dangerous situations, the struggle of competition and cooperation appears in the context, and a creative design is generated.

Human intelligence and creativity depend strongly on the collective memory of actual communities and artifacts used in those communities. A creative individual is usually thought of as working separately, but we cannot ignore the role of interaction and cooperation with others. Creative activities are generated from the interaction of individuals with their environment and others. The basis of those social interactions is for people to think, to work, and to learn with others, using tools and artifacts made in cultures.

Today various design tools and design environments, supporting the generation of creative designs based on these social interactions, have been developed. Especially it is obvious that network collaboration -- which is the method to make or allow separated subjects to collaborate beyond time and space -- is essential. It will contribute to the creative collaboration of subjects who belong to different fields, cultures, and organizations, etc.

4.5   Proposition 5: design tool

It is necessary to develop and to utilize high level systems supporting the design process which deals with complex conditions including implicit requirements in order to realize the design vision in the 21st century.

The change of design essential to our contemporary society can be explained as "from quantitative change to qualitative change", "from making to bringing up", "from design of elements to design of relations", and "from design by individual to design by group". In order to realize this change it is not enough only to construct the formal and spatial models of artifacts using CAD & CG systems. In fact it is very important to accumulate and make use of the enormous knowledge behind the physical model, and to support social interaction among heterogeneous subjects. Concretely it is necessary for us to develop and to use positively various design systems supporting functions like simulation, knowledge management, and collaboration.

For example, we can execute design simulations, including economic assessment, using virtual design environments by means of adding simulating functions to three dimensional models of artifacts, and overlaying management information on such models, and as a result we are able to develop the dynamic process of team design. Moreover there are many alternatives of design support systems such as drawing, model, internet, and workshop on the spot.

Public organizations like national and local governments should provide relevant economic and technological supports as occasions demand in order to develop and to utilize design support systems in the context of public design such as townscape design and universal design.

4.6   Proposition 6: design education and ethics

It is the user who makes the final evaluation of the quality of design, and we should consider not only the side of the designer and producer but also the side of the user in future designs. Accordingly we should promote the spread of design education and design ethics, and to supply pertinent design information actively.

We must consider not only the quality of an individual artifact but also the quality of the environmental and social system, consisting of the assembly of artifacts and natural objects in the post-industrialized society, because our lives are nurtured in the expanse of the environmental and social system. Those qualities of design are generated in the macro design process, which includes design, production and living process; the experience and aesthetic sense of the user and the mature culture of society play an important role in such macro processes. Accordingly, in order to enhance the design sense and living culture of users, it is necessary to promote the following strategies positively: promotion of design education from childhood, institutional guarantee of user participation, and supply of relevant information for users to evaluate design products.

Moreover the sense of value and world view of people must be taken into account in various contexts in future designs, and we should make an effort to spread design ethics including engineering ethics, environmental ethics, and bioethics.

4.7   Proposition 7: science of synthesis

The science of design engineering which investigates the essence of designing is the frontier of academic research to explore the method of synthesis which is required in the science of the 21st century. It should be promoted so that it can provide its research system positively.

The science of the 20th century made great progress based on the method of analysis to decompose a whole into parts, but relations among various parts are divided and a whole is often lost sight of. In contrast constructing the method of synthesis has emerged as one of the important academic researches in the science of the 21st century.

We cannot determine a unique solution of the design problem for artifact design and production. The reasons are as follows: the designing has to do not only with the analysis of design but also with the synthesis of design; moreover the design conditions change according to the culture in which the artifacts exist, and there are various methods to integrate parts into a whole. Among these is the method to grasp the interrelationships among parts and to generate an abundant whole in the design of 21st century intended for environmental and social systems.

Design engineering exploring the above-mentioned designing is in fact very exciting research, as it approaches the essential nature of human beings, including the resolution of the creative process of "abduction", and the active research which plays an important role to propose a vision of better future society such as a sustainable recycled society from a long term viewpoint.

The promotion of design engineering dealing with a method of synthesis encounters various difficulties, because it is concerned closely with the problem of value. Therefore it is positioned as a frontier science to construct the method of synthesis which is a common subject to the science of the 21st

century, and it is necessary to arrange the social research system for design engineering positively.

5.   CONCLUDING REMARKS

As explained above, the Design Engineering Section of the National Committee for Artifact Design and Production presented "The Proposal of Design Vision for Artifact Design and Production in the 21st Century" as the final report of our research activity for three years to the Japanese Government.

The proposal of design vision for a post-industrialized society is summarized as follows.

(1) We should strive to transform the qualitative concept of design for the post-industrialized society. It becomes very important not only how we make things but also what things we make in the new context.

(2) Good artifacts are to be generated in a continuous process in which designing, production, and living are closely connected. It is necessary to expand the design process of artifacts in order to include not only the making process but also the living or breeding process.

(3) Design in the 21st century should play an important role in enhancing the quality of life not only by making individual artifacts but also by improving the environmental and social system including the set of artifacts and natural objects. In this context, the object of design must be expanded from the hardware of artifacts to the software of services indispensable to the environmental and social system.

(4) The design problem is very complex, ambiguous, and unstable today. Therefore it is necessary to promote collaborative designs by various subjects in order to solve such *nasty* design problems.

(5) It is necessary to develop and to utilize high level systems supporting the design process which deals with complex conditions including implicit requirements in order to realize the design vision in the 21st century.

(6) It is the user who makes the final evaluation of the quality of design, and we should consider not only the side of the designer and producer but also the side of the user in future designs. Accordingly we should promote the spread of design education and design ethics, and to supply pertinent design information actively.

(7) Moreover it should be necessary to support academic research to explore the essence of designing, and to present the possibility to realize a better future society.

This proposal includes urgent propositions, less urgent propositions, and propositions necessary from a long term perspective. We should make a road map to realize the design vision for the artifact design and production in the 21st century. This is the important mission for the next term activity of our committee.

REFERENCES

1. The Design Engineering Section of the National Committee for Artifact Design and Production, The Science Council of Japan: "The Proposal of Design Vision for the Artifact Design and Production in the 21st Century", The Science Council of Japan, July 2003.
2. The Design Engineering Section of the National Committee for Artifact Design and Production, The Science Council of Japan: Design Engineering Symposium "The Qualitative Change of Design", The Japan Society of Mechanical Engineers, May 1999.
3. The Design Engineering Section of the National Committee for Artifact Design and Production, The Science Council of Japan: Design Engineering Symposium "The Design Vision of the 21st Century", The Japan Society of Mechanical Engineers, May 2002.

RESEARCH OF VIRTUAL PROTOTYPING FOR AUTO-BODY SHEET METAL ASSEMBLY

Wenfeng Zhu, Xinmin Lai and Zhongqin Lin
Shanghai Jiaotong University, Shanghai, P.R.China, 200030

Abstract: Virtual Prototyping (VP) technology has been studied in mechanical dynamic simulation. Aim to fit competitive market demand on auto-body R&D, this paper firstly introduces this engineering innovation to complex auto-body sheet metal assembly simulation. A VP modeling for it is presented which contains five sub-models. Firstly, it includes assembly hierarchy model, representing how those sub-assemblies are put together from different assembly level. The second is assembly behavior model that represents exclusive properties of sheet metal assembly. The third is assembly variation model describing the assembly variation. The fourth is assembly environment model, stating the real body-shop situation and finally, a data management system supporting the design data integration and collaboration. In our primary research practice, the virtual prototyping can support auto-body design and manufacturing in a cost-effective and time-reduced way.

Key words: virtual prototype modeling, sheet metal, auto-body assembly.

1.   INTRODUCTION

Concurrent Engineering (CE) is an advanced engineering philosophy that aims to build an integrated information environment on support of advanced CAX/DFX/PDM technologies so that design quality is improved and time to market is shortened. Furthermore, with integration stepping on, many manufacturing-related issues are urgent to be dealt with as early as possible that demand on system-level simulation. The ultimate development of this in terms of speed, cost and flexibility is Virtual Prototyping (VP). It includes design optimization, performance analysis and manufacturing and usability testing.[1][2] Boeing 777 's R&D is a successful case of VP whose life-cycle design, assembly and test-fly are all completed in computers. The other is Ford Vehicle Attitude Control system (VAC). By way of VP, it makes mechanical sub-system and electronic sub-system collaborate more efficiently and control optimization can be well done in even a week [3].

Within last years, automobile development has been revolutionized. Competition within the international market forces lead time to be reduced as well as contributing to a large increase of individual niche products. However, body assembly is one of the core manufacturing operation of all automotive companies. It is a complex operation consisting of welding 300~400 sheet metal parts together. It is the basis from which other parts such as the closures, engine, interior and exterior components are attached. Auto-body development ranges from initial clay modeling, concept design, detailed assembly to manufacturing planning with 4~5 physical prototype built. It is very time-costly and capital consuming, as showed in Figure-1. Obviously, this situation could not fit the urgent market need.

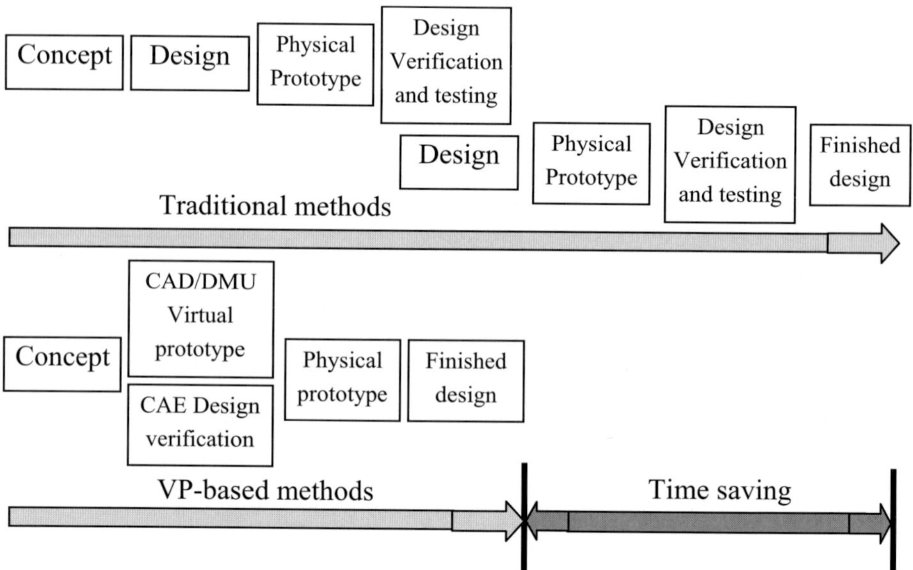

Figure-1 virtual prototyping vs traditional way

The virtual prototyping technology has been adopted for design optimization, performance analysis, assembly and usability simulation in the early development phase, thus removing the need for some or even all physical prototype[4]. It can reduce the time and cost of product development and realize auto-bogy innovative design.

In the next section, we will explain what is virtual prototyping (VP). Referring to sheet metal assembly, a VP modeling for auto-body sheet metal assembly is presented. Section 3 describes five sub-models in detail and an assembly example is introduced in Section 4. Finally, Section 5 gives conclusion and our long-term research goal in development a VP for auto-body sheet metal assembly.

2.   WHAT IS VP FOR AUTO-BODY SHEET METAL ASSEMBLY

VP technology depends on C3P（CAD/CAE/CAM/PDM）.Firstly, with help of CAD system, the parametric and solid modeling technologies have made the product geometric modification and optimization easier. Secondly, many CAE technologies have brought forward the manufacturing simulation and performance testing even before the physical prototype is built up. Thirdly, PDM system develops much more advanced and could build digital information environment and provide Product Structure Editor and release process management.

VP is based on virtual prototype and its environment [5][6]. The former includes two meanings. Firstly, it refers to virtual prototype which is a digital model built in computer and can be used to take appearance evaluation, structural, dynamic and kinematics test in place of its physical prototype. Secondly, it refers to virtual prototyping that is the process of testing and evaluation a system-level design. Both these two are generalized to be defined as VP. The virtual prototype environment focuses on the physics-based graphical simulation environment, including PDM and human interactive interface. It is reported by overseas research that VP technology can reduce time and cost by 25%.

Traditionally, auto-body sheet metal assembly design begins after the whole auto-body is divided into several parts and the boundary of auto panel has been decided [7]. It determines the position, joint type, assembly sequence and variation of the sub-assemblies and needs full scale master model measured from the initial clay model. The auto-body is called Body In White(BIW) before it is painted Since BIW 's dimensional quality and variability strongly affects the other assembly operations such as installation of closure panels , trim components, engines etc, Several BIWs have to be assembled to verify design. When design defection is found, it has no way but traces back to the beginning and leads to great hard work-over. It takes several troublesome iterations until the assembly quality is met. Obviously, based on physical prototype, this is a time and cost-consuming way and restrains the improvement of auto-body redesign. Aim to fit competitive and changeable market demand on niche automotive product, VP technology is introduced to auto-body sheet metal assembly design and a VP modeling for it is presented.

3. MODELING OF VP FOR AUTO-BODY SHEET METAL ASSEMBLY

Auto-body is composed by hundreds of compliant sheet metal parts. As to this complex and multi-level assembly work, traditional methods deal with it from single perspective and modeling from single field. However, the big opportunity to increase assembly design quality and reduce time and cost has depended on system level perspective. As a system-level modeling methodology, VP needs to build synthetic model with several sub-models that contributes lots to assembly quality. The following is the framework of VP–based modeling for sheet metal assembly presented in this paper.

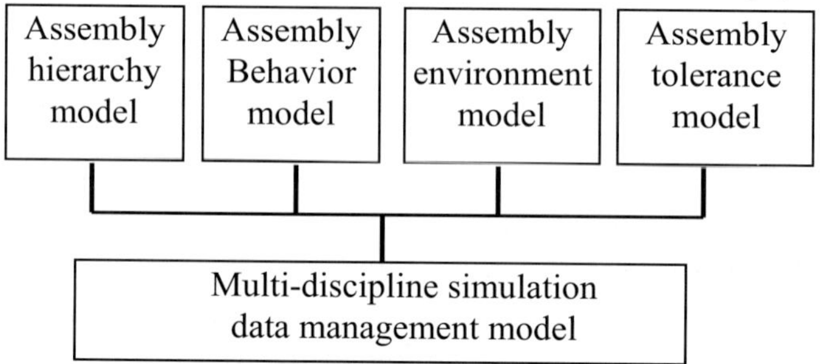

Figure-2 Model framework of VP for auto-bodysheet metal assembly

3.1   Assembly hierarchy model

As showed in Figure-3, auto-body assembly tree has an obvious hierarchical structure. The whole auto-body can be divided into several sub-assemblies and components. In body assembly design and true body shop assembly local, it takes a bottom-up way that several components are assembled to form sub-assemblies and all six sub-assemblies are then welded together to form BIW.

Assembly hierarchy model is built on the base of parametric solid CAD model of assembly sheet metal parts and their assembly constrains. Besides the dimensional and geometric constrains, it additionally defines two aspects of

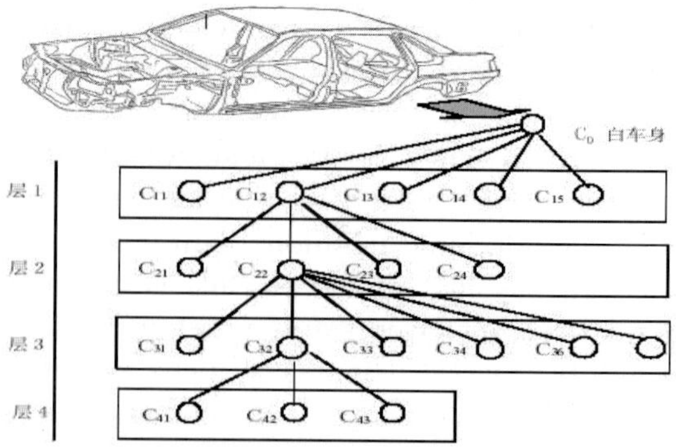

Figure-3 Assembly hierarchy of auto-body

information. One states which group and which level are some certain sheet metal assigned so that the location of all assembly sheet metals can be located in the assembly tree. The other states the assembly relationship of sheet metal locating in same group or assembly level that includes several constrains such as the relevant position, dimension, assembly joint type and assembly. These two aspects can be defined during components geometric modeling phase by re-exploring general solid modeling CAD system

3.2   Assembly behavior model

The compliant sheet metal parts used on auto-body usually has thickness of 0.6mm~1.2mm that is flexible and deformable. It has some exclusive properties that make its assembly behavior different from that of rigid body. Showed in Figure-4 is the assembly of Part 1 and Part 2. Because of sheet metal spring-back, there exists assembly clearance and it makes actual assembly position not coincide with design position. With the force of welding gun and fixture, the actual assembly point has obviously deviation from theoretical assembly point.

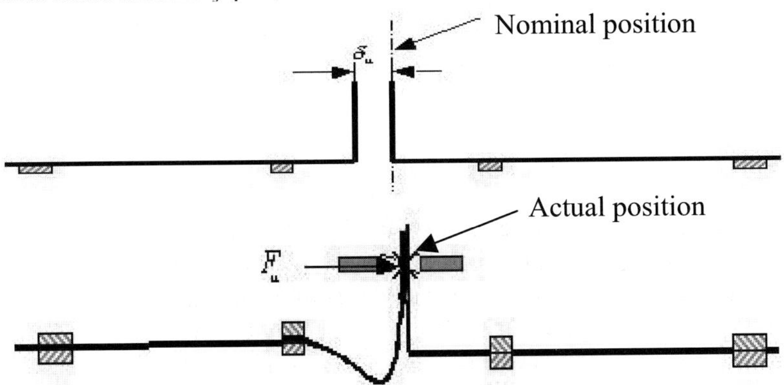

Figure-4 Sheet metal assembly behavior

Liu and Hu [8] of University of Michigan study the sheet metal assembly behavior under the assumption of little deformation and ignorance of welding thermal impact. They integrate the statistical analysis method and component dynamic model and present the Mechanical Variation model. This model sorts the joint of sheet metal assembly into three types: lap joint, butt joint, mix joint and calculates the assembly deviation on different joint type and assembly sequence. Our assembly behavior model is built on the base of their models to prescribe the effect of different joint type on assembly variation and support the algorithm used in the later assembly variation analysis

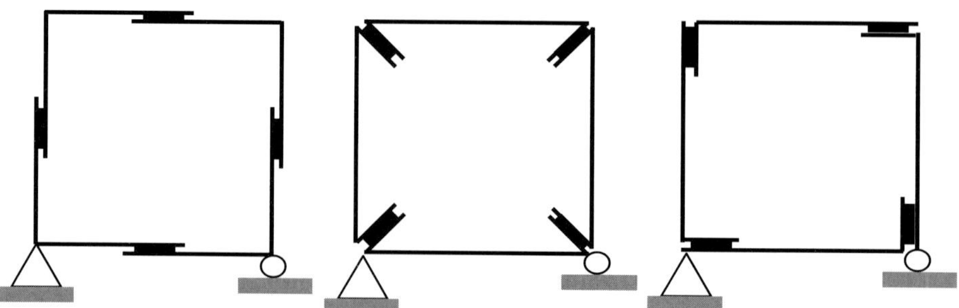

Figure-5 Three joint types(lap, butt, mix-from left to right)

## 3.3	Assembly environment model

In body shop local, sheet metals are assembled on several workstations. Since VP technology is a system-level simulation, the assembly environment modeling and simulation is one important part. It is the simulation of physical assembly environment and have two levels of modeling. One is micro-level which simulates the assembly workstation, the other is macro-level which simulates the whole assembly line. As to our research activity, only the formal is researched in this paper. It contains the modeling of assembly workstation composed by assembly robot, assembly dummy, assembly fixture and some other devices in assembly environment.

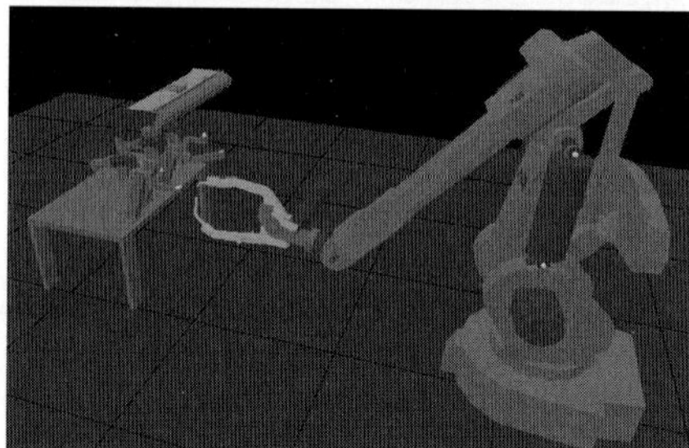

Figure-6 Physical-based graphical environment model

## 3.4	Assembly variation model

On traditional rigid body assembly variation analysis, there are some effective methods such as Worst Case method, RSS (Root Sum Squares) method, Monte-Carlo method.

The Worst Case method assumes that all the parts are assemble on its maximum or minimum dimension. As to 1D case, the assembly variation can be calculated as follows:

$$TOL_{asm} = TOL_1 + TOL_2 + \cdots + TOL_n$$

n ----numbers of assembly parts, TOL_i ----each part variation,

TOL_{asm} ----total assembly variation。

RSS (Root Sum Squares) method calculates the total assembly variation distribution parameters based on assemble parts' variation probability density function by use of a second order moment. As to 1D or 2D rigid body case, the assembly variation can be calculated as follows:

$$TOL_{asm} = \sqrt{TOL_1^2 + TOL_2^2 + \cdots + TOL_n^2}$$

n ----numbers assembly of parts, TOL_i ----each part variation,
TOL_{asm} ----total assembly variation。

Monte-Carlo method firstly builds up a probability model on parts' parameters. By use of random probability distribution function, it gets the statistical parameters of assembly variables after optimization; the assembly variation's approximation can be got. So the Monte-Carlo method is also named as stochastic simulation method.

These three methods have common ground that they all research on rigid body assembly whose variation is determined by parts' geometric and kinematics parameter. However, Takezawa[9] points out in 1980 that the theory of assembly variation accumulation based on rigid body could be applied on the auto-body sheet metal assembly. Lots of impacts during assembly process such as stamping spring-back, welding-gun and fixture force of welding can all bring into variation and they can be annihilated or enlarged at succeeding assembly process.

Figure-7 is an auto-body assembly style in 1D offset beam model which is used to deal with flexible auto-body sheet metal assembly. It is firstly presented by Liu, Lee and Hu [8] after they generalize and abstract the whole assembly process. It has taken account of the flexibility behavior of sheet metal and is used in the VP modeling of this paper.

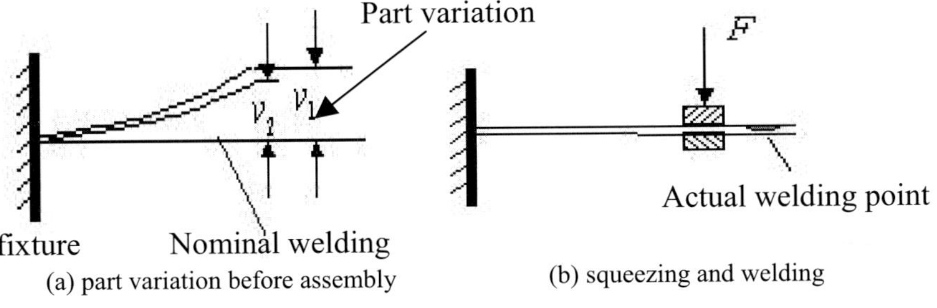

Figure-7 1D offset beam assembly model

3.5   Multi-discipline simulation data management model

Key technology of VP is the interactive and simulation of multi-discipline so as to realize the collaboration of various simulation data. All information needs an effective management. The four levels Multi-discipline simulation data management model for VP is presented as follows.

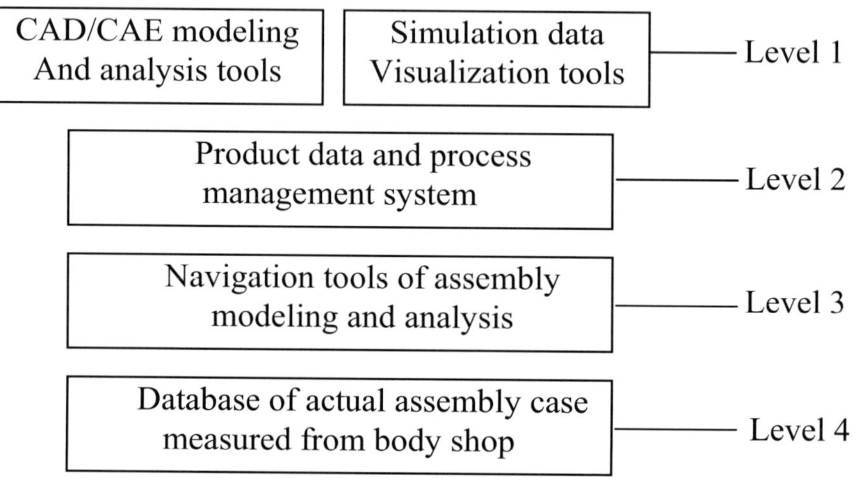

Figure-8 4-level simulation data management model

Level one contains some CAD/CAE modeling and analysis tools such as UG or CATIA for solid modeling, ANSYS for thermal-structural couple analysis. Also it has Visualization tools used as human interactive interface of simulation data. Level two realizes the simulation data and process management. It takes the advantage of PDM's tool integration and application encapsulation mechanism. Level three provides simulation application service. Since the CAD geometric model cannot be translated directly into physical model to be used for CAE analysis. It needs some modification experience that is not mastered by all engineers. So, a navigator tool that is composed by lots of successful CAE cases in auto-body sheet metal analysis is offered to engineers as wizard. Although the post-processing of simulation can take the use of contour plot and vector plot and visualization, it could not evaluate the simulation results without the comparison with actual data. Level four restores some actual assembly data case measured from body shop. It helps to make VV&A of simulation and improve simulation reliability

4.   A VP EXAMPLE OF FRONT FLOOR ASSEMBLY

As showed in Figure-9 is the exploded view of auto-body front floor. It is assembled by ten sheet metal parts (their names are in Reference [7]) and it is assembled with rear floor sub-assembly to form floor sub-assembly and finally to form Body In White with the assembly of left/right side-frame sub-assembly.

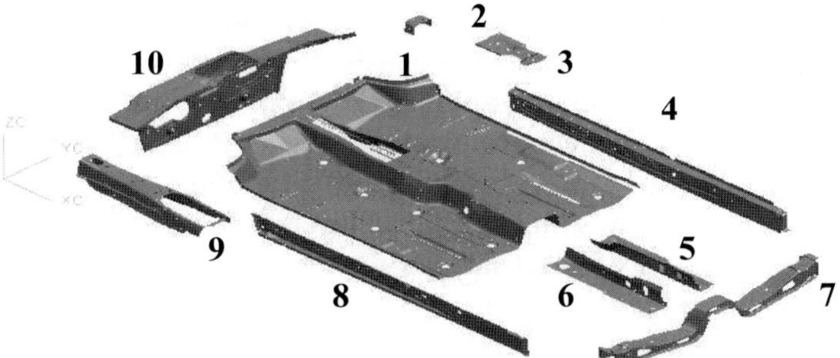

Figure-9 Exploded view of front floor assembly

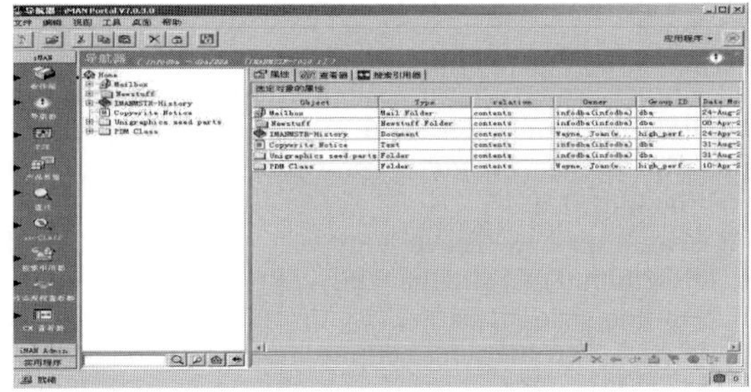

Figure-10 IMAN/portal master interface

As stated above, VP is built up on the base of many digital modeling and analysis tools. In order to build the VP for front floor assembly, UG is used for parts' CAD solid modeling and assembly hierarchy modeling. The variation analysis and assembly sequence is simulated by VIS-VSA. By re-exploring its modules of flexible body with the 1D beam model presented above, it can calculate the assembly variation of different assembly sequence and joint type. The assembly environment model is built up with DENEB/ENVISION because it provides lots of assembly units model including fixture, robot and could be used after retrieving them and giving some relevant definition. According to the Multi-discipline simulation data management, we use IMAN with its advantages of seamless integration with UG and interface to many CAE tools so that all simulation data and process can be organized effectively. Figrure-10 is IMAN/portal interface.

5. CONCLUSION

A current bottleneck in globally competitive auto-body design is that its design, creation, testing and modification need to build physical prototype. as an system-level design approach, VP allows for breakthrough in speed, cost and quality for new product design.

Sheet metal assembly design based on physical Body In White (BIW) is costly and time-consuming. In order to solve this engineering problem, a VP-based modeling is presented. It consists of several key sub-models which great affect the final product assembly variation.

On our research practice, we consider that VP-based modeling provides an integrated platform to perform assembly variation design and simulation in a cost-effective way. Although its research for sheet metal assembly is in initial stage because of sheet metal's distinctive property. It is a promising research field because of the time and cost restriction on product development from market. In our long research goal, An auto-body VP will be developed for system-level design and simulation.

REFERENCES

1. Kari Kutti, et al, Virtual Prototyping in usability testing, *Proceeding of the 34th Annual Hawaii International Conference*, 2001,3-6
2. Robert R,Ryan. *Functional Virtual Prototyping :Realization of the digital car.* Mechanical Dynamic Inc.2001
3. Chen Xiaobo, Research on Key Technology of Collaborative Simulation for Complex Product Design, Dissertation of Tsinghua University, 2003.
4. Gary Wang, Definition and Review of Virtual Prototyping, JCISE-2001-76.PDF http://www.umanitoba.ca/faculties/engineering/mech_and_ind/prof/wang/
5. Wen Sizhu, Li Min, *Research of mechanical engineering development strategy*(in Chinese), press of Tsinghua university, 2003 first edition.
6. Antonino, G. S. and Zachmann, G., "Integrating Virtual Reality for Virtual Prototyping", *Proc. Of ASME Design DETC98/CIE-5536*, Atlanta, Georgia, September 13-16. 1998
7. Zhang Yizhu, "Auto-body structure concept design and decision making based on graph theory and homogeneous transform"(in Chinese), Dissertation of SJTU, 2003.
8. S. Charles Liu, Hsin-Wel Lee and S. Jack Hu, Variation simulation for deformable sheet metal assemblies using mechanistic models, *Trans of NAMRI/SAE*, 1995.
9. Takezawa, N. An improved method for establishing the Process Wise Quality Standard, Reports of Statistical and Applied Research, Union of Japanese Scientists and Engineers(JUSE), 1980, 27(3)
10. Ma Siqun, et al, research on key technology of data collaboration supporting railway vehicle's VP (in Chinese). CIMS, Vol9, No.3.

A RETRIEVAL AGENT ARCHITECTURE FOR RAPID PRODUCT DEVELOPMENT

Stavros Dalakakis, Emil A. Stoyanov and Dieter Roller
University of Stuttgart, Institute of Computer-aided Product Development Systems, Universitätsstraße 38, D-70569 Stuttgart, Dieter.Roller@informatik.uni-stuttgart.de

Abstract: Application of agent technology in Rapid Product Development (RPD) seems to be a promising way for solving problems related to separation of services, finding an intelligent solution and synchronization of design teams. Agents are applied as tools in the context of RPD providing not only solutions for the problem but also controlling of the entire RPD Process. This paper presents a framework for building a multi-agent system operating in multicast environment. Specifically, an information retrieval system architecture within the RPD-Process. This architecture considers aspects of distributed design teams as well as modifications in the model by means of metamodel. Basic components for the design of a retrieval system are extensively described. For that, two query concepts, descriptive and navigational are examined and integrated.

Key words: Retrieval Agent, Information retrieval, Multi agent system, Multi agent system architecture.

1.   INTRODUCTION

The Rapid Product Development (RPD) as a technique for design of innovative products, is characterized by development teams distributed in time and space, with different responsibilities, background in various domains, by fast reaction to market changes and by the coordination of necessary actions for every development phase between these groups.

Currently most product development processes can be classified into two distinct groups. The common ad hoc solution where the development process is not explicitly controlled but the information flow is. For example a company, which uses CAD systems for design and process planning, may use PDM to capture and control information but the process control itself exists only on paper or in the minds of the product engineers. The second more holistic industrial design view is to integrate all the influential parameters into the product development environment. For example project management, decision management and other utilities are considered as important factors for the design of prototypes and need to be integrated with physical models and manufacturing systems. Such a system facilitates the design of the product and the control of the process, allows a clear determination of the state of the process and enables communication, cooperation and coordination. So, the second approach is supported by this project; an attempt for an integrative, holistic approach of the entire domain.

Product development is influenced from the side of computer science especially from distributed systems, knowledge representation and agent technology. Application of agent technology in RPD seems to be a promising way for solving problems related to separation of services, finding an intelligent solution, synchronization of design teams and controlling processes.

A major challenge is the selection of an adequate representation form of RPD knowledge. This is extremely difficult because of the thematic variation of the tasks that have to be done by a RPD team. Simulation, project management and design have to take and deliver knowledge to the knowledge base. This knowledge has to be exchanged by other team members in the RPD process. That is a "crossroad" for members that accumulate knowledge and for others that use this knowledge. For this knowledge storage the appropriate retrieval mechanisms have to be ensured. Storage without appropriate retrieval is becoming worthless. Therefore, additional mechanisms for supporting efficient retrieval of knowledge are required. Whereby humans are involved for main decisions as indispensable components of the system rather than just the applications that deliver or enter knowledge. For that reason the demand on means of communication, cooperation and coordination are much higher then in low intelligent product development process [1]. The chosen form for accumulating knowledge in this research project is a semantic network, a description of which is found in [2] and which will be discussed in a future paper.

In section two we present a brief review of related works. An architecture of retrieval agent is introduced in section three. In section four we discuss a specification of a proposed query language and finally a realization of the entire work is presented.

## 2.	RELATED WORK

Research approaches in the agent technology include different aspects and stretch across theoretical and logical examinations to practical implementing applications. In these are agent architectures and models, interactions protocols and language part of an agent system. The main scientific areas implicated in the design of a multi agent system are: network technologies, middleware, knowledge representation and artificial intelligence. All of them are based on the field of computer science.

The best examined model on agent architecture is the Belief-Desire-Intention Model (BDI) [3]. The BDI approach is based on practical reasoning. Another model, the behaviour model, simply reacts to the stimulus response approach. Through the combination of both approaches are the hybrid model and the layered model where the last one, combines the best alternatives of both.

In [4] is designed a metamodel for overcoming the heterogeneity of languages and architectures. To facilitate agents interoperability, the definition of terms and concepts and therefore the area of ontology, wins in importance [5].

Communication is the major characteristic for a multi agent system and by definition as "at least" criterion that has to be fulfilled for such a system [6]. That assumes a common communication language, a so-called Agent Communication Language (ACL). The Foundation of Intelligent Physical Agents (FIPA) makes available a big range of specifications around the agent technology [7] and has standardized a "message structure" [8] for ACL. Also the FIPA-ACL semantic is through the FIPA-"Communicative Act Library" (CAL) specified [9].

In the area of agent communication protocols is established the secure protocol „Light-weight Reliable Multicast Protocol" (LRMP) [10]. LRMP is used for implementing „Java Message Service", a core component for a number of commercial messaging systems [11-14] as well as for the realization of services through peer-to-peer specification JXTA [15].

The Extensible Markup Language (XML) [16] is privileged for heterogeneous environment like a multi agent system. This is a standard for the data exchange, where only the interpretation of data is defined, without to be necessary the definition of formats and interfaces. So, is offered a technological solution for the data exchange between incompatible systems. XML is most reasonable applied, if it is used for data manipulation, data transmission as well as management of semi-structured data and data integration. [17] set the question, if the relational model the unique correct paradigm illustrates. Thus, is the movement in the database theory a normal activity. The semi-structured model is the successor model in this area.

Additionally to the database management system are offered operations about data transformation, query realization, data transport and stream-based processing [18].

The XQuery approach that supplies an extended infrastructure for queries is recommended by [19]. Similar projects, for the realisation of an XQuery processor according to the specifications of W3-Consortium, are [20] named XML-DB, by [21] as Galax and by [22] as XMediator.

Interesting multi agent system approaches are already incorporated in the following projects: ABLE [23] especially in the utilization of learning strategies, in the project JADE [24] for the run-time services, in the project RETSINA [25] where the communication is realized through peer-to-peer and in the project FIPA-OS [26] with an interesting configuration approach.

3.   A RETRIEVAL AGENT ARCHITECTURE

3.1   In general

RPD projects are characterized by high complexity, variety of interrelated application domains and distributed working teams. To face these problems multi agent systems have been developed. The goal of multi agent systems is to elaborate solutions for problems by cooperation in a group of agents, by parallel work of tasks through different agents and by realization of autonomous behavior within each agent. For the definition of such groups of agent populations the term agent framework (AF) has been established. An AF enables the development of common infrastructure for agents that communicate, coordinate and collaborate inside the system boundaries with each other as well as organizing the reproduction of agents in it. In addition, other components inside each agent are exclusively responsible for "the problem to be solved" task. So, in opposite to the objects, which are encapsulating only methods, agents are privileged for encapsulating e.g. targets, knowledge or other RPD related requirements and so they are forming an abstraction for more complicated structures.

The most influential parameters for the development of a retrieval agent are identified as: the knowledge representation on which the retrieval agent will apply, the communication and coordination infrastructure inside the multi agent system and the abstraction level of the communication between application or user towards to the retrieval system expressed by the query language. All these aspects are in the following pages discussed.

3.2   The metamodel for the representation

The predefined underlying representation is decisive for the success of an information retrieval. The already chosen representation form of the RPD process for this research project is an Active Semantic Network (ASN) [2]. The representation consists of three layers. A high level RPD independent description of the system, the metamodel. The flexibility of the metamodel allows e.g. to define new relation types in an abstract manner. The second layer called Model is derived from the metamodel by the RPD team and describes the specific RPD process, for e.g. the objects, their relationships and attributes. The third layer is the ASN, which is the realization of the Model with RPD knowledge. The persistent storage of the entire knowledge of the RPD including everything around developing a product e.g. processes, construction draws, software programs, are stored in the ASN. A middleware is required to support the RPD user of application for using this knowledge. The question *how* to use this knowledge is the same as to define the functionality of the middleware. The AF is part of this middleware. That means the AF is acting orthogonal to the above representation and is not part of it. It acts at both the ASN level and at the Model level.

3.3   The model for the agent framework

The following elementary problems are considered as primary in the conception of the agent architecture: the reliability of communication into the multi agent system, redundancy of the entire system, the organization of communication among objects without knowing about the existence of each other, the perception and behaviour of the agent as well as monitoring of the internal life of the agent.

Diederich [27] has taken into account aspects of coordination and communication in an agent framework. Based on this work an architecture for totally autonomous system of software agents communicating in multicast environment is developed. The system should be redundant by means of reliable communication and constant presence of agents servicing various requirements of the RPD user, no matter of geographical location or operating system. Main components of the multi agent system are the multicast receiver and sender which form a perception module, the coordination scheme between the agents population, the agents reproduction and configuration module, instinctive and decision modules – message-event-action mapping rules formulation mechanism which forms agent's behaviour. This architecture is scalable, multicast aware with an open communication protocol.

Every agent has build-in "master agent" functionality. The master agent is the active element in the communication process, which monitors the network and keeps track of agent's status, instantiates dynamically new agents by demand. In case of absence of master agent the existing agents in the system elect one from themselves, which is to be the new master agent. The selection of included components is dependent on the specific implemented functionality and assigned problem to every agent.

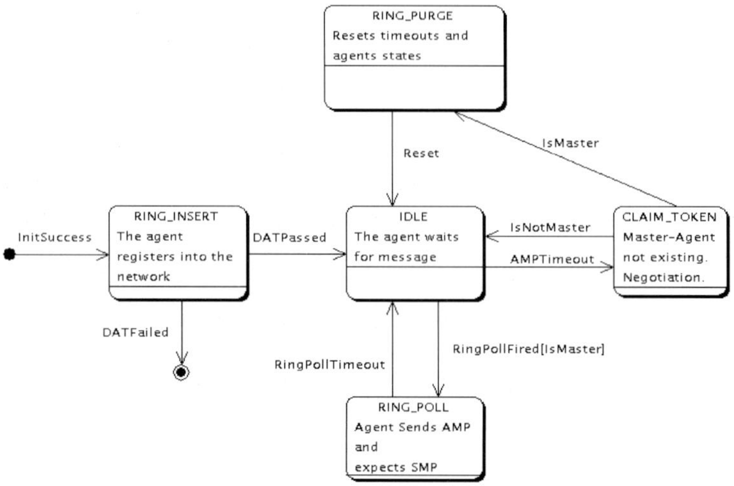

Figure 1. Agent's ring management process state diagram

The perception module is a level in the communication mechanism of an agent, which is responsible for the correct transmitting of data over reliable multicast protocol. The module consists of multicast receiver and sender components which operate with XML based messages used as transport envelope for the exchanged between agents data. As reliable multicast protocol is used LRMP [10]. The multicast technology has the advantage, in opposite to broadcast technology, to provide information exclusively to such instances that they need it. Additionally benefits are, the support of many senders in heterogeneous networks and of course reliability in the communication of distributed teams.

The described multi-agent system uses a coordination scheme to keep itself redundant and the networked components to be aware of system's current state. As a reference design is used the management scheme of the network protocol IEEE 802.5 known as TokenRing also as "Ring Management" [28]. It defines similar behaviour and is proven as efficient and reliable mechanism for keeping a network in consistent state. Directly implemented and adapted Ring Management procedures are RING_POLL,

RING_PURGE, CLAIM_TOKEN, RING_INSERT. In Figure 1 is the state diagram of the agent's ring management process. All of them are mainly based on broadcast signals, transformed into the agent implementation to single XML messages over reliable multicast protocol (LRMP).

One of the main characteristics of the software agent concept is that the agent has its own behaviour adapted to the desired goal or assigned problem. Considering this, the architecture of the described agent system was designed with the idea to make this aspect as extendible and flexible as possible. The agent reaction to what happens in the world, e.g. the multicast network, and agent's internal state is sequenced to several phases: message recognition/interception, event generation, event handling, action. These four steps are handled by two groups of components: a request processors register and an event handlers register, the first responsible for event generation and the second for event handling and decision for execution of appropriate action. Registration of multiple request processors is possible and encouraged as a way to form an instinctive behaviour – how the agent associates the visible by it world with its capabilities. Multiple event handler registration is possible as well to extend the choice of action to be executed considered with the fired event related either to internal for the agent state or to its instinctive behaviour. In figure 2 is depicted the agent's event handling.

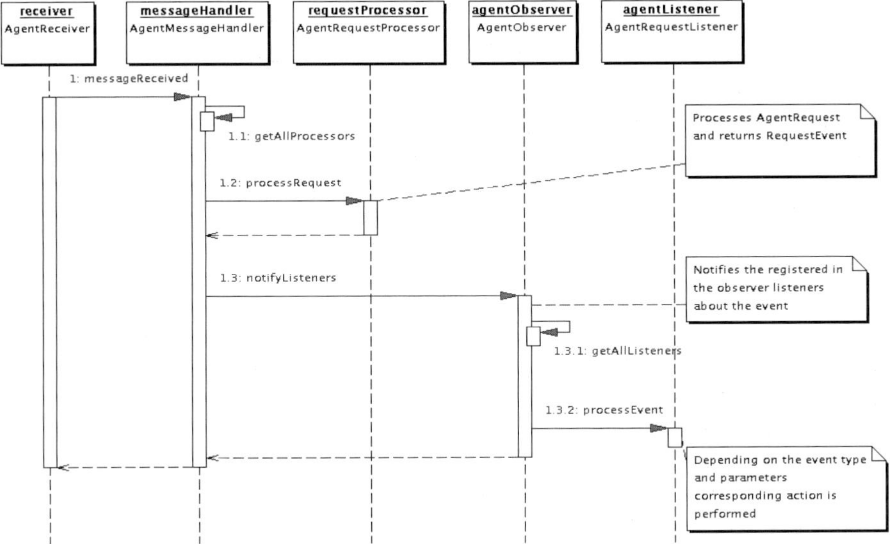

Figure 2. Event handling sequence diagram

As the described system requires redundancy and on-demand creation and enabling of agents, a common mechanism of agent instantiation is designed, both on the user's side and on the side of the agents system. A built-in agent factory provides to every agent or client application the possibility to dynamically produce a working agent from a predefined configuration set of options, including network parameters, agent types and responsible programming units. Additionally, the implemented agent architecture allows standalone components to be registered to a global data bus and monitor the internal state of the agent, thus giving the programmer easy and safe way to extend a properly working agent or client application with new functionality.

3.4   Information Retrieval

Typically retrieval ways for stored knowledge are three: browsing the knowledge, using a procedural interface and using a query language. Browsing is the target independent search in data spaces. Where the current position and the scope of the information space in not known. Procedural interface is an inflexible possibility to invoke standardized queries by configuration of their parameters. In opposed to procedural interface the specification of a query language offers a high abstractions way for formulating queries through functions and its extensions.

Main goal in this work is the consideration of semantics, represented as relations between concepts, into the retrieval process. This implicated knowledge into the relations enforces the information retrieval process with additionally knowledge. So, are yielded more precisely results. Two principles seem to be promising, the navigational and the descriptive concept.

3.5   The navigational concept

Navigation is a kind of movement in information space, where every decision about the direction is dependent from the target. In opposite, in exploration the target is to find the structure, the scope and the content of the information space. Navigation assumes interaction with the user and makes demands on the user's ability to decide. So, like in figure 4 a query consists of many part queries. After the first query is the information space provided from which the next query is set until the user reaches the result by navigation.

3.6   The descriptive concept

The user of a descriptive concept describes only the criterions of the requested knowledge. In opposite to that in the common procedural approach the stress of the question is set to *how* to find the persistent knowledge. The importance of the descriptive concept is in the formulation of the query so that is given ones the format of the results and on the other hand the query expression without to take in account the details of the knowledge storage. These ideas are incorporated in the specification of the query language in the next section.

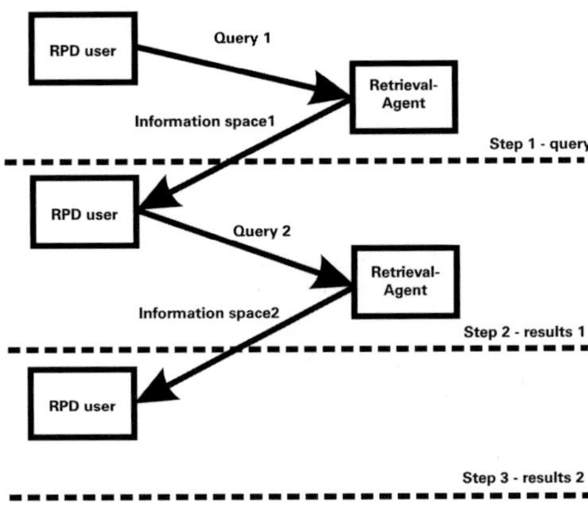

Figure 3. Query by navigation

4.   RETRIEVAL AGENT

4.1   The query language ASN-QL

The ASN Query Language (ASN-QL) has been specified considering the knowledge representation in ASN and the RPD environment. The fundamental idea in the design of this language is to link tightly the semantics within the relations into the entire retrieval process. In particular the recursive query formulation, unsharp query and the predefined structure of the query results by the user are considered. A near natural formulation of the query, as well as experience from descriptive and navigational concepts are incorporated in the design of the query language. A specification of ASN-QL in EBNF form is presented in the appendix. Figure 4 shows all the abstraction levels on which ASN-QL is built.

XQuery is an extensive query language, decoupled from the information's representation. Abstracting XQuery gives the possibility on one hand to use the existing expressions and on the other hand to extend the language. This project abstracts XQuery's ForClause LetClause WhereClause OrderByClause Return expression (FLWOR) and exploits the opportunity to define new functions. FLWOR allows to define and to iterate in a data space where the entire RPD knowledge is stored. This is in the ASN base where the FLWOR applies. Before that, the knowledge base has to be transformed into an XQuery compatible form. The FLWOR expression elaborates finally methods in sharp or unsharp form.

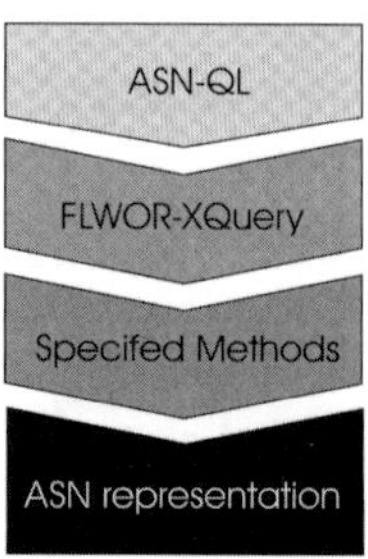

Figure 4. Abstraction levels of ASN-QL

Sharp formulation of a query is done when the user has precise knowledge about the model i.e. search for a concept through the conceptId. Unsharpness comes if different users are acting on the same model. Modelers are acting dynamically on the model. The user may not have the knowledge about the current model and therefore can't formulate the precise query. For example the modeler defined an entity as an attribute but the user supposes that it is defined as a concept, a sharp query would not find the required entity in this case but an unsharp query would provide the result. In table 1 can be seen the formulation and the call of the method through FLWOR.

Table 1. Mapping of the semanticFunction to retrieval methods

Semantic	Sharp methods	Unsharp methods
Findekonzept	GetConcept	getConceptUnscharp
hatbeziehungzu	IsAssociatedWith	isRelatedWith
Bestehtaus	isAggregatedOf	consistsOf
	isComposedOf	isRelatedWith
Istabgeleitet	IsGeneralizationOf	isRelatedWith
Istaehnlich	-	isSimilarTo

A separation in sharp and unsharp information retrieval is not a criterion for the query formulation from the user's viewpoint, because the system uses first sharp methods and if it doesn't find any results continues the search by using unsharp methods.

Following methods are designed for a sharp retrieval:

- getConcept: it has as input properties concepts and delivers one or more concepts existing in the ASN
- isAssociatedWith: it uses one or two concept descriptions for the start and target concept and delivers all exclusively associated concepts, in a specified range.
- isAggregatedOf: it uses one or two concept descriptions for the start and target concept and delivers all exclusively aggregated concepts, in a specified range.
- isComposedOf: it uses one or two concept descriptions for the start and target concept and delivers all exclusively composed concepts, in a specified range.
- isGeneralizationOf: it uses one or two concept descriptions for the start and target concept and delivers all exclusively generalized concepts, in a specified range.

In order to develop methods for retrieving unsharp knowledge we introduce the term semantic neighbourhood. Semantic neighbourhood quantifies (weights) the dependency of a relation type (existing in the knowledge base) to query. Various relations have different nearness to each other and therefore are in a different importance for a query. The semantic neighbourhood expresses the importance, from the user's point of view. The content of the query specifies the relation type to be searched. In table 2 the horizontal options express the intended search relation for one concept and the vertical options are the actual relations existing in the model. Thus, the table 2 gives the correspondence between the "to be searched" and "is modeled" relation. For example the semantic neighbourhood for the search term "association" is "low" for the modeled term "generalization" whereas it is "medium" for the terms "aggregation" and "composition".

Table 2. Weighted semantic neighbourhood for a given start concept

	Association	*Aggregation*	*Generalization*
Association	High	Low	Low
Generalization	Low	Medium	High
Aggregation	Medium	High	Medium
Composition	Medium	High	Medium

The table 2 is valid only for searching for one concept and one relation. In the case that is searched for a start concept, a target concept and a supposed transitive relation (that means two relations between three concepts, where the third concept is not known) it is estimated the semantic neighbourhood through the table 3. In the table 3 is examined only the case in which is searched for an association between the start and target concept and the first relation is given in the vertical options and the second relation is in the horizontal options. Is the structure of the model not known by the user; with the keyword "keyword" can be searched for various entities: in the concept type and name, in the attribute and value, in the relation and role. Additionally, the results relevance is incorporated into the language. For that, is given a ranking. The ranking value is calculated from the semantic neighbourhood of the target concepts.

Table 3. Weighted semantic neighbourhood for transitive relations

	Association	Generalization	Aggregation	Composition
Association	High	Low	Medium	Medium
Generalization	Low	Low	Low	Low
Aggregation	Medium	Low	Medium	Medium
Composition	Medium	Low	Medium	Medium

Following methods are designed for an unsharp retrieval:

- getConceptUnsharp: it has an input of unsharp information about an entity and delivers back all the existing knowledge stored in to ASN.
- isRelatedWith: it uses part information about the start and target concept and try to find relevant concepts through their semantic neighbourhood with the defined range
- consistsOf: this method abstracts between the relation type aggregation and composition. During the unsharp search for supposed aggregation is allowed to search also the type composition in order that possible unsharp formulated query retrieves correct results.
- isSimilarTo: it searches for similar concepts using comparison of the attributes in a vector space. The weighted for each attribute is determined by the value. The cosinus value of the concepts expresses the similarity degree.

4.2 Functionality of the Retrieval Agent

During the functionality, obtains the retrieval agent various states. Normally it starts from the waiting status as soon as a query formulation reaches it. The respectively ASN methods are called after a syntactic and semantic check of the query. The system first applies the sharp methods and in the case that no results are retrieved uses the unsharp methods. Another opportunity is given by the system if the user wishes navigation. An information space is provided to him. For this retrieving of information, two more steps are necessary: reception of ASN data in the internally agent database and processing of query in the XQuery processor. The whole process steps can be viewed in the figure 5.

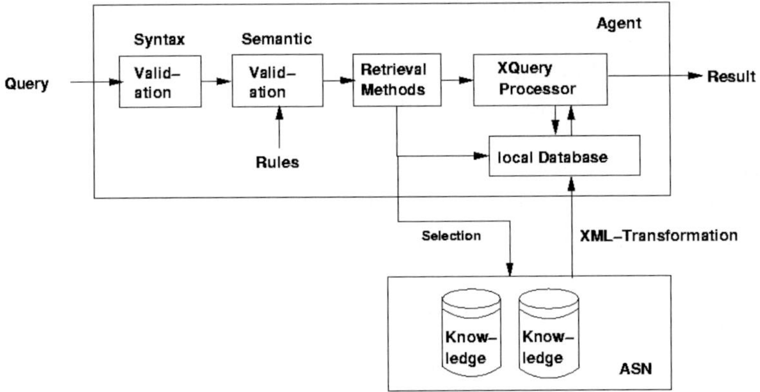

Figure 5. Processing a query

5.   REALIZATION ASPECTS

5.1   Information exchange in the multi agent system

The agent's communication between themselves and the environment is based on exchange of XML format messages over LRMP. The predefined message format has the syntax in the table 4. The message format provides necessary information for addressing agents by sender and data type.

Table 4. The agent message exchange format

```
<request sender="" destination="" reqControl="" dataType="">
   <data></data>
</request>
```

Parameter "sender" is the Unique Agent Identifier (UAI) of the sending agent. UAI is defined as a sequence of AgentName/AgentTyp/ID. The parameter "destination" is the target agent's UAI for the receiver. The value "ALL" for "destination" may be -and in the conversation protocol IS- used to address all the agents participating in the multicast environment. The parameter "reqControl" may be used in additional high-level conversation processes or used for compatibility with the earlier development of a coordination agent. The parameter "dataType" defines the content of the data body -used in the implemented protocol as identifier of the protocol command, e.g. DAT, AMP, SMP, CLAIM_TOKEN, RING_PURGE-.

5.2   System components

Distributed systems are based on the middleware model as an evolution of the client server model. Middleware systems offer the advantages of locally distributed developer teams and additionally separation of the services from elementary computer specific problems like persistence. Agents are implemented as standalone Java application. So they are fitting in a middleware implementation which is realized by containers, building on the J2EE platform from Sun Microsystems [29].

The XQuery processor is the Qexo [30]. This, GNU project is bind easily in the agents framework. The processor is extended for specific functionality. The Parser Generator JavaCC is chosen for the generation as a query parser [31].

5.3   Application

In the figure 6 is the formulation of a query and the deliver of this query to the retrieval agent.

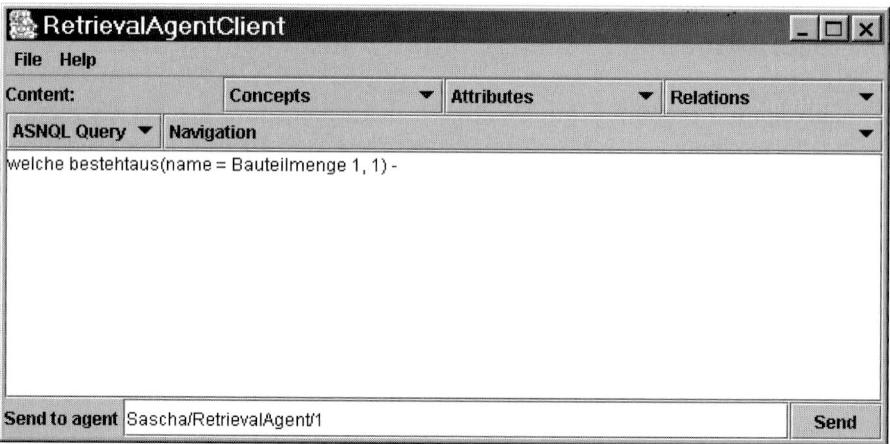

Figure 6. User interface for query formulation

The retrieved results of the query in figure 6 can be seen in the figure 7.

6.   CONCLUSIONS

Query formulations with the presented ASN-QL are more intuitively formulated then SQL queries. Retrieving knowledge by retrieval agent presuppose neither knowledge about the product or its representation nor knowledge about the development process. Additionally, we extended XQuery with commands based on the semantics of the relations.

```
Result from request [Sascha/LocalAgent/1/13]
 Sender = Sascha/RetrievalAgent/1
 DataType = query-result
Data
<result>
<concept name="Schraube 1" type="Schraube" id="190">
<searchinfo rating="0" retrievedby="bestehtaus" foundrelatedto="Bauteilmenge 1" />
<attributes><attribute name="Name">Schraube 1</attribute>
<attribute name="Dicke">1</attribute>
<attribute name="Laenge">5</attribute>
<attribute name="Material">leer</attribute>
<attribute name="Kopftyp">leer</attribute>
<attribute name="Typ">leer</attribute>
<attribute name="KostenproStueck">leer</attribute>
<attribute name="Nummer">leer</attribute>
<attribute name="Bruchfestigkeit">leer</attribute>
</attributes>
<relations><relation name="Bauteilmenge" multiplicity="one" direction="Bauteilmenge"
type="Aggregation">Bauteilmenge 1</relation>
</relations>
</concept><concept name="Schraube 2" type="Schraube" id="191">
<searchinfo rating="0" retrievedby="bestehtaus" foundrelatedto="Bauteilmenge 1" />
<attributes><attribute name="Name">Schraube 2</attribute>
<attribute name="Dicke">1</attribute>
<attribute name="Laenge">10</attribute>
<attribute name="Material">leer</attribute>
<attribute name="Kopftyp">leer</attribute>
<attribute name="Typ">leer</attribute>
<attribute name="KostenproStueck">leer</attribute>
<attribute name="Nummer">leer</attribute>
<attribute name="Bruchfestigkeit">leer</attribute>
</attributes>
<relations><relation name="Bauteilmenge" multiplicity="one" direction="Bauteilmenge"
type="Aggregation">Bauteilmenge 1</relation>
</relations>
</concept><concept name="Schraube 3" type="Schraube" id="192">
<searchinfo rating="0" retrievedby="bestehtaus" foundrelatedto="Bauteilmenge 1" />
<attributes><attribute name="Name">Schraube 3</attribute>
<attribute name="Dicke">5</attribute>
<attribute name="Laenge">20</attribute>
<attribute name="Material">leer</attribute>
<attribute name="Kopftyp">leer</attribute>
<attribute name="Typ">leer</attribute>
<attribute name="KostenproStueck">leer</attribute>
<attribute name="Nummer">leer</attribute>
<attribute name="Bruchfestigkeit">leer</attribute>
</attributes>
<relations><relation name="Bauteilmenge" multiplicity="one" direction="Bauteilmenge"
type="Aggregation">Bauteilmenge 1</relation>
</relations></concept>
</result>
```

Close

Figure 7: Query's results

                    Stavros Dalakakis, Emil A. Stoyanov and Dieter Roller

ACKNOWLEDGEMENTS

The ASN is part of the project SFB 374 supported by the Deutsche Forschungsgemeinschaft (DFG). Several Institutes from the University of Stuttgart, Fraunhofer-Gesellschaft as well as DaimlerChrysler are in this project cooperating.

APPENDIX

EBNF of the query language ASN-QL:

```
ASN-QL              = [QFunction] SemanticFunction [Restriction] [Navigation];
SemanticFunction    = ( Semantic "(" ConceptDescr "," Range ")"
                        | Semantic "(" SemanticFunktion "," Range ")„
                        | Semantic "(" ConceptDescr, ConceptDescr "," Range ")"
                        | Semantic "(" ConceptDescr, SemanticFunction "," Range ")" );
Semantic            = ("bestehtaus" | "istaehnlich" | "hatbeziehungzu„ | "findekonzept"
                        | "istabgeleitet" );
QFunction           = ("Welche" | "Wieviele" | "Wie oft " | "Welche Qualitaet"
                        | "Wodurch" | "Womit " | "Wer" );
Restriction         = AttributeDescr Operator Value { (OR | AND) Restriction} ;
AttributeDescr      = String ;
Value               = (String | Digit);
Operator            = "=" | "!=" | "<" | "<=" | ">" | ">=" ;
Navigation          = "+" | "-" ;
Range               = {Digit}-;
Digit               = ( "0 " | "1 "| "2 "| "3 "| "4 " | "5 " | "6 " | "7 " | "8 " |
  "9 ");
ConceptDescr        = ConceptDescrSharp | ConceptDescrUnsharp;
```

EBNF the sharp methods of the query language ASN-QL:

```
ConceptDescrSharp   = ( "id =" id |
                        ( "type =" Type | "name =" Name | "type =" Type "&" "name =" Name )
                        | "attribute = " Name {"&" "attribute = " Name});
    For FLWOR expression and mapping:
SharpMethods        = ("get-concept" "("ConceptDescrSharp { "," ConceptDescrSharp } ")"
                        | "isAssociatedWith" "(" ConceptDescr, Range ")"
                        | "isAssociatedWith" "(" ConceptDescr, ConceptDescr, Range ")"
                        | "isAggregatedOf" "(" ConceptDescr, Range ")"
                        | "isAggregatedOf" "(" ConceptDescr, ConceptDescr, Range ")"
                        | "isComposedOf" "(" ConceptDescr, Range ")"
                        | "isComposedOf" "(" ConceptDescr, ConceptDescr, Range ")"
                        | "isGeneralizationOf" "(" ConceptDescr, Range ")"
                        | "isGeneralizationOf" "("ConceptDescr, ConceptDescr, Range")" );
Range               = {Digit}- ;
```

EBNF the unsharp methods of the query lansuage ASN-QL:

```
ConceptDescrUnsharp = ( "type =" (Type | RegularExpression)
                        | "name =" (Name | RegularExpression)
                        | "attribut =" (Name | RegularExpression)
                            {"&" "attribut =" (Name | RegularExpression)}
                        | "Keyword =" (Keyword | RegularExpression));
RegularExpression   = ( X | Y | XY | X"|"Y | "(" X") | X"*" | X"?" | X"+" ) ;
```

```
X                       = String | RegularExpression;
Y                       = String | RegularExpression;
Keyword                 = {Char}-;
   For FLWOR expression and mapping:
   UnsharpMethods   = ( "getConceptUnsharp" "(" ConceptDescrUnsharp, RatingLimit ")"
                | "isRelatedWith" "(" ConceptDescr, Range, RatingLimit ")"
                | "isRelatedWith" "(" ConceptDescr, ConceptDescr, Range, RatingLimit ")"
                | "consistsOf" "(" ConceptDescr, Range, RatingLimit ")"
                | "consistsOf" "(" ConceptDescr, ConceptDescr, Range, RatingLimit ")"
                | "isSimilarTo" "(" ConceptDescr , Range, RatingLimit ")"  ) ;
```

REFERENCES

1. Dieter Roller, Oliver Eck, Stavros Dalakakis: "Integrated version and transaction group model for shared engineering Databases". *Data & Knowledge Engineering*, Elsevier Science B.V. **Vol. 42**, pp. 223-245, 2002.
2. Dieter Roller, Stavros Dalakakis, Oliver Eck: "A knowledge base for rapid product development". *Proceedings of the fourth international symposium on tools and methods of competitive engineering*, pp. 171-181, April 2002.
3. Georgeff, M.; Pell, B.; Pollack, M.; Tambe, M.; Wooldridge, M.; "The Belief-Desire-Intention Model of Agency". Intelligent Agents V. Agent Theories, Architectures and Languages, *ATAL'98: 5th International Workshop*, 4-7 Juli 1998 in Paris Mueller J.P., Singh M.P., Rao A.S. (Hrsg.). *Proceedings*, LNAI 1555, Springer-Verlag Berlin/Heidelberg, pp. 1-10, 1998.
4. Ferber, J.; Gutknecht, O.: "A meta-model for the analysis and design of organizations in multi-agent systems". *Proceedings of the Third International Conference on Multi-Agent* Systems: 3.-7. July in Paris, Y. Demazeau (Edited by), pp. 128-135, 1998.
5. Gruber, T. R.: "Toward Principles for the Design of Ontologies Used for Knowledge Sharing". Formal Ontology in Conceptual Analysis and Knowledge Representation (edited by) Guarino, N.; Poli, R.; Presented in the International Workshop on Formal Ontology, Padova, Italy, March 1993.
6. Genesereth, M. R.; Ketchpel, S. P.: „Software Agents". Communications of the ACM, Vol. 37 (7), pp. 48-53, 1994.
7. FIPA Abstract Architecture Specification: Foundation for Intelligent Physical Agents, 2002. URL: http://www.fipa.org/specs/fipa00001
8. FIPA ACL Message Structure Specification: Foundation for Intelligent Physical Agents, 2002. URL: http://www.fipa.org/specs/fipa00061
9. FIPA Communicative Act Library Specification: Foundation for Intelligent Physical Agents, 2002. URL: http://www.fipa.org/specs/fipa00037
10. Liao, T.,"Light-weight reliable Multicast Protocol" INRIA, Rocquencourt, BP 105 /8153 Le Chesnay Cedex, France, 2003.
11. URL: http://webcanal.inria.fr/lrmp/lrmp_v1.html
12. Hewlet Packard Message Service, Hewlet Packard. URL: http://www.hpmiddleware.com/SaIsapi.dll/SaServletEngine.class/products/hp-ms/technical_information.jsp
13. Java Shared Data Toolkit, Sun Microsystems Inc.
14. URL: http://java.sun.com/products/java-media/jsdt/
15. SwiftMQ: URL: http://www.swiftmq.com/products/index.html
16. UberMQ: URL : http://www.ubermq.com/products.html
17. JXTA: Sun Microsystems Inc. URL: http://services.jxta.org/servlets/ProjectHome
18. Extensible Markup Language (XML), XML-Standards: URL: http://www.w3c.org/XML/Core/#Publications

19. Papadimitriou, C. H.: "Database metatheory: Asking the big queries". Proceedings of the Fourteenth ACM SIGACT-SIGMOD-SIGART Symposium on Principles of Database Systems, San Jose, California / ACM Press, pp. 1-10, 1995.
20. Suciu, D.: "On Database Theory and XML". SIGMOD Record September 2001, Vol. 30, No. 3, 2001.
21. Fraunhofer - Institut Integrierte Publikations- und Informationssyteme – IPSI:
22. URL: http://www.ipsi.fraunhofer.de/oasys/projects/ipsi-xq/index_d.html
23. Oracle Corporation:
24. URL: http://otn.oracle.com/tech/xml/xmldb/htdocs/querying_xml.html
25. Lucent Technologies: URL : http://db.bell-labs.com/galax/
26. Papakonstantinou, Y.; Vinayak, B.; et. al.: "XML queries nad algebra in the Enosys integration platform". *Data & Knowledge Engineering*, Elsevier Science B.V. Vol. 44, pp. 299-322, 2003.
27. ABLE, "Agent Building and Learning Environment", IBM.
28. URL: http://www.alphaWorks.ibm.com/tech/able
29. JADE "Java Agent Development Framework", Telecom Italia Lab. URL: http://sharon.cselt.it/projects/jade/
30. RETSINA, "Reusable Enviroment for Task-Structured Intelligent Networked Agents", URL: http://www-2. cs.cmu.edu/%7Esoftagents/retsina_agent_arch.html
31. FIPA-OS, "Foundation for Intelligent Physical Agents Open Source from Nortel Networks", URL: http://fipa-os.sourceforge.net/summary.htm
32. Diederich, M. K., Leyh, J.: "Agent Based Middleware to Coordinate Distributed Development Teams in the Rapid Product Development". *Proceedings for the Conference on Product Development, 1st Automotive and Transportation Technology Congress and Exhibition (1st ATTCE)*, Barcelona, 2001.
33. IEEE 802.5: Tokenring.
34. J2EE : http://java.sun.com/j2ee/
35. Qexo: http://www.gnu.org/software/qexo/
36. JavaCC : https://javacc.dev.java.net/

COMPUTER-AIDED TECHNIQUES IN MANUFACTURING OF LARGE HYDRO TURBINE'S BLADES

Xi-De Lai, Ci-Chang Chen and Qing-Gang Li
School of Energy and Environment, Xihua University, Chengdu, 610039, China, email: greatlai@21cn.com

Abstract: Blades are one of the vital components and most difficulty in manufacturing of large hydro turbines. In order to cost-effectively and productively manufacture these kinds of blades, hydro turbine manufacturers have been trying to machine these large blades using 5-axis CNC machining techniques in recent year. To implement 5-axis simultaneous CNC machining for large blades, a series of computer-aided techniques in manufacturing of these blades have been developed. This paper presents the major computer-aided techniques that we have been developed in manufacturing of large hydro turbine blades, which include that Geometrical design of hydro turbine blades for manufacturing, Computer-aided location and the machined error evaluation by means of 3-dimensional digitized measuring, tool path generation strategy to meet requirements of enhancing machining efficiency and controlling deviation in NC machining, tool path generation and NC machining simulation by establishing a virtual NC machining environment for blades; A reasonable and feasible strategy and the systematic scheme for manufacturing for large blades by using 5-axis simultaneous CNC machining are also purposed. The developed computer-aided techniques have been successfully applied in manufacturing of both the large Kaplan and Francis hydraulic turbine blades, it shows that higher efficiency and the better surfaces finish accuracy can be achieved in machining of large blades with the right combination of these techniques.

Key words: Computer-aided techniques, Hydraulic turbine, Blade, CAM

1. INTRODUCTION

Hydro turbine blades, such as Francis and Kaplan types, are usually designed by means of detailed hydraulics methods and simulations of Computational Fluid Dynamics (CFD), or digitized from the models. Traditionally, the large size hydraulic turbine blades were machined by using manual grinding operations and inspections with sets of combined templates, which cannot satisfy for rapid developments of large hydro turbine and also cannot ensure accuracy of finished blades. Hydro turbine manufacturers have been trying to mill these large blades using CNC machining techniques in recent year [1-3]. As the surfaces of a large blade to be machined are usually from several to tens of square meters and involve bulk metal removal while machining from the casting, it tends to be non-productive to machine these kinds of blade using traditional 3-axis machining with ball-end cutters. With the advent of large size multi-axis CNC machines, an innovative 5-axis simultaneous CNC machining technology has become not only feasible but also cost effective for large parts with sculptured surfaces. As a large hydro turbine blade is closed with more than 10 sculptured surfaces that have irregular curvature distribution, and it's thickness also intensely changes, it is much difficult to mill the blades by using 5-axis CNC machining. In practice, the 5-axis CNC machining not only offers many advantages over the 3-axis machining, which include the higher productivity and improved surfaces finish, but also brings about more complex technological problems. Such as insufficient support by the conventional CAD/CAM systems, highly complex algorithms for gouging avoidance and collision detection, complex surface machining errors that result from the traditional NC programming methods, etc [4-5]. However, a reasonable strategy and systematic scheme of blade's manufacturing with the right combination of a series of computer-aided techniques can solve those complex problems in engineering and made them operational feasibility in technology. We had developed a series of computer-aided techniques in manufacturing of large blades, which mainly include: (1) Computer-aided geometrical design and 3-dimensional modeling of blades for manufacturing; (2) Computer-aided localization for position & setup by means of 3-dimensional digitized measuring of blades casting; (3) Making of machining strategy and tool path planning with the help of computer-aided geometrical analysis; (4) 5-axis machining tool path generation for different surfaces of a blade; (5) Computer simulation for machining of a large blade to verify the gouges and collisions between machine tool and blade& fixtures; (6) Computer-aided inspection and the machined error evaluation for the finished blade surfaces by means of 3-dimensional digitized measuring; With the right combination of the developed computer-aided techniques, both Kaplan and Francis hydro turbine blades can be cost-effectively and productively manufactured.

2.   MANUFACTURING STRATEGY OF LARGE HYDRO TURBINE BLADES BY USING NC MACHINING

The key to the cost effectiveness of 5-axis machining for large blades is how to make reasonable and feasible manufacture strategy and planning. By means of computer-aided techniques that we had developed, a typical blade manufacturing strategy with 5-axis CNC machining shown in *Fig.1*

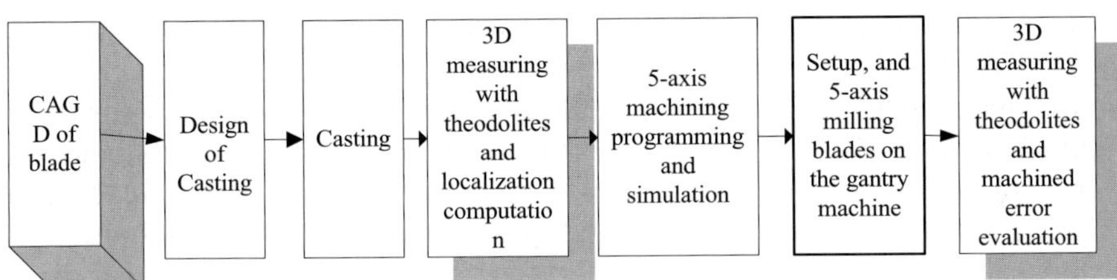

Fig. 1. A typical blade manufacturing strategy for large size hydro turbines

3.   COMPUTER-AIDED GEOMETRICAL DESIGN FOR BLADE'S MANUFACTURING

Hydro turbine blade is a very complex body closed by sculptured surfaces. Traditionally, the pattern drawing is used to represent surfaces of blade in runner design, which is not able to meet the requirements for generation of digital models in the digital design and manufacturing of hydro turbine runners [6]. For the exact geometrical modeling of a blade, it has to be transfer to the point sets along the intersection curve between blade and stream surfaces with the special developed software [6]. Then, NURBS representation is used as a unified digital model of the blade's geometrical design. In order to implement NC machining on a gentry machine, a blade is usually subdivided into more than ten sculptured surfaces [1-2]. As shown in Fig.2, a typical large Francis turbine blade can be subdivided into the surfaces of face, back, crown, band, inlet, outlet, weld preps between crown and blade, and weld preps between band and blade. Fig.3 shows a typical large Kaplan turbine blade.

4.   COMPUTER-AIDED LOTION AND EVALUATION OF MACHINE ERROR

4.1   3-Dimensional Digital Measuring for Blade's Castings and Finished Blades

Because each blade casting has different and uneven stocks distribution and there is no basis for position on blade's castings, one of key techniques in CNC machining of large blades is that how to set up correctly for each blade casting on machine in a very short time. It was difficult for the localization of a large blades casting on a machine, which was manually adjusted by means of measuring with the probe of machine. So it was arduous and time-consuming, and even allowance can't be ensured for machining of each surface. An efficient method is to use computer-aided localization technique that is based on 3D measuring each blade casting, which can compute the right spatial setup position for each blade casting by the self-developed software [1-2]. After finished machining, the finished blades need to be inspected the compliance between the designed and machined. For each blade casting or finished blade, a series of points on each surfaces of blade casting can be measured by non-contact or contact measuring methods [1-3]. The non-contact measuring instruments, such as high accuracy theodolites and laser tracker, are more convenient and suitable for large blade castings. *Fig.4* shows that a blade casting is being measured with the high accuracy theodolites (Made by Leica). In order to define the geometry of a blade casting, all surfaces of a blade casting should be measured, the theodolites are usually placed on a fixed place and a casting or finished blade has to be turn over more than one times, so all measured data should be unified in the same measuring coordinate system.

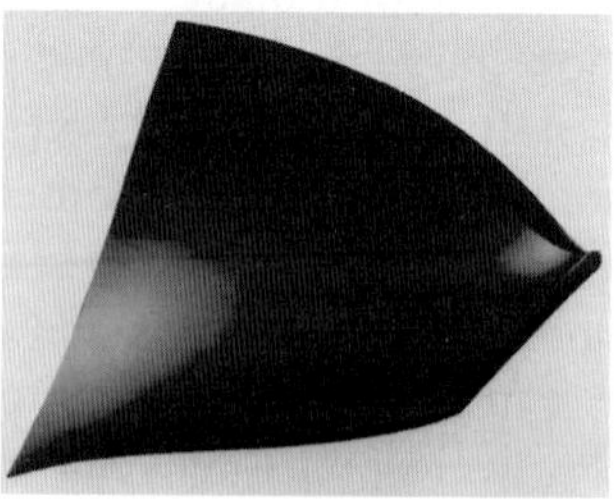

Fig.2 Francis turbine blade

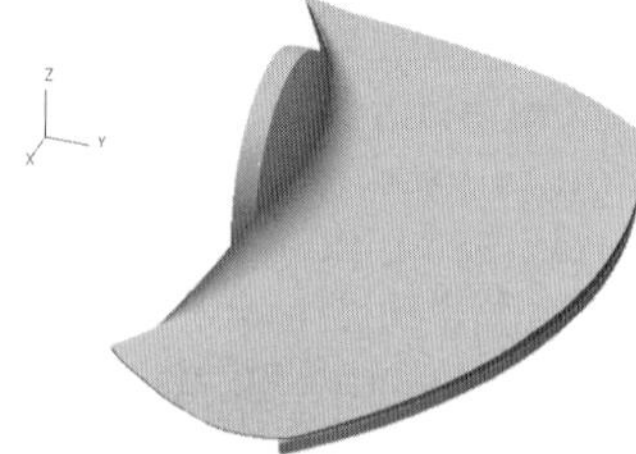

Fig.3 Kaplan turbine blade

4.2   Computer-aided Localization Computation and Setup

The localization of a blade is to align the stock surface with its CAD model, which can be done by searching the optimal Euclidean transformation between the measurement coordinate frame of the casting and the blade's CAD model coordinate frame. In the situation of computer-aided localization (CAL) for large blade, it is a hybrid location/envelopment problem in engineering. The key to solve this problem is how to quickly and accurately find the matching relation that can keep the allowance even on all surfaces between the measured blade castings and the CAD model. After we carefully studied, a new method is proposed for solving localization problem of blade, which consists of rough surface matching and accuracy surface matching. Rough matching decides the variable range of the follow-up algorithms. Accuracy matching acquires the optimal location. Then, rapid localization and clamp can be implemented. To cope with the accuracy matching, a hybrid global optimization method is adopted. It shares the advantages of both genetic algorithms (GAs) and simplex [7]. Based on the proposed method, special CAL software has been developed and used in the 5-axis machining of both the large Kaplan and Francis hydro turbine blades. With this software, we can easily and quickly search the optimal setup position for blade casting on machine [1-3]. This optimal setup position can be transformed to machine coordinate system with fixture adjusting reference points, so the setup time can be dramatically decreased to less than one hour while 5-axis machining from more than twenty hours with the probe of machine and adjusted manually. Additionally, it is more important that this method can keep the even stocks on each surface to be machined. Fig.5 shows the optimal stock distribution computation.

Fig4. 3D measuring of casting with theodolites

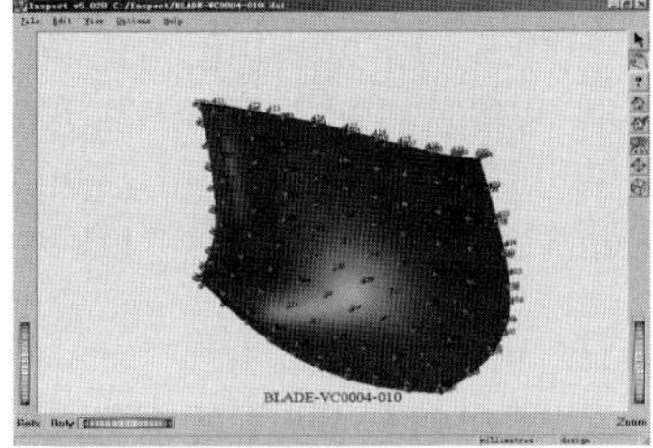

Fig.5 Optimal Computation of stocks

4.3   Computer-aided Inspection & Machined Error Evaluation

After finished machining, the finished blades need to inspect the compliance between the designed and machined. We developed special software for computer-aided inspection by comparing 3D digital measuring of finished blade with blade's CAD model based on the reference marks on the finished blade. This way of inspection is not only more effective and accurate but also cost very low than the traditional method. It has been shown that this method can be widely used in blade's manufacturing as a kind of general technique.

5.   COMPUTER-AIDED GEOMETRICAL ANALYSIS OF BLADES AND MACHINING STRATEGY

As shown in Fig.3, In order to mill the blades on an actual CNC machine, a large blade is usually subdivided into more than ten surfaces depending on the machine's configurations. Each surface has its characteristic and surface property. In order to make the reasonable machining strategy, computer-aided geometrical analysis of each surface is very helpful to decide selection of cutter diameter, cutting direction, tool-axis orientation and so on. Surface properties such as different curvatures, normal vector surfaces, highlight lines and continuity can be computed and displayed graphically by computer-aided geometrical analysis. With the help of surface curvature analysis, you can make rough selection of the diameter of cutting tools for each surface machining. For the pressure and suction face of large blades, a large diameter heavy-duty index able inserts face-milling cutter with 60 degrees rake angle can be used for rough milling, and a large diameter face-milling cutter with round inserts can be used for semi-finish and finish milling. For outlet face, a long flute end-milling cutter can be used by side-wall milling. For inlet and weld preparation surfaces, a large diameter face-milling cutter with large round inserts can be used for roughing and finishing. Those machining strategies offer a set of distinct advantages for milling large blades, which are that small scallop heights with large cross feed, improvement of tool life and better surface finish, and reduction of vibration and defection. By means of computer-aided geometrical analysis, the fairness of each surface can also be checked.

6.   TOOL PATH PLANNING AND FIVE-AXIS CNC PROGRAMMING

6.1   Tool path planning

As the surfaces of a large blade to be machined are usually from several to tens of square meters, a productive way is to mill these kinds of large blades by using 5-axis simultaneous CNC machining techniques. In order to obtain the higher productivity and improved surfaces finish, parameters of tool path planning for each surface should be carefully given by trying and verifying with combination of cutting simulation in computer. To ensure the properties of fluid dynamics, the form deviation between the design and the machined blade has to be strictly controlled [1-2]. Because the higher efficiency as well as the better surfaces finish accuracy is required in machining of large sculptured surfaces, it is very important to effectively control and reduce the machining errors in tool path planning. A way to control and reduce geometrical error in machining of large sculptured surfaces has been presented by the author [5], in which the link between the geometrical errors (non-linear error in multi-axis machining) and the parameters of tool path planning is established. Iterating tool path generation and cutting simulation in computer can optimize for tool path planning.

6.2   Five-axis CNC Programming for large blades

CNC programming is the most important task in five-axis machining of large blades. In order to do the cost effectiveness CNC programming for large blades, besides all the above computer-aided techniques will be put to use synthetically, it is necessary to combine the cutting simulation and machining process simulation during tool path generation. After tool path planning parameters is given, tool path generation can be implemented by re-development based on commercially powerful CAD/CAM software such as SDRC/Camand or Unigraphcis. The 5-axis CNC programming flow chart that we adopted is shown in Fig5.

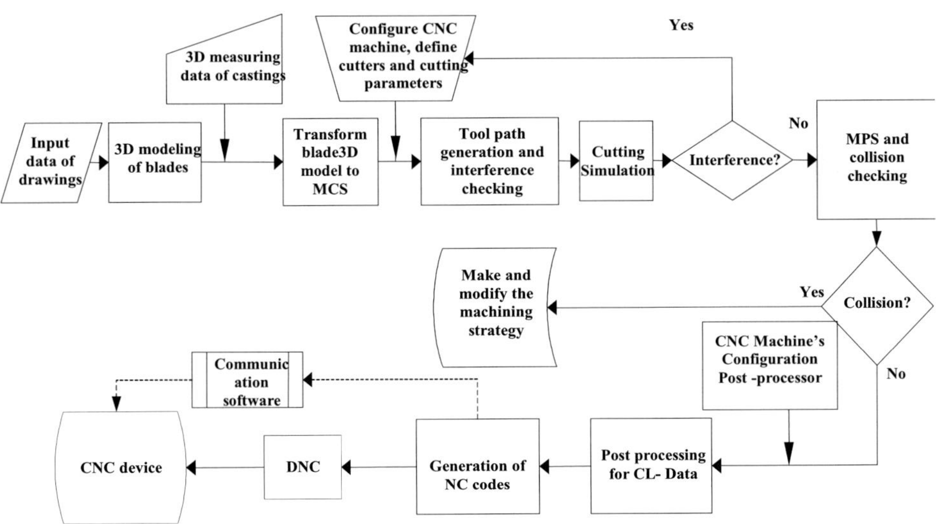

Fig.6 5-axis CNC programming flow chart for blade machining

7.   COMPUTER SIMULATION FOR MACHINING OF A LARGE BLADE

Although a lot of works, such as 3D modeling for all surfaces, machining strategy and tool path planning, have been carefully done, it is still not able to ensure that there are no problems in CNC programs of blade's machining. To avoid the over-cut and collision between tool shanks & milling head and blade & fixtures, and overload in machining, it is necessary to develop computer simulation techniques for verifying the tool path or CNC programs. Based on the tool swept body and NC-machining simulation with G-buffer method [8-9], special software for verifying and evaluating NC programs for large blades has been developed. By using the developed computer simulation software, the real machining environment of large blade and machining processing can be simulated and geometrical error of tool path can be verified, and whether collision between moving components and blade/fixtures can be checked with the simulation software. As it is an extremely complex machining process for a large blade, combining with the developed computer simulation techniques, the machining strategy and tool path planning parameters and cutting parameters can further optimized. The function of the developed software for simulation machining of large hydro turbine blades includes: (1) Tool path simulation and verification for cutting; (2) Machine processing simulation and verification for collision. Fig.7 shows a snapshot of cutting simulation for a large Francis blade, and Fig.8 shows a snapshot of machine simulation for a large Kaplan blade.

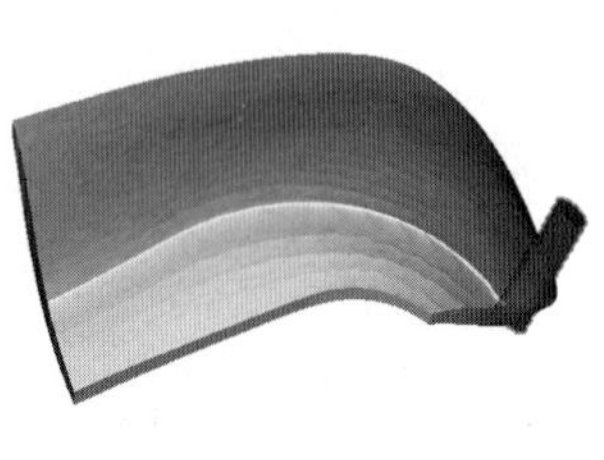

Fig.7 Simulation for cutting

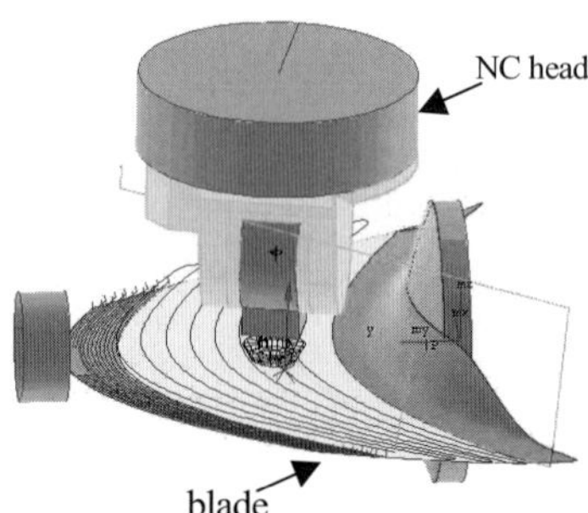

Fig.8 Simulation for machine processing

8.   ILLUSTRATIVE EXAMPLE FOR MACHINING OF LARGE BLADE

The above-mentioned computer aided techniques have been used in manufacturing of both the large Francis and Kaplan hydro turbine blades [1-2] by the authors. As an example, 5-axis machining of a large Francis blade shown in Fig. 9. This blade is for Sanxia Hydro Power Station in China and the largest Francis blade in the world so far. The blade has surface area of 40 m^2 to be machined from the casting of 30 tons to the net weight of 19 tons. The blade is closed by 11 sculptured surfaces and twisted into "X" shape in the minimum contained space of $5500mm(length) \times 4650mm(width) \times 1900mm(height)$. The blade had been machined on a large size 5 –axis gantry travel machine with $\Phi250mm$ indexable inserts face milling cutter (Flat cutter with $60°$ side angle) for roughing and semi-finishing, and with $\Phi160mm$ indexable round inserts face-milling cutter (Filleted cutter) for finishing.

Fig.9 Photo for machining of blade

9. CONCLUSIONS

This paper presents the developed computer-aided techniques in manufacturing of large hydro turbine blades, which are keys for successfully manufacturing of large blade with 5-axis CNC machines. With the right combination of the developed computer-aided techniques, large hydro turbine runner's blades of both Kaplan and Francis can be cost-effectively and productively manufactured. They are also applicable in 5-axis machining of the other large parts with sculptured surfaces

ACKNOWLEDGEMENTS

This work was supported by the National Natural Science Foundation of China (NSFC, Grant No. 59493704), and partially supported by the Key Project of "National Ninth-Five Year Plan" in China (Grant No.97-312-01) and the Science Foundation of Xihua University (Grant No.0224907). The large hydro turbine blades had been machined at Dongfang Electrical Machinery Co. Ltd. Their supports are greatly appreciated.

REFERENCES

1. Xi-de Lai: "Technology for 5-axis simultaneous CNC machining of large Kaplan hydro turbine blades". *CAD/CAM*, **Vol. 66,** pp.71-75, May 2000.   (in Chinese)
2. Xi-de LAI: "Technology for 5-axis simultaneous CNC machining of large Francis hydro turbine blades". *J. of Ordnance Industry Automation*, **Vol.61**, pp.40-43, March 2001. (in Chinese)
3. Cichang CHEN, Xi-de LAI: "Latest Progress of Design and Manufacturing for Hydro Turbine in China". *J. of Japanese Turbomachinery*, **Vol.31**, pp .193-198, March 2003,
4. C-C Lo: "Efficient cutter-path planning for five-axis surface machining with a flat-end cutter". Computer-aided Design, Vol.31, pp.557-566, Sept. 1999
5. Xi-De LAI, Yun-Fei Zhou, Ji Zhou, et al.: "Geometrical Errors Analysis and Control for 5-axis Machining of Large sculptured surfaces". *Int. J. of Adv. Manuf. Technol.* , **Vol.21**, pp. 110~118, Feb. 2003
6. Xi-De LAI, Zhen-kai Wang: "A software for Francis blade streamline surfaces transform in NC machining of large hydro turbine blades". *J. of Ordnance Industry Automation*, **Vol.45**, pp.10-14, June 1997. (in Chinese)
7. Sijie Yan, Yunfei Zhou, Xi-de LAI: "Optimization of allowance distribution on the workpieces with large sculptured surfaces in NC machining". *J. of Huazhong Univ. of Sci. & Technol.* , **Vol.30**, pp35~37, Oct. 2002 (in Chinese)
8. C.-J. Chiou, Y,-S. Lee: "A shape-generating approach for multi-axis machining G-buffer models", *Computer-Aided Design* , **Vol.31**, pp.761-776, 1999
9. D. Roth, S. Bedi, F. Ismail, et al .: "Surface swept by a toroidal cutter during 5-axis machining". *Computer-Aided Design*, **Vol.33**, pp.57-63, 2001.

USING CONTEXT KNOLEDGE BASED REASONING TO SUPPORT FUNCTIONAL DESIGN

Fayyaz Rehman and Xiu-Tian Yan
CAD Centre, Department of Design, Manufacture and Engineering Management University of Strathclyde, 75 Montrose Street, Glasgow G1 1XJ, UK, , Email: fayyaz@cad.strath.ac.uk.

Abstract: This paper describes an approach to support function based conceptual design of a mechanical artefact based component. The approach defines a broad concept of design context knowledge as the related background knowledge of design problem under consideration. Having generated the initial concepts/solutions based on the function requirements, the approach supports the conceptual design through background reasoning of functional requirements as well as generated solutions using the design context knowledge. The reasoning mechanism facilitates a designer to explore rapidly different design consequences that are likely to occur at different stages of product life cycle generated due to selection of different design alternatives. This early awareness of potential problems due to design decisions supports designer in selecting those solutions, which are less likely to cause problems at later life cycle stages.

Key words: Design context knowledge, conceptual design, decision making.

1. INTRODUCTION

The importance of design process and particularly conceptual design to the overall success of the product is crucial as once the conceptual design process has been finished, the major decisions determining the product cost and quality has been committed and fixed by selecting particular concepts/solutions as the subsequent product life cycle activities (manufacturing, assembly, use, recycle/dispose) depends on these conceptual solutions. Moreover detail design and manufacture cannot make-up for a poor or inadequate conceptual design. Decisions taken during conceptual design affect all the downstream phases of product life cycle and each design decision has downstream consequences [1]. Design context knowledge is an important source of product background knowledge and it can contribute to the generation of design consequences. This paper describes a proposed reasoning mechanism based on design context knowledge used in an on going research project in order to support the decision making at the conceptual design stage by generating and informing a designer the consequences that are likely to occur at later life cycle stages due to designer's decisions.

2.   FUNCTIONAL DESIGN

Function based conceptual design is the most abstract stage of the design process starting with required functions (what the product is to behave without saying how the product is to do it) and resulting in concepts (preliminary system configurations). The word 'function' in design is regarded as a description of the intended action or effect produced by an object [2]. Designing by functions should allow one to describe objects (which in the design context are design problems and solutions) in terms of their known intended functions. For example the function of a mechanical shaft can be described as: *a shaft transmits torque.*

Conceptual design can be defined as the transition process between three different information states: 1) a set of required functions; 2) a set of behaviours that fulfil the functions; 3) Final selected concept(s)/solution(s) that generate the behaviours [2]. Functions, behaviour and form have been identified as the major elements of information, which are manipulated in these states. The typical first step in conceptual design is to decompose the higher-level functions into sub/smaller level functions.

2.1   Function decomposition

During the functional decomposition, the functional requirements are often decomposed to a level where it is possible to identify potential means or mechanisms to realise these small sub-functions. For example, in sheet machine component design, one of the desired function requirements could be *Convert Motion,* which can be further decomposed into *Convert Rotary Motion into Translatory motion* and *Convert Rotary Motion into Rotary motion* (Figure 1).

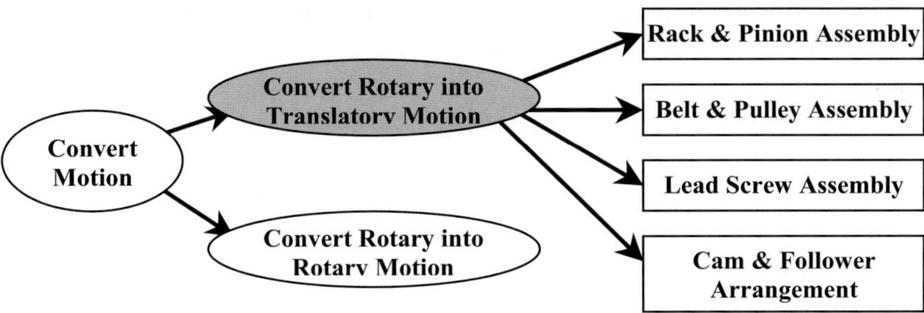

Figure 1. Function-Means association

The potential means for Convert Rotary Motion into Translatory motion from the function means mapping library could be a Rack & Pinion Assembly. Observing the product from constructional point of view gives rise to product breakdown structure (PBS). Borg et. al. [3] presented this structure as a number of elements called product design elements (PDE). PDEs are used as the key to function-oriented design in mapping PDEs to function requirements [4]. For a decomposed function structure, this research

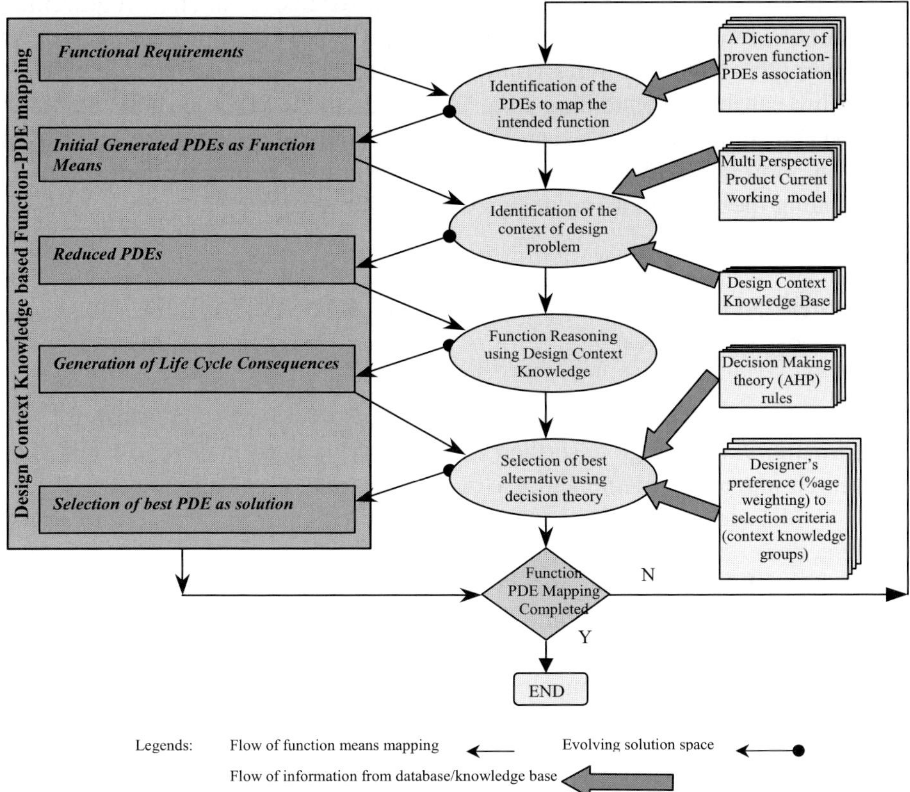

Figure2. Function to PDE(s) mapping process model

proposes the following PDE based function mapping to identify the suitable means to realise a chosen function.

2.2   Function to PDE mapping

Designing by functions or "functional design" refers to the process of generating a design solution from product function point of view, using available well-understood function-PDEs relationships to identify suitable means in the form of PDEs. For a given functional requirement, PDEs are the information carriers that allow the mapping between function requirements and physical solutions of a product.

PDEs are the vehicles to bring basic design information to the downstream product realisation phases for embodiment, detailed part design and later life cycle processes. Figure 2 shows a generic process model of Function to PDE mapping developed in this research [4]. This model can be used in any engineering design domain to generate, evaluate and select best design solution(s) keeping in view functional requirements as well as product life cycle constraints. At present this model has been successfully used in sheet metal design domain. With an extensive function requirement definition, function to PDE mapping process can usually produce a long list of alternative PDEs for a designer to consider.

This can be a demanding task if each of these PDEs is fully manually evaluated. In addition, the deadline for a design solution can be quite tight. To support effectively designers in these scenarios, this research developed a reasoning mechanism using design context knowledge.

3.   DESIGN CONTEXT KNOWLEDGE BASED REASONING

There are many uses of the word 'Context' in design, and information/knowledge described as 'Context' is used in several ways. This research intends to define and use design context knowledge in a broad term as a knowledge having information about surrounding factors and interactions which have an impact on the behavior of the product and therefore the design decision making process which results in design solutions. Therefore it can be defined as *"the related background information of a design problem under consideration"*. The next section explains how different groups of context knowledge can be used to highlight consequences that would occur in different life phases of a product while selecting a particular solution/PDE, thus providing a proactive decision support to designer.

3.1   Context knowledge categories

This research aims to use important and relevant design context knowledge to provide proactive and intelligent design support for engineering design through background reasoning. For this purpose this research identified different categories of design context knowledge These categories can be abstracted into three main groups (User related, Life Cycle related, General). This identification stems from the work done by the authors and other researchers in the areas of design synthesis for multi-X as well as product life cycle modelling [5][6][7]. It is noted that these categories of context knowledge are by no means exhaustive. There could be even more knowledge groups/categories that should be considered depending upon the nature of design problem under consideration, however in metal component design particularly in sheet metal component, these categories can be used to explore the knowledge important for consideration at conceptual design stage. These ten categories are briefly summarised:-

This category of context knowledge deals with the users of the product. Any specific requirements of the user are defined here e.g. colour preference, time impression of a product, less sharp edges, easy to handle, modular etc. Reasoning using product user requirements can help designer by gaining an insight about the user preference, which is more suitable to the user. An example could be a requirement of insulation for metal components to avoid hot contact in working environment.

Knowledge related to product material properties is essential for selecting a particular solution to a given functional requirement. It includes general material specifications of the components like strength, durability, allowable stresses, hardness etc. Timely prompting the designer using background reasoning about material properties would help designer in selecting those solutions/PDEs, which are feasible.

It is the measure/degree of fulfilling the intended function by a solution in different working environment/conditions. This also implies how much a selected solution/PDE deviates from desired behaviour due to the quality of the solution and the influence of working environment. This knowledge could be the adaptability of selected solution/PDE to different working conditions like temperature resistance, vibration resistance, and shock/impact load resistance. An example could be measure of slack in friction belt due to high temperature generated at high speed rotation of two pulleys in order to convert rotary motion into rotary motion at different axis.

Context knowledge includes preparation of components and additional items required if any during realization/manufacturing of PDEs. This type of context knowledge is normally referred as Life cycle more specifically context knowledge in the form of life phase system's constraints. Reasoning using pre production requirement information involves evaluating and comparing time and cost required incurred on pre production processes/items for different solutions. This is an important source of knowledge about the constraints that manufacture/assembly systems impose on design decisions of a solution product. Designers are often unaware of these limitations and as design decisions become more related to other factors, it is very difficult, if possible, for designers to foresee these potential decision consequences. Through the use of these Life Cycle Consequences (LCCs), it is possible to remind designers proactively the potential consequences of their decisions.

It involves knowledge about actual manufacturing/production requirements for a solution/PDE to be manufactured onto the component. This type of knowledge is important for designer not only to analyse the ease of manufacturing solutions/PDEs on the components but also to compare the cost incurred in manufacturing each of these solutions/PDEs, thus giving support to the designer in selecting low cost/easy manufacturable solutions that involve less manufacturing time and low manufacturing cost.

Post-production requirement describes if a special process is needed after manufacturing/inscribing PDE on the component. An example could be retightening of a nut in case of *Hole-fastener* as solution to a '*Assembly*' function between two products during service/use. Reasoning using this type of context knowledge generates consequences about life phase systems (Maintenance/Service) and helps designer in avoiding unintended problematic/costly consequences. The consequence in this example could be the time and cost of equipment incurred in retightening of nut. Therefore it is necessary to compare the time and cost of equipment required if any during use/maintenance/service phase of a product among all the potential solutions/PDEs to select the low cost/time solution.

Timely prompting designer about the type and cost of machine/tooling that would require to manufacture/realize a selected solution/PDE will help in making a cost effective decision as more costly and increased number of machines will add up to increased overall lead time and product cost. An example could be the use of fine blanking dies for high surface finish in punching/blanking operation of sheet metal components instead of ordinary dies which are less costly, but requires a secondary (trimming) operation to get high surface finish of product.

Quantity of product required plays an important role in selecting a particular solution/PDE to realize a particular function. The quantity of product directly affects the selection of production methods. High equipment cost could be justified if large quantity of components is to be made, due to return in profit of mass production of components. Therefore the information about quantity of product, which can be cost effectively produced is necessary for estimation at conceptual design stage to select a solution.

Time required and level of difficulty to realize PDEs vary considerably. It is therefore necessary to consider the achievable production rate of each selected means and associated cost using the selected production equipment before making a final decision to go ahead with the selected design solution to realize a functional requirement. Higher achievable production rate will not only reduce the lead-time of the product but also reduces the lab overhead costs thus reducing the overall product cost.

A selection of a PDE/solution with high degree of available quality assurance techniques helps in avoiding accidents or breakdowns due to performance of solution during use. This results into low maintenance cost as well reduced time in maintenance/repair work.

Having introduced the above ten categories of design context knowledge, design decision-making becomes much more comprehensive and holistic. At the same time, this approach may also cause many complications and potential contradictions and conflicts. This research proposed and partially developed a reasoning mechanism to solve the above problem.

3.2   Reasoning mechanism

Suitable PDEs are initially selected on the basis of desired functional requirements using dictionary of proven function-PDEs association. This dictionary can be developed by writing function-PDE mapping algorithm on the basis of knowledge available about different functions, PDEs and their relationships in literature, through experience and past case studies. Given functional requirements are used in order to identify context of design problem, which is decomposed into three groups of context knowledge requirements. These three groups are further decomposed into different knowledge categories as shown in Figure 3 and discussed in three previous sections. The number of categories in each one of three groups depends upon the nature of design problem under consideration. Thus context knowledge is used in order to classify functional requirements into different knowledge requirement categories. This whole process of converting functional requirements into different categories of knowledge requirements is encompassed as functional model in this research. The generated PDEs can be further decomposed into different attributes like *Material attributes* (Name, Physical properties), *Form attributes* (Shape, Form, Structure) and *Surface Finish* attributes (Type of Finish, Degree of Finish). This decomposition process results into Form/Structural model of PDEs/Solutions. Current working knowledge in terms of partial solution information (information generated upto the current stage of design process) is elicited from these decomposed PDEs using design context knowledge base. This current working knowledge is further decomposed into same

 Fayyaz Rehman, Xiu-Tian Yan

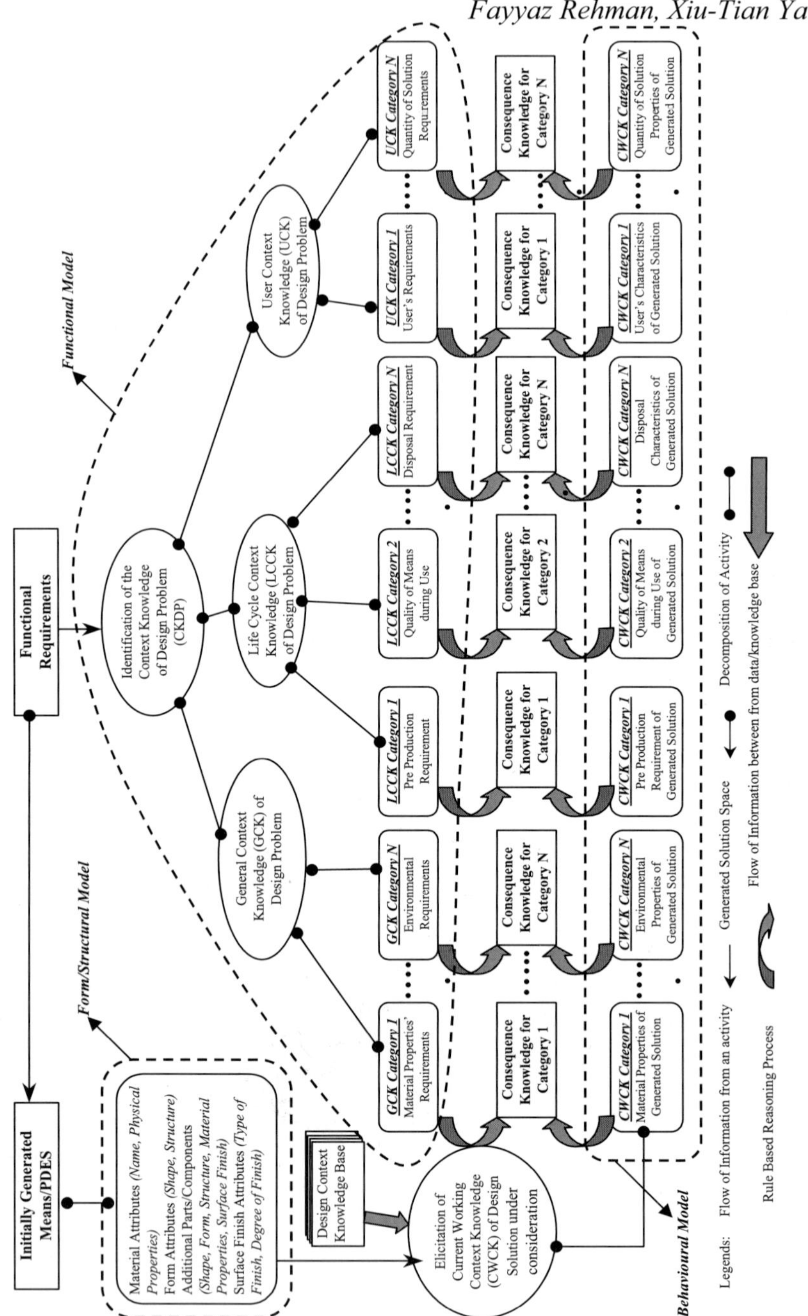

Figure 3. Design Context Knowledge based Reasoning Mechanism

number of knowledge categories as that of functional requirements under three different groups. But these pieces of knowledge are in the form of available/generated properties for each of the design solution/PDE under consideration. These categories of generated context knowledge forms the Behavioural model as behaviour of a function is context sensitive and as such, behaviour comes into play only in the context of design environment.

Having functional requirements as context knowledge in different categories as well as generated information about each solution/PDE, rule based reasoning is used to elicit the context consequences for each category. Thus potential good or bad consequences by simultaneously reasoning the context knowledge in one category and generated context knowledge of the PDE/solution under consideration for selection at the moment under the same category are highlighted for the consideration of designer.

4.   DATABASE INTERFACE FOR REASONING

The process of reasoning for the solution alternative under consideration for a particular function is repeated for all ten categories discussed in section 3.1. This situation presents this problem as a problem of combinatorial explosion and can be represented mathematically as:

$$CS = \sum_{i=1}^{i=m} F_i \forall \left\{ \sum_{j,k=1}^{j=n,k=10} \left(S_j \forall C_k \right) \right\}$$

CS denotes the consequence space, which would be generated due to the reasoning process to be performed to realise different functions of a product, *F* is a functional requirement, *and S* is solution alternative and C a context knowledge category. There can be several different pieces of information under ten different context knowledge categories identified in section 3.1. The computerisation of such vast amount of information could be a key to provide efficient support to functional design. Efficient and effective organization, storing and retrieving of this information is possible through a database interface in a computer-based support system. A schematic diagram of relational database interface adopted in this research during the reasoning process is shown in Figure 4.The flow of information towards and from database in the Figure 4 takes place in the current active session of design process. When the initial context knowledge is stored in the database the reasoning mechanism gets this knowledge and compares it with already stored knowledge for that category in the database. This comparison results in corresponding consequence knowledge for that category already stored in the database to be presented before the designer for consideration in order to anticipate problems that are likely to occur due to selection of currently considered alternative solution to the problem. Thus process of reasoning

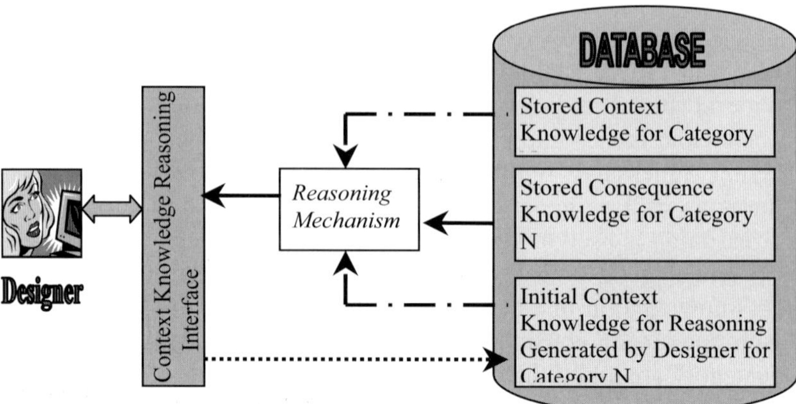

Figure 4. Database Interface during reasoning process

and data maintenance can be truthfully accomplished by employing database interface in a computer-based environment.

5.   CONCLUSIONS

Reasoning using context knowledge can further assist designers to concentrate on exploring design alternatives and generate more innovative design solutions thus reducing/eliminating the chances of redesign by considering manufacturing implications and increased costs earlier at conceptual design stage due to the selection of a particular solution. Context knowledge categories identified in this research are generic and can be used for any type of mechanical component design with a little modification. Moreover these categories show how vital this information in supporting designer's decision making at conceptual design stage.

REFERENCES

1.   Borg C J, Yan X T, Juster N, "Exploring decisions' influence on life-cycle performance to aid Design for Multi-X", *Artificial Intelligence for Engineering Design, Analysis and Manufacturing*, **Vol. No. 14**, pp. 91-113, 2000.
2.   Welch R V and Dixon J R, "Representing function, behaviour and structure during conceptual design", *Proceedings of ASME Design Theory and Methodology Conference*, **Vol. No. 42**, pp. 11-18, 1992
3.   Borg C J, Yan X T, Juster N, "Guiding component form design using decision consequence knowledge support", *Artificial Intelligence for Engineering Design, Analysis and Manufacturing*, **Vol. No. 13,** pp. 387-403, 1999.
4.   Rehman F, Yan X T, "Product design elements as means to realise functions in mechanical conceptual design", *Proceedings of 14th International Conference on Engineering Design ICED 03*, Stockholm, Sweden, AUGUST 19-21, 2003.
5.   Yan X T, Rehman F, Borg J C, "FORESEEing design solution consequences using design context information", *Proceedings of the Fifth IFP Workshop in Knowledge-Intensive Computer –Aided Design*, pp 18-33, 2002.

VIRTUAL PRODUCT DEVELOPMENT
-Solution for China's Automobile Industry

Wenli Zhang, Dieter Roller, Wenlin Chen and Hongfu Zuo
Institute of Computer-aided Product Development Systems, University of Stuttgart, IRIS Universitätstr. 38 D-70569 Stuttgar, (Wenli.Zhang,roller)@informatik.uni-stuttgart.de Dicastal Wheel MFG Co.,LTD R&D Dept China wine0335@etang.com, Nanjing University of Aeronautics and Astronautics China rms@nuaa.edu.cn

Abstract: The automobile industry has been witnessed transformed in recent years, leading automakers and their new competitors in developing countries face more fierce competitive markets. Virtual product development (VPD) are regarded as important approach to improve quality, cost/value and timeliness, In this article, the history, current and trend of solutions for automobile product process will be firstly studied, followed by a comparison between German and Chinese automobile industry. The VPD actuality in their respective automobile enterprises will be illustrated with examples such as DaimlerChrysler, First Automobile Works, Beiqi Fotor Motor, etc.. As conclusion some suggestions about VPD strategy to China's automakers will be offered.

Key words: Automobile industry, Virtual product development, CAD, CAE, CAM

1. INTRODUCTION

When many industries face the weakness and resulting uncertainty in the highly interdependent global economy, the automobile industry is also at a crossroads. On the one hand, the auto industries of developing countries have been transformed by trade and investment liberalisation policies and the global expansion of the auto industry. Governments' active promoters of the auto industry through investment incentives, local content regulations and tariffs, have brought up a new wave of automakers, which stands poised to make competitive leaps. For example in China there are domestic automakers such as First Automobile Works (FAW), Dongfong Motors and

Shanghai Automotive Industry Group (SAIC). On the other hand, in the mature, affluent markets, several of the largest established automakers and their affiliates have acutely felt that they must struggle to cope with an oversupply environment, especially while demand has to take off in emerging markets such as China and India, where huge manufacturing and assembly investments have been made. Therefore these large developing countries – notably China, India, Mexico etc. with a sizeable, growing middle-class – represent huge potential demand for private vehicular transportation, both low-cost production sites and the last remaining battlefields for automakers [1].

In such fierce competitive markets, achieving and sustaining competitive differentiation is the foremost challenge for automotive primacy and their new competitors. Virtual product development is regarded as very important approach to improve product quality, cost/value and timeliness, which are three fundamentals of competitive differentiation.

In the past decade, strategy of virtual product development (VPD) has been carefully defined to improve the product creation processes both in leading automakers and their new competitors in developing countries. There are certainly many differences between them, because of the regional, historical and culture factors. In order to offer some suggestions about VPD strategy to China's automakers, the history, current and trend of tools solutions for automobile product process will be studied. As an exemplar, virtual product development process in DaimlerChrysler will be in-depth illustrated., which followed by the study about VPD actuality in First Automobile Works, Beiqi Fotor Motor to give an overview about the VPD current status in China's automobile enterprises.

2. SOLUTIONS FOR AUTOMOBILE PRODUCT PROCESS - HISTORY , CURRENT, TREND

2.1 History

Car and aerospace industry are always the most important motivation of computer aided product development.

Before the digital revolution of the late 70's, the realization of the designer's styling concepts, the definition of body sheet metal, and the design and construction of stamping and assembly tooling relied entirely on physical models, which cost time, are very expensive, inaccurate and unrepeatable [2].

60's wire-frame 3D digital models can only describe simple geometry information, like point, line. Till 70's invention of Bezier Algorithm allowed the free form surface of auto and plane to be represented and manipulated. That was the first revolution in CAD history. Dassault's CATIA was the most successful CAD system at that time, which supports construction of free form surface. But such CAD systems were very expensive, and incapable for describing other product features like quality, center of gravity and so on. In 1979, SDRC released IDEAS, which was the worlds first solid based CAD/CAE system. But limited by hardware at that time, after 10 years such system had been widely adopted. The third revolution of CAD was the parametric design system, PTC 's Pro/E is the leading of them, which focuses on the great users of middle and small firms. To overcome the disadvantages of parametric design system, SDRC put forward advanced variation based solid modeling technology and completely rewrote whole system, in 1993, released IDEASMasterSeries [3].

In short 40 years, from 2D electronic drafting, 3D Design and Digital Mock-up, to current Virtual Product & Process, tools for the product development process which is becoming increasingly complex, has been greatly changed.

2.2   Current

The term "virtual product development" refers to the design, analysis and prove out of the product design and manufacturing processes by use of computers well before any physical products are constructed. Digital computers with powerful application software are used to define the product geometry, test the product, design the process steps, analyse and simulate manufacturing operations, simulate the ergonomics, and develop control code for the automation. Following widespread systems present state-of-the-art of current VPD.

I-DEAS NX Series form EDS is a scalable, integrated, CAD/CAM/CAE solution that supports design of new products in communication environment. Its uniquely ability is to develop digital master product models with CAE/CAD/CAM software technology. Digital master models contain advance information about the product's shape, behavior and cost long before the fabrication of costly physical models, and by providing common ground for interrelated product development tasks, it enables teams including manufacturing and production, as well as marketing, management to work concurrently. Digital master models can help designers and engineers develop and evaluate multiple design concepts so that the final product more closely matches customer expectations, as well as minimize scrap, reduce downtime and eliminate wasted or redundant operations in production process.

Instead of having individuals creating one piece of information at a time, the digital master model enables various disciplines to work together much earlier in the product development process, which can help reduce time-to-market [4].

CATIA is the product design solution offering from Dassault Systèmes. It allows manufacturers to simulate all the industrial design processes, from the pre-project phase, through detailed design, analysis, simulation, assembly and maintenance

Modular in design, the CATIA product line adapts to all customer businesses including style and form design, mechanical design, systems and equipment engineering, managing digital mock-ups, numerical control, simulation and analysis, using an open and scalable architecture. Its latest V5R12 release , CATIA offers a business to tap into and reuse its knowledge, which means shorter development times so that a business can react more rapidly to changes in the market, and revolutionary design tools like "morphing" or functional design, which generate innovation by sparking creativity [5].

Unigraphics NX from EDS gives manufacturers a unique product development environment that helps create innovative, market-leading products. From the initial concept, through product design, simulation, and manufacturing engineering, it enables digital transformation of the entire product lifecycle.

NX delivers next-generation product development through Digital Decision Making, which is founded on advanced knowledge capture tools that distill complex knowledge about products, engineering, and performance into re-usable forms. With Predictive Engineering tools - Process Wizards and Assistants, manufacturers can preload the product development process with manufacturing and systems performance knowledge to automate and optimize design decisions. The entire development process benefits from decisions that have been proven in the past [6].

PTC's Engineering Solutions for Computer Aided Styling, Design, Analysis and Manufacturing that support the entire vehicle product development lifecycle from where it starts with the Industrial Designer to Job 1. The solutions include:

CDRS – tools for the Industrial Designer

ICEM/Surf – is the premier surface modeling tool for engineers who need to accomplish reverse engineering, develop "A" Class surfaces, body-in-white and body engineering, trim and hardware as well as supporting tooling applications.

Pro/ENGINEER – is the most widely used professional 3D parametric modeling tool today. In the context of the automotive industry, it is widely used for engine design, drive train development, suspension and chassis applications

Division/Mock-up and Virtual Reality - provides digital mock-ups and electronic prototyping for virtual simulation and analysis, bringing distributed design teams into a single homogeneous review environment and supporting very large assembly review (complete vehicle definition with all component parts) [7]

Usually there are more then one sort of CAD system used in an automobile enterprise, these different systems aid one other, and usually are organized around a kernel CAD system [8]. The Table 1 shows us what some automakers have chosen for their kernel CAD system.

Table 1. Kernel CAD systems in automobile industry

Manufacturer	Nation	Kernel CAD	Date
Ford	U.S.A	I-DEAS Master Series	1995/12/19
Nissan	Japan	I-DEAS Master Series	1998/01/07
Renault	France	I-DEAS MS+Euclid	1998/02/11
Mazda	Japan	I-DEAS Master Series	1996/12/19
DaimlerChrysler	Germany	CATIA	1996/02/05
FAW	China	PTC Solutions	
Foton	China	PTC Solutions	

2.3   Trend

Now in leading automobile enterprises, 3D design and Digital Mock-up (DMU) have been almost completely applied through product process. With the vision on Virtual Product Creation: Complete virtual car, Near-physical product simulation, Realistic simulation in any situation with complete environment, Customer "designs" the product, Intelligent product model, Automatic process optimization, and so on, researchers identified respectively under these four partial objectives: DMU, Virtual Reality, Virtual Manufacturing , Product Design & Manufacturing Strategies in order to move towards the full virtual product concept, and progressive knowledge base & virtual prototype [9].

DMU focuses on a progressive enrichment and integration of the present functional DMU concept, to include increasing functional and life-related features, in an integrated way. Virtual Reality deals with improving the human interaction with the virtual product, both for development and for user oriented design and evaluation. The center of Virtual Manufacturing are means for planning and design the manufacturing plant, directly from the virtual product representation, both in terms of feasibility and required quality. Product Design & Manufacturing Strategies aims at the inclusion of best practices, and market and environment related knowledge in the virtual product representation, to be exploited in the target setting – deployment – achieving cycle, for both design and manufacturing.

In the future, as we gain a better understanding of automobile design processes, generative design software will become available for both product and process. Finite element modeling and dynamic simulation tools will reduce the cost and improve the performance of manufacturing tooling. Database tools will provide for easier access to digital designs and equipment performance histories, which are essential for a learning organization. Rapid prototyping tools will further streamline the tooling development process by eliminating steps in the manufacturing process. And on the factory floor, intelligent processes will monitor and self-correct to virtually eliminate the kinds of variability we struggle to control today. With these prospects in mind, we can expect that the next 40 years will be even more exciting than the last.

3. ACTUALITY OF VIRTUAL PRODUCT DEVELOPMENT IN GERMAN AUTOMOBILE INDUSTRY

As leading automotive manufacturers, German automobile industry has invested heavily in computer aided design and manufacturing technology to improve the safety, quality, cost, and time to market of its products. Virtual Product Development is used as an important approach to improve competitive position in the market place and strengthen the position as a premier automotive manufacturer. Here as an example the virtual automobile product development in DaimlerChrysler will be illustrated.

In the case here at DaimlerChrysler, the CATIA systems are used as kernel CAD systems [10]. For the Digital Mockup and Virtual Reality, DaimlerChrysler Research and Technology Virtual Reality Competence Center has developed a visualization and development framework – DBView, which offers real-time collision detection, physical simulation functionality, dynamic packaging applications, visualization of volume data, animation of complex finite automatons, object- and shade handling etc. advanced and complex functionality.

Using CATIA, and with the aid of other tools, such as DBView, integrated digital product development process are completed, this process chain spans the entire life cycle of the motor vehicle, encompassing design, construction, trial, production, quality and customer service right through to laying up and recycling the motor vehicle [11].

Product Concept and Design Starting in the design office, where new model vehicles are conceived, digital tools are used to model vehicle shapes. Sketching tools emulate paper, pen and paint. Modelers in virtual environment give the designer an almost true to life view of the new model.

Vehicle component design is today done entirely in CAD, computer aided design. Digital assembly of the vehicle insures that the parts will fit and function together.

By sending design information to downstream users like supplier of the electrical system DaimlerChrysler works with its suppliers. DaimlerChrysler's CATIA 3D models are used to drive suppliers' operations – from design to manufacturing, to eliminating paper, improving cycle time, cut costs and improving quality [12].

Manufacturing Tool Design and Development While the parts are being designed the virtual design and simulation of the die process for the stamped panel is also done. The 3D CAD model of panel is used for the stamping manufacturing process, the analysis tools are applied to minimize scrap in the stamping press and the number of operations within the die process. With the process feasibility established, computer based die design defines the stamping tools for making the panels. Casting configurations are customized within the CAD system. The finished design is inspected and adjusted in 3D CAD environment before build begins.

Computer based decision support tools are used extensively to capture process information prior to the actual modeling of the vehicle assembly operations, which include assembly sequence studies weld allocation and weld schedule development.

Assembly tools, like weld guns, gripper fixtures, holding fixtures and other tooling are today designed in CATIA. Once the tools are designed in 3D CAD, simulation software can be used to validate the process, cycle time, and ensure collision-free robot paths.

Digital Plant Digital tools are also used to design and simulate the plants. Assembly systems, manual workstations, conveyors, piping, and safety work envelopes are maintained within a CAD facility layout of the entire plant. Finally, because the product, process, and resource models are contained within the same CAD environment, the control code can be generated to program the automation from process and resource models.

Plant Production Computer based tools are used extensively on the plant floor. Quality data is gather using coordinate measuring machines and other measurement tools. Process control reports are generated as needed and are available to the appropriate personnel on the DaimlerChrysler Intranet.

Today's virtual product development systems, based on digital models, have transformed automobile industry. And next-generation knowledge base & virtual prototype system are coming into German automobile industry, will obviously improve the productivity and time to market.

4. ACTUALITY OF VIRTUAL PRODUCT DEVELOPMENT IN CHINA'S AUTOMOBILE INDUSTRY

As a new economy increase point for China, the automobile industry has obvious rapid growth since the country opened up to the outside world and adopted economic reforms. Now China is capable of manufacturing a complete line of automobile products, and there are total of 2,401 automotive enterprises (by the end of 2001). Apart from a few large automotive enterprises, most of them are middle or small enterprises, which benefits from the rapid growth of Chinese automobile industry, such as Beiqi Foton Motor Co., Ltd. (Foton), which was founded in 1996, but by the end of 2002, its automobile industry has covered all kinds of commercial vehicles except for passenger car [13].

For such infant enterprises with great ambition, advanced virtual product development technology are considered as a short cut to catch up the leading automobile manufacturers. Chinese automobile enterprises have invested a lot in purchasing computer hardware and software, training, researching and developing application system etc. Only PTC has sold nearly 1000 suits software to automobile manufacturers in China. These invest have already achieved great repay, for example, Foton had argued, it is the successful application of CAD, that creates Foton's miracle. But there are still many problem exist, and the difference between leading large automakers and middle or small enterprises are not trivial.

For example, FAW, which is the pioneer in CAD application in China, has applied CAD in product development process for a long time. Since 1999, it's new products were 100% designed using CAD. In these years FAW has bought over 170 suits software from PTC. With the help of other CAX systems, such as UGII , CATIA, SolidEdge, BUCLD etc. product design, analyse, verification can now be completed in an integrated CAD/CAE/CAM environment [14].

Foton can be used as an example to present the general status of virtual product development in china's automobile industry. When September 1997

CAD was firstly introduced in Foton as new technology of product development, the premier target is 2D computer aided drafting. But soon they have decided to design directly in 3D CAD systems, such as CDRS, Pro/E and UG, which firstly used as the tools for styling design and die producing. Now they focus on the more effective application of CAD/CAE Systems, try to design major vehicle components within 3D CAD system, simulate assembly process to insure that the parts will fit and function together, verify styles with digital prototypes.

Therefore for FAW and Foton completely 3D design and digital mockup are still today's focus. In China, there are still great number of small enterprises, which have just begun to apply 3D design tools.

5. CONCLUSION

In conclusion, by product development there are still a long distance between Chinese auto enterprises and leading automakers. In Chinese auto enterprises , after completely 2D computer-aided drafting, most of them are now focusing on 3D design and digital mockup in an integrated CAD/CAE/CAM environment. Only a few of them stand by virtual product and process. Contrariwise, leading automakers, such as large established German automakers are now pursuing the full virtual product development, knowledge base and virtual prototype are becoming the center in recent future. With following figure we can get an impression about it.

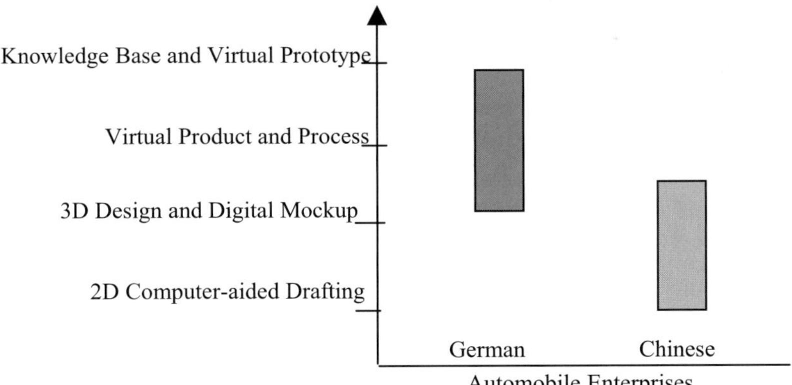

Figure 1. Solutions for Automobile Product Process in German and China

For China's automakers, a clear survey over the tools for product process is the pre-condition to make correct strategic plan to actualize virtual product development, and the experience of pioneers are valuable. But because of the large difference of development history, region, culture etc., the strategy for application of virtual product development should be carefully adjusted according to the product, finance, designer and engineer in individual enterprise. During implement of virtual product development, the kernel system must be carefully chosen, because it is very crucial to whole downstream works, such as selection of computer hardware and relative software, professional training, maintenance and so on. At last, computer hardware and software systems provide only a basis for product development, the most important factor is designers and engineers, who carry out all process. So the training is also necessary for the successful implement of VPD.

Globalization of automobile industry and the quick growth of automobile industry in developing countries bring with together opportunity and challenge to China's auto enterprises, successful application of virtual product development should be one of the solutions, which support their quickly growing. We can believe, that in such unsettled markets, all competitors, who are firm to seek their target with great ambitions, can become winners.

REFERENCES

1. Herbert K. Tay : "Achieving competitive differentiation: the challenge for automakers". *Strategy & Leadership*, **Vol.31**, pp. 23-30, NO.4 2003.
2. Daniel J. VandenBossche : "Chrysler - Virtual Manufacturing CATIA". http://www-1.ibm.com/solutions/plm/doc/content/casestudy/768669113.html
3. " Overview of CAD Technology Development History" http://www.e-works.net.cn/ewkArticles/Category29/article10221_1.htm
4. http://www.eds.com/products/plm/ideas/
5. http://www.3ds.com/en/brands/catia_ipf.asp
6. http://www.eds.com/products/plm/unigraphics_nx/.
7. http://www.automotive-technology.com/contractors/cad/ptc/index.html
8. "Trend of CAD/CAM/CAE Systems Selection in Automobile Industry"http://www.e-works.net.cn/ewkArticles/Category29/article10660.htm
9. EC Project IST-2002-37605: "The future of Virtual Product Creation - strategic roadmap". 12.2002 http://www.vip-roam.de/
10. Hans-Joachim Schöpf : "Durchgängige digitale Produktentwicklung". Virtual Product Creation 2003 – Automobile –und Motortechnische Konferenz, 30, Juli 2003, Stuttgart, Germany
11. www.daimlerchrysler.com
12. "DaimlerChrysler and UTA eliminate paper with help from CATIA". http://www-1.ibm.com/solutions/plm/doc/content/casestudy/768693113.html
13. http:/English.foton.com.cn/
14. http://www.faw.com.cn/

GRAPH BASED REPRESENTATION IN DESIGN OF MULTIFUNCTIONAL STRUCTURE

Yuemin Hou, Linhong Ji and Dewen Jin
Dept. of Precision Instrument and Mechanology, Tsinghua University, Beijing, 100084, P. R. China, hym01@mails.tsinghua.edu.cn, jilh@pim.tsinghua.edu.cn, jdw-om@tsinghua.edu.cn.

Abstract: This paper presents a design representation based on planar graph and dual graph. Substructures are represented by vertices and physical connections and signal transmission between substructures are represented by edges. By testing planarity of a graph, designers can evaluate the electronic circuit and material compatibility of embedded substructures and interconnected substructures. By generating dual graph, designers can develop the optimal layout of the electronic circuit and substructures. This graph representation allows for concurrent design and programming at both system level and subsystem level.

Key words: Design representation, graph theory, planar graph, dual graph, Multifunctional structure.

1.   INTRODUCTION

With the rapid development of Micro electromechanical system (MEMS) and spacecraft, the dramatic reductions in mass and volumes, compared to current designs, means significant economical benefit. For instance, the launch price of small satellite is around 5000-10000\$/kg, about 35%-40% of whole satellite program. On the other hand, one kilogram increase of the mass of small satellite will increase 200-300 kg of the mass of the rocket. So to reduce the mass is to reduce the cost. One solution to reduce the mass is to integrate different subsystem functions, such as signal and data transmission, thermal control, radiation and meteoroid protection, sensing, communication/surveillance, energy generation, and attitude control.

Lockheed Martin Astronautics has developed the Multifunctional Structure (MFS) concept "as a new system for spacecraft design which eliminates chassis, cables, and connectors and incorporates the electronics onto the walls of the spacecraft". [1]. Figure 1 is an illustration of MFS.

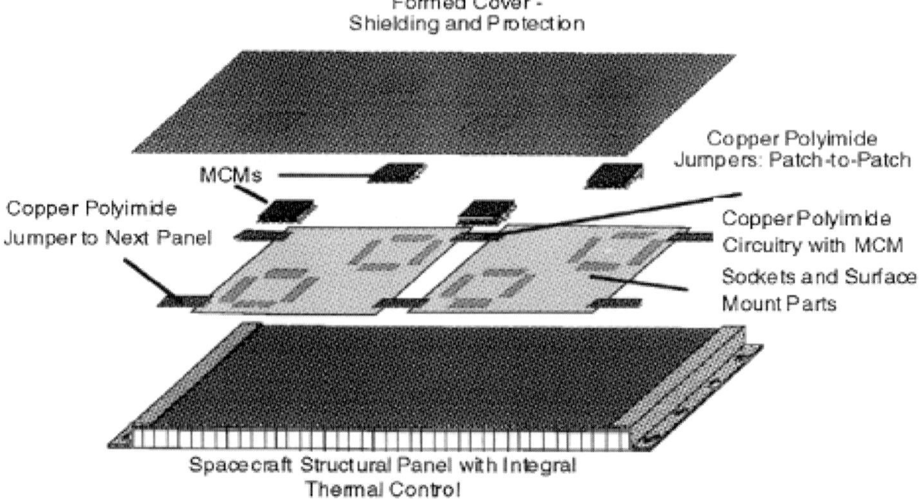

Figure 1 Illustration of MFS [1]

Some literature on MFS have been published during the past decade, most of them, to my knowledge, introduced the concept of MFS and experiment on MFS [2-4]. The approach of this paper is different with the literature published by placing emphasis on the information representation based on graph theory in the design of MFS.

Graph theory concepts have been used in conceptual design of mechanical system for decades. For example, kinematic chains and mechanisms (Dobryjanskyj, 1966), cam-modulated linkages (Pryor, 1979), automatic generation of the kinematic structure of mechanisms (Sohn and Freudenstein, 1986), gear trains and mechanism design (Olson, 1988), feature recognition of 3D solid model (Joshi an Chang, 1988), geometry in mechanical design (Chung, 1989), model of electronic, mechanical and hydraulic systems (Karnopp, 1990), transformation from conceptual design to embodiment design (Kusiak and Sczerbicki, 1993), tribological performance of mechanical system (Sharma, 1996), function of a design artifact (AL-Hakim, Kusiak, Mathew, 2000) (see [5] and reference), dynamic design verification (Deng, Britton, Tor, 2000) [6], mechanical system reliability analysis (Tang, 2001) [7], cable-membrane structures (Ivanyi, Topping, 2002) [8], qualitative description of the behavior of a system by graph describing the possible transitions between regions (Bernard, Gouze, 2002) [9], object-oriented modeling of physical systems for the design and control of machatronic system (Amerongen, Breedveld, 2003) [10]. In system analysis, graph theory was applied in analysis of large-scale systems (Kron, 1963), plastic analysis and design (Fenves and Branin, 1963, Fenves et al, 1964, 1971, 1977), formulation of dynamic motion equations (Andrews, 1971), efficient structure analysis (Kaveh, 1995, 1997), resistance graph representation of both electronic circuit and structures (Shai, 1999, 2001,) [see [11] and reference].

This paper uses planar graph and its dual graph to represent the MFS and the relationship between the comonents of MFS, aiming at providing a systematic information interface for multidiscplinary design team to benifit concurrent design of diferent discplines.

The reminder of the paper consists of following: the information representation in the design of MFS, including the core of the design of MFS, planar graph, dual graph, graph with weight; and finally a brief summary.

2.   GRAPH REPRESENTATION

2.1   The core of the design of MFS

The design issues of MFS can be classified into following groups.
1)   Miniature of electronic and optical elements.
2)   Microstructure of composite material and heterogeneous material, the distributions of material composition along the gradient direction of Functionally Gradient Material (FGM), as well as fabrication.
3)   Theory and methodology on the embryogenesis of structure.
4)   Design representation of integrating subsystems into a MFS.

The first two questions are the realm of electronics, optics, material science and so on, beyond the context of this paper. The third one is the core of design theory and methodology, but it will not be discussed here. The paper places emphasis on the forth question.

The issue of integrating subsystems into MFS can be expressed as following.
1)   Plane realization of the electronic circuit.
2)   The heterogeneous materials with embedded components can be represented as a planar graph if the materials of adjacent subsystems

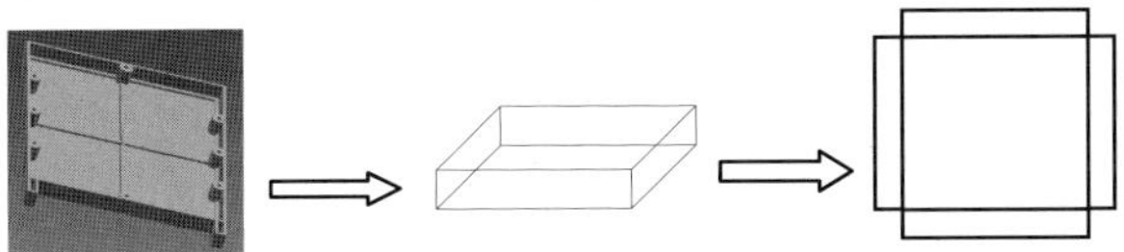

Figure 2 A modular box of a small satellite

are compatible.

Figure 2 is a modular box of small satellite weighting 50 kg, which consists of several boxes bolted together. To integrate several subsystems into a MFS, all subsystems and connections between them should be represented as a planar graph.

In general, the MFS is a panel. In case of a box, all walls of the box and the floor are allowed to place effective loads or equipments, so the box can be spread out as a plane, see figure 2. All the cables and jumpers are embedded in the composite material, and some electronic components or some pieces of piezoelectric materials, or some other components such as heat pipes are embedded in the composite material.

2.2   Information representation in the design of MFS

A design representation serves as two roles: one is a means of communication and the other is a vehicle for exploration of further knowledge of the intended artifact [12].

Graph theory provides a dimension to descript both characteristics of substructures and relationship between them.

The whole structure of MFS can be represented by a graph. Figure 3 shows a MFS panel and figure 4 shows its graph representation. Vertices represent substructures; Edges represent connections.

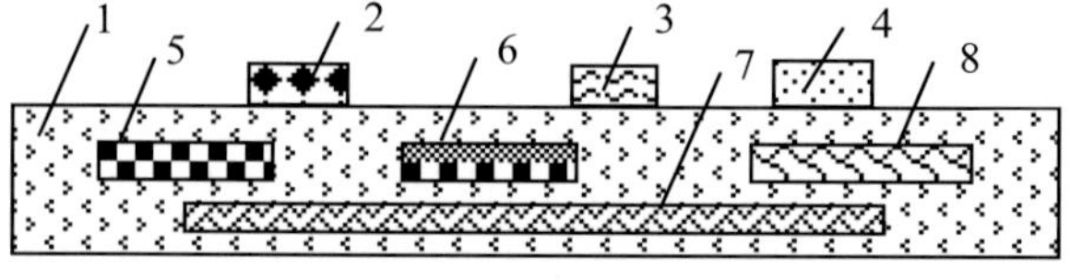

Figure 3 A MFS panel

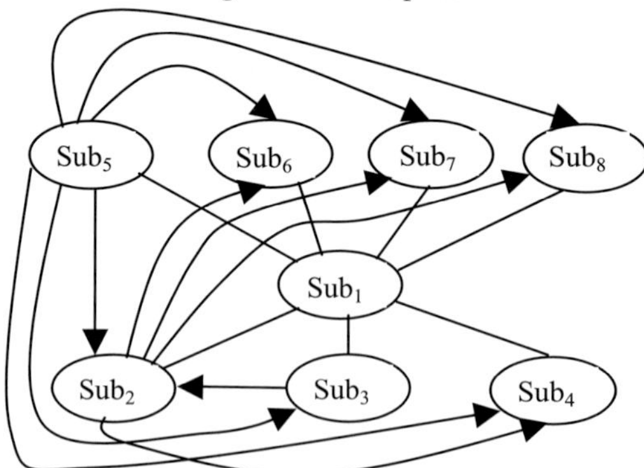

Figure 4 Graph representation of the MFS panel

2.3   Testing graph planarity

"Consider a graph drawn in the plane in such a way that each vertex is represented by a point; each edge is represented by a continuous line connecting the two points which represent its end vertices and no two lines, which represent edges, share any points, except in their ends. Such a drawing is called a plane graph. If a graph G has a representation in a plane which is a plane graph then it is said to be planar [13]. " In a word, a graph is planar if it can be drawn on a plane without edges crossing in any way.

Complex integrated circuits require several layers of circuit connections in their wiring, but it can be decomposed into a number of parts that are known to be planar [14]. On the other hand, material compatibility is represented as edge between two vertices, so all embedded substructures

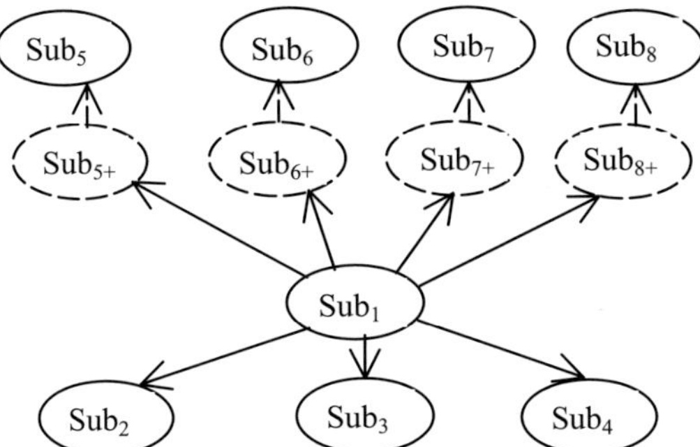

Figure 5a   Graph representation of substructures

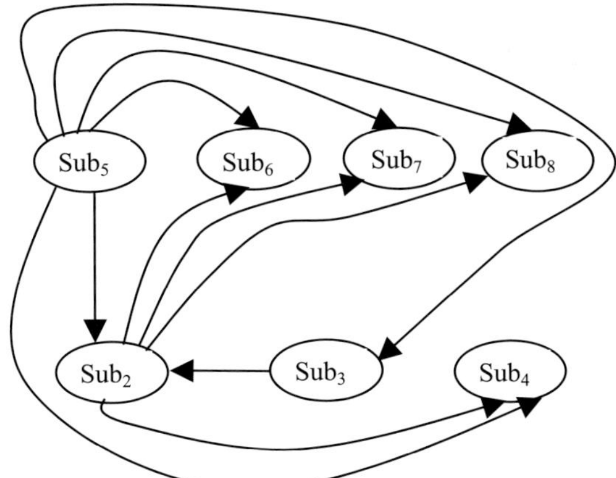

Figure 5b Graph representation of signal transmission

should be branches of a tree with base substructure as root. The base substructure is the substructure in which other substructures are embedded or to which other substructures are connected. If the materials of two substructures are not compatible, additional substructure is needed as functional grading material. Figure 5 is planar graphs of figure 4.

There are two known planarity testing algorithms which have been shown to be realizable in a way which achieves linear time $(O(|V|))$. The first algorithm starts by finding a simple circuit and adding to it one sample path at a time. Each such new path connects two old vertices via new edges and vertices. It is called path addition algorithm. The second algorithm adds in each step one vertex. Previously drawn edges incident to this vertex are connected to it, and new edges incident to it are drawn and their other endpoints are left unconnected. [13].

2.4   Duality

The graph G*(V*, E*) is said to be the dual of a connected graph G(V,E) if there is a 1-1 correspondence f: E-E*, such that a set of edges S forms a simple circuit in G if and only if f(S) (the corresponding set of edges in G*) forms a cut set in G*. [13]

Figure 6 is the dual graph of figure 5b.

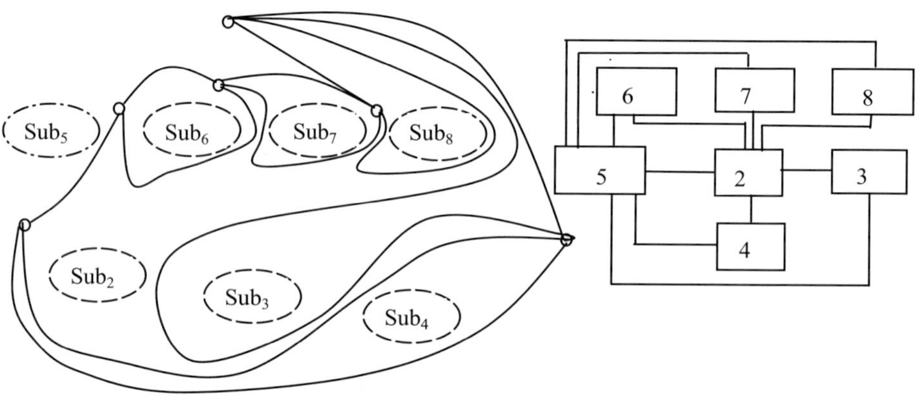

Figure 6   a. Dual graph of figure 5b.   b. Layout of electronic circuit

The faces of the dual graph represent substructure regions and edges of the dual graph represent signal transmission between two substructures. The dual graph shows the optimal layout of both electronic circuit and substructures.

2.5   Layout of the MFS

According to the dual graph, the optimal layout of the MFS panel can be achieved, see figure 7.

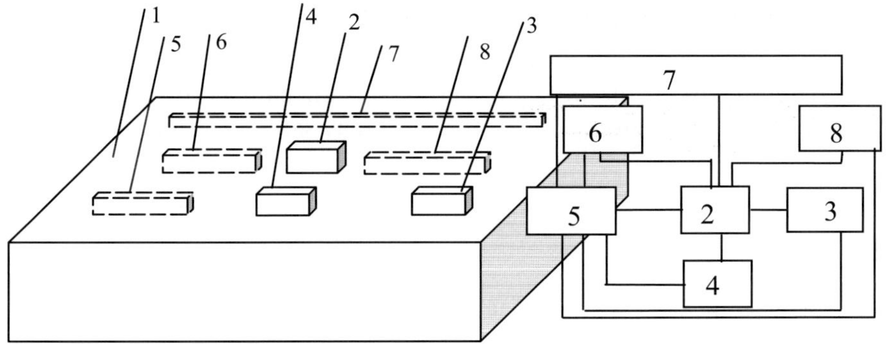

Figure 7 a. Layout of the MFS panel. b. The electronic circuit

2.6 Weighted Graph

Taking a flexible structure for example, see figure 8. Its graph representation is shown in figure 9.

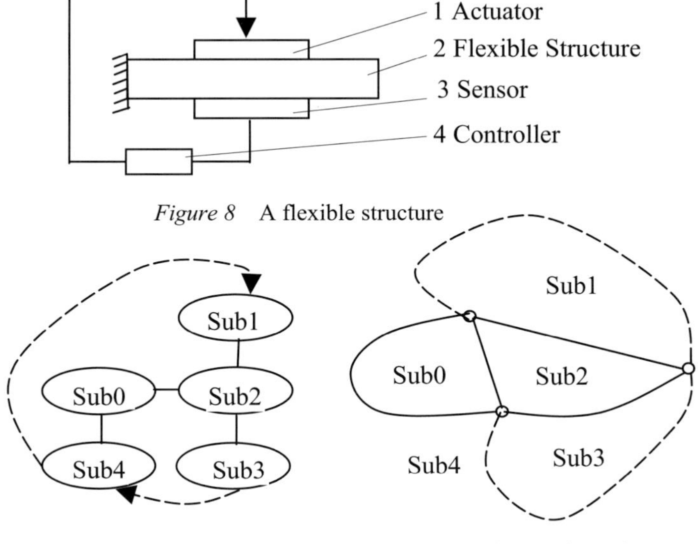

Figure 8 A flexible structure

Figure 9 *a.* Graph representation *b.* Dual graph

Assume a graph G(V,E). Let n, p, and q be the number of vertices, edges, and faces.

$$V = \{v_i, \Pr operty_i\}, i = 1, 2, \ldots, n \tag{1}$$

$$\mathrm{Pr}\,operty = \{\mathrm{Pr}\,operty_i \mid \mathrm{Pr}\,operty_i = [\tau_1, \tau_2, \cdots]\}, i = 1, 2, \cdots, n \qquad (2)$$

Where Property is the weight of vertex; n is the number of functions; τ is determined by designer; the dimension of τ_i may same or different.

$$E = \{e_l, R_l\}, l = 1, 2, \cdots, p \qquad (3)$$

$$R = \{R_l \mid R_l = [r_{ij}(Type, F, Ve, a), w_{ij}(u, I, V, B, T)]\},$$
$$l = 1, 2, \mathsf{L}, p, i = 1, 2, \mathsf{L}, n, j = 1, 2, \mathsf{L}, n \qquad (4)$$

Where T is the type of connection; F is generalized force, including force, rotation moment, bending moment; Ve is generalized velocity, including velocity, angular velocity; a is generalized acceleration, including acceleration and angular acceleration. For static connection, Ve=0.

Description of signal connection is w_{ij}. u is displacement; I is electrical current; V is Voltage; B is magnetic strength; T is temperature.

Assume G*(V*,E*,f*) is the dual graph of G(V,E).

$$V^* = \{v^*_k \mid v_k^* \leftarrow f_k, X_k^*\}, k = 1, 2, \cdots, q \qquad (5)$$

For figure 9,

$$V^* = \{v_1^*, v_2^*, v_3^*\} \qquad (6)$$

Let the subscript of f^* represents corresponding substructures.

$$v_1^* = (\forall v \in f_0^*) \wedge (\forall v \in f_1^*) \wedge (\forall v \in f_2^*) \qquad (7)$$

$$v_2^* = (\forall v \in f_0^*) \wedge (\forall v \in f_3^*) \wedge (\forall v \in f_2^*) \qquad (8)$$

$$v_3^* = (\forall v \in f_3^*) \wedge (\forall v \in f_1^*) \wedge (\forall v \in f_2^*) \qquad (9)$$

Let the subscripts of X^* represents corresponding vertices and substructures respectively.

$$X^* = \begin{Bmatrix} X_{10} & X_{20} & X_{30} \\ X_{11} & X_{21} & X_{31} \\ X_{12} & X_{22} & X_{32} \\ X_{13} & X_{23} & X_{33} \\ X_{14} & X_{24} & X_{34} \end{Bmatrix} \tag{10}$$

$$E_i^* = \{e_l^* \mid e_l^* \leftarrow e_l, R_l^* \mid R_l^* = R_l, L_l\}, l = 1,2,\cdots,p \tag{11}$$

$$\begin{aligned} E^* &= \{e_1^*, e_2^*, e_3^*, e_4^*, e_5^*, e_6^*\} \\ &= \{f_0^* \wedge f_2^*, f_1^* \wedge f_2^*, f_2^* \wedge f_3^*, f_0^* \wedge f_4^*, f_1^* \wedge f_4^*, f_3^* \wedge f_4^*\} \end{aligned} \tag{12}$$

$$f^* = \{f_i^* \mid f_i^* \leftarrow Sub_i, S_i\}, i = 1,2,\cdots,n \tag{13}$$

$$S = \{S_i \mid S_i = s_i(\Omega, m); \Omega \in R^3, m = \{E, \upsilon, \alpha_t, \alpha_d\} = \Phi(t, C, v), i = 1, n\} \tag{14}$$

$$v = \left\{ V_i/V, \sum_{i=1}^{r} V_i = 1, V_i \geq 0 \right\} \tag{15}$$

Where Ω is the topology of substructure; m is material; $E, \upsilon, \alpha_t, \alpha_d$ are Young's modulus of elasticity, Poisson ratio, thermal coefficient, electrical conductivity respectively. Material comprises microstructures, t is the type of microstructure, C is the geometry parameter, and v is volume fraction, which means the ratio of the volume occupied by the material(s) with respect to the total volume of the local region.

3. SUMMARY

The significant feature of MFS design is multidisciplinary. A team with members from design, structure, material, manufacturing, microelectronics, optics, control, communication, computer and the like is necessary to design subsystems concurrently because that the design decision in every phrase of design needs information of dynamics, thermodynamics, optics, electronics, control laws, materials and so on. Therefore, design representation must represent both systematic information and subsystem information in a way that benefit calculation, generating, coding and programming.

This paper presents a representation based on planar graph and dual graph. Substructures are represented by vertices, and physical connections and signal transmission between substructures are represented by edges. By testing planarity of a graph, designers can evaluate the electronic circuit and material compatibility of embedded substructures and interconnected substructures. By generating dual graph, designers can develop the optimal layout of the electronic circuit and substructures. This representation allows for concurrent design of a team and programming at both system level and subsystem level.

REFERENCES

1. Barnett DM, Rawal S, Rummel K. "Multifunctional structures for advanced spacecraft", *Journal of Spacecraft and Rockets*, **vol. 38, no. 2**, pp. 226-230, 2001.
2. Obal M, Stater J. "Multifunctional Structures: The Future of Spacecraft Design", *Proc. Of 5th Int. Conf. Adaptive Structures*, Sendal, Japan, Dec. 5-7, 1994, Basle, Technomic, pp. 720-734, 1995.
3. Barnett DM, Rawal SP. "Multifunctional Structures Technology Experiment on Deep Space 1 Mission", *IEEE Aerospace and Electronic Systems*, **vol. 14, no. 1**, pp. 13-18, 1999.
4. Rawal SP, Barnett DM, Martin DE. "Thermal Management for Multifunctional Structures", *IEEE transactions on Advanced Packaging*, **vol. 22, no. 3**, pp. 379-383, 1999.
5. Al-Hakim L, Kusiak A, Mathew J. "A graph-theoretic approach to conceptual design with functional perspectives", *Computer-Aided Design*, **vol. 32**, pp.867-875, 2000.
6. Deng YM, Britton GA, Tor SB. "Constraint-based functional design verification for conceptual design", *Computer-Aided Design*, **vol. 32**, pp. 889-899, 2000.
7. Tang J. "Mechanical system reliability analysis using a combination of graph theory and Boolean function", *Reliability Engineering and System Safety*, **vol. 72**, pp.21-30, 2001.
8. Iva'nyi P, Topping BHV. " A new graph representation for cable–membrane structures", *Advances in Engineering Software*, **vol. 33**, pp. 273–279, 2002.
9. Bernard O, Gouzé JL. "Global qualitative description of a class of nonlinear dynamical systems", *Artificial Intelligence*, **vol. 136**, pp. 29–59, 2002.
10. Amerongen JV, Breedveld P. " Modelling of physical systems for the design and control of mechatronic systems", *Annual Reviews in Control*, **vol. 27**, pp. 87–117, 2003.
11. Shai O. "Deriving structure theorems and methods using Tellegen's theorem and combinatorial representations", *International Journal of Solids and Structures*, **vol. 38**, pp. 8037-8052, 2001.
12. Jolion JM, Kropatsch WG, (eds.). *"Graph based representations in pattern recognition"*, New York, Wien, Springer, 1998.
13. Even S. *"Graph algorithm"*, Maryland, *Computer science press*, Chapter 7-8, 1979.
14. Tucker A. *"Applied Combinatorics"*, Third Edition, New York, JOHN WILEY & SONS INC., 1995.

PATCH UNIT METHOD FOR CAD SURFACE MODELS

Xingbo Wang and Jin-Long Shi
*College of Machatronic Engineering & Automation, National Univ. of Defence Tech.,
Changsha 410073, China; Email : xbwang@nudt.edu.cn, xbwang@doctor.com*

Abstract: Many engineering models are geometrically expressed by piecewise continuous composite-surface made up of simple smooth surface-patches in CAD modeling. This CAD model is usually too complicated for conventional numerical methods like finite element method(FEM) or boundary element method(BEM) or element free method(EFG) to perform instant simulation of many industrial requirements. This makes it very meaningful to develop numerical methods fit for CAD needs.

This paper presents a numerical method that can fast and accurately compute mechanical properties of object with composite-surface structure. The method regards a simple surface-patch on the composite-surface as a computational unit, construct a framework of the composite-surface with all the boundary areas by dividing each unit into boundary area and inner area, and adopts FEM to build the computing equations for the framework and BEM/EFM to build the computing equations for the inner areas of all the units. The method is simpler and easier than conventional numerical methods both in mesh generation and in calculation.

Key words: CAD/CAE, numerical method, composite surface, manifold

1.   INTRODUCTION

It is an essential task for a CAD designer to analyze and evaluate the mechanical behaviors of the designed model. Finite element method (FEM) and boundary element `ethod (BEM) are historically two major tools to do such analysis and evaluation. But the two methods reveal many drawbacks with the increasing requirements of contemporary industries. E.g., BEM cannot treat non-linear problem very well and FEM has fatal faults when treating problems like crack propagation, fragmentation and large deformations due to its excessive and time-consuming mesh generation. So people have been seeking better methods in the past decades.

In 1990s, Dr. Shi G H put forward a numerical manifold method (NMM) to compute rock deformation [1-2]. NMM uses two mesh systems to carry out the computation, one is called physical cover for the calculated object field and the other is called mathematical cover for constructing shape functions. Although NMM exhibits many advantages compared with FEM, its disadvantages like treating small deformation and static force problems slows down its advancing steps [3].

In an overall view, it is mesh generation that keeps FEM and the like methods from excellence. So meshless method or element free method (EFM) became a focus in the history. In 1990s, Belytschko T raised an Element Free Galerkin Method (EFGM) [4]. By a set of discrete nodes EFGM can compute the calculated field accurately and in a high precision. The method soon became a hot research point in computational mechanics [5] and now is still in study [6]. Except EFGM, EFM has many other styles such as Finite-Cover-Based Element-Free Method (FCEFM)[7] and Natural Neighbor Galerkin Method (NNGM)[8]. All the EFM have the disadvantage of low computational efficiency especially for complicated geometry domains, which is not fit for the instant simulations such as the simulation of die-forming process.

In CAD projects, people frequently need instant simulations of various kinds. A CAD model is geometrically a piecewise continuous composite-surface and this composite-surface is very complicated in many cases, e.g., the mould surface of a decorative plastic animal face. This will result in either a difficult mesh generation for FEM/NMM or a time-consuming process for EFM. Thus the methods fit for CAD projects are still in need.

This paper presents a numerical method that can fast and accurately calculate the mechanical properties on the objects with composite-surface structure. The method regards a simple surface-patch on the composite-surface as a computational unit, constructs a framework of the composite-surface with all the boundary areas by dividing each unit into boundary area and inner area, and adopts FEM to build the computing equations for the framework and BEM/EFM to build the computing equations for the inner areas of all the units. The method is simpler and easier than conventional numerical methods both in mesh generation and in calculation. In later sections we call this method patch unit method (PUM).

2. PATCH UNIT METHOD FOR CAD SURFACE MODELS

In modern geometry theory, a two-dimensional manifold is a complex surface that each of its point has a neighbourhood homeomorphic to an open set of two-dimensional Euclidean space. CAD modeling theory shows that a CAD model is geometrically a piecewise continuous composite-surface which is made up of finite simple surface-patches, as shown in figure 1.

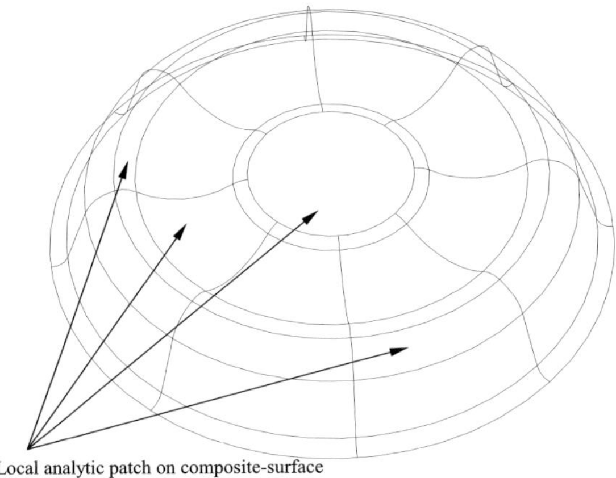

Figure 1. Surface of a CAD model is a two-dimensional manifold

It can be seen that this composite-surface is mathematically a two-dimensional manifold. Thus theoretically we can find a manifold structure on the composite-surface and develop a numerical method to compute the manifold data corresponding to the traits of the manifold structure. PUM is such a method that will be illustrated in next subsections.

2.1 Manifold Expression of a CAD Model

A surface patch U_α is such a set in 3-dimensional Euclidean space $\mathbf{E}^3$ that can be one-to-one mapped onto a planar area $D(u, v)$, namely , there exists a one-to-one mapping φ_α such that each element $X(f, g, h) \in U_\alpha$ can be expressed as:

$$\varphi_\alpha : f = f(u,v), g = g(u,v), h = h(u,v) \qquad (u,v) \in D$$

where $f(u, v)$, $g(u, v)$ and $h(u, v)$ are C^3 functions in $\mathbf{D}$.

U_α together with its coordinate mapping φ_α is conventionally denoted as $(U_\alpha, \varphi_\alpha)$, called a local coordinate system of U_α.

If the rank of the matrix

$$\begin{pmatrix} f_u & g_u & h_u \\ f_v & g_v & h_v \end{pmatrix}$$

is 2 at each point of U_α, U_α is called smooth and U_α is always supposed smooth later.

Let U_α and U_β are two patches, $< U_\alpha, x^1, x^2, x^3 >$ and $< U_\beta, y^1, y^2, y^3 >$ are their local coordinate systems respectively. If $U_\alpha \cap U_\beta = U_{\alpha\beta}$ and $U_{\alpha\beta}$ is not empty, we say there is a connection Ψ on $U_{\alpha\beta}$ between U_α and U_β, where Ψ is the mapping:

$$\Psi : x^i = x^i(y^1, y^2, y^3), i = 1,2,3 \tag{1}$$

and Ψ has inverse Ψ^{-1}:

$$\Psi^{-1} : y^i = y^i(x^1, x^2, x^3), i = 1,2,3 \tag{2}$$

Thus on $U_{\alpha\beta}$ the Jacobian matrix

$$\frac{\partial(x^1, x^2, x^3)}{\partial(y^1, y^2, y^3)} = \left(\frac{\partial x^i}{\partial y^j} \right), i, j = 1,2,3 \tag{3}$$

is non-singular.

Now we give a definition for a composite-surface.

A surface S is such a set in $\mathbf{E}^3$ that has the following properties:

1. S is covered by smooth surface patches, namely, $S = \bigcup_\alpha U_\alpha$;

2. each U_α is one-to-one mapped onto a (u, v)-plane area $D_\alpha(u_\alpha, v_\alpha)$, denoted as $\varphi_\alpha: D_\alpha \to U_\alpha$;

3. If $U_\alpha \cap U_\beta = U_{\alpha\beta}$ and $U_{\alpha\beta}$ is not empty, there exists a connection on $U_{\alpha\beta}$ between U_α and U_β and the corresponding inversible $\varphi_{\alpha\beta}$: $D_{\alpha\beta} \to U_{\alpha\beta}$ exists;

4. Suppose P, Q are two different points on S, $P \in U_\alpha$, $Q \in U_\beta$; then there exists an $\varepsilon > 0$ such that the ε-neighbourhoods of P and Q have not intersection.

If $U_{\alpha\beta}$ is a continuous curve $\gamma_{\alpha\beta}$, then $\gamma_{\alpha\beta}$ is called the joint boundary between U_α and U_β; If all $U_{\alpha\beta}$ are the joint boundaries, then S is called a composite-surface.

It is easy to see that surface S together with its coordinate systems $(U_\alpha, \varphi_\alpha)(\alpha = 1, 2, \ldots,$ finite number) form a manifold structure. Thus the surface of a CAD model, a composite-surface, is a two-dimensional manifold, and its manifold expression is as follows:

$$S = \bigcup_\alpha U_\alpha \tag{4}$$

where two adjacent patches are connected by their joint boundaries.

It is easy to get the following traits about S:

1. No matter how complex S is, each of its patch U_α has an exact mathematical (parametric) expression $U_\alpha = U_\alpha(u,v)$ and U_α is smooth. This is called local analyticity of manifold S.
2. Two adjacent patches U_α and U_β are at least C^0 connected and never overlapped. This is called global connectivity of manifold S.

2.2 Patch Unit Method for CAD-model-based Manifold

Let S be a manifold given by (4). For convenience, from now on, we think S to be a thin shell and exerted by a force F, and our task is to compute the corresponding mechanical behaviors of S.

Suppose U_α has n boundaries and denote Γ_α the set of the boundaries: $\Gamma_\alpha = \{\gamma_1^\alpha, \gamma_2^\alpha, ..., \gamma_n^\alpha\}$, then all patch boundaries $\{\Gamma_\alpha\}$ form a boundary mesh B on S. Obviously, $B=\{\Gamma_\alpha\}$.

We first take a single patch U_α as our computational unit. Although we have many methods like FEM, BEM, NMM or EFM to perform the computation for such a single patch, we'd rather adopt a little complicated way to make later computation of S simple.

To do this, let's first imagine that we cut U_α into two parts by a slice along Γ_α, then there appears a new boundary γ_i near each original boundary γ_i^α, as shown in figure 2. Denote the inner part closed by γ_i ($i=1,2,...,n$) as Ω_1 and the outer part (boundary part) as Ω_2.

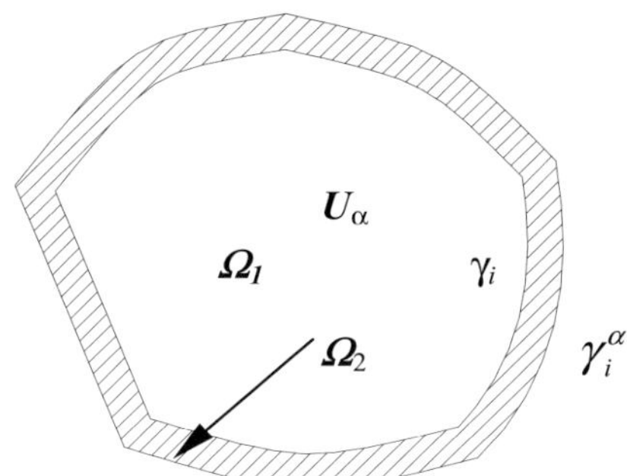

Figure 2. Patch and its boundary deformation

It is true that Ω_1 and Ω_2 will deform if U_α is under a charge or a deformation. Let $D_1(\mathbf{p})$ and $D_2(\mathbf{p})$ be the displacement vector at point $\mathbf{p}$ in Ω_1 and Ω_2 respectively, then the following must be satisfied:

$$D_1(\mathbf{p}) = D_2(\mathbf{p}) \quad \mathbf{p} \in \Gamma \quad \Gamma = \bigcup_{i=1}^{n} \gamma_i \tag{5}$$

Now we perform the computation on U_α.

2.2.1   FE-BE Blending Method

If the force F causes just a small linear elastic deformation on U_α, we use FEM to compute Ω_2 and BEM to compute Ω_1 because BEM only needs nodes on the boundary of Ω_1 and can have good result when Γ is regular. To do this, we make such a discretization on Ω_2 that the boundary is separated into n segments, and on boundary γ_i there are i_i nodes, as shown in figure 3.

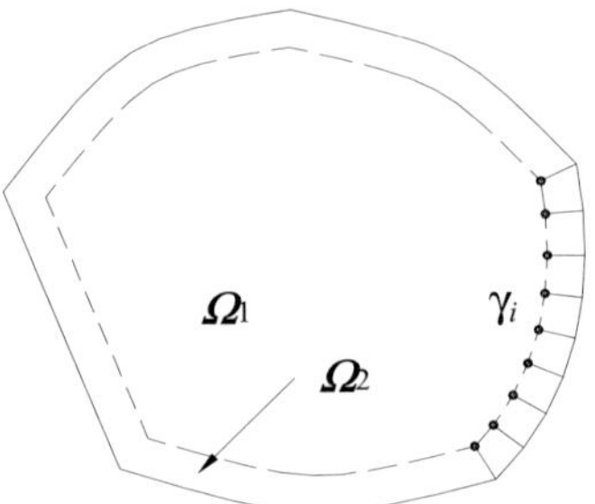

Figure 3. FEM-BEM discretization

Thus on the boundary of Ω_1, there are $N_\alpha = i_1 + i_2 + \ldots + i_n - n$ nodes. Using these nodes and by BEM process, we can have an equation on Ω_1:

$$K_{\Omega 1} D_\Gamma = F_{\Omega 1} \qquad (6)$$

where D_Γ is node displacement vector on boundary Γ and $K_{\Omega 1}$ is stiffness matrix in terms of BEM.

Using FEM process, we can have an equation on Ω_2:

$$K_{\Omega 2} D_{\Omega 2} = F_{\Omega 2} \qquad (7)$$

where $D_{\Omega 2}$ is node displacement vector on Ω_2 and $K_{\Omega 2}$ is stiffness matrix in terms of FEM.

Notice that equation (6) and equation (7) are dependent because the nodes in D_Γ are part of nodes in Ω_2. Referring to the method in Zienkiewicz's book [9], we can couple the two equations into one.

$$K_{\Omega 1 \cup \Omega 2} D_{\Omega 2} = F_{\Omega 1 \cup \Omega 2} \qquad (8)$$

FEM theory tells us (8) has a solution, hence the problem can be solved.

2.2.2   FE-EF Blending Method

If the force F causes a non-linear deformation on U_α, we use FEM to compute Ω_2 and EFM to compute Ω_1. To do this, we first discretize Ω_2 into finite elements and keep i_i nodes on boundary γ_i, then select m points on Ω_1, as shown in figure 4.

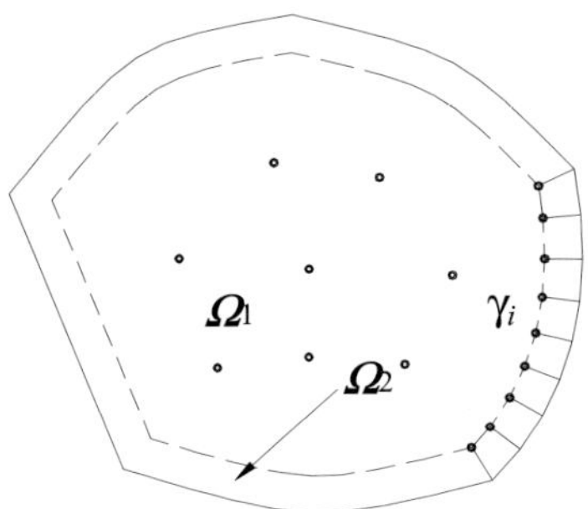

Figure 4. FEM-EFM discretization

Thus on Ω_1, there are $N_\alpha = i_1 + i_2 + \ldots + i_n - n + m$ nodes. Using these nodes and EFM process, we have an equation on Ω_1:

$$K_{\Omega 1} D_{\Omega 1} = F_{\Omega 1} \tag{9}$$

where $D_{\Omega 1}$ is node displacement vector on Ω_1 and $K_{\Omega 1}$ is stiffness matrix in terms of EFM.

Also we can have an FEM equation (7) on Ω_2 and couple it with (9) to get one total equation referring to Chen T and Saju I S [**10**].

$$K_{\Omega 1 \cup \Omega 2} D_{\Omega 2} = F_{\Omega 1 \cup \Omega 2} \tag{10}$$

The later work is to find the solution of equation (10) and here, we omit the details.

2.2.3   Blending Method for Composite Surface Model

Now we consider the computation of the total composite-surface. We first add a new boundary on each patch U_α of S to divide U_α into two parts as illustrated before. Then if U_α and U_β are two adjacent patches, the parts outside the new boundaries near the original boundaries will produce a midst area $\Omega_{\alpha\beta}$, as shown in figure 5.

It is true that the union of all midst areas will form a framework (or skeleton) of S. We call this skeleton *boundary bridge* and denote it as BB.

By definition of $\Omega_{\alpha\beta}$, we know the following formula holds:

$$BB = \bigcup_{\alpha,\beta} \Omega_{\alpha\beta} \tag{11}$$

Now it is up to the computation of BB and the inner part of each patch in terms of their new boundaries. Following the previous FEM-BEM/EFM blending methods, we can see it is better to compute BB with FEM and use BEM to compute the others if S undergoes a linear elastic deformation and use EFM if S undergoes a non-linear deformation. So the next work is to build total equations to find the solution of the problem. Since this is a redundant technical process, we here omit the details for the limitation of paper size and we will focus on the geometry continuity and mesh generation for **BB** instead.

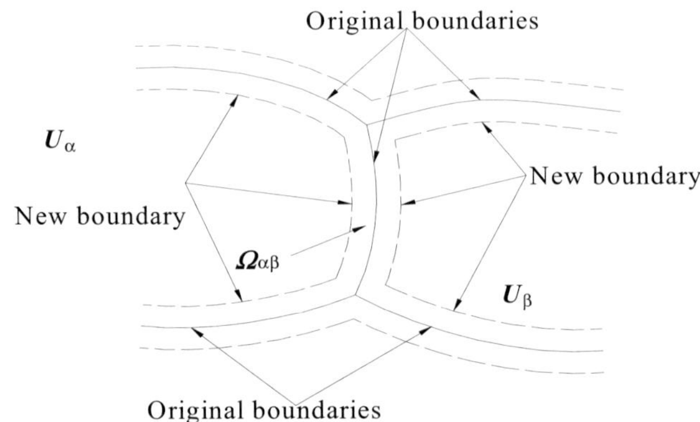

Figure 5. Midst computing area between two patches

2.2.4   Comments on Mesh Generation for Boundary Bridge

CAD modeling theory tells us that there exist kinds of geometry continuity on a composite surface. Continuum mechanics shows that different continuity on geometry raises different intrinsic mechanical properties. For example, a composite surface is totally geometric smooth if all of its patches are GC^1 or higher connected, but the surface is not totally "mechanical smooth" if there exists continuity lower than GC^2 because GC^1 connection cannot ensure deforming energy continuous at the connecting spot. This can be proved as follows.

Suppose U_α and U_β are GC^1 connected. Take two thin slices s_α, s_β from U_α and U_β respectively, and assume the two slices connected at P, as shown in figure 6.

It is true that s_α and s_β are GC^1 connected.

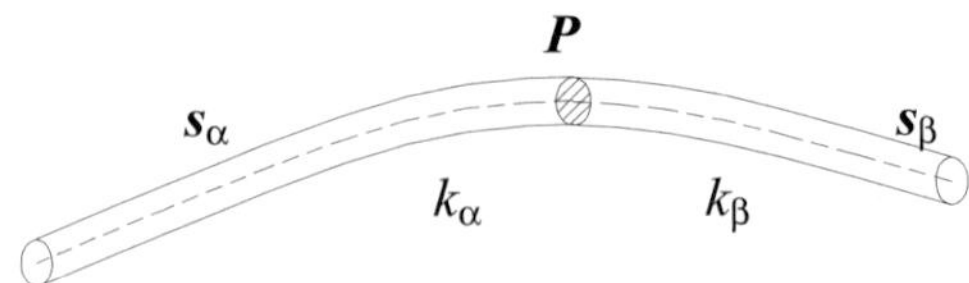

Figure 6. Geometry continuity and deforming energy

Now let k_α and k_β be curvatures near P of s_α and s_β respectively, then if s_α and s_β undergoes a elastic deformation, the corresponding deformation energy depends on the following formula [**11**]:

$$e_\alpha \approx \frac{EI_\alpha}{2}\int_s k_\alpha^2 dl \qquad e_\beta \approx \frac{EI_\beta}{2}\int_s k_\beta^2 dl \qquad (12)$$

where E is the young's modulus of the material, I_α and I_β depend on the shape of the section on s_α and s_β respectively, s is a small segment on s_α and s_β near P.

Since s_α and s_β have the same section-shape, thus $I_\alpha = I_\beta$.

Now it is proved that the deforming energy is discontinuous at the connecting line on S despite that S is geometrically smooth.

So that we should consider the geometric continuity between patches when we produce mesh on BB. In practice, we use the following rules:

i.   When two patches are GC^0 or GC^1 continuous, we select nodes on original boundary line and the two new boundaries, as shown in figure 7 case (a).

ii.   When two patches are GC^2 or higher continuous, we ignore the original boundary line and select nodes on the two new boundaries, as shown in figure 7 case (b).

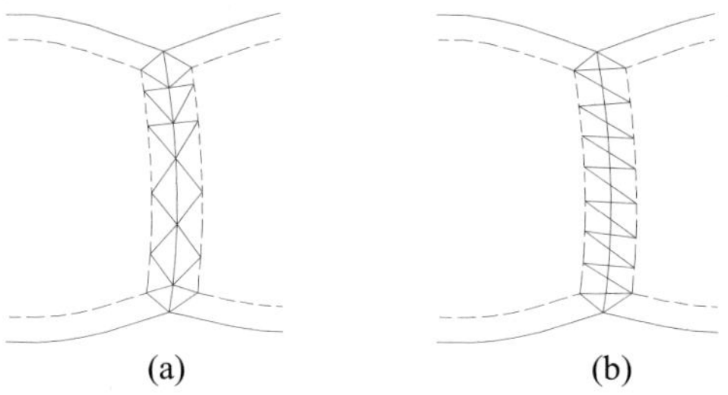

Figure 7. Geometric continuity and meshing mode

3.   SUMMARY AND CONCLUSIONS

Numerical method for CAD and industrial requirement is perpetually in need as mentioned before. PUM presented in this paper is actually a blending method of FEM, BEM/EFM. It inherits all the advantages and discards many disadvantages of the conventional methods. In conclusion, we summarize some traits of PUM.

i. PUM does not need global mesh like FEM. In PUM, only boundary bridge need meshing like FEM and this is much simpler than a global mesh.

ii. Since the whole composite-surface is decomposed into small analytic patches, computation on the inner part of patch by BEM is simpler than global BEM process.

iii. The same as above, the inner part EFM is simpler than global EFM, especially in selection of shape functions.

Reader can see PUM has many other good features but we have to stop counting for the limitation of paper size. More details and numerical examples will be shown in our later papers.

ACKNOWLEDGEMENTS

The project is supported by National Natural Science Foundation of China under grant number 50175106.

REFERENCES

1. Shi Gen Hua. "Manifold method of material analysis", *Transactions of the ninth army conference on applied mathematics and computing*, Minneapolish, Minnesoda, USA,1992.
2. Shi Gen Hua. "Manifold method", *Proc. Of the First Int. Forum on Discon. Defor. Anal. • Simu. of Discon.Media*,California,USA,1996.
3. WANG Zhiyin, LI Yunpeng. "Numerical Manifold Method and Its Development(In Chinese)", *Advances in Mechanics*,Vol.22, pp.261-266,May,2003
4. Belytschko T, et al. "Element-free Galerkin methods", *International Journal of Numeric Methods in Engineering,* Vol.34,pp.229-256,Feb,1994.
5. Belytschko T, et al. "Meshless method: An overview and recent developments", *Comput Methods Appl Mech Engrg*,Vol.139,pp.3-47,Janu.,1996.
6. Duflot M, Huang N D. "A truly meshless Galerkin method based on a moving least square quadrature", *Computations in numerical methods in engineering*,Vol.18,pp1-9,March,2002.
7. Tian Rong. *Finite-Cover-Based Element-Free Method for Continuous and Discontinuous Deformation Analysis with Applications in Geotechnical Engineering.* Ph.D. thesis, Dalian University of Technology, Nov., 2000.
8. Cueto E, Sukumar N, Cigenino J, et al. "Overview and recent advances in natural neighbor Galerkin method"*, Archives of Computational Methods in Engineering*, Vol. 10, pp:307-387, 2003.
9. Zienkiewicz O C. "Boundary Solution and FEM ", *The Finite Element Method(third edition).* Mc. Graw-Hill, New York 1977.
10. Chen T, Raju I S. "Coupling finite element and meshless local Petrov-Galerkin method for two-dimensional potential problems", *Proc 43rd AIAA/ASME/ASCE/AHS/ASC Structures, Structural Dynamics, and Materials Conference*, Denver, Colorado,AIAA-2002-1659,2002.
11. Landau L D, Lifshitz E M. "The energy of a deformed rod". Theory of Elasticity(third edition). Butterworth-Heinemann, 1997

VISUAL AIDED PROTOTYPE OF PRINCIPLE SOLUTION IN PRODUCT CONCEPTUAL DESIGN FOR CHIEF ENGINEERS

Fei Zheng, Mei Chen, Hongmei Zeng and Baoyan Duan
Xidian University, Xi'an , Shaanxi 710071, China

Abstract: Conceptual design is the main responsibilities of chief engineers, but nowadays less software tools are practical and effective for them to carry out conceptual design of new products. A Visual Aided Prototype of Principle Solution System (VAPPSS) is proposed in this paper. It uses extended tree structure, multimedia-based principle solutions and mouse-driven interface to construct a visual aided system to give chief engineers a convenient tool for conceptual design of new products. Such a prototype is fairly simple and effective in principle solutions.

Key words: Conceptual design, principle solution, tree structure, visual aided system.

1.   INTRODUCTION

It is well known that decisions made during conceptual design of a product have significant influence on the cost, performance, reliability, safety and environmental impact of a product [1]. Principle Solution is one of the key problems to be solved in the conceptual design period. After many years of studies carried out by thousands of researchers[2,3,4], it is still not clear how we make a good conceptual design in our brains. Though several design theories have been proposed, such as Comprehensive Design Methodology, Axiomatic Design, Quality Function Development and TRIZ (Theory of Inventive Problem Solving), and there are many powerful corresponding software nowadays such as UG NX, Pro/Engineering that can give effective supports for CAD/CAE/CAM, little is done to bring out practical and effective software tools to conceptual design for chief engineers in research institutes, who are taking very important roles in the conceptual design processes of new products, especially for creative products. Due to the psychological characteristics of an individual, a chief engineer is always in such a status that some engineering elements are not clearly remembered, or some important aspects are not considered, or some 3D models cannot be imaginary in the brain, which always interrupt the thinking and reasoning processes of the engineer and influence the conceptual design effects. Focus on effectively assisting chief engineers to concentrate their brains on using their intelligences and experiences to carry out the conceptual designs of inventive products, which would be very difficult or even be impossible to realize purely by computers, we attempted a Visual Aided Prototype of Principle Solution System (VAPPSS) in product conceptual design for chief engineers. The rest of our paper is arranged as follows: section 2 describes the basis treatments of the VAPPSS; section 3 gives the structure of the prototype of such a VAPPSS; section 4 concludes our results and gives further research directions of such an attempt.

2.　　BASIS TREATMENTS OF THE VAPPSS

Our VAPPSS is based on principle solutions in Comprehensive Design Methodology [2,4]. The basic idea of the VAPPSS is to let the computer remember what chief engineers cannot remember or imaginary clearly in conceptual design process of new products, and to assist them to concentrate on thinking, reasoning and solving the complex problems in their brains in a practical and effective manner. To achieve this aim, several treatments are adopted to assist the chief engineers' thinking and reasoning in conceptual design process.

1)　　Extended text-based tree structure.

Top-down and tree structure analysis method is commonly used in many aspects. File directory tree structure is a typical example of the common use of tree structure. But for conceptual design, available tree editors, which give only text information, are not sufficient and effective to assist chief engineers. So we extend a tree editor in following aspects:

* Use a text form to enable chief engineers to use it both inside and outside the tree editor freely. To minimize the input operations, space char is used to identify different levels. The amount of each level is limited to seven to fit for the memory characteristic of human brain. It is easy to process add, delete, expand, shrink or move operations in such a tree structure. Therefore, it is convenient for the engineers to express their design ideas or processes in such a manner.

* Use marks to identify different medias such as text, mathematic formula, principle draft and 3D model. For example, "\M","\F" and "\B" are marks to indicate menu, formula and 3D model, respectively. By using such simple marks, sufficient information of a principle solution can be collected and arranged in a pure text form.

2)   Multimedia-based principle solutions.

A creative design of a new product is often based on existing principle solutions which are concluded from millions of successful products in human developing history. With different physical inputs and outputs, there are many different principle solutions. A creative design is often created by combining such solutions in a suitable form. Such a process is very complex, and may not be possible to be automatically realized in computers in a short time. So we think it would be practical to give powerful assistances for chief engineers to concentrate their thinking and reasoning on solving such complex problems by arranging a suitable design flow and giving sufficient descriptions of principle solutions in a simple and agile manner.

To sufficiently describe a principle solution in a simple manner for conceptual design, a simple statement, a principle draft, a mathematic formula or even a 3D model are all necessary. Such principle solutions can be classified and stored in a database for searching and picking-up.

In order to store images and 3D models in a database, it is suitable to use hyperlink to store a link in the database to direct the correspond file to be treated. Figure 1 is an example of a principle solution in the database. Such a solution indicates that with the given input and output, a level is a selection to achieve the aim. The principle draft is stored in the file directory ".\prinGraph\gz020201.gif", the physical principle is stored in the file directory ".\prinGraph\fz020201.gif", also in image format. The example is designed to indicate a 3D model in a VRML(Virtual Reality Modeling Language) format, corresponding software functions have been realized, but the 3D model of a level drive is not built at present.

Reason(In)	Result(Out)	Physical Effort	Principle Draft	Notate	Physical Principle	Example
.....	.....	.....	.....	.....	.....	.....
Speed	Scale of Energy or Signal	Level	#prinGraph\gz020201. gif#		#prinGraph\fz020201. gif#	Level Drive
.....	.....	.....	.....	.....	.....	.....

Figure 1. A principle solution description in the database

By using the marks in the text-based tree structure and in the database stated above, it is easy to bring different forms such as mathematic formula, principle draft and 3D model in the conceptual design processes to aid chief engineers for concentrating on thinking and reasoning of complex problems in conceptual design.

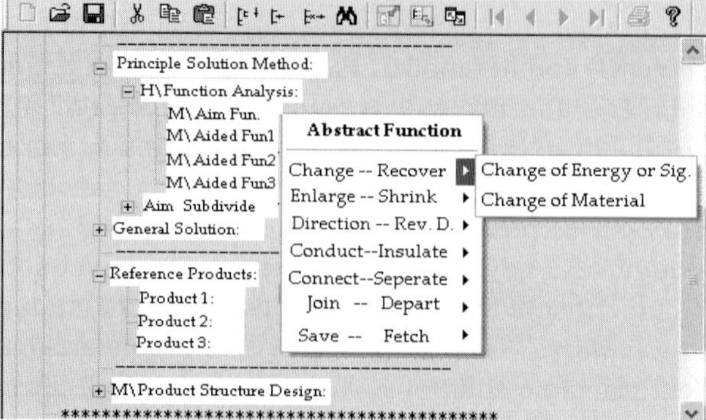

Figure 2. An operation of conceptual design in the tree structure text

3)   Mouse-driven Interface.

We use three button mouse and speed buttons to display tree structure, to fetch mathematic formula, principle draft or 3D model, to give detail description of a physical principle and to trigger sub-windows or sub-trees by just single keystroke or double keystrokes of the mouse in a tree structure. Besides expanding and shrinking functions of the tree structure, we use left keystroke of the mouse to select corresponding items, middle keystroke to modify the selected item, right keystroke to fetch 2D image marked in the tree structure text, double right keystroke to display 3D model marked in the text. Figure 2 is an example of an operation in the tree structure text.

3.   PROTOTYPE STRUCTURE OF THE VAPPSS

The prototype of the VAPPSS is realized by using VC++ language in Windows environment. It is divided into three sub-windows, all can be adjusted in size dynamically or by speed buttons.

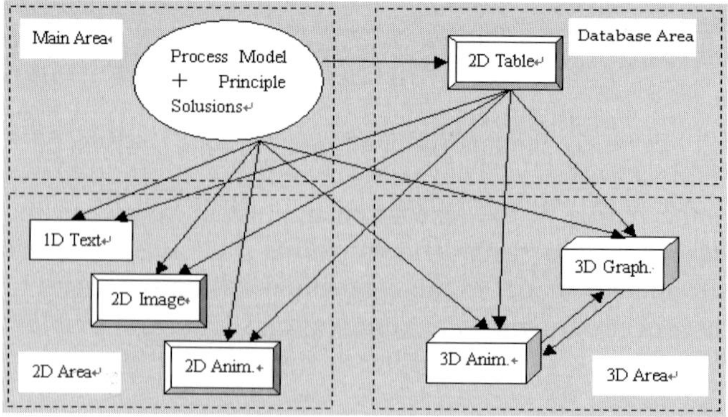

Figure 3. Frame Structure of the VAPPSS

Figure 3 is the frame structure of the VAPPSS, the arrows in the structure indicate the correlative citings and visualization effects in corresponding sub-windows. The left-up window demonstrates tree structure, which can be expanded, shrunken, added, deleted or moved conveniently just by using keystroke of mouse or speed buttons. It is also easy to trigger multimedia such as draft, image, database, 3D model and so on. The left-down window is for text explanation, 2D picture or image, which can be triggered by corresponding items in the tree structure in left-up window or in the database table in the right window. The right window displays database table or 3D model, which can be triggered by speed button or function key. In database table, items with explanations or image or 3D model can be triggered to display in the left-down window or in the triggered right window.

Therefore, with sufficient and elaborately organized medias and agile operations, the chief engineers can easily get mathematic formula, principle draft or 3D model that are convenient for computer to store and search, and then can concentrate on thinking and reasoning that are difficult for computer to carry out. It is also easy to store their design ideas in such a text tree structure. Such a prototype can be used not only to process inventive conceptual design, but also to study and record successful design ideas of available products in such a extended tree structure text format. At present, the prototype is only in Chinese form, as showed in figure 4.

4.   CONCLUSION AND FUTURE WORKS

After presences of many discussions for product developments among chief engineers from many research institutes, and after many efforts to

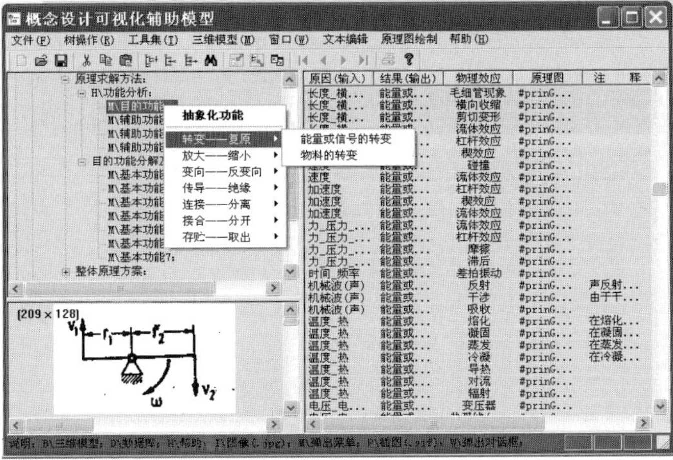

Figure 4. Prototype of the VAPPSS (in Chinese Form)

improve our ongoing software development of antenna structures, we found that there is less practical and effective software tools for conceptual design, which is the main responsibilities of chief engineers who have sufficient design intelligences and experiences and lack experiences or time in learning or using computer design software. Such design software are getting larger in size and more complex in learning but still less effective in conceptual design. So we created the idea to construct the VAPPSS for such a special and important group of people. Through some fairly simple examples, we found that such a prototype is practical and effective to give the chief engineers sufficient information of principle solutions to inspirit and steer them for inventive design.

Such a prototype is still an elementary model. Many useful modules such as TRIZ model, mechanical construct rules or mechanical transmission rules should be added. The conceptual design process model should be deeply studied and expressed in the tree structure model.

We would continue to expand the database and elaborate the layout, and would integrate it in our ongoing antenna structure software development in future works.

ACKNOWLEDGEMENTS

The study is supported by the national scientific research fund for abroad Chinese visiting scholar, the academic department, China. It is also supported by Youth Scientific Research Workstation of Xidian University, China.

REFERENCES

1. W. Hsu, B. Liu: "Conceptual design: issues and challenges". *Computer-Aided Design*, **Vol.32**, pp. 849-850, September 2000.
2. R. Koller: *Mechanical Design Methodology*, China Science Press, Beijing 1990. (in Chinese).
3. Web site: http://www.triz-journal.com
4. J. Deng, et al: *Product Concept Design*, China Machine Press, Beijing 2002. (in Chinese)

PROACTIVE SUPPORT FOR PRODUCT CONCEPTUAL DESIGN INCORPORATING MANUFACTURING SYSTEM CONSIDERATIONS

Zifang Wu[1], Xiu-Tian Yan[1], Jianhua Yang[2] and Xiansheng Qin[3]

1. CAD Centre, DMEM, University of Strathclyde, Glasgow, G1 1X, Tel:0044-141-5482374, Fax:0044-141-5520557, email: zifang@cad.strath.ac.uk
2. Manufacturing engineering research centre, DMEM, University of Strathclyde, Glasgow, G1 1XJ
3. Northwestern Polytechnical University, Xi'an, China, 710072

Abstract: The paper proposes a proactive support approach for product conceptual design considering issues related to manufacturing systems. A set of manufacturing system analytical models has been developed to estimate manufacturing cycle time for product sub-assembly. A concurrent analysis method is provided to calculate the manufacturing cost based on predicted cycle time and to analyze product quality. An overall life-cycle performance index of a design solution is evaluated to help designers make informed decisions in selecting the optimal alternative in terms of manufacturing cycle time, cost and quality for a given manufacturing system.

Key words: design support, manufacturing system, cycle time, cost, quality

1.   INTRODUCTION

With increased global competition the pressure to get quality products to market in time and at competitive cost is ever increasing. The coherent integration of design and manufacture is an important approach to achieve these objectives.

In the past decades, the area of Design for Manufacture (DFM) has come under intense investigation. DFM techniques are now an important aspect of the product design process. It incorporates the consideration of manufacturing expertise at the early design stages. Although a large number of methods and tools have been developed for manufacturability analysis and evaluation in various domains, most of them available tend to reflect the traditional sequential approach to design and manufacture by providing manufacturing feedback on completed designs rather than providing feedback as design evolves. Little attention has been given to DFM at the early stages of design.

This situation contrasts with a need evinced in a 1997 NIST study [1], which found that up to 80% of the cost of the final product is decided at the conceptual design stage. During conceptual design, important decisions are made which will influence all the subsequent product life-cycle activities (manufacturing, assembly, use, recycle, etc). Primary manufacturing costs are therefore committed and determined at this stage. Taking right decisions at conceptual stage is very critical for achieving business objectives. This underscores the importance for designers to make informed and influential decisions based on relevant manufacturing knowledge, so that they can be more confident about their designs and the intended product performance once the design solutions are passed on to manufacturing process.

More importantly, almost all the existing DFM approaches [6] lack the consideration of the impact of manufacturing systems, which are scheduled for the product, on product design. These approaches evaluate the materials, the required manufacturing processes, and the ease of assembly. In short, they evaluate manufacturing capability and measure the manufacturing cost to show how the product design could influence individual manufacturing operations, such as casting, injection moulding, and CNC milling, etc. However, they are insufficient at considering manufacturing system at the production line, factory, or supply chain level. A manufacturing system significantly influences a product design [5]. Because of the insufficient available capacity of a manufacturing system, any product design that ignores manufacturing system issues will probably lead to a longer manufacturing cycle time as queuing of components will occur at the heavily utilized resources, which means a delay in time-to-market from which a reduced market share, lost sales and profit may result. As a consequence, a product must be designed to suit the machines available. As product variety increases, product life-cycles decrease, and historical manufacturing cycle times in a factory will not be accurate enough when the product mix is different. Manufacturing system analytical models must be established to ensure that a product is designed to fit the specific facilities in a manufacturing system so that it can be manufactured quickly by exploiting capability and capacity that already exist. To overcome above limitations, this research proposes a proactive approach for product conceptual design considering issues related to manufacturing systems.

2. OVERVIEW OF PROPOSED APPROACH

Considering issues related to manufacturing systems at early design stage couldn't only focus on reducing the lead-time of a product. For alternative design solutions scheduled in a given manufacturing system, different processes and equipment, which have different utilizations, cost rates and capability data, would be selected. This would influence not only the manufacturing cycle time but also the manufacturing cost and quality of a product. As a result, this paper proposes to jointly consider the manufacturing cycle time, cost and quality of design alternatives to obtain their rankings, so that designers can be assisted in choosing the optimal solution.

The proposed approach in this research is a proactive approach for product conceptual design that can support designers in making informed decisions, through the timely provision of feedback of manufacturing consequences caused by design commitments. It calculates manufacturing cycle time and estimates manufacturing cost and quality level for design alternatives based on both available capability and capacity of manufacturing resources, and provides information on these metrics to facilitate designers to select the optimal solution against these criteria concurrently. An in-depth analysis on these metrics can also give insights to designers on the critical features and attributes of a design solution that have the most influence on design performance against design requirements. This analysis will facilitate designers to explore alternative solutions to obtain optimal design solutions.

To achieve above, three system information models are proposed for development as the base of the approach: a product model for storing the solution information of the product sub-assembly and parts, a manufacturing resource model for the available resources and a process model for generic process knowledge. On the basis of these models, a set of manufacturing system analytical models has also been developed to analyze the manufacturability of design alternatives and to predict their manufacturing cycle times.

Based on the predicted manufacturing cycle times, a concurrent analysis approach is applied to generate the estimates of manufacturing costs and to analyze quality levels in terms of product tolerance level, surface finish and consequences related to manufacturing processes. This research proposed a measurement definition for a product life-cycle performance, entitled a total product life-cycle performance index. This index is defined as a function of manufacturing cycle time, cost and quality shown above and can be calculated to evaluate design decisions. Providing timely information on manufacturing cycle time, cost and quality of a product in a given manufacturing environment early in design synthesis, during which decisions primarily revolve around controlling these three key business objectives, would be of great benefit and most practical for designers to ensure their designs to be functional, manufacturable and profitable.

3.   SYSTEM INFORMATION MODELLING

To analyze a design solution in a manufacturing system, three system models—product, process and resource—are required. These models are designed using an object-oriented classification of objects in a functional and logical hierarchy. Relevant manufacturing knowledge is applied to the objects within these models.

3.1   Product model

Product model is described by manufacturing features. It is a hierarchical structure of features and components. A component is a mechanical part that is composed of a number of features. To model a sub-assembly, a component object may have subcomponents added to it. The manufacturing process planning is dependent on this model structure. A feature in the model represents an entity to be made using a manufacturing process. Typically, a feature defines the material, shape including key dimensions and a set of process-related information, such as surface finish, interval of tolerance (IT) grade and unrelated tolerances, etc. The functional performance functions of sub-assembly, which describes the relationships between functional performances and sub-assembly features, are also added to feature so that quality metrics can be accordingly calculated.

3.2   Process model

Manufacturing is characterized by two kinds of activities: discrete parts production and the subsequent assembly of these parts to generate the finished product. The process model covers a wide variety of processes of machining and assembly operations required for machined product. A process (like machining) in the process model has a set of related sub-processes (like milling and drilling). A sub-process has a set of critical process parameters that describe its manufacturing capabilities. The sub-process capability data aggregate the best capabilities of the available equipment. To support consequence analysis of manufacturing processes for quality measure, knowledge on these consequences is also stored within this model.

4. PROACTIVE DESIGN SUPPORT METHODOLOGY

The proposed proactively supported conceptual design process model is shown as figure 1.

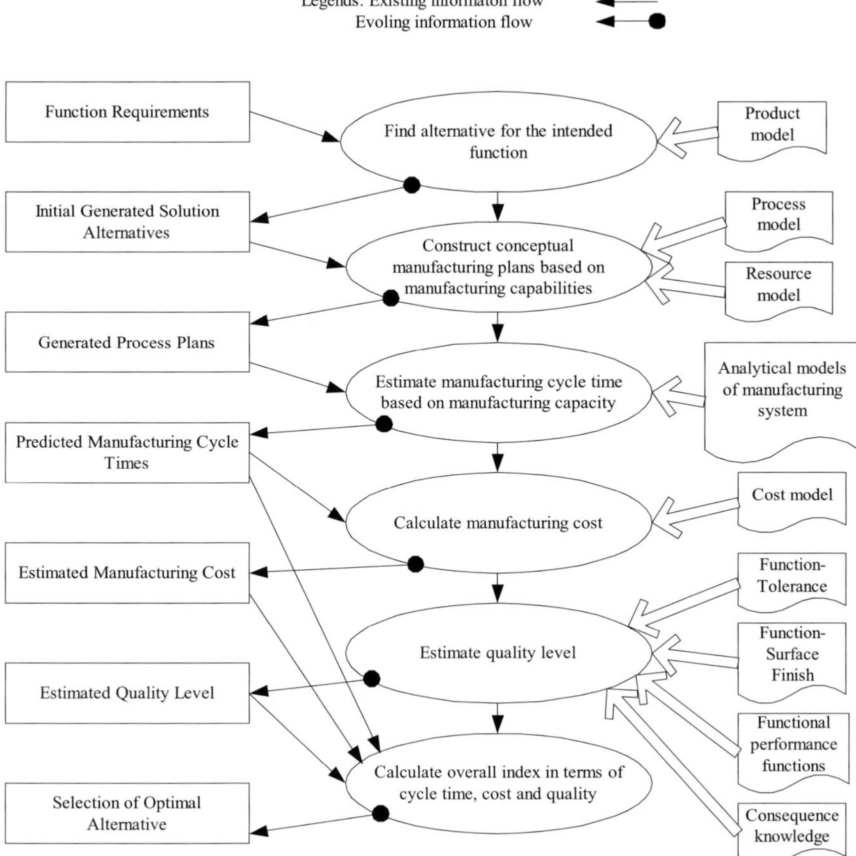

Figure 1. Proactively-supported conceptual design process model

4.1 Conceptual process planning

The function of process planning requires critical design information about the sub-assembly. The sub-assembly design alternatives are first generated by translating functional requirements into a feature-based product model. For each design alternative generated above, alternative manufacturing process plans are constructed by selecting appropriate processes and resources. For each operation in the plans, the processing times are estimated.

At the conceptual design stage, it is not possible, or indeed desirable, to create detailed plans for a product's manufacture, as the product information is incomplete and inaccurate. Therefore, the process planning can only perform at the conceptual level. The process plans may be approximate and incomplete. The necessary information is the sequence of operations, the resource required, and estimates of the time required. They may lack some details like process parameters, fixing instructions, or other operational attributes that require a detailed design to determine.

Preliminary manufacturability analysis is also conducted during the process planning through, firstly, checking that a process is capable of producing the sub-assembly feature, as designed. After the initial process selection, based on the shape producing ability, the technological checks are made to reject any processes incompatible with the feature geometry or surface roughness criteria.

The results of process planning, resources required and estimated processing times, are the basis of manufacturing cycle time calculation in the manufacturing system analytical models as follows.

4.2 Manufacturing system analytical models

Manufacturing system analytical models are used to estimate manufacturing cycle times for alternative product sub-assemblies based on the manufacturing process plans generated for each given design alternative, availability of each resource in the manufacturing system at the time the new sub-assembly is introduced, and data about the manufacturing system.

Many manufacturing system analytical models have been developed to estimate average manufacturing cycle time in a manufacturing system, such as conveyor model, queuing system model, cyclic production scheduling model, discrete event simulation model and hybrid model, etc. Some of them can provide much more information than the average manufacturing cycle time. However, most of the models require much information input that is usually not available at the conceptual design stage. The time taken for running these models is also great, which make them not suitable for quick or even real-time analysis of multiple design alternatives simultaneously to support design decisions. Queuing network model requires less information input and computational effort, but it cannot be applied to non-flow-line manufacturing system.

A school course-timetabling problem has been analyzed in this research and identified appropriate to perform the non-flow-line manufacturing system analysis. A course-timetabling problem is very similar to the manufacturing system analysis problem(see Figure 2). Assigning classes in a course-timetabling problem to different classrooms located at different department buildings can be analogous to arranging components in a manufacturing system to be manufactured on resources with different specs of different types. Although course-timetabling problem has long been used to schedule classes within a given number of classrooms and time periods in a school, this research effort is the first to apply it to manufacturing system analysis.

The advantages for applying a school course-timetabling problem to the manufacturing system analysis are as follows:

· Course-timetabling problem is a classical mathematical problem, many mature algorithms have been developed;

· Less information input is required;

· Less computational efforts are required;

· Course-timetabling algorithm can be used to solve the problem of non-flow-line manufacturing system analysis

The adaptation of course-timetabling problem to manufacturing system analysis problem is needed. The following corresponding relationships between the two are essential to realize the adaptation.

4.3 Estimate of manufacturing cost

Based on predicted cycle time using manufacturing system analytical

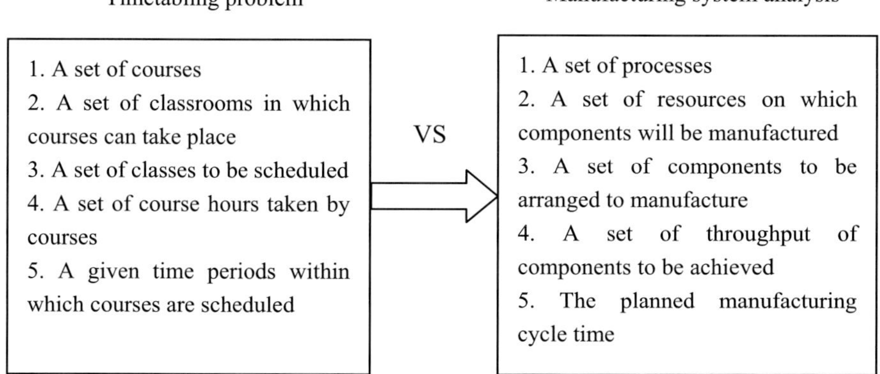

Figure 2. Adaptation relationship

models, manufacturing costs of the product sub-assembly can be estimated.

The main costs of manufacturing consist of (a) the cost of raw material, (b) the cost of machine time required, (c) the cost of tooling.

Using information about the sub-assembly volume and envelope size, material density and cost per unit mass, and process factor, i.e. the factor that allows for scrap and wastage, which varies with different manufacturing processes, the cost of raw material is calculated by equation (1):

$$Craw = (VolumeofMaterial) \times (\Pr ocessFactor) \times (Dencity) \times (UnitCost)$$

$$(1)$$

At the conceptual design stage, an accurate and detailed analysis of cost of machine time is not possible because detailed manufacturing plans have not been developed. The time to machine the sub-assembly is only derived by an approximate method based on the difference in the volume of material in the sub-assembly before and after machining. From data in well-known handbooks on cutting speeds, depths of cut and feeds, the average rate of material removal V and the time to machine a sub-assembly T can be calculated. Therefore, the cost of machine time is Cmac where

$$Cmac = T \times (CostPerUnitTime) \tag{2}$$

Tooling covers items such as jigs, fixtures, dies, moulds and patterns. Costs of tooling depend on manufacturing processes selected and the production rate. These costs can be obtained by looking up handbooks according to required dimensions, tolerances and material of the sub-assembly.

All above three costs are expressed per unit manufactured. Once they have been determined, the total cost of manufacturing is calculated according to

$$Cost = Craw + Cmac + Ctooling \tag{3}$$

4.4 Analysis of component and assembly quality

The quality of sub-assembly is measured by calculating the function and functional performance metrics, and by analyzing consequences of manufacturing processes.

Function and functional performance are two quality measures of a design from the design aspect. The satisfaction of function and functional performance requirements has important impact on a design's quality. In the present work, the function metric of quality considers four affecting factors:

dimensional tolerance, surface finish, related geometrical tolerance and unrelated geometrical tolerance. The correlation matrix of these factors and the function factors can be established by collecting from different books and by discussing with design specialists. A sample matrix of assembly-related factors is partially showed in the following table. By comparing design specifications assigned by designers with the technological attributes in the matrix, the function metric is calculated.

Table. Correlation matrix of assembly-related factors

Assembly-related factors		Dimensional Tolerances	Surface Finish
Mating Relationship	Fit	High	High
	Against	Low	Low
Fit Condition	NoFit	-	-
	Surface-Edge	High	Average
Fit Type	Locational Clearance	Low	Low
	Sliding	Low	Low

Functional performance, F(d), is a function of design variables $d=(d1,d2,...dn)^T$. For a gear design, for example, 'efficiency' is one of the functional performances, which can be represented as Efficiency$_{gear}$=f(Assembly tolerance, material, surface finish, stress,.....). The measure of functional performance often combines a number of individual functional performance functions, Fi(d) (I=1,2,...p). The overall functional performance of a design is a function of all contributing functional performance measures:

$$F(d) = f(F1(d), F2(d),...Fp(d)) \qquad (4)$$

QFD (Quality Function Deployment) method is used to define the functional performance metrics for a design.

Similarly, manufacturing consequence is another quality measure of a design from a manufacturing aspect. Manufacturing consequences are related to manufacturing processes. Welding process, for example, has consequence of residual strain, which affects the appearance quality of a component. Consequence knowledge should be stored in the corresponding process models to support this analysis of quality level. The comparable readings of above three quality measures are added together to obtain an overall quality index of a design.

4.5 Determination of overall index of design alternative

In the proposed approach, the overall index of a design alternative is determined by summing all above three measures: manufacturing cycle time, cost and quality, $TI = Icycletime + I\cos t + Iquality$ (5)

An Analytic Hierarchy Process (AHP) [10] method is used to assign weights to these three indices to obtain an overall index, which will be provided to designers to help them make better decisions on selecting an optimal design in these terms.

5.   CONCLUSIONS

Manufacturing system analysis plays a very important role in the product design. This paper presents a set of manufacturing system analytical models and a concurrent analysis method to generate estimates of manufacturing cycle time, cost and quality of design alternatives for a given manufacturing system. A novel modeling method of manufacturing system analysis is proposed. The proposed approach can assist designers in making better design decisions by providing them with knowledge on predicted manufacturing cycle time, cost and quality of their design alternatives, at the early conceptual design stage, to select the optimal solution against these criteria. Thus, the reduced time-to-market, low cost and high quality of designs can be achieved, and the redesign work later in the production phase can also be avoided. The system implementation of the approach and its case studies will be conducted shortly in the ongoing research.

REFERENCES

1.   Phillips, B. & P. Katragadda. *The ICAD System & I.C.E. Product Overviews*, NIST, 1997.
2.   F. Rehman and, X. T. Yan. "Proactive support for conceptual design synthesis using design context knowledge", *International Conference on Engineering Design ICED03*, Stockholm, August 19-21, 2003.
3.   Borg, J.C.. *Design synthesis for multi-x: A 'life-cycle consequence knowledge' approach*, Dept. Design, Manufacture, and Engineering Management, University of Strathclyde, Glasgow, 1999.
4.   Xiu-Tian Yan. "A multiple perspective product modeling and simulation approach to engineering design support". *Concurrent Engineering: Research and Application*, pp. 221-234, 2001.
5.   Jeffrey W. Herrmann, et al. "Design for production: A tool for reducing manufacturing cycle time". *DETC2000*, Maryland, September 2000.
6.   Boothroyd, G. Design for manufacture and assembly: the Boothroyd-Dewhurst's experience, *Design for X, Concurrent Engineering Imperatives*, UK, 1996.
7.   T.L. Saaty. "How to make decision: The Analytic Hierarchy Process", *Interfaces*, **vol.24**, no. 6, pp. 19-43, 1994.

AUTOMATED GENERATION OF DIMENSION CHAINS FOR AUTO-BODY DIMENSIONAL QUALITY EVALUATION

Jiangqi Zhou, Guanlong Chen, Xinmin Lai and Zhongqin Lin
Shanghai JiaoTong University, Huashan.Road 1954, Shanghai, 200030 ,tel: 021-62932125-0, fax: 021-62932125-103,e-mail: zhoujiangqi@sjtu.edu.cn

Abstract: Dimensional quality evaluation is one of essential steps during auto-body early design stage. There are plenty of elements impacting on the variation propagation of assembly process. Dimension chain can address hierarchical characteristic inherent in auto-body assembly. Traditional method based on tolerance charting technique to generate dimension chains cannot suit the case of body assembly. This paper proposed a method to generate dimension chain based on developing a joint chain model of assembly, which can be integrated with important information needed for variation analysis, such as joint types and assembly sequences. The implementation of such a procedure will contribute to the rapid evaluation of body dimensional quality in the design stage.

Key words: Variation analysis, Dimension chain, Auto-body

1.   INTRODUCTION

The construction of assembly function is one of the keys to evaluate dimensional quality of assembly structure in the process of computer-aided product and process concurrent design. The assembly function must represent the relationship among parts being assembled as well as the propagation of variation stack-up. The dimension chains directly reflect the kernel of assembly function though the task of its identification and determination is really a painstaking job for designers. For the compliant assembly, e.g. auto-body, the traditional method based on tolerance charting technique to generate dimension chains cannot be adopted by such a situation [1][2]. Auto-body product normally is made of several hundreds of sheet metal parts assembled in 55~75 assembly stations. Despite this level of complexity, the assemble process of auto-body is characterized with hierarchical layers [3]. The structures, connections, joints of body assemblies (parts), as well as the assembly sequence will substantially influence the key dimensional quality of automotive body.

In the last two decades, some researchers have developed and investigated the methodologies to predict and evaluate the variation propagation and stack-up of the auto-body assembly. S. Charles et al. studied the effects of deformation on component tolerance and developed a mechanistic variation simulation method by considering the compliance of the stamping parts and the tooling variation [4]. The method provides a more realistic simulation result. Ceglarek, D., Shi, J proposed a tolerance analysis methodology for sheet metal assembly using a beam-based model [5]. The beam-based model has the advantage of simultaneously considering the physical and functional aspect of the target assembly. Geometrical variation and fabrication errors induced by part locating layouts can be evaluated together according to their method. Ceglarek and Shi also provided an analytical tool to address the dimensional capabilities of an assembly process in the early product design stage, focusing on the interaction of part-to-part joint [6]. Bai Zhang in [7] proposed a matrix-represented assembly model and an adaptive design method considering the effects of different joint type. The critical problem still exists that how to incorporate the joint types consideration and sequence solution tightly when to evaluate different design product solution in the early design stage.

Dimension chain is a common tool used by product or process engineers when they attempt to link product functional requirements to the process of product realization. With the help of this tool, components tolerances and assembly variance can be analyzed under different constraints such as material, batch size, cost consideration and their processes to realize them. In tolerance analysis, the close link of one dimension chain is usually identified according to the significance of that dimension to the assembly. For sheet metal assembly, the close link can be determined with key product characteristic (KPC) identification involved in that assembly. The determination of constituting links will face some problems because of their high coupling with assembly process (sequence). Additionally, constituting links actually reflect the interaction between components, so the joint features of components become one of the main design considerations during setup of dimension chain. In the early design stage, it is important to quickly change design solutions and implement dimensional quality analysis efficiently.

This paper proposes a method to generate feasible dimension chains with the ultimate objective to contribute to the rapid evaluation of body dimensional quality in the design stage. The article has been organized in following fashion. Section 2 presents the graphical representation of assembly as well as the sequence and joint feature. A joint chain model is presented according to the sequence information in Section 3. Section 4 presents a procedure to obtain feasible dimension chain using graphical search algorithm. Section 5 explains the proposed method with the help of an example of an assembly. Conclusions are summarized in Section 6.

2.   GENERATION AND REPRESENTATION OF SEQUENCE

Sequence generation is one of the critical steps in the product design and process planning, and results of which highly depend on the selection of assembly model and constraints. Earlier research works show that the matrix method is the best to represent the geometrical and technological data for assembly sequence planning **[8][9]**.

Y.Z Zhang in [10] presents a procedure to automatically derive all the feasible assembly sequences for automobile body assembly based on subassembly detection and matrix method. For the auto-body assembly process, the precedence knowledge, most of which are determined by designers with their experience according to the geometric and physical constraints, are the critical constraints for designers to choose feasible assembly sequences. The precedence knowledge of an assembly can be described as a directed graph $D = \{P\ C\}$. This directed graph consists of a finite nonempty set of vertices P and a set of edges C connecting them. Each vertex represents a component identified by a number. Each edge indicates a relationship between the two linked components. To distinguish different types of relationships among the components, the "dummy connection" concept is introduced as the non-contact physical constraint between two components [7][10]. This constraint will only influence the relative assembly sequences between the two components. Fig. 1 shows an example of such a directed graph.

For the purpose of computer realization and information transformation between different design stages, the representation of a sequence S_i is expressed as bellows:

$$S_i = \{\{\{sub_1, sub_2, ...\}, sub_i, sub_j, ...\}, sub_s, sub_t, ...\} \tag{1}$$

where, $sub_s = (p_1, p_2, ..., p_l \mid 1 \le l \le n, p_l \in P)$ represent subassemblies;
$P = \{p_1, p_2, p_3, ..., p_n\}$, n is the number of parts.

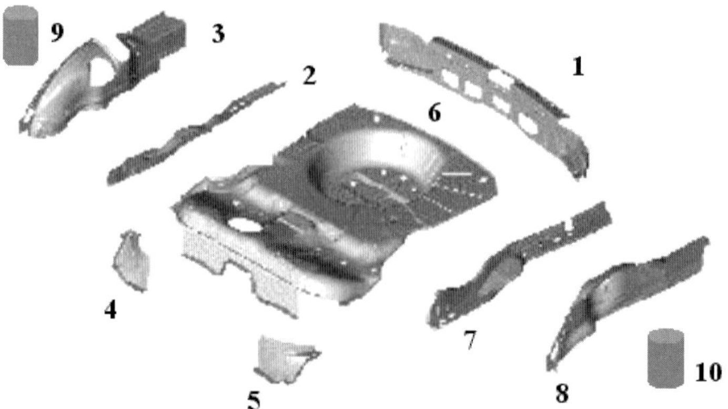

1. Inner Reinforcement,   2. Right Rail,
3. Right Wheelhouse Inner,   4. Right Extension,
5. Left Extension,   6. Rear Floor,
7. Left Rail,   8. Left Wheelhouse Inner,
9. Right Shock Tower,   10. Left Shock Tower.

(a) Major parts of a rear floor assembly

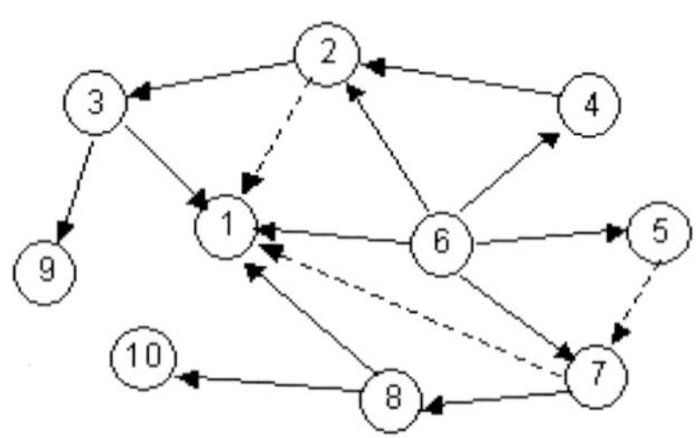

(b) Directed graph of the rear floor assembly
Figure. 1. Directed graph representation of a rear floor assembly

From Eq. (1), the hierarchical information of assembly process can be deduced by character identification. In other words, the depth of bracket pair '{}' in the expression can be mapped to the layer information of the assembly process. One of feasible sequences for the assembly in Fig. 1 is given in Eq. (2). It can be seen that subassemblies 6)4)5), 7)8)10) and 2)3)9) will be assembled firstly at three different stations, respectively, and then put together to make a subassembly of next layer. Lastly the whole assembly comprising of 10 parts is completed at last layer.

$$S_i\big|_{i=1} = \{\{\{6)4)5)\},\{7)8)10)\},\{2)3)9)\}\},1)\}$$ (2)

In brief, following information can be concluded in Eq. (2): 1) total part number, 2) subassemblies, 3) layers where subassemblies produced.

The research will use the procedure in [10] to generate feasible sequences as the prerequisite to create a joint chain of assembly.

3.   JOINT CHAIN MODEL

3.1   Joint Representation

Generally there are two types of joints used in sheet metal assembly: 1) slip joint, 2) butt joint. The concept of defined direction is used to clearly describe joints. The defined direction is the important axis of an analyzed joint, usually defined by design specifications [6] (Fig. 2). The slip joint does not introduce any constraints in the defined direction of interaction, and the butt joint does create a constraint and acts like rigid body in the defined direction. One scenario that designers have to be faced is that how to use right joint types upon right connection when a certain KPC requirement must be met. There may exist one more types of joint between any two parts, each having different defined directions. Because of the sensitivity of joint type on propagation of dimensional variation, it is the duty for designers to assign right joint type to component during design stage.

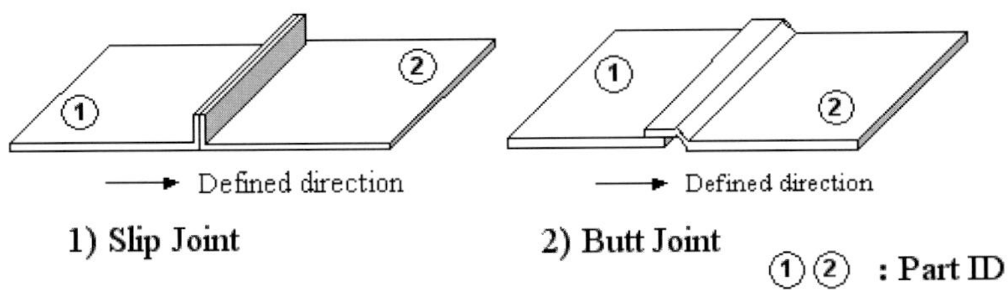

Figure 2. Joint types and their defined directions

All joint types and their defined directions between two components can be represented as a weighted graph, $J=\{P, E, W\}$, called Joint Graph (JG), where nodes P represents parts involved in joint, weight function W represents joint types and their defined directions. The weight of each edge is expressed as a two-character string. The first character stands for joint type, the second stands for defined direction.

In order to simplify the representation, we assume: 1) the coordination system for component is identical to the vehicle coordinate system, 2) the defined direction of joint is in accordance to one of vehicle coordinates. For an example, the joint and defined direction between part 1 and 2 shown in Fig. 3 is 'butt joint' and 'x axis', respectively. Both of them can be represented as the weighted edge 'bx' in JG, where 'b' is the 'butt joint' and 'x' the 'x axis' in short. So with the slip joint between part 2 and part 3, the weighted edge is 'sx'. The joint graph of a four parts assembly is depicted in Fig. 3.

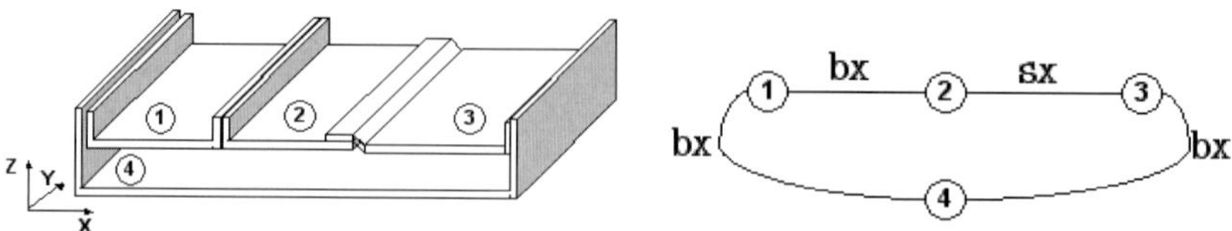

Figure. 3, Joint graph representation

Table 1, Joint Chain of sequence S_1

Layer (L)	Joints J (L)
0	Empty
1	(=part a) (=part b) (=part c)
2	(=part e)
3	
Joint Chain	JC (S_1)={J (1), J (2), J (3)}

3.2   Joint Chain Generation

The developing of joint chain (JC) begins with the assembly sequence and its joint graph. From the assembly sequence, joints that have been created at different layers (including those at the same layer) can be abstracted and recorded. Detail information about the joints can be identified from the joint graph and then integrated with corresponding layered joints. These layered joints form a joint chain, which varies with the sequence and joint types. The joint chain of exemplar assembly is listed in Table 1. In Table 1, a multi-dimension array *J(L)* is used to save joints and their defined directions created from different layers (*L*). Variable *JC (S₁)* represents the joint chain for the sequence S_1, is the result of sequencing *J(L)*.

From the view of tolerance analysis, all joints that are created at the same layer (L) should be evaluated with analysis method at the same time. The total variation produced by these joints then will propagate further to the next layer (L+1) of assembly process.

4.   GENERATION OF DIMENSION CHAIN

The candidate parts involved in one dimension chain are inherent in the directed graph of assembly. After selected the close link, all constituting links of dimension chain can be identified through following steps.

1, Produce a revised directed graph *D'* from the assembly directed graph, pruning all dummy arcs.
2, Identify two vertices according to the definition of close link, and take one of them as the start point *S*, the left as the end point *T*.
3, Search all paths that begin with *S* and end with *T* in *D'* using *modified Dijkstra algorithm* (see *APPENDIX*). While each path is searched, corresponding cut set shall be recorded as *CutSetOfPath$_i$*.
4, Check paths feasibility:
   Rule: The ith path is feasible if there exists no edges (connections) in *CutSetOfPath$_i$* that will be connected (assembled) before any one among edges in the ith path.
5, Assign detail attributions such as layers, joint types and defined directions, to every edge included in the feasible path according to JC model and specified close link.

5.   CASE APPLICATION

A rear floor assembly shown in Fig. 1 is used to illustrate the aforementioned method. The KPC of the assembly is identified as the dimension between two tower centers in Y direction. Part 9) and 10) are involved in the close link. According to the algorithm given in section 4, there are four possible paths found between part 9) and 10) in the directed graph. After feasibility verification of all possible paths, only one feasible path is left, which is shown in Table 2.

The feasible path includes expected constituting links that is essential to form a framed dimension chain with the key dimension as the close link. The final dimension chain can be produced after combined with the JC model, where some important information about joints type and layers will be abstracted. A finding is that different sequences will result in different JC models. There are four feasible assembly sequences generated for the rear floor assembly in Fig. 1, so four joint chains are obtained as a result. The final dimension chains for all sequences are listed in Table 3.

6.   CONCLUSIONS

There are a number of elements impacting on the variation propagation of sheet metal assembly process. The problem of constructing an assembly function, which may take into account these elements, has attracted intensive research efforts. From the point of view of dimension chain, this paper proposed a procedure to generate dimension chains based on the development of a joint chain model of assembly, which integrated some important information needed for variation analysis, such as joint types and assembly sequences. Graphical path search method is also proposed to find link members of a dimension chain from directed graph representation of assembly. The procedure can be implemented by software programming and the resulted design tool will contribute the rapid evaluation of body dimensional quality in the design stage.

Table 2, Generation of dimension chain of a rear floor assembly

Directed graph	$A = \{P,C\}$ (Fig. 1)
Key dimension	Dimension of distance between two tower centers in Y direction (*Start*= 9, *End*=10)
Path	$Path_1$=9, 3, 2, 6, 7, 8, 10 $Path_2$=9, 3, 1, 6, 7, 8, 10 $Path_3$=9, 3, 1, 8, 10 $Path_4$=9, 3, 2, 6, 1, 8, 10
Cut edges of path	$CutSet_1$= {(3,1), (1,6), (1,8), (6,4), (6,5)} $CutSet_2$= {(3,2), (2,6), (1,8), (6,4), (6,5)} $CutSet_3$= {(3,2), (2,6), (1,6), (6,7), (7,8), (6,4), (6,5)} $CutSet_4$= {(3,1), (6,7), (7,8), (6,4), (6,5)}
Feasible path	$Path_1$=9, 3, 2, 6, 7, 8, 10

Table 3, Final dimension chains of a rear floor assembly

Assembly sequences (S_i)	Feasible path	Joint chain	Final dimension chain (Only constituting links shown in Text)

{{{6)4)5)},{2)3)9)},{7)8)10)}},1)}		JC (S_1)	{{{(2,3)(3,9)},{(7,8)(8,10)}},{(2,6)(7,6)}}
{{{6)5)4)},{7)8)10)},{2)3)9)}},1)}	Path$_1$	JC (S_2)	{{{(2,3)(3,9)},{(7,8)(8,10)}},{(2,6)(7,6)}}
{{{6)4)5)},{7)8)10)},{2)3)9)}},1)}		JC (S_3)	{{{(2,3)(3,9)},{(7,8)(8,10)}},{(2,6)(7,6)}}
{{{6)5)4)},{2)3)9)},{7)8)10)}},1)}		JC (S_4)	{{{(2,3)(3,9)},{(7,8)(8,10)}},{(2,6)(7,6)}}

ACKNOWLEDGEMENTS

The authors would like to extend our sincere thanks to Dr. Wayne Cai (GM R&D) and anonymous referees for their valuable suggestions to improve the quality of this paper.

APPENDIX

Modified Dijkstra Algorithm:

Given a connected graph $G=(V, E)$, a weight d: E->R+, *Dijkstra algorithm* is a good algorithm to find a shortest path from s to each vertex v in V. Here we will have somewhat of *Dijkstra algorithm* modified to solve the all path problem in this study.

The all path problem can be expressed as follows:

Given an undirected connected graph $G=(V, E)$, a weight d: E->R+(here d=1) and 2 fixed vertex s and t in V, to find all paths between s and t.

Algorithm:

1. Set $i=0$, $S_0= \{u_0=s\}$, $L(u_0)=0$, and $L(v)=$infinity for v $<>$ u_0. If $|V| = 1$ then stop, otherwise go to step 2.
2. For each v in V\$S_i$, replace $L(v)$ by min$\{L(v), L(u_i)+ d_v^{u}{}_i \}$. If $L(v)$ is replaced, put a label $(L(v), u_i)$ on v.
3. Find a vertex v which minimizes $\{L(v): v$ in V\$S_i\}$, say u_{i+1}.
4. Let $S_{i+1} = S_i$ cup $\{u_{i+1}\}$.
5. Replace i by i+1. If i=$|V|$-1 then stop, otherwise go to step 2.
   These five steps above comprise of the core of *Dijkstra algorithm* [11], from which the shortest path is obtained. The following steps are designed to find other paths:
6. Record those arcs which vertices do not include s and t in the shortest path, as E (n_0, m), n_0 is the number of arcs recorded, m =1,2…
7. Cut the arcs in E (n_0, m) with option of n_0=0 to n_0 arcs, to generate G'=(V, E) from G.
8. Replace G with G' and go to step 1 to find a shortest path for G', if no path available, then go to step 7. If no G' can be generated, then stop.

REFERENCES

1. И. С. СОЛОНИН, С. И. СОЛОНИН, Li, Chunpu (Translator) *Calculation of Assembly and Process Dimension Chains*(in Chinese), Shanghai Science and Technology Literature Press, 1986.1.
2. Jianbin Xue, Ping Ji. "Identifying tolerance chains with a surface-chain model in tolerance charting", *Journal of Material Processing Technology*, **Vol123**, pp. 93-99, 2002.
3. Hu S. J. " Stream of variation theory for automotive body assembly", *Annals of the CIRP*, **Vol46** (1), pp. 1-6, 1997.

4.   S. Charles Liu, Hsin Wel Lee and S. Jack Hu. "Variation simulation for deformable sheet metal assemblies using mechanic models", *Transactions of NAMRI/SAE*, **Vol23** (5), pp. 235-240, 1995.

5.   Ceglarek, D., Shi, J. "Tolerance analysis for sheet metal assembly using a beam-based model," *ASME International Mechanical Engineering Congress and Exposition*, **DE-Vol. 94**, pp.153-159, 1997.

6.   Ceglarek, D., Shi, J. "Design evaluation of sheet metal joints for dimensional integrity", *Journal of Manufacturing Science and Engineering*, **Vol.120**, pp. 452-460, MAY 1998.

7.   Bai Zhang. *Design for the Dimensional Integrity of Automobile Body Assembles, [Ph.D. Dissertation]*. University of Michigan, Ann Arbor, 2000.

8.   Sarvananthan, V.C, Pandey, P.C. "An automated assembly sequence planning system for mechanical parts", *Proceeding of DETC98/FLEX-6015*.

9.   Dini, G., Santochi, M. "Automated sequencing and subassembly detection in assembly planning", *Annals of the CIRP*, Vol41, 1:1-4,1999.

10. Y.Z Zhang, J. Ni, Z.Q Lin, X.M Lai "Automated sequencing and sub-assembly detection in automobile body assembly planning", *Journal of Material Processing Technology*, **Vol129**, pp. 490-494, 2002.

11. http://www-b2.is.tokushima-u.ac.jp/~ikeda/suuri/dijkstra/Dijkstra.shtml.en

GENETIC ALGORITHM-BASED SCHEME GENERATION FOR MECHANICAL PRODUCTS IN CONCEPTUAL DESIGN

Rui-Feng Bo[1,2], Hong-Zhong Huang[1], Ying-Kui Gu[1] and Li-Hua Xue[1]

[1]*School of Mechanical Engineering, Dalian University of Technology, Dalian 116023, P R China*

[2]*Department of Mechanical Engineering, North China Institute of Technology, Taiyuan 030051, P R China*

Corresponding email: hzhhuang@dlut.edu.cn

Abstract: Conceptual design's intrinsic uncertainty, imprecision, and lack of information lead to the fact that current conceptual design activities in engineering have not been computerized and very few CAD systems are available to support conceptual design. In most of the current intelligent design systems, approaches of principle synthesis, such as morphology matrix, bond graph, and design catalogues, are usually adopted to deal with the scheme generation in which optional schemes are generally combined and enumerated through function analysis. However, as a large number of schemes are generated, it is difficult to evaluate and optimize these design candidates using regular algorithm. It is necessary to search a new approach or a tool to solve the scheme generation in conceptual design. Generally speaking, scheme generation in conceptual design is a problem of scheme synthesis. In essence, this process of developing design candidate is a combinatorial optimization process, viz., the process of scheme generation can be regarded as a solution for a state-place composed of multi-schemes. In this research, genetic algorithm (GA) is utilized as a feasible tool to solve the problem of combinatorial optimization in scheme generation, in which the encoding method of morphology matrix based on function analysis is applied, and a sequence of optimal schemes are generated through the search and iterative process which is controlled by genetic operators, including selection, crossover, mutation, and reproduction in GA. Several important problems on GA are discussed in this research, such as the calculation of fitness value and the criteria for heredity termination, which have a heavy effect on selection of better schemes. The feasibility and intellectualisation of the proposed approach are demonstrated with an engineering case. In this study scheme generation is implemented using GA, which can facilitate not only generating several better schemes, but also selecting the best candidate. Thus optimal schemes can be conveniently developed and design efficiency can be greatly improved.

Key words: Conceptual design; Genetic algorithm; Scheme generation; Intelligentize.

1.   INTRODUCTION

Conceptual design is the first step in the overall process of product design. But it is a dominate and creative process in which scheme is generated as a design result. For a mechanical product, its performance, quality, cost, and market to time are mainly embodied through the design scheme. Compared with the detailed design stage, more emphases are laid on innovation of a scheme which concerns the high level activities of brain in the conceptual design stage. So a large quantity of expert knowledge and intelligence are necessary in this process. Since it is a preliminary stage of design, imprecision, uncertainty and lack of design information emerge before the designers. Thus the process of scheme design can be regarded as a highly intellectualized, non-linear, and isomerous process. As a result, the current conceptual design activity in the field of engineering has not been computerized to assist designer and it is still in a manual design stage which undoubtedly has an ill effect on the design efficiency, period and quality at a large extent. The intellectualization and automatization of conceptual design has been a critical problem in the field of intelligent design.

Now in most of intelligent design systems, approaches of principle synthesis, such as morphology matrix, bond graphs, and design catalogues, are widely used to deal with the scheme generation in which optional schemes are usually combined and enumerated through function analysis. However, as a large number of combinatorial schemes are generated, it is difficult to evaluate and optimize these design candidates using regular algorithm. Hence it is necessary to search a new approach or a tool to solve the scheme generation in conceptual design.

2.   COMBINATION OF GA AND CONCEPTUAL DESIGN

In essence, scheme generation for mechanical product is a problem of combinatorial optimization and at the same time it is a scheme synthesis problem. The process of scheme design can be viewed as a global optimization of a group consisting of different mechanical schemes, viz., a process of finding the working mechanisms satisfying all functions in a mechanism database, generating numerous design candidates through different combinations, and finally selecting the best design candidate by evaluation. This optimization is also a process of searching a solution for the best candidate in a large state-place, which is usually conducted at random because of the complexity of the state-place. If this search is implemented blindly, it is inevitable to result in combination explosion. The key for the scheme design is how to find the optimal solution by minimizing the searching space and controlling the searching process adaptively according to certain rules [1]. The recent rapid advances in genetic algorithm provide a feasible tool for solving the problem of combinatorial optimization in scheme generation.

Genetic algorithm is essentially a probability search-based optimization approach by simulating the mechanisms of selection, crossover, and mutation in the process of evolution in nature. In the searching process, the knowledge for searching place can be acquired and accumulated automatically, and this searching process can be adaptively controlled using fitness function. Finally the optimal solution can be received accompanying this step-by-step evolution process. When the number of individuals consisting in a population is comparatively large, genetic algorithm is more effective than other traditional algorithms in combinatorial optimization problem. The objective of this research is to propose an approach for generating design scheme of mechanical product automatically using genetic algorithm.

Scheme generation in conceptual design is an optimization process required to satisfy certain functional requirements and design constraints. It is an iterative process of searching the best solution to satisfy some objective functions, including quality, cost, and market to time, under certain design constraints. The optimization of scheme generation in conceptual design can be described in mathematical equations, as shown in Eq.(1).

$$\min F = f(x, y, \theta)$$
$$\text{s.t. } F \in R \quad g(x, y, \theta) \le 0 \quad h(x, y, \theta) = 0 \tag{1}$$
$$F \in R, \quad (x, y, \theta) \in C$$

Where x is a continuous variable, y is a discrete variable, θ is an invariable, and each triplet, viz., $(x, y, \theta) \in C$, constitutes a solution to the problem. C represents the definition field of x, y, and θ, which can be regarded as the state-place for each solution. F is a real number which can be considered as a measurable expression of a solution's performance. f represents a mapping from a solution space to a real number field and for each decided solution, a F_i can be achieved correspondingly.

It is obvious that the objective of using genetic algorithm for optimization is to find a triplet, viz., $(x_0, y_0, \theta) \in C$, and ensure $F = f(x_0, y_0,$

θ) $\rightarrow$min. A series of solutions should be received in order to find the best candidate and better candidates in solving the multi-scheme problem.

3.   KEY TECHNOLOGIES OF GENETIC ALGORITHM-BASED SCHEME GENERATION

The following gives a case of four-station special machine tool to study some key technologies on genetic algorithm.

Four-station machine tool can accomplish four stations, viz., drilling, fan boring, reaming, and handling work piece. Before using genetic algorithm, functional decomposition for design task is performed at first and functional structure tree is acquired in Fig.1. Then we can draw the motion-converting function figure by analyzing the motion's difference of transmission chain from motor to working mechanisms, as shown in Fig. 2, where the meaning of motion-converting symbol is expressed in Ref. [3]. Through selecting the proper working mechanism for main function units, representing every function unit with y-coordinate and representing the corresponding solution with x-coordinate, the morphology matrix is constructed in Tab.1.

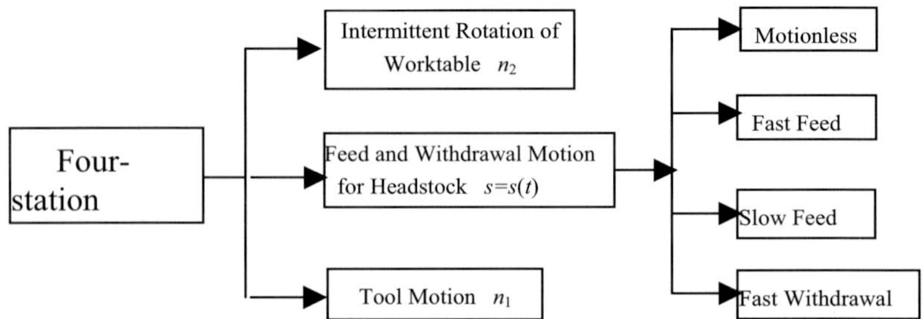

Figure 1: Functional Structure of four-station special machine tool

3.1   Scheme Coding [4]

The initial population's generation starts with individual coding. In the field of conceptual design, several coding approaches are usually adopted, including tree-structure coding based on genetic programming presented by John R. Koza [5], one-dimensional string structure coding, and two-

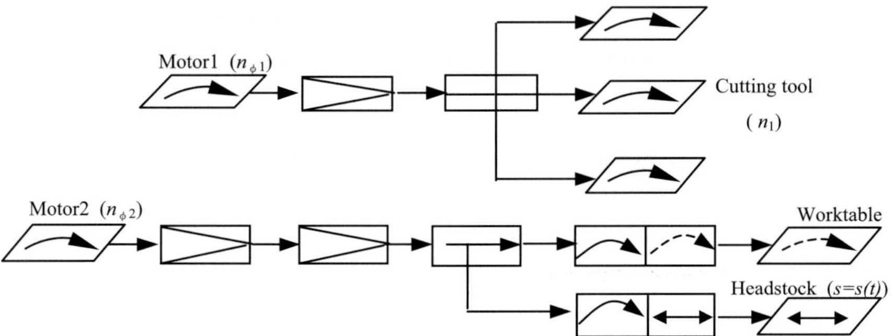

Fig.2 Motion-converting function figure

Table 1 Morphology matrix of four-station machine tool

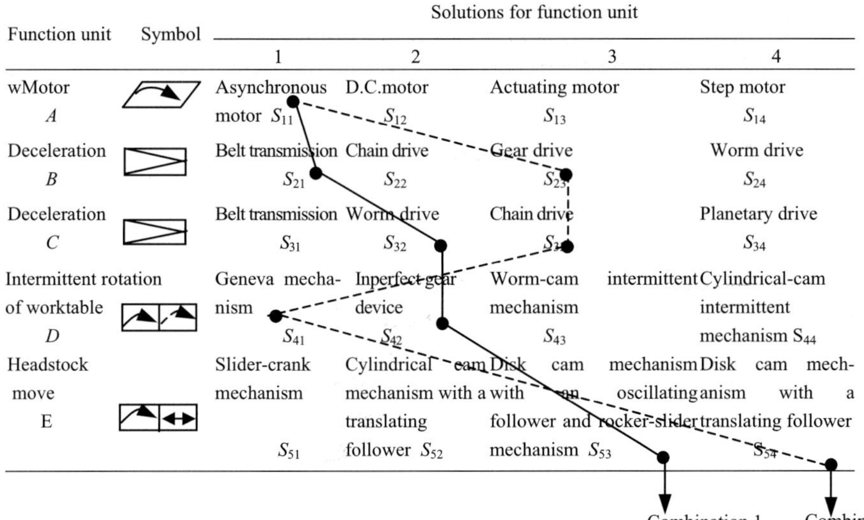

Function unit	Symbol	Solutions for function unit			
		1	2	3	4
wMotor A		Asynchronous motor S_{11}	D.C.motor S_{12}	Actuating motor S_{13}	Step motor S_{14}
Deceleration B		Belt transmission S_{21}	Chain drive S_{22}	Gear drive S_{23}	Worm drive S_{24}
Deceleration C		Belt transmission S_{31}	Worm drive S_{32}	Chain drive S_{33}	Planetary drive S_{34}
Intermittent rotation of worktable D		Geneva mechanism S_{41}	Inperfect gear device S_{42}	Worm-cam mechanism S_{43}	intermittent Cylindrical-cam intermittent mechanism S_{44}
Headstock move E		Slider-crank mechanism S_{51}	Cylindrical cam mechanism with a translating follower S_{52}	Disk cam mechanism with an oscillating follower and rocker-slider mechanism S_{53}	Disk cam mechanism with a translating follower S_{54}

dimensional structure coding, etc. Binary string structure coding based on morphology matrix is introduced in this research. In the morphology matrix shown in Tab.1, there are five major function units and each possesses four optional function solutions. A series of scheme individuals can be acquired after any solution for every function unit are combined using different combinatorial ways and the number of combinatorial schemes is 1024. Two of them are shown in Tab.1. It is noticeable to remain the physical or geometrical consistency among combinatorial solutions, that is to say, solution's combination must satisfy certain constraint conditions. In the process of individual (chromosome) coding, every solution (gene) is coded firstly. Gene coding is expressed with two digits binary string which can nicely represent four optional solutions to every function unit, viz., 00, 01, 10, and 11. In Table 1, combination 1 (chromosome 1) and combination 2 (chromosome 2) are represented with binary strings as 0000010110 and

0010100011 respectively. After generating the best scheme, its coding can be decomposed into corresponding solutions to function units using the opposite decoding approach.

3.2   Population initialization

Genetic algorithm is an optimization approach based on population which is composed of an amount of individuals. Accordingly, a certain number of scheme individuals are required to constitute an initial population. In this research, a certain number of individuals are generated according to constraint conditions. Then the individuals with better fitness are selected from them and fed into the initial population. This process is continuously performed until the number of individuals in initial population arrives at scheduled number. As a result of it, initial population of a certain scale is constructed through generating a number of binary coding strings.

3.3   Calculation of fitness

In genetic algorithm, fitness is an important index to evaluate the performance of individual. It is correspondent to objective function of optimization model and the calculated fitness has an effect on heredity and variation of population. In this research, the value of fitness is calculated using an approach of evaluating the scheme individual quantitatively. For the reason that the scheme to realize the total function is constructed with consistent function units, the fitness of individual equals weighted sum of the fitness of each gene, as calculated in Eq.(2)

$$F_i = \sum_{j=1}^{m} G_{ij} \cdot W_j = W_1 G_{i1} + W_2 G_{i2} + \cdots + W_m G_{im} \tag{2}$$

In Eq.(2), F_i is the fitness value of the ith individual (scheme) where $1 \le i \le l$ and l is the number of schemes, G_{ij} is the fitness value of the jth gene (function unit) of the ith individual where $1 \le j \le m$ and m is the number of genes. W_j is the weight of the jth gene which represents the importance of different function units and

$$\sum_{j=1}^{m} W_j = 1 \tag{3}$$

It is apparent that the calculation of scheme individual's fitness is based on the calculation of function units' fitness, which can reflect satisfaction of the mechanism to function unit. The fitness value of function unit is calculated by Eq.(4)

$$G_j = \sum_{k=1}^{n} f_j(V_{jk}) \cdot W_{jk} \tag{4}$$

Where G_j is the fitness value of the jth function unit, V_{jk} is the kth evaluating index where $1 \leq k \leq n$ and n represents the number of evaluating indexes, these evaluating indexes involves elementary function index, working performance, dynamical performance, economy index, compactness index and customer's special requirement, W_{jk} is the weight value of correspondent index where

$$\sum_{k=1}^{n} W_{jk} = 1 \tag{5}$$

and $f_j(V_{jk})$ is the evaluating value giving by experts.

The satisfaction degree (Fitness) of functional unit is determined by expert knowledge which is easily acquired, expressed, and accumulated to construct a database. For instance, in the equation of G_3 (*Answer* (Geneva mechanism), Intermittent rotation) $=0.8$, 0.8 is a fuzzy number which represents satisfaction of Geneva mechanism to intermittent rotation.

3.4   Genetic Operators

In the scheme generation of conceptual design, there are three dominating operators, including reproduction, crossover, and mutation [2].

1) Reproduction

The objective of reproduction is to improve the average fitness of population and embody the evolution principle of the survival of the fittest in nature. The mode of selection adopted in this research combines proportional fitness assignment and roulette wheel selection. After ranking of individual's fitness, coding strings with better fitness are reproduced to the next generation by being selected from population of father generation which can ensure the accomplishment of global search with a high searching efficiency.

2) Crossover

The objective of crossover is to generate new individuals to embody the superiority of recombination in biology. Double-point crossover is used here, and depicted as follows.

Double-point

A : 000|00101|10   crossover   A' : 000 01000 10

B : 001|01000|11 $\longrightarrow$ B' : 001 00101 11

Here, two scheme individuals are selected at random, then two exchanging position are generated at random, and next the correspondent binary digits belonging to two strings between these two positions are exchanged. As a result of it, two new individuals are generated. In the example of four-station machine tool, crossover process is based on the foundation of morphology matrix and it belongs to a plane-crossover mode. Suppose there are two scheme individuals in which individual A is 0000010110 and individual B is 0010100011, the crossover process is expressed as above and can be shown by morphology matrix, as illustrated in Fig.3.

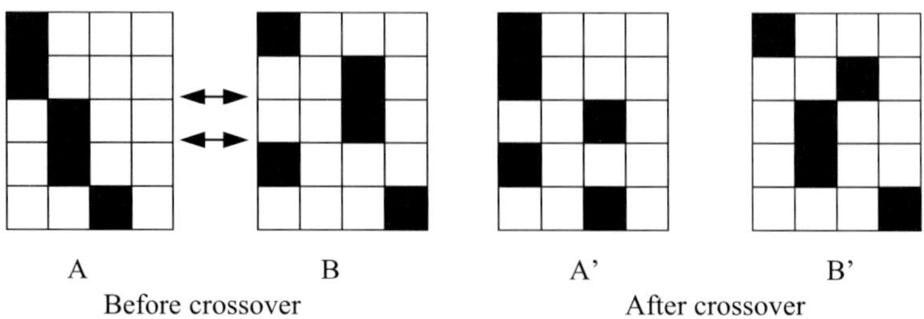

Figure.3: Crossover

3) Mutation

The objective of mutation is to enhance the diversity of new individuals in population by simulating the gene variation in biology, viz., the variation of population is accomplished by selecting other solution for function unit in solution set to replace the current solution. In the process of mutation, an integer is generated at random at first to determine the position of mutation, and then a random integer is then generated to represent the value of mutation. This mutation is performed within a certain bound, and random of mutation decreases at some extent. Furthermore, a gene's mutation will generate more than one result.

3.5   Criterion for Heredity Termination

In the process of scheme generation, there are two terminating heredity criterions for GA, including satisfaction degree criterion and the maximum generation criterion. After a new population is generated by selection, crossover, and mutation, the individuals of this population are evaluated to decide whether the evaluating result satisfies design requirements or iterative number exceeds the maximum generation. If the result doesn't satisfy these conditions, the process of genetic operation is repeated until the satisfactory result is acquired. Generally speaking, a group of individuals need to be acquired by GA, and these individuals are ranked in terms of the fitness value of each. Thus more than one optional scheme is generated. Then the designer is required to determine the best scheme for further development by the mode of man-machine conversation. In the case of four-station special machine tool, the number of total schemes is 20 (l=20), the number of function units is 5 (m=5), and each function unit possesses 4 solutions. The number of evaluating index can be decreased to only 2, including the index of satisfaction to main function and economy index (k=2). The pcross is equal to 0.90, the pmutation is equal to 0.05, and the maximum of evolving generation is 150. The evolving process is repeated until the average fitness of population remains consistent. One of the schemes for four-station machine tool is shown in Fig.4 and the combination of mechanisms can be acquired by decoding, expressed by $S_{11}+S_{21}+S_{32}+S_{42}+S_{53}$.

4.   CONCLUSIONS

Scheme generation of mechanical product in conceptual design is a dominant part for mechanical design. In this research scheme generation is implemented using GA, which can facilitate not only generating several better schemes, but also selecting the best candidate. Thus scheme can be generated automatically and design efficiency can be improved. However, the proposed approach is still in the stage of basic research by which only some simple design problem can be solved. For those complicated design problems, this approach isn't perfect enough to deal with them for the reason that this approach is based on the implementation of function units which

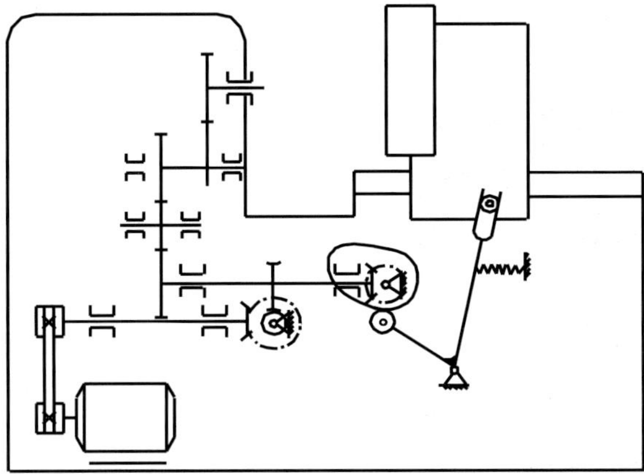

Figure 4: One scheme for four-station machine tool

restricts the innovation of design scheme. Therefore, much effort is still required for the application of this approach in the field of engineering design.

ACKNOWLEDGEMENTS

We gratefully acknowledge the financial support of the National Science Foundation of China (under the contract number 50175010), the Excellent

Young Teachers Program of the Ministry of Education, China (under the contract number 1766), the National Excellent Doctoral Dissertation Special Foundation, China (under the contract number 200232), and the Natural Sciences and Engineering Research Council of Canada.

REFERENCES

1. Shu Q., Hao B, "An Automatic Conceptual Design System for Mechanical Products". *China Mechanical Engineering* **13**: pp. 1676-1678,2002.
2. Zhou M., Sun S. *Genetic Algorithm: Principle and Application.* Beijing, National Defence Industry Press,1999.
3. Zou H., Fu X., Zhang C. *Mechanism and Machine.* Beijing, Higher Education Press, 2002.
4. Huang H. Z., Zhao Z., Guang L. "Intelligent Conceptual Design Based on Genetic Algorithm". *Journal of Computer -aided Design & Computer Graphics* **14**: pp. 437-441, 2002.
5. Koza J. R. *Genetic Programming: on The Programming of Computers by Means of Natural Selection.* Cambridge, The MIT Press, 1992.

RESEARCH ON PRODUCT MODELING TECHNOLOGY FOR DISASSEMBLY

Weixiang Guo, Guangfu Liu, Zhifeng Liu and Haihong Huang
*School of Mechanical & Automobile Engineering, Hefei University of Technology, Add: P.O.
Box 2,193 Tunxi Road, Hefei, Anhui 230009, China, E-mail: guowayxiang@etang.com*

Abstract: Disassembly analysis is a very complex process that needs the supports of much information and many methods. The contents of disassembly analysis are illustrated; to give a strong foundation for the disassembly analysis, an integrated model is introduced in this paper that integrates the 3D assembly model, the Hierarchical-Network Model and an open database; a case study is given to show that this method is operable and reasonable in the end.

Key words: Disassembly, integrated model, 3D assembly model, the Hierarchical-Network Model, database.

1. INTRODUCTION

With the increasing environmental and economical concerns, more and more nations have highly focused on the recycling of waste products. One of the important researches for products recycling is products disassembly analysis [1]. Much information is required in this research, so a product disassembly model is necessary to support the whole process of disassembly analysis.

Up to now, a lot of papers have given different methods on the modeling for disassembly. These include Directed Graph Model [2], Undirected Graph Model [3], And/Or Graph Model [4], Petri Net Model [2] etc. However, it is difficult for most of these models to support the whole process of disassembly analysis, because those mostly were built only to support one aspect of disassembly analysis, and contained very limited information.

An integrated model is introduced in this paper, which is the integration of the 3D assembly model, the relational model, the hierarchical model and an open database. It can provide effective help for the disassembly analysis.

2.   PRODUCT DISASSEMBLY

The different recycle strategies can be adopted for the end-of-life products in order to gain the maximal recycling benefit and good environmental influence [5]. Figure 1 describes the whole process of product recycling analysis. The general principle of product recycle is to obtain the maximal profit, reuse the parts and recovery the material furthest, and produce the least waste.

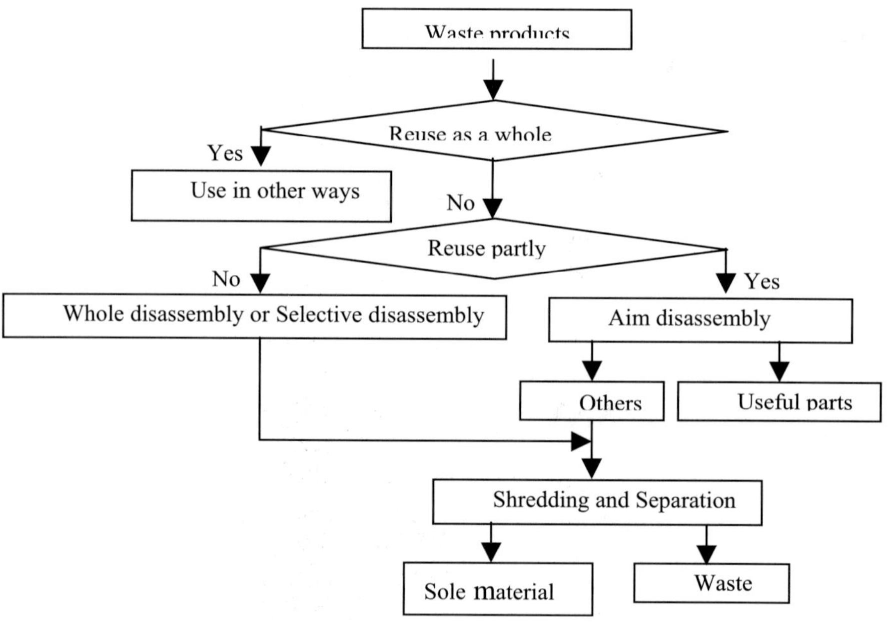

Figure 1. Process of product recycling

Disassembly aims to dismantle the products to parts or components according to the requirement of recovery. Disassembly is one of the most important steps in the product recovery. Effective disassembly analysis will make the disassembly process more reasonable and scientific. And disassembly analysis is also very important to the recovery decision-making. Reasonable disassembly analysis can help to decide which level the product should be disassembled and recycled. The disassembly analysis mainly includes the several following contents:

- The generation and evaluation of disassembly sequence. Some reasonable disassembly sequence should be gained, and the optimal should be selected by the evaluation system.
- The analysis of disassemblability. Such as the disassembly reachability, the number of fasteners and the level of standardization will be taken into account to valuate the disassemblability of the products, and the improvement advices will be given.
- Research on disassembly strategies. The combination of economics and environment will be considered to decide the level of product disassembly, so the maximum synthetical benefit will be gained.
- Disassembly simulation. The process of disassembly will be simulated according to the result obtained from the disassembly decision-making and the disassembly sequence planning, so the disassembly process can be previewed.

3. AN INTEGRATED MODEL FOR DISASSEMBLY

The integrated disassembly model is the precondition and foundation of disassembly analysis. Figure 2 is the description of integrated disassembly model. The 3D parts database is constructed; the 3D product assembly model is built; the Hierarchical-Network Model is gained automatically according to the assembly model; and a disassembly information database is provided to support the whole process of disassembly analysis.

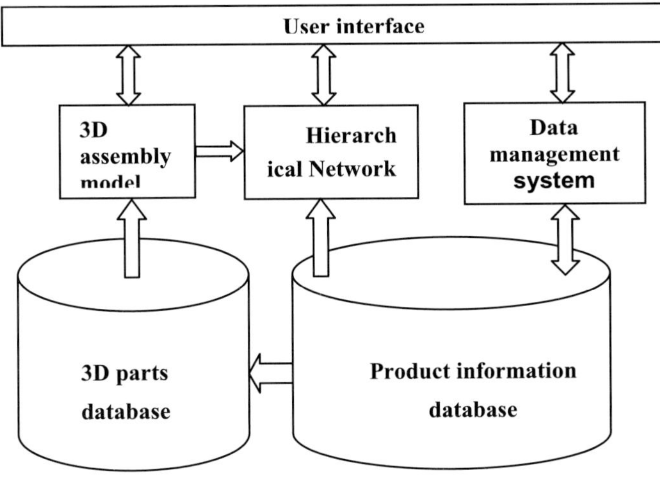

Figure 2. The integrated model.

3.1   3D Assembly Model

The 3D assembly modeling is vital in the integrated disassembly model. The geometrical forms and structure are presented through the 3D assembly model; the analysis of disassembly interference and the dynamic simulation have to depend on the 3D assembly model. The 3D modeling includes both part modeling and assembly modeling.

The part modeling process includes 3D modeling and information input. The 3D part model is the carrier of basic part information and the basic element of assembly model. The design software UG provides many ways including Solid, Surface, Wire frame and modeling based on the feature to help build the 3D model. A lot of information of parts such as type, weight and material should be considered in the disassembly analysis. However, the information can't be gained directly from the 3D assembly. Therefore the information has to be given to the part model by man-machine way. UG also provides the attribute method, and the necessary information can be supplied to the part model through the attribute dialog. The information can be attained in the application program by the API function UF_ATTR_read_value () provided by UG. These finished part models will be stored in the part database, and any of them can be picked up to compose the assembly model when needed.

In the assembly model, all the parts data are only the reflection of the real part model. The assembly model and the part model are related to each other, and any one has been modified, the other one will be renewed simultaneously. The constraints and connections in the real products can be represented by the logical mate, align and center and so on in the process of assembly modeling. To express the relations among different components in the assembly distinctly, the different parts or components can be displayed by different colors. The parts can be assembled to some subassemblies firstly in accordance with the relations of mate and function among them, and step-by-step, the product assembly model can be built in the end. The hierarchical assembly model makes the disassembly analysis easier.

3.2   Hierarchical-Network Mode

3.2.1   Undirected Network Graph

Generally, Graph E includes two sets V and E:

$$G = (V, E)$$

Where V is the finite non-vacuous set of nodes, and E is the finite set of edges that haven't direction.

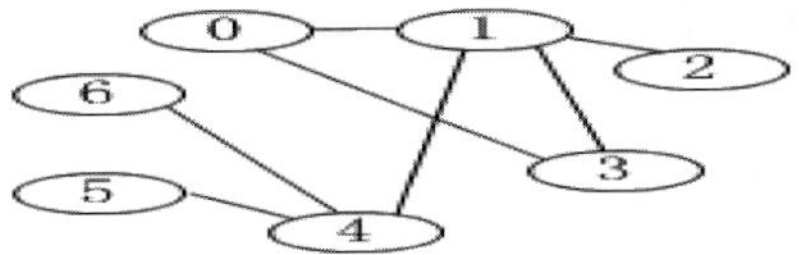

Figure 3. The Undirected Network Graph

The Undirected Network Graph $G=(V, E)$ can show the relations among the components of a product. The parts and components can be represented by nodes $V=\{v_1, v2, \cdots, v_n\}$, where n is the number of components. The relations among them can be represented by edges $E=\{e_1, e_2, \cdots, e_m\}$, where m is the number of edges. If two components v_i and v_j $(i \neq j)$ have the connective relations, then (v_i, v_j) $\square E$. Figure 3 is an Undirected Graph for a product including 7 components. There may be more than one mates between two components, but there is only one edge between them in the Undirected Network Graph. Therefore, the edges in the Undirected Network Graph just represent the existence of mate, not the specific number of mates.

Adjacent Matrix can be used to represent the connective relations among components in the Undirected Network Graph. Suppose $G=(V, E)$ is a Undirected Network Graph containing n nodes, the Adjacent Matrix for G can be presented as follows:

$$\begin{bmatrix} e_{11} & e_{12} & e_{13} & \cdots & e_{1n} \\ e_{21} & e_{22} & e_{23} & \cdots & e_{2n} \\ \vdots & \vdots & \vdots & \cdots & \vdots \\ e_{n1} & e_{n2} & e_{n3} & \cdots & e_{nn} \end{bmatrix}$$

$$e_{ij} = \begin{cases} 1 & \textit{component i and j have the connective relations , and } i \neq j \\ 0 & \textit{otherwise} \end{cases}$$

The nodes contain the information such as name, weight, type and material of components; and the edges contain the information such as name of components connected by the edge, connective types and disassembly vectors [5].

In a sense, disassembly is the reverse course of assembly, therefore the data like connective relations and disassembly vector can be gained by analyzing the assembly model. These data can be gained automatically by some API functions provided by soft ware.

3.2.2　Hierarchical-Network Model

The Hierarchical-Network Graph is a set T comprising one or more nodes; there is only one root node, the other nodes can be separated to m sets: $T1$、$T2\cdots Tm$, which don't intersect with each other, and every one of them is called a sub-graph. If two nodes have affiliation between them, one of them is the child of the other one.

The Hierarchical-Network Graph is composed of cell Undirected Network Graph. There are three different nodes in the Hierarchical-Network Graph. The node is a root node if it has no father node, and there is only one root node in a Hierarchical-Network Graph; the node is a free node if it has no child node; the node is a middle node if it has not only father node but also child nodes. There are also two kinds of edges: one represents the connective relations; the other represents the affiliation, while it doesn't belong to any cell Undirected Network Graph, and it only points to the sub-graph from its father node.

Cell Graph (Cell Hierarchical-Network), nodes and edges are the three elementary inscapes, and three classes are defined for them. It not only helps to construct the Hierarchical-Network Graph, but also embody the thought of OODM (Object Oriented Data Model).

Node definition:

```
class NodeInfo
{
char *ComponentName;
tag_t componentID;
double ComponentWeight;
char *ComponentType;
char *ComponentMaterial;
}
```

Edge definition:

```
class EdgeInfo
{
NodeInfo   *node1st;
NodeInfo   *node2nd;
int Constraintsnum;
double  disassemblyVector[3];
}
```

Cell Graph definition:

```
class GraphInfo
{
  int    graphId;
  NodeInfo  *parentNodeInfo;
```

```
    int    nodeNum;
       NodeInfo *nodes;
    int    edgeNum;
    EdgeInfo  *edges;
    char  *ProductName;
}
```

The Hierarchical-Network Graph shows the hierarchical relations and connective relations among the components, and it is also the carrier of product disassembly information. The Hierarchical-Network Graph can be constructed based on the finished 3D assembly model. All the hierarchy information of product has been saved by the UG in the course of assembly, and the API functions used to pick-up the information have been provided by UG, so the Hierarchical-Network Graph can be constructed in the environment of VC++ with the help of the API functions.

3.3   Disassembly information database

The disassembly information database should includes all the data related with the disassembly as follows:

(1) The original design information. The design attributes of product including product lifetime, material types, product structure, dimension, weight etc. are very important to disassembly analysis. These attributes have been decided in the stage of design.

(2) Basic Component Information. The basic information includes the component type, shape, weight and material etc.

(3) The Usage and Maintenance Information. Many uncertain factors such as different work environment and different users would change the product a lot; even the structure of product would be changed. The disassembly performance will be changed by these situation. So the disassembly information should include the use time, use environment and maintenance information etc.

(4) Material Information. All the material factors including the chemistry, physics, price and the influence to environment should be considered in the process of disassembly. So there must be a list of the materials that are used usually in the products.

(5) Disassembly Time. The disassembly time is a very important evaluation criterion of disassembly performance. The disassembly time can be gained by the normal disassembly operation to relative product.

(6) Disassembly Energy Consumption. There are not only fastener joints but also chemic joints such as welding and gluing in the normal product. The consumption of energy is different with the different join types.

(7) Economic Information. The economic information includes labour cost, tool cost, and material recovery cost etc. The economics performance of disassembly can be calculated effectively by the information, and provides an important foundation for disassembly decision-making.

(8) Disassembly Tools. The choice of disassembly tools is a key factor of disassembly, so the information about the disassembly tools should be included in the database.

There are so many data in the database that the hierarchy is adopted to build the disassembly database. The hierarchy is a typical inverse tree. There is only a root record; every other record has a father record; if the record has no child record, it's a leaf record. Table 1 shows the format of the database that including serial number, property name, property capability and property value. The serial numbers reflect the hierarchy of the records, and the serial number can be adjusted automatically according to the addition and deletion of records. The property name is the description of content in the database. The property capability shows the number of the child records under the current record, the value of property capability will change automatically while adding and deleting the child records. Only the leaf records have the property value, and the property value of others is null.

Table 1. Disassembly Information

Serial number	Property name	Property capability	Property value
000000	Disassembly information	10	
010000	Material information	8	
011000	Material price	253	
011100	Copper price	0	8200
011200	Aluminium price	0	6900
……	……		……

In the table of database, every column has the same variable type. However, the variable type in the property value column should be different according to different types of properties. All the property values are defined as the float type, so that it is convenient to store and calculate. Some kinds of property values like price, time, dimension and weight can be defined by float type directly, but some property values like disassembly difficulty and part forms can't be defined directly. So there must be a transition table that can convert the other types of property values to float type. For example, the different numbers can represent different disassembly difficulties.

The relation between the Hierarchical-Network Graph and disassembly database has been built, and the useful information can be picked up from the database according to the key words provided by the Hierarchical-Network Graph in the process of disassembly analysis. The researchers and the managers can update and complement the information at any moment by the database management system. They can also search some useful information in the database when they need.

4. CASE STUDY

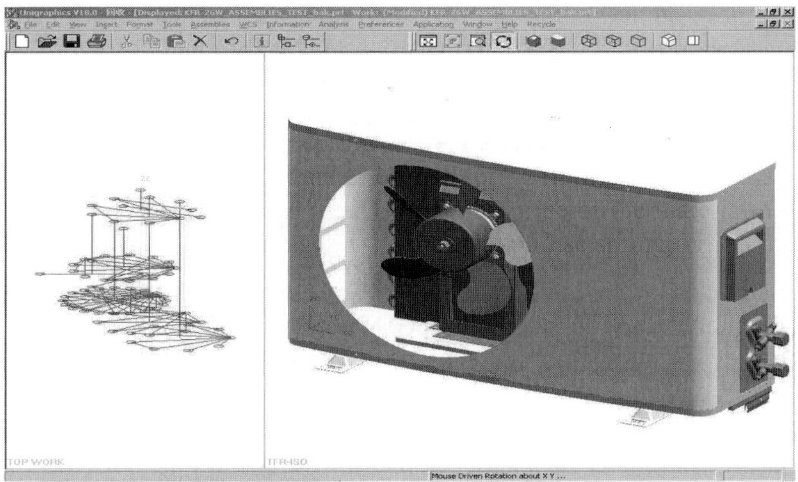

Figure 4. Green Product Development System

We have developed the GPDS (Green Product Development System) as shown in Figure 4. The 3D assembly model and the Hierarchical-Network Graph of an air-conditioner are illustrated.

Before disassembly, we must make the decision that which parts or components should be separated from the products and which are not necessary to be separated. Generally this can be judged by the attributions including the value, toxicity, and reuse of components. With help of the integrated model, all the attributions are considered synthetically, thus the level of disassembly can be attained.

The interference check-up, disassembly sequence generation and optimization, disassembly simulation have been realized with the help of the integrated disassembly model. At the same time, the disassembly performance of the product can be attained with the disassembly analysis. The disassembly performance analysis can provide an information feedback for the designers to modify their design plan.

A lot of information is needed in the whole process of disassembly analysis. The information provided by the disassembly database ensures the rationality and correctness of disassembly analysis.

5.   CONCLUSIONS

The disassembly analysis, as a complicated process, includes many research contents, which needs an integrated model to support. The disassembly model established by some other researchers usually supports only one aspect of disassembly analysis, while the model presented in this paper can provides support in many fields of the disassembly analysis.

ACKNOWLEDGEMENTS

Sponsors of this research include the National Natural Science Foundation, China (Granted No. 50375044) and the Excellent Youth Foundation of Anhui Province and the Anhui province Natural Science Foundation (Granted No. 01044105). The authors wish to thank all the collaborators from the Institute of GD&M Engineering in HFUT for their useful discussions and contributions associated with this work.

REFERENCES

1.   Liu Guangfu, Liu Xueping, Liu Zhifeng *et al*. " Analysis of Disassembly and recycling property on mechatronic product", *Mechanical Engineer*, **No.1**, pp.13~16, 2001.
2.   Tang Ying, Zhou Mengchu, Zussman Eyal *et al*. "Disassembly modeling, planning and application: a review", *Proceeding of the 2000 IEEE International Conference on Robotics & Automation*, San Francisco, California ,USA, April 22-28, 2000, pp. 2197-2202 .
3.   Zhang H C, Kuo T C. "A graph-based disassembly sequence planning for EOL product recycling", *IEEE/CPMT Int'1 Electronics Manufacturing Technology Symposium*, Austin ,TX,USA, October, 1997, pp. 140-151.
4.   Pnueli Y,ZussmanE. "Evaluating the end-of-life of a product and improving it by redesign", *Int. J. Prod. Res.*, **Vol. 35** No.4, pp.921~942, 1997.
5.   Wang Shuwang, Liu Zhifeng, Liu Guangfu. "Study on the recycling technology of mechatronic products", *Chinese Journal of Mechanical Engineering*, **Vol.38** No.12, pp.76~81, 2002.

MULTIOBJECTIVE OPTIMIZATION-BASED MANUFACTURABILITY EVALUATION FOR PART STRUCTURE

Shan Jiang[1], Yuxin Wang[2] and Yiming Yang[1]

1 School of Mechanical Engineering, Tianjin University, Tianjin, China 300072, E-mail: js73@eyou.com
2 School of Mechanical Engineering, Tongji University, Shanghai, China 200092, E-mail:creative@263.net

Abstract: Manufacturability evaluation for part structure scheme is a multi-objective problem. Some factors that affect evaluation result can't be calculated quantitatively. Different sub-objective functions are built up for different measure factors. In order to unify dimension, sub-objective functions are normalized linearly. Analytic Hierarchy Process (AHP) method is used to confirm weighting coefficients of the evaluation ingredients. Weighting factors and normalized sub-objective functions are taken into evaluation function. With the value, the evaluation result can be received, and the optimized scheme of part structure can also be gained.

Key words: manufacturability evaluation, multi-objective optimization, AHP, part.

1. INTRODUCTION

Concurrent engineering (CE) has taken Design for Manufacturing (DFM), Design for Assembly (DFA) and Computer Aided Engineering (CAE) one step further by prescribing simultaneous design of a product by engineers of different specialties. Manufacturability evaluation system can make the CE process more simplified. These systems are intended to provide manufacturing feedback to the designer.

The issue of improving manufacturability through design was first addressed by N.P.Suh et al [1], who defined a set of axioms to be followed during design. The same approach was also suggested in Scarr [2], where 62 rules and guidelines for automated assembly were presented.

The quantitative measures of manufacturability that have appeared in literature have very limited use. Several cost equations were appeared in Boothroyd[3], where the cost of material, machining, tools used for machining and some other fees are considered. Through the calculation, the total cost of manufacturing can be gained. N.J.Yannoulakis[4], who described the rotational features by means of two primitive elements: the generalized taper and the generalized arc. Trajectory-specific parameters, velocity-specific parameters and other parameters are determined by the shape of feature. And cost equation was used to calculate the machining time. But this method could only be used for the rotational parts. Shoji Arimoto[5] built a Machining-productability Evaluation Method system(MEM), which divided shape features into 20 elements. Cost evaluation was realized through the mapping between the elements and the machining cost.

The methodology for classification and cost evaluation of machining form features and relationships among features was presented in Chang-Xue (Jack) Feng et al [6]. Machining form features are classified as simple and complex. The unit manufacturing cost is determined by two major factors: the manufacturing activities and the corresponding time required to perform these activities. The feature-based evaluation of manufacturing cost is formulated as the shortest path problem, and a mathematical model as well as an algorithm is presented to determine the minimum cost design alternative. The impact of tolerances and the sequence of operations were not considered in the paper.

Evaluation has relationship with part structure, manufacturing facilities and machining time [7]. Part structure involves the shape, tolerance and surface quality of the product. So manufacturability evaluation is a concept with several factors. Some of them can be measured quantitatively others can only be evaluated qualitatively. Build a reasonable measure model is the basis for the manufacturability evaluation.

The paper is organized as follows. Section 2 will build several sub-objective functions. After linearly normalized, the functions' dimension can be unified. Section 3 is dedicated to an introduction of how to use AHP to calculate the weighting factors. Section 4 is about how to build a unified function by using sub-objective functions and the weighting factors. Section 5 gives an example, which is used to validate that the system is effective. Finally, some conclusions and perspectives for future work are outlined.

2. ESTABLISHMENT OF MULTIOBJECTIVE FUNCTION

2.1 Establishment of multi-objective function

The design space of part structure is a hyperspace. Every dimension denotes an element that affects the lifecycle of the part. The character of the multidimensional design space determines that the evaluation criteria are affected by many factors. These criteria have the following attributes:

1. There are several factors that affect the result of measurement, so local optimum solution often appears. That is: when one factor up to the best, the others present the worst.
2. Several criteria can be measured quantitatively but others can only be evaluated qualitatively.
3. Some of them have no direct relationship with the economic evaluation result.

Of all the attributes mentioned above, an appropriate model should be built for the evaluation.

Because the design of part structure has been affected by many factors, and these factors have no relationship with each other. In this paper, a multi-objective methodology is used to evaluate the scheme. First of all, several sub-objective functions are built. After normalized and unified, a uniform function is created. With this function we can know which scheme of part structure is the best. This function is:

$$\min f[f_1(X), f_2(X), \cdots, f_q(X)]$$
$$s.t. \quad g_j(X) \geq 0 \qquad (j = 1, 2, \cdots, p) \tag{1}$$

In this research, the objective of evaluating function is: using manufacturing resources as fewer as you can. Because the fewer manufacturing resources are used the less cost is expended. That is:

$$\min f[f_1(X), f_2(X), f_3(X), f_4(X), f_5(X)]$$
$$s.t. \quad g_j(X) \geq 0 \qquad (j = 1, 2, \cdots, 5) \tag{2}$$

where

$$f_1(X) = J_i^N = \sum_{e=0}^{p} j_i^e$$ -machining *No .i* structure scheme use J_i^N kind of

machining tools.

$$f_2(X) = D_i^N = \sum_{e=0}^{q} d_i^e \text{ -machining } No.i \text{ structure scheme use } D_i^N \text{ kind of}$$

cutting tools.

$$f_3(X) = K_i^N = \sum_{e=0}^{r} k_i^e \text{ -machining } No.i \text{ structure scheme use } K_i^N \text{ kind of}$$

fixtures.

$$f_4(X) = L_i^N = \sum_{e=0}^{s} l_i^e \text{ -machining } No.i \text{ structure scheme use } L_i^N \text{ kind of}$$

measures.

$$f_5(X) = T_i = \sum_{e=0}^{u} t_i^e \text{ - total machining time of machining } No.i \text{ structure}$$

scheme is T_i .

Normalize linearly for equation (2), and receive

$$\min f(X) = \min \sum_{i=1}^{q} w_i f_i(X) \tag{3}$$

Where, $\sum_{i=1}^{q} w_i = 1, w_i \geq 0, i = 1,2,\cdots,q$

Machining tool, cutting tool, fixture, measure and machining time are evaluation factors. Since they have different dimensions and the results are great disparity, they cannot be put into one equation. So in order to eliminate the gap among them, the dimensions of sub-objective functions should be unified. That is:

$$\overline{f}_i(X) = \frac{f_i(X) - \alpha_i}{\beta_i - \alpha_i} \tag{4}$$

$$f(X) = \sum_{i=1}^{q} w_i \overline{f}_i(X) \tag{5}$$

Where, $\overline{f}_i(X)$ - sub-objective function that is deal with dimensionlessly

w_i - weighting factor of sub-objective function

$$\alpha_i \leq f_i(X) \leq \beta_i (i = 1,2,\cdots,q) \tag{6}$$

the result of each sub-objective function can be transformed into 0~1, so the functions can be unified.

3. APPLICATION OF AHP IN CONFIRMING THE WEIGHTING FACTOR

After unification of sub-objective functions, effect caused by the evaluation criteria will be weakened. Because weighting factor means the importance of each sub-objective function in the unified function, it will become more and more important. In the manufacturability evaluation, some criteria cannot be measured quantitatively. In this paper, AHP method is used to confirm the weighting factors. The strength of this approach is that it organizes tangible and intangible factors in a systematic way, and provides a structured yet relatively simple solution to the decision-making problems. In addition, by breaking a problem down in a logical fashion from the large, descending in gradual steps, to the smaller and smaller, one is able to connect, through simple paired comparison judgments, the small to the large.

3.1 Modeling multi-attribute problem using AHP

The problem can be structured as Fig.1, where the first level is goal. The second level is the criteria, which include the amount of manufacturing resources when machining the part. Such as machining tool, cutting tool, fixture, measure and cutting time. The third level is the sub criteria, which contain the shape feature, general dimension, tolerance and the roughness that will affect the amount of using manufacturing resources. The bottom is the alternative part structure schemes.

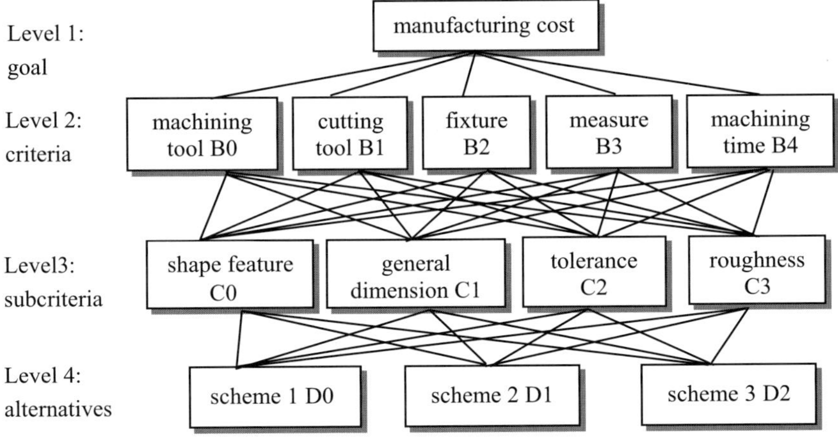

Figure1. The hierarchy model of multi-attribute problem

3.2　Construct pair-wise comparison matrix

In order to solve the problem, comparison matrices are constructed according to the hierarchy model as Figure.1. By using the relative scale measurement shown in Table.1, a set of pair-wise comparison matrices (size n×n) are built up for each of the lower levels with one matrix for each element in the level immediately above. The pair-wise comparison are done in terms of which element dominates the others.

$$A = \begin{bmatrix} \dfrac{W_1}{W_1} & \dfrac{W_1}{W_2} & \cdots & \dfrac{W_1}{W_n} \\ \dfrac{W_2}{W_1} & \dfrac{W_2}{W_2} & \cdots & \dfrac{W_2}{W_n} \\ \vdots & \vdots & \vdots & \vdots \\ \dfrac{W_n}{W_1} & \dfrac{W_n}{W_2} & \cdots & \dfrac{W_n}{W_n} \end{bmatrix} = (a_{ij})_{n \times n} \tag{7}$$

where, $W_1, W_2, \cdots W_n$ are relative weights amongst elements.

Table 1. Pair-wise comparison scale for AHP preferences [8]

Numerical rating	Verbal judgments of preferences
9	Extremely preferred
8	Very strongly to extremely
7	Very strongly to preferred
6	Strongly to very strongly
5	Strongly preferred
4	Moderately to strongly
3	Moderately preferred
2	Equally to moderately
1	Equally to preferred

The significance of the eigenvector derived from the above comparison matrix is the degree of relative importance amongst the elements. At the same time, the maximum eigenvalue can be derived. This value can be used to determine the strength of consistency amongst comparisons. The method of derivation is as follows:

$$K = AW = \begin{bmatrix} \dfrac{W_1}{W_1} & \dfrac{W_1}{W_2} & \cdots & \dfrac{W_1}{W_n} \\ \dfrac{W_2}{W_1} & \dfrac{W_2}{W_2} & \cdots & \dfrac{W_2}{W_n} \\ \vdots & \vdots & \vdots & \vdots \\ \dfrac{W_n}{W_1} & \dfrac{W_n}{W_2} & \cdots & \dfrac{W_n}{W_n} \end{bmatrix} \begin{bmatrix} W_1 \\ W_2 \\ \vdots \\ W_n \end{bmatrix} = \begin{bmatrix} nW_1 \\ nW_2 \\ \vdots \\ nW_n \end{bmatrix} = nW \tag{8}$$

where, W is the eigenvector for A matrix, and n is the maximum eigenvalue.

Having made all the pair-wise comparison, the consistency is determined by using eigenvalue λ_{max} to calculate the consistency index, CI as follow:

$$CI = \frac{\lambda_{max} - n}{n - 1},$$ where n is the matrix size. Taking the consistency ratio CR of CI can check consistency judgment: $CR = CI/RI$, RI is the appropriate value in Table 2. The CR is acceptable unless it exceeds 0.10. If it is more than 0.10, the judgment matrix is inconsistent. To obtain a consistent matrix, judgments should be reviewed and improved.

Table2. Average random consistency (RI) [8]

Size of matrix	1	2	3	4	5	6	7	8	9
Random consistency	0	0	0.58	0.9	1.12	1.24	1.32	1.41	1.45

4.   CHOSEN OF THE BEST ALTERNATIVE

Several kinds of structure schemes can be gained after the embodiment design. Use the CAPP system, we can statute the amount of manufacturing resources for machining each structure scheme: $f_1(X)$, $f_2(X)$, $f_3(X)$, $f_4(X)$, $f_5(X)$ Put them and weighting factors $\sum_{i=1}^{q} w_i = 1$, $w_i \geq 0$, $i = 1,2,\cdots,q$ into the equation (5). The best scheme will appear according to the result.

5.   EXAMPLE

Use the box part of a kind of reducer as an example. There are three alternatives as in Figure.2 that can meet the requirement. Weighting factors for criterion elements are in Table 3. The statistics for manufacturing resources is in Table 4. After unifying, the value is in Table 5.

Table3. Weighting factors for criterion elements

A	B_0	B_1	B_2	B_3	B_4	W	consistency judgment
B_0	1	4	5	6	7	0.4841	$\lambda_{max} = 5.1493$
B_1	1/4	1	4	5	6	0.2359	$CI = 0.0373$
B_2	1/5	1/4	1	4	5	0.1246	$RI = 1.12$
B_3	1/6	1/5	1/4	1	4	0.0909	$CR = 0.0333$
B_4	1/7	1/6	1/5	1/4	1	0.0645	

Table 4. Statistics of manufacturing resources

	$f_1(X)$	$f_2(X)$	$f_3(X)$	$f_4(X)$	$f_5(X)$
Alternative 1	3	11	3	2	310.9
Alternative 2	3	8	3	2	264.8
Alternative 3	3	8	3	2	262.7

Table 5. Unified value for statistics of manufacturing resources

	$\bar{f}_1(X)$	$\bar{f}_2(X)$	$\bar{f}_3(X)$	$\bar{f}_4(X)$	$\bar{f}_5(X)$
Alternative 1	0.2222	0.5263	0.2222	0.1111	0.6210
Alternative 2	0.2222	0.3684	0.2222	0.1111	0.5287
Alternative 3	0.2222	0.3684	0.2222	0.1111	0.5244

Taking value W in Table 3 and $\bar{f}_1(X) \sim \bar{f}_5(X)$ in Table 5 into the equation (5), then

$$f_1(X) = 0.3096$$
$$f_2(X) = 0.2663 \qquad (9)$$
$$f_3(X) = 0.2660$$

According to the above value, a conclusion can be drawn that the third scheme is the best, which can be machined more easily. The second is similar with the third one, and the first is the worst.

Figure2. Three kinds of box part schemes for a reducer

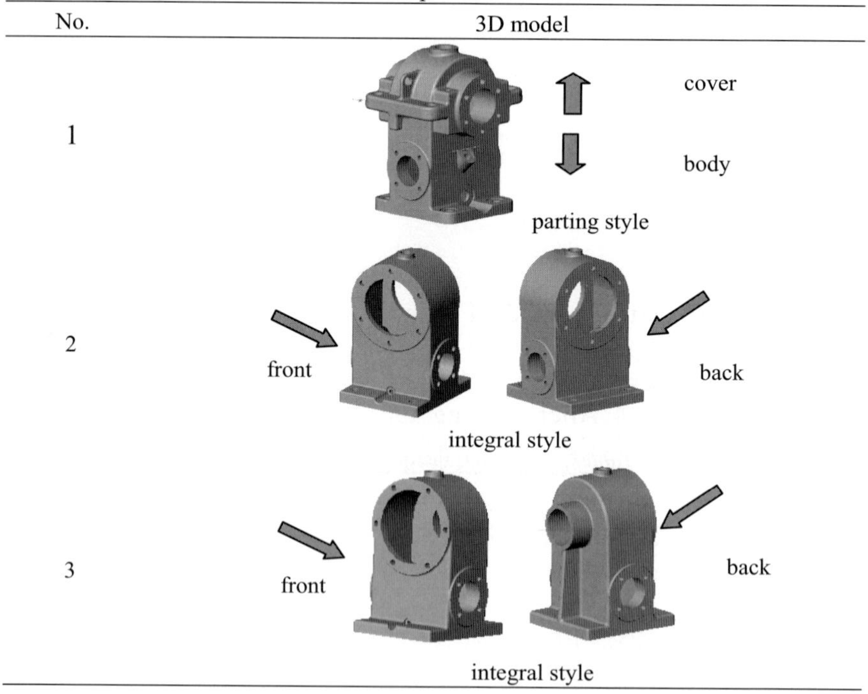

No.	3D model

1

2

3

6.   CONCLUSIONS

Multi-objective optimum method is used to evaluate the part structure schemes. There are some conclusions as following:

1. Several sub-objective functions are built. After unified and normalized, these functions can be compared with each other.
2. With the use of AHP method in confirming the weighting factors, non-quantified elements can be quantified by numerical values to indicate a decision's priority after evaluation and a mathematical process. A decision maker can reach the choice of evaluation alternatives in a very short time.
3. Box parts are used to validate the effectiveness of this method.

ACKNOWLEDGEMENTS

This research is sponsored by the project fund supported by the ministry of education, Research on the exploitation of automatic system for 3D virtual mechanism.

REFERENCES

1. Suh NP. *The Principle of Design*, New York, Oxford University Press, 1990.
2. Scarr AJ. "Product design for Robotic Automated Assembly", *Proc. of the IEEE international Conference on Robotics and Automation*, San Francisco, CA, USA, April, 1986, **vol. 2**, pp. 7-10.
3. Boothroyd G, Dewhurst P. "Product Design for Manufacture and Assembly", *Computer Aided Design*, **vol. 26 no.7**, pp. 165-173, 1994.
4. Yannoulakis NJ. "Quantitative Measures of Manufacturability for Rotational Parts", *Transaction of the ASME Journal of Engineering for Industry*, **vol. 11, no. 2**, pp. 189-198, 1994.
5. Shoji A. "Development of Machining-Producibility Evaluation Method", *Annals of the CIRP*, vol. 42, no.1, pp. 43-56, 1993.
6. Chang XF, Kusiak A. "Cost Evaluation in Design With Form Features", *Computer Aided Design*, **vol. 28, no.9**, pp. 679-885, 1996.
7. Wang XK, Bai G. "Design and Implementation of DFM System for Stamped Products", *Journal of Tsinghua University (Science & Technology)*, **vol. 37, no.11**, pp. 26-29, 1997.
8. Saaty TL. *The analytic hierarchy process*. New York. McGraw Hill, 1980.
9. Kamal M, Al-Subhi, Al-Harbi. "Application of the AHP in project management" *International Journal of Project Management*, **vol. 19, no.1**, pp. 19-27, 2001.
10. Lin ZC, Yang CB. "Evaluation of machine selection by the AHP method". *Journal of Materials Processing Technology*, **vol. 57, no.5**, pp. 253-258, 1996.

A VIRTUAL PRODUCT FAMILY TO SUPPORT DESIGN FOR CUSTOMERS

Yan Li, Shujuan Li and Shaokun Wei
School of mechanical & instrumental engineering, Xi'an University of Technology, Xi'an, P.R.China, 710048, jyxy-ly@xaut.edu.cn

Abstract: This paper proposed virtual product families' method (VPF), This VPF not only supports product development phase, but also supports CAPP, CAM and assembly all of life cycle. In this paper, we only discuss how to meet customers' needs in product development. For uncertain customers' needs, fuzzy logic is an adequate way of modeling this imprecision. For verified customers' needs, this can be precisely defined, the product or component is constituted by selecting component or attributes. The neural net is adopted to obtain the product or component of customers required. Above research results constitute a VPF model. The protocol system is being developed in networks. Finally, a lathe is used as a case study to show the effectiveness of the proposed method in this paper.

Key words: Mass Customization, VPF, Customers' Needs Analysis, Fuzzy Logic, Neural Net.

1. INTRODUCTION

Along with customer's requirements diversification and dynamic, product renovation speed is rapider. Product development's dilemma is exacerbating difficulties. In order to achieve high customer satisfaction, better quality, low cost, shorter time to market, and many people put forward many viewpoint and method, such as:

Al-sultan and Fedjki (1997) use the viewpoint of the process relationship between parts and machines to identify the similarities amongst the parts.

And by the method of cluster analysis by the k-means algorithm, they deal with the relationship matrix and successfully divide the parts into families.

S.Bin et al (2001) put forward web-based product family development using Artificial neural network.

Kao and Moon (1998) consider the viewpoints of the features of the parts as the distinguish rules for the similarities amongst the parts. And put parts into families with a feature-based associated method.

Zhu Jiacheng et al (2001) supported virtual development center dynamic organization and product data management techniques supports for individuated product develop.

Suresh et al. (1999) identified the similar sequence of operations in parts as the viewpoints. And with the ART (Adaptive Resonance Theory) neural network, they create the part families.

Javier and Otto (2000) find the right mixes of module instances to use in each of the desired variants, as well as determine the targets for the instances that need to be newly created while product family is designed with this different module platform.

All these works can work well in the product family development with different methods and different viewpoints for identifying the similarities, but some limitations are also in them:

- Limited scale for the products number. Most of the methods especially the typical method, cluster analysis, can only deals with limit number of the products; this makes the method become limit to some extent.
- Limited viewpoints. Most of the works have just considered the products with one viewpoint. This is suitable for the typical part products, but for complex product, only the single view is not enough.
- Simple object. Most of the above works are used to work for the part family development. But for the general work of the product family development, these methods are needed to modify.

This paper proposed virtual product families' method (VPF), it is a generic product families architecture (PFA) representation in terms of representing multiple views of product families. This VPF not only supports product develop phase, but also supports CAPP, CAM and assembly all of life cycle. But in this paper, only discuss how to meet customer requirements in product development. It provides customer chances to participate in product design conveniently or directly; and conveys customer demands to product's performance and feature, dispatches these performance and feature to all product design phrases, establishes product life-cycle numeral model.

As well known, the product has some special attributes or characters different from the other products, no matter what VPF or basic product. Base on above attributes, product or component-function shion matrix is produced. For verified or empirical customer requirements, this can be precisely defined; the neural net is adopted to obtain the product of customer required. For uncertain customer requirements, fuzzy logic is an adequate way of modeling this imprecision. Above research results constitute a VPF model.

This paper is arranged as follows. Section 2 analyses and processes customers' needs, section 3 presents the hybrid simulation of fuzzy logic and neural net the web-based system, section 4 discusses a case study, section 5 obtains conclusion.

2. ANALYZING AND PROCESSING CUSTOMERS' NEEDS

During the development of new product, at first, market investigate and research must be done in order to understand customers' needs and ensure that these needs are not distorted in the later design, manufacturing and assembly. The first step, to understand customers' needs exactly, is the most important step to assure quality in manufacturing process and improve the degree of customers' satisfaction. In the process of acquirement of customers' needs, some instances often meet with which can be classified two groups approximately. Firstly, certain function and certain dimension parameters can be described in customers' needs. Faced to such instance, customers' needs can be acquired and confirmed by means of the selection of certain parts and components or module customisation. Secondly, customers' needs are expressed that manoeuvrability is convenience, precision is high and so on, and these cannot be described with data exactly. In this paper, the method of fuzzy logic will be adopted to deal with them. The key to set up VFP model is to analyze and process these customers' needs. Here, the customers' needs processing are divided into two sections: analyzing customers' needs and process customers' needs.

2.1　Describing and analyzing customers' needs

Each product has its own attributes and these attributes can be described. According to different attributes, enterprises can establish each product's attributes tree.

Fig 1 is the product's attributes tree in an enterprise manufacturing machine tools. The tree comprises machine tool's all attributes.

In case having the products attributes tree, any requirement that customers put forward always can be searched relevant item from it. In the product's attributes tree, as well seen, there be two instances. Firstly, for instance, common horizontal lathe can machining shaft, hole and screw etc, its maximal machining diameter is 400mm and so on, these which can be definitely described by data, are named as certain factors. Secondly, such as operability and maintainability, these, which are unable to be definitely described with data, go by the name of uncertain factors. Aiming at different factor, different method should be adopted.

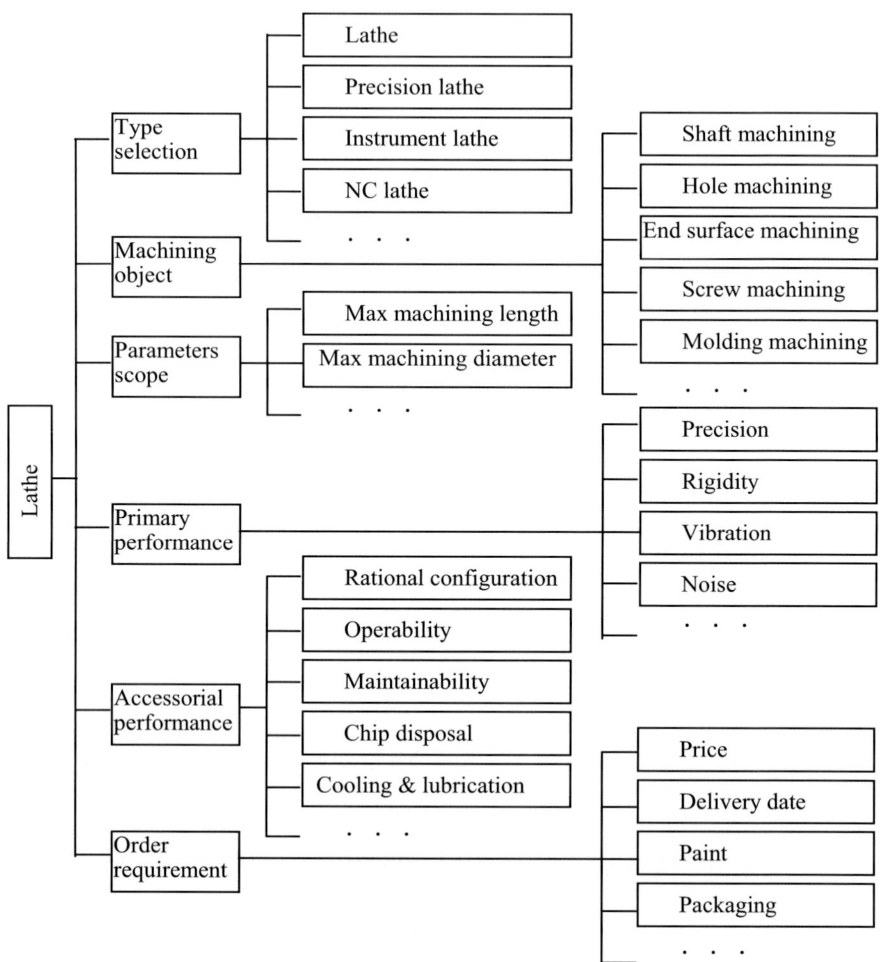

Figure 1. The product attributes tree of the lathe

2.2　Processing customers' needs

For certain customers' needs ,which are able to be described with data exactly, we can divide into levels; for example when the length of the track is not more than 750mm, certain part and component can be acquired by selection. For uncertain customers' needs, which are uncertain semantics in course of customers' needs description such as higher precision, lower noise etc, the fuzzy logic method is introduced for data description. Performance items of the machine tool processing results are shown in table1.

Table 1. Performance items of the machine tool processing results

Semantic description	Data expression	Machining precision	Rigidity (kN/mm)	Vibration (μm)	Noise (db)
best	1	≤IT5	23.25~30	<5	>85
good	0.75	IT5~IT6	16.5~23.25	5~10	75~85
generic	0.5	IT7~IT9	9.75~16.5	10~15	65~75
inferior	0.25	IT9~IT11	3~9.75	15~20	50~65
worst	0	IT11~IT13	<3	20~25	<50

Machine tool configuration is deployed with selectable parts and components. Therefore, the selection and the employment of product parts and components become a key. Assuming that a product P is composed of some parts and components, for any part or component $C_i (i = 1,2,...,n)$, n is the number of parts or components. k is the type of C_i, and each type is provided with m attributes $(j = 1,2,...,m)$, and expressed by A_{ji} respectively. According to above, the function matrix of product virtual parts and components families can be constructed. The row expresses attribute values of C_i, and the column expresses the family of C_i. As far as the track of the machine tool is concerned, the value of each element in the matrix is acquired by experimental formula, experimentation and existent knowledge. For example, Comparing the wearability of cast-iron track with that of weld track, getting 0.8 and 0.6 respectively, that expresses the wearability of cast-iron track is better. Other attributes value can be acquired by the same method. Table2 is the function-attribute matrix demonstration of the track family obtained by analysis and calculation.

Table 2. The function-attribute matrix demonstration of the track family

Type / Attributes	Cast iron track	Weld track	Marble track	Cast- iron adhibited plastic	Weld adhibited plastic
Rational configuration	0.6	0.8	0.4	0.6	0.8
Wearability	0.8	0.6	0.7	0.9	0.9
Heat distortion	0.7	0.2	0.9	0.8	0.4
Vibration resist	0.75	0.45	0.55	0.85	0.65
Economy	0.6	0.8	0.4	0.55	0.75
Precision maintenance	0.7	0.5	0.9	0.8	0.65

Values obtained from the function-attributes matrix of parts or components are prepared for fast selection of parts and components and meeting customers' needs in the follow NN processing. After such data processing, parts and components can be selected by the means of NN.

2.3   Nerve Network processing

NN is a calculating system composed of a mass of nerve cells co-adjacent greatly. The knowledge it describes is expressed by the mode of the interconnection among nerve cells, the importance of each input to nerve cells is memorized by the mode of right, the information processing is to respond to input by right dynamic change, and network learning and identification lie on the dynamic evolvement process of inter right series of each nerve cell. The hybrid model of fuzzy logic and NN about components selection is founded as shown in fig2.

Based on the foregoing analysis, the NN model can be founded after all certain and uncertain elements about customers' needs are processed, as follows.

$$P = f(C_1, C_2, ..., C_i, ..., C_n)$$

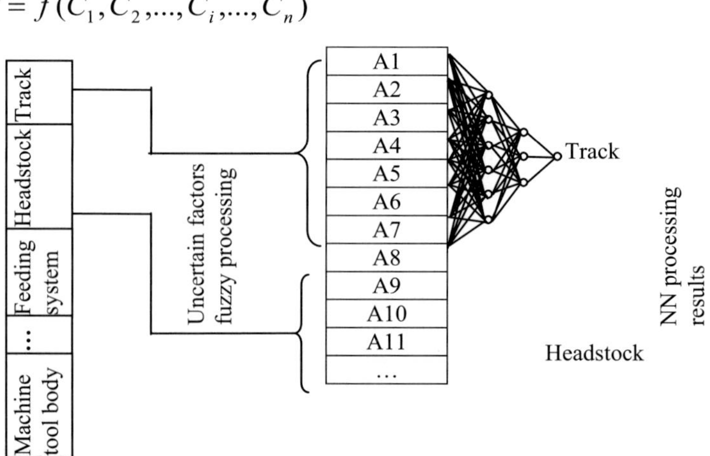

Figure 2. The hybrid processing results of fuzzy logic and NN

In the formula, product P is NN output, while $C_1, C_2, ..., C_n$ is NN input. During confirming each part or component, the fuzzy logic method is adopted to processing uncertain elements. Furthermore, NN is also adopted to confirm each part and component which can confirm a product finally. The product is possibly existent, also possibly the deployed part and component or product which constitutes the virtual product family or the virtual family of parts and components.

3. THE MODEL OF VFP-BASED DESIGN FOR CUSTOMERS

According to the analysis and processing of customers' needs in section 2, the Web-based VFP model can be set up as fig 3. In this model, at first customers input their own needs at the browser, whether these needs are certain or uncertain, then submit this page. Secondly based on customers' needs, the system processes them automatically. For certain needs, system can choose relevant parts and components, whereas for uncertain ones, it

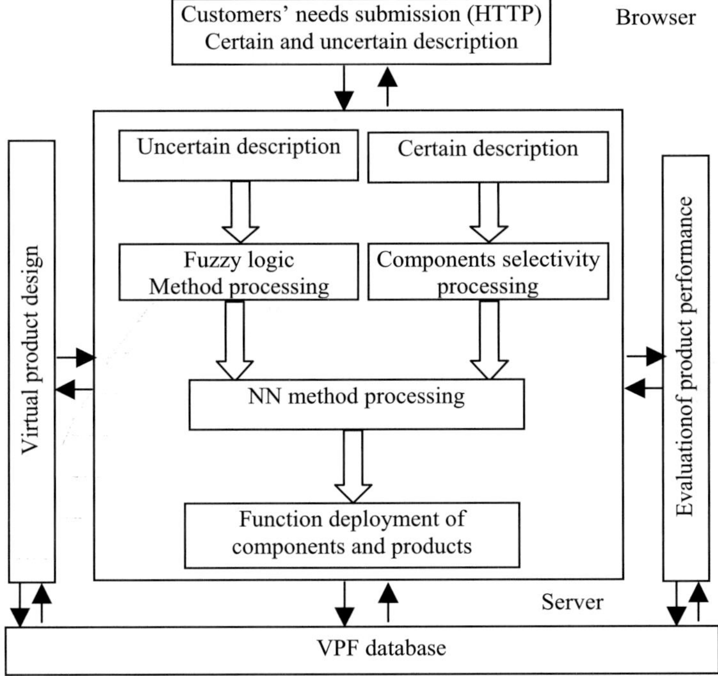

Figure 3. Web-based VFP implementation model

will process data according to customers' semantics, form definite data values, then all these together enter into NN to be processed and the final part and component or the final product can be acquired. So the result is fed back to the browser page as the reference for customers. To the selected part and component or the product that isn't existent, the later design process should be the design of the virtual part and component or the virtual product according to the processing results. The chosen product or part and component also should be evaluated based on product performance, then its results are saved into the VFP database. The VFP database embodies parts and components attributes and product attribute trees of all products as well as virtual product trees and the architecture of virtual parts and components tree.

4.   A CASE STUDY

Based on customers' needs in fig 1, VFP model Web-based in fig 3 is adopted, the machine tool selection list is achieved, as shown in table 3.

Table 3. The machine tool selection list

Item	Result
Machine tool type	Horizontal lathe
Track	Cast-iron saddle track
Track length	750mm
Max machining diameter	600mm
Principal axis revolution scope	25~1600rpm,　16 grade
Feeding scope	0.08~1.2mm/r,　32 grade
Transmission type	General motor＋gear＋lead screw
Cooling type	Water
Tailstock	No
Tool post	4 working positions
Reference price	75 thousand　(¥)
Delivery date	2.5 month
Paint	Green $312^{\#}$

5.   CONCLUSION

In the process of Mass Customization Production, it is principal and key to acquire exactly, analyze, and process customers' needs, in order not only to satisfy customers' needs rapidly, but also improve the quality of product design and manufacture. In this paper, design for customers is achieved by the means of VFP model under Networks. Among customers' needs, for factors which are certain and able to be described with data, the method of selection and deployment in parts and components families is adopted to constitute product. On the other hand, for those which are uncertain and unable to be expressed with data, the fuzzy logic method is used to process semantics into data; further NN method is adopted to process produced data, and meet customers' needs well. Compared with the traditional design for customers, in this model, customers can participate in product design directly, which not only ensure that customers' needs can be expressed exactly, and help to implement the later design process, but make customers browser the product customized by themselves rapidly and directly.

ACKNOWLEDGEMENTS

Results proposed in this paper have been developed in research projects supposed by the Natural Science Fund-aided project (Research No: 2000C31) of ShaanXi Province as well as the Science-Tech fund-aided project (Research No: 00JK232) of ShaanXi Provincial Education Department.

REFERENCES

1. Joseph P.B.. "Mass Customization Products and Services". *Planning Review*, **Vol.6**, Jul/Aug 1993.
2. Charles T., Mosier, MahMood F.. "Simultaneous identification of group technology machine cells and part families using a multiple solution framework". *International Journal of Computer Integrated Manufacturing*, **Vol. 9, No. 5**, pp402-416, 1996.
3. Javier P. Gonzalez-Zugasti, Otto K. N.. "Modular platform-based product family design". *Proceeding of DECT'00, ASME Design Engineering Technical Conferences and Computers and Information in Engineering Conference*, 2000.
4. Jiao J. X., Tseng M. M.. "An Information Modeling Framework for Product Families to Support Mass Customization". *Annals of the CIRP*, **Vol.48**, Jan 1999.

THE RESEARCH ON INTEGRATED SYSTEM OF MECHANICAL PRODUCT INTELLIGENT DESIGN AND MANAGEMENT DECISION SUPPORT IN CE ENVIRONMENT

JingXin Chen,[1,] Jing Xu[1,] Hong Li,[1] JiPing Zhou,[1] and Lan Cai[2]
1.Yangzhou Univ, Yangzhou, China 225009, jxchen@yzu.edu.cn,
2.Jiangsu Univ, Zhenjiang, China 212013

Abstract: From the views of optimising the process of product design and promoting the reliability and time utility of the management decision and improving the competitive ability of the product in the market, the theories and methods of the integration of mechanical product Design (MPD) and Management Decision Support (MDS) are studied, and the related structural model of the integrated system are also built up with the theories of concurrent engineering and system engineering and decision support. Combining the theory of knowledge engineering with the Computer Aided Design technology, the characteristics of knowledge based intelligent design (KBID), the methods of acquirement and expression of design knowledge and the implementing methods of intelligent design are researched. Also the key technologies of intelligent design such as the three-dimensional modeling technology based on features, and the methods of rule-based and case-based reasoning are studied. Finally the application of integrated system of knowledge based intelligent design and management decision support (KBID&MDS) of Disc Spring in CE environment is developed using the method of intelligent design, the method of evaluation and decision based on analytical hierarchy process and the distributed system architecture technology. And the integration of the intelligent design of disc spring structure and its machining process, technology equipment and evaluation and decision in technology and economy is implemented.

Key words: Mechanical product, system integration, knowledge based intelligent design, management decision, concurrent engineering.

1. INTRODUCTION

In order to adapt to the various competition, the decision support using OR/MS (Optional Research/Management Science) has been the hit of current research and development. Although this kind of decision in some extend is scientific and systematic, the data collected by decision support system lacks of accuracy and time utility because of no reliable support from technical system. As a result the decision provided by the support system lacks of reliability and effectiveness.

In the article, the idea of effective management decision has been put forward that it takes the key factors of the product's competency: cost, quality and time as the targets of decision, and it makes prompt response and adjustment and make reasonable decision according to the changes of the market and customers, using the prompt and accurate data with the systematical and scientific methods. With elevation of the product competency from the market, the established models are becoming more and more complicated, and depend on the data more and more stronger. If the models are not based on the technical data of the product, the accuracy and time utility of the data can't be guaranteed. So the effective management decision must be constructed on the base of the planning and design of the product technical scheme. So building up the integrated system of product technology scheme and management decision and implementing the optimization of product technology and economy schemes has great practical significance for the improvement of the market adjusting and competing ability of the enterprise.

2. THE STRUCTURAL MODEL OF INTEGRATED SYSTEM OF MPD&MDS IN CE ENVIRONMENT

2.1 The connotation of the integration

The effective management decision support system emphasizes on the accuracy and time utility of the product data, and also emphasizes on the rationality and effectiveness. The integration of mechanical design system and decision support system is a system in multi-platform and multi-region. It has the following advantages:

Firstly the integration of the mechanical product design system and the management decision system is the integration of the technologies, the processes and the information in the concurrent environment, and it is a system that supports the all phases of product design and the management decision using kinds of computer aided technologies. Secondly The mechanical product design system and management decision system are the two systems in the CE environment, which are separated and related one another. Meanwhile the mechanical product design system can work out the design destinations of all the phases of product design procedure according to the strategic intentions from the decision layer, and it also provide the decision support system with the data related to the decision, and it still can change the product design with respect to the comprehensive technological and economical evaluation and the opinions of the customers. The management decision support system can make the comprehensive technological and economical evaluation and decision of the product's technological scheme and economic scheme.

2.2   The general function and structure of the integrated system

According to the above discussion, the integrated system should have three functional modules. There are the module of mechanical product technical scheme design which provide the related technical parameters and schemes to the management decision; the module of data transaction which transforms the technical data of the technical scheme design subsystem into the economical data needed by the evaluation and decision; the module of

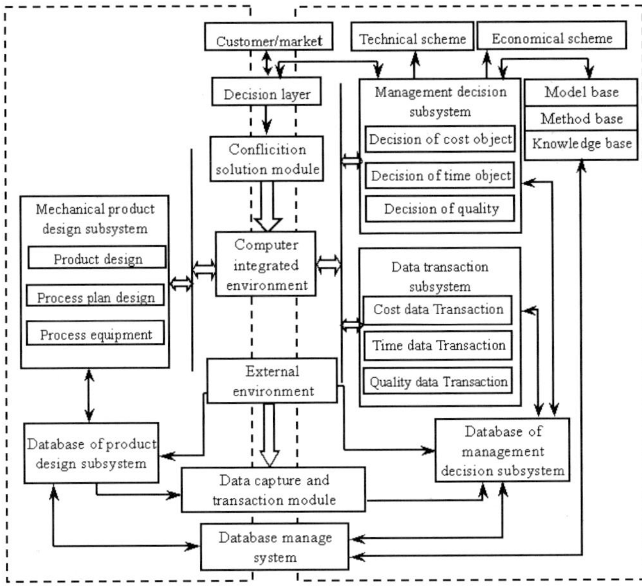

Figure1. The general function and structure of the integrated system in CE environment

evaluation and decision to the economical scheme which comprehensively evaluates the economical scheme.

The general function and structure of the integrated system of mechanical product design and management decision support is constructed as shown by Figure.1. In the CE environment this system uses the technical data from the design subsystem as the data support and then reconstructs the technical data structure into data structure needed by management decision. The evaluation and decision subsystem makes the comprehensive economical evaluation and decision, and acquires corresponding technical and economical schemes, using the related decision analyzing and evaluating methods in the model base, method base, and knowledge base, according to the information such as product technical information, economical information, market information and so on.

3. KBID OF MECHANICAL PRODUCT

3.1 The characteristics of KBID

The KBID can more clearly embody the features of product and more adjust to the development of current design[4], comparing with the traditional product modeling method. Its main features are: a) the engineering model is constructed on the product itself and the entire design procedure; b) the product design knowledge is made up of the engineering standards of design, analysis and manufacture, the geometrical and non-geometrical constraints, which drives the product model, and the knowledge is the driving power; c) it can automatically modify the product model according to the acquired and integrated design knowledge, and improve the ability of self-adjusting; d) it can express the design experience, standard and idea of multi-form in multi-area in visible method, and store them into the knowledge base, and become visible knowledge for further reuse.

3.2 The implementation methods of KBID

The key to implementation of KBID is to use design knowledge to drive the product geometric model.

3.2.1 Parameter driving of the geometric model

The system parameters of the CAD software can transfer the dimensional constraints into the characteristic parameter of the product geometry, which can directly drive the geometry. The user parameters must connect with corresponding system parameter with formula and form kinds of knowledge expressions to drive geometry.

3.2.2 Table driving of the geometric model

Based on the parameter driving geometric model, table can store a series of parameters, which can be stored in file or in database. The connection between table and parameters must be founded, and then the change of table record can change the value of parameter, there for to drive the geometry. This method is especially fit for the standard part and series part.

3.2.3 Rule-based reasoning driving of the geometric model

In the method, the design guideline, standard, theory and experience can be expressed in form of IF-THEN. The corresponding rules base is constructed to store the mass rules, which can be used to guide the product design and control product structure and also drive the product model.

3.2.4 Case based reasoning driving of the geometric model

The method is more suitable for the kind of experience design in the design area, which uses the former successful design case as its references[5]. The reasoning machine maybe an embedded program in CAD software programmed with script language, and also can be external reasoning application, which can communicate with CAD software through OLE Automation technology and use the reasoning result to drive product model.

3.3   The key technologies of KBID

3.3.1   3.3.1 the three-dimensional modeling technology

The three dimensional parametric modeling is one of the key and basic technology in KBID.

When modeling, first the features of the product need to be analyzed, then the characteristic parameters are extracted from the real product. Here the features of the product are the geometric features, material features and process features and so on. The characteristic parameters are extracted from the product key features, which are expressed in letters or code. The extracted characteristic parameters are transferred into the parameters of the model, the parametric model is presented with the extracted parameters.

3.3.2 the compatibility and complementarity of design model

In order to support the entire process from concept design, detailed design to the process and equipment design, the different modeling technologies of the design procedure should be compatible with each other[2].

The entire parametric feature based model is one of the key for the compatibility and complementarity, which can present the structural hierarchy and reflect the assembly constraints relations of the product, and can ensure the dimensional coordination of the parameters in the parts and components of the product. The expert rules for the machining process scheme and equipment process scheme are also another key for the compatibility and complementarity.

3.3.3 the reasoning methods based on rule and case

With the object-oriented method of knowledge expression and the rule and case based reasoning method, the product knowledge base is constructed combining with the experience and theories of the product design. Finally the rule and case based reasoning of the design model is implemented. And the parametric model with compatibility and complementarity can change with the variation of the design constraints, and generate the new product and equipment three-dimensional model, which is the final key technology to implement the rapid design and intelligent design supporting the whole design procedure.

4. THE EVALUATION AND DECISION METHODS BASED ON ANALYTICAL HIERARCHY PROCESS

From the view of product design, the price, cost and time are the key factors[6]. Using the key factors of product competitive ability as the evaluation factors, the hierarchy model for decision is constructed. The highest level of the hierarchy model is the total object of decision, scheme decision. The layer bellower of the highest contains the standards and its specific index of the total object. Here cost, quality and time are considered in this layer. The lowest layer consists of the kinds of schemes to be ordered. Here are the three schemes to be compared, and the analytical hierarchy model for decision of the three schemes is shown as Figure.2.

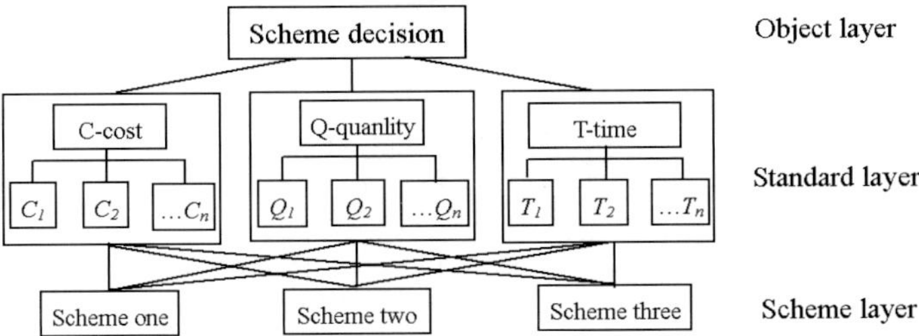

Figure2. The analytical hierarchy model of the technical schemes

After the hierarchy model constructed, the evaluation level of elements, which belong to the same layer should be confirmed according to a certain element in the up layer. The every two elements of the same layer are compared with each other, and then the relative importance of every element is determined, and the judging matrix is constructed. Then the elements belonging to the current layer are ordered in response to the relative importance aiming at some of the up layer element and the coincidence of the matrix is checked. If the coincidence is unsatisfied, the modification is applied to the matrix till it has the satisfied coincidence in order to ensure the reliability of the ordering result. After this, through the comprehensive importance calculation the order of the three existing schemes are obtained. At last, the coincidence of the final result is checked. If it is satisfied, the decision can be performed in response to the final result.

5. THE DEVELOPMENT OF INTEGRATED SYSTEM OF KBID&MDS OF DISC SPRING

5.1 Developing environment of system and its structure

In the network environment based on TCP/IP, the prototype system is developed with the tools such as object-oriented language Delphi, CAD software CATIA V5 and the database management system SQL Server.

The system adopts the three-tier distributed structure, which is shown in Figure.3.

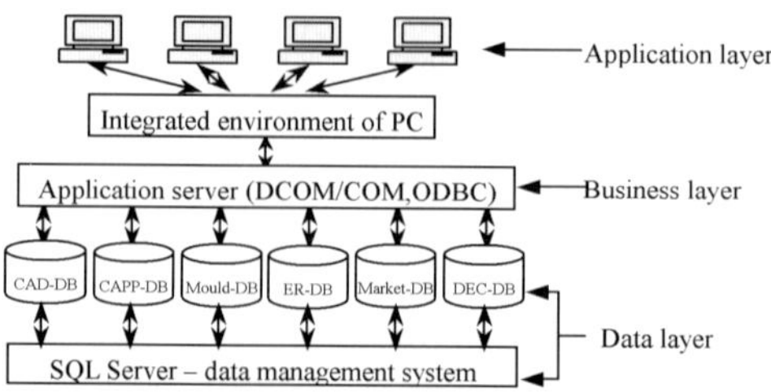

Figure.3 The distributed system structure

The business layer uses the multi-tier application server and the distributed component model to connect with the application layer, and it also connects with the data layer through the opening database connecting and data engine.

5.2   The KBID system of disc spring

5.2.1 KBID of disc spring structure

Disc spring is a simple revolution object. With the analysis of the disc spring, its main geometric parameters are selected as the outer diameter-D, the inner diameter-d, the thickness-t and the inner cone height-h.

There are two other important characteristic parameters: material and sort. Normally the disc spring has two types of material 60Si2MnA and 50CrVA. The sort parameter indicates the existence of support plan feature of the disc spring. Through the above analysis, the sketch is drawing on the plan thought the revolve axis. In the sketch, the characteristic parameters are expressed by the dimensional constraints and some of the angular constraints, which are shown in Figure.4.

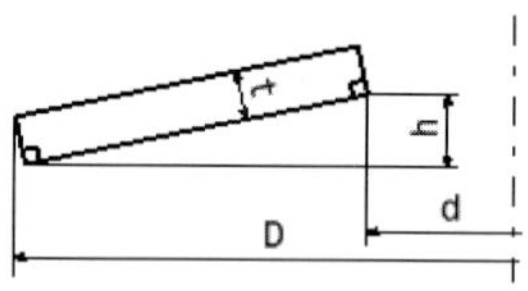

Figure4. The sketch of the disc spring

5.2.2 KBID of disc spring stamping process

Disc spring is mainly produced by the punching and stamping process. The process scheme is related to the disc spring's thickness. From the long-term experience of the factory, disc spring has four typical processes. When the thickness is not more than 1mm, the process is made up of the punching stamping and forming compound step, heat treatment and so on. When the thickness is higher than 1mm and not more than 3mm, the process has the punching and stamping step, the lathing step, the forming step, the heat treatment and other steps. When the thickness is from 3mm to 6mm, the process consists of the punching and stamping step, the lathing step, the forming step, the heat treatment and other steps. When the thickness is more than 6mm, the process includes the punching step, stamping step, lathing step, forming step and so on.

The rule set for the disc spring process scheme is built according to the thickness of disc spring and the experience knowledge of the factory. Through the rule based reasoning, the intelligent process scheme design of disc spring is implemented.

5.2.3 KBID of disc spring mould [7]

First of all, analyze the structural hierarchy of the mould and setting up the assembly template should be done. For disc spring with different thickness, it has different punching and stamping and forming process plan and different mould structure. Normally the layout of disc spring mould has the single die, progressive die and compound die. The single die has three types, which are punching die, stamping die and forming die, and the progressive die has the punching and stamping die, and the compound die has two types, which are punching stamping and forming die and the punching stamping die. From the view of mould's structure, the die is made up of the part of protrude, the part of concave, the part of die set and the auxiliary part which are separated with each other and contains their selves parts. In order to make the rapid variant design of the mould, the assembly template is founded, which not only contains all the components and parts, but also clearly embodies the structural hierarchy and assembly hierarchy.

Because the mould assembly template is the first step for mould design, it could not have the complete parts. Therefore the virtual parts are introduced into replace the real part in the mould structure for building the assembly constraints.

Then refine the assembly template and parameterise all the parts. In the elementary assembly template, the parts are not designed in detail and the virtual parts only have some construction elements. In this stage, all the parts are refined and parameterised. After this the assembly template is illustrated according to the assembly constraints and the three dimensional parametric model is generated and stored into the mould case base.

Construct the design knowledge base of mould design. Mould design is a kind of experienced design work, which not only needs the former design experience, but also needs the support of mass knowledge in mould design field such as kinds of formulas and tables. How to deal with these experience and rules becomes the premise of the mould automatic design.

Form the above analysis of disc spring process scheme, the rule set for mould structure reasoning can be set up, the reasoning network can be constructed as figure5.

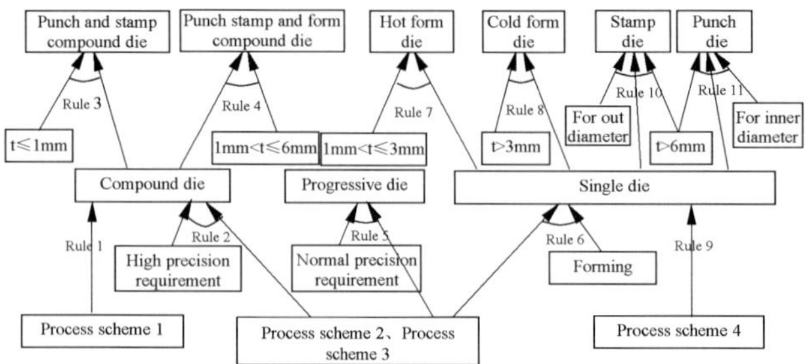

Figure5. The reasoning network of mould structure

Finally all the design rules and checks are stored into the rule base and can be used in later design.

Rule-based and case-based reasoning implement mould intelligent design. When the mould of disc spring is designed, the designer can select the mould structure using the rule-based reasoning according to the input information of disc spring. Then through the cased-based reasoning, the similar case of mould structure is searched out. And with further rule-based reasoning and the related detailed design, the complete mould structure is finished.

5.3   The data transaction system

Using different technical parameters, the intelligent design system of disc spring works out different technical schemes through the knowledge based reasoning methods, so the quality-Q, the cost-C and the time-T of these schemes are different with each other. The data transaction system can transform the technical data from intelligent design system into the quality-Q, the cost-C and the time-T for the management decision system.

The quality of disc spring is determined according to the customer's request and the practical and technological condition by the designer. Then the time of delivery and cost depend on the quality of disc spring.

The time of delivery mainly depends on the machining time of disc spring.

The producing cost-C_p is the critical factor influencing the total cost of disc spring.

Take the producing cost as an example, C_p is made up of the material cost-C_m, the worker's salary-C_w, and the energy consumption-C_e and the depreciation and maintenance of the equipments-C_d:

$$C_p = C_m + C_w + C_e + C_d \tag{1}$$

The material cost is:

$$C_m = \pi \cdot th \cdot (D^2 - d^2) \cdot \rho \cdot P_m \cdot \eta / 4 \tag{2}$$

In the expression(2):

th is the thickness of disc spring (mm), D is the outer diameter of disc spring (mm), d is the inner diameter of disc spring (mm), ρ is the density of the material of disc spring (kg/mm3), P_m is the price of the material of disc spring (Yuan/kg), η is the usage of the material of disc spring;

As to the calculation of worker's salary:

$$C_w = P_w \cdot t / 60 \tag{3}$$

P_w is the price of per working time (yuan/h),
t is the total machining time (min).
And to the energy consumption and depreciation and maintenance cost:

$$C_e = K_e \cdot C_w \tag{4}$$

K_e is the ratio of the Ce and Cw, which is decided by the process.

$$C_d = K_d \cdot C_w \tag{5}$$

K_d is the ratio of the Cd and Cw, which is decided by the process.

According to the related technical parameters in the design system, the data transaction system can calculate the total producing cost by means of the formulas from (4-1) to (4-5).

5.4   The management decision system based on analytical hierarchy process (AHP)

In order to get the different quality, the technical schemes are also different, and the cost and time of these schemes are different too. Suppose that there are two technical schemes A and B. Schemes A's quality, cost and time are Q_a, C_a and T_a; Schemes B's quality, cost and time are Q_b, C_b and T_b.

Then according to the calculated total cost and time of different technical schemes in the data transaction system, the management decision system will calculated with the combination of the qualitative and quantitative methods. It makes the comprehensive evaluation and decision using the multi-object AHP in section3 and obtains the best technical scheme. The confirmation of corresponding evaluation system and relative importance and the construction of judgment matrix and the coincidence of the matrix will not be expatiated.

6. CONCLUSION

Through the research on the theories and methods of the integration of mechanical product design and the management decision, and the development of integrated system of KBID&MDS of disc spring, the concept design, structure design, process design, mould design and management decision support are optimally planned. The comprehensive planning implements the integration of the mechanical design and management decision support in CE environment. It has provided the methods for the optimization of the product design procedure and the enhancement of reliability and time utility of management decision and realizes the comprehensive optimization of technological and economical schemes. It has also put forth effective ways to improve the enterprise's competitive ability.

REFERENCES

1. Boxiong Lan, Jiayao Pan, Xiaoyan Li. "The Decision Support System for Enterprise Management under CIMS Environment", *System engineering theory and practice*, pp.78-84 March 2000.
2. Fei Liu, Xiadong Zhang. *Manufacturing system engineering*, National defence industry publishing house. China 2000.
3. ChenJingxin, LiHong, XuJing, ZhouJiping, CaiLan. "Knowledge-based Intelligent Design and Its Application", *Agriculture mechanical Journal*, **vol.34**, pp.109-112, July 2003.
4. Bravo-Aranda G, Hernandez-Rodriguez F. "Knowledge-based system development for assisting structural design", *Advances in Engineering Software,.***30**, pp.763~774. December 1999.
5. Sehyun Myung, Soonhung Han. "Knowledge-based parametric design of mechanical products based on configuration design method", *Expert System with Applications.* **Vol.21**, pp. 99-107. August 2001.
6. Nanua Singh. "Integrated product and process design: a multi-objective modeling framework", *Robotics and Computer Integrated Manufacturing* **Vol 18 (2):** pp157~168,2002.
7. ChenJingxin, LiHong, XuJing, Guoming Lu: "Means and Technology of Intelligent Design about Disc Spring Die under CATIA V5 Environment", *Mould Industry*, **Vol.271**, pp.12-16, September 2003.

A STUDY OF DECISION-MAKING ON PROCESS METHODS SELECTION BASED ON GA

Qing-Jun Lee and Tan-Cheng Xie
Electromechanical Engineering College, Henan University of Science and Technology, 471003, liqingjun123@163.com or Xietc@mail.haust.edu.cn

Abstract: Genetic algorithm (GA) was applied to the decision-making of process methods for components' feature. The optimal or near-operation choices for the production requirements were obtained by using GA. A new sort of fitness function was studied and defined.

Keywords: Genetic algorithm, Process Link, Optimization, Decision-making.

1.   INTRODUCTION

With the development and application of artificial intelligence technology in many fields particularly in mechanical manufacture engineering field, it is to necessary and reasonable to develop intelligence Computer Aided Process Planning (CAPP) systems. Fuzzy logic, Artificial Neutral Network (ANN) and Genetic Algorithms (GA) are three main artificial intelligence technologies, which have been researched and explored by many researchers.

According to final requirement for surfaces of component features, a CAPP system can select directly process method, which can be looked up in process method tables. Based on an expert system, adaptable method can be selected by using the pattern, "if … then …"[1]. There are also several other methods including Multi-criteria decision-making etc. But all those methods have the following shortcomings: strict requirement of manufacture resource, poor flexibility, few decision-making parameters, and complex judgments. To improve the intelligence of decision-making, many experts studied these methods. **[2-4]**.

The essential character of GA technology can overcome the shortcomings of classical methods. Firstly, intelligence of GA is self-organized and self-adapted; Secondly, GA's process logic reflects a kind of character of gene in the evolvement of nature. Thirdly, GA's multiple solutions can supply many similarly solutions to users. Fourthly, GA's robust character makes it more practical and efficient. Finally, GA has the most possibility to obtain the method of global optimization.

## 2.	PROCESS METHODS DECISION-MAKING

## 2.1	Gene coding

According to the technologic knowledge and the design requirements of components, every process method is expressed as a gene chromosome. For instance, to process a hole, based on Mechanism Process Manual, about 20 different process methods can be obtained, each of which are coded. Also, to simplify GA operation, the mapping module is developed to map decimal system into binary system and the contradictory. Hole gene chromosomes can be valued in the integral field from one to twenty because the number of whole process methods is less than twenty. For example, "Drill" can be coded to "1", "Drill-Bore" can be coded to "2" and so on. There are 17 codes of process methods. In the hole process methods system, a five digits binary number is needed to encode every one of 20 methods, just like table 1.

Table 1. Gene coding of Process methods for Hole feature

Process methods	Gene coding	
	Decimal system	Binary system
Drilling	1	00001
Drilling + Rough Reaming + Finishing Reaming	3	00101
Drilling+ underreaming + Reaming	6	00110
Rough Boring （or underreaming） + half-finishing boring （or finishing underreaming) +finishing boring （or Reaming） +finishing boring with free floating block	12	01100

Process parameters and manufacture conditions that are related to process method decision were also encoded into an adaptable station. For example, "roughcast is mild steel" was encoded into "2". These codes may seem the same as those of process methods, but they are in a different field from process methods gene.

2.2　Design of fitness function

Fitness function is to estimate each gene's ability to fit the environment. The best method is the one with the highest fitness value. Therefore, to design an adaptable fitness function is a key in using GA.

According to Mechanics Process Manual and experts' experience, a conception named nearness was defined to express the distance between components' designing requirement and final quality of every method. The conception will be brought after several definitions as following.

Defining 1 : Components' requirements and manufacture conditions of components feature are defined as Process Factor Set (PFS), expressed by U in equation (1),

$$U = \{X \mid X = IT, Ra, MC,...\} \tag{1}$$

Where, U is PFS; IT is machining precise grade; Ra is Roughness value requirement; MC is several other machining conditions, which including roughcast type, feature dimension and other factors. Elements of PFS can be changed according to practical requirement of the designer.

Defining 2: Every element of PFC is a number set, which is defined as Manufacture Condition Set (MCS) expressed by X in equation (2).

$$X = \{x \mid x \in [x_{min} \ x_{max}]\}, c_{max} = x_{max} - x_{min} \tag{2}$$

Where, X is MCS; x_{min} and x_{max} is the minimal number and the maximum number in MCS; c_{max} is the length of MCS.

Defining 3: The final quality of components can obtain by using the process method coded into "n" must be a child set of MCS, which is defined as Adaptable Set of n (ASN), expressed by equation (3).

$$X^{(n)} = \{x \mid x \in [x_1 \ x_2]\} \tag{3}$$

Where, $X^{(n)}$ is ASN; x_1 and x_2 is the minimal number and the maximum number of the range.

Following relations in equation (4) can be concluded by the three defines :

$$X^{(n)} \subseteq X \in U$$
$$x_{min} \leq x_1, x_2 \leq x_{max} \tag{4}$$

Components' designing requirements that user enters can be encoded into a number set named Enter Set (ES), expressed by u in equation (5).

$$u = \{x \mid x = it, ra, mc, ...\} \tag{5}$$

Where, "u" is ES; "it" is machining precise grade; "ra" is Roughness value requirement; "mc" is several other machining conditions, which including roughcast type, feature dimension and other factors.

Based on those defining, two different arithmetics of Nearness were considered as Equation (6) and Equation (7).

$$f(x,X) = 1 - \frac{\left| \frac{1}{2}(x_1 + x_2) - x \right|}{c_{max}} \tag{6}$$

$$f(x,X) = 1 - \frac{\min\{|x_1 - x|, |x_2 - x|\}}{c_{max}} \tag{7}$$

Where, f(x,X) is Nearness of x (ES) and X (MCS); other signs are similar with defining 2 and defining 3.

Equation (6) means to compare user enter with Adaptable Set of n, which obtain an absolute value. Then Nearness can be gained by calculating between the absolute value and the length of MCS (c_{max}). Where nearer is x_0 to the mid of X, the higher is the Nearness. Though equation (6) expresses an arithmetic that is simple to calculate and similar with thinking process of man, the arithmetic can't obtain a satisfying result when the bounding of Adaptable Set of n is bigger or one ASN overlaps another.

However, Equation (7) gains a satisfying result. Equation (7) is a sort of arithmetic that considers the boundary of ASN and only accepts the boundary value (x_1 or x_2). That is to say, Nearness only relates to the boundary value that is nearer to x. After comparing data many times, Equation (7) was selected.

So, Fitness function of ES to PFS is defined as equation (8).

$$f(u,U) = \sum_{i=1}^{m} k_i f(u(i), U(i)) \tag{8}$$

Where, u(i) is the element of u indexed by "i"; U(i) for the element of U indexed by "i"; ki for the weight of element of U. Process designer may select suitable value for the weights according to different cases. Commonly, the weights were normalized with Equation (9).

$$\sum k_i = 1, k_i > 0 \qquad (9)$$

In case of the hole feature characterized by "roughcast is alloy steel", "Process precise is IT7", and "Roughness value is 6.3", a Enter Set can be gained as following,

$$u = \{7, 6.3, 2\}$$

Calculation of fitness of several Process methods is showed in table 2. In case of X is machining precise grade (IT), Adaptable Set of "1" can be obtained as $X^{(1)}$ in following equation,

$$X = \{x \mid x \in [0,13]\}$$

$$X^{(1)} = \{x \mid x \in [11,13]\}$$

Nearness of x_0 to x_1 of IT can be calculated as following equation,

$$\left. \begin{array}{l} x_{min} = 0, x_{max} = 13 \\ x_1 = 11, x_2 = 13 \\ x_0 = 7 \end{array} \right\} \Rightarrow f(7, IT) = 0.6923$$

Similarly, Nearness of Ra and MC also can be obtained and Fitness of gene "1" can be calculated as following equation,

$$\left. \begin{array}{l} f(7, IT) = 0.6923 \\ f(6.3, Ra) = 0.748 \\ f(2, MC) = 0.8333 \end{array} \right\} \Rightarrow f(u, U) = \frac{1}{3}(0.6923 + 0.748 + 0.8333) = 0.7579$$

Where, the weights of IT, Ra and MC were selected with the same value according to process knowledge. Table 2 shows that gene 6's fitness is the highest, that is to say, the process method which gene 6 is the best result.

Table 2. Calculating of fitness of several Process methods

Gene encoding of Process methods	Machining precision $IT = \{x \mid x \in [0,13]\}$			Roughness value $Ra = \{x \mid x \in [0,25]\}$			Some rest machining condition $MC = \{x \mid x \in [1,7]\}$			Fitness $f(X_0,A)$
	X_1	X_2	X_0	X_1	X_2	X_0	X_1	X_2	X_0	
1	11	13	7	12.5	25	6.3	1	1	2	0.7579
3	7	8		0.8	1.6		1	1		0.8817
6	8	9		1.6	3.2		2	2		0.9330
12	6	7		0.4	0.8		4	4		0.8156

2.3   Initiating colony and GA operation

In process methods decision-making of hole feature, four genes were selected as the initial group of GA operation at random. Every gene is a single process method that was encoded into a binary number of five digits.

Selection operation is used to aggregate the group to a higher fitness field, while mutation operation is used to aggrandize the multiplicity of the group. In process methods decision-making of hole feature, roulette selection [5-6] and one-point crossover was used. The probabilities of selection, crossover and mutation are selected according to the case.

2.4   Rule of operation stopping

There are two parameters to control GA operation's stopping. The first parameter is "M", the biggest repeated epoch, at which GA stops operating. "M" can be selected according to the number of process methods. The second one is "d", the lower ascending rate of Best Fitness of the GA group, at which GA stops operating. In hole process application, "M" is "50" and "d" is "0.01".

3.   INSTANCE

A hole feature process methods decision-making system based on GA was developed. The system is running well in the platform of Window 2000 with Matlab 6.1[7]. The main user interface can be seen like figure 1.

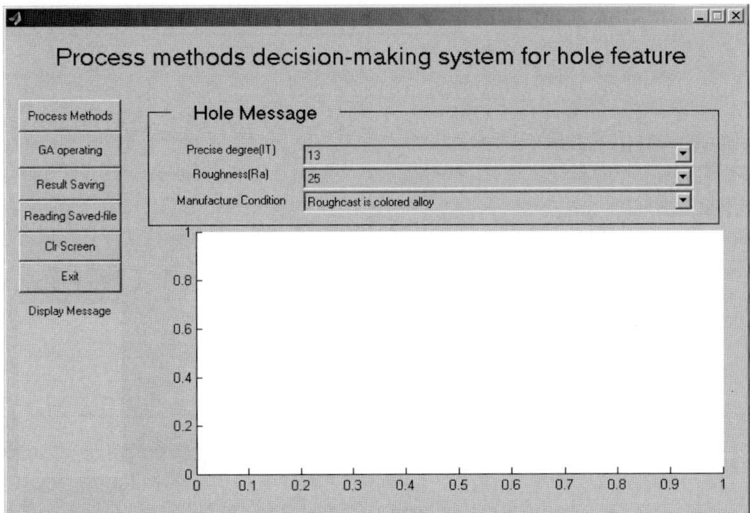

Figure 1. Main user interface of Hole Process method System based on GA

A hole feature of a component of YT company, characterized by "the roughcast is colored alloy", "Machining precise grade is IT7", and "Roughness value is 6.3" was tried an experiment with. After several epochs' GA operation, a decision was made by the system, which can be described as "Rough Boring (or underreaming) + half-finishing boring (or finishing underreaming) +finishing boring (or Reaming) +finishing boring with free floating block". The calculating process also was plotted as figure 2. Compared with the real process plan for the component, the result is satisfied.

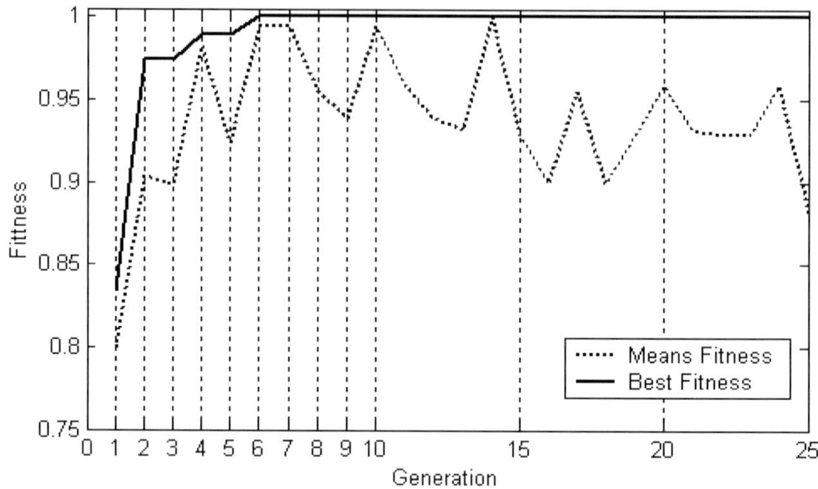

Figure 2. Changing of fitness of hole process method decision using GA

4. CONCLUSION

Process method selection by using GA is a new method to be an addition to original theory of process decision-making. GA can overcome errors of decision-making from common process, thereby optimize process planning and improve intelligence of decision-making. If further research is done, it is possible to realize GA optimization of process method and make CAPP system more intelligent.

REFERENCES

1. Wang Xiaoping, Cao Liming: *GA – Theories, Application and Realization*, Xi'an China ,Xi'an Jiaotong University Press, ,2002.
2. Li Qingjun, Liu Jun: "Study on CAPP development tools intelligence module". *Modern Manufacturing Engineering*, **Vol 6**, June 2003.
3. Zhao Liangcai: *CAPP system design*, Beijing China, China Machine Press, 1994
4. Dong Jiaxiang: Intellective CAPP development tools, Beijing China, National Defiance Industry Press, August 1996.
5. Chen Guoliang: *GA and its application*, Beijing China, People's Tele Press, June 1996
6. Yasuhiko Kitamura (Japanese): *Artificial Intelligence*", Beijing China, Science Press, February 2003
7. Wen Xin: *Matlab Ann application design*, Beijing China, Science Press, July 2000

CAD MODELING OF STRAIGHT BEVEL GEARS

Wei Wei, Keke Zhu, Luling An and Laishui Zhou
*Nanjing University of Aeronautics and Astronautics, 29th Yudao Street, Nanjing, China,
210016,Tel: 86-25-84892570 Fax: 86-25-84611802,meewwei@nuaa.edu.cn*

Abstract: In this paper, the foundation equations for gear modeling are presented. Based
on ACIS geometric engine, gear-modeling software is developed. The solid
model of a straight bevel gear is generated.

Key words: CAD, straight bevel gear, involute, modeling.

1.   INTRODUCTION

Straight bevel gears have been widely used in automobiles, tractors, motorcars, and a variety of mechanical equipment. Compared with cylindrical gears, their manufacturing is more difficult. There are mainly two types of fabrication for straight bevel gears, i.e. generating cutting and precision die forging. The latter has the advantages of high productivity, low cost for volume production.   The electrode for form forging moulds is needed with high quality, and their manufacturing becomes the most important factor. In general, they are fabricated with gear shapers.

When the gear shaper are used to machine the straight bevel gears, the adjusting processes of tool parameters are quite complex and several test machining procedures have to be made, which is inconvenient to machine such workpieces as precision die forging gear electrodes with multiple types and small amount. The above problems can be solved through modeling of straight bevel gears with CAD/CAM technology and machining with high-speed milling. The production cost can be reduced and efficiency be enhanced by replacing special machine and tool with general machine and tool. It is the foundation of realizing the above ideas to establish the modeling of straight bevel gears.

In this paper, the foundation equations for gear modeling are presented. Based on ACIS geometric engine, gear-modeling software is developed. The solid model of a straight bevel gear is generated.

2. SPATIAL MAP EQUATIONS OF OUTSPREAD PLANE TO BACK CONE OF STRAIGHT BEVEL GEAR

It is quite difficult to machining the straight bevel gears with spherical involute tooth faces. Therefore, their engagement is often approximately substituted by that of the equivalent cylindrical gears on the respective back cones. In such a way, the tooth faces of the straight bevel gears are formed. Therefore, the spatial map equations of the outspread plane to back cone must be deduces in order to obtain the tooth face equations of the straight bevel gears (Fig.1).

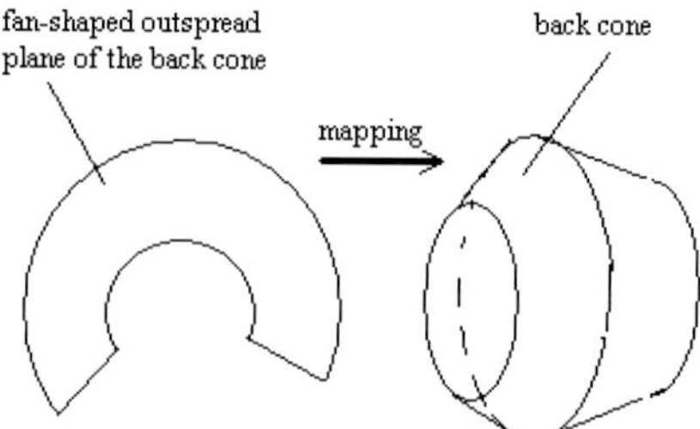

Fig.1 Mapping of outspread plane to back cone

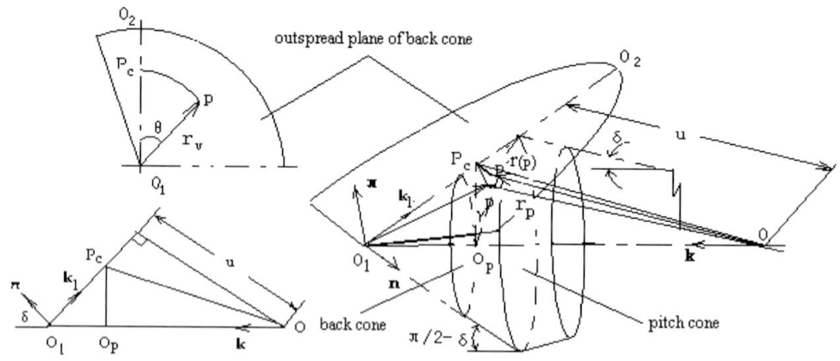

Fig.2 Vectors of spatial mapping of outspread plane to back cone

As shown in Fig.2, O and O_1 are the apexes of the gear pitch cone and back cone respectively, δ is the pitch conic angle, and u the pitch conic distance. The unit vector of O O_1 is k; the normal vector of the back cone plane is π, and k_1 is the unit vector of O_1O_2: $n = k_1 \times \pi$ and

$$k_1 = \pi \times n$$

Since $\pi = k \cos \delta - n \times k \sin \delta$

Then $k_1 = k \sin \delta - n \times k \cos \delta$ $\qquad$ (1)

For a point P on the back cone plane, we have $O_1P = r_v, OP = r_p$. The arc through point P on back cone plane intersects with O_1O_2 at point Pc. Let

$$r_v \cdot n = a, \quad r_v \cdot k_1 = b \qquad (2)$$

From equation (1) we have

$$r_v = a n + b \ k_1 = a n + b k \sin \delta - b n \times k \cos \delta$$

$$r_p = u k \sec \delta + a n + b k \sin \delta - b n \times k \cos \delta$$

$$= a n + (u \sec\delta + b \sin\delta)k - b n \times k \cos\delta$$

Then,

$$a = r_p \cdot n \text{ and } b = -(r_p, n, k) \sec \delta \qquad (3)$$

Here $(r_p, \ n, \ k)$ is the vector product of r_p, n and k.

$$O_1P_c = k\sqrt{a^2 + b^2} \sin\delta - n \times k\sqrt{a^2 + b^2} \cos\delta$$

Let θ be the angle between vectors O_1Pc and O_1P, then

$$\theta = arctan \frac{a}{b}$$

Let L be the length of arc PPc, then
$$L = |r_v|\theta = \sqrt{a^2 + b^2} \ arctan \frac{a}{b}$$

When the outspread plane is against upon the back cone, let P' be the corresponding point of P. And we have $OP'=r(p)$. Make a plane through point Pc and perpendicular to the cone axis. It intersects with the cone axis at Op, and then

$$\boldsymbol{OO}_p = \left(u\sec\delta - \sqrt{a^2+b^2}\,\sin\delta\right)\boldsymbol{k} \tag{4}$$

$$\boldsymbol{O}_p\boldsymbol{P}_c = -\sqrt{a^2+b^2}\,\boldsymbol{n}\times\boldsymbol{k}\cos\delta \tag{5}$$

Let γ be the angle between OpPc and OpP'. The arc length PcP' equals to that of PPc, that is

$$\mathrm{L} = \left|\boldsymbol{r}_v\right|\gamma\cos\delta = \left|\boldsymbol{r}_v\right|arctan\frac{a}{b}$$

Therefore

$$\gamma = \sec\delta\,arctan\frac{a}{b} \tag{6}$$

From equations (5) and (6), it can be obtained :

$$\boldsymbol{O}_p\boldsymbol{P}' = \boldsymbol{O}_p\boldsymbol{P}_c\cos\gamma - \boldsymbol{k}\times\boldsymbol{O}_p\boldsymbol{P}_c\sin\gamma$$

$$= \sqrt{a^2+b^2}\,\boldsymbol{n}\cos\gamma\sin\left(\sec\delta\,arctan\frac{a}{b}\right) -$$

$$\sqrt{a^2+b^2}\,\boldsymbol{n}\times\boldsymbol{k}\cos\delta\cos\left(\sec\delta\,arctan\frac{a}{b}\right) \tag{7}$$

From equations (4) and (7), the spatial mapping equation of the outspread plane of the back cone can be obtained:

$$r(\mathrm{p}) = \boldsymbol{OP}' = \boldsymbol{OO}_p + \boldsymbol{O}_p\boldsymbol{P}'$$

$$= \sqrt{a^2+b^2}\,\boldsymbol{n}\cos\delta\sin\left(\sec\delta\,arctan\frac{a}{b}\right)$$

$$+ \left(u\sec\delta - \sqrt{a^2+b^2}\,\sin\delta\right)\boldsymbol{k}$$

$$- \sqrt{a^2+b^2}\,\boldsymbol{n}\times\boldsymbol{k}\cos\delta\cos\left(\sec\delta\,arctan\frac{a}{b}\right) \tag{8}$$

Where a and b are defined by equation (3) ; δ — pitch conic angle.

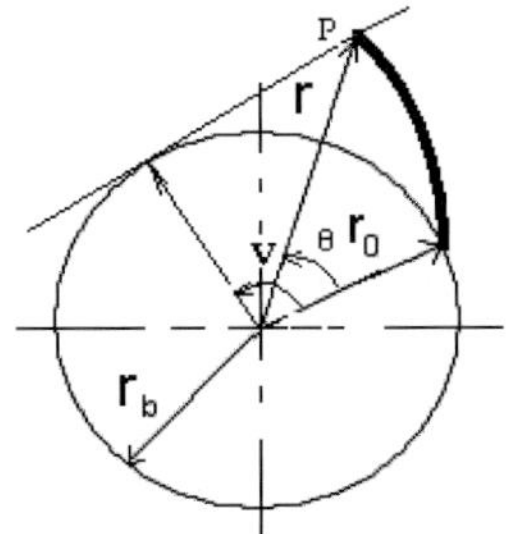

Fig.3 generation of involute

3.   TOOTH FACE EQUATIONS OF STRAIGHT BEVEL GEAR BASED ON EQUIVALENT CYLINDER GEAR

As shown in Fig.3, when a line rolls on the base circle, the involute is generated, which is the track of a point on the line. Its equation can be obtained as follows:

$$r(v) = (cos\,v + v\,sin\,v)r_0 + (sin\,v - v\,cos\,v)\pi \times r_0 \tag{9}$$

Where r_b is the radius of the base circle; r_0 is the base vector of the involute; v is the angle between **OP** and r_0; and π is the normal vector of the current plane.

From equation (9), the following equation can be obtained:

$$|r(v)| = r_b\sqrt{v^2 + 1} \quad v = \frac{\sqrt{r^2 - r_b^2}}{r_b} \tag{10}$$

Especially, let r_p and r_a be the radii of pitch circle and addendum circle, and v_p and v_a the relevant parameters, then

$$v_p = \frac{\sqrt{r_p^2 - r_b^2}}{r_b}, \quad v_a = \frac{\sqrt{r_a^2 - r_b^2}}{r_b} \tag{11}$$

If θ is the angle between r and r_0, then $\cos\theta = \dfrac{cos\,v + v\,sin\,v}{\sqrt{v^2 + 1}}$ (12)

The tooth thickness on the index circle is

$$s = \left(\frac{\pi}{2} + 2\tan\alpha + \tau\right)m \tag{13}$$

Where χ --- Radial modification coefficient
         α --- Pressure angle
         τ --- Tangential modification coefficient
         m ---- Module

The slot width between the teeth on index circle is

$$e = m\pi - s = \left(\frac{\pi}{2} - 2tan\alpha - \tau\right)m \tag{14}$$

The index circle diameter of the equivalent gear is

$$dpv = 2u\tan\delta \tag{15}$$

Where   u-----tooth width of the bevel gear
         Δ ----conic angle of the bevel gear

As shown in Fig.4, the outspread plane of back cone contacts tangentially on OO_1, which is the center of the tooth slot. The central angle relating to the slot width between the teeth on index circle is $2\omega p$, then

$$\omega_p = \frac{e}{d_{pv}} = \frac{(\frac{\pi}{2} - 2\chi\tan\alpha - \tau)m}{2u\tan\delta} \tag{16}$$

The index circle radius of the equivalent gear is

$$r_{vp} = u\tan\delta \tag{17}$$

The base circle radius of the equivalent gear is

$$r_{vb} = u\tan\delta\cos\alpha \tag{18}$$

From equations (11), (17) and (18), we have

$$v_p = \sin\alpha \tag{19}$$

As shown in Fig.4, B is the base point of the involute, and the index circle intersects with O_1O_2 at A, then

$$O_1B = r_0 = \frac{r_{vb}}{r_{vp}}\left[cos(\omega_p - \theta_p)O_1A + sin(\omega_p - \theta_p)\pi \times O_1A\right]$$

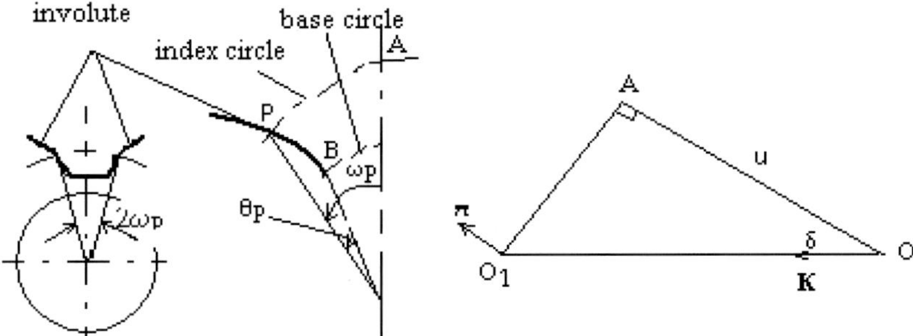

Fig.4 mapping of the slot between teeth to back cone

$$= u\,cos\alpha\left[cos(\omega_p - \theta_p)(\pi - sec\,\delta\mathbf{k}) + sin(\omega_p - \theta_p)\pi \times (\pi - sec\,\delta\mathbf{k})\right]$$

$$= u\,cos\,\alpha\left[cos(\omega_p - \theta_p)(\pi - sec\,\delta\mathbf{k}) - sin(\omega_p - \theta_p)sec\,\delta\,\pi \times \mathbf{k}\right] \quad (20)$$

$\omega\,$p and $\ \theta\,$p are defined by equations (16) and (12) respectively.

From equation (11), we have

$$\mathbf{r}_0 = \mathbf{O}_1\mathbf{B} = -u\,cos\alpha osa[t\;sin\delta\,cos(\omega_p - \theta_p)\mathbf{k} + tan\delta\,sin(\omega_p - \theta_p)\mathbf{n}$$
$$+ sin\delta\,cos(\omega_p - \theta_p)\mathbf{n} \times \mathbf{k}\,] \quad\quad (21)$$

$$\pi \times \mathbf{r}_0 = -u\,cos\alpha\,sec\delta ec\delta[(\omega_p - \theta_p)\pi \times \mathbf{k} - sin(\omega_p - \theta_p)\mathbf{k} +$$
$$cos\delta\,sin(\omega_p - \theta_p)\pi]$$
$$= u\,cos\alpha osa[t\;sin\delta\,sin(\omega_p - \theta_p)\mathbf{k} - tan\delta\,cos(\omega_p - \theta_p)\mathbf{n}$$
$$+ sin\delta\,sin(\omega_p - \theta_p)\mathbf{n} \times \mathbf{k}\,] \quad\quad (22)$$

The involute equation on the tangent plane of the back cone can be obtained from equations (21), (22) and (9) as follows:

$$\begin{aligned}
\boldsymbol{r}_{\text{binv}}(v) = &-(cosv + v\,sinv)u\,cos\alpha os\alpha [t\,sin\delta\,cos(\omega_p - \theta_p)\boldsymbol{k} \\
&+ tan\delta\,sin(\omega_p - \theta_p)\boldsymbol{n} + sin\delta\,cos(\omega_p - \theta_p)\boldsymbol{n} \times \boldsymbol{k}] \\
&+ (sinv - v\,cosv)u\,cos\alpha os\alpha [t\,sin\delta\,sin(\omega_p - \theta_p)\boldsymbol{k} - tan\delta \\
&cos(\omega_p - \theta_p)\boldsymbol{n} + sin\delta\,sin(\omega_p - \theta_p)\boldsymbol{n} \times \boldsymbol{k}] \\
= &\, u\,cos\alpha os\alpha\{t\,sin\delta in\delta [(\omega_p - \theta_p)(sinv - v\,cosv) - cos(\omega_p - \theta_p) \\
&(cosv + v\,sinv)]\boldsymbol{k} - tan\delta an\delta [(\omega_p - \theta_p)(cosv + v\,sinv) + cos(\omega_p - \theta_p) \\
&(sinv - v\,cosv)]\boldsymbol{n} + sin\delta in\delta [(\omega_p - \theta_p)(sinv - v\,cosv) - cos(\omega_p - \theta_p) \\
&(cosv + v\,sinv)]\boldsymbol{n} \times \boldsymbol{k}\}
\end{aligned} \tag{23}$$

If the cone distance u is taken as a variable, then the involute conical surface equation (outspread) is obtained from the above equation.

$$r(u,v) = \boldsymbol{OO}_1 + \boldsymbol{r}_{\text{binv}}(v) = u\,sec\,\delta\,\boldsymbol{k} + \boldsymbol{r}_{\text{binv}}(v) \tag{24}$$

Let $a(u,v) = -r(u,v)\cdot\boldsymbol{n}$, $b(u,v) = (r(u,v),\boldsymbol{n},\boldsymbol{k})sec\,\delta$

That is

$$\begin{aligned}
a(u,v) = &\, u\,cos\alpha\,tan\,\boldsymbol{\delta}[sin(\omega_p - \theta_p)(cosv + v\,sinv) + \\
&cos(\omega_p - \theta_p)(sinv - v\,cosv)]
\end{aligned} \tag{25}$$

$$\begin{aligned}
b(u,v) = &\, u\,cos\alpha\,tan\,\delta[sin(\omega_p - \theta_p)(sinv - v\,cosv) - \\
&cos(\omega_p - \theta_p)(cosv + v\,sinv)]
\end{aligned} \tag{26}$$

$$\sqrt{a(u,v)^2 + b(u,v)^2} = u\,cos\,\alpha\,tan\,\delta\sqrt{1 + v^2} \tag{27}$$

$$\frac{a(u,v)}{b(u,v)} = \frac{sin(\omega_p - \theta_p)(cos\,v + v\,sin\,v) + cos(\omega_p - \theta_p)(sin\,v - v\,cos\,v)}{sin(\omega_p - \theta_p)(sin\,v - v\,cos\,v) - cos(\omega_p - \theta_p)(cos\,v + v\,sin\,v)} \tag{28}$$

Finally, the tooth face equation of the straight bevel gear based on the equivalent cylindrical gear of the back cone is obtained from equations (26) and (27):

$$r_{\text{inv}}(u,v) = \sqrt{a(u,v)^2 + b(u,v)^2}\, \boldsymbol{n}\cos\delta\,\sin\left(\sec\delta\,\arctan\frac{a(u,v)}{b(u,v)}\right)$$

$$+ \left(u\sec\delta - \sqrt{a(u,v)^2 + b(u,v)^2}\,\sin\delta\right)\boldsymbol{k}$$

$$- \sqrt{a(u,v)^2 + b(u,v)^2}\,\boldsymbol{n}\times\boldsymbol{k}\cos\delta\,\cos\left(\sec\delta\,\arctan\frac{a(u,v)}{b(u,v)}\right)$$

$$= u\cos\alpha\,\sin\delta\,\sqrt{1+v^2}\,\sin(\sec\delta$$

$$\arctan\frac{\sin(\omega_p-\theta_p)(\cos v + v\sin v) + \cos(\omega_p-\theta_p)(\sin v - v\cos v)}{\sin(\omega_p-\theta_p)(\sin v - v\cos v) - \cos(\omega_p-\theta_p)(\cos v + v\sin v)})\boldsymbol{n}$$

$$+ u\sec\delta\left(1 - \cos\alpha\,\sin^2\delta\,\sqrt{1+v^2}\right)\boldsymbol{k} - u\cos\alpha\,\sin\delta\,\sqrt{1+v^2}$$

$$\cos(\sec\,\arctan\frac{\sin(\omega_p-\theta_p)(\cos v + v\sin v) + \cos(\omega_p-\theta_p)(\sin v - v\cos v)}{\sin(\omega_p-\theta_p)(\sin v - v\cos v) - \cos(\omega_p-\theta_p)(\cos v + v\sin v)})$$

$$\boldsymbol{n}\times\boldsymbol{k}$$

$$(29)$$

4. STRAIGHT BEVEL GEAR MODELING

In addition to the tooth face, the dedendum conic face and the transition faces constitute the face of a whole gear tooth. They connect with each other smoothly, as shown in Fig.5. With the conventional modeling method, first the equations of the dedendum conic face and the transition faces are deduced. Then through intersection and trimming the multi-face model of the bevel gear is generated. In this paper, after the equations are deduced, the intersecting curves are determined. Then the topological structure is built on basis of topological relations. On basis of ACIS, one of the powerful geometric engines in the world, a solid model of straight bevel gear is generated from the above faces generated and their relevant

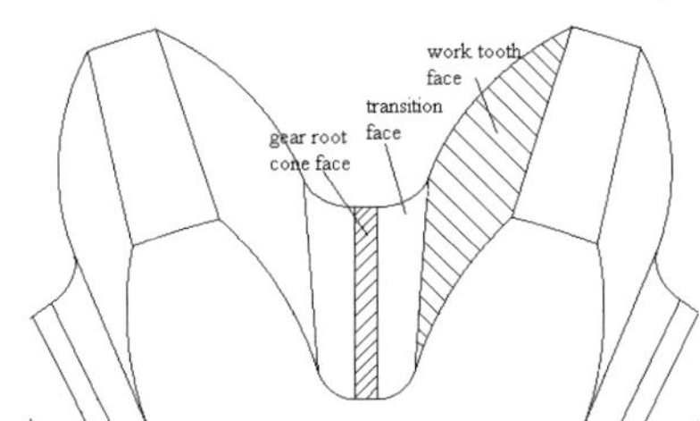

Fig.5 three pieces of surfaces of a straight bevel gear

Fig.6 Solid model of a straight bevel gear

intersecting curves. Fig.6 shows the solid model of a straight bevel model generated in the software developed by the authors.

The above method presented by the author is suitable to the modeling of Revacycle straight bevel gear. The method has been proved feasible by experiments.

5.   CONCLUSIONS

To solve the problems of machining straight bevel gears with general-purpose NC machining centers, the gear CAD models should be established. In this paper, the tooth face equation of the straight bevel gear based on the equivalent cylindrical gear of the back cone is deduced, which provides the foundation for the modeling of the straight bevel gears. Furthermore, the solid models of the gears are generated based on ACIS, with which the machining tool path can be generated by using such software as MasterCAM, UG, and etc. The method presented by the authors is suitable to the modeling of other gears.

REFERENCES

1.   Litvin, F.L., *Gear geometry and applied theory*, Chicago, Prentice-Hall, 1994.
2.   Darle W Dudley. *Gear Handbook*, New York, McGRAW-Hill Book Company INC. 1962.
3.   Sandro Barone, Leonardo Borgianni, Carlo Culla, "Marco Pieve Digital simulation procedures to create 3D solid models of gear geometries", *XII ADM International Conference*, Italy Sept. 5 -7, 2001

THE RESEARCH OF A NEW GEOMETRIC CONSTRAINT SOLVER

Chunhong Cao[1] , Wei Fu[2] and Wenhui Li[1]

1:College of computer science and technology, Jilin University,Changchun 130012, P.R.China,Chunhongcao_li@163.com

2: computing center Zhangzhou Teachers College, Zhangzhou, 363000, P.R.China

Abstract: GQA combines genetic algorithm with quantum computing. Instead of binary, numeric or symbolic representation, we introduce qubit chromosome representation. GQA is used in the process of geometric constraint solving in order to get the solution sequence. The proposed algorithm is offered in a quantum spirit, to allow analogue quantum computers to solve global optimization problems. GQA can solve problems in polynomial time instead of the exponential time required by classical algorithm. The experiment indicates that GQA is good for the fast solving for general geometric constraint problem.

Key words: Geometric constraint solver   geometric decomposing   genetic quantum algorithm   qubit chromosome

1.   GEOMETRIC CONSTRAINT SOLVING

Geometric constraint solving approaches are made of three approaches: algebraic-based solving approach, based solving approach and graph-based solving approach. One constraint describes a relation that should be satisfied. Once a user defines a series of relations, the system will satisfy the constraints by selecting proper state after the parameters are modified. The idea is named model-based constraints.   Constraint solver is a segment for the system to solve the constraints.

Geometric constraint problem is equivalent to the problem of solving a set of nonlinear equations substantially. In constraint set every constraint corresponds to one or more nonlinear equations, all the unattached geometric parameters in the geometric element set constitute the variable set of nonlinear equations. Nonlinear equations can be solved by Newton-Raphson algorithm. When the size of geometric constraint problem is large, the scale of set of equation and variable of nonlinear equation is very large. It will be difficult for the efficiency and stability to meet the requirement of interactive design. In order to solve problems hereinbefore, we must decompose the geometric constraint system into a series of small ones. Constraint decomposing is based on the fact: not all the geometric variables x_i (i=1...n) exist in a certain constraint by a obvious formula, in other words a usual geometric constraint system is a relax coupling system that can be seen from the strong sparsity of its structure matrix.

We can decompose large constraint problem into a series of small problems that can be solved step by step. The subsidiary problems and their solving sequence constitute a new geometric constraint-solving problem. In many conditions, subsidiary problem can be solved by geometric method instead of numeric overlap. The constraint solving process based on the geometric constraint decomposition consists of two phases:(1) decomposing phase and (2) execution phase. [1]

Geometric problems defined by constraints have an exponential number of solution instances in the number of geometric elements involved. Selecting a solution instance amounts to selecting one among a number of different roots of a nonlinear equation or system of equations. The problem of selecting a given root was christianized in as the Root Identification Problem. We introduce Genetic Quantum Algorithm (GQA) to solve the root identification problem by searching in the solution space automatically. The user specifies the intended solution instance by defining a set of additional constraints or predicates on the geometric elements that drive the search of genetic quantum algorithm. [2]

Geometric solving is an approach that resolves the point, line and circle using geometric information according different geometric constraint patterns. Even in the well-constraint pattern, the character of multi-solutions still exists. We adopt two basic principles: (1) topology relation invariability, (2) change least.

2. GENETIC QUANTUM ALGORITHM (GQA)

2.1 Genetic Algorithm

Genetic Algorithm is brought forward by Holland and developed to a superposition self-adaptation enlightened probability search algorithm. Practical worker favours this algorithm because of the robust, global best quality, juxtaposition quality and high efficiency. It applies evolution operation in a group of genes that code searching space. In every generation of genetic algorithm, the algorithm can search different areas of parameter space. Then it focuses on the highest part of expectation value in the solution space.

Supposing there is a problem to be optimized $F=f(x, y, z)$ $F \in R$, x, y, $z \in D$

In the formula, x, y, z are independent variables, which can be number, logic variable, even any symbols. Every $(x_i, y_i, z_i) \in D$ constitutes a solution of the problem. So D is considered defined field, or the constraint condition, or the solution space made of the solutions that must satisfy the constraint conditions. F is a real number belonging to real number field R, and can be considered a measure to the quality of every solution. Function f shows a mapping from solution field to the real number field.

2.2 Quantum Computation

Now the classical computer is realized by binary system, that is to say a bit represents 0 or 1. Quantum computer is not restricted by classical physical world. Quantum computer is a system that can fulfil computing task following quantum mechanics theory. It codes information with quantum states and the basic union to storage quantum information is a quantum double-state system (or to say two dimensional Hilbert space) named qubit. We can consider the quantum computer as a circuit made up of a series of quantum gates. We can suppose the circuit is composed of n logic gates and acts on m qubits. A qubit is a two-state $\{0,1\}$ system made of microcosmic particle. A qubit can be 0 or 1 state, also their superposition state. The qubit superposition $|\psi\rangle$ can be expressed: $|\psi\rangle = p_0|0\rangle + p_1|1\rangle$, here p_0 and p_1 are both pluralism to represent the swing of basic sates 0 and 1. $|p_0|^2$ and $|p_1|^2$ represent the probability of the basic state 0 and 1 of the system respectively and can be induced to 1, i.e. $\sum|p_i|^2=1$. When we measure the superposition $|\psi\rangle$ of quantum system, the state of the quantum register is in $|0\rangle$ state in probability of $|p_0|^2$ and $|1\rangle$ state in probability of $|p_1|^2$.[3]

The traits of quantum computer are[4]:

(1)The input and output states of quantum computer are common superposition usually not orthogonal;

(2)The transforms in the quantum computer are all the possible unitary. After getting output state, the quantum computer measures the output state and gives the computational result.

The substantial characters of quantum computer are superposition and entanglement. The transformation for every superposition part is a classical computation. All these classical computations can be finished simultaneous and superposed according to a certain probability swing then can get the final result of quantum computer. The computation is named after quantum juxtaposition computation. The quantum juxtaposition computation can improve the efficiency of quantum computer greatly and make it finish the task which classical computer cannot do. The entanglement trait has been well used in the entire quantum hyper speeding algorithms [5].

A qubit string of the length m represents a linear superposition of solutions to the problem. The length of a qubit string is the same as the number of items. The i-th item can be selected with the probability $|\beta_i|^2$ or $(1-|\alpha_i|)^2$. Thus, a binary string of the length m is formed from the qubit string. For every bit in the binary string, we generate a random number γ from the range [0.. 1]; if $\gamma > |\alpha_i|^2$, we set the bit of the binary string. The binary string x_j^t, j=1,2...n, of P (t) represents a j-th solution to the problem. For notational simplicity, x is used instead of x_j^t in the following. The i-th item is selected for the problem if $x_i=1$, where x_i is the i-th bit of x.[6]

Genetic Algorithm with qubit representation has a better characteristic of diversity than classical approaches, since it can represent superposition of states. Only one qubit chromosome is enough to represent eight states, but in classical representation at least eight-chromosomes (000) , (001) , (010) , (011) , (100) , (101) , (110) , (111) are needed. Convergence can be also obtained with the qubit representation. As $|\alpha_i|^2$ or $|\beta_i|^2$ approaches to 1 or 0, the qubit chromosome converges to a single state and the property of diversity disappears gradually.

2.3   Genetic Quantum Algorithm

The GQA can be written as follows:
Procedure GQA
Begin
$t \leftarrow 0$
initialise Q (t)
make P (t) by observing Q (t) states
evaluate P (t)
store the best solution among P (t)

```
while  ( not ccase condition )  do
  begin
      t←t+1
      make P (t) by observing Q (t-1) states
      evaluate P (t)
      update Q (t) by U (t)
      store the best solution among P (t)
  end
  end
```

Definition. QUBIT   a unit of information used in quantum computing. It is distinct from an ordinary bit in that it can encode a superposition of values.

In the step of initialise Q (t), α_i^t 和 β_i^t, i=1, 2,... , m, of all q_j^t , j=1, 2,...n means that one quantum chromosome, $q_i^t|_{t=0}$ represents the linear superposition of all possible states .The next step makes a set of binary solutions, P (t), by observing Q (t) states, where P (t) $=\{x_1^t, x_2^t,...,x_n^t\}$ at generation t. One binary solution, x_j^t , j=1 , 2 , ... , n, is a binary string of the length m, and is formed by selecting each bit using the probability of qubit, either $|\alpha_i^t|^2$ or $|\beta_i^t|^2$, i=1 , 2 , ... , m of q_i^t. Each solution x_j^t is evaluated to give some measures of its fitness. The initial best solution is then selected and stored among the binary solutions P (t).

Genetic Quantum Algorithm is a searching method of superposition and it approaches the best solution instead of equal to exactly, so we should set down a cease condition. Most common methods are to set down genetic generation. In this paper we use the method of controlling deviation. Once the difference of target function value between from the genetic quantum algorithm and the actual is less than an allowed value, the algorithm ceases.

In the process of updating Q (t), qubit chromosome is updated by some appropriate quantum gates U (t). We will discuss the quantum gate in detail in the following.

It should be noted that some genetic operators could be applied, such as mutation and crossover that can make the probability of linear superposition of states change. But as GQA has diversity caused by the qubit representation, there is no need to use the genetic operator. If the probabilities of mutation and crossover are high, the performance of GQA decreased notably.

Figure 1 shows the application of the genetic quantum into geometric constraint solving.

Geometric constraint decomposing and geometric constraint solving are the kernel processes and Genetic quantum algorithm（GQA）is used to construct the solving sequence. Then we solve the geometric constraints according to the solving sequence in turn.

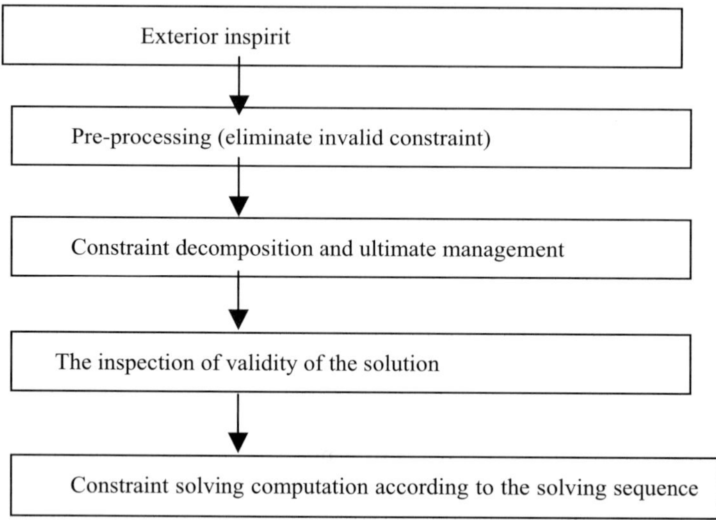

Figure1. Applying the genetic quantum algorithm into geometric constraint solving

The following features[7] represent an initial attempt to characterize a methodology for the design and development of genetic quantum algorithm.

(1) GQA is probabilistic.
(2) The problem should be expressed in a numerical in a numerical form, and if it is not, then a method should be employed to convert it into such a form.
(3) The initial configuration should be determined.
(4) The ceasing condition should be defined concisely.
(5) The problem should be amenable to being divided into smaller sub-problems.
(6) The number of universes required should be identified.
(7) Each sub-problem is assigned its own universe.
(8) Computations in the different universes occur in parallel.
(9) There must be some form of interaction between all of the universes. The interference must either yield a solution, or new information for the universes to utilize in locating a solution.

2.4 Quantum Gate

A quantum computer is equivalent to a quantum Turing. How to construct a quantum computer? It has been proved theoretically [8] that quantum Turing is equivalent to a quantum logic circuit, so we can form a quantum computer by the combination of quantum logic gates. Quantum logic gates can be distinguished by one bit, two bits and three bits etc logic gates according the number of input bit. An XOR gate of two bit and a gate operating on one bit arbitrarily can constitute a universal quantum gate set.

The quantum gate is an I/O equipment whose input and output are separate quantum variable (such as spin) [9][10]. Its operating input variable is expressed in a unitary function, viz. the state vector from $|\psi\rangle$ to $|\psi'\rangle = U|\psi\rangle$. As we know, to a digital computer, if there were linear logic gates such as NOT gate and COPY gate, as well as nonlinear logic gates such as AND gate, it will be able to finish arbitrary logic or arithmetic task. For the quantum computing, it is also true. Barenco [11] in the Oxford has proved it true. In 1982 Fredkin and Toffoli brought forward a reversible computing logic gate (for short Frendkin gate, namely Fredingate)[12], which is an equipment possessing three input lines and three output lines. This is shown in Figure 2. One of lines is input line (assuming it is line a) and its logic state has not been changed when passing the gate. If the bit of controlling line is

Figure2 Frendkin quantum logic gate

0, then the bit of other two lines is invariable; and if the bit of controlling line is 1, then the bit of other two lines should be exchanged.

We can get Frendkin gate by basic one bit and two bits quantum gates. [13] In this paper we adopt Frendkin gate.

3. APPLICATION INSTANCE AND RESULT ANALYSIS

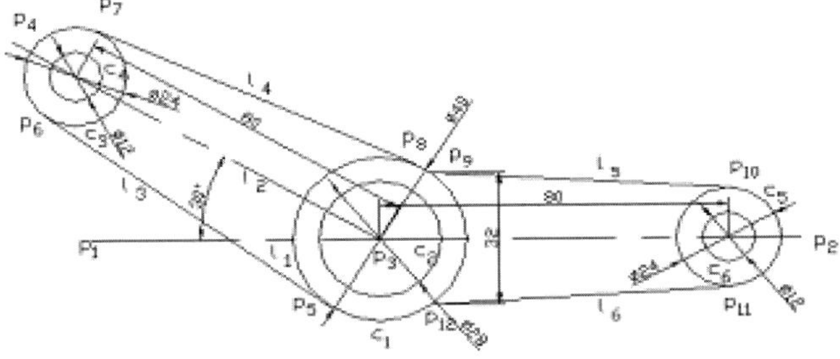

Figure 3. A design instance

$$p_9 \rightarrow p_{12} \rightarrow c_1 \rightarrow l_1 \rightarrow p_3 \rightarrow p_{11} \rightarrow c_5$$

$$l_6 \rightarrow p_{10} \rightarrow l_2 \rightarrow c_3 \rightarrow l_4 \rightarrow p_7 \rightarrow p_8$$

$$l_3 \rightarrow c_4 \rightarrow p_5 \rightarrow p_6 \rightarrow c_2 \rightarrow c_4 \rightarrow c_6$$

$$l_5 \rightarrow p_1 \rightarrow p_2 \rightarrow p_4$$

Figure4. The final solving sequence

In Figure 4, there 24 geometric elements, the freedom degree is 54 and the constraint degree is 49. We change DIST_PP (p_{c1} , p_{c3} , 80) to DIST_PP (p_{c1}, p_{c3} , 60), VDIST_PP (p_9, p_{12}, 32) to VDIST_PP (p_9, p_{12} ,30), ANGLE_LL (l_1, l_2, $\pi/6$) to ANGLE_LL (l_1, l_2, $\pi/4$). If we use 49 nonlinear equations to superpose in order to solve 54 position variables corresponding to 24 geometric elements, the solving will be very inconvenient. Here we use GQA to decompose the constraint problem. The solution sequence decomposed by genetic quantum algorithm is as follows:

We realize GQA starts with a random search and this shows the initial quantum chromosome can include all solutions with the same probability. Then the probability of the cases adjacent to the solution increases. It changes into a local search step by step after 50 generations. Finally, the probability of the qubit chromosome converges to the best solution. That is, GQA starts with a global search and changes automatically into a local search.

The final solving result is as follows:

We can get the instance 2 in the same way as figure 6.

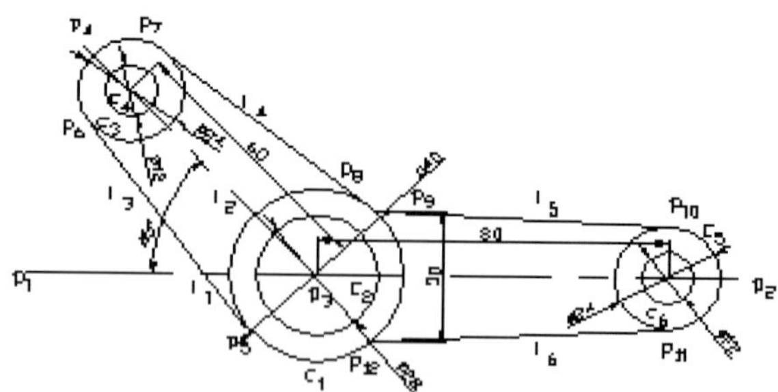

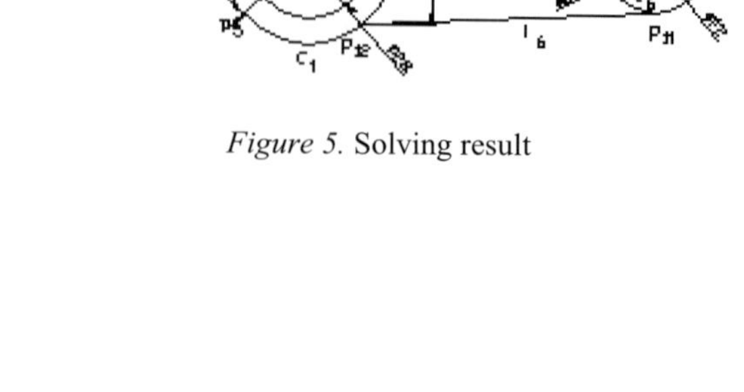

Figure 5. Solving result

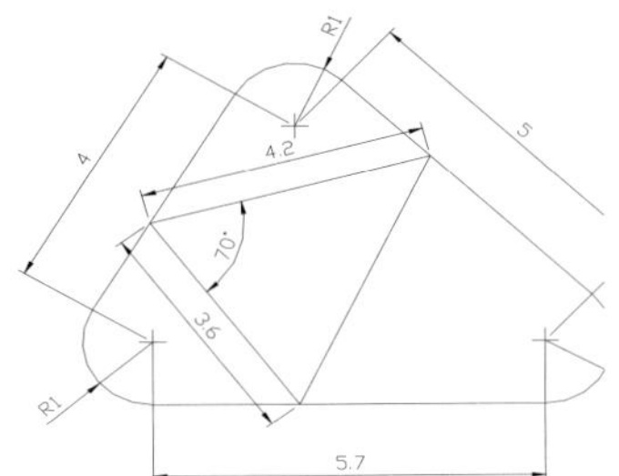

Figure6. (a) original figure

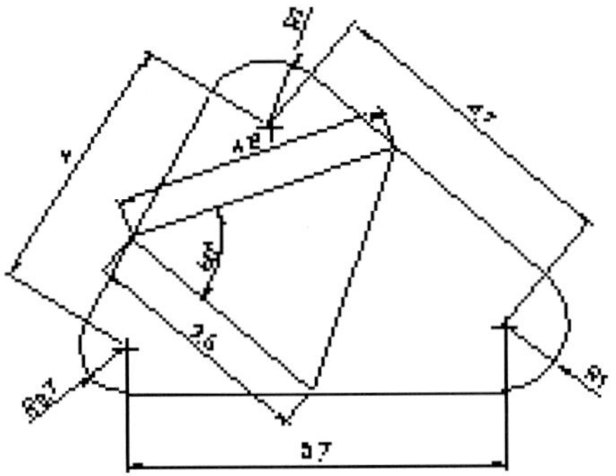

(b) changed figure

4. RESULT AND DISCUSSION

The genetic quantum algorithm can finish decomposing the geometric constraints by getting the sequence of geometric elements. The geometric constraint solver has the function of flexible parameterisation. User can design convertibly by modifying dimension parameters or adding or deleting geometric constraints. Combining genetic algorithm and quantum computation, it is a new idea by itself. Now we apply it into the geometric constraint solving and the experiment indicates that GQA is good for the fast solving for general geometric constraint problem.

The advantage of quantum computation is represented in the quantum parallel disposal. Quantum computers can solve problems in polynomial time instead of the exponential time required by classical computers. The proposed algorithm is offered in a similar spirit, to allow analogue quantum computers to solve global optimization problems. Quantum parallel computation and quantum simulation all make use of quantum entanglement. Unfortunately, quantum entanglement cannot be preserved in practical system. Because we only adopt the quantum theory to design algorithm not adopt the real quantum computer to do experiment, it does not have an effect on the computation result.

ACKNOWLEDGEMENTS

This research has been supported by National Nature Science Foundation of China. (authorizing number: 69883004)

REFERENCES

1. Yuan Bo, *Research and Implementation of Geometric Constraint Solving Technology*, Doctor dissertation of Tsinghua University, 1999
2. R. Joan-Arinyo, M. V. Luzon, and A. Soto "Constructive Geometric Constraint Solving : A New Application of Genetic Algorithms", September, 2002
3. Zhong Cheng, Chen Guoliang *Quantum computation and its application,* GuangXi University Transaction (natural science edition) March, 2002 26(1)
4. S. Lloyd, *Science* 1993, (261) : 1569 ; S. Lloyd , *Science* 1994, (263), 695
5. D. Divincenzo, *Science,* 1995, (270) : 255
6. Kuk-Hyun Han , Jong-Hwan Kim *Genetic Quantum Algorithm and its application Combinatorial Optimization Problem,* 2000
7. Ajit Narayanan *Quantum computing for beginners,* IEEE 1999
8. A. Yao, in *Proceedings of the 34th Annual Symposium of Foundation of Computer Science. (IEEE Computer Society, Los Almamitos, CA),* 352, 1993,
9. D. Deusch, *Proc. Roy. Soc* London Ser, 73, A425, 1989.
10. A. Barenco et al. , *Phys. Rev,* 3457, A52, 1995.
11. E. Fredkin and T. Toffoli, *Int. J. Theor. Phys. ,* **21,** 219, 1982.
12. Pothen A and Fan C J. "Computing the block triangular form of a sparse matrix". *ACM Trans.Mathematical Software,* 16(4): 303-324, 1990.
13. H. F. Chau and F. Wilczek, *Phys. Rev. lett. ,* 74, 75, 1995

MODELING OF COM-BASED VIRTUAL FMS[1]

Guo-Fu Ding, , Kai-Yin Yan,, Zhao-Xue Gao and Yi-Shen Zou
Advanced Design & Manufacturing Technology Research Institute, School of Mechanical Engineering, Southwest Jiaotong University, P.C.CHINA, 6 10031, Dingguofu@163.com or ding_guofu@yahoo.com.cn

Abstract: Virtual flexible manufacturing system(Abbrev. FMS) is composed of NC machining equipment, material handling system and computer control system, which can effectively adjust manufacturing scheme according to variation of product task and production environment to optimize manufacturing process plan. Based on structure analyses of flexible Manufacturing System, this paper builds the framework of virtual flexible manufacturing system. On the basis of research, some components which are constructed with the method of object-oriented, are used to plan the virtual Flexible Manufacturing System according to different work piece and its machining process, with which we can evaluate our machining plan. Equipments of FMS are also divided into several functional parts, which are modeled from geometric and kinematic method. Under the different condition of machining process, the actions of different equipment components, which are triggered with multi-timer style and harmonized with each others in good order, are also tested and verified in OPENGL-based virtual manufacturing environment, the result shows that we can obtain a satisfied process plan.

Key words: Virtual manufacturing; FMS; Computer simulation; Computer Graphics; Component Object Model

1.   INTRODUCTION

Virtual Manufacturing technology is a new way and hot spot simulating the complexity of structure, production procedure and its product in mechanical manufacturing system, lots of studies have aimed at it[1]-[5], and achievements have made. FMS is made up of NC equipment, material

[1] This work is supported by Scientific Development Funds of SWJTU

handling facilities and computer control system, etc.. to study FMS by using virtual manufacturing, two aspects need to be done, one is simulating the functions of equipment distributing in this system in a right form, the other is how to plan the production procedure and obtain the exact evaluation of this FMS. Here presents a novel component-based virtual FMS plan, and simulating their working procedure in OPENGL-based developing platform, which is different from the other scheme.

2.   THE FRAMEWORK OF VIRTUAL FMS

Figure 1 shows a typical flexible manufacturing system, in which warehouse, fork truck, stock, conveyor, all kinds of CNC or DNC, etc., are planned according to the product line. A complicated process network is also constructed where nodes are named as the units composed of machining equipment and other machinery. To map it into the computer, some abstracts must be done, the geometrical model of these nodes can be simplified. However, their functions and actions should be rightly modeled. So a virtual flexible manufacturing system can be described as a framework of figure 2.

For certain production task, to ensure the reliability of process evaluation, product plan in virtual manufacturing must be consistent with that of realistic manufacturing.

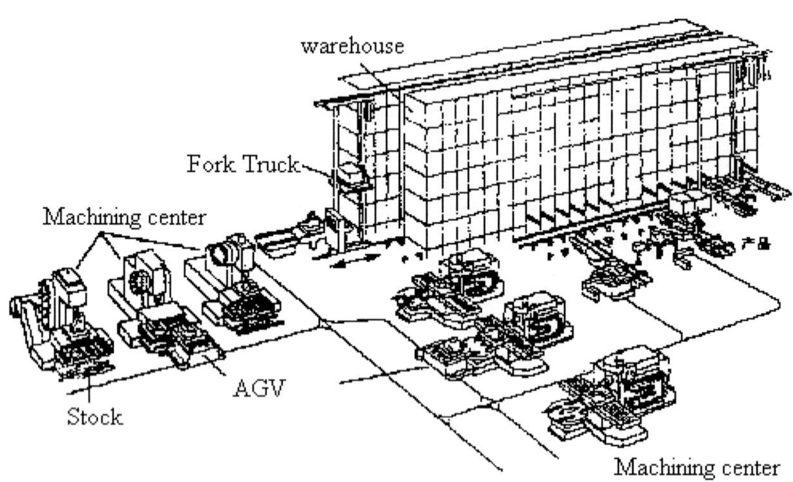

Figure 1. The framework of FMS

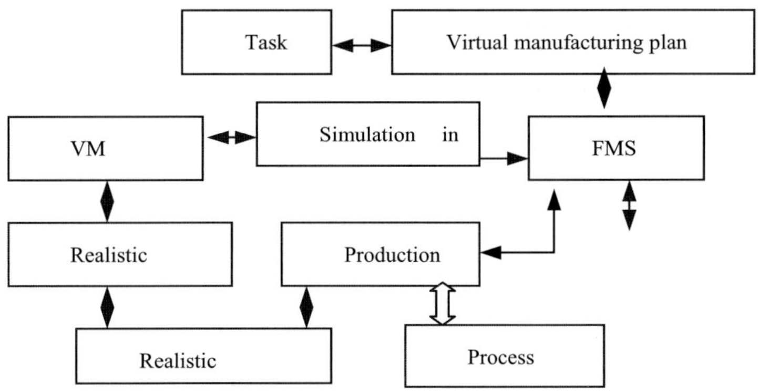

Figure 2. The framework outline of virtual FMS

3.   MODELING OF COM-BASED VIRTUAL FMS

To plan a FMS, first, a rational production layout should be executed, which is on the basis of product process. a typical unit in FMS, has public characteristics such as input/output, reuse, etc.. It resembles a programmable object in object-oriented engineering. So, abstract object that has its own functions, actions and properties, can be done in all kinds of units of FMS, some private and public properties and functions can also be classified in these objects, interfaces can be used to define output/input among objects. More deeply, these objects can be constructed by using Component Object Model (abbrev. COM), which have better characteristics of encapsulation, reuse, etc.. Each COM can be an instance, they include lathe object, milling machine object, machining center object, robot object, storage object, stock object warehouse object, fork truck object, conveyor object, etc.. These objects will be modeled in three aspects: solid geometrical modeling, kinematic modeling and property modeling. A working flow of virtual FMS composed of these COMs is described in Figure 3

In this figure, the solid lines represent information flow sent out from center computer, and the dot lines represent material flow, which is in charge of transporting materials controlled by information flow, for example, work piece. All of these components make the production activity in a good order

4.   MODELING OF DIFFERENT COM

Virtual equipment model resembles realistic physical one in terms of appearance, function and action. So the cutting modeling, solid geometrical

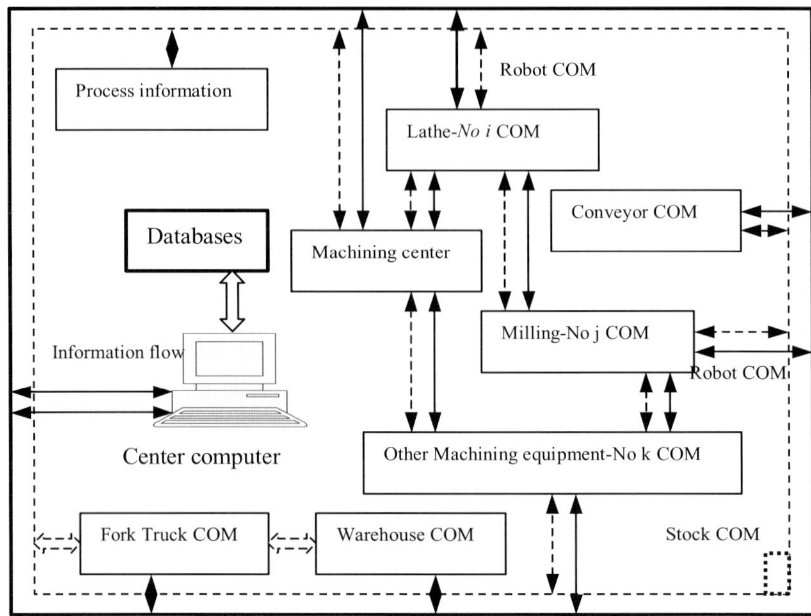

Figure 3. COM Cooperation in FMS

modeling and kinematic modeling should be considered, the first one can be studied in future plan, here gives detailed research to the last two.

4.1   Geometrical modeling

Geometrical model is a characteristic description of realistic physical model, it presents basic shape information such as bed, workbench, spindle, etc.. By using some CAD software, Pro-E, UG or Solid works, each model is divided into several components according to their functions. For a lathe, it

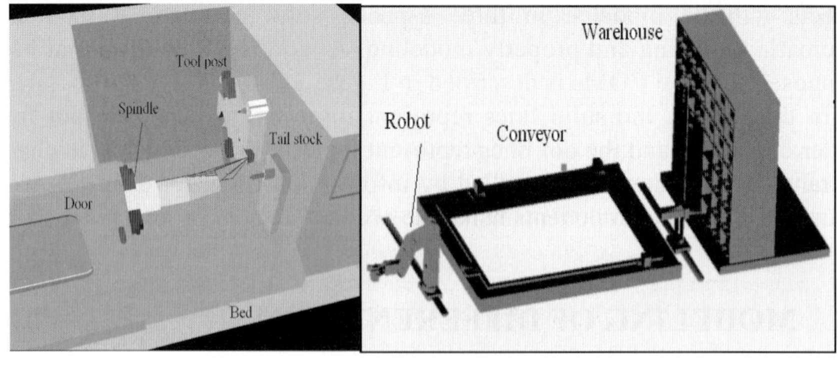

(a) Virtual lathe   (b) Virtual material handling system
Figure 4. Virtual models

is made of bed, tool-post, spindle, tapered center, carriage, saddle, set screws and side wards, etc.. they are converted into OPENGL-based virtual environment, precise coordinates will be obtained, series of transformations are implemented in order to assembly them into a virtual lathe in realistic structure. So do the other equipment, milling machine tool, robot, warehouse, stock, conveyor, etc.. Figure 4 is a list of virtual models.

4.2 Kinematic modeling

Once material flow and information flow are input, virtual control unit will be triggered, the equipment and its components will move according to given instructions.

The movement of each equipment is on the basis of its own kinematic property, and is planned. For a machining tool, when planning the movement, NC principles must be obeyed, and machining coordinate system, the machine tool origin and machining origin can be given. For simulation of NC interpolation such as arc and line etc., the step, the end of this cutting, and the decision of interpolation direction, will also be confirmed.

The movement of each component is triggered with timer, multi timers

(a) Milling a contour   b) Lathe a turning part with conic surface
Figure 5. Working procedure simulation for Lathe and Milling machine tool

will be used to harmonize with the other components. In a machine tool, for rotation of spindle, single axis movement or multi axes co-movement, robot picking and placing, material loading/unloading, etc., different timers can be defined, so do other components of equipment.

As an example, the kinematic functions of lathe and milling machine tool, a working procedure can be simulated like Figure 5.

In Figure 5(a), the machine tool is milling a contour with cylindrical surface. In Figure 5(b), the tool is lathing a turning part with a conic surface, and the NC codes are listed as follows:

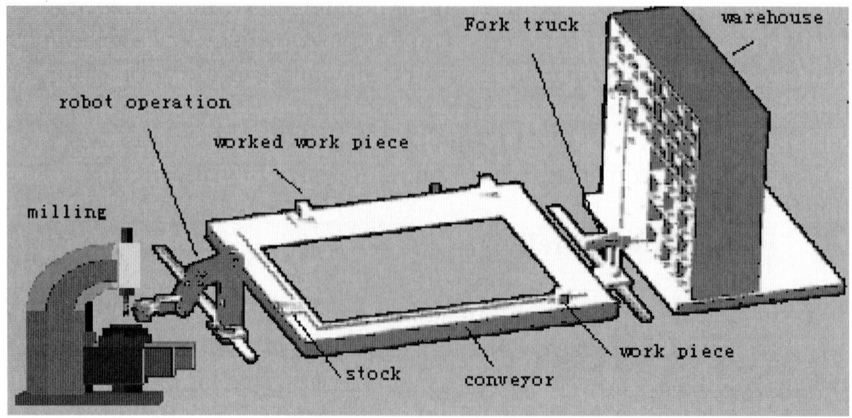

Figure 6. Simulation of Virtual FMS

Table 1. NC codes

Milling	Lathing
N01 S200	N01 S10 X-80
N02 G00 X85 Y140	N02 X-100
N03 G01 Z120 F500	N03 X-50
N04 G01 X-65 F500	N04 X-190 T06
N05 G01 X-15 Z-50 F500	N05 Y100
N06 G01 Z-120 F500	N06 X-150 Z70 T04
N07 G01 X55 Z-120 F500	N07 Y200
N08 G02 X85 Z-90 I0 K-30 F200	N08 S-1000 F50 T02
N09 G01 Z0 F500	N09 X-10 Y50 Z40
N10 G00 X0 Y0	N10 X10 Y20 T05
	N11 X-100 T03

5.   RESULTS AND CONCLUSIONS

After modeling, these objects are saved in unplanned virtual FMS. To complete a working simulation, first, analyses the working task, then select modeled objects to plan a rational layout, at last, according to material flow and information flow, distribute simulating time and define timers. After all is ok, it is controlled by computer center and the simulation starts. The procedure can be represented in Figure 6.

All of these results have been testified in OPENGL-based virtual environment, they show that the idea is reliable and the scale of virtual FMS can be extended and reuse because all of the models are on the basis of component object models, the system will be very useful to future process planning and decision research.

REFERENCES

1. Shuka C., Nazquez M. Chen F. F.. "Virtual Manufacturing :an Overview". *Computers & Industrial Engineering*, **Vol.31**, pp.79~82, January, 1996
2. Onosato M. "Development of a Virtual Manufacturing System by Integrating Products Model and Factory Models". *Annals of CIRP*,1993,42
3. Kimora F. "Product and Process Modeling as a Kernel for Virtual Manufacturing Environment". *Annals of CIRP* ,**Vol.42**, pp.147~150, 1993
4. Yuan Qingke, Zhao Rujia. "Virtual Manufacturing System. China Mechanical Engineering", **Vol.6**, pp.10~13, April, 1995
5. Yan Junqi, Fan Xiumin, Yao Jian. "The Architecture and Key Techniques for VM System. China Mechanical Engineering", **Vol.9**, pp.60~64, 1998 (in Chinese)
6. Han Xiangli, Yang Gang, Xiao Tianyuan. "Virtual Manufacturing and Its Application in CIMS Engineering Research Center". *High Technique Letters*. pp.1~6, January,1999 (in Chinese)

EDUCATING THE GLOBAL DESIGNER

William J Ion, Andrew Wodehouse, Neal Juster, Hilary Grierson and Angela Stone
Design, Manufacture and Engineering Management
University of Strathclyde, 75 Montrose Street, Glasgow G1 1XJ
e-mail: w.j.ion@strath.ac.uk

Abstract: Distributed design teams place a far heavier reliance on communication and collaboration than conventional collocated teams and as a consequence require design team members to acquire and develop an enhanced collaborative skill set. Given that distributed design teams are likely to remain as a dominant feature of product development for the foreseeable future there is a need to ensure that engineering graduates are equipped with the necessary knowledge and skills. This paper describes work carried out and the University of Strathclyde and elsewhere in the development of a curriculum for distributed design.

Key words: Design Education, Collaborative Working in Design

1.   INTRODUCTION

The design and development of new products for the global marketplace requires engineers to perform in internationally situated teams. Modern communication technologies such as virtual environments, digital libraries, shared workspaces, video and audio conferencing and email are increasingly being used to enhance performance by supporting information creation and sharing. Distributed design teams place a far heavier reliance on communication and collaboration than conventional collocated teams and as a consequence require design team members to acquire and develop an enhanced collaborative skill set [1]. Given that distributed design teams are likely to remain as a dominant feature of product development for the foreseeable future there is a need to ensure that engineering graduates are equipped with the necessary knowledge and skills. It is therefore necessary for design and engineering students to learn to work in distributed teams by utilizing cutting edge information management technologies and experiencing the social and cultural challenges of working at a distance. This paper will describe work undertaken at the University of Strathclyde that addresses these issues.

2.   GLOBAL DESIGN

2.1   What is Global Design?

Advances made in the computing world and in particular the expansion of the Internet have been key factors in the increased use of distributed design teams in the past few years. Another key factor has been the growth of the global market fuelled by demand for technologically advances products. The knowledge and skills required to develop and manufacture these products rarely resides within a single location leading to the need to establish distributed and possibly global design teams. Additionally, recent approaches within the industrial world such as Supply Chain Management, Collaborative Product Commerce, Change Management, the increase in outsourcing and the concept of the extended enterprise have promoted the growth of distributed work. As an example, Siemieniuch [2] reports that the percentage of an automobile being outsourced has risen from 5-15% in 1989 to 40-80% in 1997.

2.2   Distributed Team Collaboration

A distributed design team is a multidisciplinary team containing the core skills required to undertake a project. A typical team would consist of a variety of engineers (e.g. design, mechanical, electrical, manufacturing), marketing, finance and other personnel. These individuals would interact synchronously or asynchronously supported by a 'collaborative toolkit'. To fully understand the requirements for this toolkit it is worth considering the nature of the interaction between team members in the design process in more detail.   Design is a collaborative process involving communication, negotiation and team learning. Efficient communication is critical to achieving better co-operation and co-ordination among design team members.   Fruchter, [3] has made the following observations on conventional design team communication methods :

- Designers record background information and results of reasoning and calculations in private notebooks;
- Information in the form of text, calculations, graphics and drawings is captured in paper or computer based forms. Unfortunately, much of the design intent in a design dialogue is lost because it is partially documented. The final decision tends to be recorded but much of the interaction and developmental thinking of a design discussion is not;
- The process of identifying shared interests within a design team is ad-hoc and based on participants imperfect memories and retrieval of available documents. This error-prone and time consuming process rapidly leads to inconsistencies and conflicts;
- Meetings are usually the forum in which inconsistencies are detected and resolved before a project can progress. Telephone conversations are also used to resolve conflict and inconsistencies within the design decision-making process as and when they occur. Discussion of graphic or numerical information by telephone, fax etc. is difficult and leads to misunderstandings/misinterpretations and eventually increased product cost;
- Time delays in fax and telephone based communication between design team members can cause significant delays in project timescales. Additionally, these communication methods tend not to develop a team ethos or a common sense of ownership of design decisions.

The majority of communication within the design process is of an asynchronous nature with information being relayed between design partners in a sequential manner. For certain forms of design communications this means of operation is not satisfactory and direct interaction between design participants, in synchronous mode, is necessary. [4]

2.3 The Concept of a Designer's Shared Workspace

The shared workspace concept encapsulated both the physical and software environment that a designer would work within as a part of a distributed team. In physical terms this would consist of a networked computer workstation plus ancillary input devices (cameras, scanners etc) at each team member's location. In software terms the workspace would include all the tools necessary for effective collaboration at all stages of the design process. The tools required for effective collaboration can be identified through examining the information generation and sharing needs within the design process shown in Table 1[4]:

Table 1 Information and the Design Process

DESIGN STAGE	EXAMPLES OF INFORMATION GENERATED / SHARED
Market	Reports on current market situation, meeting notes, brainstorming notes/sketches
Specification	Specification documents, internal and client meeting notes, brainstorming, notes/sketches, general communications
Concept Generation	Brainstorming meeting notes/sketches, meeting notes, sketches, drawings, rough calculations, general communications
Concept Evaluation	Internal and client meeting notes, sketches, drawings, rough mock-ups and physical models, cost evaluation calculations, general communications
Detail Design	Meeting notes, detailed drawings and design calculations, final costing calculations, 3-D solid models, detailed mathematical and numerical models, general communications
Manufacture	Final product documentation - manufacturing drawings, bills of materials, test specifications
Sell	Sales presentations, demonstrations, photographs artwork, presentation graphics etc.

From this table it can be seen that a designers shared workspace system would be required to support both synchronous and asynchronous communication of numerical, textual, graphic, audio and physical information (e.g. models). To support this a shared workspace for design would require the following facilities: data conferencing; video and audio conferencing; application sharing; information storage and management tools. Figure 1 represents the shared workspace concept graphically.

3.   EDUCATIONAL ISSUES

3.1   The Educational Context

Incorporating distributed design into the curriculum involves addressing a number of diverse concepts. These concepts can be conveniently classified as relating to either the technological or social environment within which the design will take place.

Experiments with the teaching of collaborative design using virtual and managed learning environments, conducted at the University of Strathclyde over a period of 6 years [5], have identified a number of barriers to successful implementation. These barriers, listed below, identify some of

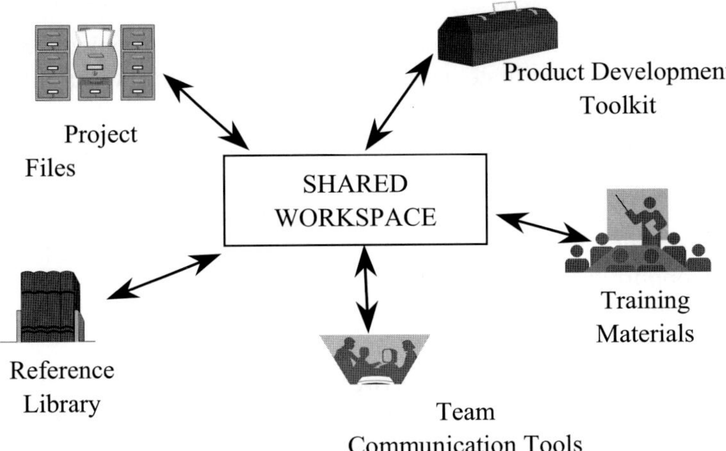

Figure 1 The Shared Workspace Concept

the issues that must be addressed in integrating collaborative design into the curriculum.

Technology/Environmental Barriers
- Unfamiliar software environment
- Inadequate software
- Lack of technical training and support
- Poor physical environments
- Difficulties in conveying design ideas

Social barriers to communication
- Not knowing collaborators in advance
- Not having clearly defined roles and responsibilities
- Misunderstandings
- Frustrations of asynchronous collaboration
- Conflicting institutional cultures and methods
- Differing student backgrounds and personalities

3.2   Coaching

High performance design engineering teams are composed of autonomous learners, who can independently determine and pursue their learning goals and content. The nature of design activity requires them to act that way; designing is context dependent and open-ended, and therefore, does not revolve around a specific body of information or knowledge. This poses a problem for design education since teachers cannot predict in advance what students will decide to learn. Coaching, rather than didactic teaching, has proved to be effective in addressing that problem [6]. Expert coaches guide and facilitate rather than try to specify what information should be used.

The educational paradigm shift from teaching to coaching requires students to have access to as wide a range of information as possible. In most cases, much of that information lies outside the students' immediate domain. Virtual learning environments incorporating digital libraries provide an excellent opportunity for extending the range of information available to design students.

3.3   Design Knowledge Framework

The design knowledge framework shown in Figure 2 was developed at the Centre of Design Research at Stanford University. [6] It is a convenient way of representing the interactions between a design team, coaches and the product development activity. An understanding of the interactions in the framework can lead to a better understanding of the educational issues within collaborative design projects.

The framework makes a distinction between formal and informal aspects of practice and knowledge. The instructor, product development history and product development process are considered to be predominantly formal elements. Coaches, teams and product development practice are considered to be informal elements. The arrows represent the 'acquisition' or 'co-generation' of product development knowledge.

Further consideration of the framework in a learning context leads to the identification of three fundamental learning mechanisms associated with the activity, referred to as 'learning loops'. The three learning mechanisms are

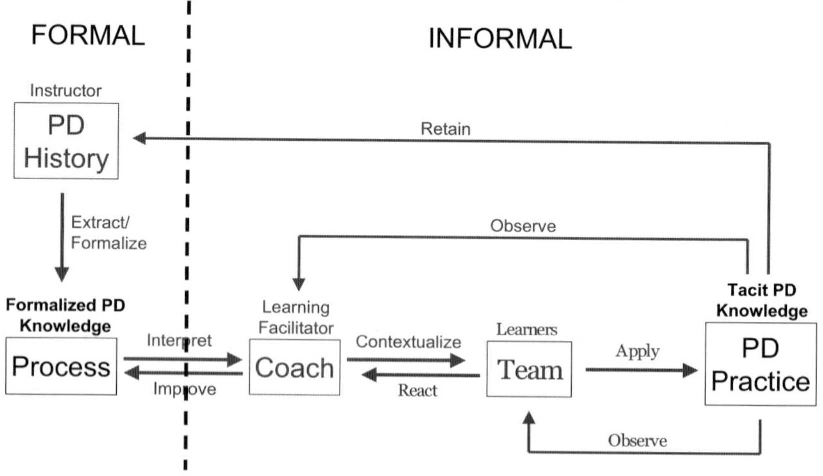

Figure 2 Design Knowledge Framework [6]

shown in Figure 3 and may be described as follows;

- **Learning Loop 1 – Designing**: Teams apply the product development process contextualized for them by coaches in their design practice. They utilize the information embodied in the process, and in doing so, generate new information.
- **Learning Loop 2 – Coaching**: Coaches observe the design practices of teams, and use the understandings they gain in contextualizing the product development process for them. Based on the needs of teams, coaches selectively extract information from the product development process and present it to the teams in a meaningful way.
- **Learning Loop 3 – Capturing, Indexing, and Publishing**: Instructors retain a history of the new knowledge generated during design practice, and extract new elements from it in order to improving the product development process. Instructors manage the capture, indexing, and publishing of the new information that teams generate in loop 2 in the form of a product development process.

Within these learning loops, students would conduct the process of team based design engineering.

Collaborative Design Projects at the University of Strathclyde

Having discussed collaborative design in a global context and some of the associated educational issues we shall now turn our attention to reviewing work carried out at the University of Strathclyde over the past nine years.

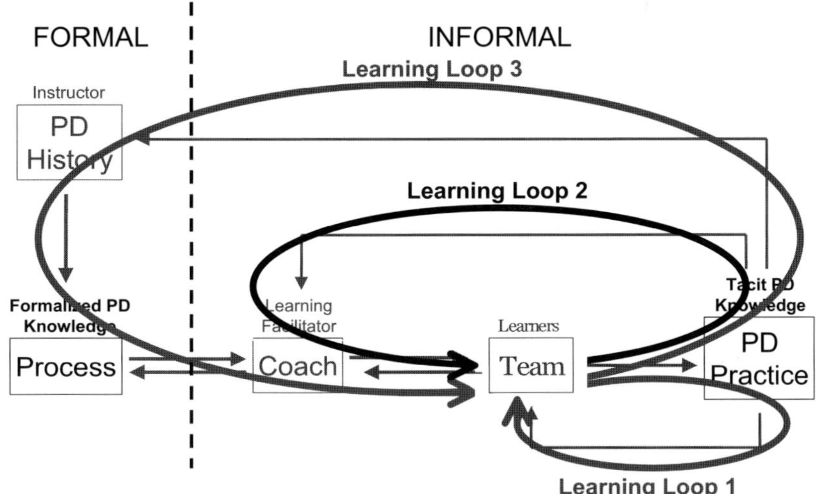

Figure 3 Learning Loops [6]

Clyde Virtual University (CVU)

The University of Strathclyde has a history of educational development in shared design project workspaces and managed learning environments. This dates back to 1995 when the Clyde Virtual University (CVU) was established by five higher education institutions in West Central Scotland. CVU is an Internet based virtual learning environment containing online courses within a virtual *Lecture Theatre,* a text-based discussion capability within a *Virtual Café,* backup textual materials in the online *Library* and web-based assessments in an *Assessment Hall.* [7].   This virtual university infrastructure was used as a conceptual model for the academic staff development program in new technologies at Strathclyde University [8] and has inspired a number of associated projects.

The Clyde Virtual Design Studio (CVDS)

The Clyde Virtual Design Studio (CVDS) was developed in 1997/98 by the University of Strathclyde and the Glasgow School of Art as a part of CVU [9]. CVDS is a computer based design project environment capable of being operated both locally and across a distance using communication techniques such as video conferencing and white boarding. The main objective in building the design studio was to provide a platform for investigating the feasibility of using computer networks as a means of communication in educational design projects. The design specification for CVDS reflected the communication needs of design project work and the constraints imposed by the educational environment. The key elements of the work included the provision of a comprehensive range of facilities within the environment to allow teams to collaborate - these teams could be separated geographically or be located at the same site.

The ICON (Institutional Collaboration over Networks) Projects

CVU and CVDS provided the supporting environment and stimulus for a number of collaborative design projects including ICON. ICON (Institutional Collaboration Over Networks) began in June 1997 [10]. The students came from the Product Design Engineering course at the University of Strathclyde and the joint course of the same name at Glasgow School of Art/Glasgow University. The main objectives of the ICON project were to ascertain the practicality of facilitating Internet based remote collaborative design projects for students, to investigate the impact of the technology on the effective teaching and practice of the design process and to encourage diversified learning. ICON utilised the environment provided by the CVDS.

The VIDEEO Project

The Clyde Virtual Design Studio provides a supportive environment for product development complete with design and communication tools. However it does not incorporate company and business constraints; these need to be addressed separately by the design team. To address this problem the Virtual Development Enterprise for Europe (VIDEEO) project [11], funded by the Socrates ODL program was started in 1998. The University of Strathclyde, through the department of Design, Manufacture and Engineering Management (DMEM), was one of eight educational institutions from across Europe, with backgrounds in engineering and business, collaborating on the project. The VIDEEO environment provides an educational and training tool for not only designers but also for other disciplines with an involvement in the design process such as marketing and manufacturing. The environment was created utilising modern computer and network technology as appropriate to provide a complete teaching and learning experience for the participants. The VIDEEO project represents a further step in the development of a fully integrated product development environment for education.

The Client Project

The knowledge gained from the ICON and VIDEEO projects has assisted in the development of the CLIENT project undertaken by the Faculty of Engineering at the University of Strathclyde in session 2001/02. This project set out to explore how mobile computing facilities (in the form of laptop computers and wireless LAN) could best be used to support team-based learning. Student teams working on product development project were each issued with a laptop computer. A central project server was established providing; project information, access to project resources/data and a web based file management system. Students were, therefore, able to access all project resources at any time from any location either via the mobile laptop, fixed desktop commuters or central server. Staff and industry based project sponsors were also able to access or supplement easily the student project data if required.

The DIDET Project

The projects described above focused on different aspects of the complete shared workspace environment. The early work focused on technological aspects. The performance and reliability of Internet based collaborative technologies such as video and data conferencing now readily accommodates the needs of distributed design teams at a cost affordable by most educational institutions. Recent work has therefore focused on enhancing the collaborative shared workspace environments available to student design teams and in particular to addressing issues relating to information storage and retrieval. The DIDET project [12]focuses on this particular aspect.

The DIDET (Digital Libraries for Distributed Innovative Design Education and Teamwork) project started in the first quarter of 2003 and is funded by the Joint Information Systems Committee (JISC) in the UK and the National Science Foundation (NSF) in the US [13]. It is a collaboration between the University of Strathclyde in Glasgow, UK and Stanford University, CA, USA. The goal of DIDET is to *enhance student learning opportunities by enabling them to partake in global, team based design engineering projects, in which they directly experience different cultural contexts and access a variety of digital information sources via a range of appropriate technology.*

The DIDET project is attempting to develop a test bed to integrate digital library capabilities into the collaborative shared workspace environment. This test bed will be used to teach students state of the art team based engineering design process skills augmented by information archiving and retrieval skills. The nature of the state of the art engineering design processes that will be used will require the students to interact with and generate large amounts of information. Being able to quickly ascertain the nature of the relevant design information, to determine if it exists, and to access it and utilize if it does exist, accelerates and improves the design process. Also, for the creation and reuse cycle to continue, it is imperative that, when new information is created, it is converted to appropriate digital formats, indexed, and rapidly made accessible on the World Wide Web. The DIDET project is attempting to inculcate in students the ability to utilize digital libraries technologies and information handling behaviors, thus not only transforming design teams into autonomous and effective learners but also improving their design performance.

4.   CONCLUSION - A CURRICULUM FOR COLLABORATIVE DESIGN

In considering how to incorporate collaborative design into the curriculum the educator must consider all the issues discussed above. Our work at the University of Strathclyde to date has clearly demonstrates that it is simply not sufficient to provide a distributed team of students with a set of collaboration tools and expect them to be able to work together effectively. Indeed, it is likely that this would be counter productive as the frustrations encountered by the students would impact negatively on their learning. In setting the curriculum consideration must be given to enhancing students' understanding of the following in a global collaborative context;

- collaborative technologies e.g. audio and video conferencing, shared whiteboards, digital libraries, groupware.
- socio/cultural issues e.g. how to conduct distributed meetings, the effects of the reduced level of gestures etc., time zone effects
- the nature and operation of the distributed design process e.g. the roles of synchronous and asynchronous communication
- information access, generation and management e.g. storage and retrieval strategies, formal and tacit information
- project management skills

In addition, the educator must give consideration to where and when in the curriculum such material should be incorporated and to the pedagogical approach to be adopted. Our experiences would suggest that the required knowledge and skills development should be embedded into the curriculum from first year onwards with practical experience of full scale distributed design projects taking place in the latter years. As previously discussed, the recommended pedagogical approach is based on coaching combined with an understanding of the learning loops shown in Figure 3.

REFERENCES

1. MacGregor S. *Describing and supporting the distributed workspace – towards a prescriptive process for design*; PhD thesis University of Strathclyde, 2002
2. Siemieniuch, C. E. and M. Sinclair, "Real-time collaboration in design engineering: an expensive fantasy or affordable reality?"; *Behaviour & Information Technology* **18(5):** 361-371, 1999.
3. Fruchter, R. "Interdisciplinary communication medium in support of synchronous and asynchronous collaborative design"; *Proceedings of the First International Conference of Information Technology in Civil and Structural Engineering Design (ITCSED)*; University of Strathclyde, Glasgow. 14-16th August, 1996.
4. Ion W J and Neison, A I. "The use of shared workspaces to support the product design process"; *Proceedings of the International Conference on Concurrent Engineering;* University of Rochester, Detroit, 1997.
5. Sclater N, Grierson H, Ion W J, MacGregor S P. "Online collaborative design projects: overcoming barriers to communication"; *International Journal of Engineering Education,* **vol. 17, no.2,** pp189-196, 2001.
6. Eris O and Leifer L. "Facilitating product development knowledge acquisition: Interaction between the expert and the team". *International Journal of Engineering Education* special issue on the Social Dimension of Engineering Design, 2002.
7. Whittington C D and Sclater N. "Building and Testing a Virtual University";. *Computers and Education* **30**, 41-47, 1998
8. Littlejohn A H and Sclater N. "The virtual university as a conceptual model for faculty change and innovation"; *Journal of Interactive Learning Environments 7, 209-226,* 1999.

9.   Ion, W.J., Thomson, A.I., Mailer, D.J. "Development and evaluation of a virtual design studio; Proceedings of EDE '99", September 1999, 163-172, 1999.
10.   Campbell L, Ali-MacLachlan I, Thomson A I, Ion W J, MacDonald A S. "Institutional collaboration over networks – ICON: A comparison of two collaborative design projects"; *Proceedings of CADE '99*, University of Teeside, April 1999
11.   Thompson A I, Ion W J, Temple B K, Allan M, Kernohan N, Davidson S; "A Virtual Development Enterprise for Europe"; paper number DETC20001/IED21207 *in Proceedings of the ASME Design Technical Conferences, Pittsburgh, Pennsylvania*, on CD, 2001.
12.   http://dmem1.ds.strath.ac.uk/didet/introduction.htm
13.   http://www.jisc.ac.uk/iundex.cfm?name=programme_dlitc

WEB EDUCATION PORTAL FOR THE TEACHING OF MECHANICAL DESIGN AND MANUFACTURING
Use of XML language

Didier Remond[1], Alexandre Toumine[1], Benoît Eynard[2], Nadège Troussier[3], Alain Daidie[4], Jean-Pierre Devaujany[1], Samuel Gomes[5] and Lionel Roucoules[2]

[1] *Institut National des Sciences Appliquées de Lyon, Bâtiment Jean d'ALEMBERT, 8, rue des Sciences, F.69621 VILLEURBANNE CEDEX, France. Didier.Remond@insa-lyon.fr*
[2] *Université de Technologie de Troyes, 12 rue Marie Curie, BP 2060, F.10010 TROYES CEDEX, France.*
[3] *Université de Technologie de Compiègne, BP 60319, F.60203 COMPIEGNE CEDEX, France.*
[4] *Institut National des Sciences Appliquées de Toulouse, 135, Av. de RANGUEIL, F.31077 TOULOUSE CEDEX 4, France.*
[5] *Université de Technologie de Belfort-Montbéliard, F.90010 BELFORT CEDEX, France.*

Abstract: In the field of design and manufacturing in mechanical engineering, it is difficult to give an overview of lecturing steps because of the splitting up of required knowledge and the expertise level of the teachers. According to the OPALYS project objectives, we present the application of an approach of documents sharing enabled by the use of XML language. We detail the chosen conformity structure for the various kinds of documents. We also detail the choices that were made in order to create links between the documents, while the authors did not know exactly their integration context. The main encountered difficulty is related to the XML tools that did not yet become mature enough for not expert users. The developed demonstrator is then presented in its main functionalities and highlights the relevancy of our approach.

Key words: ICT, Design, Manufacturing, Mechanical Engineering, XML, Teaching Document.

    D. Remond, A. Toumine, B. Eynard, N. Troussier, A Daidie, J. P Devaujany, S. Gomes and L. Roucoules

1. INTRODUCTION

This paper aims at presenting the work of several teachers from several universities, all faced with the same common problem: the sharing and transfer of lecture material for engineer's education in mechanical design and manufacturing.

The training of students leads to a splitting up of knowledge. The aim, in this paper, is to create and develop a web education portal in order to enable the link and the assimilation of the various viewpoints. Therefore, a software demonstrator has been developed. Some lecture documents enable to describe quite specific aspects of the training such as practical works, capitalization pages, etc. As each author can (must) not know the whole content of the lecture material, it is necessary to create links in a systematic way between the documents, via key words. For its rigor in the description of the data structuring and the freedom for the output formats, the documents use XML language; however they are written by non expert (in data processing) what leads to problems of control of technology to be selected.

The paper is structured in four sections. First, the context of the study and targets to be reached are described. Second, our approach for training documents sharing is presented. This approach is based on a web portal dedicated to mechanical design and manufacturing. Third, the use of XML language in order to structure the contents of the training is pointed out. Finally, the way to use the numerical educational support is shown.

2. IDENTIFIED NEEDS AND TARGETS IN THE FRAMEWORK OF OPALYS PROJECT

Mechanical design and manufacturing is featured by various skills that require a good knowledge in several fields [1-2-3-4]. The whole lectures provided by each university are usually specific and directly linked to a partial view of the design or manufacturing tasks. A great difficulty encountered by the students is then the necessity to integrate the whole design and manufacturing process and to adopt a global viewpoint. Therefore, it could be useful to use of information and communication technologies to achieve a shared web education portal. This portal should enable each teacher or student to position himself in relation to the whole context. What can help them to understand their own specificities?

Otherwise, the concepts addressed in design and manufacturing is often linked to a context. The design choices depend on the environment which can be material, cultural, related to a field of industrial activities, related to a level of knowledge, etc. It is then important to show students that the suggested process is not single and generic but is related to the environment.

The design process changes are also constrained by the fast evolution of used technologies and tools. The fundamental bases remain the same (for example as regards the embodiment of mechanical systems) but the applications evolve quickly. Then, it is difficult to keep a good quality level for education, and to be relevant regarding the industrial tools.

The OPALYS project also aims at the capitalization of student and teaching works, carried out during several years. Data structuring enables to present the same lecture material with various approaches for each university. Moreover, the evolution of several works during few years in order, for instance, to illustrate the dependency of the process according to the context and to the tools evolution.

In summary, the objectives of the OPALYS project are to provide to teachers a system enabling to share, develop and re-use lecture material via an web education portal [5]. Each teacher should both share lecture material with others teachers and link this to existing one, and that, without knowing the already existing material lecture.

The student should follow a teaching course recommended by the teacher without losing the possibility of seeking and of seeing related illustrations.

The whole lecture material must be linked via a characterizing semantic tool. The suggested approach is, when developing the lecture material, to specify the necessary elements to integrate each material in a broader and not well known context. Our application field (mechanical design and manufacturing) is well relevant regarding the problems, especially in the current teaching context (dissemination and contextualizing of the approaches). Thus the aim is to have a teaching resources base usable by teachers or students, all these resources being inter-connected either on a semantic level or by a chronological order related to a teaching approach.

In other words, the OPALYS project is not a simple database of lecture material without links and without common structure, neither a complete on line course, nor a semantic tool for automated indexing of contents. The objective is to develop a set of lecture material structured in the same way and enabling to have a large range of approaches regarding the same subject.

The aim of this paper is to present the choice of the data modelling language and, more in detail, the data structuring. This data structuring is defined both in the document and in a database of the web education portal centralizing documents. In this context, this paper presents the lecture material structuring process and the issued results.

         D. Remond, A. Toumine, B. Eynard, N. Troussier, A Daidie, J. P Devaujany, S. Gomes and L. Roucoules

3. CONTEXTUALIZING AND SHARING OF THE DOCUMENTS

In the above described context, it seems interesting to specify a common data structure enabling the exchange and capitalizing digital lecture material. A web education portal has been developed to share numerous lecture material. The features detailed bellow require a strong structuring of the material available on the web portal. The strong constraints, in terms of semantic modelling, have some effects on the way of working and of defining the lecture material:

–   The first of these constraints lies in the need to formalize the process of design and manufacturing by using the same terminology and especially by confronting two viewpoints. On the one hand, the "company" viewpoint describes the exchanges between several departments. On the other hand, the "product" viewpoint presents the product lifecycle. In our approach, these two viewpoints, illustrated in figure 1, must exist together, at the same time and should characterize the lecture material. To summarize, the need to formalize the process has led to specify the structuring key elements and to position the lecture material according to the "company" and the "product" viewpoints.

–   The need of several viewing level of the knowledge according to the wished deepening drives to the second constraint. Whether a "curious" level (informational and basic reading) or a "technician" level (enables to replay a process according to a well defined process) or a "engineer" level (enables the understanding, the appropriation and the adaptation to different situations). The level addressed by the lecture material is provided by the author when developing the document.

–   A third constraint is introduced by the kind of documents or process illustrated by the lecture material. In our case, several aspects have been implemented such as the case study, the course, the capitalization, etc. The different kind of documents defined is described in following sections.

Moreover, it is absolutely necessary to be able to link the various generated documents without asking the author to know the whole lecture material. The use of keywords, defined by the author, enables the link, but introduces the problem of the relevance and consistency of the used keywords. This point will not be detailed here; it seems quite obvious that the proposal can be largely improved by current indexing technologies.

However, the authors have to define these keywords and the minimal required knowledge to enable the integration in the set of lecture material. This means the specification of keywords shared by all authors. These keywords must be relevant regarding the lecture material. We preferred that way instead of defining of hypertext links into lecture material: links being able to highlight the importance degrees between the various documents.

Lastly, it is absolutely necessary to allow the author defining a wished teaching approach according to the specificity of his teaching and the use case. Thus the author is free to define the viewing sequence of the lecture material. This new constraint must also be taken into account in the document while enabling the modification and use of the lecture material by any one.

These various constraints must be easily handled by the authors, that limits the development solutions and the choices made for structuring and use this kind of modelling. The following section describes the made choices to develop a relevant and useful demonstrator.

4. CONTENTS AND DATABASE STRUCTURING

In this kind of approach we need to highlight the descriptive elements (meta-data) which must remain in the document and the elements referring to the management, implemented on the web portal in a database (indexing, development, search and links). This section describes the choice of the language used to specify the document structure. Furthermore, it presents how to modify a document structure.

4.1 Choice and advantages of XML language

In the above defined context, the XML language enables to specify a structured document [6] and to ensure the consistency of the lecture material regarding predefined standards. It allows modifying display and presentation in a contextual way. It also allows filtering the data according to the lecture material nature. Moreover, the development of a data server in HTML appears to complex for an evolving management of the lecture material and for an easy use of the author who proposes this material. Then, the creation of the lecture material based on XML language [7] has been chosen. This requires from the author to provide the wished data and to homogenize the teaching fields of the lecture material.

The conformity to standards requires a DTD description (Document Type Definition) detailed in the following section. The use of the DTD ensures the author to provide a well structured document which will enable the integration on the web portal. This specific point seems essential in order to easily integrate some lecture material regarding a given context and in order to allow them to be efficiently shared.

The choice of XML language leads the author to use some software that still under development or slightly mature. This kind of software offers a lower maturity level compared to traditional office software. The constraints introduce by this lack of maturity lead this software useless for the authors. Several software have been tested (see section 5) without fully met of the needs. The contribution of XML is interesting when using a tool for structuring the description of the lecture material. It also enables to several authors to develop and share lecture material.

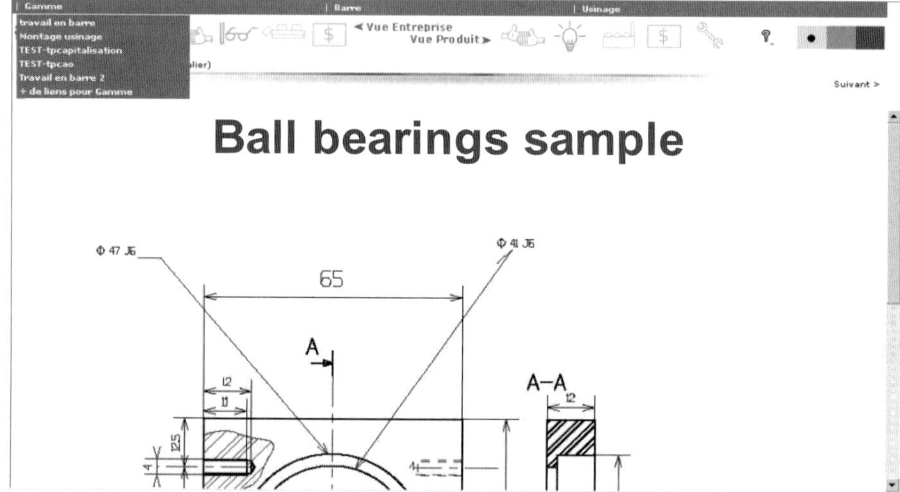

Figure 1. Sample of the Student Graphic Interface.

4.2   Description of DTD

In order to describe a DTD structure, figure 1 illustrates the Student Graphic Interface of the material lecture.

At the frame top part of this figure, the structuring approach can be seen. In this part, two kinds of viewpoints (« company » and « product ») are provided. The knowledge level (« curious », « technician » or « engineer ») appears using a colour indicator (in the top left of the frame). The description of keywords enables the linkage among the whole material lecture (the blue scroll down menu in the top left corner). It also enables the following a specific teaching process. Moreover, the viewing of the location in the web portal and of the table of contents of the document is also able.

We will use the sample of a capitalization documents in order to clarify description of a DTD in the following sections. It allows showing all the structuring features.

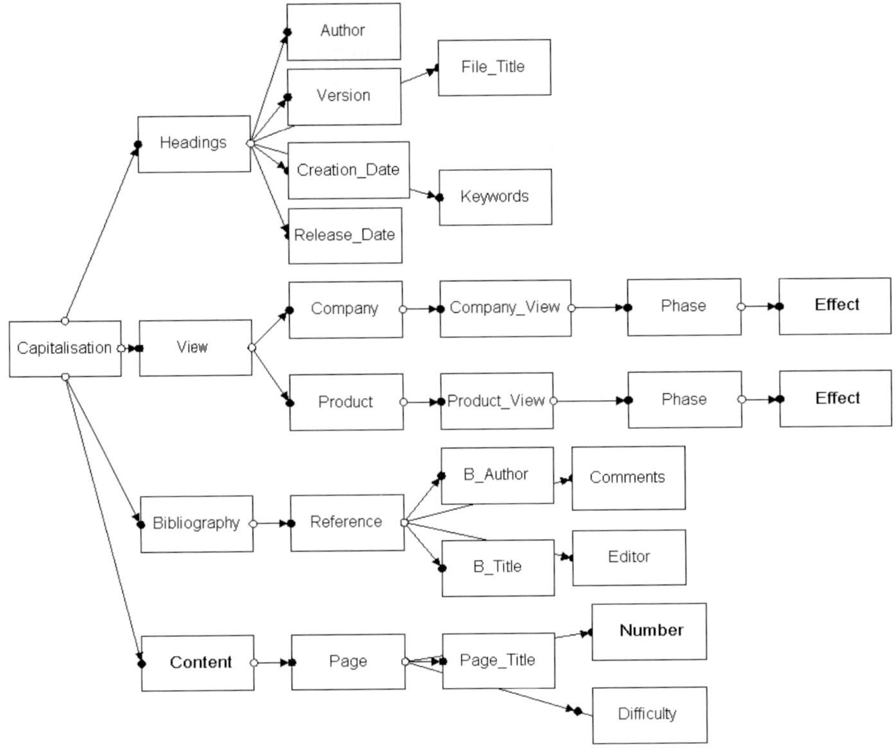

Figure 2. Higher part of the Capitalisation DTD.

4.2.1   Higher part of the description of DTD

The higher part of the document is structured as described in figure 2.

For each document, feature like the metadata is include in the heading part and can be filled in to fit the LOM standards under specification [8]. In the heading part, the keywords used to put some links among documents and the descriptions of the two viewpoints ("company" and "product") are also provided. An optional part concerned with the bibliography can be increased using additional data taken part from the standards but has not been yet developed.

In the part "content" (content), several pages can be included and identified by a number to indicate the order that defines a given teaching process. A title and a knowledge level can also be defined according to the student concerned with.

To sum up, a page can be defined as a displayable lecture material (on a screen or other) and is featured by a knowledge level and a number for ordering.

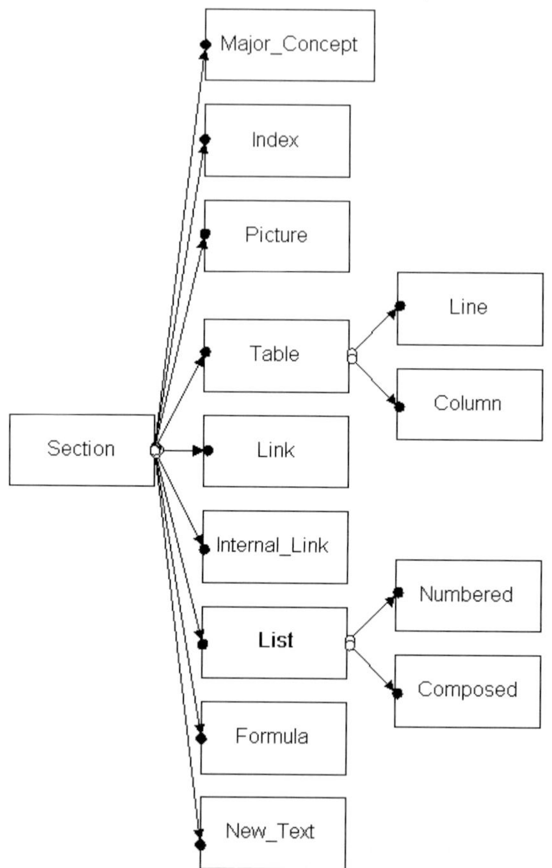

Figure 3. Lower *part of Capitalisation DTD.*

4.2.2   Lower part of the description of DTD

The lower part of the DTD (Figure 3) is concerned with the display data. This part shows our wish to not consider the display specification and to let it at the author choice. The author is just required to use three different raised texts (for instance in order to highlight three different importance levels).

This lower part allows the author to present a content of just one page and proposes some predefined formats. The XML language presents a great interest in order to match the display of some documents, carried out by several authors. The display can be modified using XSLT specifications [9].

4.2.3   Intermediate part of the description of DTD

The most interesting part of the DTD is the intermediate one. It enables to structure and organize the information of each lecture material. In the current case, the lecture material is concerned with capitalization sheet which aim at summing up some engineering design mistakes made, each years, by students achieving projects. The content of the card is structured using a DTD (Figure 4). A mechanical part must be addressed, and is presented via a picture and a title. The content is breaking down in topics for engineering design and in production planning for manufacturing.

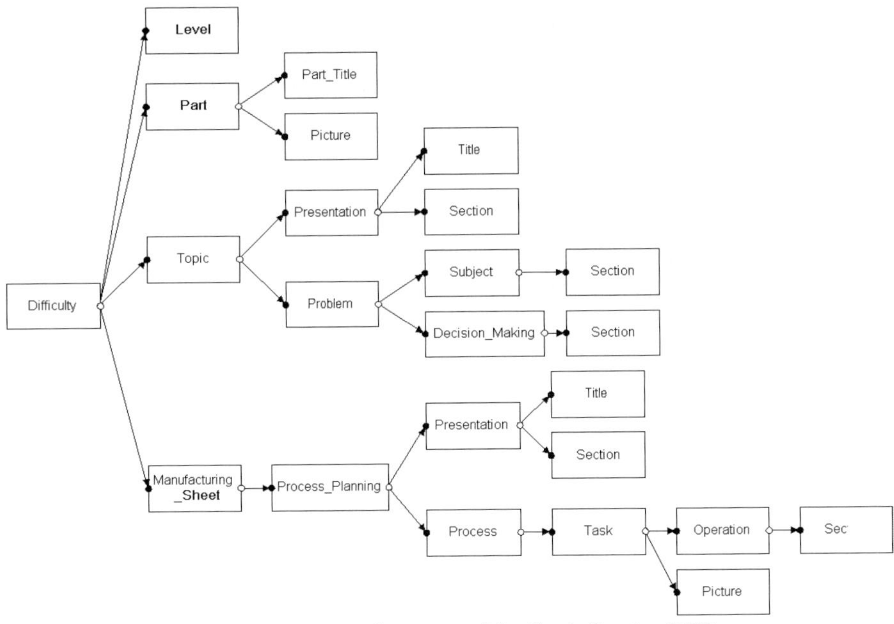

Figure 4. Intermediate part of the Capitalisation DTD

The topic is described by a title and a section (mainly text, pictures, etc.). Then it is composed of a problem breaking down itself in a problem subject and some made decisions or specified solutions (there are numerous possible solutions for a same problem subject).

In the same approach, the process planning is composed of a context presenting part (a title and a section) and a part describing of each planning with its tasks scheduling. Each one is specified by a basic operation and a picture.

The context presenting part of document will structure the way of fulfilling the "Capitalization sheet". According to this structure a consistent presentation of the Capitalization sheet" will be ensured. Then various authors will be able to write concurrently the sheet.

4.3 DTD modification

The definition of new type of document will be enabling based on the reuse of the existing DTD structure in two parts: according to the higher and lower parts described above. The document structure in those two parts is generic and then there are no needs of modification. The macro description of document and the document form do not change. But the own structure of each document is specified in the DTD intermediate part. This description should be based on existing documents. It also requires the agreement of each author. Then each department or authors group is able to structure their documents as they want. They are able to structure the document as they wish but integrating the chosen description of the design-manufacturing lifecycle and the decided form.

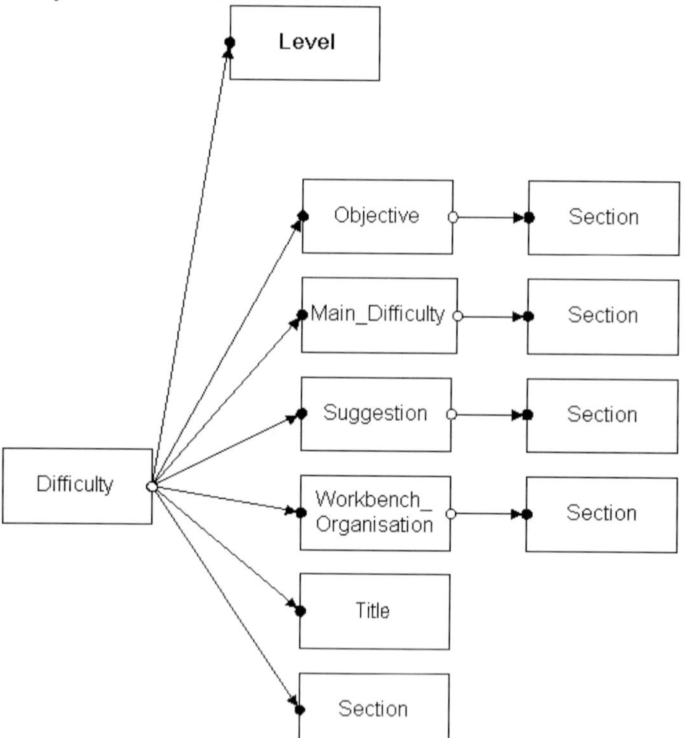

Figure 5. Intermediate part of the DTD for subject of Student Practice Works.

The figure 5 describes the intermediate part of a document corresponding to the subject of Student Practice Works.

The main difficulty is the extraction of a structure based on existing documents and enough finalized. This very important step should be carefully carried out while the DTD modification leads to the complete redefinition of all documents.

5. IMPLEMENTATION AND USE OF THE WEB EDUCATION PORTAL

The implementation of that kind of sharing and capitalisation of lectures material has required the development of a demonstrator based on a web portal integrating the display and the data vault for uploading the material. This development has been mainly based on PHP language and MySQL database [10 – 11]. The graphical user interface is presented in figure 1. The whole contain of this paper describes the functionalities and features of this display detailed section 4.2.

Regarding the uploading of lecture material, the chosen approach allows to validate the document based on a DTD. It also allows specifying a set of key words. They should be proposed in the list of key words existing on the web portal. The indexing of documents and the creation of links is then carried out in the database. The figure 6 presents a step of the uploading process of material in the web portal.

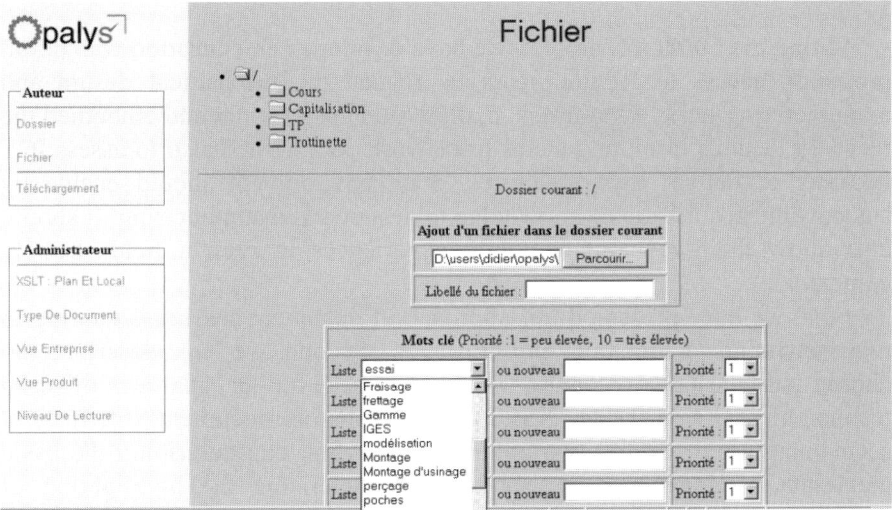

Figure 6. Workspace for uploading of lecture material

The definition of HTML web pages uses the PHP module and then provides an easy access based on all kind of browsers for the viewing of the lecture material. This viewing uses the XSLT specifications associated to each kind document.

The first definitions of materials highlight the useless of writing tools of XML documents. Currently they are not enough use and are not user friendly oriented for non expert people. The tested tools are close to templates allowing the capture of markup or close to office software allowing the users to not fulfil to the DTD specifications. In the first case, these tools are considered not user friendly oriented by authors who were not trained to that kind of tools. In the second case, it has been difficult to specify the structure or the extract the whole required information for the specification of a structure XML document. The intermediate tool does not currently existing while available ones are quickly evolved. Concerning the specification of structuring part it is required to keep the expert tools because that work will certainly remain in the expert area.

The use of the demonstrator in an education context has not been currently tested but several students have contributed to the definition of the lecture material. They seem to strongly appreciate this education positioning closer to a real industrial context quite according to our education domain of mechanical engineering.

6.   CONCLUSION

Within the OPALYS project, we have developed an education web portal providing access to lecture materials related the mechanical design and manufacturing fields. Our aim is to allow students to apply and embodied the received lectures in more general framework. We wished also to assess ICT in order to deliver expert lectures. Such environment should enable by students the acquisition of knowledge in design and manufacturing. It should also allow them to capitalize their experiences in such a field strongly contextual.

First, we have presented our approach of document sharing according to two viewpoints "Company" and "Product". Second, we have described the chosen structuring of lecture material based on an analysis of used technologies (XML, DTD, XSLT). Based on this document structures it seems possible to extract two common parts: first one describing the basic meta-data of documents; second one characterizing the highlighting of lecture material. The structuring of lecture material requires a huge analysis work of already existing material. To conclude, we have detailed the implementation and use of such an education web portal.

Our objective to develop some web lecture material, that the authors have no ideas of using context of documents, seems globally to be reached. The demonstrator allows linking by key words the indexed documents.

Our main difficulty has been to find some simple and efficient tools enabling non expert authors to produce some lecture material using XML format. The web portal demonstrator should grow in order to get a critical size in terms of lecture material. Then the demonstrator will be really usable by students and teachers. Based on the flexibility of the web portal, it is ease to increase the number of documents according to the future needs.

ACKNOWLEDGEMENTS

The authors wish to thank MM. S. Giroud and P. Will for their work in the development of the demonstrator web portal.

REFERENCES

1. Pahl G., Beitz W.: *Engineering Design: a Systematic Approach*, Springer-Verlag, London. 1996.
2. Ullman D.G.: *The Mechanical Design Process*, McGraw-Hill, New York, 1992.
3. Boothroyd, G.: *Product design for manufacture and assembly*, Marcel Dekker, New York, 1994.
4. Schey, J.: *Introduction to manufacturing Processes*, McGraw-Hill Co., Singapore, 1987.
5. Troussier, N., Daidié, A., DeVaujany, J.P., Eynard, B., Gomes, S., Rémond, D., Roucoules, L., Toumine, A.: "Exemple d'application des TIC pour un projet de création de support pédagogique". *Proceedings of the International Symposium on Technologies of Information and Communication in Education for engineering and industry – TICE 2002*, pp. 381-382, Lyon, France, November 13-15, 2002.
6. Bonneau, S., Kohl, T., Duckett, J., Williams, K., Tennison, J.: *XML Design Handbook*, Wrox Press, 2003
7. Delestre N., Frénot S., Mottelet S., Vayssade M.: "Distributed PolyTeXML Une nouvelle plateforme de partage d'items didactiques". *Proceedings of the International Symposium on Technologies of Information and Communication in Education for engineering and industry – TICE 2002*, pp. 149-156, Lyon, France, November 13-15, 2002.
8. LOM. Draft Standard for learning Object Metadata. IEEE P1484.12/D6.1.18., 2001.
9. Sussman, D., Kay, M.: *XSLT Programmer's Reference*, Wrox Press, 2000.
10. EasyPHP, http://www.easyphp.org/, 2003.
11. MySQL, http://www.mysql.com/, 2003.

FRAMEWORK FOR MECHATRONICS SYSTEMS DESIGN
- dimensions in current education

Lars Hein
*Technical University of Denmark (DTU), Building 404, DK-2800 Kgs. Lyngby, Denmark,
tel +45 45932522, fax +4545932529, email LH@MEK.DTU.DK*

Abstract: The teaching of Mechatronics Systems Design requires in its planning stage a conscious selection of how the multi-dimensional theme of Mechatronics shall be covered. Also, with the growing number of dedicated consecutive courses in Mechatronics Systems Design, it becomes increasingly important to be able to understand and articulate how each particular course covers the subject. This paper discusses eight of the more important dimensions that define the Mechatronics space.

Key words: Mechatronics, Design, Education, Framework, Methodology, Dimensions.

1.   EDUCATION IN MECHATRONICS

The teaching of Mechatronics Systems Design requires in its planning stage a conscious selection of how this multi-dimensional theme shall be covered. As is frequently the case in the teaching of Mechatronics Systems Design one course in will follow another, in order to adequately cover the field, and these consecutive courses must be tailored in such a way as to build on, respectively prepare for, the pervious and the subsequent course. Thus, it becomes increasingly important to be able to understand and articulate how each particular course covers the subject. Recent years has seen the emerging of some important patterns, which together may be said to define the space of Mechatronics Systems Design. It should be emphasized though; that the many more dimensions exist than those dealt with in this text, and that the conscious selection of the dimensions used to define the mechatronics space in any given context is important.

2. DIMENSIONS IN MECHATRONICS

In order to define for any practical purpose the space of Mechatronics Systems Design, more than one dimension in necessary. On the other hand, taking into account all the possibly relevant dimensions would certainly render too complex pattern for practical purposes. In this context the following eight dimensions and their respective grading are considered:

- Design Process
- Product
- Methodology
- Technology
- System
- Proficiency
- Domain
- Organizational frame

According to the situation, dimensions dealing specifically with for example Industrial Design, Environmental aspects, or Economy could also be given preference. However, these are not dealt with here.

2.1 Design process

An important aspect of the teaching of Mechatronics Systems Design is the extent to which the students get to carry through the total design process from conception, through design, to the final implementation and operation of the system. It is generally acknowledged that the true understanding of how decisions throughout the design process influences the end result is best arrived at by having the student(s) follow the process through. However, this is a dimension with severe time (and sometimes also economical) implications, and often a compromise as to how far the process is carried has to be made. It is reported also ([1,4]) that there is a great motivation effect in the last stages of the process:

Conceive

From the basic need situation or problem formulation: to decide on the system boundaries, to determine the basic systems requirements, and to develop basic ideas, concepts and working principles

Design

To carry out the engineering design including the detailed design, with considerations of materials, production processes, sales, distribution, etc.

Implement

To manufacture and/or establish the mechatronics system in its environment, if relevant also counting the human resources interface

Operate

To operate the system according to its intended use, and to learn about any shortcomings of the basic idea, the design and/or the implementation.

The realization of the importance of this process and its effects upon the learning outcome has led to the initiation of an international cooperation where several leading engineering schools in the United States and Europe has formed a collaborative to conceive and develop a new vision of engineering education — the CDIO Initiative [2].

2.2   Product

The Mechatronics system may be treated as either a functional entity in its own right, as a product, or as part of a product family:

Device

The functional entity as it is derived from the purely functional specifications, measured solely against functional performance goals.

Product

The system seen as a product; serving the purpose of the company as a generator of revenue and of market position. Optimization parameters may cover competitive edge, economy, etc.

Multiproduct

The system seen as part of a larger product family. This may include topics like product variants, modularisation, product architecture, product family master plan (PFMP), and product planning.

2.3   Methodology

The methodology used in the teaching of Mechatronics Systems Design may vary over a broad spectrum, depending on the focus of the course. Except for the cases where the subject is totally omitted, the spectrum could be described as follows:

Creativity

A methodology based on simple creative methods and on the procedure for general problem solving.

Design Methodology

The full design methodology, as it may be pertinent to the individual domains in the systems design (se 2.7). There is a number of sources for design methodology. However, methodologies for the individual domains (mechanics, electronics, and software) are not very well coordinated in the literature. For design methodology in the mechanical domain see [4].
Meta-tools
A number of meta-tools has been developed and deployed. These are predominantly aggregates of tools from design methodology, set up to target more or less specific issues:

- *Design for Manufacture (DFM)*
- *Design for Assembly (DFA)*
- *Design for Quality (DFQ)*
- *Design for Environment (DFEn)*
- *etc*

Generally, these tools refer to a product development framework like integrated product development [3,6,7]. (For an example on DFM, see [8].)

2.4   Technology

One of the very apparent dimensions of Mechatronics Systems Design is the often very complex technologies involved. However, a specific course in Mechatronics Systems Design may or may not take on the task of teaching these aspects.
No technology
Where no specific technologies are taught or used as the basis for systems design.

Process & Materials Technology
Where the technologies involved are related to the manufacturing process.
Complex Technologies (product technology)
Where the course may cover product technologies, such as specific IT technologies, opto-electronics, Micro- or nano-mechanics, etc.

2.5   System

The systems dimension may range from simply introducing the mechatronics system as a 'device', to be designed according to some narrow specification, and to being part of a complex socio/technological system:
Device
The bare functional entity as it is composed as just so many physical parts, to be designed according to specification.

Electro/mechanical system

To see the entity as part of a larger electro/mechanical system with which it is in some kind of interaction, often involving a human operator.

Socio/technological system

To see the mechatronics system as part of a larger socio/technological system with which it is in some kind of interaction, with implications on society, and on the environment.

2.6   Proficiency

An important dimension is the skills aspect: to what depth the teaching and learning shall penetrate, and how far the developed skill shall reach:

Know of

To know that the subject exists.

Know about

To know what the subject is about, and what the importance of it is in a given context. To know also when the current level of understanding is insufficient for the task at hand.

Command

To be able to command the skills involved, and to be able to base the design task on those skills.

Implement/teach

To have such an insight and understanding that one may transfer the understanding to, and develop the skills of others.

Add to our understanding

To be able to develop our understanding and command of the topic further, This clearly is the ambition when undertaking a PhD in Mechatronics Systems Design.

2.7   Domain

The three classical domains of Mechatronics Systems design are the Mechanical, the Electronics, and the Software (or IT) domains. Some of the education in Mechatronics Systems design covers but one of these, whereas others deals with more:

Mono-domain
Where only one of the Mechanical, the Electronics, and the Software / IT) domains are taught.

Duo-domain
Where two of the domains are taught or used as basis for the systems design. This would then presumably be Mechanical/Electronics or Electronics/Software, as it is until now only very special incidents where Mechanical/Software combination is present, without electronics.

True M-E-SW
To teach a combination of all the three domains. This is a very demanding, if also very worthwhile, task. The difficulties at present lies in the apparent incompatibility of the design tools and procedures inherent in the different domains. This however, is a topic of further research.

2.8   Organizational frame

The teachings may focus on the individual, may focus on the design team, or may take into account the whole of the company (or even of the extended company):

Individual
The individual seen as the primary agent in the systems design process.

Project/team
The project team with its individual competencies (M, E, SW, ...) and with the different roles for project manager, specialists, designers, etc..

Company
The individual and the team as seen in the context of the company, with the team communicating with many different departments, teams, and individuals throughout the systems design process.

The Extended Company

The team and the systems design process seen in the light of 'the extended factory'; where the team may be communicating around the globe with other team members, or with transnational departments or sub suppliers in the company.

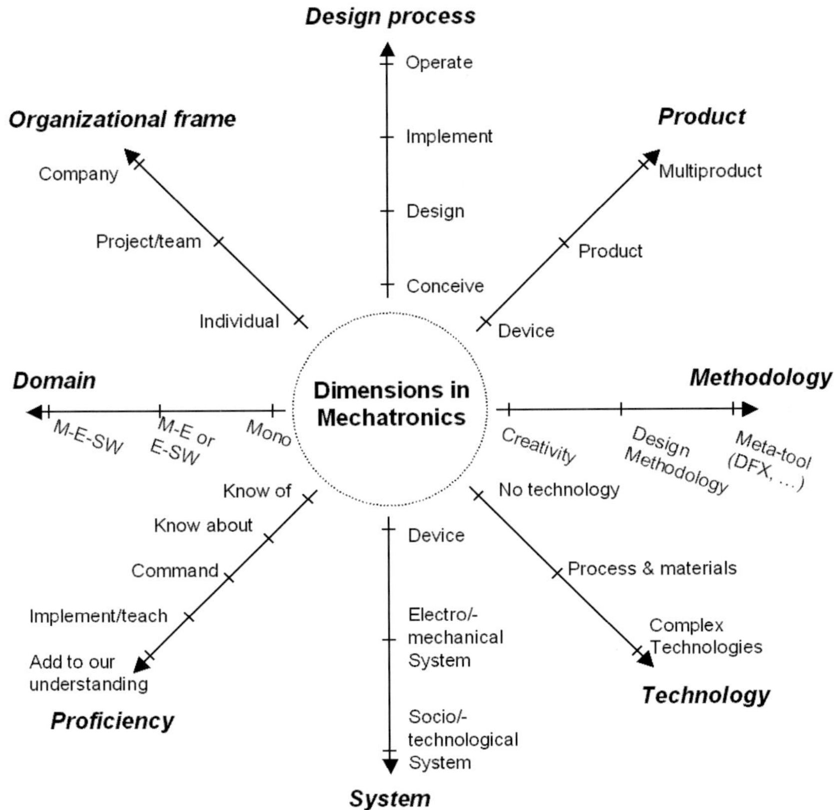

Figure 1. Eight of the central dimensions in Mechatronics with corresponding instances of their grading.

3. MAPPING MECHATRONICS SYSTEMS DESIGN

According to the dimensions listed above, a complex descriptive pattern for the 'space' of Mechatronics Systems Design is suggested (figure 1).

It is further suggested that these dimensions may be a useful framework for depicting any given course, or any given education in Mechatronics Systems Design.

At the Technical University of Denmark, a newly established M.Sc. engineering education in 'Design & innovation' [**5**] has a course (200 hours) in Mechatronics Design in its third semester. In figure 2 it is shown how this specific course maps out in the framework.

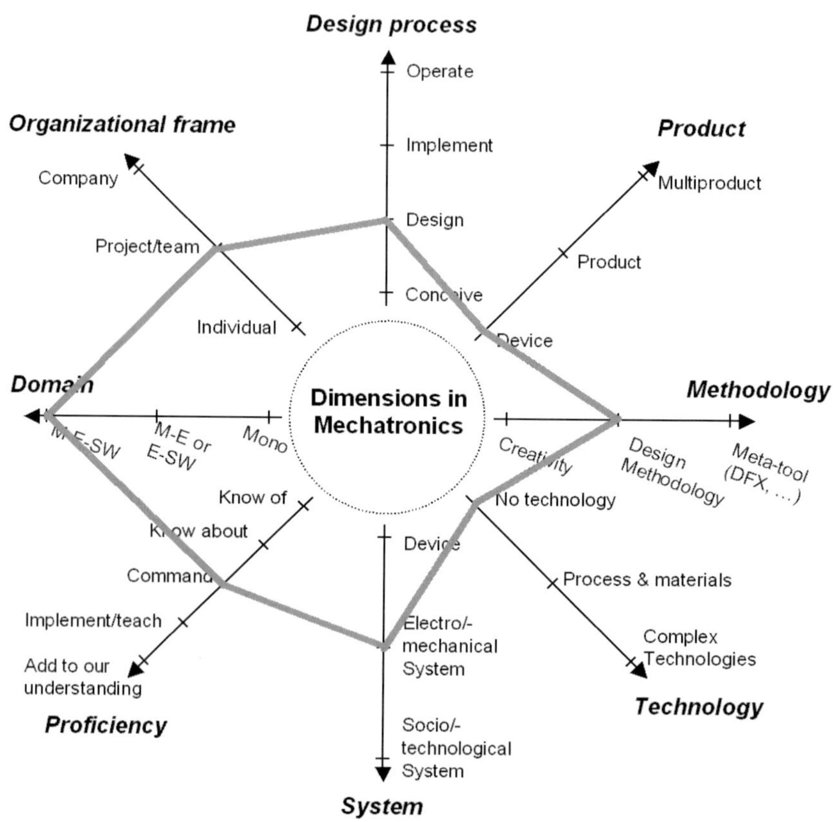

Figure 2. Example: The 200 hours course 'Mechatronics Design', in the third semester in the M.Sc. education 'Design & innovation' at the Technical University of Denmark.

4. CONCLUSION

A framework for the mapping of courses in Mechatronics System Design is suggested, the aim of which is to facilitate the planning and execution of single or multiple courses. It is then demonstrated how one of the current new courses at the technical University of Denmark 'Mechatronics Design' maps onto the framework.

The dimensions which constitutes the component parts of the framework, while fully operational for the indented purpose, are by no means fully explored as of yet. Future research by the group at our university will comprise:

- Further development and coordination of the design tools and procedures inherent in the different domains of M, E, and SW.
- Further development of understanding, tools, and procedures for treating the Mechatronics Systems in a socio/technological context.
- Further development of understanding, tools, and procedures for taking on multi product development of mechatronics systems.
- Further understanding of the implications in an educational context of the Conceive-Design-Implement-Operate mode.

REFERENCES

1. Hein L. "The Danish Approach to Mechatronics", *IEEE seminar 'Mechatronics in Scandinavia'*, 1996.
2. Berggren K-F, Brodeur D, Crawley E, Ingemarsson I, Litant W, Malmqvist J, Östlund S. "CDIO: An International Initiative for Reforming Engineering Education". *World Transactions on Engineering and Technology Education,* **Vol. 2 No. 1,** 2003.
3. Andreasen M, Hein L. *"Integrated Product Development"*. London – Berlin – New York, IFS/Springer, 1987
4. Tjalve E. *"Systematic Design of Industrial Products"*. Copenhagen, The Institute for Product Development (reprint), 2003.
5. Hein L. "Mechatronics at DTU – synergy in Research, Education, and Development", *3rd European Workshop on Education in Mechatronics,* Copenhagen, 2002.
6. Hein, L, Andreasen M M. "Integrated Product Development - a New Reference System for Methodical Design". *Serie WDK (Workshop Design-Konstruktion) 10, Proceedings of ICED'83,* **vol.1.** Zürich, Heurista 1983.
7. Cantamessa M, Duffy A, Hein L, and Rimmer D. "Design Co-ordination in New Product Development". *Proceedings from the International Conference on Engineering Design (ICED'99),* Munich, Germany, 24-26 August, 1999.
8. Fabricius F.: *"Design for Manufacture"*. EUREKA Famos, 1994

COLLABORATIVE AND REMOTE DESIGN OF MECHATRONIC PRODUCTS
Case studies based on student projects

Benoît Eynard[1] and Samuel Gomes[2]

[1] LASMIS, Troyes University of Technology, 12 rue Marie Curie, BP 2060, F.10010 TROYES CEDEX, France, Tel Tel: +33 3.25.71.58.28, Fax: +33 3.25.71.56.75 email: benoit.eynard@utt.fr

[2] SeT Laboratory, Belfort-Montbéliard University of Technology, rue du château, F.90010 BELFORT CEDEX, France, Tel: +33 3.84.58.30.06, Fax: +33 3.84.58.30.30, email: samuel.gomes@utbm.fr

Abstract: The paper presents collaborative design experiences in a multi-site and remote work organization.. These projects involve student-teams in mechatronic product design. Based on a web CSCW, the students have to experiment collaborative design approach with all its difficulties: project management without meeting face to face, the complexity of the data exchange as well as that of communication and sharing know-how between mechanical and electrical engineering. Finally, this work aims also to study, analyze and assess the benefits and the limits of such approach and computer support.

Key words: Collaborative Design, Mechatronic Product, Product Data Management, Web-Based System, Design Education.

1. INTRODUCTION

In today's challenging global market, companies must innovate to survive. Business innovation must occur in all dimensions—product, process, and organization— to improve competitiveness and business performance. To differentiate themselves, enterprises must capture, manage, and leverage their intellectual assets. This can best be accomplished through proper application such as Computer Supported Collaborative Work (CSCW) approach or Product Lifecycle Management (PLM) that addresses the needs of the extended enterprise [1]. This kind of strategic business approach helps companies achieve its business goals of reducing costs, improving quality, and shortening time to market, while innovating its products, services, and business operations. These new collaborative design methods and tools tend to introduce, very early in their design projects, various dimensions such as the technical, human, organizational, social, and economic points of view. Software tools supporting such collaborative engineering are required to help design team members in their tasks especially when they are remotely carried out.

This paper presents a collaborative design projects involving student-teams in a multi-site and remote work organization. These design projects are the result of collaboration between two French Universities of Technology. A CSCW, based on Web technologies [2] and linked to other Computer Aided X (CAD, FEM, etc), was used to help design team members to manage the project, product, process and activities (ergonomics) data. This tool, called ACSP for "Atelier Coopératif de Suivi de Projet" in French [3], was developed in order to support a product development process based on advanced approaches such as integrated product development [4], concurrent engineering [5] [6] or collaborative design [7]. This latter and the ACSP environment have been experimented on a design project for mechatronics products. Finally, this work aims to study, analyze and assess the benefits and the limits of collaborative design and based on CSCW in remote projects after 3 years of experience.

An overview of the ACSP environment is presented in section 2. Section 3 detail the mechatronics design project organization and issues. Finally, section 4 concludes and sets up further works.

2. ACSP: A CSCW FOR COLLABORATIVE DESIGN

ACSP is a web-based environment enabling the collaborative and remote activities of the design team members [2]. The ACSP user module is broken down into four main sub-modules managing data from the project, product, process and activity design domains [3].

Each of these design domains can be examined from several aspects (or models) in interaction, as defined in the previous approach. According to object oriented concepts [8] and system theory [9], we chose to develop three aspects in each design domain:
– a functional aspect, which describes the main objectives and goals of the system,
– a structural aspect, defining the system elements and architecture,

– a dynamic aspect, which describes the chronological behavior of the system.

In this approach, other design aspects such as physical or geometric models are directly linked to the structural aspect of the system. For example, applied to the product design domain, this kind of association generates functions commonly found in PDM (Product Data Management) or PLM (Product Lifecycle Management) systems [10].

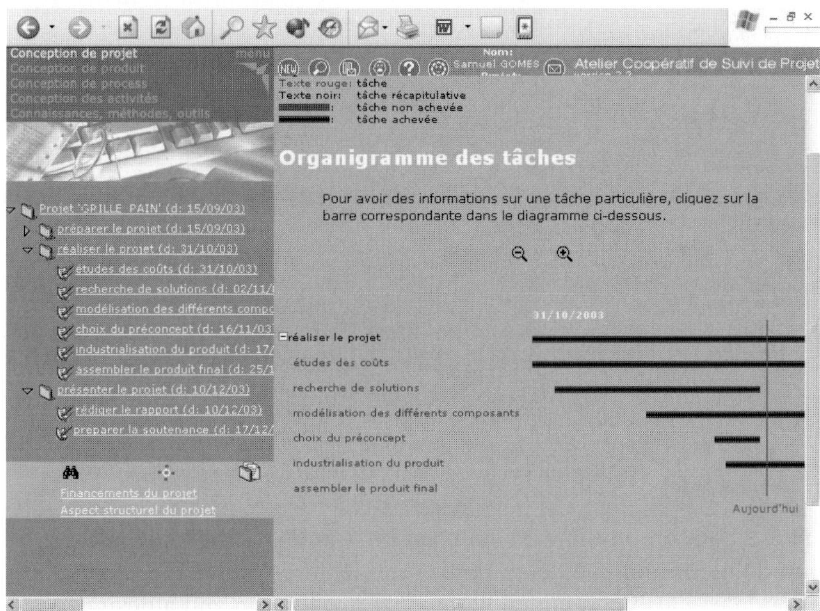

Figure 1 : ACSP display describing an example of project task planning integrated into the ACSP environment

For example, various types of data integrated into the ACSP environment can be displayed:

– project data, such as human and material resources (structural aspect) or tasks planning, as illustrated in Figure 1 (dynamic aspect),

– product data, such as product breakdown including the various product components (structural aspect) linked to CAD files (geometrical aspect) or functional specifications (functional aspect) available in different situations in the product's life cycle (Figure 3),

– process data, such as whole manufacturing process including the different machines (structural aspect) linked to CAD files (geometrical aspect) or production engineering specifications describing various manufacturing, maintenance, recycling, etc. tasks (dynamic aspects).

– activities data, such as various Human-Machine-Environment interactions in different life situations (structural aspect), multimedia documents describing dynamic sequences like video-recorded data from human work activities or virtual films (dynamic aspect).

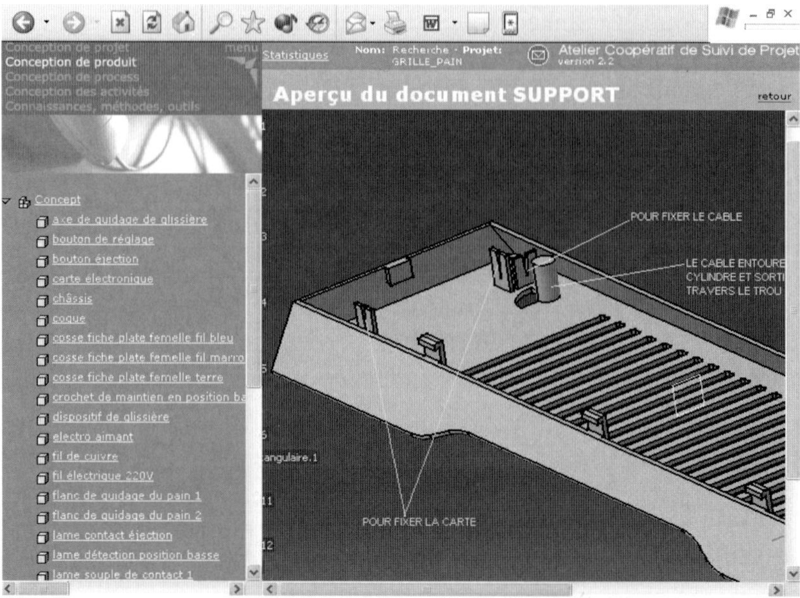

Figure 2 : ACSP display describing a geometric model linked to the product structure breakdown integrated into the ACSP environment

These data are completed with internal and external interactions in the design domains and even communication features (email, forums, etc.). The user module is also completed with an administration and a designer activity analyser module.

The administration module includes several features for managing projects, design contributors, specific company needs, etc., such as creating, modifying, deleting, storing and archiving data with the Data Base Management System included in ACSP.

The designer activity analyser module has been defined to carry out research works in contextual design process modeling within the field of concurrent engineering. These research works use the designer activity traceability when designing with this CSCW environment. Traceability analysis features, showing how designers are applying the proposed methodology, are available in the ACSP environment.

From a technical point of view, ACSP can be defined as an asynchronous CSCW based on a Data Base Management System (DBMS) connected to various Computer Aided X: CAD, CAM, FEM Solvers, etc. (Figure 3). ACSP is available as a Web Server with security layers managing user access [10]. The system has a client-server architecture available for heterogeneous environments (NT, Unix, Mac, etc.).

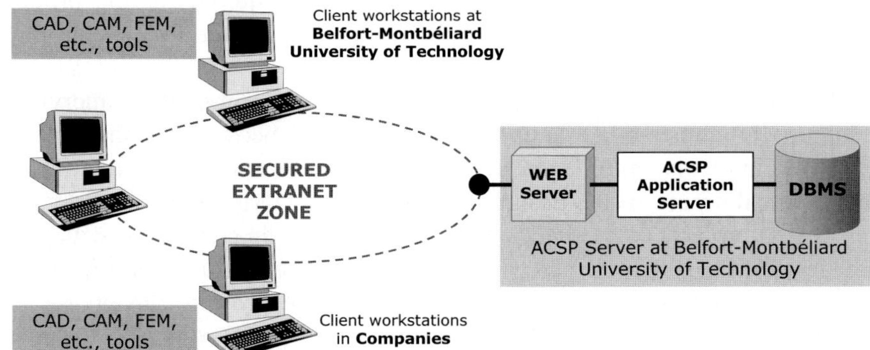

Figure 3 : ACSP client-server architecture connected to a database through a Web Server

3. MECHATRONICS DESIGN PROJECT ORGANIZATION AND ISSUES

As three years ago the Belfort-Montbéliard and Troyes Universities of Technology having already carried out an ACSP environment based projects in local site, our aim with the mechatronics design project was to assess the feasibility of carrying through a multi-site and remote organization. Supervisors decided on such a configuration in order to confront students with collaboration difficulties and project management without meeting face to face, etc. Then they had to carry out a real collaborative and remote design project. The objective was also to show them the difficulties linked to the complexity of the data exchange as well as that of sharing know-how. The students were confronted with the complexity of a concurrent and collaborative design project. Remote configuration reinforced this complexity. Last the subject dealing with mechatronics introduced another challenge that was the communication and how-know exchange between mechanical engineering and electrical engineering students.

3.1 MECHATRONIC PRODUCT DESIGN PROJECTS

Over the last decades the complexity of products increased, amongst others because of the application of electrical and software components. The increased product complexity also affects the number of people that are required to develop a product. Companies apply multidisciplinary teams to facilitate close co-operation and the exchange of preliminary information in the early stages of the design life cycle. These multidisciplinary teams consist of representatives from different phases of the product lifecycle (Marketing, Ergonomics, Engineering, Manufacturing, Maintaining, Recycling, etc.) and from different technologies required for the product (mechanical, electro-mechanical, chemistry of materials, electronics, software, etc.).

Regarding the large range of mechatronic products supervisors intentionally chose quite simple design subject because of the already existing complexity introduced by the chosen organization, considering the geographical spread of the design team members. During the two last years, five design projects of mechatronic products have been carried out with good and unsatisfactory issues: electric mower, toaster, electric drilling machine, electric juice extractor, and aero generator. These projects run on several phases of the product lifecycle (from functional specifications to technical and geometrical definition, including the manufacturing operations constraints), and they all apply a collaborative design framework:

- multidisciplinary teams (8-9 students in mechanical and electrical engineering, 2 supervisors per project), managed by a student project manager, divided on 2 engineering sites apart 300 km,
- 10 hours of engineering per week for each person, on a 4 months period,
- ACSP environment linked to various other software tools (CAD, FEM, etc.) that have helped to share, exchange, capitalize and re-use project, product, process and activity design data.

Two projects only are presented, on the basis of result obtained at the end of the project. These two projects consist in developing an electric juice extractor and a toaster while respecting the following constraints:

- adaptability with respectively various bread and citrus fruits sizes,
- have a good ergonomics (comfort, simplicity and effectiveness),
- reduce strongly the risks, in particular for the children, while considering thermal risks for the toaster and mechanical risks for the juice extractor,
- have a new and original style,
- have a reliability comparable with the other commercial products,
- reduce the costs to -15% compared to competition,

Lastly, the products will have to fulfil the standards, in particular, the food standards.

3.2   ISSUES OF THE DESIGN EXPERIENCE

At the end of the project period, students present the results of the product development process integrating the different points of view: Marketing, Ergonomics, Engineering, Manufacturing, Maintaining, Recycling, etc.

The following Table 1 gives a summary of the number of data recorded in the ACSP environment at the end of 2 projects. These data are classified according to the 3 previous design fields (Project, Product, Process) and on the communication aspect. These first results show that even if the data projects are rather close (number of project members, tasks and project documents), the number of data created in the Product and Process design domain is twice more important in the Toaster Project.

		TOASTER PROJECT	JUICE EXTRACTOR PROJECT
Project domain	Project members	10	11
	Project tasks	32	32
	Project documents	36	22
	Total data for Project domain	**78**	**65**
Product domain	Product and part data	300	94
	Functional data	100	91
	Product and Part documents	118	83
	Total data for Product domain	**518**	**268**
Process domain	Process data	23	5
	Process documents	18	0
	Total data for Process domain	**41**	**5**
Communication	Forum messages	150	62

Table 1 : Number of data available in the ACSP environment at the end of the 2 projects

This table shows an intensity of action and of formal communication (150 messages compared to the 62 forum messages recorded in the ACSP database) higher in the toaster project than in the juice extractor project.

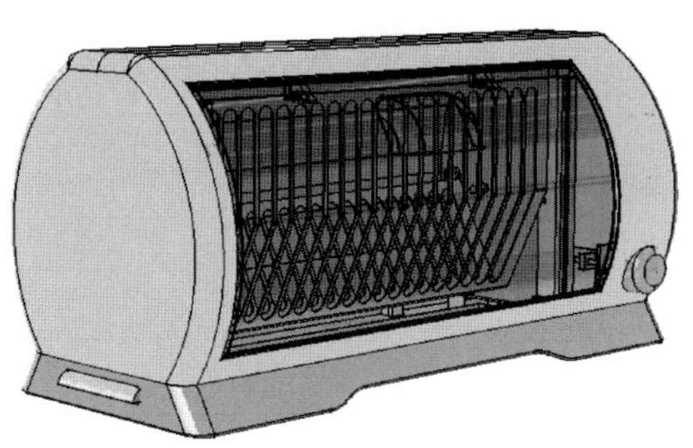

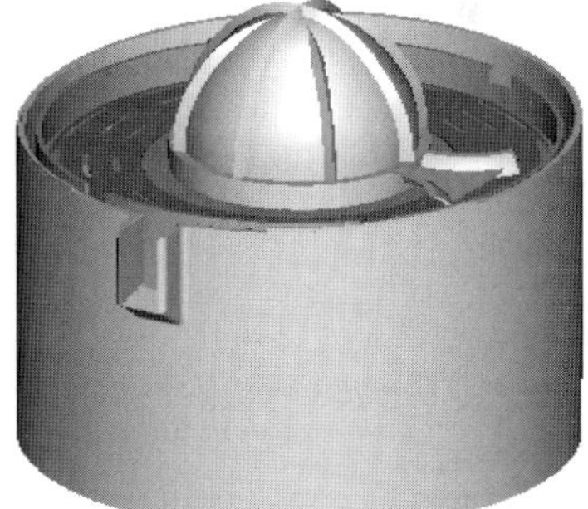

NEW TOASTER CONCEPT

ELECTRIC JUICE EXTRACTOR CONCEPT

Figure 4 : Geometric models of the concepts presented by the students at the end of the projects

If the final designed products are considered (Figure 4), the first project team has designed a really new concept of toaster with a full correspondence

to the initial constraints detailed previously. This concept integrates 33 different parts:
- 19 mechanical parts (glass door, grids, crumbs collector, etc.),
- 10 electro-mechanical parts (electromagnet, connectors, etc.),
- 4 electronic parts : (quartz tubes, potentiometer, electronic card).

The second project team has designed a quite classical concept of juice extractor with a weak correspondence to the initial constraints. This concept integrates 22 different parts:
- 16 mechanical parts (bowl, gears, axes, screws, etc.)
- 6 electro-mechanical parts (electric motor, pushbuttons, etc).

In order, to manage research activities in collaborative design process modeling, ACSP environment is used as an experimental research tool to analyze designer activity in the above mentioned design projects. The first quantitative and qualitative results from the ACSP designer activity analyser module relative to these projects are now available. For example, Figure 5 confirms the difference of work intensity between the 2 projects. Insofar as, for an equivalent number of connections, the collection of information and the actions carried out are 2 times more important from one project to another.

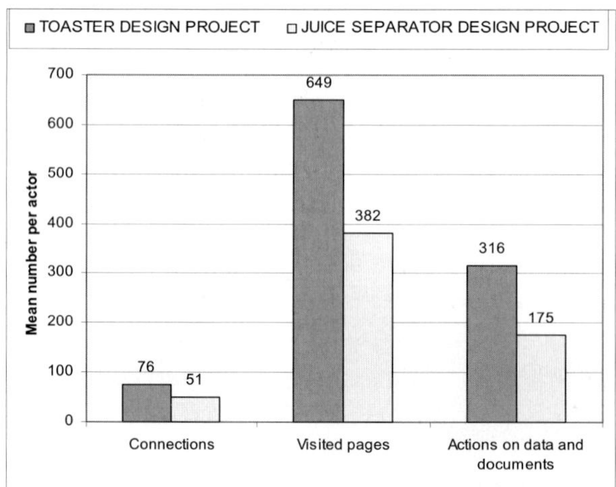

Figure 5 : Comparative assessment of number of connections, of visited pages (information gathering) and of action on design data (create, modify and delete)

As illustrated in Figure 6, it is possible to analyze the time evolution of the mean rate of red pages per connection (information gathering) compared to the rate of action (create, modify and delete / project, product and process data) for the 2 studied projects.

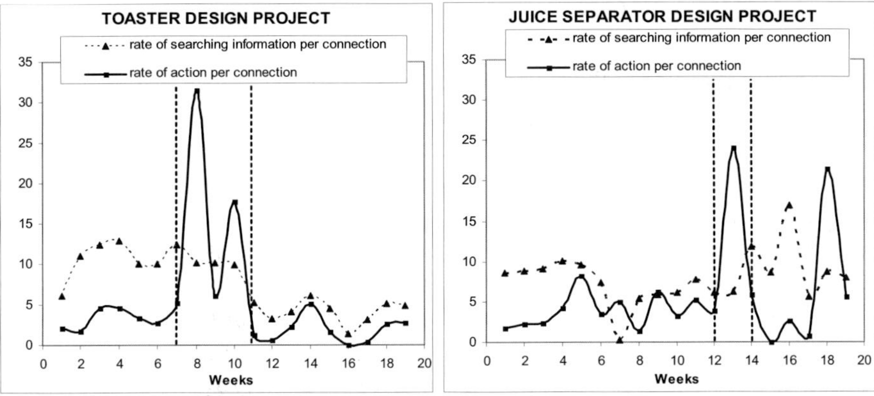

Figure 6 : Time evolution of the mean rate of visited pages per connection (information gathering) compared to the mean rate of action per connection (create, modify and delete / project, product and process data) for the 2 studied projects

Through these 2 diagrams, it is possible to identify three main phases during the project development cycle:

— a familiarization phase, in which design team members discover at the same time the purpose of the project and the detailed ACSP functionalities. This step can be considered as an observation step where designers collect more information than act on the design data (create, modify, delete data). This phase runs from week 1 to week 6 for the toaster design project, and from week 1 to week 11 for the juice extractor design project,

— a capitalization phase, in which design team members create and modify a high number of design data per connection on their project. This phase can be considered as a period of intensive activity where designers contribute to filling the ACSP database. In our case, this phase runs from week 7 to week 11 for the toaster design project, and from week 12 to week 14 for the juice extractor design project,

— a regulation phase, in which design team members manage the design data capitalized previously. They make some adjustments to data, integrating the interactions between the design domains to achieve the final goal of the project. During this phase, designers usually connect to gather information and perform less actions on design data than in the previous capitalization step. This last phase runs from week 12 to week 19 for the toaster design project, and from week 15 to week 19 for the juice extractor design project, with an activity peak during the last 2 weeks.

4.   CONCLUSION

In this paper, an experience of collaborative design projects has been presented. These projects involve student-teams in a multi-site and remote work organization, using a web-based CSCW. The results obtained at the end of the mechatronic design project are not uniform. Concerning the first analyzed design project (toaster), all seems well to have been held, with a great integration of the different technologies (mechanical, electro-mechanical, electronics, etc.). On the other hand, the juice extractor design team members have some organizational problems during their project, insofar as, they began the effective work very late in the project. They carried out late changes in their product architecture and in the project due to the lack of coordination and communication at the early steps of the project. These late changes result in delays in the project which are also more expensive than early changes [1]. As it is well known, a collaborative design requires frequent face-to-face and bilateral communication of preliminary information instead of late release of complete information (which is the case in a sequential process). After discussion with the concerned students, the conclusion is the juice extractor team project has applied a none optimized sequential development process.

REFERENCES

1.   Helms RW. *Product Data Management as enabler for concurrent engineering – Controlling the flow of preliminary information in product development*, PhD Thesis of Technical University of Eindhoven, 2002.
2.   Zhang. S, Shen W., Gheniwa H. "A review of Internet-based product information sharing and visualization", *Computers in Industry*, 2004, in press.
3.   Gomes S, Sagot JC. "A concurrent engineering experience based on a cooperative and object oriented design methodology", *Integrated Design and Manufacturing in Mechanical Engineering*, Dordrecht, Kluwer Academics Publisher, pp. 11-18, 2002.
4.   Andreasen MM, Hein L. *Integrated product development*, London, Springer-Verlag, 1987.
5.   Prasad B. *Concurrent engineering fundamentals* - **vol. 1**, Englewood Cliffs, Prentice-Hall, 1996.
6.   Sohlenius G. "Concurrent engineering", *Annals of the CIRP*, **vol. 41**, pp. 645-655, 1992.
7.   Shen W. "Editorial of special issue on knowledge sharing in collaborative design environment", *Computers in Industry*, **vol. 52**, pp. 1-3, 2003
8.   Graham I. *Object oriented methods*, Addison-Wesley Publishing Company, 1994.
9.   Le Moigne JL. La théorie du Système Général - théorie de la modélisation, Paris, PUF, 1977.
10. Liu DT, Xu XW. "A review of web-based product data management systems", *Computers in Industry*, **vol. 44**, pp. 251-262, 2001.

COMPARISONS OF CHINA'S EDUCATION IN SCIENCE AND TECHNOLOGY TO THE DEVELOPED WORLD

Xincai Tan and Xiu-Tian Yan
CAD Centre, DMEM, University of Strathclyde, 75 Montrose Street, Glasgow, G1 1XJ, UK.
Email: xincai.tan@strath.ac.uk;

Abstract: Education in science and technology (EST) has played a critical role in state construction and economic development in China. This paper presents the facts of education in China and it aims to establish a correlation between the reported economic development and education progress. Through statistical approach and associated techniques it gives an insight to the effect of EST on economical development based on case studies. Comparisons of relevant data have been carried out between China and globe. It is believed that China will make greater efforts to develop EST to match the WTO's challenges.

Key words: China, economic development, education, science and technology.

1.   INTRODUCTION

The People's Republic of China (China) plays more and more important role in the world economy since she became the 143rd member of the World Trade Organisation (WTO) [1]. Splendid achievements in science and technology in China have been achieved nurturing of the state leaders, some world-class scientists and engineers, and promoting the state's economy. Education in science and technology (EST) has a very significant influence on construction and modernization of China. There have been a number of investigations into education [2] and economic development [3,4] in China. Few reports are found dealing with the effects of EST on the national administration and development. Thus this paper presents comparisons of China's educational facts with international data; and discusses the effects of EST on the state leaders, export and import, as well as industrial productions; and then considers WTO's challenges to China's EST; and finally concludes with some suggestions for China's EST.

2.   CHINA'S EDUCATION

Fig.1 shows a comparison of educational data between China and the lower-middle-income group (LMIG) [5] of the world in the year 2002, as an octagon with eight main factors. For illiteracy, percentage of population age 15 and over in China is about 14%, while the average percent of the LMIG 13%. Urban population in China is relatively lower than the average of the LMIG, that is, percentage of urban population to total population in China is 38%, whereas that of the LMIG is 49%. Trained teachers in primary schools in China are up to 97.39% and the average value for the LMIG is 94.6%. Primary teacher-pupil ratio in China is the same as for the LMIG, 1:21. Public expenditure per student for tertiary education in China is much higher, although those for primary and secondary are relatively lower.

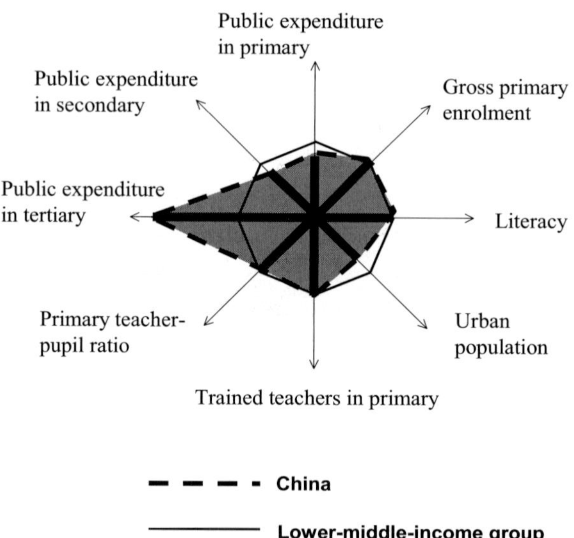

Fig.1 A comparison of China's education with the Lower-middle-income group in the world for year 2002. Data sources in [5].

In China, public expenditures per student for primary, secondary and tertiary are 6.1%, 12.1% and 85.8% of Gross Domestic Product (GDP) per capita, respectively; and those corresponding average values for the LMIG are 7.6%, 16.1% and 40.1%. The gross primary enrolment ratio in China (106.4%) is somewhat higher than that for the LMIG (102.6%), where both values are greater than 100% as a result of grade repetition and entry at ages younger or older than the typical age at that grade level.

Similar to western countries, such as the UK [6], EST starts with primary and secondary schools. China's educational system has been institutionalised as a comprehensively complete educational system comprising:

(a) **Preschool education**. Mainly in kindergartens, before the age of six.

(b) **Primary**. 6 years of schooling. During this period, pupils have to take fundamental mathematics and simple science and technology in addition to other curriculum such as Chinese, social science.

(c) **Secondary**. Divided into two parts: 3 years for junior, and 3 years for senior. Students of junior secondary begin to learn science curriculum including mathematics, chemistry, physics, biology etc.

(d) **University**. 4 or 5 years for Bachelor degrees, and 2-3 years for non-degree specialised courses. Issued by the Ministry of Education, the nationwide unified university-entrance examinations fall into two categories: the humanities, and the science and technology.

(e) **Postgraduate education**. Divided into two parts: 2.5 – 3.5 year for Master degrees, and 3 – 5 years for Doctorate degrees. After completing a first degree, a graduate may apply for further education for a Master degree. Although doctorate degrees are research degrees similar to western countries, the candidates who wish to take the studies usually have to complete their Master degree studies and pass the entrance examinations. The nationwide entrance examination system has been adopted for the postgraduate education, including Master training and Doctorate training programmes.

(f) **Post-doctoral**. 3 – 5 years. This is a job-based training programme, specially designed for recent Ph.D. graduates to focus on a specialized area mentored by experts approved by the Ministry of Education. Special funding is available to support a post-doctoral candidate to set up a research project.

China has the biggest educational system in the world in terms of the numbers of both schools and students. From 2002, official data [7] shows that in mainland China there were 1.17 million schools, 318.79 million students, and 15.81 million members of educational staff. The population involved in educational field was 26.2% of total population of the nation.

A comparison of estimated absolute number (in thousands) (AN) and number of students per 1,000 population (NSPTP) of school enrolment for China with the world, Europe and Northern America in 1996 can be seen in Table 1. The NSPTP for China are quite lower than Europe and Northern

America for various levels. For example, the numbers of students enrolled to tertiary education were 29.59 and 53.57 per 1,000 population for Europe and Northern America, respectively. That for China, however, only 0.79 per 1,000 population.

Table1 A comparison of estimated absolute number (in thousands) (AN) and number of students per 1,000 population (NSPTP) of school enrolment for China with the world, Europe and Northern America in 1996. Data source in [8,9].

		Population, all ages	Enrolment			
			All levels	Primary	Secondary	Tertiary
World total	AN	5,767,443	1130,667	659,106	386,386	85,175
	NSPTP		196.04	114.28	66.99	14.77
Europe	AN	728,561	137,522	46,866	69,103	21,554
	NSPTP		188.76	64.33	94.85	29.58
Northern America	AN	299,250	67,122	26,736	24,355	16,031
	NSPTP		224.30	89.34	81.39	53.57
China	AN	1,223,890	50,103	25,295	23,842	966
	NSPTP		40.94	20.67	19.48	0.79

For China's higher education, students in EST are always more than those studying social science. A comparison of percentages of bachelor's degrees awarded in science and technology for China and some selected countries and selected years is shown in Table 2. Percentages of all EST degrees and only engineering degrees in China were in the highest rank, 64.05% and 40.66%, respectively; followed by Korea (36.7 % and 25.9%), Germany (33.52% and 21.38%) and Japan (23.50% and 21.00%).

Table 2. A comparison of percent of bachelor's degrees awarded in science and technology for China and some selected countries and selected years. Data source in [8,9].

Country	China	Canada	Germany	Japan	Korea, Republic of	Nether-lands	UK	US
Year	2000	1999	1999	1990	1999	1999	1999	1999
All EST degrees, %	64.05	19.68	33.52	23.50	36.66	16.47	28.93	17.43
Engineering, %	40.66	7.47	21.38	21.00	25.88	11.56	12.72	6.92

Fig.2 shows the numbers of students for higher education in China between 1978 – 2001 at home and abroad. The numbers of students in China both graduated from and enrolled to Universities for degrees' education are shown in Fig.2a. Since resumption of examinations for various levels of education in 1978, the numbers of students both graduated and enrolled have totally been increased yearly. During the period of 1949 to 1965, however, China adopted former Soviet model of higher education, and there were some students enrolled to and graduated from Universities, but education endured dilemmas due to oscillating policies of Chinese Government. Whereas between 1966 and 1976, various types of vocational-technical schools and universities had stopped educational performance.

In China, students in EST have mostly studied in native China. In the mean time, the Chinese Government has regularly sent selected students and scholars abroad for university and advanced studies. Most of these students have been in EST. The total number of students sent abroad between 1950 – 1963 was 9594 [11]. Of these students, 8357 students were sent to the former Soviet Union, 273 to the German Democratic Republic, 238 to Czechoslovakia, 160 to Poland, 88 to Hungry, 75 to Romania, 68 to Bulgaria, and 17 to other European and Asian countries.

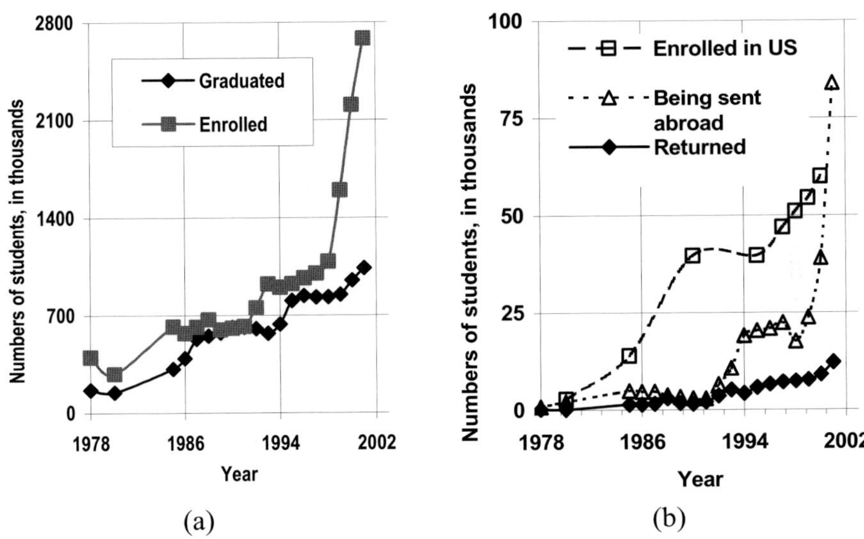

Fig.2 The numbers of students for higher education in China between 1978 – 2001 at home and abroad. Data sources in [8, 9].
(a)   Both graduated from and enrolled to Universities for degrees' education;
(b)   (i) Students from China enrolled in U.S, (ii) students sent by Chinese Government to study abroad, and (iii) official educated returnees to China.

Since "Reform and Opening Up" in China in 1978, more and more students have been sent to the developed countries, including United States, Japan and west European countries for graduate studies. Official numbers [9] of students for both being sent abroad and returned between 1978 – 2001 are summarized in Fig.2b. In general, the numbers of such students increased yearly, and have significantly increased since 1999. The number of students sent by the Government to study abroad was increased to 12,243 in the year of 2001. At the same time, self-fund students studying abroad have also followed the rend and increased since 1978. For example, students enrolled in U.S. alone between 1978 – 2000 were more than those sent by the Government.

3.   EFFECTS OF EST

EST plays a critical role in the rapid development of state economy and construction of a modern China. This has been reflected by the high numbers of prominent figures appearing in the who's who. For example [12], a former President of PR China, Jiang Zemin, was majored in electrical engineering; and Yuan Longping known as "father of hybrid rice" was specialized in agriculture.

Fig.3 shows effects of EST on state administration and industrial production. Currently, most of the Chinese state leaders, administrators and provincial governors are of EST background, as shown in Fig.3a. Among 25 members of the Political Bureau for the Sixteenth National Congress of the Communist Party of China (CPC), there are 18 members with a background of EST. Ninety percent of the State Councillors including Premier and Vice Premiers had EST. Of Chairmen and Vice Chairpersons for both the Tenth National People's Congress (NPC) and the Tenth Chinese People's Political Consultative Conference (CPPCC) (members of 16 and 25 respectively), 69% and 56% had EST, respectively. The ministers and the governors of provinces with background of EST are 58% and 52%.

The General Secretary of the CPC and President of PR China, Hu Jintao, the Chairmen of both the NPC and the CPPCC, Wu Bangguo and Jia Qinglin, and Premier Wen Jiabao, were all educated in science and technology. Due to their educational background, the current state leaders are all very keen and able to develop industries and state economy. This is a stunning contrast to Mao Zedong's policy of political struggles by launching of the disaster Great Proletariat Cultural Revolution, who had no education in Science and Technology.

As an example of key economic indicators, Fig.3b shows the output of motor vehicles between 1971 and 2001, and crude steel production between 1978 and 2003 in PR China. The motor vehicle manufacture and crude steel production have increased exponentially since 1980s when skilled workers with EST had been trained out. As shown in Fig.2, since 1980s, number of students graduated from and enrolled to universities has increased steadily. It is these graduated who have been the most important labour force to promote industrial productions.

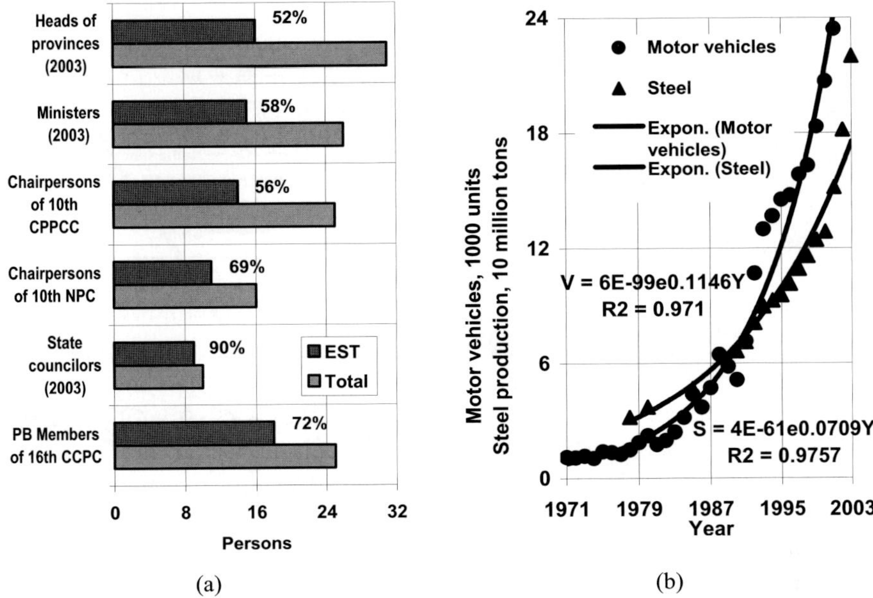

Fig.3 Effects of EST on state administration and industrial production.
(a) Most of the state administrators having background of EST. Data source in [12, 13].
(b) Output of motor vehicles between 1971 and 2001, and crude steel production between 1978 and 2003 in PR China. Data sources in [9, 14, 15].

China's industries and economy would be impossible to rise so rapid in exponent without those workers trained in EST. The development of economy in China does significantly depend on EST. Fig.4 shows export and import of PR China between 1952 – 2003. The tendencies of export and import plotted in Fig.4a are very similar to the industrial production in Fig.3b. It is that both export and import increase significantly after 1980s when a number of skilled workers with well trained in EST have been working for the state construction. By comparison of selected eight exporters in the world between 1980 – 2002, as shown in Fig.4b, China's export rate increases steadily during the period, but its magnitude has been quite low.

The scientists and engineers in China who once obtained EST abroad have made their great contributions to the state construction and development, as well as education. Some of them have achieved notable achievements. Fig.5 shows percentages of the academicians with EST at various counties for both (a) the Chinese Academy of Science (CAS) since 1955 (1026 members in total) and (b) the Chinese Academy of Engineering (CAE) since 1994 (700 members in total).

                                                          Xincai Tan and Xiu-Tian Yan

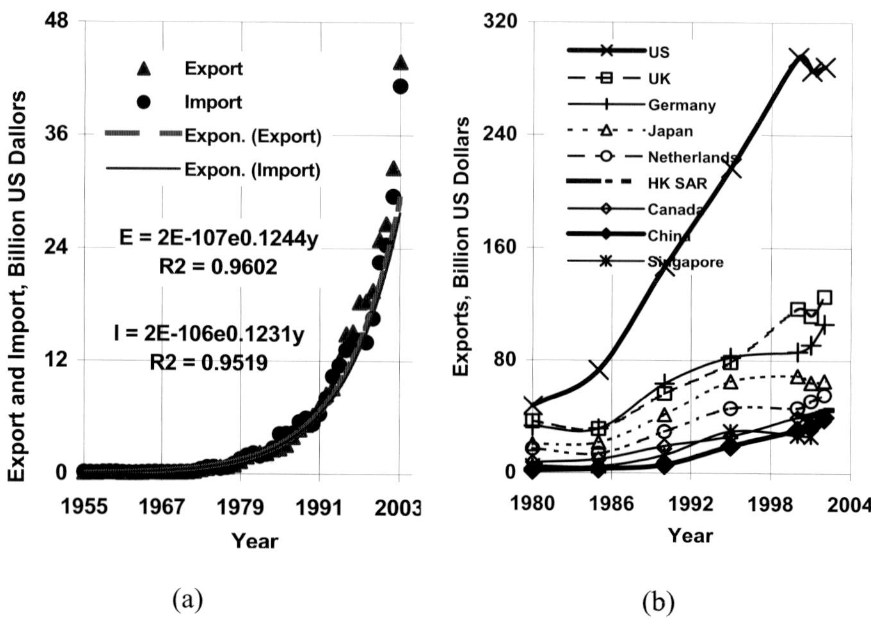

Fig.4 (a) Export and import of PR China between 1952 – 2003; and (b) a comparison of China's export 1980 - 2002 with 8 major export countries/regions in the world, the United States, the United Kingdom, Germany, Japan, the Netherlands, Hong Kong (China, SAR), Canada and Singapore. Data sources in [16-19].

These academicians are considered as the top scientists and engineers in their fields in the country. Among the members of CAS in Fig.5a, more than a half are of their EST in native China, about a quarter of members have studied/research-experienced in the United States. The number of the foreign countries for CAS members obtained EST are up to 19 countries. Among the members of CAE in Fig.5b, more than 60% members are of EST in native China, 13.71% of the members are of their EST in the former Soviet Union, 12.43% in the United States. The studied-in countries were 17. Needless to say, modern science and technology in China have been combined with the world's science and technology, especially of the United States, the former Soviet Union and European countries.

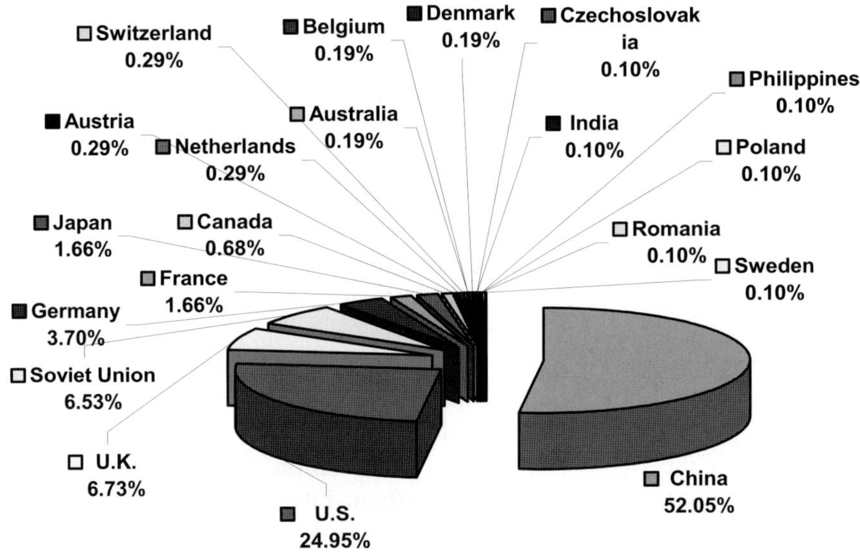

(a) Percent of CAS members once-studied in various countries

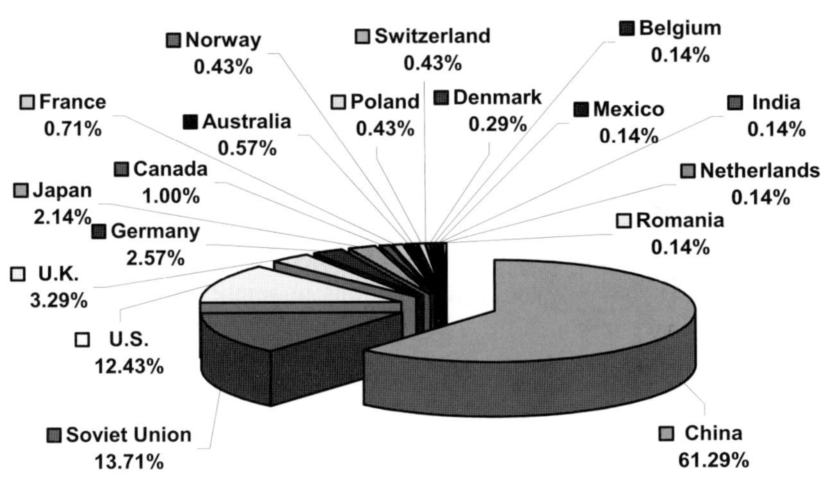

(b) Percent of CAE members once-studied in various countries

Fig.5 Percents of Members with EST/research experience in various countries for (a) CAS between 1949-2003, total members 1026; and (b) CAE between 1994 – 2003, total members 700. Data sources in [21, 22].

4. WTO'S CHALLENGE

For China in the WTO, the twenty-first century will be full of opportunities and challenges. Scientific and technological advances will not only change our lifestyles, but also the methods by which we produce goods. With its 1.3 billion consumers, China will offer new market opportunities in sectors such as agriculture, industry, telecommunications, insurance and finance. The present level of technical skills will not be sufficient to sustain China's future competitiveness as the industrial sector adopts advanced modern technologies, and as technologies themselves become more skill-intensive.

In order to increase China's overall national strength and international competitiveness pronouncedly, China needs workers with well EST at home and abroad to manoeuvre advanced equipment, and to manage state and companies. In the fierce international competition, China should not only retain its comparative advantages in labour-intensive industries, but also develop technology-intensive industries. China's state enterprises need a great number of workers educated in science and technology to innovate the current "out-of-date" equipment and technologies, and to develop and employ advanced equipment and technologies.

5. CONCLUSIONS

In China, people educated in science and technology have played their important role in state administration, economic development and industrial production. Modern science and technology in China have been combined international science and technology. It has been being useful for China to learn from and exchange with various nations in science and technology to develop and manage national industries and state economy.

As a WTO's member, China is getting successful in economy progress. China's modernization in industrial, science and technology require more and more skilled workers in EST, thus we suggest in China:

(a)   Education inputs of primary and secondary education relatively lower than LMIG, more inputs should input later on.

(b)   Tertiary enrolment was quite low in number of students per thousand population, therefore college/university education in size should be expanded.

Acknowledgement: This paper has been produced with the financial assistance of the European Community. The views expressed herein are the authors of University of Strathclyde and can therefore in no way be taken to reflect the official opinion of the European Commission.

REFERENCES

1. Magarinos, CA, Yongtu, L and Sercovich, FC (Editors). "China in the WTO: The Birth of a New Catching-Up Strategy", *United Nations Industrial Development Organization*, Palgrave Macmillan, New York, 2002.

2. Tsang, MC Education and national development in China since 1949: "Oscillating policies and enduring dilemmas", *China Review 2000* (Editors: L. Chung-ming and J. Shen), the Chinese University Press, pp.1-26, 2000.

3. Hu, A and Hu, L and Chang, Z. "China's economic growth and poverty reduction", on: http://www.imf.org/external/np/apd/seminars/2003/newdelhi/angang.pdf. 2003.

4. Wong, J and Ding, L. *China's economy into the new century*, World Scientific Publishing, Singapore, 2002.

5. The World Bank Group. Data & statistics, on: http://www.worldbank.org/data/.

6. Department for Education and Skills (UK). National Curriculum, on: http://www.nc.uk.net/home.html. 2004.

7. Ministry of Education of PR China Statistics of education, 27 February, on: http://www.moe.edu.cn/stat/tjgongbao/report_2002.doc, 2003.

8. National Center for Education Statistics (USA) (2004). Digest of education statistics, on: http://nces.ed.gov/programs/digest/

9. National Bureau of Statistics of China. Statistical Communiqué 2002, 28 February, on: http://www.stats.gov.cn/tjsj/ndsj/index.htm. 2003.

10. World Bank Group. China: World Bank approves two new projects for China, September, http://www1.worldbank.org/education/news.asp. 2003.

11. Song, J. Generations of studies abroad for hundreds years, Science and Technology Daily (in Chinese), 12th September 2003.

12. People's Daily, http://english.peopledaily.com.cn/

13. PRC Government Homepage, http://www1.cei.gov.cn/govinfo/english/default1e.shtml.

14. International Iron and Steel Institute. World Steel in Figures (2003 edition), http://www.worldsteel.org/

15. Interfax, http://www.interfax.com/com?id=5684105&item=China.

16. Ministry of Commerce of the People's Republic of China on: http://www.moftec.gov.cn/jinchukou.shtml, 2004.

17. MacDougall, C. Policy Changes China's Foreign Trade Since The Death of Mao, 1976-80, *China's New Development Strategy*, (Editors: J. Gray and G. White), Academic Press, London. (1982).

18. Denny, DL. "Recent Developments in the International Financial Policies of the People's Republic of China", *China's Changing Role in the World Economy* (Editors: B G Garth et al.), Praeger Publishers, New York, pp163-186. 1975.

19. UNCTAD, *Handbook of Statistics* 2003. http://stats.unctad.org/restricted/eng/ReportFolders/Rfview/Explorerp.asp.

20. Stavis, B and Gang, Y "Babcock and Wilcox Beijing Company Ltd", *China Business Review*, July-August, 10-12, 1988.

21. Chinese Academy of Science , 2004. http://www.cas.ac.cn/.

CASE STUDY OF COLLABORATIVE KNOWLEDGE MANAGEMENT IN THE EARLY PHASES OF PRODUCT DEVELOPMENT

James Gao[1], Hayder Aziz[1], Paul Maropoulos[2] and Wai M.Cheung[2]
*1 Enterprise Integration, Cranfield University, UK. *j.gao@cranfield.ac.uk*
2 School of Engineering, University of Durham, UK.

Abstract :   This paper reports on a case study of collaborative product development and knowledge management platforms for Small to Medium Enterprises. It has been recognised that current Product Lifecycle Management (PLM) implementations are document oriented, have a non-customisable data model and inter enterprise integration difficulties. To overcome these, an ontological knowledge management methodology using semantic web standards was added to a PLM and an open-source alternative. The client-server shortcomings have been overcome using a de-centralised architecture. This is implementable at low cost, the scale and accessibility of the system increases in line with user numbers.

1.   INTRODUCTION

Tools such as product data management (PDM) and its offspring product lifecycle management (PLM) enable collaboration within and between enterprises [1]. Large enterprises have invariably been the target of software vendors for development of such tools, resulting in large centralised applications. These are beyond the means of small to medium enterprises (SME). Even after these efforts had been made, large enterprises face numerous difficulties with PLM [2]. Firstly, enterprises evolve, and an evolving enterprise needs an evolving data management system. With large applications, such configuration changes have to be made at the server level by dedicated staff. The second problem arises when enterprises wish to collaborate with a large number of suppliers and original equipment manufacturer (OEM) customers. Current applications enable collaboration using business-to-business (B2B) protocols. However, these do not take into account that disparate enterprises do not have unitary data models or workflows. This is a strong factor in reducing the abilities of large enterprises to participate in collaborative projects.

The enterprises' prime asset, its knowledge, is not managed coherently thus perpetuating an 'invisible' limit on the company's knowledge, based on the impulse of knowledge workers' recollections of previous experience. In addition, the problems for inter-enterprise collaboration are also inherent when enterprises with different level of detail and different nomenclature collaborate. SME's have up to now been left out of developments in PDM/PLM, and knowledge management in spite of forming the majority of the world's engineering community. This negatively affects the ability of SME's to manage their knowledge, reuse existing expertise, and collaborate with other SME's and larger enterprises. This weakness affects not only the SME's but also the large enterprises that SME's work with. Smaller engineering enterprises lack the infrastructure and manpower for complex solutions.

This study has focused on the experiences of automotive and discrete machining companies in integrating their product realisation cycles with their supply chains and their customers in real-time, and also to enable the companies to make rapid appraisal and cost-estimation for their customers, from simple concept designs. The system has to be able to reuse the company's existing base of knowledge and to push the manufacturing knowledge higher up into the design chain to reduce the need for costly and time consuming reworks and engineering changes.

Based on the above set of problems facing industry, the authors have drawn up three core requirements to codify the needs of SME's in managing their product and process information i.e.

(I)   Enabling project managers and all knowledge workers to have access to the functionality and to create and manage knowledge within their domain according to the agreed nomenclature and ontological representation;

(II)   Information created has to be in a form that can be queried, reused and transformed into new representations through the use of rules and agents; and

(III)   Enabling the real-time collaboration between SME's, and larger partners, by facilitating the fast and cost-less construction of VE's;

To test the requirements and try to meet the objectives the authors took an evolutionary approach by assessing the current technology and methods, and then constructing three new example applications. The as-is technology used was a web-based commercial PLM system. The three test scenarios constructed are: (i) Modification of the PLM system for flexible and customisable data model; (ii) Implementation of a functionally equivalent system based on open source tools and (iii) The implementation of a peer-to-peer (P2P) based system.

Lihui et al [3] wrote an extensive survey of collaborative design systems, and highlighted eight areas as having scope for development including: System architecture for web-based collaborative design, Collaborative Conceptual Design Modelling and Data Sharing, Conceptual Design Selection, Knowledge Management in Collaborative Environments. Beckett [4] discussed the topic of communication and understanding in VE's between unfamiliar participants. He discussed the best tools and standards to apply in VE settings for collaboration and knowledge management and has an overview of various methodologies and applications. Camarinha-Matos [5] on a similar note, reviewed current trends in VE developments, and conclude that there is a need to develop a generalised framework for VE's, to enable harmonisation, international collaboration and rapid deployment.

Kim [6] developed a 'Distributed open-intelligent PDM' system, which adopts ISO standard STEP, whilst offering standard PDM functions. A dynamic and flexible workflow model is implemented. This could greatly enhance the flexibility of the system. Goh [7] proposed the STEP workflow management facility as a rule based workflow for PDM that is compliant with the Workflow Management Coalition (WfMC) guidelines. As part of the proposal, an object oriented data model driven system to store STEP entities is described. Zha [8] proposed a STEP based application to manage the entire product lifecycle. The information that is not already defined in STEP is modelled in EXPRESS. The system is focused primarily on assembly mating features and does not consider the machining requirements of each component. Zhou[9] tried to solve the problems of functional design knowledge management within a platform neutral setting through the definition of STEP models with the addition of semantics and the use of Artificial Neural Networks to aid the selection process. The authors' implementation used express data models and is accessible through a web portal. Vasara [10] proposes ARACHNE, the adaptive network strategy to enable integration between 87 enterprises, the authors highlight the benefits of peer to peer networks to achieve synergy between collaborators. Their developed methodology called RosettaStone enabled many-to-many integration between enterprises using a three-tier architecture.

To overcome the above mentioned problems, a new methodology was devised. In order to achieve the management of knowledge and integration with downstream applications the STEP standard was adopted. Its advantages are that it is a mature and internationally agreed standard. However, it has the disadvantages of not being deployed widely. The main reason is the cost associated with compatible tools as well as the lack of heterogeneous collaboration tools on the web. For this reason the format used for sharing the data has been changed from STEP Part-21 to the semantic web formats to enable the seamless management and sharing of knowledge over the web and enable heterogeneous systems to query and infer knowledge from the system.

2.   KNOWLEDGE REPRESENTATION METHODOLOGY

The authors define knowledge as the semantically complete definition of a domain's information that is both machine readable and interpretable. Thus a knowledge base contains an ontology which defines the classes and their relationships, instances or objects that form the domain 'data', meta-data that constrains the data within a particular domain (transforming it into information), and Universal Resource Identifiers (URI) that allow the global identification and contextual interpretation of the information. The solution provided is a user modifiable object-oriented ontology for managing all information in the product development process as distinct objects within the PLM systems and tied together with URI's. This enables the system to create reports that mimic the 'layout' of static documents but rely on a single source of up-to-date knowledge, eliminating duplication, and enabling the user to modify sub components and assemblies of geometric models without retrieving the complete model.

The resource description framework (RDF) has been used in this application as the format for storing the knowledge base, as opposed to eXtensible mark-up Language (XML). The reasons for this choice are the extra flexibility and 'machine-understandable' format of RDF graph triple model as opposed to the simple 'machine-readable' XML based mark-up vocabularies. In effect, any RDF-parser can derive the semantics and context from the URI and metadata attached to every instance. RDF has been modelled in this instance using the Protégé KBS and visualisation was made with ontoviz and tgviz. The ontology was sub-classed from the base data model of the PLM system for flexibility, ease of deployment and to make use of the lifecycle and workflow functions offered by the PLM systems. Integration with the p2p system was through a Protégé plugin.

4 based automatic process planner by Sharma [11] to generate plans from concept designs. Aggregate process plans for assemblies are generated using Cheung's [12] process planner. The STEP AP-224 feature models are defined in Protégé and use the Java expert system shell (JESS) to define the rules and functions of the standard and the constraints these impose enable

error-checking for the user during the definition of features. However, these constraints can only be applied to individual entities within the STEP model and cannot enforce any constraints between entities/features. Due to the size of the standard, only a subset of AP-224 has been translated. The definition of the Express (STEP) data types is also contained in the Clips interface. This is very flexible as it allows for the inclusion of STEP data types to other components within the ontology on a need basis without having to have any expertise in Express or any other programming language. The above enables the mixing of feature and meta-data information in the knowledge base, meaning that users can access the information stored in STEP models using queries and RDF parsers. This integration at low level between the geometric, feature and 'meta-data' within a single environment is intended to reduce repetition and errors, and also enables the reuse of all the data created during the conceptual design process.

3. COLLABORATION ON DISTRIBUTED DESIGN

One of the first problems of collaboration is trying to understand what the other says and means. The previous section showed the methods used to create the data models. This development is interrelated to the problem of collaboration and the authors overcame one of the perennial problems of collaborative design between enterprises of different sizes and complexity, that is, mapping between low and high content data models which results in irretrievable loss of information from the high data model. This problem cannot be overcome traditionally by creating a mapping from one data model to another. Instead the authors have sought to create a universal project oriented ontology that can be created, shared and used by all parties collaborating in an enterprise in real-time. This eliminates the problems faced when low-end suppliers collaborating with advanced enterprises face integration issues. Using the building-block ontologies and STEP standards, all enterprises, small and large, have access to enterprise features. With this problem out of the way, a number of mechanisms to establish the collaboration and manage the knowledge were investigated.

Business to Business (B2B) integration is the traditional method for companies to collaborate. For example, integrating 15 companies together on a one-to-many basis where a single repository manages the project knowledge and workflows would need 14 separate mappings. To empower the individual enterprises within the collaborative environment and decentralise the system would require many-to-many i.e. 94 B2B integrations!

3.1   Application Service Providers

Application Service Providers (ASPs) can be set up in two ways, either by a large 'controlling' enterprise, or through independent third party hosting. These services intend to provide the same utilities as enterprise level systems but in a non-enterprise specific service. ASP's offer project and product data management vaults where the administrator can customise the third-party portal for their own use. Advantages include Reduced cost for the enterprise as maintenance and backup is delegated. Increased opportunities if customers and VE initiating enterprises seek out partners through the portal. This is a "Democratic" system where no one enterprise controls the server and data. It sets down de-facto standards for data exchange, to which other enterprises in the same domain will adhere to in order to join the network of enterprises.

There are of course some fundamental disadvantages to the use of ASPs for product development, and these include: The bandwidth and server bottleneck problem associated with centralised services. The security fears of intellectual property rights being compromised. The potential risks of downtime and data losses in an "uncontrollable" environment and the liability issues associated with it. The difficulty of creating direct interfaces from the enterprise system to the ASPs portal. The exact functionality required may not be available from the "generic" ASP.

There are already some ASPs operating in the automotive and aeronautical sector enabling supply companies to interact and bid openly for contracts with OEMs and then manage the project/product information on the portal. However due to the disadvantages highlighted above, the authors sought to find a third way. Whilst traditional client-server systems can operate in a collaborative manner, for example over a LAN or Internet, they are not truly distributed as they are centralised.

3.2   Peer to Peer Systems

Peer to Peer (P2P) applications address the needs of de-centralised organisations to collaborate and share knowledge regardless of geographical location. The principle of P2P has been around for a long time, and is today implemented in a number of applications such as instant messaging and file-sharing (www.GNUtela.com). There are already a number of P2P PDM in existence. Primarily aimed at the lower end of the market. The two commercial applications are AutoManager workflow from Cyco (www.cyco.com) and Columbus from Oasys Software (http://www.oasys-software.com/). The latter is available for free, and aimed at AutoCAD users within the construction sector. However they are crude solutions relying on the underlying file system and adding some "meta tags" to files for version control. As an example of what can be achieved, Alibre is a P2P CAD/PDM and collaboration tool in one. It uses the STEP standard and combines low cost and fast configuration.

The advantages offered by P2P applications are (i) no single point of failure, the network is alive as long as one peer is on-line, (ii) distributed sharing of bandwidth storage and processing power, so the system becomes more powerful as more users attach, (iii) lower running cost due to the lack of servers or high bandwidth central nodes, as well as (iv) maintaining individual control of the shared knowledge. P2P groups can be used to create profiles of the peer, and also more importantly of the peer's list of contacts within different domains. These profiles can be used within the network to search for and assess people's competences, interests, and memberships of trusted groups, and can aid in the construction of new relationships based on commonalities and third party assessments. There have been a number of issues that reduce the performance of the system using pure P2P architecture. The lack of indexing and routing services in P2P degrades the peer discovery and query functions. In order to leverage the advantages of client/server systems with the independence and interoperability of P2P systems a hybrid system where "super peers" act as peers to the extended P2P network and as a server to the enterprise's internal peer network is used. In addition rendezvous peers can be assigned to manage some of the peer information assigned to particular peer nets or projects. This hybrid has been shown to have the best potential for high-performance de-centralised services.

3.3   Inter-enterprise Communication Architecture

Since no two enterprises are the same, the idea of using XML based messaging for inter-enterprise collaboration is not easy as the two company schemas have to be mapped to each other. However, a new methodology developed and applied in this instance has been to utilise the open standard and open source philosophies. In order to achieve the speedy interoperability a standard has to be set for basic messaging. Some like STEP PDM implement the complete information structure of the engineering enterprise. Others like JuxtaPose (JXTA) implement only the "messaging" components. The choice of standards to use depends on the level of "standardisation" that all enterprises can adhere to. In the view of the authors only a very low level subset of all enterprises can be "standardised". This small subset should exist only as a medium to enable communication, identification and access control management. All other aspects of the collaborative data environment are enterprise specific (although constructed from a subset of ontological components).

This project created a two tier architecture for communications, speeding up of the integration process is further enhanced through the use of open standard semantic web enabled ontology formats for storing the information, plus the use of open-standards in addition to company ontologies to define those ontology components.

4. IMPLEMENTATION

4.1 Commercial PLM System

Windchill is a traditional document management tool, whilst it can store all manner of data and make revision controls. It does not however have an intelligent method of containing and persisting information in an object oriented format. To alleviate this, the functionality of the system has been extended to include the management of knowledge from Protégé. The integration with Protégé was made using the workflow engine. Windchill is inherently centralised, overly complex to set up and requires a long period of time for customisation. The system's main strengths are in the workflow tools. However, even here there is the problem of lock-in. The workflow and any customisation carried out cannot be reused on another system. In addition Windchill has an application layer for inter-enterprise communications, and inherent in its weakness is the B2B paradigm which has been elaborated previously. The RosettaNet standard implemented can decrease the amount of customisation needed. In addition, unlike the ontological 2-tier system, Windchill's data models are neither portable nor standardised.

The data model can be modified by an expert who has to model, program, compile, update the database and integrate the code into Windchill before it can be used. This does not provide for the flexibility and ease of use for modelling ontologies that a project manager needs. The system, implementation and running costs are very large and only practical if enterprises share it between each other over a long period of co-operation.

4.2 Open Standard and Open Source tools

Open source applications are widely deployed, and the engineering/manufacturing sector can benefit greatly from leveraging the available source code for easily customised applications. Two applications used in this project are the Protégé ontology editor and Sun's JXTA peer to peer protocol. They offer the best usability, portability and cost/performance capability within their niches, and when allied to a scalable open source database such as SAP-DB, can offer enterprise level performance for zero capital expenditure and low customisation cost. Moreover they free the customer from software vendors and the vagaries of obsolescence. Other open-source tools used in this work include OPEN-CASCADE (STEP modeller) and the openflow engine that forms the heart of the open source PLM.

In the application the bottom layer consists of the database MySQL. On top of this layer are the two open source gateways, Apache for serving static and PHP based web pages and the Tomcat servlet engine for Java server pages (JSP) based applications. These three layers (database, web server and application server) form the server side of the system. Figure 1 illustrates the three-tier architecture.

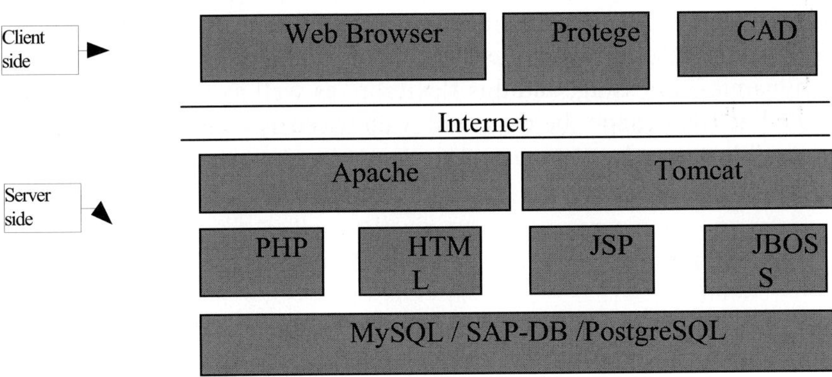

Figure 1: High-level view of open source and user interface architecture

On the client side, the user has three main applications i.e. (1) the web browser through which interactions with the PLM system are carried out, (2) the Protégé Java applet that allows the user to query and manage the knowledge base and (3) a CAD system to enable the user to create and manipulate the STEP based models held in the PLM system.

The above systems are the evolution of product development management systems from document centric PLM systems to knowledge centric, intelligent systems of the future. However, it had been discovered during the course of the project that centralisation, by its very nature, is an inhibiting factor for inter-enterprise collaboration as the only methods available for collaboration in such an environment are B2B custom integration and the centralised web-enabled or ASP models. Portals in this instance would simply be classified as B2B integrations as each data source has to be separately mapped into the portal.

4.3   Peer to Peer Implementation

The back end consists of the open source SAP-DB database with the Java database connectivity (JDBC) connector to Protégé. Connectivity is achieved using an open source implementation of JXTA open standard P2P network protocol (www.jxta.org). The choice was made because JXTA implements a unique but anonymous identification mechanism for peers and for rendezvous peers. As well as "advertisement" implemented for all peers that give information about the peer to other peers. Rendezvous peers can act as managers for peer groups and store the peer advertisements for the group for distribution to other P2P networks. An extension to enable RDF queries and ontologies to be shared over P2P is used to share the knowledge base. Queries, project management of collaborative groups, group chat and instant messaging are readily implemented by the JXTA protocol. The systems' settings enable enterprises and users without static addresses to collaborate using dynamic addressing, and this flexibility as well as the users' ability to work off-line (that cannot be done with web based systems) empowers users in all possible network scenarios.

5.   EXAMPLE

The enterprise's processes were split into lifecycle states and defined in workflow processes. Due to the complexity of the overall processes, only the conceptual design phase was modelled into the ontology and workflows. The sequence of processes for this phase are:
- Customer request for quotation is submitted including requirements and associated verification methods;
- Generation of specifications through analysis of earlier customer requests and testing against the new verification methods;
- Creating concept design options from the specifications by pattern matching against earlier concepts developed with similar specifications; and
- Selection of concept and utilising the verification methods;

User interaction through the interface as shown in Figure 2 was straightforward. Workflow sequences acted as widgets guiding the user through different configurations.

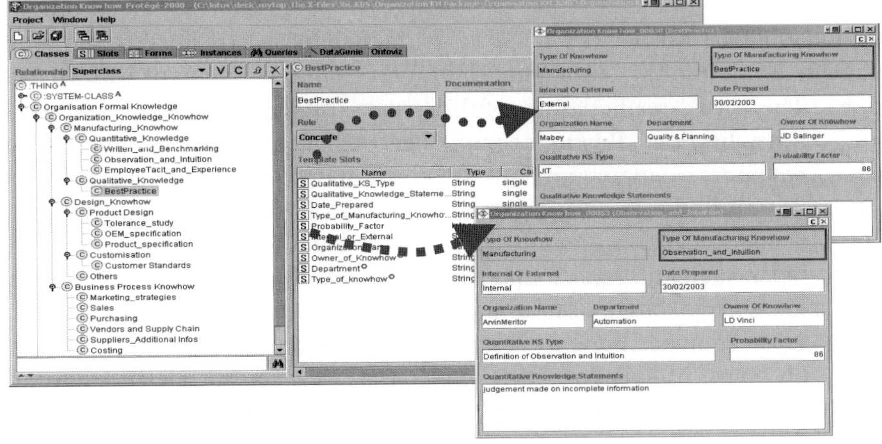

Figure 2 Protégé user interface for knowledge acquisition

The retention plate and four item assembly for the door latch (see Figure 3) was modelled for the entire lifecycle. In addition the full project data including over 300 parts were modelled into the Bill of Material. New components are entered interactively and work seamlessly throughout the lifecycle. The application proved very simple and intuitive to use. The collaboration tools use proven technologies by SUN and worked seamlessly for many-to-many connections.

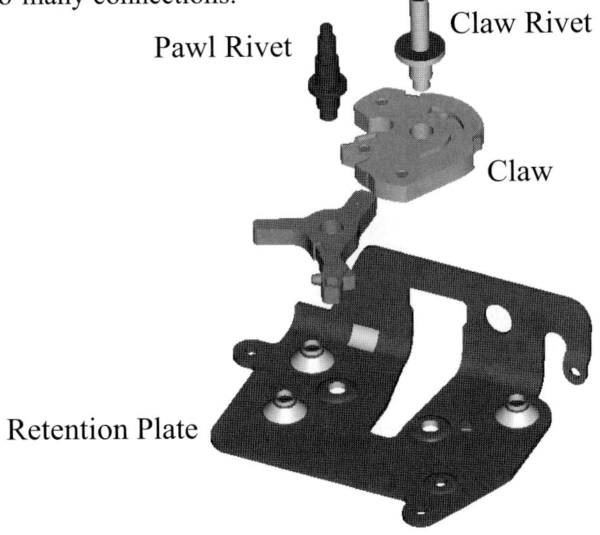

Figure 3 Latch assembly view. (Courtesy ArvinMeritor)

6. BENEFITS

- The new methodology fulfils the following criteria for Decentralisation: Reduce centralised bottlenecks in bandwidth and resources, empowerment of collaborators within networks to "control" the knowledge they create, manage intellectual property rights and enable domain professionals to tailor the system.
- Open standard: reduce interoperability issues for product and project knowledge, easier set up for inter-enterprise collaboration.
- Open source: Elimination of software licence costs, a solution to the problem of vendor lock-in in the long term, elimination of unnecessary complexity and freedom to modify the application.
- Semantic Knowledge: Eliminating the ambiguous context of the knowledge. Efficient query and retrieval mechanisms, intelligent agents to function on context aware information.
- Platform and application independence: Enable the enterprise to concentrate on its work and not be tied in to any vendor, rapid migration to future applications.

7. CONCLUSIONS

The methodology described is a suitable solution for collaborating enterprises especially SME's, to create manage and reuse their knowledge, collaborate easily and without expense. There is a lot of work in progress, and final conclusions can only be drawn once the knowledge and communication protocols are integrated more completely.

ACKNOWLEDGEMENTS

The authors would like to thank the UK Engineering and Physical Science Research Council (EPSRC) for its financial support. The industrial collaborators ArvinMeritor, LSC Group, Mabey & Johnson and PTC Corporation.

REFERENCES

1. Gao; Aziz; Maropoulos; Cheung, "Application of product data management technologies for enterprise integration". *International Journal of Computer Integrated Manufacturing* **2003** 16, 491-500.
2. Aziz; Gao; Maropoulos; Cheung, "A design environment for product knowledge management and data exchange", *Methods and Tools for Co-operative and Integrated Design*, **2003** (ed) Serge Tichkiewitch and Daniel Brissaud, published by Kluwer Academic Publishers.

3. Lihui; Weiming-Shen; Xie; Neelamkavil; Pardasani, A. "Collaborative conceptual design - state of the art and future trends". *Computer-Aided Design* **2002**, 981-996

4. Beckett, "Determining the anatomy of business systems for a virtual enterprise". *Computers in Industry* **2003**, *51*, 127-138

5. Camarinha-Matos; Afsarmanesh, "Elements of a base VE infrastructure". *Computers in Industry* **2003**, *51*, 139-163

6. Kim; Kang; Lee; Yoo, "A distributed, open, intelligent product data management system". *International Journal of Computer Integrated Manufacturing* **2001**, *14*, 224-235

7. Goh; Koh; Domazet, "ECA rule-based support for workflows". *Artificial Intelligence in Engineering* **2001**, *15*, 37-46

8. Zha, Du A "PDES/STEP-based model and system for concurrent integrated design and assembly planning". *Computer-Aided Design* **2001**, *In Press, Uncorrecte,*

9. Zhou; Chin; Xie; Yarlagadda, "Internet-based distributive knowledge integrated system for product design". *Computers in Industry* **2003**, *50*, 195-205

10. Vasara; Krebs; Peuhkuri; Eloranta, "Arachne--adaptive network strategy in a business environment". *Computers in Industry* **2003**, *50*, 127-140

11. Sharma; Gao; Bowland, "Implementation of STEP Application Protocol 224 in an automated manufacturing planning system". *Journal of Engineering Manufacture (Part B)*. **2003**

12. Cheung; Maropoulos; Gao; Aziz, "Knowledge-enriched Product Data Management System to Support Aggregate Process Planning", *1ˢᵗ International Conference on Manufacturing Research*, Glasgow **2003**. Advances in Manufacturing Technology XVII, Ed. Y Qin and N Juster, 253-258

A DEVELOPMENT ON MANUFACTURING INFORMATION COMMUNICATION IN AGILE MANUFACTURING WITH STEP AND XML*

Xiaoli Qiu and Hong Yi
Department of Mechanical Engineering, Southeast University, P.R.China, zip:210096,Email: xliqiu@sina.com

Abstract: In Agile Manufacturing (AM), it is a problem how to cooperate between the different applications with heterogeneous information format through the Internet. A new theory is put forward in this paper. It is concerned with a CAD system collaborated with another CAD partners based on STEP and XML in AM through the Internet. The STEP standard and the XML conception are introduced to support the system, as well as the EXPRESS and DTD for the corresponding standard. Technology on how to integrate STEP and XML as the neutral formal model is presented detailed. As a result, the theory and the framework of the system based on STEP and XML is demonstrated.

Key words: XML, STEP, Agile Manufacturing, Computer Aided Design.

1.   INTRODUCTION AND BACKGROUND

As we transition into the twenty-first century, the speed of the Internet has inspired manufacturers to offer products faster and with more unique customer specifications. The marketplace has become truly global. The rapid change in technology has lead to the need for far greater cooperation within and between firms. No company can have all the required skills and knowledge. In high-tech areas it is very often the small and virile organizations that develop and harness the latest advances. It is just not

* **Project 70102010 supported by NSFC**

possible for one firm to have everything it takes to fully meet a customer's need. Thus, the AM is developed to adapt to the situation.

Firms cooperated in the AM are called virtual corporations. They may physically locate at different places. These virtual corporations are opportunistic alliances of core competencies across several firms to provide focused services and products to meet the customers highly focused needs [1]. In the AM, virtual corporations may be skilled in different stage of the lifecycle of production. These virtual corporations use different software varies from CAD, CAPP, CAM, to CAE, etc. Each application has different software for its purpose. Certainly, they have different format output data. In this environment of heterogeneous systems, which is used at different stages of the product development lifecycle, a crucial issue is the data communication and exchange between these systems. The STandard for the Exchange of Product model data (STEP) is an evolving international standard for the representation and exchange of product data[8].

A new problem is, applications cannot obtain data from STEP files on the net. The possible solution to this problem is to provide the XML based services for STEP data translation through the Internet. It is fortunate that the ISO has developed a new standard called XML Representation Methods for EXPRESS-Driven Data in November 1999. This standard provides an analysis for the representation of EXPRESS-driven data using XML syntax: the late binding approach being developed as Part 28 of STEP (ISO 10303) [3].

In this paper a STEP and XML based AM system is introduced. This paper focuses on how to use XML and STEP to exchange product data between different CAD software. The paper is organised as follows. Section 1 provides some background on the development on CAD, the STEP international standard (IS) which is currently being developed. An explanation of the XML and the Internet is detailed in section 2. The technology and the architecture of the system, and the integrating of STEP, XML, Internet, CAD in AM are detailed in section 3. An example testify the possibility of the realization of the integration is illustrated in section 4. Lastly, conclusions are presented in section 5.

2.   THE NEED OF DATA EXCHANGE WITH STEP AMONG CADS IN AM

Computer-aided design (CAD) is the technology concerned with the use of computer systems to assist in the creation, modification, analysis, and optimization of a design[6]. More recently, the emergence of a new manufacturing paradigm called Agile Manufacturing (AM) has made concept of cooperation between different engineering activities to a higher degree. In AM, parts, components, and products, which constitute a product, are developed by different "partner" virtual companies' different CAD system. So, product data like design, manufacturing, maintenance and disposal are not only shared between various departments within a company, but also shared between various "partner" companies of a virtual enterprise [5].

One CAD application needs the product data such as the geometric information, assembly process information and the structure tree of the product, which is all output by other CAD systems. In fact, each "partner" virtual company may have its own CAD system different from other "partner" virtual companies' in the AM. Thus, different virtual companies share data and information about their products through a standard form of exchange, the representation of product data in multiple views through a standard exchange format for multiple manufacturing applications has become very essential. The shared data and information must describe the product uniformly, precisely, and unambiguously. Moreover, these data are considered to be divorced from CAD systems where they are designed and output. In this context, a CAD system can share data from multiple CAD systems through the Internet.

The STEP standard has emerged as the means of neutral form for data exchange between companies[7]. STEP, a key international product data technology, provides an unambiguous, computer sensible description of the physical and functional characteristics of a product throughout its life cycle. In fact, most nowadays CAD systems can export parts or product as STEP model. As a result, STEP allows companies to effectively exchange information with their partners all over the world, as well as internally.

STEP provides a neutral computer-interpretable representation of product data throughout the life cycle of a product independent of any particular system. The most important aspect of STEP is extensibility. STEP is built on

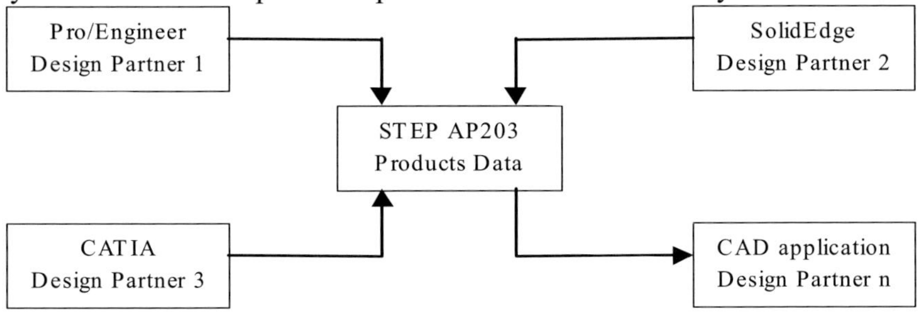

Fig.1 Data Exchange between CADs with AP203

a language that can formally describe the structure and correctness conditions of any engineering information that needs to be exchanged. This language is called EXPRESS. The function of EXPRESS is to describe information requirements and correctness conditions necessary for meaningful data exchange. The EXPRESS language can document constraints as well as data structures. These formal constraints are a explicit correctness standards for the digital product data.

Faster design times, better communication and longer lasting data are the benefits of STEP. And because it is open and extensible, it can be sure to meet the design and manufacturing needs well into the next century. EXPRESS has a range of rich data structures, including an advanced form of inheritance. But when EXPRESS in STEP is used to describe the information requirements of many "partner" virtual companies' applications such as another CAD application on the Internet, how can the information in the EXPRESS files be distributed? Luckily, XML can help working on it. In this paper, an important STEP part is concerned. It is STEP Part 28, which defines specifies use of the XML to enable the transfer of both schemas and data specified using the EXPRESS information specification language. [3].

3. THE NEWEST DEVELOPMENT ON XML AND DTD

XML is the universal format for data transformation on the Web. It is developed by the World-Wide Web Consortium (W3C) from 1996 and it is a W3C standard since February 1998. XML is a set of rules, guidelines, or conventions for designing text formats for data such that the files are easy to generate and read, they are unambiguous, and they are extensible. It aids in putting information on the Web and in the retrieval of that information.

XML is a markup language for documents containing structured information. Structured information contains both content (words, pictures, etc.) and some indication of what role that content plays. XML specifies neither semantics nor a tag set. In fact XML is really a meta-language for describing markup languages. All of the semantics of an XML document will either be defined by the applications that process them or by style sheets. XML was created so that richly structured documents could be used over the web. It is a family of technologies because it is defined by a number of related specifications [4].

DTD, short for Document Type Definition, is a mechanism to describe the structure of documents. The DTD is the original modeling language or schema for XML. The DTD defines the constraints on the structure of an XML document. It declares all of the document's element types, children element types, and the order and number of each element type. It also declares any attributes, entities, notations, processing instructions, and comments in the document [2].

The purpose of a Document Type Definition is to define the legal building blocks of an XML document. It defines the document structure with a list of legal elements. With DTD, the XML files can carry a description of its own format with it. With a DTD, independent groups of people can agree to use a common DTD for interchanging data. Applications can use a standard DTD to verify that the data received from the outside world is valid.

4.   THE INTEGRATION OF XML AND STEP IN AM

4.1   Output STEP model by different CAD

The emerging paradigm of Agile Manufacturing has imposed additional requirements of ''neutral format'' so that products data can be readily shared among multiple partners of a virtual enterprise. The STEP has emerged as the means for neutral form exchange of product related data. STEP also allows dynamic sharing of data between different systems through the standard data access interface SDAI. Several APs have already been developed to support different kinds of engineering applications and quite a few are currently being developed for various other manufacturing processes. Currently, the most widely used AP is the IS AP203 which is meant for representing design and configuration management information [7].

Many CAD/CAM systems, such as Pro/Engineer, SolidEdge and ACIS, support STEP AP203. In Agile Manufacturing discussed here, there are many application systems. The data exchange is happened among these applications. They all need data with AP203. Fig.1 depicts an application scenario for such a case, in which four designers use different CAD systems to complete the design of various components of the same product.

4.2   Data exchange by XML and STEP in Web

The structure of XML file can be defined in a Document Type Definition (DTD) file. This file is used to check that the provided XML file is valid XML document. It serves the same purpose for XML file as EXPRESS schema serves for STEP part 21 file.

The XML value of each root entity instance shall be developed by traversing the XML DTD and finding the EXPRESS entity or EXPRESS attribute instance that corresponds to each element or attribute in DTD. If there is no corresponding EXPRESS entity or attribute instance for an XML element or attribute then the XML value for that XML element or XML attribute shall be empty. An error shall be reported if the XML data generated from the EXPRESS data do not conform to the rules described by the DTD [3].

One of the ideas introduced in this paper is after providing a schema independent DTD, which can be used for sharing product data regardless on what schema it is based on, the CAD application can get products data produced by CAD modules through the Internet. In on going effort STEP community is developing a standard ISO 10303-28 (Part 28) "Product data representation and exchange: Implementation methods: XML representation of EXPRESS-driven data" where a more sophisticated approach has been taken to address this issue.. This work as converting STEP files into XML is being done. FirstSTEP EXML (http://www.pdml.org/exmlintro.html) and ST-Repository (http://www.steptools.com) can provide the converting. And there's STEPml which is a library of XML specifications -- DTDs and/or XML Schemas -- for product data, based on information models from STEP.

4.3   The realization and the framework of the system

Fig.2 presents the architecture of the AM system based on XML and STEP. In AM, the "partner" virtual company (partner y) CAD application can login on the net. The "partner y" obtains product data from the Internet. These data are output by the designing partners (partner 1, partner 2, ..., partner n), and are stored with the XML format in the Web server.

The designing partners may cooperate with each other to design different parts or components of a product with their CAD software. Different CAD software can output the designing results as STEP files with AP203. Then these STEP files are converted into XML files with an translator according to the STEP Part28 standard.

These XML files converted from STEP files are saved in the web server. The CAD application can browse these XML files and get the product data provided by the different design partners. After the "partner y" finishes its processing, advice may be given to the design partners through the Internet. Thus the design partners can correct their design so as to fit to the application.

The designing partners may cooperate with each other to design different parts or components of a product with their CAD software. Different CAD

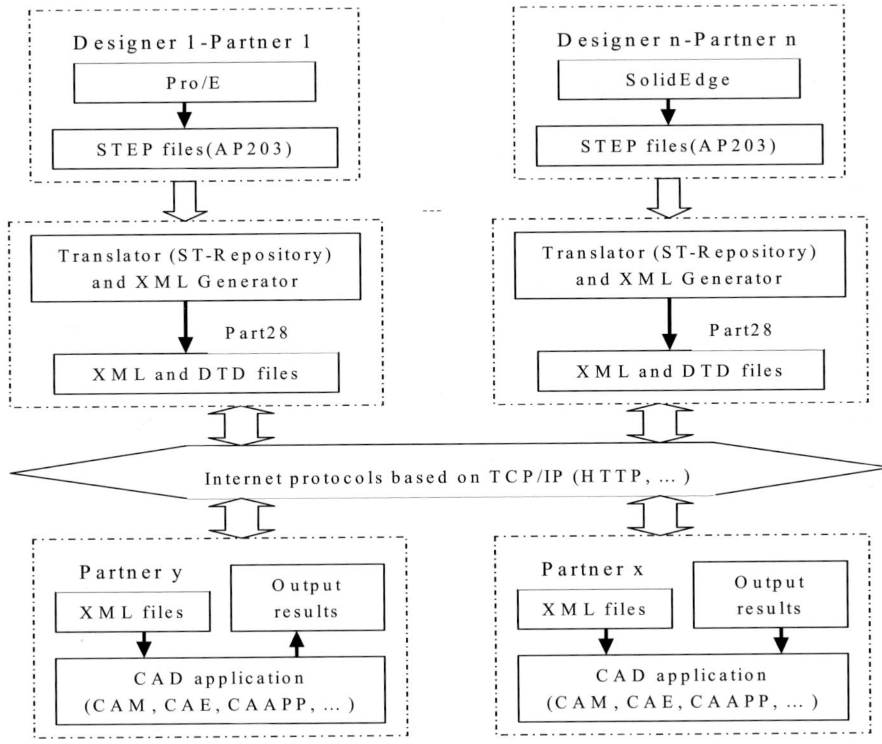

Fig. 2. Architecture of the Agile Manufacturing system

software can output the designing results as STEP files with AP203. Then these STEP files are converted into XML files with an translator according to the STEP Part28 standard.

5. DEMONSTRATION OF THE INTEGRATION OF CAD, STEP AND XML WITH EXAMPLE

Fig.3 shows a test part created using Pro/Engineer for the testing of the integration and converting between STEP and XML. The part name is "prtpart.prt". Pro/Engineer has the function to export the part as STEP model. The name of the STEP file is "prtstep.stp". So firstly, the converting from CAD part to STEP file is realized by the CAD system itself. Fig.4 shows some of the STEP file.

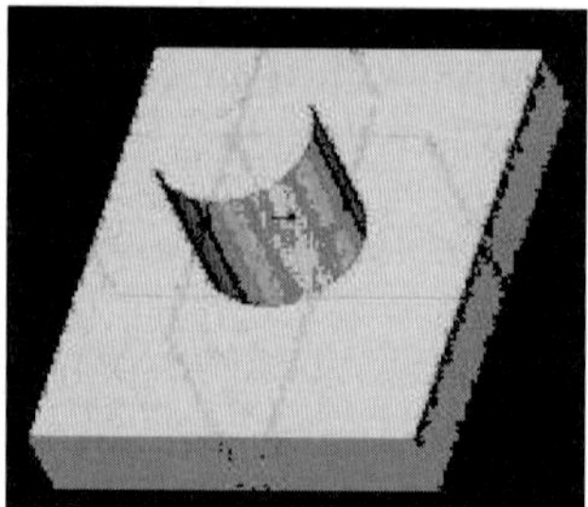

Fig.3. Sample part created by Pro/Engineer

Then a translator is applied to convert the STEP file into XML model. In this example, ST-Repository (http://www.steptools.com) is used to make this converting. An XML file named "prtstepxml.xml" is generated by the translator. Some information of this file is shown in Fig. 5.

---

1. STEP FILE
ISO-10303-21;

HEADER;

FILE_DESCRIPTION((''),'1');

FILE_NAME('PRT0001','2000-11-10T',('Administrator'),(''),

Fig. 4. Part of the STEP file

---

2. XML FILE
<?xml version="1.0" ?>
-    <!-- STEP/XML translator developed
-    by STEP Tools, Inc. using ST-Developer -->

Fig. 5. Part of the XML file

After the XML file is published to the web, the partners all over the world can get the part data from the XML. Then those partners can use this information to finish their applications, or inversely translate it into STEP model, and then reuse the design result of the part with another CAD software. There may be more and more tools realising the translation for different purposes with the development of XML and STEP. However, it is outside the scope of this paper.

6.   CONCLUSION

As the designers and the CAD applications have different CAD software, a neutral format file containing product data must be applied. STEP is the standard model. In AM, these neutral data must be obtained easily and conveniently by the partners through the Internet, so XML is introduce to publish the product data to the web. In some case, EXPRESS or SCHEMA for STEP is familiar to DTD for XML. One of the latest significant developments in STEP is the recent agreement to provide mapping to XML. This new technology is rapidly becoming the preferred method for complex data access on the Web. The flexibility and growing availability of commercial Web/XML tools with STEP will greatly increase the sharing of information across disciplines, with universal access. In the paper an example is demonstrated to show that the integration of STEP and XML is possible in the AM.

REFERENCES

1.   B. H. Maskell, "An Introduction To Agile Manufacturing", http://www.maskell.com.
2.   B. Marchal, "XML by Example", Que, 1999.
3.   ISO TC184/SC4, "ISO/WD 28, XML representation of EXPRESS-driven data", ISO 10303 Editing Committee.
4.   Jerry Banks, John S. Carson II, Barry L. Nelson, David M. Nicol, *Discrete- Event System Simulation*, 3th Edition, Prentice Hall, 1999.
5.   John M. Usher, "A STEP- Based Object- Oriented Product Model for Process Planning", *Computers In Engineering*, Vol. 31, No. 1/2, pp. 185-188, 1996.
6.   K. Lee, Principles of CAD/ CAM/ CAE Systems, Addison- Wesley, 1999.
7.   M. P. Bhandarkar, R. Nagi, "STEP- based feature extraction from STEP geometry for Agile Manufacturing", *Computer in Industry* 41, (2000) 3-24, ELSEVIER.
8.   Y. Zhang, Ch. Zhang, H. P. (Ben) Wang, "An Internet based STEP data exchange framework for virtual enterprises", *Computers In Industry,* 41 (2000)51-63, ELSEVIER.

WEBCAPP: COLLABORATIVE INTELLIGENT PROCESS PLANNING SYSTEM

Jitender Rai[1] and S.S.Pande [2]

[1] Sr. Research Engineer, Mechanical Engg. Department, Indian, Institute of Technology Bombay, Powai, Mumbai – 400076, INDIA,   Email: jiten1pm@iitb.ac.in

[2] Professor, Mechanical Engg. Department, Indian Institute of Technology, Bombay, Powai, Mumbai – 400076, INDIA, Corresponding author. Email: s.s.pande@iitb.ac.in

Abstract: This paper reports the design and implementation issues of WebCAPP: an intelligent Process Planning system for prismatic machined components. WebCAPP has been designed to function with Palantir, the web based collaborative part modeling environment indigenously developed. Clients from any location can collaboratively create CAD part models and submit them to the WebCAPP server to carry out tasks such as setup planning, GA based operation sequencing, CNC code generation and simulation. CNC program can be transmitted to a remote CNC machine for telemanufacturing. WebCAPP was tested for number of components and was found to generate optimal process plans in a consistent manner.

Key words: Collaborative CAD/CAM, Genetic Algorithm, Optimal Process Planning.

1.   INTRODUCTION

Manufacturing industries worldwide face several challenges due to shorter product life cycles, reduced time to market and need for collaboration among geographically distributed designers, suppliers and manufacturers for global product development. Though the CAD/CAM tools developed in the last two decades significantly shortened product development cycle, they are suitable for standalone operation with proprietary software [1]. Researchers worldwide are thus focusing on the development of web based collaborative CAD/ CAM systems to enable clients at different geographical locations to collaboratively create CAD part models, analyze them, generate optimal process plans and finally enable *telemanufacturing* at a remote site.

A need thus exists to design and develop such comprehensive systems.

2.   LITERATURE REVIEW

Last two decades witnessed intense research focused on various CAD/CAPP/CAM issues related to feature based product modeling and automated process planning [1]. Today feature based part modeling has become de facto industry standard. Compared to these works on stand-alone systems, scant research work is reported on the development of collaborative systems for part modeling and manufacturing. Important work relating to web based CAD and CAPP systems are surveyed here.

For collaborative part modeling, web based systems like webCAD by Kim et. al. [2], CSM by Chan et. al. [3], webSPIFF by Bidarra et. al.[4], have been reported. These systems provide limited modeling capabilities and user friendliness. Most of them produce simplistic part models suitable for display purposes.

Extensive research on standalone CAPP systems has been reported following Variant, Generative and Expert system techniques for specific part shapes and shop conditions [1,6]. Automated setup planning and process sequencing form the heart of any CAPP system. Strategies such as rule based by Sabourin et. al. [1], generative approach by Patil and Pande [6] have been reported. These are however domain specific and often shop-based.

Recently genetic algorithms are being applied to various process-planning issues like setup planning, operation sequencing, cutting parameter selection etc.[7,8,9]. These systems focus on specific shop/part domain, and have ignore the precedence constraints during the operation sequencing. They thus, do not optimize tool path and the number of tools used.

This paper reports the design and implementation issues of a web based collaborative system (WebCAPP) for collaborative CAD part modeling, Intelligent process planning and transfer of CNC code to a remote client CNC machine for telemanufacturing.

3.   SYSTEM ARCHITECTURE

Fig. 1 shows the modular architecture of WebCAPP. Clients at different geographical locations can create part model in collaborative environment using Palantir [5]- the feature based collaborative part modeler indigenously developed by us. WebCAPP server will carry out intelligent process planning and CNC code generation. The code can be transferred over the network to a remote CNC machine for telemanufacturing the part.

WebCAPP comprises of three main modules viz. communication module, web based CAD modeler (Palantir) and the CAPP module.

4.   COMMUNICATION MODULE

It handles the collaborative sessions and makes the facilities offered by the CAD and CAPP modules available over the WWW. It consists of an Authentication Server, Session Manager for each session and a Client connection on the server side and a server connection on the client side for each client. The Authentication Server is responsible for the setting up and closing of a collaborative session. The Session Master starts the session into which any number of authorized clients can join. Strict token passing is incorporated to ensure concurrency and synchronization [5]. In a collaborative session, all clients get to see the model but revision is based on the token passing. Session manager holds the final rights to accept the revisions. The collaboration module has been implemented using the Java with Native Interface to interact with Palantir. Client modules are implemented as applets, which can be run from any browser.

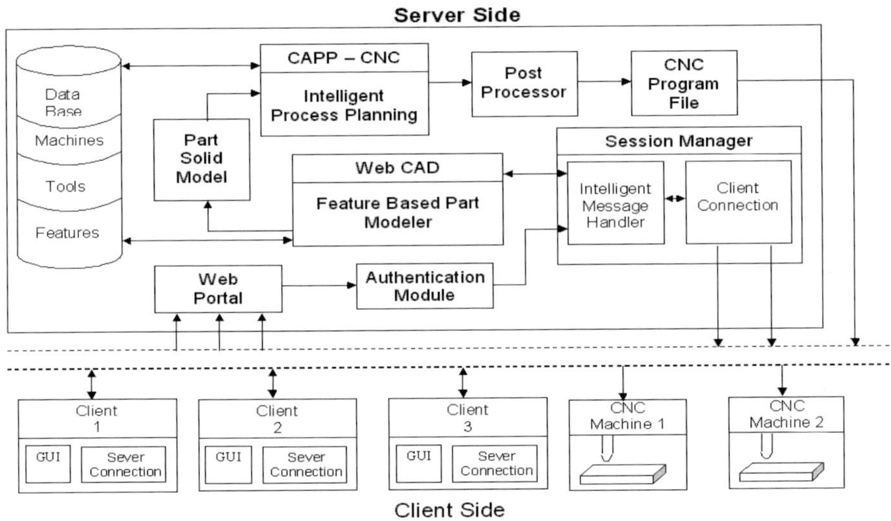

Figure 1 Modular Architecture of WebCAPP

5.   "PALANTIR" – FEATURE BASED MODELER

Palantir is based on the client –server architecture and supports feature based modeling (FBM) of complex prismatic parts. The feature taxonomy implemented in the FBM comprises of standard 2.5D features families such as Holes, Pocket, Slot, Step, Patterns etc. and 3D features like Lofted pockets and freeform (NURBS) pockets which gives the user capability to create a very wide range of complex prismatic components.

It has been implemented using ACIS 7.0 solid modeling kernel. The FBM is responsible for issues like virtual part synthesis, geometry handling, part and feature data validation, storage etc and is installed on the server.[5]

The GUI of Palantir (Fig. 2) has been designed to include a command prompt, toolbox, set of menus and panel for visualization of the model. These have been designed to give the look and feel of a stand-alone CAD systems so that the user feels comfortable with them.

In a collaborative session, all clients see the CAD model all the time but the token holder is authorized to carry out revisions/editing on the model. Exact geometry (ACIS) model resides on the server. Client can see Wire frame or faceted B-Rep model (with/without features) as per their needs and the available network speed. Java 3D is used to display B-Rep model. Users can save part models on the server in different formats viz. SAT, STEP, STL and our native (PRT) format for interfacing with other software.

6.   INTELLIGENT PROCESS PLANNING

WebCAPP functionally comprises of various sub modules viz. CAD file processing, setup planning, GA based operation sequencing, CNC code

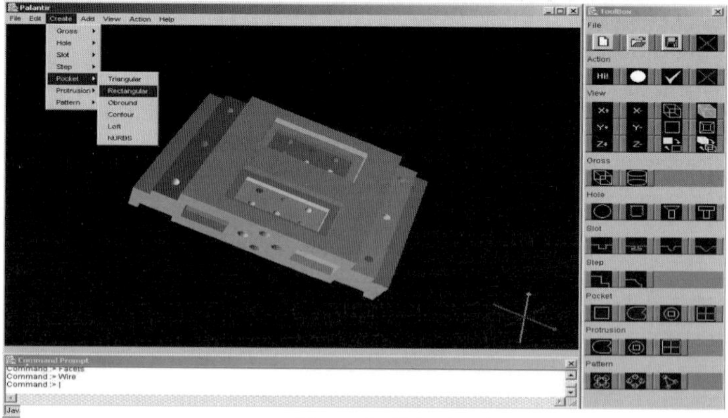

Figure 2 GUI of Palantir

generation and simulation. These modules are implemented on the server using C++ programming language.

6.1 CAD file Processing

The CAD part model received from the client is processed to extract various part feature attributes like geometry, dimension, tolerances, location (datum) etc. OOPS has been used for the representation of data for part features and associated machining processes.

6.2 Setup Planning Module

Setup planning is an important module in WebCAPP, which plans the number & sequence of set-ups and the processes performed in each setup. Setup planning is carried out in two stages. Initially part features are grossly grouped into six set-ups (+/-X, +/-Y, +/-Z) based on their Tool Access Directions (TAD). The total number of set-ups and features per setup are then minimized by following some knowledge rules written into the system. Two typical rules are enumerated.

Rule 1: Include edge-based features like Slot and Step in a setup in which the depth of cut along the TAD is minimum.

Rule 2: Follow strictly the Geometric Tolerance bindings and datum precedences between the features for sequencing the set-ups.

For the component shown in Fig. 2 WebCAPP automatically generates and sequences three set-ups for TADs of –Z, –Y and +Y. (Table 1)

6.3 Operation Sequencing using Genetic Algorithm

Operation sequencing aims to determine proper order between selected operations to minimize time to manufacture the component. In this research work, Genetic Algorithm based sequencing strategy has been developed for determining the optimal sequence of machining operations. Fitness function criteria like minimum tool travel distance and minimum number of tool changes have been employed. Compared to traditional optimization and search procedures like Dynamic Programming, GAs search the optima globally & more exhaustively and are thus expected to yield more optimal process plan. Important steps in the GA based operation-sequencing module include Problem Representation, Population Initialization, Reproduction and Fitness Evaluation. These are explained one by one.

6.3.1   Problem Representation- Encoding

The initial and important step in developing GA is Encoding – i.e. to represent the problem solutions (operation sequence) as a chromosome. Fig. 3 shows a typical component having 22 features to be machined. WebCAPP automatically generates Feature Precedence Graph (FPG) (Fig. 4), from the CAD model file. In the FPG, vertices represent features and the edges represent the precedence relations between them. The problem of operation sequencing is essentially the ordering of vertices in a FPG. Since part features have many precedence constraints, the traditional representation scheme i.e. directly encoding chromosomes using feature IDs (e.g. 1,2,4,16…) may generate infeasible solutions. Encoding strategy proposed by Chiung Moon et. al. [10] for the TSP with precedence constraints is used in the present work. To derive all feasible solutions, Random Priority Assignment technique is used where priority strings having length equal to the number of machining features are generated and mapped with a string containing vertices of the given FPG.

6.3.2   Population Initialization

The initialization of the solution population in the matting pool is done randomly by generating number of chromosomes equal to the population size. Each digit of the chromosome string represents priority of the gene and ranges between one and the number of features. Topological Sort and random priority assignment technique used in the present work always ensure the creation of valid (legal) population.

6.3.3   Reproduction

To create the next generation (offsprings), genetic operators such as <u>Selection</u>, <u>Crossover</u> and <u>Mutation</u> are employed on the chromosomes in the mating pool.

Selection: Parent chromosomes are selected for reproduction from the mating pool based on their fitness values, using the roulette wheel strategy. Elitist approach is used to ensure passing best chromosomes to the next generation. 'Good' results achieved so far are thus, not lost from the evolution process.

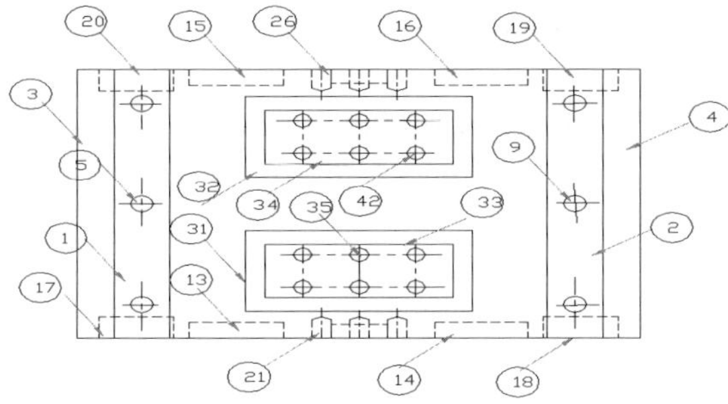

Figure 3 CAD drawing for part shown in Fig. 2

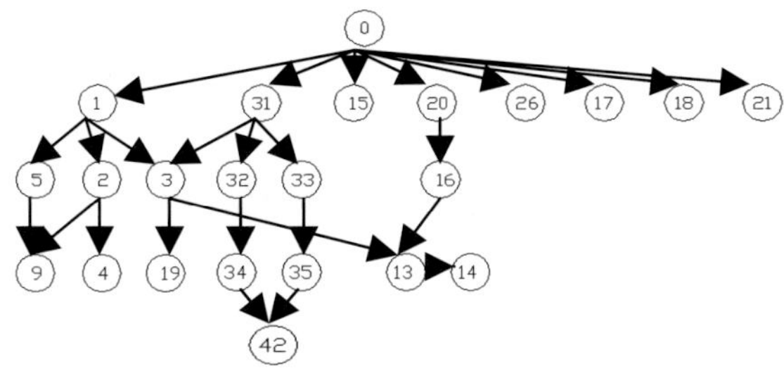

Figure 4 FPG for part shown in Fig. 3

Crossover: Crossover operator operates on two parent chromosomes (P1, P2) to generate two new offspring (O1, O2). It generates a random crossover point and reorders genes of the parent chromosomes after the crossover point according to the order of the other parent to get more diverse offsprings (Fig. 5). 4 and 6-point crossover operators developed in the present work were found to be very effective than the single point operators for handling complex operation sequencing problems having many features.

Mutation: Mutation is used to explore some of the unvisited points in the search space to prevent premature convergence. In our work, 6 -point mutation operators is developed which generates six mutation sites randomly on a single parent to swap the corresponding genes.

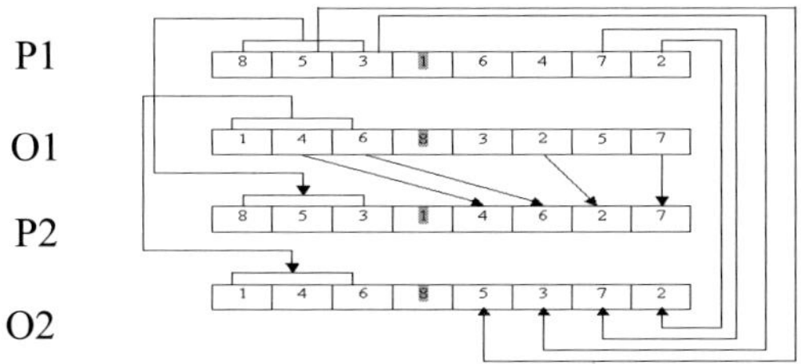

Figure 5 Single point Crossover Operator

6.3.4   Fitness Evaluation

Each chromosome in the population is evaluated using a measure of fitness. In the present work, two criteria viz. *minimum tool travel distance* and *minimum number of tool changes* have been chosen to minimize production time of a component. Quantitative measures of Tool travel distance (TTD) and Tool change distance (TCD) are computed by using feature geometric and location data available in the CAD part model.

The fitness function (f_{chr}) for each chromosome is expressed as under.

$$f_{chr} = w_1 * TTD + w_2 * TCD$$

The weightage factors w1 and w2 can be chosen by the user depending upon the specific user shop preferences for optimization.

6.4   CNC code Generator and Simulator

After operation sequence planning for a setup, three functional activities viz. Tool path planning, CNC code generation and graphical simulation are carried out.

For feature wise Tool path planning, generic machining strategies are written considering the tool allocation and cutting practices (spiral/zigzag). The CL data file so generated is postprocessed to produce Controller specific CNC code in G and M format. Setupwise CNC programs are generated and stored on the server, which can be downloaded by the clients for review, storage and transmission to a remote client CNC machine. In WebCAPP clients can validate the CNC code using VRML-based 3D graphics simulation.

WebCAPP enables a client to directly transfer the CNC code over the network to the controller of a CNC machine at other site. It thus, enables the telemanufacturing of parts on a remote client CNC machine.

7.   IMPLEMENTATION AND TESTING

Various modules of WebCAPP were implemented using PhP, HTML, C++ and Java and the server was set up on a Pentium IV PC with 256 MB RAM. WebCAPP was extensively tested from different Intranet locations considering clients with low to high-end PC configurations, using different OS like Win 98/2K, Unix etc.

WebCAPP was extensively tested by modeling prismatic components with varying complexities taken from industries as well as NIST part library in the collaborative manner. Process Planning algorithms were tested using components needing multiple set-ups each having as many as 20-25 features.

8.   CASE STUDY

The part illustrated in Fig. 3, is to be manufactured from a block on a 3 axis-CNC milling machine. It has 22 machining features with 18 precedence constraints.

The component was modeled in Palantir and transferred to CAPP module automatically. Three set-ups were planned for this part. GA algorithm performed about 20 generations with a population size of 10, single point crossover with probability 0.9 and 6-point mutation with probability 0.6. Both minimum tool travel distance and minimum number of tool changes were optimized. Table 1 depicts optimal operation sequence for Set-ups 1[TAD= -Z], 2[TAD= -Y] and 3[TAD= +Y]. The CNC code generated by WebCAPP was transferred over the network to a remote 3 axis CNC milling center with FANUC controller available in CAM lab, IIT Bombay and actually used for machining the part.

Table 1: Optimal Operation Sequence for Set-ups 1,2 and 3

Setup 1(-Z)			
Feature ID	Name	Operation	Tool ID
1	RectangularStep (1)	Step Milling	2
2	RectangularPocket(31)	Pocket Milling	2
3	RectangularPocket(32)	Pocket Milling	2
4	RectangularStep(2)	Step Milling	2
5	RectangularStep(4)	Step Milling	1
6	RectangularPocket(33)	Pocket Milling	1
7	RectangularPocket(34)	Pocket Milling	1
8	RectangularStep(3)	Step Milling	1
9	RectangularArray(35)	Drilling	4
10	RectangularArray(42)	Drilling	4
11	RectangularArray(5)	Drilling	6
12	RectangularArray(9)	Drilling	6

Setup 2(-y)			
1	Circular Array (26)	Drilling	3
2	Rectangular Slot (20)	Slot Milling	5
3	RectangularPocket(15)	Pocket Milling	5
4	RectangularPocket(16)	Pocket Milling	5
5	Rectangular Slot (19)	Slot Milling	5
Setup 3(+y)			
1	Circular Array (21)	Drilling	3
2	Rectangular Slot (17)	Slot Milling	5
3	RectangularPocket(13)	Pocket Milling	5
4	RectangularPocket(14)	Pocket Milling	5
5	Rectangular Slot (18)	Slot Milling	5

9.   CONCLUSIONS

WebCAPP was extensively tested in a collaborative manner for a large variety of prismatic parts with different degrees of complexities, number of features and tolerance binding constraints. It was found to generate optimal and consistent CNC codes very quickly.

The collaborative Intelligent CAD/CAPP/CNC system reported in this paper would be very suitable for global product design and manufacturing scenario needing 'anytime, anywhere' strategy.

REFERENCES

1. Laurent Sabourin and Francois Villeneuve, "Omega an Expert CAPP System, Advances In Engineering Software", Vol. 25, pp 51-59, 1996.
2. Kim, J.H. et. al., "Design for Machining over Internet", *Design Engineering technical conference (DETC) on Computer Integrated Engineering* Paper Number DETC'99/CIE9082, September 1999, Las Vegas, NV, 1999.
3. Chan, S.C.F., "A solid modeling library for the World Wide Web", *Computers Networks and ISDN Systems*, **30**, 1853-1863, 1998.
4. Rafael Bidara et al, "Collaborative Modelling with features", *Proceedings of DETC'01*, 2001, *ASME Design Engineering Technical Conferences*, Pittsburgh, Pennsylvania, September 9-12, 2001.
5. Niranjan, A.S., *Collaborative Intelligent Product Modeling and Manufacturing*, Master's thesis, Indian Institute of technology, Bombay, 2003.
6. Lalit Patil and S.S. Pande, "An Intelligent feature-based process planning. systems for prismatic parts", *International Journal of Production Research*, **Vol. 40**, pp 4431-4447, 2002.
7. Turkay Dereli, I. Huseyin Filiz, "Optimisation of process planning function by genetic algorithms", *Computer & Industrial Engineering*, **Vol.36**, pp 281-308, 1999.
8. B. Awadh, Nsepehri and O. Hawaleshka, "A computer-Aided Process-planning Model Based on Genetic Algorithm", *Computers Ops Res.* **Vol. 22**, pp 841-856, 1994.
9. M.S. Shunmugam, S.V. Bhaskara Reddy and T.T. Narendran, "Selection of optimal Conditions in multi-pass Face-milling using a Genetic Algorithm", *International journal of machine Tools & Manufacture*, **Vol. 40**, 2000, pp 401-414
10. Chiung Moon, Jopngsoo Kim, Gyunghyun Choi and Yoonho Seo, "An Efficient Genetic for the traveling salesman Problem with Precedence constraints", *European Journal of Operational Research*, **Vol. 140**, pp 606-617, 2002.

COORDINATION COMPONENT BASED ON PETRI NET FOR COLLABORATIVE DESIGN

Feng Zhou, Hong-Zhong Huang and Xu Zu
School of Mechanical Engineering, Dalian University of Technology Dalian, Liaoning 116023, P. R. China E-mail:dlut_zhouf@163.com;hzhhuang@dlut.edu.cn; zu_xu@163.com

Abstract : Considering the lack of efficient coordination of task interdependencies in the collaborative design system, the temporal and resource coordination mechanisms for respective problems are established based on Petri Nets. The whole system could be expanded as a Petri Nets for simulation and analysis. Architecture of reusable and pluggable components is also introduced to implement such mechanisms. The presented approach offers strong support to evaluate the performance of the collaborative design.

Key words: Petri Net, coordination mechanism, temporal interdependencies, resource management, component

1.   INTRODUCTION

With the fast development of computer technology and Internet, we have witnessed the rapid establishment of virtual society and realized the feasibility of remote interaction transcending the geographic location and time constraint. It is necessary for designing of complex artifacts and systems that the cooperation of multidisciplinary design teams which using multiple sophisticated commercial and non-commercial engineering tools such as CAD tools, modeling, simulation and optimization software, engineering databases, and knowledge-based systems. And this has been viewed by researchers and industry engineers as the key for reducing cycle times and improving product quality and reliability [1,3].

The specialists and scholars from the entire world have design and developed many models or environments to meet the requirements mentioned above. Wang introduced a CAD and CAM integrated system to shorten the product development cycle to the greatest degree and rapidly respond to the unceasingly changing market requirement [14]. Some models was established based on VE (Virtual Environment), Java, etc technologies to access, explore and collaborative through the Internet [1]. These models or environments have met those requirements and improved the efficiency of design and development in a certain extent. But none of them have the effective method to coordinate the interdependent tasks in the large the system..

Shirmohammadi and Georganas presented an architecture that support tightly coupled collaborative tasks to be performed efficiently in the design system [12]. But the rigidity of the collaborative protocols restricts its ability of collaboration in the complex system.

In this paper, we establish the coordination mechanisms for temporal and resource interdependencies respectively, and encapsulate the two kind of mechanisms into the coordination components to coordinate the task components which represent the actual tasks in the collaborative design and development system.

2.   RELATED CONCEPTS

2.1   Designing

The process of design is an act that involves members of many professions. Collaborative design is that teams of designers, engineers and manufacturers from several areas and diverse geographical locations work together over networks. [6,15]

As shown in Figure1, the kind of activity is "loosely coupled

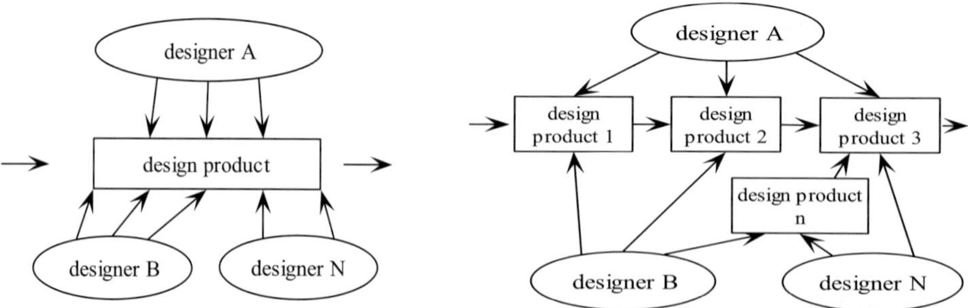

Figure 1. Loosely coupled collaborative design   Figure 2. Tightly coupled collaborative design

collaborative design". It could be well coordinated by a social protocol and participant's abilities even without any explicit coordination mechanism.

Figure 2 shows the activities are what we called "tightly (close) coupled collaborative design". In this kind of activity, one task depends on one another to start, to execute or to end, so it is necessary for the kind of activity that sophisticated coordination mechanisms. During the process of collaborative design, each participant hopes every interdependency task succeed and every other partner succeed too. However, they are not always harmonious. So there must be coordination between these interdependency tasks to ensure the whole systems execute successfully [8].

2.2   CSCW

In collaborative design systems, individuals or individual groups of multidisciplinary design teams usually work in parallel and separately with various engineering tools, which are located on different sites, often for quite a long time. At any moment, individual members may be working on different versions of a design or viewing the design from various perspectives, at different levels of details. Computer supported collaborative work (CSCW) in Design is concerned with the development of such environments.

CSCW systems offer a tool that could potentially enhance the productivity and effectiveness of teams. The primary goals for applying computer supported cooperative work (CSCW) technology include cost reduction, space optimization, and improved performance, effectiveness, and satisfaction. There are also more specific goals, such as improving group cohesiveness and diminishing the influence of dominant figures [5,6].

2.3   Component

A component is a reusable software package or application. A standard application has two main parts, the implementation and the data. The interaction between two applications is through the database. In fact this interaction is limited to the data sharing. A component-based application provides operation services containing the data operation and method operations in its implementation [11].

Cockburn, etc examines and records the major causes of component failure, and provides four principles of component design to encapsulate the problems and guide the designers. The four principles are maximization personnel acceptance, minimization requirement, minimization constraints and external integration. This is the basic principles for establishing all kind of component [2].

3. COORDINATION MECHANISMS

The design and development of a product involve many tasks and these tasks could be decomposed into many subtasks. We can classify these tasks into two main classes, those independent and those interdependent. Independent task has no or little relation with others and therefore can implement separately. Experience tells us that most tasks belong to the latter which interact with others in one or several aspects. Interdependencies are divided into two types, temporal and resource dependencies [7,10].

3.1   Temporal dependencies

Temporal dependencies establish the execute order for the tasks. Figure3 shows the temporal dependencies between two tasks [8].

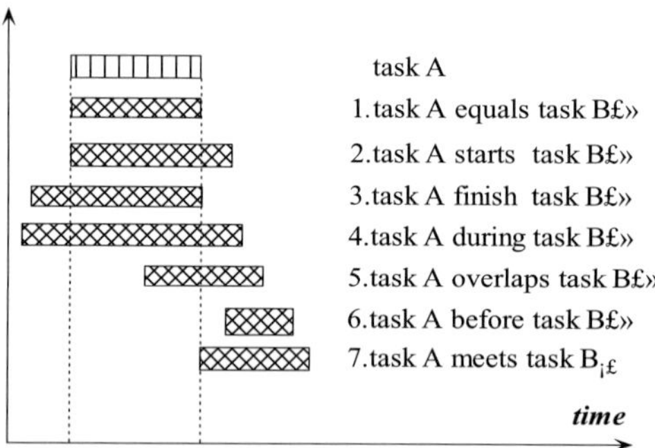

Figure 3. Temporal interdependencies

1. Task A equals task B: Task A and task B start together and have the same time interval.
2. Task A starts task B: Task A and task B start together, but have the different time interval.
3. Task A finishes task B: Task A and task B finish together, and have the different time interval.
4. Task A during task B: Task A is totally contained in task B.
5. Task A overlaps task B: Task A starts before task B, which starts before the end of Task A.
6. Task A before task B: Task A happens before task B, and they do not overlap.
7. Task A meets task B: Task A happens before task B, which starts immediately after the end of Task A.

These conditions include the possibility that may happen between two tasks. And what we could do is to establish corresponding mechanisms to coordinate these dependencies.

3.2   Resource management

The resources needed for tasks are managed by one or more resource managers. Resource managers control the allocation of resources to tasks. Resource management interdependencies are complementary to temporal ones and may be used in parallel to them. This kind of interdependency deals with the distribution of resources among tasks. Three basic resource management dependencies are defined here [8,13].

Sharing: A limited number of resources may be shared among several tasks. It represents a common situation that occurs, for example, when several designers edit a product drawing.

Simultaneity: A resource is available only if a certain number of tasks request it simultaneously. It represents, for instance, a machine that may only be used with more than one operator.

Volatility: Indicates whether, after its use, the resource is available again. For example, a printer is a non-volatile resource, which a sheet of paper is volatile.

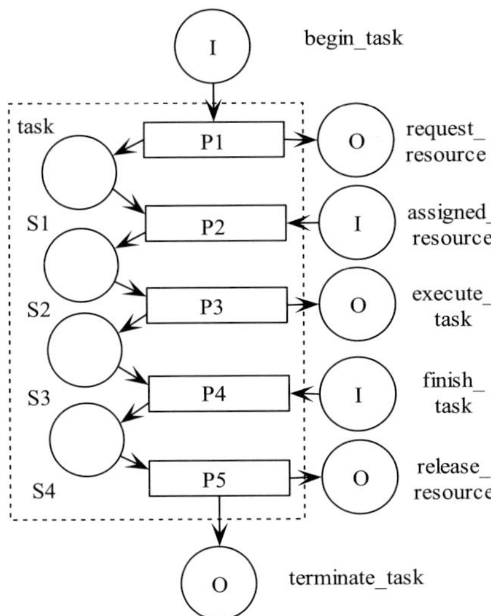

Figure 4. Structure of task [13]

3.3   Coordination mechanisms

In this context, the coordination is defined as the act of managing interdependencies tasks performed to achieve a goal. The coordination is a highly dynamic process because of the renegotiation during a collaborative effort. Without the coordination mechanisms, the participant may get involved in conflicting or repetitive tasks [9].

One coordination mechanism only can solve one kind of corresponding problem, so in the large systems for designing the complex product, many coordination mechanisms need to establish to solve the different kinds of potential problems.

The coordination mechanisms modeled for the temporal and resource management are based on the classical Petri Nets. Figure4 shows the structure of a task which is based on the Petri Nets and Figure5 shows the structure of resource manager [13]. As shown in Figure4, each task has a dependency on another task with five transitions (P1, P2, P3, P4 and P5) and four places (S1, S2, S3 and S4). The places request_resource, assigned_resource and release_resource connect the task with resource manager. The places execute_task and finish_task connect the task with the temporal coordination mechanisms. Consequently both the temporal and resource interdependencies are coordinated.

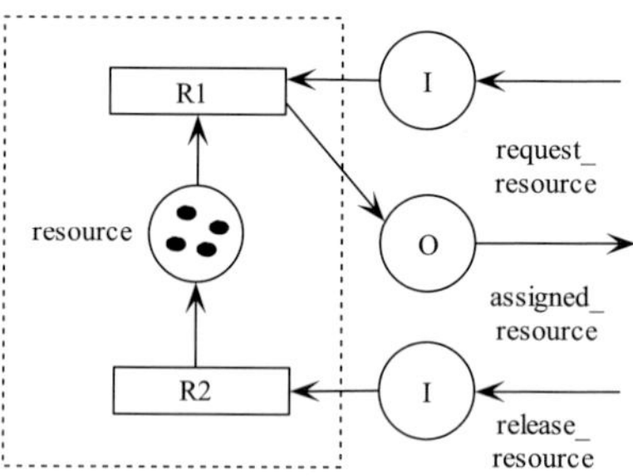

Figure 5. Structure of resource manager [13]

The proposed coordination environment includes three distinct hierarchical levels, workflow, coordination and execution [9]. The whole design system is delineated at the workflow level. All tasks are assigned to different participants, at the same time, temporal and resource interdependencies between the tasks and participants are established. Under the workflow is coordination level where the interdependent tasks are coordinated by the coordination components. All tasks of the system are actually executed at the execution level.

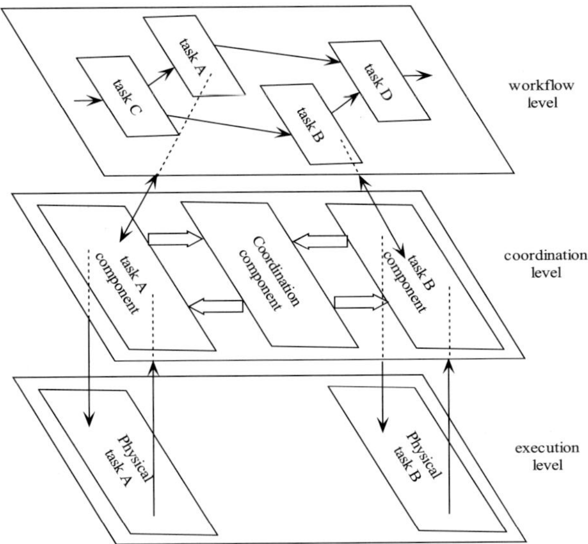

Figure 6. Three levels of collaborative design system

4.   COORDINATION COMPONENT

Figure6 shows the components involved in the coordination interdependencies between the two tasks in the three levels scheme. At the coordination level, there are three components, a coordination component

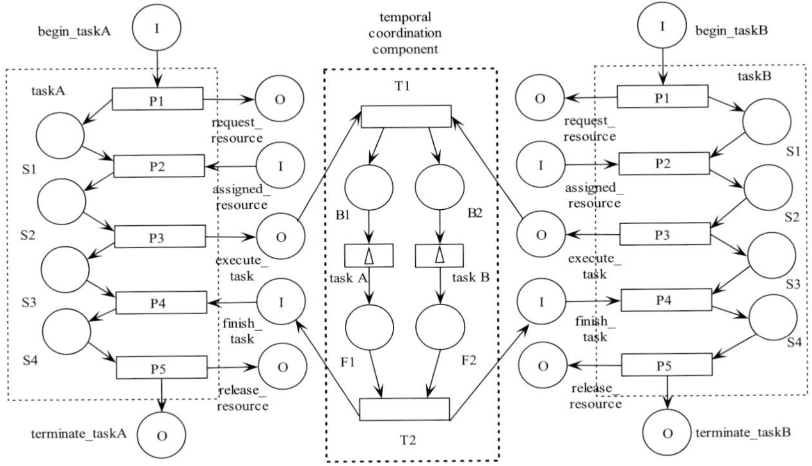

Figure 7. Coordination mechanisms for task A equals task B

and two task components. The two task components represent task A and task B respectively and maintain the task's schedule. The coordination component implement the modeled coordination mechanisms, both for temporal and resource management dependencies.

In order to coordinate temporal relations which were shown in Figure 3, several coordination mechanisms are established. Figure 7 presents the model of coordination mechanism for the temporal relation task A equals task B. Between the two task component task A and task B is temporal coordination component which is ensure both the beginning and finishing of task A and task B is simultaneity. Transition T1 ensures the simultaneous beginning of the two tasks and transition T2 ensures the simultaneous finishing. Transitions task A and task B in the coordination component are non-instantaneous transitions. This means they are the actual execution of the two tasks in the execution level. Different coordination mechanisms are established to coordinate respective interdependencies shown in the Figure 3.

In Figure8, between the two tasks is the resource coordination component which manages the resource between task A and task B. This figure shows the sharing, one of the basic resource management dependency defined above. The place resource contains three tokens which represent the available resource. According to conditions of the input places request_resource and resource, transition R1 will determine whether assigns the corresponding resources to the task which has send the request. Transition R2 reclaims the resources which have been released.

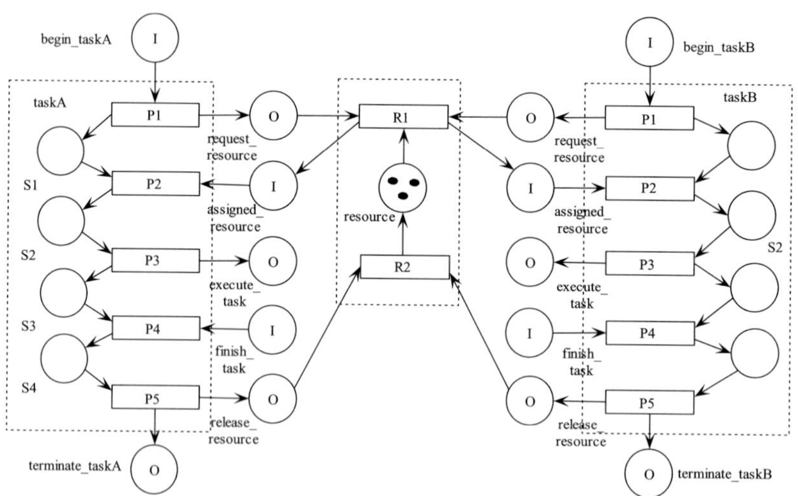

Figure 8. Resource manager for two tasks sharing resources.

In the design and development system, the coordination component was established to coordinate the interdependency tasks. Temporal and resource coordination mechanisms, as two main mechanisms of the coordination component, were encapsulated into the component to implement their functions. So in the coordination component, besides the interaction between the coordination and tasks, there are interactions between these two mechanisms to ensure the component's efficient work. Because we have established all kinds of coordination mechanisms for the resource and temporal interdependencies, the coordination component could coordinate the CSCW system automatically and efficiently.

5. CONCLUSION

In this paper we have presented an approach for tasks coordination in CSCW system. Task interdependencies include temporal and resource interdependencies, and there are several kinds of relations between tasks in two kinds of interdependencies. Different coordination mechanisms were established respectively according to these conditions. For all of these mechanisms were based on the Petri Nets, the whole system could be expanded as a Petri Nets for simulation and analysis, therefore some potentially problems in the system could be found in advance. The approach that encapsulated the two mechanisms into the component which called coordination component makes the coordination in a modular and pluggable way and the whole system have the more flexibility.

In the future, we will combine the CSCW system with VE (virtual environment). The system will be more complex, the coordination mechanisms should be developed to meet the new requirements of the system. We will consummate the mechanisms to improve the ability of coordination, and at the same time improve the computing ability of system to reduce the chance of deadlocks.

6. ACKNOWLEDGMENTS

This research was partially supported by the National Natural Science Foundation of China under the contract number 50175010, the Excellent Young Teachers Program of Ministry of Education under the contract number 1766, the National Excellent Doctoral Dissertation Special Foundation under the contract number 200232, and Taiyuan Heavy Machinery (Group) Co., LTD, China.

REFERENCES

1. Benford S, Snowdon D, Brown C, *et al.* "Visualizing and populating the Web: Collaborative virtual environment for browsing, searching and inhabiting Web space", *Computer Networks and ISDN Systems*, **29**, 1751~1761, 1997.
2. Cockburn A, Jones S. (1995) "Four principles of groupware design". *Interacting with Computers*, 7(2), 195~210.
3. Kan H Y, Duffy V G, Su Chuan-jun. An Internet virtual reality collaborative environment for effective product design. Computers in Industry, 45, 197~213, 2001.
4. Kamel N N, Davison R. "Applying CSCW technology to overcome traditional barriers in group interactions". *Information & Management*, **34**, 209~219, 1998.
5. Monplaisir L. "An integrated CSCW architecture for integrated product/process design and development". *Robotics and Computer Integrated Manufacturing*, **15**, 145~153, 1999.
6. Kvan Thomas. "Collaborative design: what is it?". *Automation in Construction*, **9**, 409~415, 2000.
7. Raposo A B, Cruz A J.A, Adriano C M, etc. "Coordination components for collaborative virtual environments". *Computers & Graphics*, **25**, 1025~1039, 2001.
8. Raposo A B, Magalhaes L P, Ricarte I L M, etc. "Coordination of collaborative activities: a framework for the definition of tasks interdependencies". *Proceeding of 7th International Workshop on Groupware*. Darmstadt: IEEE Computer Society, pp. 171~179, 2001.
9. Raposo A B, Magalhaes L P, Ricarte I L M, *et al.* "Petri Nets Based Coordination Mechanisms for Multi-Workflow Environments". *Computer Systems Science and Engineering*, **15(5)**, 315~326, 2000.
10. Ricarte I L M, Raposo A B, Magalhaes L P, "Coordination in collaborative environments-a global approach". *7th International Conference on Computer Supported Cooperative*. Rio de Janeir,pp. 25~30, 2002.
11. Rosenman M, Wang Fu-jun. (2001) "A component agent based open CAD system for collaborative design". *Automation in Construction*, **10**, 383~397.
12. Shirmohammadi S, Georganas N D. "An end-to-end communication architecture for collaborative virtual environments". *Computer Networks*, **35**, 351~367, 2001.
13. van der Aalst W M P, van Hee K M, Houben G J. "Modeling and analyzing workflow using a Petri Net base approach". *Proceeding of 2nd Workshop on Computer-supported Cooperative Work, Petri net and related formalisms*, pp. 31~50 , 1994.
14. Wang Hui-fen, Zhang You-liang "CAD/CAM integrated system in collaborative development environment". *Robotics and Computer Integrated Manufacturing*, **18**, 135~145, 2002.
15. Xu X W, Liu Tony. "A web-enabled PDM system in a collaborative design environment". *Robotics and Computer Integrated Manufacturing*, **19**, 315~328, 2003.

DYNAMIC DATA EXCHANGE IN THE COLLABORATIVE DESIGN BASED ON INTERNET ENVIRONMENT

Tianhong Luo, Xiaoan Chen, Lihong Lin, Bing Zhang and Zhihong Xian
Chongqing University, China, The State Key Laboratory of Mechanical Transmission, Chongqing University, China, 400044, email: lthcqu@sohu.com

Abstract: The SQL Server is a powerful database management system. This paper analyzes the features of data for collaborative design based on Internet, researches the data transformation, data transmission and data transaction of collaborative design, and develops the system configuration and organization mode of dynamic data exchange for Internet-based collaborative design, in which, STEP serves as the product data exchange standard. Moreover, SQL Server is the kernel database of the system, so that each of the CAD systems can synchronously access the SQL Server database, share and exchange the design information. The STEP- SQL Server serves as the organizing mode of dynamic data exchange for the collaborative design based on Internet, realizing the real-time exchange and consistency of the dynamic data among the cross-platforms and CAD systems.

Key words: collaborative design; dynamic data exchange; STEP; SQL Server.

1.   INTRODUCTION

In the collaborative design based on Internet environment, the product and associated information are visually represented to the team members, what's more, the product data and information can be real-time transmitted among the members. The system is structured to allow team participants to easily access to the relevant product data and information. The characteristics of data for collaborative design can be concluded as followed. (1)The categories of data product are various, such as graphics data and image data which present the product model, documents, charts and formulas and so on. (2) Product data exchanges are very much frequently. Product design is a creative collaborative work which agglomerates the collective wisdoms. In order to successfully design product, members at different space must real-time exchange data and information among the designers. (3) It is the more strict consistency of product data. The members among the teams are correlative and dependent each other, if any data are changed, the correlative data must accordingly change, so that the consistency of data must consider when we design Web database. (4) The concurrency accessing the product data is very much frequently.

According to the characteristics of data management and collaborative design based on Internet, this paper develops the dynamic data exchange mechanism for collaborative design based on Internet, which adopts the multi-database system in which the SQL Server is the kernel database under Client/Server structure, according to the design processing and the change of design objects, in term of certain data model, rationally presents the data storage and abstract, the data access and data consistency, to meet the collaborative design based on Internet.

This paper is organized as follows. Section 2 presents the system framework of dynamic data exchange for the collaborative design based on Internet, defining the function of general components such as data exchange interface, database management system, and communication mechanism etc. Section 3, expresses the dynamic data exchange mode for collaborative design based on Internet. Section 4 expresses the realization process of dynamic data exchange model for collaborative design based on Internet. Section 5 concludes the paper.

2.   SYSTEM STRUCTURE

WWW is an ally of server and clients, supports the HTTP protocol in the Internet. It is an open standard and can be performed at different platforms. Because the dynamic exchange data may be the documents, graphics, images and sounds, clients must lunch appropriate browsers to the locative address. The client-server model describes communication between service consumers (clients) and service providers (servers). This model allows for exchanging messages interactively at different sites. The system structure of dynamic data exchange based on Internet is shown as Fig.1.

In the structure, server maintains the product information that is accessed by the clients, transfers the data into an appointed data format. The clients can be those that are using a CAD system or those that view the product data through WWW browser. For those clients that use CAD systems, the data

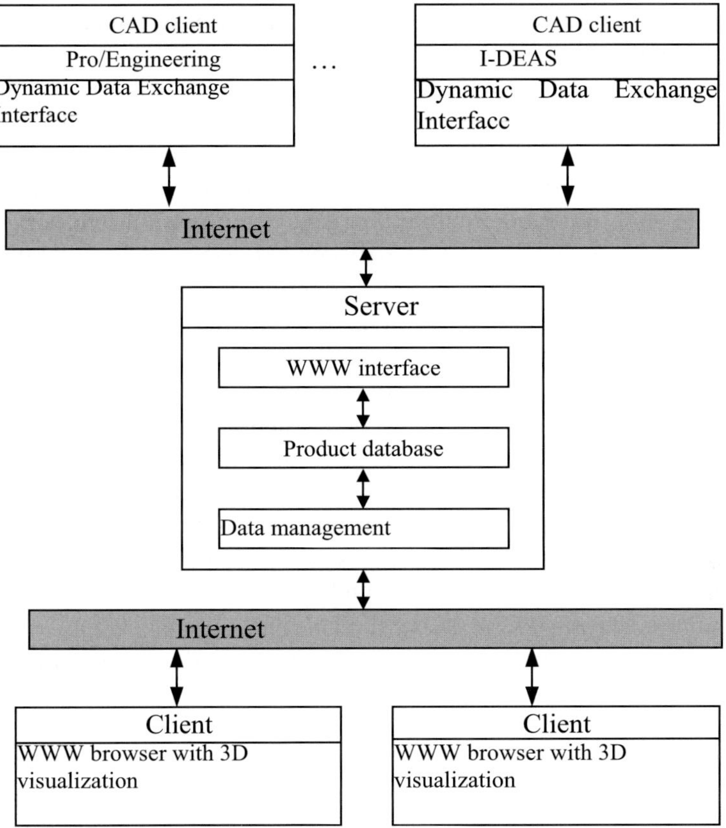

Fig.1 system structure of dynamic data exchange

from the CAD systems are translated to and from the product data standard before being transmitted to and from the server. Communication is carried out using the Internet. Centralized design is no longer necessary, however, at the phases of conception design and detail design, design information can be transmitted between the server and clients, and clients can view and operate the design, at the same time, bring forward the modification suggests and measurements [1]. By this mean, clients can successfully design product to meet the collaborative design. Designers may operate at locative sides by using themselves CAD systems. The CAD data first are transferred into a standard data format and stored into the product database, and other CAD/CAM designers and process planners can access the data via WWW.

The system structure is component of data exchange interface, server, WWW browser and Internet communication protocol. Server is the kernel of the system, including in WWW interface, data management system and product database. In the system model, data management system is component of multi-database management, super data management and data exchange interface.

The multi-database management system is a close-coupling mode, manages the design information of various design phases, and provides the valid, reasonable and credible design environment for clients. SQL Server works as the kernel database to store the crucial data. CAD system databases servers as child-database to store the local data, in which, to certain design phase, data organization and management are relatively reliable and independent based on certain transformation strategy. For close-coupling mode, clients access the data in SQL Server, what's more, it help real-time transmit the information between the kernel-database and child-databases. The transformation strategy of variant databases guarantees the manoeuvrability among the CAD systems, under the condition of uninvolved in the detail of design, processes the data and manages the data.

Super data management is component of edition control, configuration management, distributed access, concurrency control and the data safety. Edition management based on hierarchy network model can meet the demand that clients store, read and abstract the historic information. Configuration management stores the complicated objects, saves the design state to be convenient for design, feedback and adjusting the priority. Distributed access and concurrency control guarantee the consistency of the data[2].

Data exchange interface is the kernel interface for collaborative design, helps system to realize the kernel functions. System automatically transfers the design data model into interior data model, in which child-system can abstract the data needed. Clients provide the HTTP viewer, which responses the interior structure of database. Users can faster view and access the database, understand the change of design condition in time, edit the database and feed back the result to clients. The data exchange of variant child-systems becomes transparency, simple, reliable and safety, so as to predigest the process of colligating system, slow the complexity of maintaining the system. Data exchange interface, including the application interface and exchange transmission interface, provides the services of transparent data exchange among the variants of network, platforms and databases, which simples the synthesis of application system. Application exchange interface provides the exchange standard between the application operation and transmission, including the file catalogue interface and API interface, whose contents are typical symbol of application data, address of application transmission, priority of application transmission. Exchange transmission interface provides the exchange standard between the exchange layer and communication layer, which includes typical symbol of data exchange, address of exchange transmission, priority of exchange transmission.

WWW clients use the WWW viewers, such as NSCA Mosaic, software Explorer, Netscape Navigator, to access these data. The viewers help clients send requests, launch them to the server, and return the results to users. Then, the server receives the request, validate the request, gain the data, and send them back the applicants. WWW browser integrates the traditional network service and multimedia viewer, so that clients can obtain the various of services by the intuitionistic graphic transaction interface.

The characters of system are shown as followed.

(1)The system can adapt to the complicated network environment, the various Internet protocol and communication manners, has the multi-client interfaces and the cross-platforms mechanism.

(2)The system has dynamic expansibility to all sorts of network structures, because any changes of the network structure don't influence the exchange system.

(3) Data exchange and application transaction are independent each other. The data exchange is foreign to the contents and formats of data.

(4) The system has better interface mechanism for application system, provides appropriate encapsulation API interface for application system.

(5) The middleware of the product guarantee the safety, reliability, and are independence of the network environment.

3.   THE MODE OF DYNAMIC DATA EXCHANGE

This paper adopts the STEP-SQL Server structure to establish the organization mode of dynamic data exchange for collaborative design based on Internet, in which the distributed C/S serves as the sustainment platform of the system, shown as fig.2. The data organization mode of SEP-SQL Server is based on Client/Server, in which, the SQL Server is the kernel database lied on the Server, manages the vital information, such as the relationship of databases, the system information and the common data. The kernel database harmonizes all the child-databases by information transmission.

SQL Server is a relationship database management system based on the Client/Server, which transmits requests and responses between Clients and server by Transact-SQL language. SQL Server uses Client/Server structure to divide the tasks into the tasks on the server and the tasks on the clients. The application on the clients provides the data for a client and more one client, which can run not only the clients but also the server that manages the database and assigns the practicable server resource. SQL Server can be

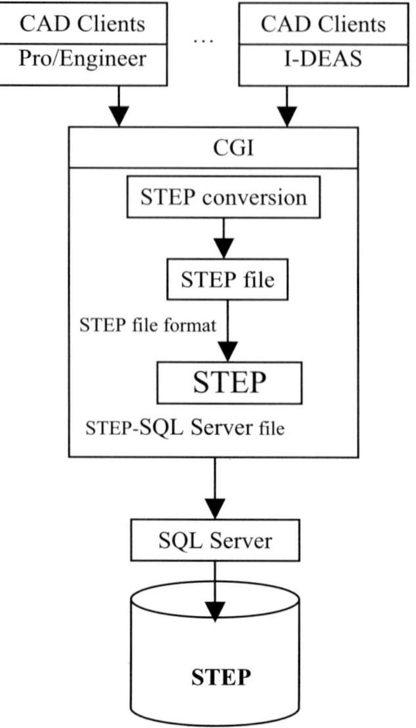

Fig.2 the STEP —SQL Server mode

operated at different platform, so the server of SQL Server may be run at the Windows NT or Windows 9x, while, the environment of Client may be the windows NT, Windows 9x, Windows 3.x, the third platform or Internet viewer. SQL Server Integrates the Windows NT employs the many functions of the NT, the Microsoft Index Server. SQL Server includes the MSSQL Server 、 SQL Server Agent and Microsoft Distributed Transaction Coordinator (MSDTC). When SQL Server system sets up the replication function, the server produces a distributed database which notes all the data and other information during the replication. Data transform server (DTS) provides the functions of output, input, transmitting data between Microsoft SQL Server and ODBS, OLE DB, or among the text files.

STEP manages the product data via uniform product data model and data management software, what' s more, different systems can directly exchange information one another, so it is a data exchange standard and expression standard toward product data define. STEP can integrally present product data and support wide application fields, which includes in each phrase of the life-cycle of product. Moreover, it is a neutral mechanism and independent of any CAX system. What's more, it has many realization fashions, not only is applicable of neutral files, but also supports the data exchange among the modes[3]. STEP presents the product information model by EXPRESS. EXPRESS is a formalization information modeling language, which extracts the functions and characteristic from many formalization express language and programming language. EXPRESS language may be understood by designers and computers, which improves the intellect of designers, can create the application via computer explain mechanism.

Common Gateway Interface (CGI) is a standard, which is used to define the communication manner between the Web server and exterior program. CGI can make exterior application to create HTML file, image and the other files. However, the transaction fashion of server is same with the transaction fashion of un-exterior files. So, CGI application creates static contents and the dynamic contents. CGI provides an interface between computer program and HTTP protocol or WWW server, which is alternation interface of human-computer. Server can provide the alternant sites to replace the old static documents and images, via CGI application. CGI is often compiled by PERL、C/C++、DELPHI. The most main trait of common gateway interface (CGI) application is that it can be written by any computer language, and be run at any platform, as long as operation according with the CGI criterion. CGI has many important functions, such as dealing with HTML list, establishing un-static files on Web page, dealing graphics and image files, searching through Web page, establishing documents independent of platform, founding exchange application, dynamically creating page, editing documents according to clients requirement.

Child-databases of CAD systems lie on the Clients, manage own data in the field. Child-databases of CAD users often use own DBMS. Moreover, data organization and management are relatively independent, correlative with the certain design phase. The STEP geometry schema provides explicit representations for shape or geometric entities of products including geometric items, topological items, and geometric representation models. The geometric representation models are compositions of various geometric items and topological items [4]. CAD clients in the collaborative design have own data format. In order to accurately exchange data among different CAD systems, a data exchange standard must be established to express the data of different CAD systems. The process of CAD system dealing with STEP files is shown as followed. Former processor of STEP transfers data from certain CAD system (A) into a standard data format according with STEP criterions, then, behind processor reads STEP files, and transforms standard data into the certain data which are understood by other CAD system (B)[5].

Any STEP physic files can be transformed and stored in the SQL Server. The contents of STEP- SQL Server are the 3D product design data that may come from different CAD systems, such as CATIA, I-DEAS, AutoCAD, Pro/Engineer and so on. The CAD systems provided the interface module to send their data to STEP physic files[6-7]. In order to load the data into SQL Server, system develops a STEP Loader which reads SQL Server files and creates new geometry models. Resultant of STEP file in the SQL Server can be view and read via WWW.

In order to data safety of the system, we develop a safety measure from two different hierarchies. In principle, clients who are authorized can only amend the data. Firstly, setting different priority according to edition state, system automatically controls the edition operation according to edition state. Secondly, setting display priority for the data of the server, by setting passwords, clients in the item have different dealing priority to the data of the server.

4.   THE PROCESS OF DYNAMIC DATA EXCHANGE

In the fields of collaborative design based on Internet, the data involved in collaborative design are often whose computer aided process of the life-cycle such as CAD, CAPP, CAE, CAQ and CAM (namely CAX), creates and applies the data. Dynamic data is a subset of product data, however, the dynamic data exchange is the kernel of CAX, because dynamic data exchange can be carried through at the different CAX, the different editions and different product types of same the CAX, even during produce assembly and parts. Data exchange is importance of collaborative design, and is the keys of data integration, data share, data real-time exchange and data consistency.

We make some defines shown as followed.

Define 1. Dynamic data transaction includes Insert,   Delete and Update.

Define 2. The data types exchanged are the 3D geometric graphics, images, documents and design parameters, which are created in CAD systems, such as CATIA, I-DEAS, AutoCAD, Pro/Engineer.

Define 3. Exchange objects are the data operated by exchange transaction, not all the data.

Define 4. The operation systems may be the different platforms, such as Windows, Unix.

The dynamic data exchange mechanism and organization mode for collaborative design based on Internet developed by author can succeed in data real-time exchange and share among the members. The steps are shown as followed.

(1)CAD system clients request the server for the data needed exchange (including shared information, graphics, documents, etc) via data exchange interface.

(2) After receiving the request, the server validates the request by multilevel safety authentication and data checkout mechanism, so that the illegitimate requests are refused, contrarily accept them.

(3)The server transforms the data into STEP standard format.

(4)DBMS stores the STEP data in the SQL Server database.

(5)The clients needed the data can abstract the STEP data from SQL Server database. After the STEP data are processed according to data processing information or themselves logic, the data can be reverted the entity model.

We can generalize the advantages of the dynamic data exchange system for collaborative design based on Internet.

(1)The dynamic data exchange system has the exclusive integer mode, and is a close-coupling multi-database system in which SQL Server database is the kernel, can realize the graphics and data real-time exchange between clients and server, among the clients.

(2)The system adopts C/S structure, which is a multi-clients system, and has good expansibility.

(3)SQL Server is the kernel database, each of CAD system clients can use MDBS interface and primary application interface to access the SQL Server database, share and exchange the information. CAD systems define output mode of database, and divide databases into private data and share data which build-up integer mode.

(4)The system has multilevel safety authentication and data check-up mechanism, has the characteristic of non-repetition transmission, non-pretermission transmission and wrong transmission, has data compress encrypt interface, which guarantees the good safety and reliability of the system.

5.   CONCLUSION

The shifting from the traditional design and manufacturing paradigm to a new, virtual and agile model is globally observed. The traditional model characterizes the limited information sharing, static organization structure, and almost no cooperation, Whereas, the new model exhibits information sharing, collaboration, and dynamic organization. Collaborative design based on Internet is an innovative paradigm for product development, which integrates widely, distributed engineers for virtual collaboration. Dynamic data exchange is the key for collaborative design based on Internet. The paper, according to the traits of the collaborative design based on Internet, develops dynamic data exchange structure and mode among different CAD systems at different platforms. Dynamic data exchange operates the data changed during product design, which reduces the load of Internet transmission, improves the efficiency of data exchange.

ACKNOWLEDGEMENTS

The work described herein is funded by the National Natural Science Foundation, China (Grant No. 50175113, 59875087).

REFERENCES

1. Mitchell, James E. "Using the World Wide Web in an Architectural Engineering (AE) Design course", *Proceedings of the 1998 International Computing Congress on Computing in Civil Engineering*, Boston, MA, USA, Oct. 18-21, 1998, pp. 129-132
2. Liang Zhongming. "Dynamic data exchange between CAD and spreadsheet for mechanism design". *Proceedings of the 1995 Annual ASEE Conference*, Anaheim, CA, USA, Jun. 25-28, 1995, pp.1762-1766.
3. Moorthy, Shreekanth. "Integrating the CAD model with dynamic simulation: simulation data exchange". *1999 Winter Simulation Conference Proceedings (WSC)*, Phoenix, AZ, USA, Dec. 5- 8, 1999, pp.276-280.
4. Bhandarkar, Manqesh P. "STEP-based feature extraction from STEP geometry for Agile Manufacturing", *Computers in Industry*, **4(1)**, p 3-24, 2000.
5. Han Soonhung, Choi Young. "Collaborative engineering design based on an intelligent STEP database". *Concurrent Engineering Research and Applications*, **10(3)**, pp.239-250, 2002.
6. Yeh S C, You C F. "Combining EXPRESS and UML to implement a STEP-based system". *International Journal of Computer Applications in Technology*, **15(1)**, pp.98-108, 2002.
7. Gu P, Chan Km. "Product modeling using STEP". *Computer-Aided Design*, **27(3)**, pp.163-180, 1995.

THE IMPACT OF DOCUMENTATION AND REFLECTION ON STUDENT LEARNING IN ENGINEERING DESIGN
A concept development project in the digital domain

A. Wodehouse, H. Grierson, W.J. Ion, N. Juster and A.S. Stone
Department of Design, Manufacture and Engineering Management (DMEM), University of Strathclyde, James Weir Building, 75 Montrose Street, Glasgow, G1 1XJ. Email: andrew.wodehouse@strath.ac.uk

Abstract: As product development teams become global in scale, more of this process is carried out in the digital domain. This paper examines the impact of basing a student design project in this environment, and in particular how the increased documentation and reflection afforded by this impacts upon student learning. The mechanisms for achieving this included templates, information repositories and video presentations. It was found that a shared information resource had an impact on concept direction and that although students found critical reflection on their design process difficult, that the increased documentation of a digital repository encouraged more transparent working practices.

Key words: Documentation, reflection, design, learning.

1.   BACKGROUND

Throughout the conceptual design phase of product development, high volumes of information are acquired and processed by the designer in the rapid production of ideas [1], as indicated by Fig. 1. Indeed, it has been suggested [2] that designers spend 24% of their time sourcing or locating relevant information or knowledge. Student engineers, particularly, often find the shift from information gathering to sketching and idea generation difficult as the initial design boundaries specified in the product design specification (PDS) are lost. This can be addressed by storing the information more effectively, ensuring that it is constantly updated, checked and reflected upon. As a result, the final design is more likely to meet the requirements of the PDS.

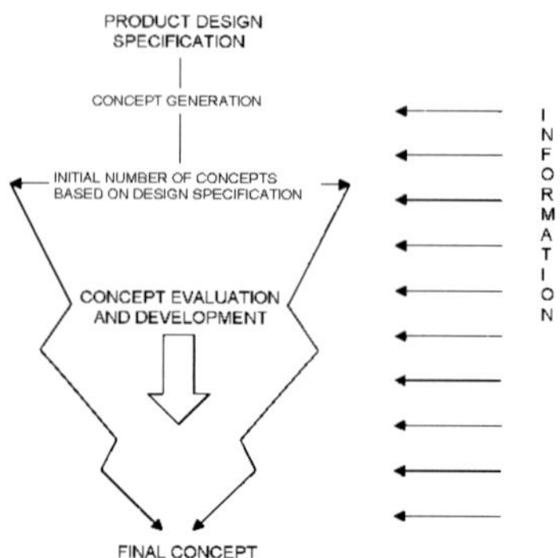

Figure 1. The concept design process (after Pugh [1])

This project seeks to address this issue through the creation of a digital information repository by the students themselves, and a series of checkpoints throughout the design process to encourage student reflection during the free-flowing stage of conceptual design. The aim is to help the designer store information in a manner which allows quick retrieval, and to structure design tasks so that awareness of where they are in the process and the next steps to take are more readily apparent. Computer tools lend themselves well to this increased interaction with information: work in digital libraries has shown that storing information digitally can improve teamwork and understanding [3] and several systems have been developed to provide support in terms of information [4, 5].

This study is part of a larger investigation on 'Digital Libraries in the Classroom' funded by JISC/NSF. The project entitled 'Distributed Innovative Design, Education and Team working' has partners at the Universities of Strathclyde (UK) and Stanford (US).

2.   RESEARCH QUESTIONS

By asking students to undertake a high level of project documentation during the course of a design project, it was possible to ask them to form a critical overview of their information utilisation and concept development process as the project progressed. This paper is therefore primarily concerned with evaluating the impact of this upon their learning experience, and whether it justifies the additional work involved for the student. The key questions addressed are:

- Did reflecting on the information resources gathered in the early stages of the project affect the concept development path?
- Did documenting all project information and storing it in a single digital repository affect working patterns or enhance the learning experience?
- Did documenting work help students to reflect critically on their design process at the end of the project?

3.   DESCRIPTION OF WORK

The test bed was a group of 3rd year Product Design Engineering students who were asked to work in teams of four to rapidly design a can crushing device for soft drinks. The project was organised and run over 6 weeks using TikiWiki (6), an open-source groupware product, in addition to a weekly studio session. The ten teams of four students were each asked to use TikiWiki as a digital repository and collaborative tool. Each team was provided with a team workspace where they could upload images and files,

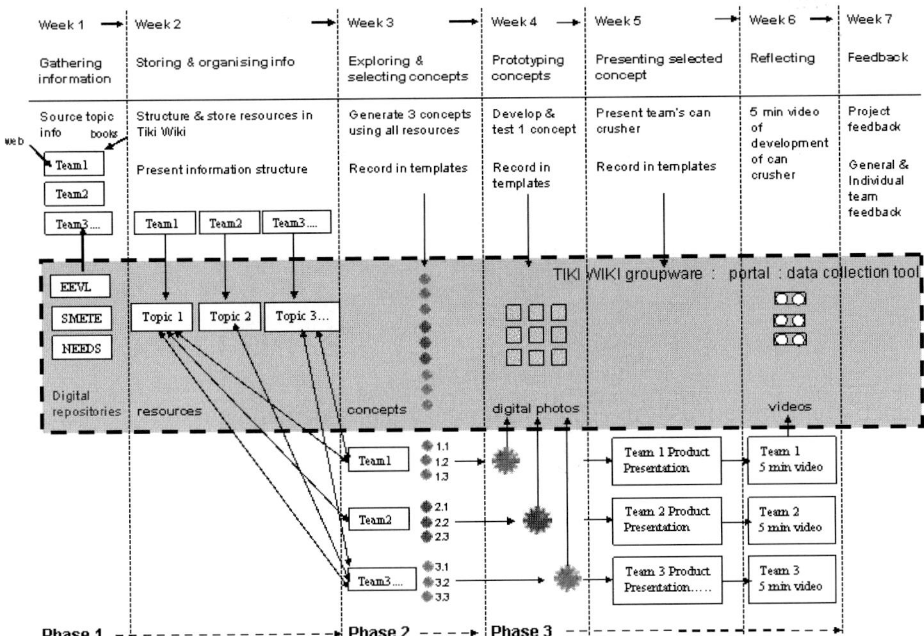

Figure 2. The project structure of the can crusher project.

and create Wiki pages (similar to web pages). Within the Wiki pages this information could be linked and organised into hierarchies.

The work was split into 3 main areas: information gathering (Phase 1), concept generation (Phase 2) and concept development (Phase 3). Figure 2 illustrates the flow of information documentation and key reflection points for students during the project.

During Phase 1, the teams created ten resource sites on topics relating to a can crusher (recycling, market, user environment, mechanisms, aesthetics, ergonomics and safety). They uploaded these to the shared groupware/team workspace and then created an organizational framework for these resources using structured and interlinked Wiki pages.

At the beginning of Phase 2 the group were given access to the information resources created by the other teams (it became a class workspace) and asked to develop three concepts for the can crushing problem with specific references to the created resources. All these concepts were documented and stored digitally using TikiWiki templates. These templates required concepts to be produced at certain project milestones by each team. Most did this by either scanning or digitally photographing sketchwork and embedding the images in the templates.

The final stage of the project (Phase 3) was concept development. Students were required to build a proof-of-concept model and reflect on their design. Each group completed a TikiWiki template outlining their chosen concept, and gave a presentation to discuss the concept's merits. In Week 6 the teams had to reflect on their design process through a video presentation (Fig. 3).

Figure 3. Team 6's reflective presentation in the television studios.

4. RESULTS

In this section, the results have been broken down chronologically. The techniques used to gather this information included staff observation and discussion with students in class, group critiques on information searching, observation of team pages and files, brief weekly questionnaires and online polls, and 5 minute reflective video presentations.

4.1 Gathering background information

The students were asked to make a presentation after Phase 1 on the information they had gathered and the way they had structured it for use in the concept generation stage. The process of creating and organising resources demanded student to think critically about information, its hierarchy and inter-relationships. The TikiWiki graph function (Fig. 4) gave the students a clear understanding of relationships between critical resources. However, they were only encouraged to use this after creating their repository, when it could have provided useful feedback as they developed them. Better teams tended to make the prioritising information transparent, showing clearly how it related to can crushing devices and would help form useful design knowledge; poorer teams simply pasted up text which was unedited and did nothing to inform their design concepts.

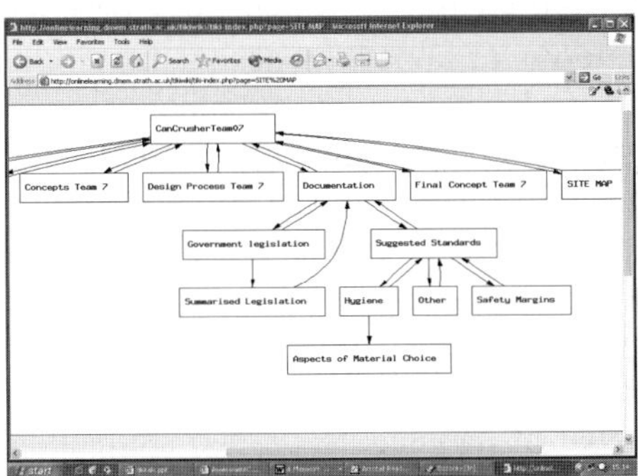

Figure 4. Team 7's site map.

Making resource collections can directly contribute to student learning as students have to interact with the information they have found [7], and improved focus on the major design issues associated with the topics certain teams researched was observed. For example, Team 4 researched mechanisms and produced a final concept with an appreciable more sophisticated crushing mechanism than other groups, and Team 7, which had examined safety, produced an automatic design which reduced any risk to the user.

4.2   Uploading background information

When students were uploading information into TikiWiki, they were required to enter certain metadata, including keywords and descriptions for each file. It has been shown that this process in itself can aid reflection and help students understand how the information relates to the overall design process [8]. However, it is often not immediately apparent to the designer how metatagging will be of use and students found the use of metadata difficult for several reasons: too long, too confusing and not used to them were some of the comments. This was reflected in the relatively poor use overall of keywords and file descriptions. Some thought has been given to creating a new and interactive interface [9] which would make the process quicker, more convenient, and more visual. Additionally, it would be desirable to relate the information more closely to the design process. Work has already been carried out on implementing a new time management module to TikiWiki, called TikiTok. By tying the uploaded information to this chronologically, another dimension of information management can be implemented.

4.3   Documenting design information

The methods used to capture design information (sketchwork, models, meeting minutes) were mainly scanning and photographing, which were equally popular. In terms of impact on design activity itself, it didn't seem to interrupt the 'flow' of idea generation too much, but additional technologies, such as A3 scanners for capturing large-format sketch work, have been identified for future purchases. DMEM provides digital cameras, scanners, electronic whiteboards and access to laptop computers within its design studios, as this is seen as preferable to 'media rooms' which attempt to document the design process in its entirety but at the expense of realistic working conditions.

Providing a degree of choice encouraged students to use their discretion when recording their work. For example, some students typed up the results of brainstorming, whereas others simply photographed their output.

4.4   Storing and retrieving design information

The teams were given three templates in the form of editable Wiki pages during the course of the project as a fixed format to deliver their design concepts. This comprised an image of the concept, a text description, keywords and references to the resources that informed that concept. The purpose of this was to create a uniform hand-in and to save student effort on formatting. Students appreciated this simplicity, and it offset the extra effort required in the documenting of design work as when compiled at the end of the project, the templates also formed the basis of the overall project report.

Additionally, all concept templates were posted on the site and the whole class had access to them during their concept development in Phase 3. Given that each team produced three initial concepts, this meant that there were a lot of "failures" which students could reflect on and learn from while continuing to develop their chosen concept.

Having all information in the same place was extremely useful as it allowed teams to work together in different places at different times. There was evidence that this happened, with students logging on to the TikiWiki site at unusual hours at weekends. There were, however, clear peaks of activity as project deadlines approached!

Students were encouraged to show Wiki links for their concepts, i.e. to indicate which information sources they had found most useful in informing particular concepts. This showed some results in that the better concepts tended to have the links completed, but can also attributed to the fact they were the better groups.

4.5   Reflecting through video presentations

Finally, students were videotaped (5 mins) talking about their overall process. The presentations were supported by the use of project documentation in the TikiWiki environment. Teams created a script which tended to formalise their design thinking, but often this was carried out retrospectively, i.e. they tried to make what they did fit an imagined 'ideal' process rather than understanding why their design took a particular direction. This is similar to the concept of "airbrushing" discussed by Lloyd et al. [10] in relation to students' videos of their design work.

The teams had to present their concepts in a critique in Week 5. There was a 1-week gap before they had to do a reflective video presentation in Week 6. Between these two presentations several teams had taken on board comments on their concept and sought to address them. Many teams tried to force-fit retrospectively their work to an "ideal" design process rather than talking about the real issues and problems they faced.

5.   DISCUSSION

In this section, the three main research questions will be addressed in turn:

5.1   Reflecting on information

Many high quality resources were generated as a result of the project, and it is hoped that a selection from these resources can be harvested to benefit future cohorts of students. Creating their own information resource and being asked to present their findings forced students to think more hierarchically, and to prioritise the important issues. However, not all teams had structured their Wiki pages as well as might have been expected. Some students had not sufficiently considered how others might use their resources; others were reluctant to review and edit information. The teams which had prioritised and organized hierarchically the information they had found were the teams which went on to produce the better concepts.

In retrospect, insufficient preparation had been given to students in the creation of keyword and descriptive metadata. Therefore, adding keywords as they uploaded information had little reflective value- the words chosen seemed random and often inappropriate. Also, it became apparent that the structured Wiki pages (which could be browsed) had undermined the value of metadata as a tool for searching. One member of staff commented that 'ideally students would have used browsing mode [of Wiki pages] for inspiration and keyword searching for targeting'.

5.2   Single digital repository

Observation of team concepts, and the presentations by students, confirmed to teaching staff that all teams had used resources sourced by other teams to inform their concepts. Analysis of TikWiki concept pages revealed that 50% of the teams referenced resources in their templates and that the best concepts were those generated by teams that had interacted with a wider range of resources. This was also evidenced in the assessment reviews; teams that appeared most knowledgeable were those that reported having browsed resources in the early stages of concept generation and/or had having regularly revisited these materials to further develop their concepts. Staff reported that 'the shared resources had helped improve the concept designs compared to previous years'. However, although many high quality resources were generated as a result of the project, students could have made even more use of these to inform their concepts.

Further work is required to monitor exactly how the information resources formed affect particular concepts and this is being attempted in a current project through the use of more detailed and descriptive project templates.

5.3   Reflecting critically on design process

The emphasis on documenting design work in the digital domain meant it was easier to keep track of the work which was carried out by students. Encouraging more transparency in their work - they often tend to internalize their thinking - made it easier both for reflecting on their process and for staff to assess, which then allowed more tailored feedback to be provided.

The students found it difficult to reflect critically on the path their concept development had taken. This was evident in the tendency to attempt to retrospectively make their process fit an imagined ideal rather than addressing the key issues of why their design took a particular direction. An example of this was that most groups submitted a controlled convergence matrix [1] despite in being inappropriate for the level of detail of their concepts and ineffectual in helping to choose the correct way forward.

There was evidence that the topic-specific nature of the information resources created initially informed their final concept, e.g. the mechanisms group had a strong mechanical bent to their design. Having access to the resources and concepts created by other teams, however, encouraged flexibility in their thinking. On occasion, and particularly after critiques and staff feedback, students realised errors in their design and altered their thinking accordingly. One group, however, showed no flexibility in their thinking and ended with the weakest concept of the class. By designing the templates to encourage this type of flexible reflection, rather than force-fitting their design process retrospectively, students could deliver more robust concept designs.

6.   CONCLUSION

This paper reflects on the first project implementation of TikiWiki groupware, which will be the key tool used by the University of Strathclyde in its efforts to create global design projects for engineering design students in the future. The effect of recording work digitally on student learning has been monitored specifically to allow changes to the system before it is used on a global level.

It was found that a shared information resource had an impact on conceptual design work, and that although students found critical reflection on their design process difficult, the increased documentation required by a digital repository encouraged more transparent working practices.

7.   FUTURE WORK

TikiWiki is open source, allowing many modifications to be made to the system to suit particular projects. It is anticipated that this optimisation will continue apace. New visual interactions and gaming techniques are envisaged as methods of making the process of searching, retrieving and using information within a structured design process. Alternative ways of capturing design work will be considered, and the template system for design documentation will be enhanced, expanded and integrated with information management tools to create a system which works in the background to help and support the student designer through the product development process.

ACKNOWLEDGEMENTS

Thanks to our Strathclyde colleagues Andrew Lynn, Allison Littlejohn, David Nicol and Lou McGill for their input.

REFERENCES

1.   Pugh, S. Total design: integrated methods for successful product engineering, Addison-Wesley, 1990.
2.   Crabtree, R.A., M.S. Fox and N.K. Baid "Case Studies of Coordination Activities and Problems in Collaborative Design", Research in Engineering Design 9: pp. 70-84. 1997.
3.   Nicol, D.J. and I. MacLeod, , "Using a Shared Workspace and Wireless Laptops to Improve Collaborative Project Learning in an Engineering Design Course", *Computers and Education.* 2004, under review.
4.   Rodgers, P.A., A. P. Huxor and N. H. M. Caldwell.. "Design Support Using Distributed Web-Based AI Tools", *Research in Engineering Design.* 11: pp. 31–44, 1999.
5.   TikiWiki (2004). Home page: http://www.tikiwiki.org (R 1.6), [Accessed 11/10/03]. University of Strathclyde homepage: http://onlinelearning.dmem.strath.ac.uk/tikiwiki/tiki-index.php
6.   Denard, H., (2003). E-Tutoring and Transformations in Online Learning, Interactions, Summer term 2003, 7(2), School of Theatre Studies, University of Warwick. http://www.warwick.ac.uk/ETS/interactions/vol7no2/denard.htm
7.   Grierson H., D. Nicol, A. Littlejohn and A.J. Wodehouse. "Structuring and Sharing Information Resources to support Concept Development and Design Learning", Networked Learning Conference, Lancaster, UK, 2004.
8.   Wodehouse, A.J. and D.A. Bradley, "Computer Tools in Product Development", International Conference on Engineering Design 2003, Stockholm, Sweden, The Design Society, pp. 333-334, 2003.
9.   Lloyd, P., Valkenburg, R., McDonnell, J. "The truth about designing: Conclusions from the Video Assisted Learning in Design (VALiD)", *International Conference on Engineering Design 2003*, Stockholm, Sweden, The Design Society, pp. 377-380, 2003.

ALUMINOTHERMIC REDUCTION OF ZIRCONIA

Xiaoli Zhe, C A Dioka and A Hendry
Materials & Metallurgy Group, Department of Mechanical Engineering, University of Strathclyde, Glasgow, UK [3]

Abstract: Results are presented for the reaction of aluminium metal with zirconia ceramics under (reducing) sintering conditions. Aluminothermic reduction is a form of self-sustaining high-temperature synthesis (SHS). During reaction the principal phase formed is zirconium monoxide which is shown to have a rock-salt cubic structure and a unit-cell dimension of 0.46258nm.

When a yttria partially-stabilised zirconia is used as starting material, composite structures of ZrO, alumina and cubic zirconia are produced which are the result of reduction reactions which proceed through the liquid metal phase. The results are interpreted through X-ray diffraction, microstructural and phase equilibria considerations which show that the sintering gas also plays a role in reaction through the oxygen partial pressure.

The results allow calculation of an approximate value of the free energy of formation of zirconium monoxide ($-294,500 J.mol^{-1}$) which is consistent with the reported values for the monatomic carbide and nitride of zirconium and with other similar transition metal carbide, nitride and monoxide series.

Key words: ceramics, composites, reactions

1.   INTRODUCTION

Cubic zirconium dioxide is known to be an oxygen ion conductor[1] and an important phase in toughenable cubic/tetragonal partially stabilised zirconia structural ceramics[2]. Thus there are features of the behaviour of zirconia which are important in both functional and structural ceramic applications. In addition, zirconia is also used as a corrosion-resistant high-temperature material in slag-band inserts in submerged entry refractory tubes in one continuous casting of steel[3]. The technology of zirconia is well understood in each of those respects but the development of specific properties depends on the processing route and in many cases these are complex and expensive. There is a need therefore to develop for industrial ceramics new methods of producing tailored zirconia microstructures which do not require high-cost materials or complex production routes. This applies in particular to zirconia for functional use. The present work describes the development of such materials and microstructures for sensor applications.

The origins of the present paper lie in the application of self-sustaining high-temperature synthesis (SHS) in the form of aluminothermic reduction to the formation of controlled morphology zirconia composites. Previous work has shown[4,5] that a simple approach of powder processing and control of ceramic/gas interaction during sintering can produce microstructures with combinations of functional properties and good high-temperature mechanical strength.

The technological applications are speculative but the preliminary work[6] has shown that microstructures which can be used as sensors are possible. For example, in nitride-bonded silicon carbide tubes for combustion an insert of the appropriate zirconia composite with ionic oxygen-conducting character could be used as the electrolyte to monitor combustion conditions inside the tube. The material has sufficiently good strength and is resistant to thermal shock but can also be incorporated into the tube structure by conventional isostatic pressing and firing during which the aluminothermic reduction forms the required microstructure in situ. However, before any such applications can be considered a full knowledge of the parameters controlling the development of microstructure must be established.

2.   EXPERIMENTAL PROCEDURE

To produce a highly reduced zirconium oxide, commercial aluminium powder (Goodfellow Metals) with mean particle size of 30μm and purity 99.9% and tetragonal yttria-stabilised zirconia powder (Zirconia Sales U.K. Ltd.) with mean particle size of 0.5μm and purity 99.7% were used as reactants. Up to 20w% (53.9 mol%) of commercial aluminium was mixed with 80w% Y-stabilised zirconia powder (Table 1) and ball milled with alumina balls and alcohol in a propylene jar for one hour. The dried mixture was isostatically compressed for sintering at pressure of 350MPa into pellets of diameter 20mm and thickness 5mm.

Control samples of aluminium with pure monoclinic zirconia were prepared in a similar manner.

Table 1. Compositions

(a) Composition & phase of powders (mol%)			
Powder	**Additive**	**ZrO_2**	**Phase**
ZrO_2	-	>99.8	Monoclinic
Y-ZrO_2	$3Y_2O_3$	97	Tetragonal

(b) Starting composition of composites (wt%)		
Sample	**Al-metal**	**Ceramic**
20Z	20	80 ZrO_2
5Y	5	95 Y-ZrO_2
10Y	10	90 Y-ZrO_2
20Y	20	80 Y+-ZrO_2

Hot press sintering (HP) was conducted in air using an induction heated facility with graphite tooling. A boron nitride (BN) powder bed was employed to prevent direct reaction between the green pellet and the graphite die in a standard hot-pressing assembly and this arrangement also provides a quasi-isostatic pressure during hot pressing. The temperature was controlled manually and monitored by an optical pyrometer. The reading taken from the surface of the graphite die was corrected to a real temperature inside the die with accuracy of $\pm10°C$. After heating to sintering temperatures within 0.5h, samples were held for 0.5h under a quasi-isostatic pressure of 70 bar and then cooled to below 1000°C in 0.5h with the pressure removed.

X-ray diffraction data were obtained using a Philips step-scanning diffractometer with Cu Kα radiation and a Ni filter using a step size of 0.1° and a counting time of 20 seconds. The microstructure of the composite was examined using a CAMECA SX100 electron-probe micro-analyser (EMPA) with EDS (energy dispersive spectoscopy) and WDS (wavelength dispersive spectroscopy) capability for the light elements O, N and C.

3.   RESULTS

General view of phases in the composites

X-ray diffraction results (XRD) of sample 20Z (pure zirconia) hot-pressed between 1000°C and 1500°C for 0.5h, show that the only Al-containing compound is α-Al_2O_3 and no metallic Al can be detected in the sample. This clearly indicates that Al reacted totally to form alumina during

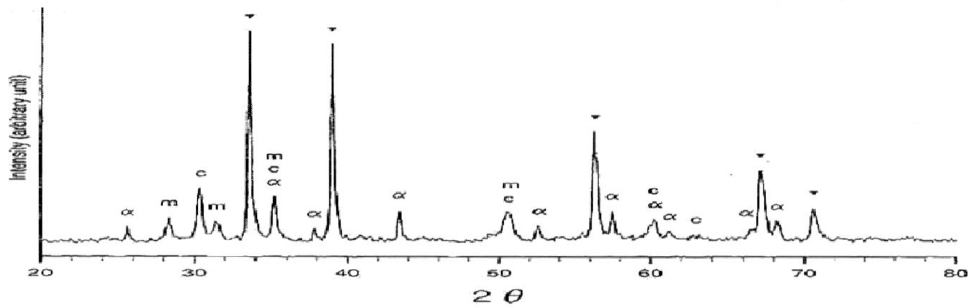

Figure 1. XRD pattern of composite sample 20Z hot-pressed at 1500°C.

sintering. Figure 1 is a representative XRD pattern of a sample hot pressed at 1500°C. There are two modifications of zirconia present, monoclinic and cubic phases. The former is untransformed zirconia from the starting materials but the latter is unusual in that the cubic phase does not usually occur in a sintered sample without a stabilising oxide additive (as discussed later). The major phase is marked by (∇) is similar to ZrC and ZrN in the intensity of the peaks which are characteristic of a rock-salt structure but different from the latter in cell dimension, which is close to ZrO(s). To identify this ZrO-like phase, a more precise diffraction pattern with scan step of 0.01° and internal standard KCl was taken from the same specimen for further indexing of reflections and refinement of the lattice constant.

Indexing and refinement of cell dimensions for the ZrO-like phase
The corrected peak positions with accuracy better than 0.02° from specimens of 20Z were used to carry out indexing and refinement of the lattice constant. The refined cubic unit cell of $a = 0.46258$nm is significantly different from 0.4577nm for ZrN and 0.4693nm for ZrC (ICDD cards 35-753 and 35-784). A similar result was obtained ($a = 0.46289$nm) for this specimen by a Hagg-Guinier camera technique[7]. The only similar literature evidence of a rock-salt structure is zirconium monoxide ZrO, reported as a calculated pattern with $a = 0.4620$nm in ICDD Card 20-684. This substance was generally believed not to be stable as a pure solid phase[8].

Phases in Y-stabilised zirconia composites
Phases in 20w% Al-Y/ZrO$_2$ (20Y) hot-pressed at 1500°C are shown by

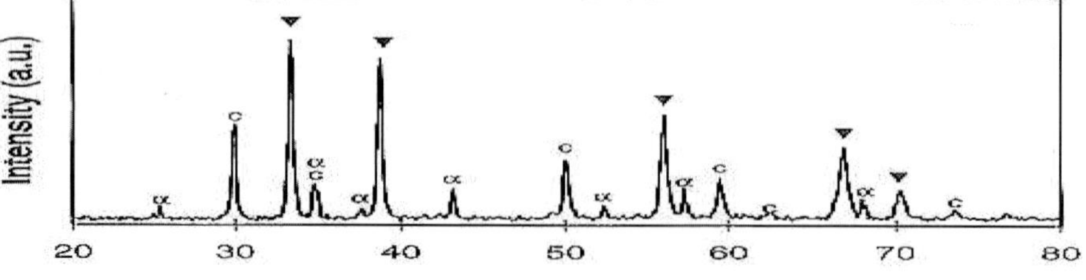

Figure 2. XRD pattern of composite sample 20Y hot-pressed at 1500°C

XRD in Figure 2. The major phase is again a ZrO-like compound with cell dimension of 0.46250nm which differs only slightly from that in 20Z sintered at the same temperature. The only Al-containing phase is α-Al$_2$O$_3$ and no Y-Al garnet was found in the samples contrary to expectation. This indicates that yttria remains in solid solution in the ZrO$_2$ phase as a stabiliser and although the starting ceramic is tetragonal ZrO$_2$, there is only the cubic modification after sintering.

For different amounts of Al additive, the composition of phases of Y-stabilised zirconia composites hot-pressed at 1400°C and 1500°C is shown in Table 2. In contrast to composites with pure zirconia there is a strong tendency to form cubic ZrO$_2$ even at the lower temperature. The ZrO-like phase is not formed in conditions where temperature is <1400°C and Al additive is <10wt.%.

Table 2. Phases in Y-composites hot-pressed at 1400 and 1500°C

Temperature (°C)	5Y	10Y	20Y
1400	c + α	c + α	c + ∇ + α
1500	c + α	c + ∇ + α	∇ + c + α
Note: ∇ --- ZrO-like, c--- c- ZrO$_2$, m--- m-ZrO$_2$ and α--- α-Al$_2$O$_3$; appear in high to low sequence against the most intensive peak in each phase.			

Microstructure

An optical micrograph of the general appearance of the composites is shown in Figure 3. This appearance is typical of both hot pressed and of pressureless sintered samples and shows roughly spherical regions which reflect the shape and size of the original aluminium metal particles which have subsequently melted and reacted during sintering. Within these regions there is finer microstructural detail which is discussed below. The matrix is typical of sintered zirconia with a relatively fine grain structure and associated porosity.

EPMA data in the form of area plots have been used to identify the elemental distribution in the microstructure. Figure 4 shows elemental maps of the distribution of Al, Zr and O in the reacted metallic particles. It is clear that zirconium and oxygen have been incorporated into the aluminium metal resulting in a eutectic-like structure of mixed zirconium and aluminium oxides and with a region around the outer periphery which is high in alumina and lower in zirconia.

4.   DISCUSSION

It is clear from the results of X-ray diffraction studies that the phases formed by aluminothermic reduction of zirconia depends on both the nature (stabilised or pure material) of the zirconia used and the conditions of sintering.   In the absence of a stabilising additive (ytrria) the product of reaction hot pressing is primarily the ZrO rock salt phase with some residual monoclinic zirconia from the starting material and also some cubic zirconia. All of the aluminium metal has been converted to aluminia.   When yttria-stablised zirconia is used the composition after hot pressing is similar but no monoclinic zirconia is present indicating that additional stabilisation by yttria has produced only cubic zirconia.   In all cases however the microstructure is "composite"; that is it is a heterogeneous distribution of phases.

In order to assess the phase distributions which might be expected from such reactions the phase diagram (plotted in atomic concentration) shown in Figure 5 has been assembled. It is necessary to plot in this way rather than the more usual oxide compositions in order to take account of the variable valency of zirconium between ZrO and ZrO_2.   The diagram shows the stoichiometric composition (S) which is given by the reaction,

$$2Al + 3ZrO_2 \rightarrow Al_2O_3 + 3ZrO \tag{1}$$

and which corresponds to an aluminium concentration in the reactant mix of $12.7^{w}/_{o}$. Thus the present experiments with $20^{w}/_{o}Al$ have an excess of metal for this reaction as represented by the composition, E_o. Further, from analysis of the X-ray diffraction results (Figure 1) the products of reaction lie within the triangle ZrO - ZrO_2 - Al_2O_3.   There is therefore an inconsistency between the starting and final phase compositions thus defined which can only be explained by a net increase in the amount of oxygen in the reaction system as the ratio of the metallic elements remains the same. This can be represented either as a chemical reaction or by plotting on the ternary phase diagram. Construction of such a diagram (Figure 5) is achieved by taking quantified diffraction data for the proportions of phases from a sample of composition 20Z sintered at 1500°C.   From the data in Figure 1 the product of reaction has the phase composition $46^{w}/_{o}$ ZrO, $33^{w}/_{o}$ Al_2O_3 and $21^{w}/_{o}$ ZrO_2 and this is plotted (in appropriate composition units) on Figure 5, (E_s).   The change in oxygen concentration from the starting to the final composition is given by E_oE_s on the diagram. There is no gain or loss of the metallic elements and this is confirmed by the extrapolation of E_oE_s which passes through the oxygen (O) corner of the diagram.   The increase in oxygen content during sintering is from 49.6at% (21.6wt%) in the starting mixture to 59.4at% (28.8wt%) in the sintered sample and this is assumed to come from oxygen in the sintering (hot-pressing) atmosphere.

The equivalent chemical reaction (equivalent to the point E_s in Figure 5) is then,

$$2Al + 2ZrO_2 + 0.8O_2 \rightarrow Al_2O_3 + 1.4ZrO + 0.6ZrO_2 \qquad (2)$$

(Given the accuracy of determination of the phase composition by diffraction and hence the precision of point E_s, Equation 2 has been rounded to the first decimal place of numbers of moles). In principal, from such an equation an oxygen partial pressure could be calculated for reaction of the pure phases and which could then be compared to the expected po_2 for the sintering system. However no data is available for the free energy of formation of ZrO and so the calculation can not be done in that way. Conversely, if a value of po_2 is calculated for the hot-pressing conditions used in these experiments then an estimate of the stability of ZrO can be made as follows.

The hot-press system consists of carbon tooling which at the start of the experiment is in air ($21\%O_2 + 79\%N_2$). On heating it is assumed that all of the oxygen contained in the heated enclosure is converted at 1500°C to carbon monoxide

$$2C + O_2 \rightarrow 2CO \qquad (3)$$

and thus 2 moles of CO are created from every mole of oxygen with the nitrogen remaining unchanged. The relative partial pressures are then (assuming the heated enclosure to be a closed system), $p_{CO} = 0.35$atm $p_{N2} = 0.65$ atm.

The equivalent partial pressure of oxygen can then be calculated from Equation 3 and the free energy of formation of carbon monoxide[9], $p_{02} = 2.2 \times 10^{-17}$ atm.

Putting this value into Equation 2 and using the relevant free energy data for that set of compounds in their pure state[9], the free energy of formation of ZrO at 1500°C is,

$$\Delta G_O (ZrO) = -294{,}500 \text{ J.mol}^{-1} \ (-70{,}460 \text{ cal.mol}^{-1})$$

This value is consistent with the values of free energy of formation of the corresponding carbide and nitride of zirconium and with that of other transition metal carbide, nitride and monoxide series[9].

The above analysis of Figure 5 and Equation 2 gives the overall changes which result from the experimental conditions used. However it does not provide a mechanism by which the reaction takes place and by which the observed microstructure forms; this is somewhat more speculative.

It may reasonably be assumed that just as the oxygen pressure in the sintering system is controlled by the external gas atmosphere, so the number of moles of oxygen in addition to those in the zirconia starting compound (Equation 2) arise from gas (air) within the compact pellet. As the temperature rises and metallic aluminium melts then reaction with the gas takes place to form a thin skin of aluminium oxide around the metal particles. This is exothermic but at this stage is limited in extent. It is known from thermal analysis data[6] that a marked exotherm occurs at about 1000°C when the major reaction of aluminium metal occurs with residual gas and, more importantly, above 1200°C in the reduction of the zirconia ceramic powder matrix. As this strongly exothermic reaction proceeds both zirconium and oxygen are incorporated locally in metal particles to form Al-Zr-O mixed oxide phases which when cooled are seen as the $Al_2O_3 - ZrO_2$ "eutectic" structure of Figures 3 and 4. Although the areas corresponding to the prior metal particles have a eutectic oxide appearance and the phase composition is consistent with this (Figure 5) it would require temperatures locally in excess of 1710°C (Reference 10) to form a true eutectic. The strongly reducing action of the aluminothermic reaction reduces the matrix of zirconia grains to the monoxide (ZrO) phase which is confirmed by X-ray diffraction (Figure 1) as the major phase. It is emphasised that these deductions are speculative but they are consistent with the diffraction and microstructural evidence.

5.   CONCLUSIONS

It has been shown that the aluminothermic reduction of zirconia can be used to produce composite microstructures with a heterogeneous distribution of phases. Under suitably reducing conditions the principal phase is zirconium monoxide which is formed together with alumina and cubic zirconia. The results show that reaction of aluminium metal with zirconia powder proceeds with incorporation of oxygen from the sintering gas at low partial pressure and that both zirconium and oxygen are incorporated into the former aluminium particles as reaction proceeds. The resulting microstructure is consistent with a mechanism in which dissolution of zirconium and oxygen in the (liquid) metal at sintering temperatures leads to formation of a reduced matrix of zirconium monoxide.

The free energy of formation of zirconium monoxide has been estimated as -294,500 $J.mol^{-1}$ which is consistent with the values for similar transition metal monoxides. The structure of ZrO is a rock-salt cubic unit cell and the unit-cell dimension is 0.46258nm.

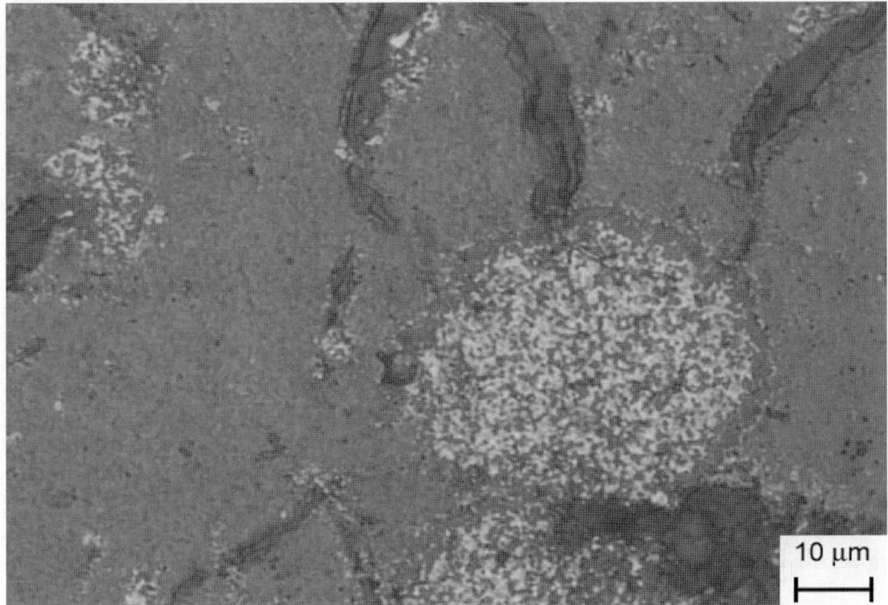

Figure 3. Optical micrograph of the reacted microstructure

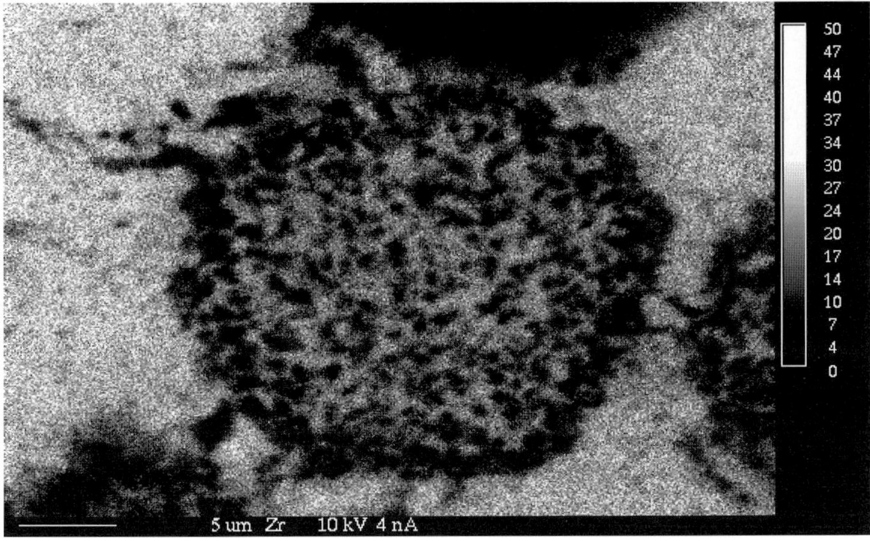

Figure 4(a) EPMA zirconium concentration map

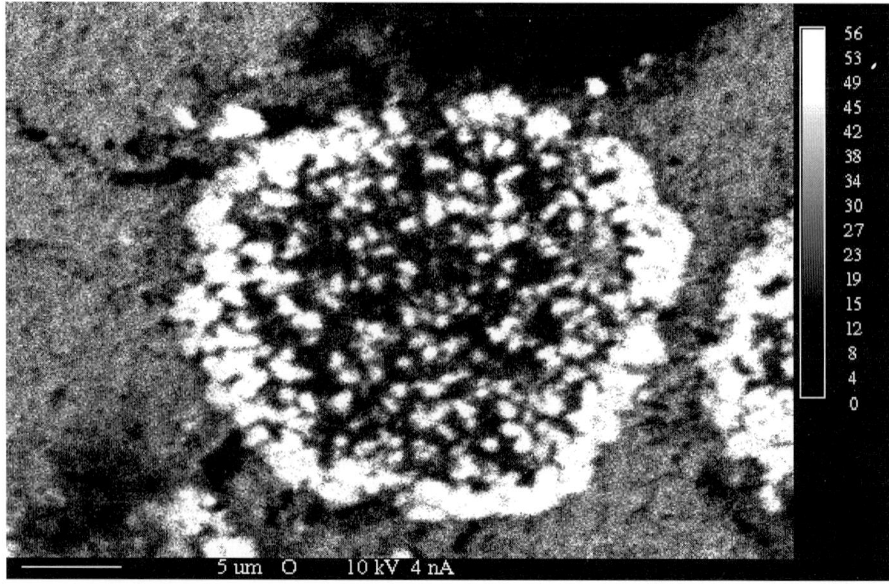

Figure 4(b) EPMA oxygen concentration map

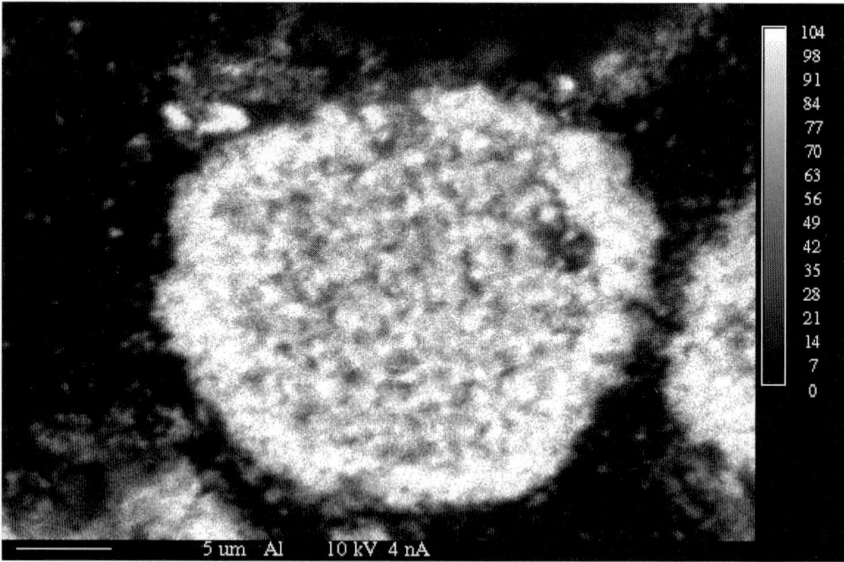

Figure 4(c) EPMA aluminium concentration map

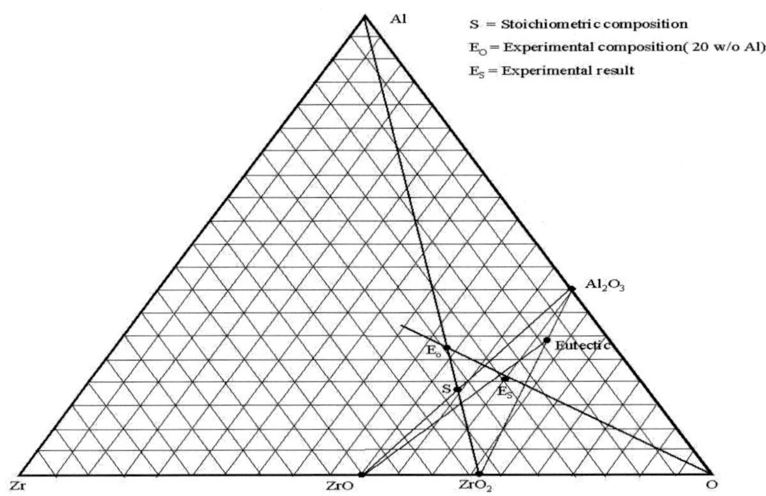

Figure 5 Ternary composition diagram of the Zr-Al-O system (atomic concentration)

REFERENCES

1. R Stevens, "An introduction to zirconia", *MEL Publication* **No 113**   (Magnesium Elektron; Manchester, England), 1983.
2. EP Butler, "Transformation toughened zirconia ceramics", *Materials Sci &   Tech*, **1**, 417, 1985.
3. JR Singh, RB Poeppel, JJ James & JJ Picciolo, "Microstructural development   of refractory composites with improved fracture toughness", in Microstructure & Properties of Refractories, Eds JR Singh & S Banerjee, Key Engineering Materials, **88**, (Trans Tech Publications; Aedermansdorf, Switzerland) p103, 1993.
4. X Zhe & A Hendry, "In situ synthesis of hard and conductive ceramic composites   from Al and ZrO_2 mixtures by reaction hot pressing", *J Mater Sci Letters*,   17, 687, 1998.
5. X Zhe, LM Watson & A Hendry, "Formation of meta-stable zirconium   monoxide during thermite reaction hot pressing" *in Proc of 4th Int Conf on SHS, Ed   MA Rodriguez*, (ICV; Madrid, Spain), 1997.
6. X Zhe, "Novel zirconium oxide-based ceramic composites", PhD Thesis, University   of Strathclyde, 1999.
7. K Liddell, (1996), University of Newcastle upon Tyne, Unpublished work.
8. RJ Aschermann, SP Garg & EG Rauh, "High temperature phase diagram of   the   system Zr-O", *J Am Cer Soc*, **60**, 341, 1977.
9. OA Kubaschewski, "Metallurgical Thermochemistry", (Pergamon Press; London) 1979.
10. "Phase Diagrams for Ceramists", Eds EM Levin & HF McMurdie, (American   Ceramic Society; Columbus, Ohio, USA), Figure 4377, 1975.

MODELING THE OXIDATION AND CORROSION BEHAVIOR OF 3D CARBON FIBER REINFORCED SIC MATRIX COMPOSITES FROM 400 TO 1500 DEGREES C

Qingfeng Zeng, Laifei Cheng, Litong Zhang, Xingang Luan, Xiaowei Yin and Yongdong Xu
State Key Laboratory of Solidification Processing,Northwestern Polytechnical University, Xi'an Shaanxi 710072, P.R. China, Email: qfzeng@mail.nwpu.edu.cn

Abstract: SiC coated three-dimensional textile carbon fiber reinforced SiC matrix composites (3D C/SiC) are promising materials for high temperature from 400 to 1500degreesC, oxidizing and corroding environment. In different temperature range from 400 to 1500degreesC, the oxidation and corrosion mechanisms of 3D C/SiC are quite different at different pure or mixed atmosphere including oxygen, water and sodium sulfate. The quality of SiC coating, which can be characterized with deposition times, thickness and grain size, is a key factor that effects on the weight change of the composites working in an environment. The environmental effects on the weight change of the composites have been studied through a series of oxidation and corrosion experiments. Quantitative models were developed based on the mechanism analyses and experimental data, which can predict the weight change of composites with different SiC coating in pure and mixed gaseous atmosphere from 400 to 1500degreesC. Oxygen and water are the controlling factors below and above 950degreesC, respectively. Na_2SO_4 causes a relatively great weight loss of C/SiC above 1300degreesC. Four layers of SiC coating with the total thickness of 80 microns can improve the oxidation and corrosion resistance of 3D C/SiC.

Key words: Carbon composites, Oxidation, Modeling.

1. INTRODUCTION

Owing to their lightweight, high strength, and temperature capability, 3D C/SiC composites are promising candidate ceramic matrix composites for applications at high temperature up to 1500degreesC in the oxidizing and corroding environment. Lots of achievements have been received in processing, microstructure characterization, property tests and applications of this kind of materials(Naslain 1999, Naslain 1998, Xu et al. 1998 and Trabandt et al. 1999). Manufacture and property tests of CMCs are costly and time-consuming. Modeling the properties evolution of materials in working environment can help to understand the environmental behavior of materials, find out environmental controlling factors, and reduce testing cost.

Weight change of 3D C/SiC is one of the important parameters used for environmental performance characterization. Halbig and Cawley (2000) modeled the oxidation behavior of one-dimensional textile carbon fiber reinforced SiC matrix composites (1D C/SiC) with finite difference method (Fig.1 (a)), but it is very difficult to develop an analytical model for 3D C/SiC with this method. Because the textile structure and the distribution of SiC matrix in a 3D carbon fiber braiding preform (Fig.1(b)) are too complicated to be modeled, it is a big obstacle to model the diffusion of the mixed gases (H_2O, O_2, and Na_2SO_4) in the composites and their chemical reactions with 3D C/SiC.

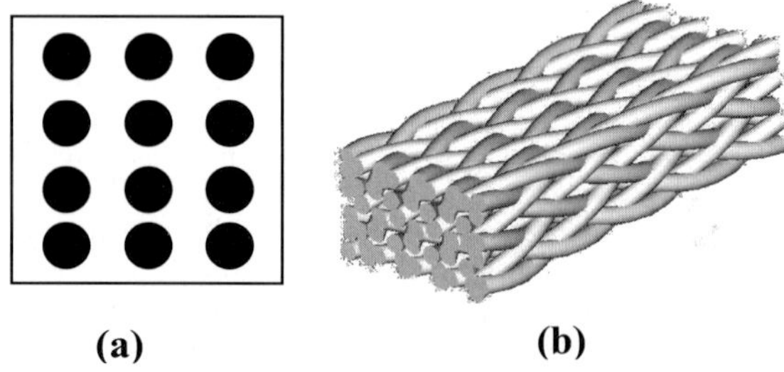

(a)   (b)

Figure 1. (a) 1D C/SiC model used in Ref. (Halbig and Cawley, 2000), black solid circles represent carbon fibers; (b) a 3D carbon fiber braiding preform

Cheng et al. (2001, 2002a and 2002b) reported their work on the modeling the effects of pure O_2 or Na_2SO_4 on the oxidation and corrosion behavior of 3D C/SiC. This paper mainly discussed modeling the effects of mixed gases on the oxidation and corrosion behavior of 3D C/SiC with different layers of SiC coating. When these models were used to predict the weight change of a real sample in an environment, the predicted values were not always satisfied with the experimental values very well, because the quality of the SiC coating was not always the same as that of the samples used to develop the models, although the real sample was coated with the same layers of coating. A mathematical method was proposed for solving this problem in this paper.

2.   EXPERIMENTAL PROCEDURE

T-300 carbon fiber from Japan Toray were employed to braid three-dimensional performs by four-step processing with the fiber volume fraction of 40-45%. The preforms were deposited with a pyrolytic carbon interlayer and a SiC matrix by low-pressure chemical vapor infiltration (LPCVI) process at 950degreesC to obtain C/SiC composites. Multilayer SiC coating was prepared on the surface of C/SiC composites by chemical vapor deposition (CVD) process at 950degreesC. The thickness of each SiC coating is about 20 microns.

Environmental experiments were conducted in different oxidizing and corroding environments at temperatures from 400 to 1500degreesC. The pressure of oxygen was in the range from zero to 21kPa; the pressure of water vapor was in the range from zero to 21kPa; the concentration of Na_2SO_4 was in the range from zero to 300ppm.

The microstructure of the fracture surface was observed with a scanning electron microscope (SEM, JEOL 840).

3.   MODELING

3.1   Effects of a pure gas on the environmental behavior of 3D C/SiC

The weight change of 3D C/SiC in the oxidizing and corroding environment obey Arrhennius relation (Cheng et al.,2001):

$$\Delta W_i = A_i(t)\left(1-\exp\left(-B_i(t)T_i^{n_i(t)}\right)\right), \quad i=1,2,3,... \tag{1}$$

Here, ΔW_i is a weight change caused by mechanism i, %; T_i is temperature, K; t is time, h; A_i, n_i and B_i are functions of t and the pressure of atmosphere.(1) is a monotonic function, and the total weight change ΔW_{tot} can be expressed by the sum of ΔW_i, like in (2). More details can be found in Ref. (Cheng et al., 2001, 2002a and 2002b).

$$\Delta W_{tot} = \sum_i \Delta W_i \tag{2}$$

3.2   Effects of mixed gases on the environmental behavior of 3D C/SiC

Microcracks that formed in the SiC matrix and coating on cooling from processing temperature can render the highly reactive carbon constituents more accessible to oxygen in oxidizing environment. Based on the experimental results, there were intense coupling effects of O_2 and H_2O on each other, while there were no prominent coupling effects of Na_2SO_4 on both O_2 and H_2O because of the low partial pressure of Na_2SO_4 vapor. Consequently, the weight change of C/SiC in a mixed gaseous atmosphere can be expressed by (3).

$$\Delta W = \Delta W_{(S \neq 0, O = 0, H = 0)} + \Delta W_{(S = 0, O \neq 0, H \neq 0)} \tag{3}$$

Here, S, O and H represent Na_2SO_4, O_2 and H_2O respectively. For simplicity, if the pressure or concentration of a gas is zero, it will not be written in the equations. For example, (3) can be written as (4) (similarly hereinafter).

$$\Delta W = \Delta W_{(S \neq 0)} + \Delta W_{(O \neq 0, H \neq 0)} \tag{4}$$

$\Delta W_{(S \neq 0)}$ can be modeled according to the method mentioned in section 3.1, since it models the effects of pure Na_2SO_4 on the corrosion behavior of 3D C/SiC. The second item of (4) can be expressed as (5).

$$\Delta W_{(O \neq 0, H \neq 0)} = \frac{1}{2}(k_1 + 1)\Delta W_{(O \neq 0)} + \frac{1}{2}(k_2 + 1)\Delta W_{(H \neq 0)} \tag{5}$$

Here,

$$k_1 = \frac{\Delta W_{(O \neq 0, H \neq 0)} - \Delta W_{(H \neq 0)}}{\Delta W_{(O \neq 0)}} \tag{6}$$

and

$$k_2 = \frac{\Delta W_{(O \neq 0, H \neq 0)} - \Delta W_{(O \neq 0)}}{\Delta W_{(H \neq 0)}} \tag{7}$$

k_1 represents the coupling effect of O_2 on H_2O, that is, how much difference between $\Delta W_{(O\neq 0, H\neq 0)}$ and $\Delta W_{(H\neq 0)}$ caused by 1% change of $\Delta W_{(O\neq 0)}$; k_2 represents the coupling effect of H_2O on O_2, that is, how much difference between $\Delta W_{(O\neq 0, H\neq 0)}$ and $\Delta W_{(O\neq 0)}$ caused by 1% change of $\Delta W_{(H\neq 0)}$.

Both $\Delta W_{(O\neq 0)}$ and $\Delta W_{(H\neq 0)}$ can be modeled according to the method mentioned in section 3.1, since they model the effects of pure O_2 and H_2O on the oxidation and corrosion behavior of 3D C/SiC respectively.

Based on experimental data, k_1 and k_2 are shown as (8) and (9).

$$k_1 = \left(0.0069t + 0.9939\right)^{-10} \tag{8}$$

$$k_2 = \left(0.0016t + 0.0024\right)^{-0.2} \tag{9}$$

Here, t is time, h. Consequently, the coupling model of ΔW should be

$$\Delta W = \Delta W_{(S\neq 0)} + \frac{1}{2}(k_1 + 1)\Delta W_{(O\neq 0)} + \frac{1}{2}(k_2 + 1)\Delta W_{(H\neq 0)} \tag{10}$$

3.3   Standardizing the quality difference of the real SiC coating

Fig.2 shows that the oxidation and corrosion resistance of C/SiC can be improved by deposition several layers of SiC coating. In Fig.2, $\Delta W_{SiC\ layer=1}$ represents the weight change of C/SiC with one layer of SiC coating. The weight change of C/SiC with different layers of coating can be expressed with (11).

$$\Delta W = k_{coating} \Delta W_{SiC\ layer=1} \tag{11}$$

Here, $k_{coating}$ is a function of the number of coating layer and it can be obtained by fitting the experimental data.

The quality of the SiC coating, however, is not always the same as that of the samples that used to develop the models, although they had the same layers of coating. And the quality of a coating is too complicated to characterize quantitatively, while the number of coating layer is a simple characterization parameter. In fact, the information of the coating's quality has already been included in the coating layer. Before the predicting tasks, the coating's quality must be standardized as following steps:

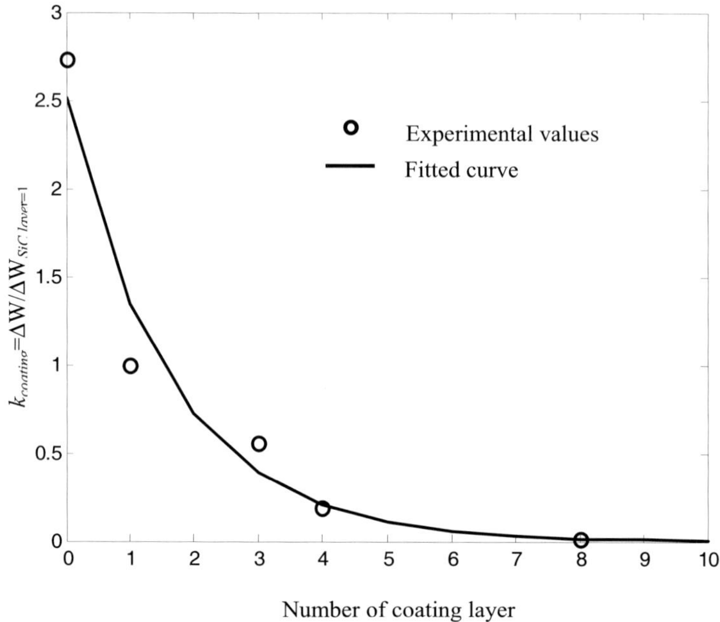

Figure 2. The relation curve of normalized weight change with coating layer

1) setting the models' environmental parameters the same as that of the predicting tasks;

2) searching a proper layer so that the predicted values approach to the measured values.

In most cases, two experiments, at a lower and a higher temperature, are enough for a standardizing task. Weight change analyses can be carried out by using the new layer in other environments and at other temperatures.

4.   RESULTS AND DISCUSSIONS

4.1   The effect of coating layer on the environmental behavior of 3D C/SiC

The oxidation and corrosion resistance can be improved by increasing coating layer (Figs.3 and 4). In the case of one layer of coating, it shows that

1) at the temperature from 400 to 700degreesC, weight loss is caused by the reaction of carbon phases with oxygen;

2) at the temperature from 700 to 950degreesC, weight loss decreased with the sealing of the cracks in matrix and coating;

3) at the temperature from 950 to 1200degreesC, weight loss is caused by the reaction of carbon phases with H_2O and a small quantity of SiO_2 formed;

4) at the temperature from 1200 to 1400degreesC, weight loss decreased with a mass of SiO_2 formed and the oxidizing and corroding gases are difficult to diffused through the SiO_2 film;

5) above 1400degreesC, weight loss is caused by the vaporization of SiO and the reaction of carbon phases with mixed gases, oxygen, water vapor

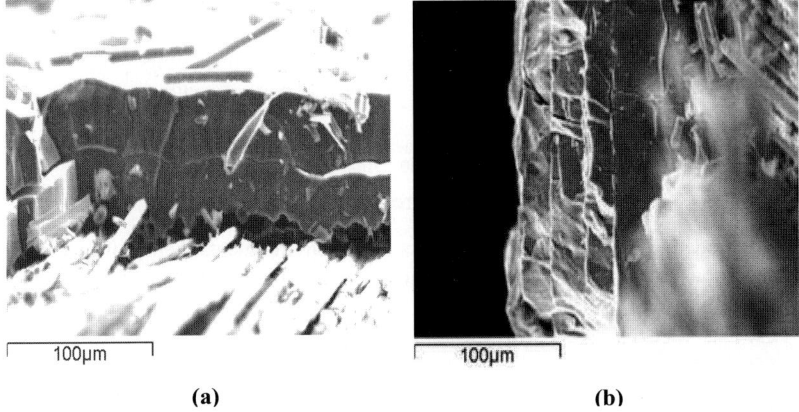

Figure 3. SEM images of 3D C/SiC with different SiC coating layer
Carbon phases are more accessible to oxygen with 2 layers of coating (a) than
that with 4 layers of coating (b).

and Na_2O_4 vapor.

As Fig.4 shows that in the cases of two and four layers of coating, the environmental behavior is the same as that in the case of one layer of coating, and the weight loss decreased a lot compares with that in the case of one layer of coating. The weight loss is decreased a lot when the C/SiC is coated with four layers of SiC coating whose total thickness is about 80 microns.

4.2   The effects of pure and mixed atmosphere on the environmental behavior of 3D C/SiC

Fig.5 shows the corrosion behavior of 3D C/SiC in a Na_2SO_4 atmosphere. Surface corrosion causes a slight weight gain below 1200degreesC; carbon phases corrosion and the vapor of SiO causes a relatively large weight loss above 1300degreesC.

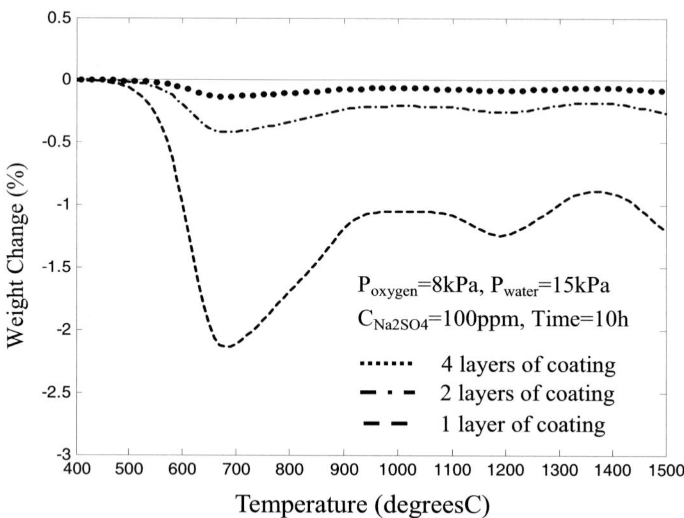

Figure 4. Modeling the effect of coating layer on the environmental behavior of C/SiC
The oxidation and corrosion resistance can be improved by increasing coating layer.

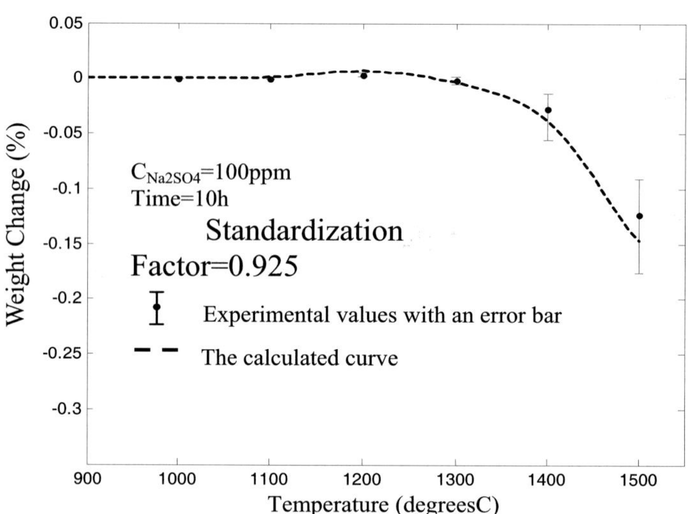

Figure 5. Modeling the effect of Na_2SO_4 on the corrosion behavior of 3D C/SiC
There is a slight weight gain below 1200degreesC and a relatively large weight
loss above 1300degreesC.

Fig.5 shows the oxidation and corrosion behavior of 3D C/SiC in an O_2
and H_2O mixed atmosphere. 700 and 1200degreesC are two dangerous points

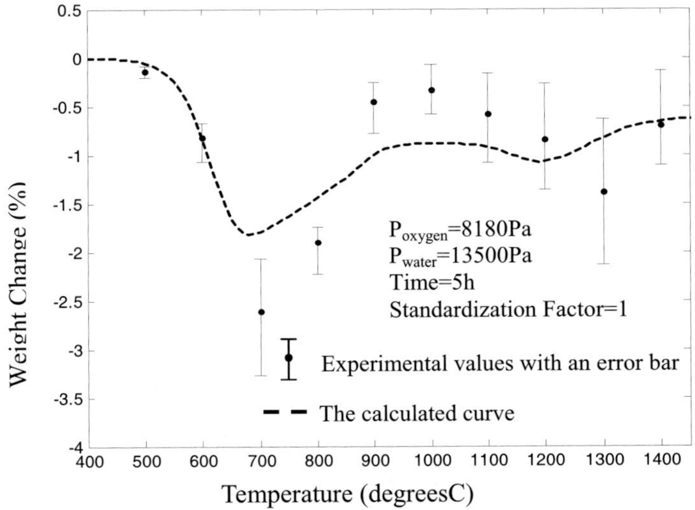

Figure 6. Modeling the effect of O_2 and H_2O on the oxidation and corrosion
behavior of 3D C/SiC with 2 layers of SiC coating
700 and 1200degreesC were two dangerous points in the low temperature and
high temperature ranges respectively, where were relatively large weight loss.

in the low temperature and high temperature ranges respectively, where are
relatively large weight loss. Oxygen is the controlling factor below
950degreesC and water is another controlling factor above 950degreesC.
Because the processing temperature is 950degreesC, the microcracks in the
matrix and coating at a temperature below 950degreesC are wider than that
at a temperature above 950degreesC. Carbon phases are more accessible to
oxygen below 950degreesC than that above 950degreesC. Because Na_2SO_4
vapor causes a relatively great weight loss of C/SiC above 1300degreesC,
Fig.6 is different from Fig.5 above that temperature.

5.   CONCLUSIONS

Based on mechanism analyses and necessary experimental data,
mathematical models were developed to model the environmental effects on
the oxidation and corrosion behavior of 3D C/SiC.

Mathematic models can be used to predict the weight change of 3D
C/SiC in practice by standardizing the SiC coating quality. The calculated
values accorded with the experimental values quite well.

The oxidation of carbon phases caused a large weight loss below
950degreesC--the processing temperature. Weight loss decreased with the
sealing of microcracks in the matrix and coating above the processing
temperature.

Oxygen is the controlling factor below 950degreesC and water is another controlling factor above 950degreesC. Na_2SO_4 causes a relatively great weight loss of C/SiC above 1300degreesC.

The oxidation and corrosion resistance of 3D C/SiC is improved by increasing the SiC coating up to 4 layers, about 80 microns.

REFERENCES

1. Cheng L.F., Y.D. Xu, L.T. Zhang and X.W. Yin "Effect of Carbon Interlayer on Behavior of C/SiC Composites with a Coating from Room Temperature to 1500degreesC". *Materials Science and Engineering* A 300(1-2): pp. 219-225, 2001.
2. Cheng L.F., Y.D. Xu, L.T. Zhang and X.G. Luan (2002a), "Corrosion of a 3D-C/SiC Composite in Salt Vapor Environments". *Carbon* **40(6)**: pp. 877-882.
3. Cheng L.F., Y.D. Xu, L.T. Zhang, Q.F. Zeng and X.G. Luan, "Factorization Method for Failure Mechanism Analysis of C/C and C/SiC Composites". *Science and Engineering of Composite Materials* **10(1)**: pp. 45-49, 2002b.
4. Halbig M.C. and J.D. Cawley. *Modeling the Environmental Effects on Carbon Fibers in a Ceramic Matrix at Oxidizing Conditions*. NASA-TM-2000-210223, 2000.
5. Naslain R. (1999), "Processing of Ceramic Matrix Composites". *Key Engineering Materials* 164-165: pp. 3-8.
6. Naslain R. "The Design of the Fibre-matrix Interfacial Zone in Ceramic Matrix Composites". *Composites Part* A 29(9-10): pp. 1145-1155, 1998.
7. Trabandt U., H.G. Wulz and T. Schmid. "CMC for Hot Structures and Control Surfaces of Future Launchers". *Key Engineering Materials* **164-165**: pp. 445-450, 1999.
8. Xu Y.D., L.T. Zhang, L.F. Cheng and D.T. Yan. "Microstructure and Mechanical Properties of Three-dimensional Carbon/silicon Carbide Composites Fabricated by Chemical Vapor Infiltration". *Carbon* **36**(7-8): pp. 1051-1056, 1998.

DIRECTIONAL BEAM BASED APPROACH APPLIED TO THE CHOICE OF MACHINING DIRECTION FOR A SCULPTURED PART
APPLICATION FOR FINISHING PROCESS

Yann Quinsat, Laurent Sabourin and Grigore Gogu

LaRAMA, IFMA and University Blaise pascal, BP 63175 Aubière, e-mail: yann.quinsat@ifm

Abstract: Within the framework of the manufacture of sculptured parts in three axes (foundry moulds, plastic injection and stamping die), CAD/CAM software currently offers a broad range of operations associated with increasingly powerful tool-path generators. However, there is no methodological guide that facilitates an optimized choice of the machining process (roughing, finishing) and the adjustment of the various parameters of the tool-path generators. The objective of this research is to provide a tool to assist in the choice of application of machining strategies for sculptured parts. We decided to focus on the machining strategy choice in the finishing process. This operation determines the final quality of the part. The objective of this work is to provide a methodology for machining strategy choice in the finishing process based on the criteria of surface roughness.

Key words: Machining Strategy, Tool-path, Roughness, Finishing Process.

1. INTRODUCTION

The sculptured parts are defined from parameterised surfaces. These surfaces have an orientation and curvatures, which depend on the considered point. Moreover, they may have discontinuities of order n. From a functional point of view, there are four types of standardized specifications (dimensional, shape defect (ISO 1101), surface roughness (ISO 4287), visual aspect (ISO 8785). To manufacture sculptured parts in three axes (foundry moulds, plastic injection and stamping die), CAD/CAM software currently offers a broad range of operations. These are associated with increasingly powerful tool-path generators [11-12] and numerous research tasks are undertaken [8-15]. However, there is no methodological guide that facilitates an optimized choice of the machining process (roughing, finishing) and the adjustment of the various parameters of the tool-path generators.

The objective of this research is to provide a tool to assist in the choice of application of machining strategies. We propose a definition of a machining strategy as well as the description of the parameters.

We decided to focus on the machining strategy choice in the finishing process. This operation determines the final quality of the part and the respect of the design department specifications. The concept of directional beam will be introduced.

2. MACHINING STRATEGY : DEFINITION AND DESCRIPTION

The purpose of this first part is to specify all the terms and concepts associated with the definition of a machining strategy. After comparing the various definitions in process planning [1], we introduce a definition of a machining strategy [13]: A machining strategy is a methodology used to generate a series of operations, with the aim of producing a given form. It makes it possible to link a machining process with a machining feature.

We develop this concept for the finishing process of a sculptured part.

2.1   Machining strategy of finishing process

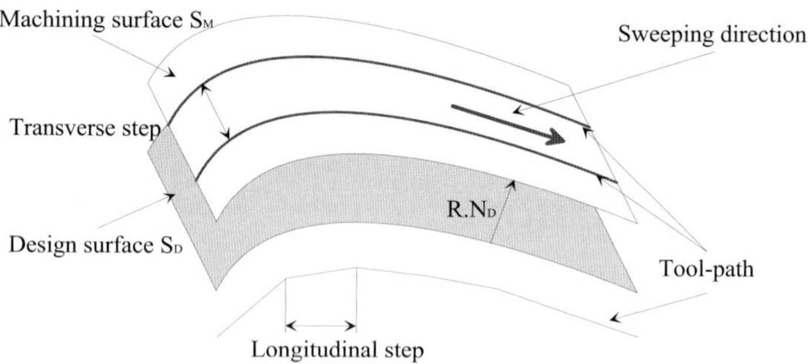

Fig 1: Description of the tool-path construction

A machining strategy for the finishing process is a methodology used to generate an operation with the aim of producing a geometrical feature in its final form. The choice of machining strategy is dependent on a number of factors e.g. shape defect, surface roughness.

It makes it possible to characterize the tool, the cutting conditions and the adjustment parameters of the tool-path generator. The main parameters are [9]:

- The machining direction (sweeping).
- The transverse step.
- The longitudinal step.

2.2   Existing research

Within the framework of an optimal machining strategy choice, we distinguish two research directions. The first is interested in the local properties of the surface in order to define a global choice. Chiou et al. [3] propose for the 5 axis machining of complex parts, to locally define the machining directions allowing a material removal rate increase. The tool-paths are then building by connecting the different points by a curve tangent to these chosen directions. Feng et al. [7] define for each point, the machining direction allowing to maximize the feed rate while respecting the form defect. They analyze the variation of the cutting forces according to surface (approximated locally by a plane) and the machining direction. Chen et al. [2] define a machining direction allowing the greatest material removal. This direction is defined by the greatest engagement of the cutting edge for a tolerance defined a given surface. For each point, they show that this direction corresponds to the projection of the surface gradient on an orthogonal plane to the tool axis. The purpose of this various work is always to define a machining direction. This direction optimises the machining time by increasing the feed rate or the material removal, but there is no methodology linking the geometry of the part and the optimal choice of the machining strategy parameters.

Other studies directly define a single machining direction for the part. This direction optimises the machining. Lee et al. [10] study the influence of the machining direction on the surface roughness and the vibrations of the part in the case of a thin plate. Ramos et al. [14] describe the influence of the strategy machining (sweeping mode and direction) on the surface roughness for the machining of a boat propeller. The principal disadvantage of these general studies is that they focus on particular cases and cannot be applied to more general cases. The authors do not give methodology to define a machining strategy from A to Z.

Now we introduce the concept of directional beams which allows a link between the parameters of the machining strategy and the surface geometry at each point.

3.　DIRECTIONAL BEAMS

3.1　Fitness function

The machining strategy which optimises the machining time while respecting the design department's specifications (shape defect, surface roughness) will be the most powerful one. The total machining time is calculated by $T_{tot}=\sum_{trajet}\int_{trajeti}ds/V_f$. As the feed rate is difficult to foresee along the tool-path, this total time is hard to compute. The fitness function is a function which makes it possible to link the machining direction at a point and the cutting parameters to the improvement of the machining time. This function is described by:

$$\left(\begin{array}{ccc} (D,S_M,P_c) & \rightarrow & \Re \\ (d,M,P) & \mapsto & G_P(d,M,P) \end{array} \right)$$

D is the set of machining directions and d a direction in D. M is a point on surface S_M. Finally, P_C represents the set of cutting parameters (f_z, transverse step) and P is an element of this set. This function G_P expresses the performance at point M on surface S_M of a machining direction d for parameters P.

A high value of this fitness function G_P gives a low total cutting time, conversely, a low value of G_P leads to an increase in cutting time.

Now we introduce the concept of directional beams which allows a link between the parameters of the machining strategy and the surface geometry at each point.

3.2　Definition of the directional beams

For a given point M on the surface and fixed cutting parameters P, it is possible to define directions d which maximize the fitness function G_P. The maxima of G_P is noted:

$$G_{pmax}(M,P)=\max_{d\in D}(G_P(d,M,P))|_{(M,P)\ fixed}$$

From this maximal value, the machining direction d which enables us to obtain this maximum value, are gathered in a set $D_{max(M,P)}$. It is defined by:

$$D_{max(M,P)}=\{d\in D,\ G_P(d,M,P)=G_{pmax}(M,P)\}$$

We define the concept of value proximity by the calculation of a relative difference and the definition of a coefficient

Directional Beam: A directional beam is a set of machining directions for a given point M and cutting parameters P. The fitness function G_P is close (with a coefficient α) to its maximal value G_{pmax} for all directions.

$$Fd(M,P)=\{d\in D,\ G_{pmax}(M,P)-G_P(d,M,P)\leq\alpha.\ G_{pmax}(M,P)\}$$

These beams enable to introduce flexibility in the choice of machining direction. They describe the concept of proximity between two machining directions.

If two points M_1 and M_2 present two beams ($Fd(M_1,P)$ and $Fd(M_2,P)$) non-disconnected, they could be machined with a common machining direction. This direction guarantees a machining time close to an optimal value in M_1 and M_2.

3.3   Machining feasibility criterion

The objective is to define a criterion of machining feasibility of the surface S_M. We must be able to establish whether optimal machining direction for this surface. We decided to limit ourselves to the choice of a direction for parallel plane machining.

To determine a machining direction for all points S_M, it is sufficient to calculate the beam intersection. We note:

$$I = \bigcap_{M \in S_M} F_d(M,P)$$

- If $I \neq \{\phi\}$ then there is a machining direction allowing the machining of the surface close to the optimal fitness at each point. It is sufficient to choose a direction in I.
- If $I = \{\phi\}$ then there is no machining direction allowing to machine the surface close to the optimal fitness at each point for a rate α.

If the beam intersection is not empty, the selected direction guarantees that for each point the fitness function is close to its maximum. But there is no guarantee that the direction chosen is the best one. It is necessary to decrease α in order to diminish the directional beams.

3.4   Algorithm to define choice

In order to determine a machining direction which makes it possible to minimize α for any point we propose the algorithm described in figure 2. This algorithm begins with a high value of α, then decreases this value in order to reduce the number of elements at the intersection. If the intersection becomes empty then the step chosen to decrease rate α is reduced and the process starts again. This algorithm thus makes it possible to minimize α and to determine a machining direction close to optimal (G_{pmax}) at any point.

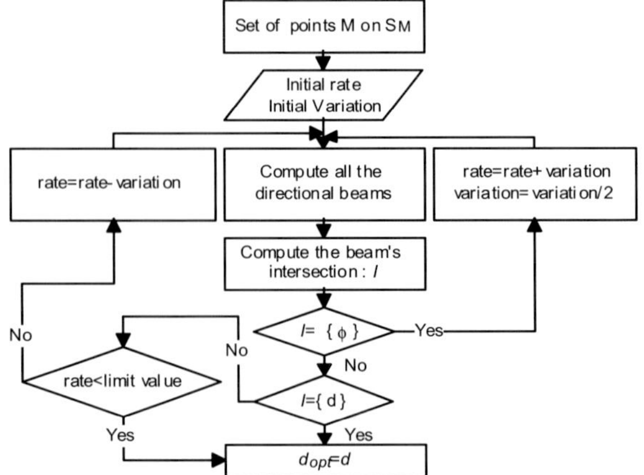

Fig 2: Algorithm for defining choice of machining direction

4.   CASE OF SURFACE ROUGHNESS CONSTRAINT

The objective of this section is to apply the definitions and concepts set out in the case of the finishing operation of complex parts by ball end mill. The performance function will be developed under the constraints of surface roughness.

4.1   Surface roughness criteria

It has been shown [13] that the pattern obtained by a ball end mill encompasses numerous spherical segments. Currently, there is no standardized criterion to describe this pattern. Moreover, traditional criteria of characterization (Ra, Rt defined in ISO 4287, [4-5]) do not allow to highlight the coupling between the feed rate and the transverse step. To define the surface roughness corresponding to this pattern, we use the surface criterion S_z [6] defined in the standard draft ISO 12085. This parameter corresponds to the height deviation between the lowest and the highest points of the surface. For a spherical segment of Ro radius, and dimension f_z and p, the maximum height is at one of the vertices of the rectangle defined by f_z and p. We express S_z by:

$$S_z=((f_z^2+p^2)/(8.R_o))$$

4.2 Construction of the fitness function under the constraint of surface roughness

From the criterion highlighted in the preceding paragraph, we sought to determine the best choice of machining direction under the constraints of surface roughness

4.2.1 Parameter definition

The surface is defined by its parametric form $S_D(u,v)$. Hence the perpendicular vector n is defined by:

$$n(u,v) = \frac{\dfrac{\partial S(u,v)}{\partial u} \times \dfrac{\partial S(u,v)}{\partial u}}{\left\| \dfrac{\partial S(u,v)}{\partial u} \times \dfrac{\partial S(u,v)}{\partial u} \right\|}$$

For a given point on the surface, the machining direction is noted as d with $d=(cos\theta,sin\theta,0)^T$. θ defines the orientation of d from the x axis in the (x,y) plane. Locally, the tool-path is assumed to be defined on the tangent plane of the surface. The transverse direction d_T is orthogonal to the plane $P_n=(d,n)$. We choose $d_T=((d\times n)/(\|d\times n\|))$. Now, it is possible to define the vector n' by $n'= d_T \times d$.

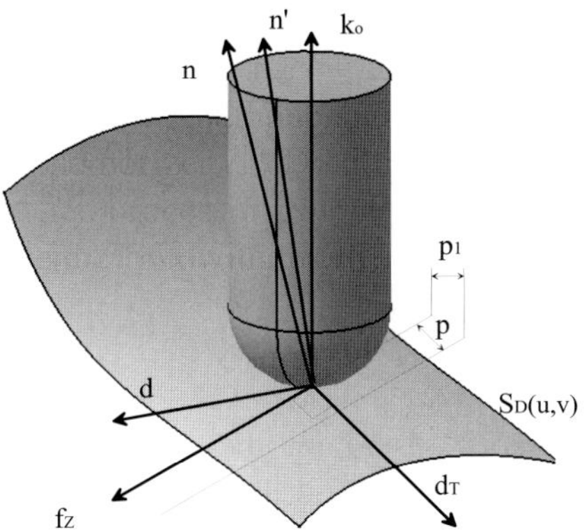

Fig 3: Parameter description

We can distinguish two cases:

- n' is oriented outside the part. n' and k form an acute angle (because the surface has no undercut). In this case, β can be defined by $n'.k=cos\beta$.
- n' is oriented inside the part. n' and k form an obtuse angle, in this case $n'.k=cos(\pi-\beta)=-cos\beta$.

Generally, the expression of β is $\beta=cos^{-1}(|n'.k|)$.

4.2.2 Fitness function

We consider here that the transverse step selected out of the whole of the part is p_1. The effective height of pattern S_z was defined in the preceding paragraph. At the considered point M (locally the surface is approximated by a plane) $p=p_1/\cos\beta$. We can derive:

$$S_z=((p_1{}^2/\cos^2\beta+f_z{}^2)/(8.R_o))$$

Replacing β in the equation, the calculation of the acceptable step at the point considered M becomes:

$$p_1^2 = \left(8.S_Z.R_O - f_z^2\right)\left(\left(\frac{d\times n}{\|d\times n\|}\times d\right).k\right)^2$$

The total machining time is calculated by $T_{tot}=\sum_{\text{tool-path}}\int_{\text{tool-path\_i}}(ds/V_f)$. The reduction in the machining time forces an increase in the rate of material removal. However, locally, this material removal rate is proportional to $A=p_1.f_z$. The model set up [13] shows that the patterns on manufactured surfaces have a projected surface equal to $p_1.f_z$. If the lengths of the tool-path are equivalent, the total time will decrease as factor A increases. Locally the optimal machining direction makes it possible to maximize the transverse step (p_1) while respecting the constraint of surface roughness S_z. A fitness function respecting this objective could be defined by:

$$G_p(d,M,P) = \left(8.S_Z.R_O - f_z^2\right)\left(\left(\frac{d\times n}{\|d\times n\|}\times d\right).k\right)^2$$

5. APPLICATION

Our method has been tested on the surface defined in figure 4. This surface is contained in a cube of $200\times200\text{x}100$ mm³. The tool used is a ball-end mill 10 mm diameter. The feed rate (f_z) used is 0.05 mm/tooth. The algorithm ends up in a direction where $\theta=45°$.

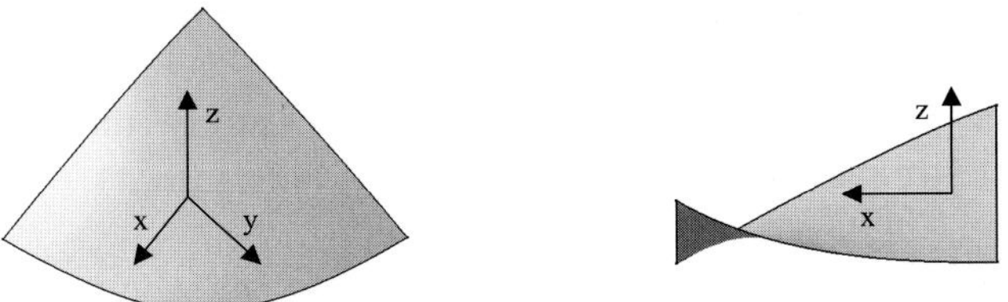

Fig 4: Description of the test surface

In order to check this result, tool-paths have been programmed on CAM software (Catia V5). The feed rate used was 2000 mm/min (f_z =0.1 mm/tooth and N=10000 tr/min). For these cutting conditions, time keepings on machine tool (HURON KX15 with a Siemens 840D) were carried out. The results in figure 5 confirm a notable reduction in machining time (20%).

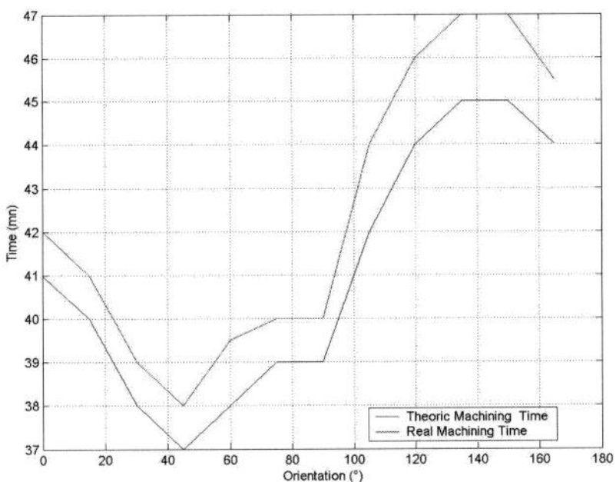

Fig 5: Time machining comparison

6.   CONCLUSIONS AND PERSPECTIVES

In this paper we propose a tool for the choice of machining direction. First of all, we specify the concept of machining strategy, as well as the main parameters. In the second part, the concept of directional beam was introduced. This beam allows to define a set of directions close to an optimal value of a fitness function. A choice at algorithm based on the directional beams guarantees an optimization of the fitness function for any point of the surface. Then, the fitness function is linked to the constraint of surface roughness and its dominating parameters p and f_z. The algorithm developed is correlated with time keeping's on a machine tool. The results obtained guarantee a good robustness of the method.. The results obtained show that an optimized choice of machining direction reduces the machining time by 20%. Another application of this work is to be able to cut up the surface according to optimal machining direction in the case of non-connected beams. Now, we are working to define new fitness functions. These functions will take into account the kinematics of the machine tool (feed rate acceleration). Now, our next objective is to apply this method to the manufacture of industrial parts.

REFERENCES

1. A. BERNARD, "Usinage tridimensionnel d'outillage de topologie complexe : Analyse des contraintes de production et contribution à l'optimisation du processus d'usinage", PhD Thesis ECOLE CENTRALE PARIS/LURPA/ENS CACHAN 1989.
2. Z. CHEN, Z. DONG, G. W. VICKERS, "Automated Surface Subdivision and Tool Path Generation for 3 1/2 1/2 axis CNC Machining of Sculptured parts", *Computer in Industry*, 2003,**vol 50**, pp 319-331.
3. C.- J. CHIOU, Y.S. LEE, "A Machining potential field approach to tool path generation for multi-axis sculptured surface machining", *Computer Aided Design*, 2002, **vol 34**, pp 357-371.
4. W. P.DONG, P. J.SULLIVAN, K. J. STOUT. "Comprehensive study of parameters for characterizing three dimensional surface topography I: Some inherent properties of parameter variation", *Wear*, **volume 159**, 1992, pp 161-171.
5. W. P.DONG, P. J.SULLIVAN, K. J. STOUT. "Comprehensive study of parameters for characterizing three dimensional surface topography II: Statistical properties of parameter variation", **Wear**, **volume 167**, 1993, pp 9-21.
6. W. P.DONG, P. J.SULLIVAN, K. J. STOUT. "Comprehensive study of parameters for characterizing three dimensional surface topography IV: Parameter for characterizing spatial and hybrid properties", *Wear*, **volume 178**, 1994, pp 45-60.
7. H.-Y FENG, N. SU, "Integrated tool path and feed rate optimization for the finishing machining of 3D plane surfaces", *International of Machine Tools and Manufacture, 2000*, **vol 40**, pp 1557-1572.
8. T. KIM, S. E. SARMA, "Tool-path generation along directions of maximum kinematic performance; a first cut at machine-optimal paths", *Computer Aided Design*, **vol 34**, 2002, pp.453-468.
9. C. LARTIGUE, E. DUC, C. TOURNIER, "Machining of Free-Form Surfaces and Geometrical Specifications", *Proc Instn Mech Engrs*, 1999, **vol 213**, pp 21-27.
10. C.M. LEE, S. W. KIM, K. H. CHOI, D. W. LEE, "Evaluation of cutter orientations in high-speed ball end milling of cantilever-shaped thin plate", *Journal of Material Processing Technology*, 2003, **vol 140**, pp 231-236.
11. S. C. PARK, B. K. CHOI, "Tool-path planning for direction-parallel area milling", *Computer Aided Design*, 2000, **vol 32**, pp. 17-25.
12. S. C. PARK, "Tool-path generation for z-constant contour machining", *Computer Aided Design*, 2003, **vol 35**, pp. 27-36.
13. Y. QUINSAT, L. SABOURIN, G. GOGU, "Help for sculptured surface machining strategy choice : application to finishing process of sculptured surface", *Conference on Mechanical Design and Production*, Cairo, Egypt, January 4-6, 2004.
14. A.M. RAMOS, C. RELVAS, J.A. SIMOES, "The influence of finishing milling strategies on texture roughness and dimensional deviations on the machining of complex surface", *Journal of Materials Processing Technology*, 2003, **vol 136**, pp.209-216.
15. C. TOURNIER, E. DUC, "A surface based approach for constant scallop height". *International Journal of Advanced Manufacturing Technology*, 2002, **vol 19**, pp. 318-324.

A DISTRIBUTED PRODUCTION CONTROL SYSTEM ON A COMPONENT-BASED SOFTWARE DEVELOPMENT AND OPERATION FRAMEWORK

Hiroyki Sawada[1], Norio Matsuki[2] and Hitoshi Tokunaga[3]
National Institute of Advanced Industrial Science and Technology (AIST), Japan, 1-2-1 Namiki, Tsukuba, Ibaraki 305-8564, Japan, h.sawada@aist.go.jp[1], matsuki.n@aist.go.jp[2], tokunaga.h@aist.go.jp[3]

Abstract: Systematization and digitisation of manufacturing processes by introducing information technology (IT) tools is regarded as a key approach for increasing manufacturing companies' competitiveness. In order to support them to build manufacturing software applications for their own use, we have developed a component-based software development and operation framework, named "MZ-Platform". In this paper, we show its architecture, functions and advantage of reducing time and cost of building manufacturing software system through an example of a production control system working among control and manufacturing sections of a company.

Key words: Component-based development, manufacturing software, network application

1.   INTRODUCTION

In recent years, manufacturing companies are being challenged to bring new products of high quality and low cost to a market in shorter time and they are also subject to increasing pressures to reduce the cost and lead time. Systematization and digitisation of manufacturing processes by introducing information technology (IT) tools is regarded as one of key approaches for resolving the above issue and increasing manufacturing companies' competitiveness. In the last three decades, computer technology has been significantly improved and a lot of useful IT tools are now available, such as CAD. However, costs of purchasing and maintaining those IT tools and training engineers and operators for them, are a heavy burden to manufacturing companies, especially when the IT tools are used among distributed sections via network. In addition, manufacturers should meet a big challenge when they try to customize the IT tools for their own use.

Conventional IT tools such as CAD systems are often too big software packages for manufacturers, especially small and medium-sized enterprises, to manage, maintain and customize. In order to overcome those difficulties, we employ a component-based software development approach [1]. That is, we provide manufacturers with a set of software components that carry out a few limited functions. Each software component consists of data, methods and interfaces to the other software components. Users develop an IT tool by integrating those small software components into a large software system on our framework named "MZ-Platform" [2]. This component-based approach has a big advantage over a conventional software development manner from a viewpoint of customisation and maintenance. Users can easily modify and improve functions of an existing IT tool only by replacing and adding appropriate software components, if necessary. In this paper, we introduce the architecture and functions of MZ-Platform, and show its advantage of reducing the costs and time of building manufacturing software applications through an example of a production control system working among distributed sections of a company via network.

The structure of this paper is as follows. Section 2 gives the outline of MZ-Platform and describes the way of developing and operating network software applications. In Section 3, we show a production control system developed on MZ-Platform in cooperation with a manufacturing company.

2.   MZ-PLATFORM AND NETWORK SOFTWARE APPLICATIONS

2.1   Outline of MZ-Platform

MZ-Platform helps users in developing software applications that work in a fully event-driven manner. It is based on Java and JavaBeans technology. Figure 1 shows the architecture of MZ-Platform.

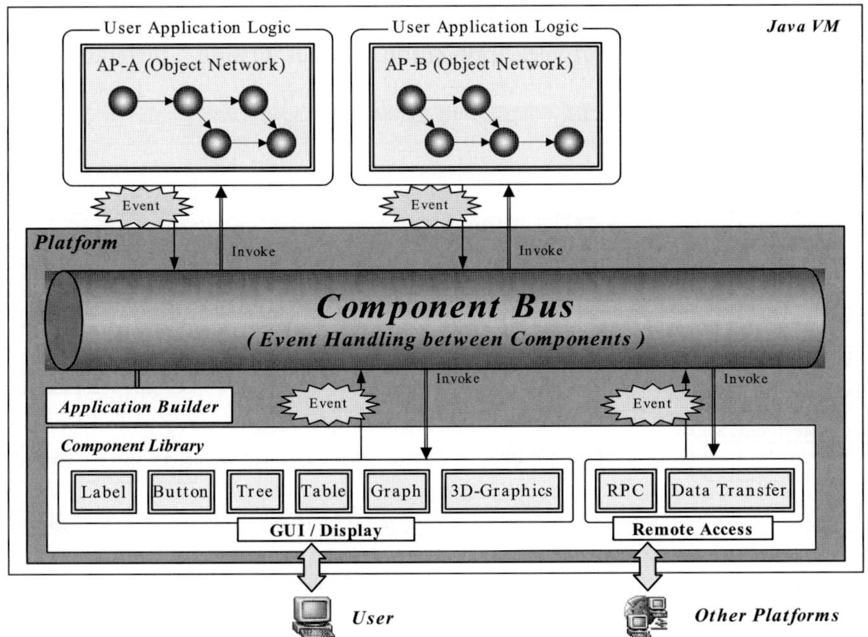

Figure 1. Architecture of MZ-Platform

When users develop a software application, first of all, they select necessary components from the *Component Library*. Each component is implemented as a Java Bean, and carries out a few limited functions. Currently, MZ-Platform provides more than one hundred components including GUI, graphic display and remote access components. Users can develop and register a new component with the Component Library. MZ-Platform also provides template files for development of users' own components.

After selecting components, the users define an interaction between those components on the *Application Builder*. The components work in a fully event-driven manner under the control of the *Component Bus*. When one component passes an event to another component via the Component Bus, it executes a designated method. It is often the case that events may be propagated one after another. The users define the paths of event propagation and invoked methods on the Application Builder.

Figure 2 shows an interface window of the Application Builder. Rectangles and lines represent components and event-propagated paths respectively. Events are passed from rectangles on the left side to those on the right side along the lines. MZ-Platform provides twenty kinds of events, such as a mouse event and an action event, generated on different occasions.

Names of events and invoked methods are displayed above the lines. In this way, users can build a software application without writing a program.

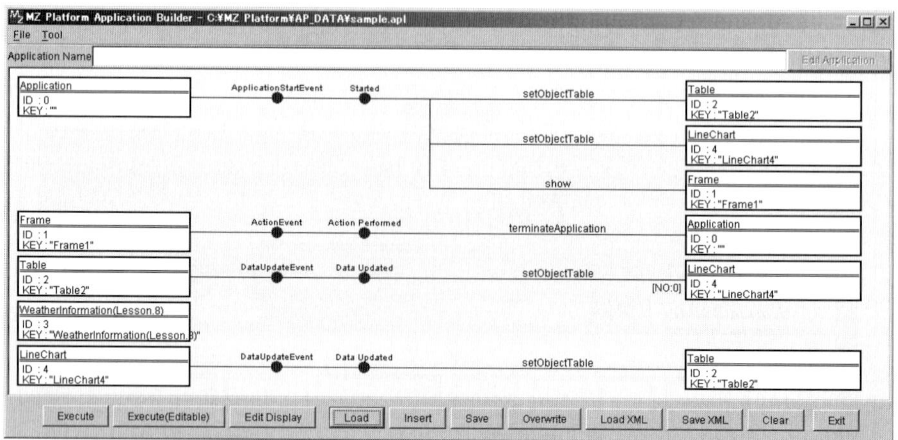

Figure 2. Application Builder

The following two subsections describe the remote access mechanism between distributed MZ-Platforms, and development and operation of network software applications on MZ-Platform.

2.2   Remote access mechanism

MZ-Platform provides functions of data cooperation between distributed MZ-Platforms for development and operation of network software applications. Figure 3 shows the overview of the remote access mechanism.

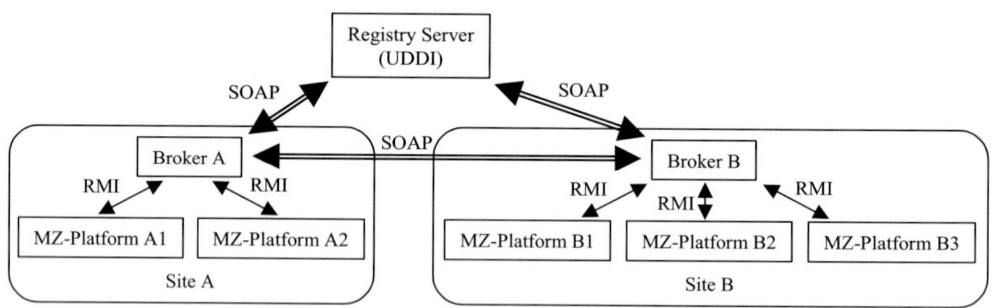

Figure 3. Remote access mechanism

There is one registry server that holds information about MZ-Platforms and their components and accessible methods within the network. The registry server uses UDDI (Universal Description, Discovery and

Integration) protocols [3]. Each local site has a broker. It manages communication between MZ-Platforms in the local site and with the registry server and the other brokers. SOAP (Simple Object Access Protocol) [4] is used for the communication between the registry server and the brokers, and Java RMI (Remote Method Invocation) is used for the local communication.

When one MZ-Platform calls a method provided by another MZ-Platform, it firstly accesses the broker on its local site. The broker makes a message in SOAP format, inquires of the registry server the destination, and then sends the message. The broker that received it invokes the designated method and returns the result in the same way.

This mechanism is hidden from users. That is, users can use the data cooperation functions without recognizing the above mechanism.

2.3 Network software applications on MZ-Platform

When users build a network software application such as a distributed database system, it is quite important to maintain consistency of data held at different sites. MZ-Platform provides a simple and unique mechanism to keep data consistency. Figure 4 shows its overview.

All the MZ-Platforms, both the server and clients, have the same data sets. When a user tries to add/update/remove data, the command is sent to the server without changing the local data set. The server collects those commands, changes its own data set, and then sends all the clients the same commands in the same sequence. In this way, all the data sets held by all the MZ-Platforms are kept consistent.

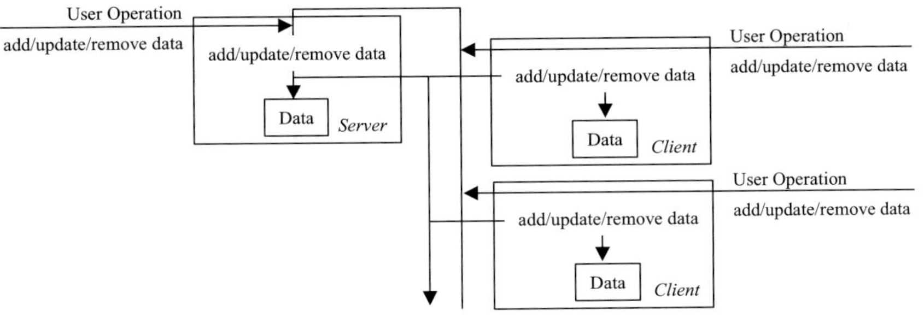

Figure 4. Maintenance of data consistency

A unique feature of this mechanism is that the server is not fixed. When an application starts on one of MZ-Platforms, it tries to find out the server. If the server exists, it replicates the data set of the server and works as a client. If the server does not exist, it begins to work as a server. This mechanism gives MZ-Platform an advantage of constructing a robust network application over a conventional server-client system in which a server has to be determined beforehand. Even though some trouble occurs over a server, the network application can work appropriately.

The above data consistency maintenance mechanism is provided as a set of functions of one component. Users can use this mechanism by installing the component.

3. DISTRIBUTED PRODUCTION CONTROL SYSTEM DEVELOPED ON MZ-PLATFORM

We have developed a production control system on MZ-Platform in cooperation with a precision machinery company, Daiya Seiki Co., Ltd., Japan. This system works among control and manufacturing sections of the company via network. It helps workers in notifying production plans, collecting results and comparing plans with results for their next production planning.

Section 3.1 introduces the production process of the company. After that, Section 3.2 describes the developed production control system.

3.1 Production process

Figure 5 shows the production process model of Daiya Seiki Co., Ltd.

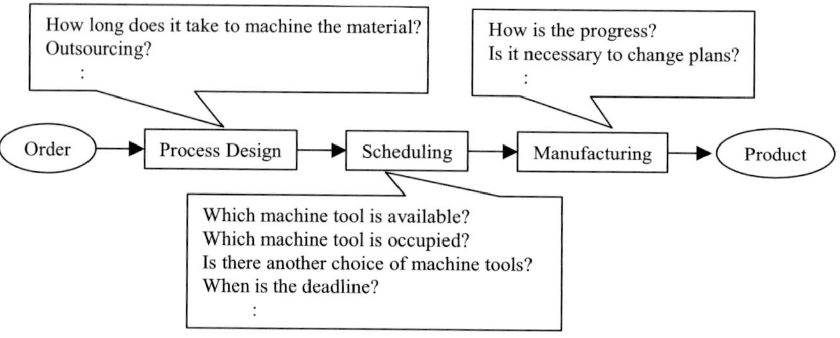

Figure 5. Production process model

When the company receives an order of products, first of all, a manufacturing process of the products is designed. After that, for each stage of the manufacturing process, an appropriate machine tool is assigned and the task is scheduled. Then, materials are manufactured into the products. During each manufacturing process, a worksheet is attached to the materials and semi-manufactures. Workers are required to write records including workers' name, date of machining and machining time on the worksheet after they complete their tasks.

At each stage during the production process, the following information is necessary.

1. Process Design
   In order to estimate the necessary machining time, process designers need machining time required for the same/similar materials/products in past jobs.
2. Scheduling
   In order to select appropriate machine tools and decide schedules, workers in charge of scheduling have to know loads on machine tools.
3. Manufacturing
   In order to manage manufacturing processes, workers in charge of process control need to know progress of manufacturing tasks.

Principally, the above information should be fed back from the manufacturing processes. In particular, it is necessary to get information about progress of the manufacturing tasks in real time.

Conventionally, since the task progress data were recorded on the worksheets attached to materials and semi-manufactures, it was very difficult, even impossible, to obtain such information immediately. In addition, since most of the other data such as the schedules were recorded and stored as paper documents, it is quite difficult to extract necessary information for process designing and scheduling.

The objective of our cooperative work is to overcome the above problem by constructing a production control system that enables every section of the company to share the same information about production process with all the others.

3.2 The distributed production control system developed on MZ-Platform

Figure 6 shows the overview of the production control system developed in this cooperative work. This system is based on the data cooperation mechanism of MZ-Platform (Section 2.3), and works among different sections with sharing the same data via network. Therefore, the workers in different sections can share the same production information only by starting this application on their PCs.

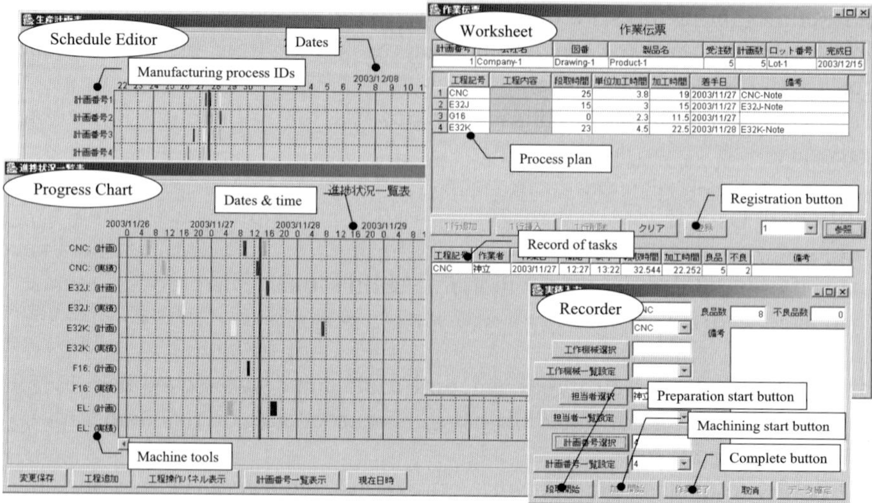

Figure 6. Overview of the production control system

The production control system consists of four main windows: *the Worksheet, the Schedule Editor, the Progress Chart* and *the Recorder*. For each order, process designers write a manufacturing process plan on the Worksheet and register it by clicking the registration button. Then, the new manufacturing process plan is added to the Schedule Editor, which shows all the manufacturing process plans in the form of Gantt chart.

The Schedule Editor can invoke two sub-windows that show two different kinds of charts of machine tool loads. Figure 7 shows them.

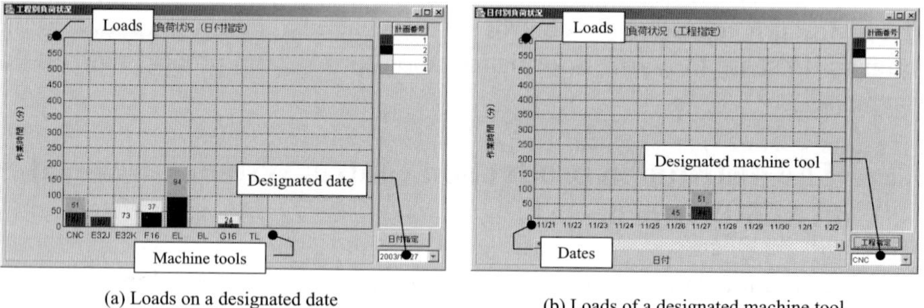

(a) Loads on a designated date   (b) Loads of a designated machine tool

Figure 7. Charts of machine tool loads

These charts show machine tool loads on a designated date and those of a designated machine tool respectively. Workers in charge of scheduling can schedule tasks by referring to these charts so that all the manufacturing processes work efficiently.

Workers in the manufacturing section record the information about their tasks on the Recorder in a quite simple way. When they start preparation of machining, they click the preparation start button. After finishing the preparation, they click the machining start button and start machining. When the machining is completed, they click the complete button. The time of preparation and machining is automatically calculated and recorded. Workers' names can be recorded by using a bar code reader, if they are held in the database of the system.

The information of tasks recorded on the Recorder is sent to the Progress Chart and the Worksheet immediately. The Progress Chart shows the plans and results of all the tasks in the form of Gantt chart. It gives the workers in charge of process control the information of task progress graphically and intuitively. Based on the information, they can decide whether it is necessary to change production plans.

The lower part of the Worksheet shows the detailed information about each completed task: machining date, starting and finishing time, and preparation and machining time. Process designers can access all the records only by selecting manufacturing process ID from the pull-down menu.

We are now using this system experimentally among a few sections of the company. We plan to collect feedbacks from the workers to improve functions of this system as well as MZ-Platform.

4.   CONCLUDING REMARKS

We have developed a component-based software development and operation framework MZ-Platform. Our aim is to encourage manufacturing companies, especially small and medium-sized enterprises, to increase their competitiveness by introducing IT tools.

MZ-Platform enables users to develop and operate software applications working on both local and remote sites for their own specific needs without writing any program or considering network mechanisms. This means that manufacturers with insufficient knowledge about programming can develop software with necessary and sufficient functions for themselves, and that they can reduce the costs and burdens of software development and maintenance.

Actually, when Daiya Seiki Co., Ltd. asked some software vender to estimate for the development of this kind of production control system, it was answered that a few hundred thousand dollars and several months would be needed. On the other hand, in our cooperative work, one of authors, who is not an expert programmer, has completed the first version of the production control system within two months by using MZ-Platform. This illustrates the effectiveness of MZ-Platform to reduce the costs and burdens associated with software development, operation and maintenance.

Currently, we operate MZ-Platform only on Windows OS. Since MZ-Platform works on Java VM, it should also work on the other operation systems such as Linux and Solaris. We are planning to check the behavior of MZ-Platform on the other operation systems, and to create new application areas of MZ-Platform.

ACKNOWLEDGEMENTS

We would like to Mr Yuji Oguchi and the staffs of Daiya Seiki Co., Ltd., for their advices and fruitful discussions.

REFERENCES

1. Cmkovic, I. and Larsson, M.: Building Reliable Component-Based Software Systems, Attech House, 2002.
2. Matsuki, N., Tokunaga, H. and Sawada, H.: "A Component-Based Software Development and Execution Framework for Cax Applications". *Proc. AED 2003*, G1.2, 2003.
3. http://www.uddi.org/
4. http://www.w3.org/TR/SOAP/

JOB-SHOP SCHEDULING USING IMMUNE ALGORITHM BASED ON DYNAMIC EVALUATION

Jianjun Yu, Shudong Sun and Ganggang Niu
Institution of System Integration and Engineering Management
Northwestern Polytechnical University, XI'AN 710072
npu_yjj@163.com   npu_yjj@sohu.com

Abstract:
The job-shop scheduling problem (JSP) is NP-hard. Traditional algorithms have their features and disadvantages. The powerful system processing capabilities of the immune system provide rich metaphors for its artificial counterpart. As a result, immune algorithm has emerged, and gradually been applied to many engineering practices. Due to the stubborn nature of the JSP, this paper initially brings forward a dynamic evaluation based immune algorithm (DEIA) as an attempt to solve JSP. By simulations of FT10x10 benchmark problem and comparisons with other algorithms, the proposed immune algorithm proved to be efficient in solving JSP.

Key words:
Dynamic evaluation based immune algorithm; job-shop scheduling; FT10x10; information entropy

1.   INTRODUCTION

The job shop scheduling problem (JSP) is NP-hard [1] and has continuously challenged computational researchers. Exact methods [2-6] have been successful in solving small instances. Problems of big dimension are usually considered to be beyond the reach of exact methods [7]. For such problems there is a need for good heuristics. Surveys of heuristic methods for the JSP are given in [8, 9]. These include dispatching rules reviewed in [10], the shifting bottleneck approach [11, 3], local search [12], simulated annealing [13], tabu search [14, 15], and genetic algorithms [16]. A comprehensive survey of job shop scheduling techniques can be found in Jain and Meeran [17]. Each algorithm has its feature and disadvantages: Tabu Search reaches a decision rapidly, although usually to the local optimum; finding a clear stopping criterion for Tabu Search is impossible; GA reaches the global optimum with more calculation; and the process for reaching the global optimum is dependent on the initial conditions.

Owning to immune system's powerful information processing capabilities, such as feature extraction, pattern recognition, learning, memory, and its distributive nature, artificial immune system (AIS) suggests new solutions to engineering problems including JSP. AIS methods primarily use three immunological principles including the immune network theory, the mechanisms of negative selection, and the clonal selection principles. The objectives and constraints are expressed first as the antigen inputs. The antigen and antibody are treated as the objective and the feasible solution of a conventional optimization method. The immune operators including crossover and mutation are then processed to produce antibodies in a feasible space.

As a typical optimization instance, the traveling salesman problem (TSP) spontaneously becomes all-important application object. N-TSP was solved by introducing two memory mechanisms which highly improved the search efficiency [18-19]. TSP size between 30 and 100 was simulated separately by immune method and others, and simulation results showed excellent performance of the immune method [20-21]. Immune algorithm can also be applied to solve multi-objective, multi-modal and combinatorial optimization problems. It has solved a nonlinear and multi-objective optimization problem in optimal-switching operation [22] and a scheduling problem with two sub problems including variable batch sizing and sequencing [23]. Besides, many algorithms based on some immune mechanisms were also brought forward and applied to engineering optimization [24-26]. As an attempt in scheduling problem, Flow-shop scheduling of small dimension has gained satisfactory result by immune algorithm [27-28]. Furthermore, immune algorithm was also applied in other optimization fields [29-32].

Due to the stubborn nature of the JSP and limitations such as the lack of convergence efficiency in the existing immune algorithm models, this paper proposes a dynamic evaluation based immune algorithm to solve JSP, and the simulation results of FT10x10 prove its efficiency. The contents of this paper are as below: section 2 puts forward the dynamic evaluation based immune algorithm; section 3 sets up the model of the job-shop scheduling problem; section 4 applies the dynamic evaluation based immune algorithm to job-shop scheduling problem; section 5 shows the simulation experiment and the results analysis; section 6 is the conclusion.

2.   DYNAMIC EVALUATION BASED IMMUNE ALGORITHM

2.1   Steps of the algorithm

Step 1: Recognition of antigen (problem)

Analyze problems and find out the characteristics of the solution, and then encode the antibody as described in Fig. 1

Step 2: Production of antibody group (solution group)

When iterating for the first time, the primary solution group consists of N antibodies generated randomly in the solution space and N_k antibodies picked up from the memory library. Here, N_k is the antibodies' number in the memory library. If there are no antibodies in the memory library, $N + N_k$ antibodies are randomly generated from the solution space to form the primary solution group. In the following iterating, N new antibodies are generated by the operation of selecting, crossover, and mutation based on the former group. Then these antibodies, together with the N_k memory antibodies form the new solution group.

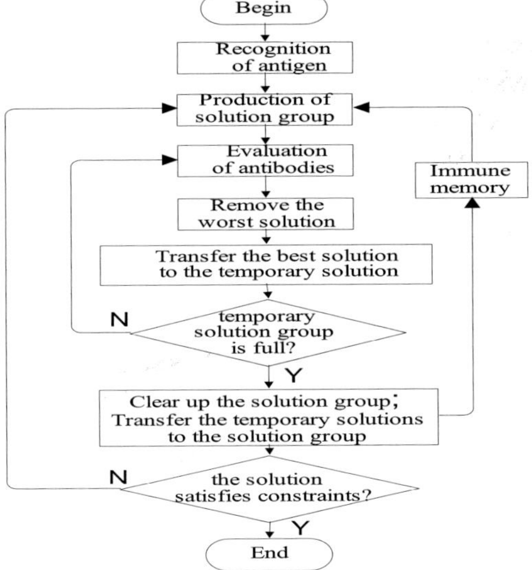

Figure 1 The flow of the dynamic evaluation based immune algorithm

Step 3: Evaluation of antibodies

In order to make a global searching, we should limit the antibodies with quite a big size, and at the same time increase the antibodies with a small size. Consequently the evaluating standard is defined by the antibody's expected reproduce rate e_v which lies on the affinity and the concentration, and the affinity is defined by the information entropy.

The affinity between two antibodies is used to show the degree of association between antibodies. The affinity between antibody v and antibody w can be represented as below:

$$ay_{v,w} = \frac{1}{1 + H(2)} \qquad (3)$$

Where: H (2) is the information entropy between antibody v and antibody w. Similarly, the affinity between antibody and antigen is defined as below:

$$ax_v = \frac{1}{(1 + opt_v)} \qquad (4)$$

Where: opt_v shows the degree of the diversity between antibody and antigen.

In the paper, the concentration of antibodies shows the size of an antibody and the resembling antibodies together, and is defined as below:

$$C_v = \frac{1}{N} \sum_{w=1}^{N} ac_{v,w} \qquad (5)$$

$$\text{where :} \qquad ac_{v,w} = \begin{cases} 1 & \alpha \cdot a_v \le a_w \le \beta \cdot a_v \\ 0 & otherwise \end{cases}$$

α and β are adjustable coefficient between 0 and 1, generally we choose the same value for both of them. The expected reproduce rate of an antibody is defined as below:

$$e_v = \frac{ax_v \prod_{w=1}^{N} (1 - ae_{v,w})}{\sqrt{C_v}} \qquad (6)$$

$$\text{where :} \qquad ae_{v,w} = \begin{cases} ay_{v,w} & \alpha \cdot a_v \le a_w \le \beta \cdot a_v \\ 0 & otherwise \end{cases}$$

Step 4: Transfer of the best solution and removal of the worst solution

This step corresponds to the proliferation and suppression of antibodies. In case of the slow convergence speed caused by the elimination of the less optimal solution with a higher concentration in later calculation, we introduce a new group: the temporary solution group, whose capacity is N. By eliminating the antibody with the lowest expected reproduce rate and transfer the antibody with the highest rate to the temporary solution group according to the descending sequence of expected reproduce rate of the primary antibodies, we can cut down the two extremenesses of antibodies. Then proliferate and suppress the antibody group in a dynamic way time after time.

Step 5: Fullness judgment of the temporary solution group

If the number of antibodies in the temporary solution group is below N, turn to step 3, and reevaluate the remaining solution space. This process is necessary to show the facility and the dynamic evaluation because of the elimination of the two extremenesses of antibodies. When the temporary solution group is full, algorithm turns to the next step.

Step 6: Clearing up of the solution group; and transfer of the temporary solutions to the solution group

Since the temporary solution group stores the less optimal solutions selected in a dynamic way, it is necessary to clear up the solutions in the solution group, and transfer the temporary solutions to the solution group to form the father generation group. At the same time, put the first N_k antibodies picked up from the temporary solution group into the memory library. In the end, clear up the temporary solution group.

Step 7: Satisfaction judgment of the solution to constraints.

If the solution satisfies constraints, algorithm ends. Otherwise, algorithm turns to step 2.

2.2 The characteristics of the dynamic evaluation based immune algorithm

(1) By introducing the temporary solution group, the evaluation, proliferation and suppression of antibodies take more times in a dynamic way, which updates the solution group and improves the convergence speed of the algorithm in later calculation.

(2)Unify the processes of generating the primary antibodies and generating new antibodies in step 2 (Production of solution group), which makes the algorithm uniform and orderly.

(3)In this algorithm the memory library is used from the beginning to the end, which helps to accelerate the convergence speed.

(4)The calculation of antibodies' concentration is based on antibodies whose fitness degree is not a certain value but between a certain interval and the parameters in the calculation can be adjusted agilely, which makes for limiting antibodies with a high concentration.

(5) The calculation of antibody's expected reproduce rate is based on the non-linear combination of affinity and concentration, which shows better the effect to the expected reproduce rate for each factor.

(6) Information entropy is more efficient in evaluating the affinity of antibodies than other methods such as: objective functions (questing the maximum of objective functions), reciprocal of objective functions (questing the minimum of objective functions), Euclidean distance, and Hamming distance.

3.   THE MODEL OF THE JSP

The JSP model is described as below:
Objective functions:

$$F = \min(\sum_{i=1}^{N_O} w_i F_i)$$ (7)

$$F_1 = \max(e_{i,j} \in \forall wp_{i,j})$$ (8)

$$F_2 = \sum_{m=1}^{N_M} \sum_{i=1}^{J_m-1} z_{i,i+1}^{(m)}$$ (9)

$$F_3 = \sum_{i=1}^{N_P} \sum_{j=1}^{L_i-1} (s_{i,j+1} - e_{i,j})$$ (10)

$$F_4 = \sum_{m=1}^{N_M} K_m \cdot P_m$$ (11)

$$\vdots$$

subject to:.

$$s_{i,j+1} \geq e_{i,j} \quad (12) \quad , \quad s_{i,1} \geq 0 \quad (13)$$

$$z_{i,j,m} = \begin{cases} 1 & \text{the } j\text{-th operation of product } i \text{ is processed on machine } m \\ 0 & \text{otherwise} \end{cases}$$ (14)

$$\begin{cases} s_{i,j} \geq e_{p,q} \\ z_{i,j,m} = z_{p,q,m} = 1 \end{cases} \quad \text{or} \quad \begin{cases} s_{p,q} \geq e_{i,j} \\ z_{i,j,m} = z_{p,q,m} = 1 \end{cases}$$ (15)

$$\sum_{m=1}^{N_M} \sum_{j=1}^{J_m} z_{i,j,m} = L_i$$ (16)

$$z_{i,j}^{(m)} = \begin{cases} 1 & \text{proculct type of operation } wm_{m,i} \text{ and } wm_{m,j} \text{ is different} \\ 0 & \text{otherwise} \end{cases}$$ (17)

$$t_{i,j} = PZ_i \times tu_{i,j}^{(m)} \quad (18), \quad e_{i,j} = s_{i,j} + t_{i,j} \quad (19)$$

$$nl_m = \sum_{i=1}^{N_P} (PZ_i \cdot Q_{i,m}) \qquad (20), \qquad P_m = \begin{cases} 1 & nl_m \geq nlu_m \\ 0 & \text{otherwise} \end{cases} \qquad (21)$$

$$Q_{i,m} = \begin{cases} 1 & \text{product } i \text{ is processed on machine } m \\ 0 & \text{otherwise} \end{cases} \qquad (22)$$

where :

F_i : i-th object function, w_i : weight of F_i

N_O : mumber of object functions, m : machine m

$wp_{i,j}$: j-th operation of product i

$t_{i,j}$: process time of $wp_{i,j}$

$wm_{m,j}$: j-th operation of machine m

J_m : operation number of machine m

$s_{i,j}$: start time of $wp_{i,j}$, $e_{i,j}$: completion time of $wp_{i,j}$

$tu_{i,j}^{(m)}$: processing time per one unit of $wp_{i,j}$ on machine m

nl_m : load of machine m

nlu_m : upper limit of load of machine m

K_m : coefficient of penalty of machine m

P_m : penalty function of machine m owning to overload

Q_m : subsidiary function, N_M : number of machine

N_P : number of variety of products

PZ_i : size of product i

L_i : number of operations of product i

4.   TO SOLVE JSP BY DEIA

The antibody constructed here as described in figure 2, doesn't only have one segment, but M segments. Here, each segment corresponds to a machine. The number in each gene locus represents the product number, and

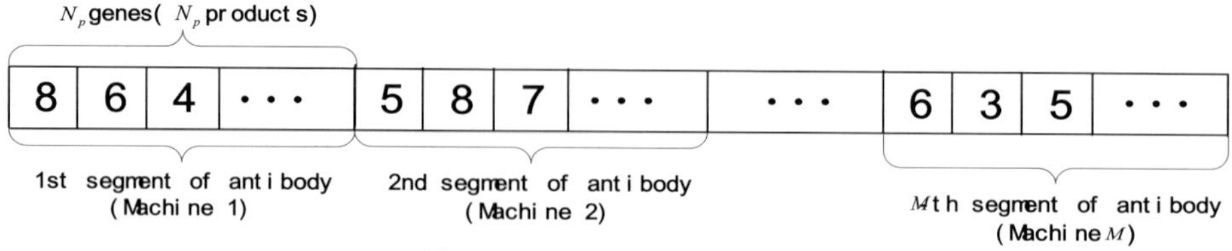

Figure 2 Antibody encoding

the sequence of the number represents the processing sequence of the products. So the m -th segment shows the processing sequence of the products on machine m . The first segment shows the processing sequence on machine 1 is : product 8, product 6, product 4, ..., product N_P . In addition. Considering the complexity of JSP, we only choose the minimization of the maximum processing time F_1 as the objective function F.

5.   SIMULATIONS

This paper separately applies simulated annealing (SA), immune algorithm（IM）, and dynamic evaluation based immune algorithm(DEIA) to simulate the well-known FT10x10 (10 jobs and each job including 10 operations) problem put forward by Fisher and Thompson in 1963, and makes a comparison of the results. The simulation curves are shown separately in figure 3, figure 4, and figure 5. From the simulation results, we see that the optimization of the FT10x10 problem can be achieved in the 500th iteration using the dynamic evaluation based immune algorithm, which proves the efficiency of the algorithm. Table 3 shows the optimal processing sequences and the processing time.

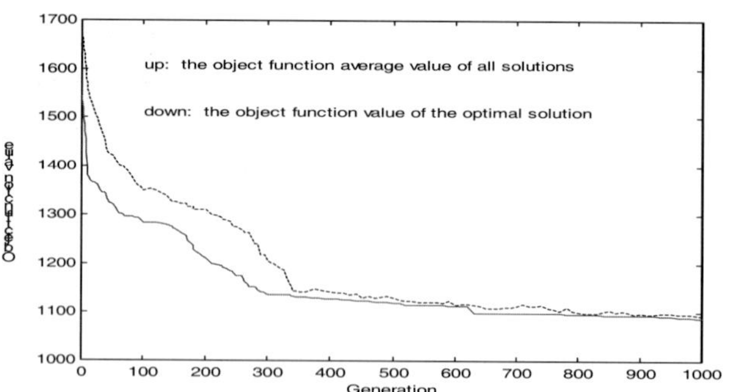

Figure 3 Results of running 1000 generations by SA

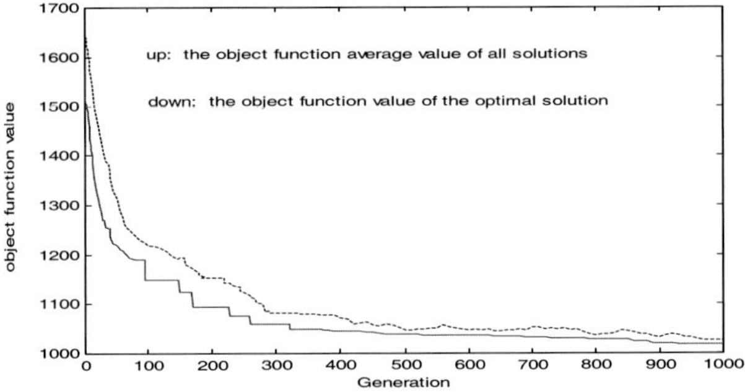

Figure 4 Results of running 1000 generations by MA

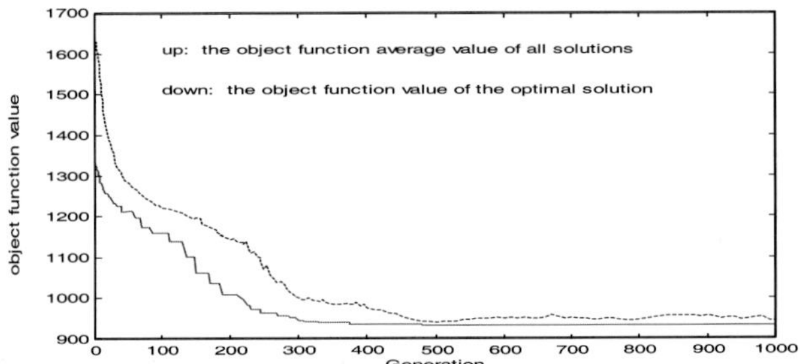

Figure 5 Results of running 1000 generations by DEIA

6.   CONCLUSIONS

This paper proposed a dynamic evaluation based immune algorithm, and constructed the antibody and the immune operator applicable to JSP. By introducing the temporary solution group, the evaluation, proliferation and suppression of antibodies take more times in a dynamic way, which updates the solution group and improves the convergence speed of the algorithm in later calculation. Besides, the memory library is used from the beginning to the end, which also helps to accelerate the convergence speed. Simulation results of the FT10x10 problem shows that the proposed algorithm not only is feasible, but also avoids prematurity, and has a higher convergence speed. This is a totally new algorithm in solving job-shop scheduling problem.

ACKNOWLEDGEMENTS

The authors gratefully acknowledge the financial support for this research from National Natural Science Foundation of China (Grant No. 69984004) and from National Hi-technology Development Program of China (Grant No. 2001AA412150, 2003AA411110).

REFERENCES

1. J. K. Lenstra and A. H. G. Rinnooy Kan. "Computational complexity of discrete optimization problems". *Annals of Discrete Mathematics*, **4**:121{140, 1979.
2. D. Applegate and W.Cook. "A computational study of the job-shop scheduling problem. ORSA" *Journal on Computing*, 3:149{156, 1991.
3. P. Brucker, B. Jurisch, and B. Sievers. "A branch and bound algorithm for the job-shop scheduling problem". *Discrete Applied Mathematics*, **49**:105{127, 1994.
4. J. Carlier and E. Pinson. "An algorithm for solving the job-shop problem". *Management Science*, 35:164{176, 1989.
5. J. Carlier and E. Pinson. "A practical use of Jackson's preemptive schedule for solving the job-shop problem". *Annals of Operations Research*, **26**:269{287, 1990.
6. B. Gi_er and G. L. Thompson. "Algorithms for solving production scheduling problems". *Operations Research*, **8**:487{503, 1960.
7. S. Binato, W.J. Hery, D. Loewenstern, and M.G.C. Resende. "A GRASP for job shop scheduling". In P. Hansen and C.C. Ribeiro, editors, *Essays and surveys on metaheuristics*. Kluwer Academic Publishers, 2001.
8. E. Pinson. "The job shop scheduling problem: A concise survey and some recent developments". In P. Chr_etienne, E.G. Co_man Jr., J.K. Lenstra, and Z. Liu, editors, *Scheduling theory and its application*, pages 277{293. John Wiley and Sons, 1995.
9. R. J. M. Vaessens, E. H. L. Aarts, and J. K. Lenstra. "Job shop scheduling by local search. INFORMS" *Journal on Computing*, **8**:302{317, 1996.
10. S. French. *Sequencing and scheduling:An introduction to the mathematics of the job-shop*. Horwood, 1982.
11. J. Adams, E. Balas, and D. Zawack. "The shifting bottleneck procedure for job shop scheduling". *Management Sciences*, 34:391{401, 1988.
12. R. J. M. Vaessens, E. H. L. Aarts, and J. K. Lenstra. "Job shop scheduling by local search. INFORMS" *Journal on Computing*, **8**:302{317, 1996.
13. P. J. M. Van Laarhoven, E. H. L. Aarts, and J. K. Lenstra. "Job shop scheduling by simulated annealing". *Operations Research*, **40**:113{125, 1992.
14. E. D. Taillard. "Parallel taboo search techniques for the job shop scheduling problem". *ORSA ournal on Computing*, 6:108{117, 1994.
15. E. Nowicki and C. Smutnicki. "A fast taboo search algorithm for the job shop problem". *Management Science*, **42**:797{813, 1996.
16. L. Davis. « Job shop scheduling with genetic algorithms". In *Proceedings of the First International Conference on Genetic Algorithms and their Applications*, pages 136{140. Morgan Kaufmann, 1985.
17. A. S. Jain and S. Meeran. "A state-of-the-art review of job-shop scheduling techniques. Technical report", Department of Applied Physics, Electronic and Mechanical Engineering, University of Dundee, Dundee, Scotland, 1998.
18. Endo, S., Toma, N. & Yamada, K. (1998), "Immune Algorithm for n-TSP", In *Proc. of the IEEE SMC'98*, pp. 3844-3849.
19. Toma, N., Endo, S. & Yamada, K. (1999), "Immune Algorithm with Immune Network and MHC for Adaptive Problem Solving", *In Proc. IEEE SMC'99*, **4**, pp. 271-276.
20. Liu Kesheng, Cao Xianbin, Zheng Haoran, Wang Xufa. "Solving TSP Based on Immune Algorithm", *Computer Engineering*, **Vol.26**, No.1, January 2000.
21. Zhou Hui, Li Ling, Hu Jiaqing. "An immune algorithm for traveling salesman problem", *Journal of Changsha University of Electric Power(Natural Science)*, **Vol.18, No.2**, May 2003-12-7
22. C.H.Lin, C S Chen, C.J.Wu and M.S.Kang. "Application of IA to optimal switching operation for distribution-loss mini and loading balance", *IEEE Proc.-Gener. Transm. Distrib.*, **Vol. 150.No.2**. March 2003

23. Kazuyuki Mori, Makoto Tsukiyama and Toyoo Fukuda. "Adaptive scheduling system inspired by immune system", *IEEE International Conference on Systems, Man, and Cybernetics*, 1998, pp.3833-3837.

24. Leandro N. de Castro, Femando J. Von Zuben, "learning and optimization using the clonal selection principle",*IEEE Transation on Evolutionary Computation*, **Vol.6, no 3**, pp.239-251,2002

25. Leandro Nunes de Castro, Fernando J. Von Zuben "the Clonal Selection Algorithm with Engineering Applications", In Workshop Proceedings of GECCO, pp. 36-37, *Workshop on Artificial Immune Systems and Their Applications*, Las Vegas, USA, July 2000.

26. Huang Xiyue, Zhang Zhuhong "multi-objective optimization immune algorithm based on immune response principle and its application", *Information and Control*, **Vol.32, No.3**, June 2003.

27. Yang Jianguo, Ding Huiming, Li Peizhi, "Immune Scheduling Algorithm for Multi-object Flow-shop Problem", *Machine Design and Research*, **Vol.18, No.4**, Aug 2002.

28. Tan Zhiyang, "Application of Artificial Immune Algorithm for Flow-shop Problem", *Computer engineering and applications*, in 2002.

29. Shyh-Jier Huang. "Enhancement of thermal unit commitment using immune algorithms based optimization approaches", *Electrical Power and Energy Systems 21* (1999) 245–252 246

30. Gao Jie. "The Application of the Immune Algorithm for Power Network Planning", *System Engineering Theory and Practice*, **No.5**,May 2001

31. Qi Xia, Chen. Senfa, Huang Kun, Zhou Zhengguo. "Study on the Logistics Distribution VRP Based on Immune Algorithm", *China Civil Engineering Journal*, **Vol.36, No.7**, july 2003.

32. Jang-Sung Chun, Jeong-Pil Lim, and Hyun-Kyo Jung. "Optimal design of synchronous motor with parameter correction using immune algorithm", In *Proceedings of the 1997 IEEE International Electric Machines and Drives Confeence Record*, pages TB2/2.1-2/2.3.

33. Ge Hong, Mao Zongyuan, "Improvement for Immune Algorithm", *Computer engineering and application*, 2002.

34. H. Fisher and G. L. Thompson. "Probabilistic learning combinations of local job-shop scheduling rules". In J. F. Muth and G. L. Thompson, editors, *Industrial Scheduling*, pages 225{251. Prentice Hall, Englewood Clis, NJ, 1963.

PREVENTIVE MAINTENANCE AND JOINT BUFFER INVENTORY FOR PARALLEL MACHINES :
An Approach Based on Experimental Design

Valerio Boschian[1], Nidhal Rezg[1] and Gilles Cormier[2]
1 AGIP/GIPM, Ile du Saulcy 57045 Metz-Cedex, FRANCE
2 Faculté d'ingénierie, Université de Moncton, Moncton, NB, CANADA
boschian@enim.fr, nrezg@loria.fr, cormieg@umoncton.ca

Abstract: This paper presents an approach for jointly optimizing buffer inventory and preventive maintenance. Due to the complexity of the mathematical model, experimental design is used in order to determine the optimal values of the decision variables, namely, the maintenance periodicity and the inventory level. We have also proposed a decomposition method for the case of parallel machines. By including the value of the scale coefficient beta in our cost equations, we have determined a model able to adapt itself to numerous machines.

Key words: preventive maintenance, inventory control, simulation, optimisation, experimental design

1.   INTRODUCTION

In order to reach their productivity, profitability and development objectives, industrial enterprises have to exert control over their production equipment. Hence, maintenance activities must allow the latter to perform the required production function at an optimal total cost (standard AFNOR X 60-000, 010).

In the "Just in Time" philosophy, which aims to eliminate inventory (Abdulnour et al., 1995), the production units operate such that they receive the raw materiel just prior to their utilization. This system can only be efficient in a context where both workers and production equipment are very reliable. A production system is nevertheless subject to numerous uncertainties, for instance, non identical cycle times, breakdowns, varying repair times, production shortages, etc... All these uncertainties encourage the use of safety stock, which can be less expensive than a loss of production (Groenevelt et al., 1995) and the optimization which contributes to the profitability objective.

Computer simulation allows us to test hypotheses (Claver et al., 1996) particularly when the complexity of the situation precludes an analytical approach (Ouali 1999).

Under such circumstances, experimental design and simulation can also be combined, see for instance Gharbi et al., (2003) who used this approach to compute a period to perform preventive maintenance.

In this paper, the capacity of the production line may at time be insufficient to satisfy demand, the decision thus being taken to increase the production capacity by using machines belonging to other production lines. The objective of this study is to reduce total cost through the optimisation of the decision parameters, that is to say, the maintenance frequency of each machine and the level of the ending inventory. In order to achieve this, we propose a decomposition method which entails allocating demand to subsets of machines in proportion to their production capacity.

This paper is organised as follows. Section 2 presents a description of the problem at hand and emphasize the challenge posed by applying a straight analytical method to jointly model maintenance and production. To overcome this difficulty, we propose the use of simulation. Section 3 presents the simulation model. Section 4 presents a method to solve the problem, which entails formulating the cost function based on the experimental design technique. Section 5 describes a method for decomposing the production system. While the conclusions and the future research opportunities are discussed in section 6.

2.　PROBLEM DESCRIPTION

2.1　Notations

MTBM : Mean Time Between Maintenance
MTTR: Mean Time To Repair
$M_{p(i)}$: preventive maintenance cost of machine i, per occurrence
$M_{c(i)}$: failure repair cost of machine i
R(t): production unit reliability function

F(t) : probability distribution function of the production system lifetime.

2.2   Analysis of the Model

Our study concerns a production system at first consisting of a single machine, but with the possibility of incorporating other machines in nearby production lines, if demand justifies it and at a certain setup cost. This system manufactures only one type of parts which form the ending inventory S from which demand d is satisfied. Demand d is assumed to be stationary (constant mean and variance). The machine can work at a pace U_{max}, which is superior to the average demand and allows for the accumulation of the ending inventory.

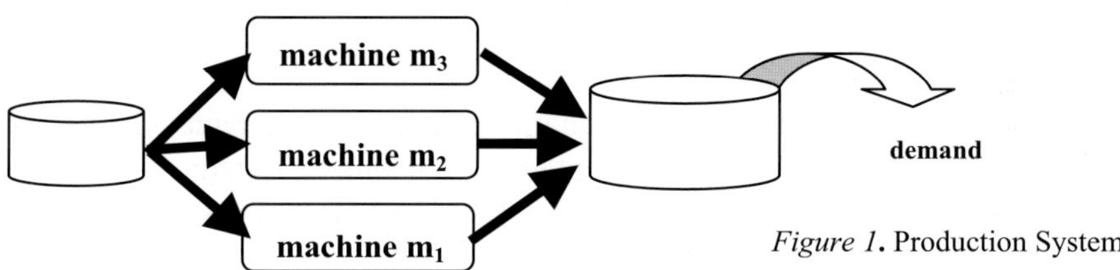

Figure 1. Production System

Production machines are subject to failure according to a probability distribution. The failure of the production machines immediately interrupts production and, in case of insufficient ending inventory, leads to unfulfilled demand.

2.3   Profile of the Ending Inventory

The profile of the ending inventory consists of three phases (Fig. 2) :
- During the first phase, the machine produces at a rate U_{max} in order to constitute an inventory of size h. We assume that the machine, being as new, will not breakdown. The value s(t) of the inventory at time t is $(U_{max}-d)*t$. This will be used to satisfy the demand during the period of preventive maintenance.
- During the second phase, production is equal to the demand d.
- During the third phase, the machine is inoperative and undergoes preventive maintenance. The value s(t) of the inventory therefore decreases so as to satisfy the demand d and can actually became negative of a shortage arise.

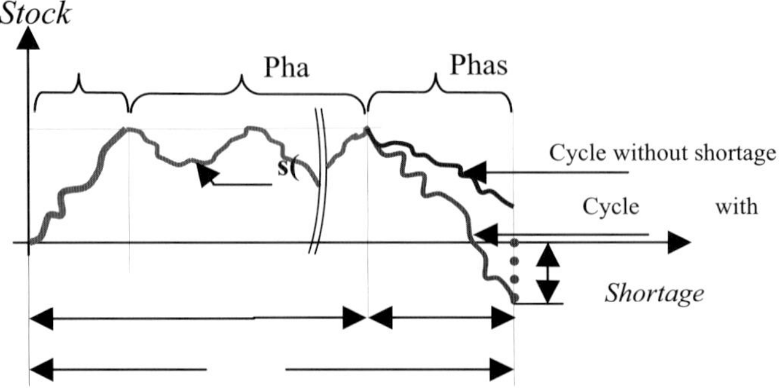

Figure 2. Variation of the inventory over time

2.4 Analytical model for joint maintenance and inventory management

The integrated model for maintenance and inventory management under study combines the rules imposed by a maintenance strategy based on the age of the equipment (Barlox et al., 1965), which consists of performing corrective maintenance after a failure has occurred or preventive maintenance after m time units without a breakdown, with an inventory policy either with or without shortages. Let $\Pi_{(h,m_i)}$ be the total average cost of maintenance and inventory operations per time unit over an infinite horizon. Then

$$\Pi\left(h, m_i\right) = \frac{\delta_{c/p}(h, m_i) + \sum_{i=1}^{n} M_{p(i)}R(m_i) + \sum_{i=1}^{n} M_{c(i)}F(m_i)}{E\left[T_{cyc}\right]}$$

Where $\delta_{c/p}(h, m_i)$ represents the average total cost of inventory as a function of h and m_i.

In the case of constant demand, this cost is difficult to calculate, whereas it is impossible to do so in the case of uncertain demand. The latter situation will hence be dealt with via simulation and experimental design.

3. SIMULATION MODEL

We have used the Promodel simulation software to model and simulate the foregoing strategies.

3.1   Simulation Scenario

The simulation model was developed following a three stage method, as described below.

Stage 1 : *Setting up a physical model*. The physical model represents the configuration of the production system to be studied. It defines the operations sequence of the products and describes their corresponding routing through the production system. The latter consists of three manufacturing centres, each of which is supplied by a beginning inventory and produces a product. Product is placed in an output buffer where the ending inventory accumulation (if $h_i < h_{maxi}$) and from which customer demand is fulfilled.

Stage 2 : *Programming of the decision rules*. The next stage consists of programming the decision rules which govern the model reflecting the inventories maintenance policies.

Stage 3 :*Scenario analysis and optimisation*. Taking into account the analytical models previously derived, the total operating cost is a function of the age *m* of the machine and the level h of the inventory S. The software (*Promodel*) allows to generate simulation scenarios which combine different values of h and m while taking into account the model parameters such as the probability distributions of breakdown and repair time, the cost associated with corrective and preventive maintenance, the holding cost, the shortage cost, etc...

3.2   Numerical Example

We now describes an example of the proposed simulation model assuming the following data :

- Time to failure probability distribution, F(t): Weibull $W(\alpha_i, \beta_i)$ with $\alpha_1 = \alpha_2 = 2$, $\alpha_3 = 1.5$, $\beta_1 = \beta_2 = 200$ and $\beta_3 = 100$
- Repair time probability distribution : Exponential $E(\delta_i)$ with mean=10 time units, for each i.
- Preventive maintenance time probability distribution : Exponential $E(\varepsilon_i)$ with mean=5 time units, for each i.
- Preventive maintenance cost M_p is 300 monetary units;
- Failure repair cost M_c is 2000 monetary units;
- The holding cost C_s and the shortage cost C_p are respectively 3 and 250 monetary units per time unit;

The demand d has a normal distribution with mean 4 time units and a standard deviation of 0.5 time units.

3.3   Conclusion

The simulation model described in this section allows for breakdowns during the inventory building phase and is thus realistic. We can then obtain experimental values of the total cost by varying the different parameters, which will become the starting point for our experimental design..

4.   GLOBAL SOLUTION METHOD

4.1   Introduction

As we have alluded in section 3.1, the analytical model is complex and even impossible to solve on account of the characteristics of the demand d. We are hence using simulation to derive a cost model, based on experimental design (Sado and al., 1991) (Boschian and al., 2003). Our aim is to express the total cost in terms of our four decision variables, namely the m_i's and h.

4.2   Problem modeling

Let Y denote the system response and X_i the independent variables, such as X_{mi}, X_h, their combinations or functions of them, where $Y=f(X_i)$. The model can alternatively be written as

$$Y = \alpha_0 + \sum_{i=1}^{k} \alpha_i . X_i + \sum_{i=1}^{k-1} \sum_{j=2.(j>i)}^{k} \alpha_{ij} . X_i . X_j$$

where α_i represent the coefficients which must be determined.

First, we have verified that the cost function is not a first class model. Then, we shall use a factorial plan completed with centre points. Preliminary runs made by fixing h and varying T (and vice versa) allowed us to select the domains of interest for each parameter. The cost function was found to have a quadratic form, h varying between 1 and 49 and T between 10 and 390. According variables h and m_i are reduced to the following levels :

X_i	Different levels of reduced variables		
	Level -1	**Level 0**	**Level 1**
h	1	25	49
m_i	10	60	110

Table 1. Correspondence between real and reduced variables

By varying the four reduced variables between their three possible values, we have obtained 81 experimental results, gathered together in a matrix [Y] so that $^t[Y] = [y_1, y_2, ..., y_{81}]$.

The model coefficients are calculated by multilinear regression on the entire set of 81 experiments. The system of equations we have to solve is:

$$[Y] = [X][A] + [E]$$

Where [X] = the regression matrix
$^t[A] = [\alpha_0, \alpha_1, ..., \alpha_q]$ the vector of coefficients, to be determined
$^t[E] = [e_1, e_2, ... e_{81}]$ = the vector of the distances

The sum of the squared distances is written $[^TE].[E]$, which will be minimal in connection with the coefficients if $\dfrac{d[^TE][E]}{dA} = 0$

The coefficients α_i are obtained as follows :

$$[A] = ([^TX].[X])^{-1}.[^TX].[Y]$$

We deduce from there the following mathematical model :

$$Y = 109 - 10{,}6X_h - 68{,}4\ X_{m1} - 71\ X_{m2} - 160\ X_{m3} + 4{,}3X_h.X_{m1} + 4{,}3X_h.X_{m2} - 4{,}7X_h.X_{m3} + 15{,}6X_{m1}.X_{m2} + 22X_{m1}.X_{m3} + 18{,}4X_{m2}.X_{m3} + 22{,}5\ X_h^2 + 71{,}8X_{m1}^2 + 71{,}1X_{m2}^2 + 157X_{m3}^2$$

4.3   Significance tests of the coefficients

We shall verify if the above coefficients are significant by computing the distances ei (or remainders) and applying Student's test.

By applying this test, we establish that the first class interactions between the stock level and the periodicities of the maintenance are insignificant. We also suppress these coefficients from the vector A and the corresponding columns in the reduced matrix. We deduce the following mathematical model :

$$Y = 97{,}3 - 18{,}5 * X_h - 67 * X_{m1} - 75.6 * X_{m2} - 81.5 * X_{m3} + 8{,}87 * X_{m1} \cdot X_{m2} + 10{,}9 * X_{m1} \cdot X_{m3} + 9{,}14 * 4X_{m2} \cdot X_{m3} + 25 * X_h^2 + 70{,}4 * X_{m1}^2 + 80{,}5 * X_{m2}^2 + 86 * X_{m3}^2$$

4.4 Model Validation

We have tried to compare the experimental values y_i of the total cost obtained with our simulation model and the calculated values γ_i. Given $e_i = y_i - \gamma_i$,

$$\Delta Y\% = \frac{\sum_{i=1}^{81} |e_i|}{\sum_{i=1}^{81} y_i} = \frac{984}{22060} = 4.4\%,$$

which reveals that the variation between the costs obtained by experimentation is very low in comparison with that obtained by calculation.

4.5 Results

The optimal values of X_h (denoted X_h^*) and of the X_{mi} (denoted X_{mi}^*) are given by solving the following systems differential equations:

$$\left. \frac{dY}{dX_h} \right|_{X_h = X_h^*} = 0 \qquad\qquad \left. \frac{dY}{dX_{mi}} \right|_{X_{mi} = X_{mi}^*} = 0$$

$$\Rightarrow X_h^* = 0.43,\ X_{m1}^* = 0.4213,\ X_{m2}^* = 0.4263 \text{ and } X_{m3}^* = 0.4305$$

The aim here is to express the total cost as a function of h and m_i using the transformation: $X_h = \frac{h-25}{24}$ and $X_{mi} = \frac{mi-60}{50}$. By executing variable substitutions : $(X_{hi}, X_{mi}) \Rightarrow (h_i, m_i)$, we obtain $h^* = 36.8$ units, $m_1^* = 142$ time units, $m_2^* = 144$ time units and $m_3^* = 146$ time units.

4.6 Conclusions

Through experimental design, we have determined a mathematical model of total cost. The validity of this second order equation has been verified with appropriate statistical tests. The reduced matrix includes nevertheless a great number of parameters such that the number of required experiments is important.

5.   DECOMPOSITION METHOD

5.1   Method Principle

We shall decompose the global demand into elementary demands, which can be satisfied separately. Each machine i will satisfy an elementary demand d_i, determinal as a function of its maximum capacity $U_{max(i)}$, that is,

$$d_i = \frac{U_{max(i)}}{\sum_{i=1}^{n} U_{max(i)}} * d$$

For each machine i, we shall simulate a virtual inventory S_i (of level h_i) assigned to each machine. Under these conditions, we set the following :

$$\frac{dCost\_of\_Global\_Machine}{dh} = \sum_{i=1}^{3} \frac{dCost\_of\_Elementary\_Machine\_i}{dh_i}$$

$$\frac{dCost\_of\_Global\_Machine}{dmi} = \frac{dCost\_of\_Elementary\_Machine\_i}{dmi}$$

The level h of the ending inventory S will be determined by

$$h = \sum_{i=1}^{n} h_i$$

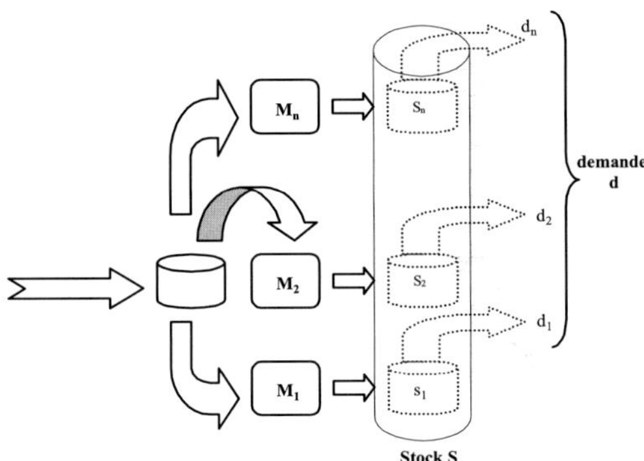

Figure 3. Decomposition of the demand

5.2   Solution Using Experimental Design

Let us now apply the method of the previous section to each machine individually. For each machine i, we make the hypothesis that the real model can be approximated with a mathematical model of the form :

$$Y = \alpha_0 + \sum_{i=1}^{k} \alpha_i.X_i + \sum_{i=1}^{k-1} \sum_{j=2.(j>i)}^{k} \alpha_{ij}.X_i.X_j$$

X_i	Different level of reduced variables		
h	1	26	51
m_i	10	100	190
beta	80	100	120

Table 2. Correspondance between real and reduced variables

In this experiment, we introduce the variable beta, the scale parameter of the Weibull law, which varies here between 80 and 120. By varying the three reduced variables between their three possible values, we have obtained 27 experimental results y_i, aggregated in a matrix $[Y]$, such that ${}^t[Y] = [y_1, y_2, \ldots, y_{27}]$. The model coefficients are calculated by multilinear regression on all 27 experimental results. The system to be solved is formulated as $[Y]=[X][A]+[E]$

We obtain, for machine i, the following result :

$$Y_i=58 + 8*X_{hi} - 67.9X_{mi} - 1,4X_{beta} + 5,7X_{hi}.X_{mi} + 8,16X_{hi}.X_{beta} -$$
$$0,44X_{mi}.X_{beta} + 10,9\,X_{hi}^2 + 70,9X_{mi}^2 + 0,86X_{beta}^2$$

For machine 1 (beta=80, X_{beta}=-1)

$$Y_1=60,26 - 0,16*X_{h1} - 67.46*X_{m1} + 5,7*X_{h1}.X_{m1} + 10,9*\,X_{h1}^2 +$$
$$70,9*X_{m1}^2$$

$$\left.\frac{dY}{dX_{h1}}\right|_{X_{h1}=X_{h1}{}^*}=0 \Rightarrow X_{h1}{}^* = -0.110, \quad \left.\frac{dY}{dX_{m1}}\right|_{X_{m1}=X_{m1}{}^*}=0 \Rightarrow X_{m1}{}^* = 0.480$$

For machine 2 (beta=100, Xbeta=0): $X_{h2}{}^*=$ -0.49 $X_{m2}{}^*$=0.4989
For machine 3 (beta=120, Xbeta=1): $X_{h3}{}^*=$ -0.8933 $X_{m3}{}^*$= 0.5176
By executing variable substitution :$(X_{hi}, X_{mi}) \Rightarrow (h_i, m_i)$, we obtain
$h_1{}^*= 23.1$ units $h_2{}^*=13.2$ units $h_3{}^*=3.6$ units
$m_1{}^*=143$ time units $m_2{}^*=144$ time units $m_3{}^*=145.9$ time units

We can thus determine the cost function of machine i in terms of h and m_i.

5.3   Conclusions

	Global Model	Decomposed Model			$\sum_{i=1}^{3}$	relative error
		Machine 1	Machine 2	Machine 3		
h	36.8 pieces	23.1	13.2	3.6	39.9	8,4%
m1*	142 units	143				0,7%
m2*	144 units		144,1			0,07%
m3*	146 units			145,9		0,07%

Table 3. Results Comparison

We observe that the sum of the inventory obtained using the decomposition model is approximately equal to the inventory calculated by the global method. The relative error totals 8,4%, which can be explained by the simulation tool. As for the maintenance periodicity m_i, we note the near perfect correspondence of the results between the global and decomposition methods. Indeed, the number of trials can be less important for large matrix sizes. With machines having coefficients of identical scales, the experiments realized in section 4.2 would have needed only 2 reduced variables (X_m, X_h), and this in turn would have reduced from 27 to 9 the number of necessary experiments.

These encouraging results are an indication that we could safety extrapolate from 3 to n machines in parallel production lines.

6.   CONCLUSIONS AND FUTURE RESEARCH

In this paper, we have presented an approach for jointly optimizing inventories and maintenance policies.

We have used experimental design in order to determine the optimal values of the decision variables, namely, the maintenance periodicity and the inventory level. We have also proposed a decomposition method for the case of parallel machines. By including the value of the scale coefficient beta in our cost equations, we have determined a model able to adapt itself to numerous machines. This technique can be extended to n parallel machines and to the case of variable demand. Our future objective is to adapt this method to machines having different production capacities.

REFERENCES

1. Abdulnour G., Duddek R.A and Schmitt M.L., "Effect of maintenance policies on just-in-time production system", *International Journal of Production Research*, **vol. 33**, 1995, p.565-585.
2. Boschian V., Rezg.N. « Utilisation des plans d'expériences pour une approche intégrée de la maintenance préventive et gestion des stocks », 5ème congrès international de GI, Québec, octobre 2003.
3. Barlow R.E. and Porschan F., *Mathematical theory of reliability*, Jhon Willey&Sons, New-York, 1965.
4. Claver Jean-François, Jacqueline Gélinier, Dominique Pitt, Gestion de flux en entreprise, Editions HERMES, 1996.
5. Gharbi A., Beauchamps Y., Andriamaharosoa S.: Stratégie de maintenance préventive de type age : approche basée sur l'intégration de la simulation et des plans d'expériences... to appear.
6. Groenevelt Harry, Pintelon Liliane and Seidmann Abraham, *"Production Lot sizing with machine breakdown"*, Management Science, **Vol. 38, n°1**, January 1995, pp. 104-123.
7. Ouali Mahamed-Salah, Ait-Kadi Daoud, Gharbi Ali « A simulation model for opportunistic maintenance stategies » 7th IEEE (ETFA'99), Barcelon, Spain.
8. Sado Gilles, Sado Marie-Christine « Les plans d'expérience », AFNOR TECHNIQUE 1991, ISBN :2-12-450311-1.

DESIGN AS PRODUCTIVE SCIENCE
Conception And Implication

Yuemin Hou, Linghong Ji and Dewen Jin
Dept. of Precision Instrument and Mechanology, Tsinghua University, Beijing, 100084, P.R China, hym01@mails.tsinghua.edu.cn, jilh@pim.tsinghua.edu.cn, jdw-om@tsinghua.edu.cn.

Abstract:　Philosophy of design concerns concept and language, while the design methodology concerns design process. We focus on the nature of design as the key to understanding design process. This paper commences by an overview of design and designing for further analysis. The fundamental idea of this paper is that design is productive science, which is founded upon the science classification of Aristotle. Proceeding with this idea, the core of design is analyzed and the implication of design as productive science is discussed. A brief summary concludes this paper

Key words:　Productive science, Aristotle, philosophy of design, artefact.

1. INTRODUCTION

What is the core of design? What distinguishes design with other disciplines?

One way of stating these questions precisely is to look to philosophy for help. The contributions in this field may be catalogued into philosophy of design. The philosophy of design, according to Per Galle, is characterized as the pursuit of insights about design by philosophical means [1]. Love further distinguishes philosophy of design from design philosophy: the latter is associated more with the philosophical study of design method, whereas the former concerns what design is in general [2].

There are a great deal of literatures contribute to design studies, most of which could be catalogued into design methodology. Also there emerges a voluminous bunch of papers about "the philosophy of design", most of which concern the philosophical concept of design. I would like to cope with design methodology from philosophy angle aiming at new insight into the character of design process that go unseen otherwise. This paper narrows the discussion to design as productive science, see figure 1.

In the reminder of the paper, we first present a brief overview on design

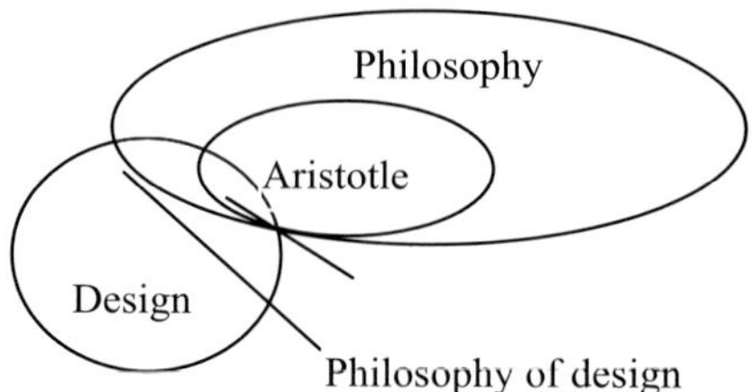

Figure 1 Design as productive science

and designing. Then we philosophise design in light of Aristotelian science classification. Thirdly we discuss the triple nature of designs and the core of design as productive science. The paper is concluded with a brief summary.

2.   DESIGN AND DESIGNING

To clarify the meaning of design, we first distinguish design and designing. There are a variety of definition about design and designing, just list some:

Design:

a) A noun referring to specification or plan for making a particular artefact or for undertaking a particular activities [3].
b) Bridging function and structure using knowledge [2].
c) Design is a socially mediated activity [4].
d) To transform requirements, more generally termed function, that embodies the expectations of the purposes of the resulting artefact, into design description [5].
e) Design is a purposeful human activity in which cognitive process are used to transform human needs and intend into an embodied object [6].
f) Turning ideas into reality [7]].
g) A complex problem-solving activity [7].
h) Finding the best possible solution to a specific program [7].
i) Intellectual attempt to meet certain demands in the best possible way [7].
j) The creative and enabling between deficiencies in man's human condition and their alleviation through the understanding and application of natural laws [7].

Designing:

a) Non-routine human activity leading to the production of a design [5].
b) Primary human function on a par with thinking and feeling [8].
c) A particular type of action, which, in turn, is described in terms of plans, intentions, and practical reasoning [3].
d) Larger process of plan design, which aims at establishing the intended use of the artefact in question [1].
e) Social process awash in uncertainty and ambiguity [9].
f) A goal-oriented, constrained decision-making, exploration and learning activity that operates within a situation that depends on the designer's perception of the situation and results in the description of a future engineering system [8].

To sum up, design is thinking on linking function and entity, whereas designing is the action to embody the thinking, including graph, calculation etc. In other words, design is referred to conceptual design, while designing is referred to detail design. The detail design is the embodiment of the conceptual design, a quantities description of design thought.

To explore design further, we can classify design into different groups. Gero called them routine and non-routine designing [10]. The first "produces designs that are some minor variation of existing designs, and the second produces designs that are noticeable different to existing designs". Further, he defined non-routine designing as innovative designing and creative designing: the former means that "when the context that constrains the available ranges of the values for the variables is jettisoned so that unexpected values become possible", the latter means that "when one or new variables is introduced into the design". Love narrows designing to non-routine process [3] and design as novel design.

Broadly to say, design can be catalogued into three groups, namely simulation, replication, and creative design. When a new solution is found or created to bridge function and structure, this kind of activity can be called creative design; when this solution is used with minor variation, it might as well be called replication; when the principle of the solution is applied to realize different functions, this kind of activity is called simulation.

3. DESIGN AND SCIENCE

An important viewpoint on design is that design is artificial science [11]. In terms of Simon, design is any activity planned to change reality, including social activities, programming and the like. The two crucial issues of design are: (1) utility theory and decision theory to make reasonable choices among alternatives: (2) optimization methodology to evaluate existed alternatives.

This opinion reminds of Popper's conjecture/analysis method. 'There is no more rational procedure than the method of trial and error." [12]

In light of Simon, artificial science focuses on interface of inner environment and outer environment. How to bridge inner and out environment? The question can be expressed in detail as following. How to establish the relationship between function description and structure description? How to describe structure in terms of function description? How to transmute function conception to structure conception? How to bridge intention to physical object? And what are the methods designers used? And what roles they play in design?

Simon stressed the role of decision in design. However, science on man-made things has its own place in philosophy.

3.1   Science classification

The genesis of science is philosophy, and philosophy originates from Greece. According to Aristotle, there are three classes of science, viz. natural science (phusikos episteemee), productive science (poieetikee episteemee), and practical science (praktikos episteemee) [13-14]. Natural science is defined to trace laws that govern nature by human being's wisdom, and nature is the object being studied. In productive science, "the principle of movement is in the producer and not in the product, and is either an art or some other faculty." In practical science, "the movement is not in the thing done, but rather in the doers." [13-14]

What does Greek word "poieetikee" means? In contrast with generating (genesis) and thought (noeesis), poieetikee means making.   And "all makings proceed either from art or from a faculty or from thought" [13-14].

Poieetikee is translated as productive in English. Merriam-Webster's Collegiate Dictionary gives productive the following definitions [15]:

1: having the quality or power of producing especially in abundance; 2: effective in bringing about; 3 a: yielding results, benefits, or profits; b: yielding or devoted to the satisfaction of wants or the creation of utilities; ….

The American Heritage Dictionary of the English Language gives productive the following explanation [16]:

From Greek *poiētikos*, creative, from *poiētēs*, maker, from *poiein*, to make.

In Chinese, poieetikee is translated as creation and making [13].

To sum up, the term "productive" means creative making in the context of productive science.

As stated in the second section, design is thinking how to create things to realize certain functions. If all thought is either practical or productive or theoretical, as physics must be a theoretical science, so design must be productive science.

In terms of Aristotle, "Every science which is ratiocinative or at all involves reasoning seeks certain principles and causes for each of its objects, more or less precise." "For each of these marks off a certain class of things for itself and busies itself about this as about something existing and real-not however qua real; ..."[13-14]

Science is to explore laws that govern nature, in other words, to find explanation for natural phenomenon. By analogy with the role of science, to explore laws that link function with entity should be the essence of design. Simply to say, design as productive science should to explore new explanation (cause) for bridging function and structure. Hence, only creative design is possible to be considered as science since it provides a new explanation. But even if creative design may not be qualified as productive science if it fails to explain the linkage between the function and structure with necessary accuracy and generalization, where creativity in design and productive science branch away.

3.2   Procedure of Science research and design

Science research is to conjecture a procedure P, starting from cause, passing P step by step, ending at result. The physical explanation is a procedure in mechanical design, which consists of several steps embodied by structure, i.e. a mechanical system. Through the procedure, certain predicted result will come true.

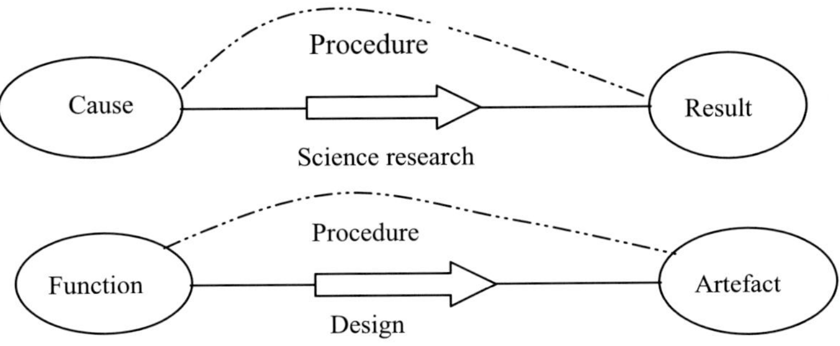

Figure.2 Design vs. science

To compare science with design will benefit design research. See figure 2. The start point and end point are reverse in science and design. Gero expressed the idea as "designing is the opposite of the traditional scientific explanation" [5]. Scientists first discover some phenomenon, namely result, then conjecture the cause and verify it, while designers first determine function, namely result, then, construct the environment, namely cause.

The procedure would be new theories in science, which can interpret a class of phenomena, and no evidence violates the explanation up to now, so it is generalization, such as relative theory. On the other hand, the procedure in design would be new approach to entity, through which certain natural principle works. The entity is an instance, such as steam turbine, satellite, but design procedure is generalization from function to structure. Generalization is one of criteria to partition off typical design from productive science. The other criterion is, as discussed above, a new and better explanation.

4.   DESIGN AND ARTEFACT

4.1   The dual nature of artefact

Although the design is process oriented, the result is an artefact, whose function is the first criterion to justify design.

Man-made things are called artefacts, or technical products, which come from something and come to be some other thing resulted from faculty or technology [13-14].

Peter Kroes defines technical artefacts as following: technical artefacts are objects with a physical structure consciously designed, produced and used by humans to realize its function [17].

Technical artefacts have dual nature: on the one hand they are physical objects that may be used to perform a certain function, on the other hand, they are intentional objects since it is the function of a technical artefact that distinguishes it from physical (natural) objects and this function has meaning only within a context of intentional human action [17].

Although artefacts are man-made things, they are physical entity once they are made. Therefore, they comply with natural laws. What designers can do is to build an environment, viz. constrains, in which certain result will come out. Will the result be predicted precisely? The answer is "no". Quantum mechanics tell us that single atom cannot be interpreted by cause and result, it has to be described by statistics instead. Why? Maybe it comes of that our knowledge of nature is deficiency, or that the laws humans have generalized take effect only on ideal condition or in ideal object, or that there is no natural laws but statistical inference to natural phenomena. No matter what reason it might be, the fact is that no exact description of cause and result is possible at present so that any physical object only can be controlled in high probability, or their states can be predicted in high probability [18].

To sum up, artefacts have dual nature, intentional object and physical object. Design is to build an environment, viz. physical object, to create condition under which some natural laws would work so that intentional result occurs in high probability. Design is to find and build connection between phenomenon and physical environment. Here, connection means new explanation, function is the description of intention, and physical environment is materialized by structure.

4.2   The triple nature of design

As previously discussed, the result of design is an artefact, and artefacts have dual nature: they are both physical objects and intentional objects. As physical objects, like other natural things, their performances and characteristics result from natural laws, why and how they occur is unknown more or less, so research on them falls on the realm of natural science. As intentional objects, on the other hand, their performances come true only under control of humans who design an environment and input movement or force into it so that it results in intended performance. To describe the process is so called design theory and methodology. To design such an environment locates in the domain of productive science. Besides, design is to reflect natural principles though man-made environment, such as wind wheel, pendulum and so on. In this sense, design belongs to the domain of practice design.

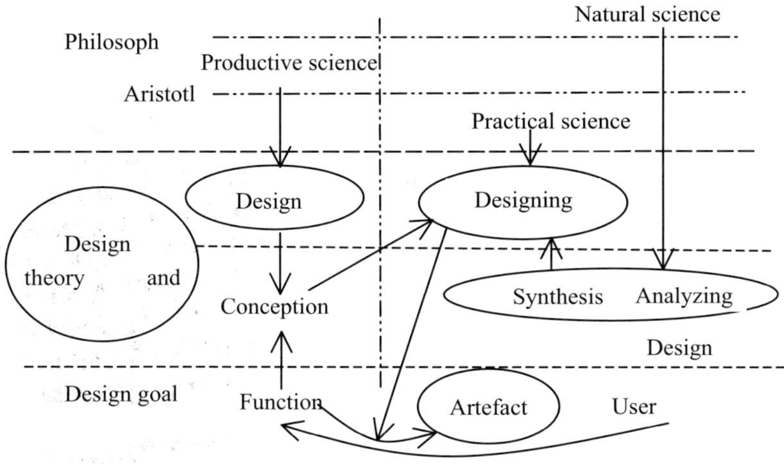

Figure 3 Design and designing

The triple nature of design makes design traverse three hierarchies. As show in figure 3, philosophy locates in the first hierarch, and there are three class of science according to Aristotle: natural science, productive science, and practical science. Design theory and method locates in the second hierarch, encompassing design and designing. Design goal occupies the third hierarch, which includes user specification, function, and artefact. Design in the context of this paper is regarded as productive science, while designing as practical science, and analysis and synthesis are either in the domain of natural science if some new explanation is generalized, or in the domain of practical science when only natural laws are applied.

4.3　　Implication of design as productive science

Kant claimed that human being cannot comprehend the essence of nature by pure wisdom [19], as a result, conjecture and analysis is the possible bridge from function to entity. The Popper's method of conjecture/analysis [12] is a generalization of all design activities: first conjecturing some structure, then analysis whether it will yield the result required. However, when structures surrogate theories in C/A model, the methodology of science is reduced to design method because it lacks exactness to map function and structures. And this kind of methodology results in that the iteration of revising and calculating costs a great amount of computer time.

As stated in previous sections, the goal of design is artefact, but design research as productive science should provides explanation on linking structure and function as accurate as possible, and the explanation is generalization [20].

So, the core of design as productive science is meticulously mapping every part of artefact and functions, and, of particularly important, exploring rules that serve as descriptions of mapping them. This thought leads to a function-driven methodology, which means deducing substructures from function based on induction rules step by step, namely, from local to whole body. In a word, structure arises progressively (namely epigenesis). Therefore, structure, material, control, and drive should be designed concurrently and interdependently.

Then a question arises: does there exist rules linking function and structure? No direct link between function and structure exists according to Gero, instead, there exists link between structure and behaviour [10]. Behaviour is clearly defined as how something acts in response to its environment [6]. And behaviour can be described as property of structure [20]. Simon considered that the relation between structure and behaviour could be generated by input data and output data, which is so called scientific discovery [11].

Since the property of organism is what artifacts pursue to possess, it is natural to explore the rules by analogy with biological life. According to Kant, each part of organism fits together not by a plan outside them, instead, by force inside them [19]. Paley had expressed the thought this way: "it is the suitableness of these parts to one another; first, in the succession and order in which they act; and, secondly, with a view to the effect finally produced" [21]. Dawkins claims "Biology is the study of complicated things that give the appearance of having been designed for a purpose" [21]. By analogy with embroygenesis and neurogenesis, rules of generating structure could be generated for description from function to structure.

Nowadays, Laser-aided manufacture technology, designed materials [22-24], smart material, MEMs, smart material and multifunctional structure provide much broader space for design possibility than traditional materials and manufacture technology can do. So, mapping every part of artefacts and functions meticulously is realizable both in material technology and in manufacturing technology. Literature [20] advanced a framework of bridging function and structure based on the thought of this paper.

5.   SUMMARY

There are three classes of science: natural science, productive science, and practical science in light of Aristotle. The triple nature of design renders design procedure to encompass the three classes. To distinguish between different classes of design will benefit to find the core of design and to frame the design methodology for different class. This paper focuses on design as productive science by philosophizing the nature of design and analyzing its implication. The core of design as productive science is meticulously mapping every part of artefact and functions, and, of particularly important, exploring rules that serve as descriptions of mapping them. This design thought implies epigenesis of artefact, namely, structure arises progressively. Therefore, structure, material, control, drive and the like should be designed concurrently and interdependently.

ACKNOWLEDGEMENTS

We would like to thank professor Linda Schmidt of the Department of Mechanical Engineering of University of Maryland for her valuable comments on the earlier version of this paper.

REFERENCES

1.   Galle P. "Philosophy of design: an editorial introduction", *Design Studies*, **vol. 23, no. 3,** pp. 211-218, May 2002.
2.   Love T. "Constructing a coherent crossdisciplinary body of theory about designing and designs: some philosophical issues", *Design Studies*, **vol. 23, no. 3**, pp. 345-361, 2002.

3.   Love T. "Philosophy of design: a metatheoretical structure for design theory", *Design Studies*, **vol.21, no.3**, pp. 293-313, 2000.
4.   Liddament T. "The Computationalist Paradigm in Design Research", *Design studies*, **vol. 20, no. 1**, pp. 41-56, 1999.
5.   Gero JS. "Computational Models of Innovative and Creative Design Process", *Technological Forecasting and social Change*, **vol. 64, no. 2**, pp. 183-196, 2000.
6.   Rosenman MA, Gero JS. "Purpose and function in design: from the socio-cultural to the techno-physical", *Design Studies*, **vol. 19, no. 2**, pp. 161-186, April, 1998.
7.   Kim H, Querin OM, Steven GP. "On the development of structural optimisation and its relevance in engineering design", *Design studies*, **vol. 23, no. 1**, pp. 85-102, 2002.
8.   Galle P. "Design as intentional action: a conceptual analysis", *Design Studies*, **vol. 20, no. 1**, pp. 57-81, 1999.
9.   Bucciarelli LL. "Between thought and object in engineering design School of Engineering", *Design Studies*, **vol. 23, no. 3**, pp. 219-231, 2002.
10. Gero JS. "Creativety, emergency, and evolution in design", *Knowledge-Based System*, **no. 9**, pp. 435-448, 1996.
11. Simon HA. "The science of artificial", Cambridge, The MIT press, Second edition, 1982.
12. Bamford G. "From analysis/synthesis to conjecture/analysis: a review of Karl Popper's influence on design methodology in architecture", *Design Studies*, **vol. 23, no. 3**, pp. 245-261, 2002.
13. Aristotle. "Metaphysics", Ed. Miao Li Tian, Beijing, Renmin University of China Press, 1996. (In Chinese)
14. Aristotle. "Metaphysics", Written 350 B.C.E, Translated by W. D. Ross, website: classics.mit.edu/Browse/browse-Aristotle.html.
15. "Merriam-Webster's Collegiate Dictionary", Deluxe Audio Edition, 2000.
16. "The American Heritage Dictionary of the English Language: Fourth Edition", 2000.
17. Kroes P: "Design methodology and the nature of technical artifacts", *Design Studies*, **vol. 23, no. 3**, pp. 287-302, 2002.
18. Reichenbach H. "The Rise of Scientific Philosophy", Berkeley, University of California Press, 1954, Chinese translated version, Shanghai, Commercial Press, 1983.
19. Brooke, JH. "Science and Religion", Cambridge, Cambridge University Press, 1991. Chinese translated version, Shanghai, Fudan University Press, 2000.
20. Hou YM, Ji LH, Jin DW. "An approach to Bridge Function and Structure", *Journal of the Asian Design International Conference*, **vol. 1**, abstract in pp. 48, full paper in CD-ROM (pp. 1-11), Tsukuba, Japan, Oct. 14-17, 2003.
21. Bateson P. "Design, Development and Decisions", *Strd. Hist. Biol. & Biomed. Sci*, **vol. 32, no. 4**, pp. 635-646, 2001.
22. Strang G, Kohn R.. "Optimal Design in Elasticity and Plasticity", *International Journal for Numerical Methods in Engineering*, **vol. 22**, 183-188, 1986.
23. Olson GB. "Beyond Discovery: Design for a New Material World", Calphad, **vol. 25, no. 2**, pp. 175-190, 2001.
24. Sanjay R, et al. "Representation of heterogeneous objects during design, processing and freeform-fabrication", *Materials and Design*, **vol. 22**, pp. 185-197, 2001.

OPTIMIZATION OF CUTTING PARAMETERS FOR PERIPHERAL MILLING BASED ON MACHINING STABILITY

Yong Chen[1] and Xiong-Wei Liu[2]

[1]Institute of Advanced Manufacturing Technology, Huaqiao University, Quanzhou, 362011, China, Email: chenyong@hqu.edu.cn [2]Department of Aerospace, Automotive and Design Engineering, University of Hertfordshire, UK[3]

Abstract: On the basis of a comprehensive and improved nonlinear mathematical dynamic cutting force model, a computer simulation model for peripheral milling process in time domain is developed with consideration of regenerative chatter. Variable-step numerical integral algorithm based on explicit forth-order Runger-Kutta formula and recurrence algorithm are used to solve the second-order dynamics equations to ensure convergence and validity of the solution. The visual simulation results, including the dynamic cutting forces and the vibration between the cutter and workpiece as well as their power spectra exhibit the same trends as those obtained from cutting trials. Comparing the simulation results under different cutting conditions, a set of optimal process parameters of milling operation are obtained in terms of the minimum of amplitudes of the vibration between the cutter and workpiece.

Key words: Peripheral milling; cutting force; process parameters; optimal design

1.   INTRODUCTION

Peripheral milling is widely used for the machining of profiled components in aerospace, automotive and die manufacturing. The mechanism of milling process is very complex due to the occurrence of the periodical and intermittent cutting process, semi-closed-form machining and chip thickness variation. The cutting force deflects the workpiece and cutter and these deflections have a significant influence on the geometric and dimensional errors for the parts machined. As a result, the machining instability, or self-excited chatter, occurs due to the interaction between the cutter and workpiece. Unless avoided, machining with the presence of chatter leads to poor surface finish and excessive tool wear even breakage. It is essential to model the multi-variable milling process and select a set of optimal process parameters, including the spindle speed, radial and axial depths of cut, and the feed rate. These are the most important factors that influence the peripheral milling accuracy and surface finish.

Efforts have been made in the last decade to predict analytically machine stability in milling. As a fundamental mechanism of the chatter vibrations, Tlusty's chatter stability theory has been widely used by researchers due to its simplicity and straightforward [1]. Considering the time varying forces and chip loads, as well as the nonlinearity in the process, Tlusty and Ismail presented numerical algorithms in predicting the chatter stability lobes in milling [2]. Sridhar presented the first basic two dimensional chatter stability formulation in milling using a simple example, although they did not propose a mathematical solution due to complexity of periodic delayed differential equations [3]. Minis and Yanyshevsky presented the first analytical prediction of chatter stability by iteratively solving the two dimensional dynamic milling problem[4]. Altintas and Budak presented the two dimensional chatter stability theory and solved the chatter stability directly without any iteration, and showed that the two dimensional chatter stability in end milling can be predicted as accurately as in time domain solution. However its simulation is limited only in a shorter computation time period [5]. Fuh and Chang presented a dimension-accuracy model for the peripheral milling using response surface design under the appropriate cutting conditions [6]. At present, most of the research derive the machining stability through the estimation of the cutting forces which involving various factors. However, as the primary model in all these mentioned articles, the analytical model of the cutting force did not include the size effect of uncut chip thickness, the influence of the effective rake angle and the chip flow angle together.

In this paper, based on a comprehensive and improved theoretical nonlinear dynamic cutting force model presented in the papers [7-8] with consideration of above factors, a second-order ordinary differential equation of machining dynamics for peripheral milling process is presented [9]. Variable-step numerical integral algorithm based on explicit forth-order Runger-Kutta formula and recurrence algorithm are used to solve the dynamics equation to ensure convergence and validity of the solution. Then a computer simulation model for peripheral milling process is developed. A series of simulation and analyses in time domain with specific module parameters of machining system are conducted to investigate machine tool vibration by adopting different process parameters of milling operations and geometrical parameters of cutter. Comparing the simulation results under different cutting conditions with those of the cutting trials, a set of optimal process parameters of milling operation are obtained in terms of the minimum of amplitudes of vibration between the cutter and workpiece.

2.   ESTABLISHMENT OF COMPUTER SIMULATION MODEL

2.1   Theoretical model of dynamic cutting force

Since cutting force is one of the most important outputs of milling process, which directly influences the quality of machined parts, developing an accurate and practical cutting force model is essential. A comprehensive and improved nonlinear theoretical dynamic cutting force model for peripheral milling process is briefly summarized here based on the Liu's work [7,8]. In particular, the size effect of uncut chip thickness, the influence of the effective rake angle and the chip flow angle that are usually missed in most existing models of the same sort, are included in the proposed model.

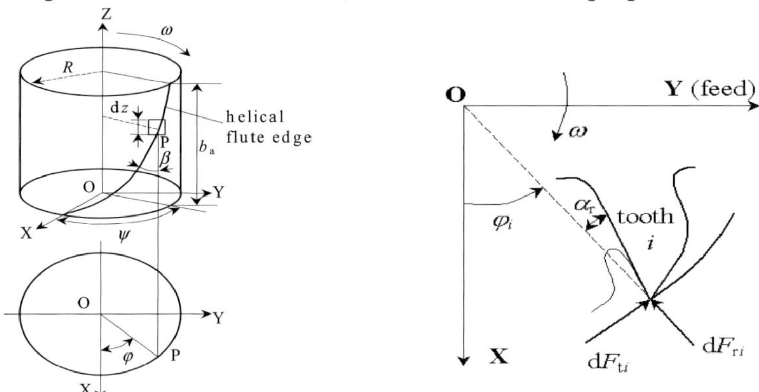

(a) Geometric model of helical cutter   (b) Differential tangential and radial foroces
Figure 1. Cutting force model of peripheral milling

A typical peripheral milling cutter with helical flutes is shown in Figure 1. The milling cutter is axially digitized with small elements with a uniform differential height of dz in the z-direction. In this figure, R is the cutter radius, ω is the angular velocity of the spindle, β is the helix angle of the cutter, φ_i is the position angle of a point on the cutting edge of the ith helical flute, α_r is the radial rake angle. Within each slice, the cutting action for an individual tooth is modeled as a single point oblique cut. The cutting forces acting on the helical flute's rake face are dependent on the uncut chip thickness. The axial cutting force isn't considered in this case.

The differential cutting forces of the tooth in tangential and radial directions with a height dz for the rotational position φ on the rake face are calculated from:

$$\begin{bmatrix} \mathrm{d}F_{ti}(\varphi_i) \\ \mathrm{d}F_{ri}(\varphi_i) \end{bmatrix} = \begin{bmatrix} K_s \\ cK_s \end{bmatrix} [R\cot\beta]\, t_i(\varphi_i)\, \mathrm{d}\varphi \tag{1}$$

where c is the force ratio, $t_i(\varphi_i)$ is the instantaneous uncut chip thickness of the tooth element, K_s is the tangential cutting force coefficient, which has the same meaning as the total energy per unit volume u.

Considering the size effect of uncut chip thickness and the influence of effective rake angle, there is:

$$K_s = u_0(1 - \frac{\alpha_e - \alpha_{e0}}{100})(\frac{t_0}{t_i(\varphi_i)})^{0.2} \tag{2}$$

where u_0 is the initial total cutting energy per unit volume, α_e (in degrees) is the effective rake angle, α_{e0} (in degrees) is the initial effective rake angle and t_0 is the initial uncut chip thickness.

Once the tangential and radial cutting forces are identified, cutting forces are projected to the Cartesian coordinate system as follows:

$$\begin{bmatrix} \mathrm{d}F_{ix} \\ \mathrm{d}F_{iy} \end{bmatrix} = \begin{bmatrix} -\sin\varphi_i & -\cos\varphi_i \\ \cos\varphi_i & -\sin\varphi_i \end{bmatrix} \begin{bmatrix} \mathrm{d}F_{ti}(\varphi_i) \\ \mathrm{d}F_{ri}(\varphi_i) \end{bmatrix} \tag{3}$$

The details of calculation of cutting forces in feed (Y) direction and normal (X) direction can be found in the papers [7,8]. The conclusions are briefly summarized here as follows:

For down-milling, assuming $u' \approx K_s(\sin\varphi_i)^{0.2}$, there is:

$$\left\{ \begin{aligned} \mathrm{d}F_{ix} = &-u'R\cot\beta[(f_t + y_c - y_{c(-T)} - y_w + y_{w(-T)})(\sin^{1.8}\varphi_i + c_1\sin^{0.8}\varphi_i\cos\varphi_i) \\ &+ (x_c - x_{c(-T)} - x_w + x_{w(-T)})(\sin^{0.8}\varphi_i\cos\varphi_i + c_1\sin^{-0.2}\varphi_i\cos^2\varphi_i)]\mathrm{d}\varphi_i \\ \mathrm{d}F_{iy} = &\,u'R\cot\beta[(f_t + y - y_{(-T)} - y_w + y_{w(-T)})(\sin^{0.8}\varphi_i\cos\varphi_i - c_1\sin^{1.8}\varphi_i) \\ &+ (x - x_{(-T)} - x_w + x_{w(-T)})(\sin^{-0.2}\varphi_i\cos^2\varphi_i - c_1\sin^{1.8}\varphi_i)]\mathrm{d}\varphi_i \end{aligned} \right. \tag{4}$$

(2) For up-milling, assuming $u' \approx K_s[\sin(-\varphi_i)]^{0.2}$, there is:

$$\begin{cases} \mathrm{d}F_{ix} = u' R \cot\beta [(f_t + y_{c(-T)} - y_c - y_{w(-T)} + y_w)(\sin^{1.8}(-\varphi_i) - c_1 \sin^{0.8}(-\varphi_i)\cos\varphi_i) \\ \qquad + (x_{c(-T)} - x_c - x_{w(-T)} + x_w)(-\sin^{0.8}(-\varphi_i)\cos\varphi_i + c_1 \sin^{-0.2}(-\varphi_i)\cos^2\varphi_i)]\mathrm{d}\varphi_i \quad (5) \\ \mathrm{d}F_{iy} = u' R \cot\beta [(f_t + y_{c(-T)} - y_c - y_{w(-T)} + y_w)(\sin^{0.8}(-\varphi_i)\cos\varphi_i + c_1 \sin^{1.8}(-\varphi_i)) \\ \qquad + (x_{c(-T)} - x_c - x_{w(-T)} + x_w)(-\sin^{-0.2}(-\varphi_i)\cos^2\varphi_i - c_1 \sin^{0.8}(-\varphi_i)\cos\varphi_i)]\mathrm{d}\varphi_i \end{cases}$$

where f_t is the feed rate per tooth, y_c, y_w and $y_{c(-T)}$, $y_{w(-T)}$ are the dynamic displacements between the cutter and workpiece in the Y direction, for current and previous tooth passes at the position φ_i of the ith tooth, respectively. Same meaning is applied in the X direction. In Equations 4 and 5, dynamic displacement due to regenerative chatter in the Y and X directions are included in instantaneous chip thickness.

Equation 4 and Equation 5 represent an elemental cutting force acting on one tooth in one slice for two different cutting styles respectively. Then the total cutting forces for the rotational position φ in X and Y directions can be calculated by integrating Equation 4 or Equation 5 by a small amount $\Delta\varphi$ along the axial depth of cut and digitally summed for all cutting flutes which are in contact with the workpiece.

2.2 Principle of the computer simulation

As described models in several previous works, the structural dynamics model for the peripheral milling process in the paper is describes as a two-degree-of-freedom spring-mass-damper vibratory system in the two mutually perpendicular directions X and Y, as shown in Figure 2.

For the general peripheral milling process, its dynamics equation at a discretional time t is described as follows:

$$\begin{cases} F_x(t) = m_x \ddot{x}(t) + c_x \dot{x}(t) + k_x x(t) \\ F_y(t) = m_y \ddot{y}(t) + c_y \dot{y}(t) + k_y y(t) \end{cases} \tag{6}$$

where F_x, F_y are the total cutting forces acting on the tool, m, c, k are the mass, damping and stiffness of either tool or work-piece at the contact zone in the feed or normal directions, x, y represent relatively vibratory displacement in the cutter-work piece system in the normal and feed directions respectively.

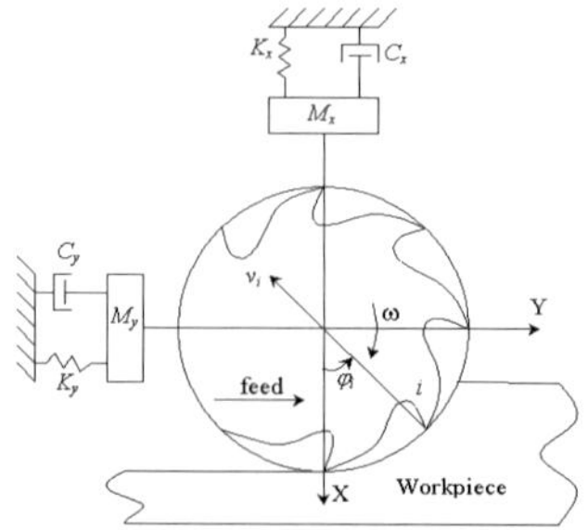

Figure 2. Structural dynamic model for the peripheral milling process

In milling, considering regenerative chatter, the cutting forces excite vibrations in the normal (X) and feed (Y) directions periodically, causing dynamic displacements Δx and Δy respectively. As a result, the cutting forces and regenerative displacements have a closed-loop interaction.

It is assumed that T is the tooth passing interval and G_{xx}, G_{xy}, G_{yx} and G_{yy} are the direct and cross transfer functions in the normal and feed directions of the machine tool/work-piece structure. Superscript c and w mean cutter and workpiece respectively. Dynamic displacements between cutter and work piece in the Laplace domain can be expressed as:

$$\begin{Bmatrix} \Delta x_c(s) \\ \Delta y_c(s) \end{Bmatrix} = (1 - e^{-sT}) \begin{bmatrix} G_{xx}^c(s) & G_{xy}^c(s) \\ G_{yx}^c(s) & G_{yy}^c(s) \end{bmatrix} \begin{Bmatrix} F_x(s) \\ F_y(s) \end{Bmatrix}$$

$$\begin{Bmatrix} \Delta x_w(s) \\ \Delta y_w(s) \end{Bmatrix} = (1 - e^{-sT}) \begin{bmatrix} G_{xx}^w(s) & G_{xy}^w(s) \\ G_{yx}^w(s) & G_{yy}^w(s) \end{bmatrix} \begin{Bmatrix} F_x(s) \\ F_y(s) \end{Bmatrix}$$

In practical milling process, the direct and cross transfer functions of cutter and work piece in the normal and feed directions identified by experimental modal analysis technique can be described as:

$$G_{xx}^c(s) = \sum_{j=1}^{l_{cx}} \frac{\alpha_{cxj}s + \beta_{cxj}}{s^2 + 2\zeta_{cxj}\omega_{ncxj}s + \omega_{ncxj}^2} , \quad G_{yy}^c(s) = \sum_{j=1}^{l_{cy}} \frac{\alpha_{cyj}s + \beta_{cyj}}{s^2 + 2\zeta_{cyj}\omega_{ncyj}s + \omega_{ncyj}^2}$$

$$G_{xx}^w(s) = \sum_{j=1}^{l_{wx}} \frac{\alpha_{wxj}s + \beta_{wxj}}{s^2 + 2\zeta_{wxj}\omega_{nwxj}s + \omega_{nwxj}^2} , \quad G_{yy}^w(s) = \sum_{j=1}^{l_{wy}} \frac{\alpha_{wyj}s + \beta_{wyj}}{s^2 + 2\zeta_{wyj}\omega_{nwyj}s + \omega_{nwyj}^2}$$

where s is Laplacian, α_j, β_j are mode coefficients in the corresponding directions. l is the modal number, ω_n and ξ are the natural frequency and structural damping ratio respectively and can be evaluated with damping ratio and stiffness C_x, C_y, K_x, K_y derived from modal analysis. The expression are given as follows: $\omega_{nx} = \sqrt{k_x/m_x}$, $\xi_x = 0.5c_x/\sqrt{k_x m_x}$,
$\omega_{ny} = \sqrt{k_y/m_y}$, $\xi_y = 0.5c_y/\sqrt{k_y m_y}$.

2.3 Algorithm of simulation

In Equation 6, each term is a second-order dynamical ordinary differential equation. It is impossible to solve results with ordinary algorithm. Discrete recurrence algorithm used to the solution in the past easily results in distortion of the final results because errors of the solution are amplified with gradual increase of steps in the simulation[9].

To obtain accurate results, variable-step numerical integral algorithm based on explicit forth-order Runger-Kutta formula and recurrence algorithm are used in the solution work to ensure convergence and validity of the solution. Furthermore, application of these algorithms in the solution model greatly reduces the time-consuming and improves the accuracy of the mathematical solving work.

2.4 Experiment verification

On the basis of the mathematical model of cutting force and the solution algorithms, a computer simulation model for peripheral milling is developed, involving almost all the parameters of milling process and cutter geometry. To verify the effectiveness of the simulation model, a series of simulation are conducted with the given set of cutting conditions that are same as experimental conditions, shown as Figure 3. The simulation results and the experimental results of the cutting forces are illustrated with the smooth curves and rough curves respectively. In Figure 3, it shows that there is a reasonably good agreement between the experimental results and the simulation results. The simulation results of vibratory displacement will be presented in the following.

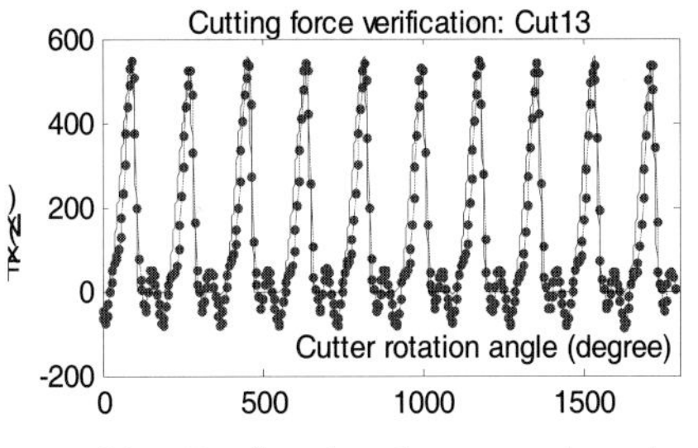

(a) Predicted and measured cutting forces in normal direction

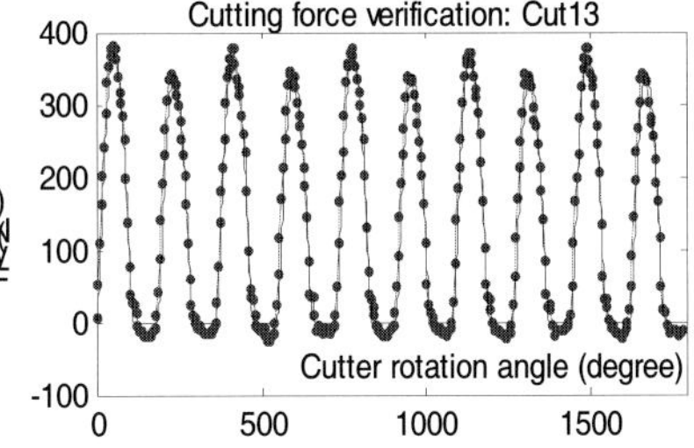

(b) Predicted and measured cutting forces in feed direction

Figure 3. Predicted and measured cutting forces. Cutting conditions: helical end mill, rake angle:5° , Cutter diameter:20mm, helix angle:30° .Spindle speed :1592 rpm, feed rate per tooth:0.05mm, axial depth of cut: 5.03mm,up milling, slotting with fluid. carbide cutter with 2 flutes, work piece material: EN8

3. OPTIMIZATION OF CUTTING PARAMETERS

It is well known that the machining stability of peripheral milling is primarily influenced by the spindle speed, the axial and radial depth of cutter, and the feed rate. The current practice of milling process is to choose overly conservative cutting parameters in order to achieve the specified parts accuracy and to avoid excessive tool deflections, which usually leads to low metal removal rate and low productivity. Based on the simulation model, a set of simulation with various cutting conditions are conducted for the optimization of the milling process and cutter geometry. Initial cutting conditions, cutter geometry and modal parameters of the milling system in the normal and feed directions obtained from experimental modal analysis are summarized in Tables 1 and 2.

Table 1 Initial cutting conditions and cutter geometry parameters

Work piece material	Cutter material	Numbers of Helix flutes	Helix angle
Aluminium alloy:7075-T6	High-speed steel	4	30˙
Cutter diameter	Effective rake angle of cutter	Feed rate per tooth	Spindle speed
φ25.4mm	5˚	0.1mm	2400rpm
Axial depth of cut	Radial depth of cut	Cutting mode	
4mm	12.7mm	up milling with fluid	

Table 2 The modal parameters of machine tool in X and Y directions

Modular mass of machine tool	$m_x = m_y = 110g$	Modular stiffness of machine tool	$k_x = 7.7 \times 10^6 \; \text{N}/_m$ $k_y = 8.1 \times 10^6 \; \text{N}/_m$
Nature frequency of machine tool	$\omega_{nx} = 264.58 rad/s$ $\omega_{ny} = 271.36 rad/s$	Damp ratio of machine tool	$\xi_x = \xi_y = 0.019$

3.1 Effect of spindle speed on machining stability

The main cutting frequencies of peripheral milling include the spindle frequency (SF), tooth passing frequency (TPF), which equals to the tooth number times SF, and their N-times frequencies. These cutting frequencies depend on the spindle speed. If any of these frequencies is close to the natural frequency of the cutting system, the vibration between the cutter and workpiece will be amplified and may be converted to a chatter vibration. Simulation results of the vibration are shown in Figure 4(a) and (b) at spindle speeds of 2000rpm and 2600rpm respectively, in which the first case describes an unstable cutting process, and the second one indicates a chatter free cutting process.

3.2   Influence of axial depth of cut

The amplitude of the vibration between the cutter and workpiece will increase if increasing the axial depth of cut. When the axial depth of cut is larger than the limit axial depth of cut, the chatter will occur, as shown in Figure 5(a) and (b). The conclusion accords with the most previous research results.

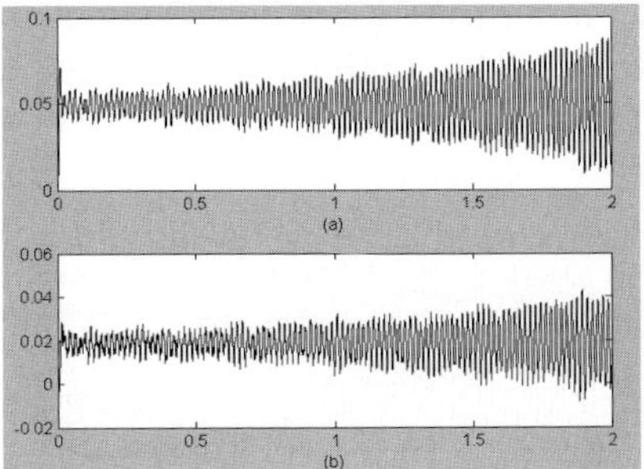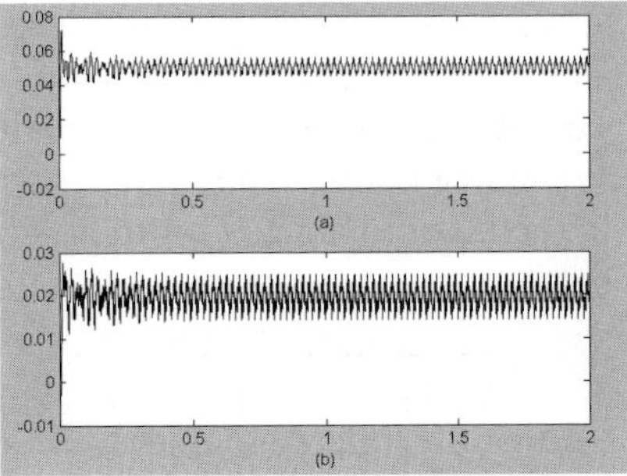

(a) vibration at a spindle speed of 2000rpm (b) vibration at a spindle speed of 2600rpm
Figure 4 Vibration in the normal and feed directions in the cutter/workpiece system at different spindle speed

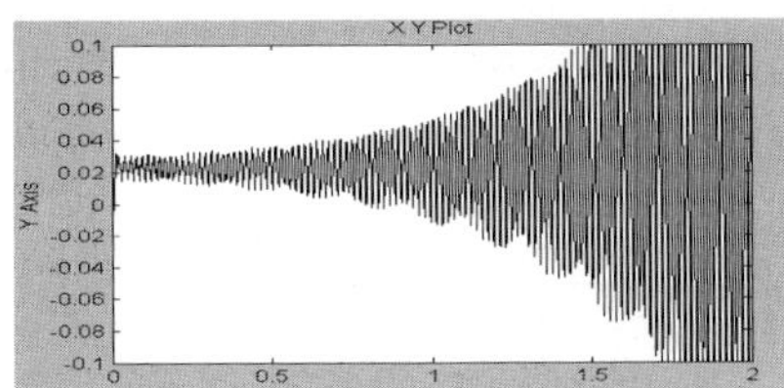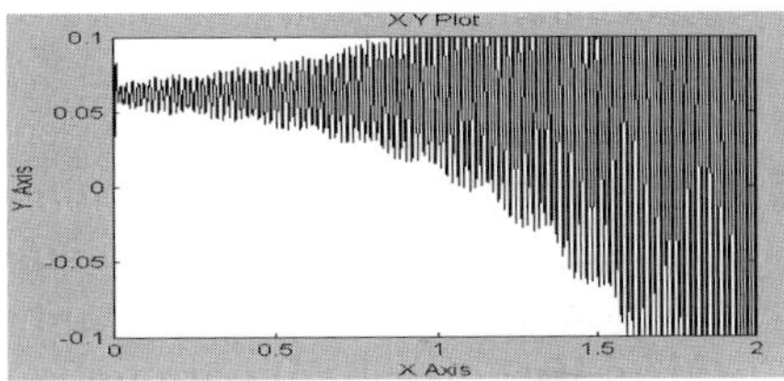

(a) vibration in the feed direction   (b) vibration in the normal direction
Figure 5 Vibration in the cutter/workpiece system at axial depth of cut 5mm

3.3   Effect of the number of cutter teeth on machining stability

The amplitude of the vibration between the cutter and workpiece with a 5-flutes cutter is nearly double as that with a 4-flutes cutter, as shown in Figure 6(a) and (b). Similar instance is seen in milling with a 3-flutes cutter.

3.4   Effect of helix angle on machining stability

The amplitude of the vibration between the cutter and workpiece increases a little as the helix angle of the cutter decreases. There has opposite trend with increased helix angle.

Using the above approach, the optimal process parameters of milling operation are obtained in terms of the minimum of amplitude of the vibration between the cutter and workpiece with comparison of a series of simulation results under different cutting conditions. The optimal parameters are presented in table 1 and the resultant vibrations and their power spectra are shown as Figures 7 and 8.

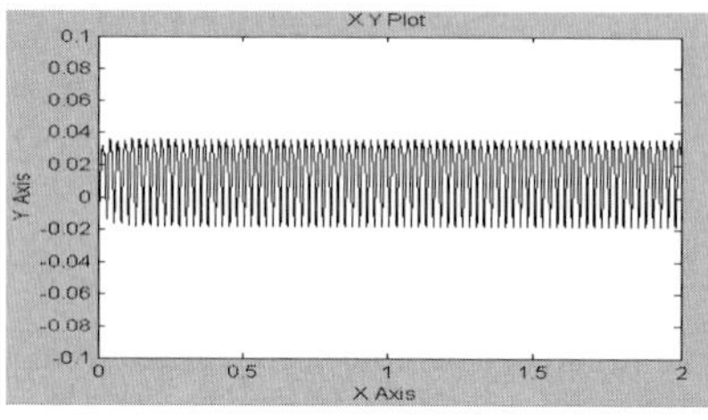

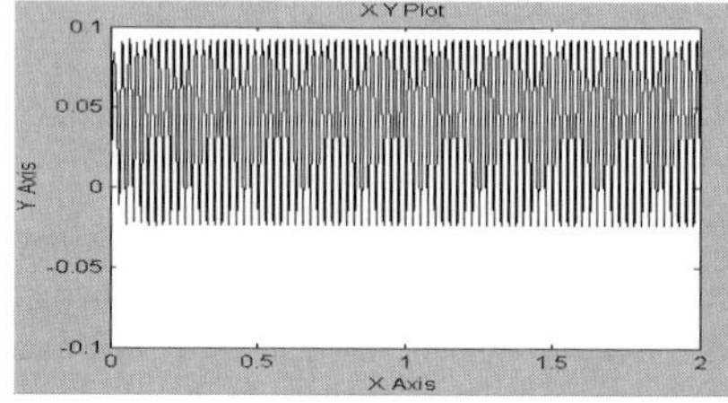

(a) vibration in the feed direction   (b) vibration in the normal direction
Figure 6 Vibration in the cutter/workpiece system with cutter with 5 flutes

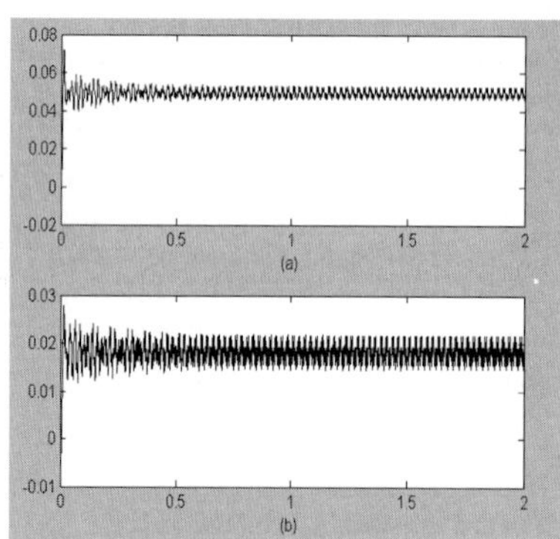

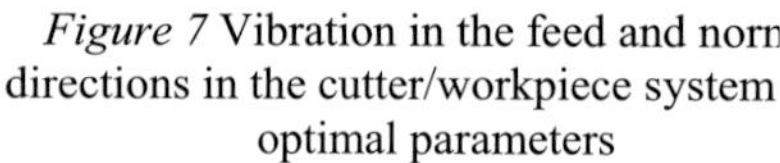

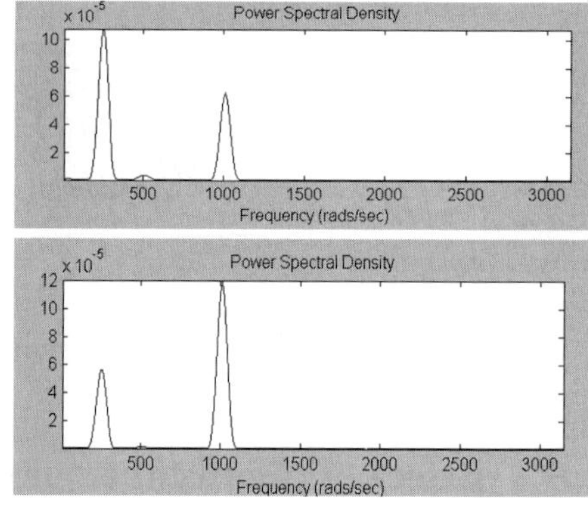

Figure 7 Vibration in the feed and norm directions in the cutter/workpiece system optimal parameters

Figure 8 Power spectral density of vibration in the feed and normal directions in the cutter／workpiece system with optimal parameters

4.   CONCLUSION

The accuracy of peripheral milling is generally determined by the cutting parameters and cutter geometry parameters. Based on the machining dynamics simulation model developed in the paper, the minimum of amplitudes of the vibration between the cutter and workpiece can be estimated efficiently and used to select the milling conditions for the desired accuracy in the peripheral milling process. As a valid research platform, the computer simulation model and the dynamic cutting force model can be used to substitute the expensive cutting trials, optimize the cutting parameters, predict and eliminate chatter. There is also potential capability for the design of machining tool structure and cutting tools with the optimized cutter geometry.

REFERENCES

1. Koenigsberger F, Tlusty J. *Machine Tool Structures*. Pergamon Press, 1967.
2. Tlusty J, Ismail F. "Basic Nonlinearity in Machining Chatter", *Annals of the CIRP*, **Vol.30, no.1**, pp.299-304, 1981.
3. Sridhar R, Hohn RE, Long G.W. "A Stability Algorithm for the General Milling Process", *Trans, ASME Journal of Engineer for Industry*, **Vol.90, no.2**, pp.330-334, 1968.
4. Minis I, Yanushevsky T, Tembo R, Hocken R. "Analysis of Linear and Nonlinear Chatter in Milling*"*, *Annals of the CIRP*, **Vol.39, no.1**, pp.459-462, 1990.
5. Altintas Y, Budak E. "Analytical Prediction of Stability Lobes in Milling", *Annals of the CIRP*, Vol.44, no.1,pp.357-362, 1995.
6. Fuh KH, Chang HY. "An Accuracy Model for the Peripheral Milling of Aluminium Alloys Using Response Surface Design", *Journal of Materials Processing Technology*, **Vol.72**, pp.42-47, 1997.
7. Liu XW, Cheng K, Webb D, Luo XC. "Improved Dynamic Cutting Force Model in Peripheral Milling – Part I: Theoretical Model and Simulation", *International Journal of Advanced Manufacturing Technology*, **Vol.20, no.9**, pp.631-638, 2002.
8. Liu XW, Cheng K, Webb D, Longstaff AP, et al. "Improved Dynamic Cutting Force Model in Peripheral Milling – Part II: Experimental Verification and Prediction", *International Journal of Advanced Manufacturing Technology*, (in press).
9. Chen Y, Liu XW. "Dynamic Simulation for Peripheral Milling Process Based on Matlab/Simulink Environment", *Journal of Huaqiao University*, **Vol.24, no.3**, pp. 168-173, 2003.

A REMOTE TECHNIQUE SUPPORT SYSTEM OF ENGINEERING EQUIPMENT

Huan-Liang Li, Qi Zhang, Cheng-Xian Yang and Song-Hui Guo
Department of Mechanical Engineering of Engineering Institute of Engineering Corps, PLA. University of Sic. & Tech., Nanjing 210007 China

Abstract: According to the characteristics of engineering support, three kinds of wireless info-net model of remote technique support for equipment are built based on the Satellite Mobile Communication Technique (SMCT), the IEEE802.11b Standard and the Bluetooth Technique. The Computer-mounted state-monitoring system of equipment detects the running parameters and faults of the equipment; the GPS satellite receiver device calculates the position of the equipment automatically; on the base of assuring the transport speed, the wireless data transport system gets rid of the limitations of traditional wire net; the Technique Support Center (TSC) supplies the remote users with powerful technique support. On the ground of the situation of Engineering Equipment Technique Support (EETS), the Remote Technique Support System of engineering equipment (RTSSEE) is developed to realize the remote fault diagnosis for engineering equipment. Video-Audio Conference System is established to support the info communication under its operation. Therefore, the digital level of the EETS is greatly improved.

Keywords: engineering equipment, remote technique support, wireless communication, digital.

1.   INTRODUCTION

Engineering equipment is playing an important role in economic construction and country defence engineering. The level of EETS is relate to the maximum affectivity of the engineering equipment to be achieved or not, the mission to be finished timely and accurately or not. With the development of computer technique, communication technique, and network technique, the EETS is stepping towards information, intelligent control, and integration. It has become an essential project to research RTSSEE and to pursue developing strategies to enhance the digital level of EETS.

The techniques used in the system include sensor technique, computer video, image process, signal process, auto-control, etc. According to the characteristics of engineering equipment and engineering support, the paper establishes three kinds of wireless info-net hardware model of remote technique support for equipment respectively based on the SMCT, the IEEE802.11b Standard and the Bluetooth Technique. On the ground of the situation of EETS, the software of RTSSEE is designed and programmed to realize the remote fault diagnosis for engineering equipment. The Video-Audio Conference subsystem can make the users exchange their video and audio information timely. Similarly some other subsystems to assure the equipment remote technique support are described in detail as follows.

2.   THE HARDWARE STRUCTURE OF RTSSEE

The hardware structure of the system is composed of four parts: the computer-mounted state-monitoring system of equipment, GPS satellite location signal-received device, wireless data transport system （WDTS）, and the TSC[1-2].

2.1   The Computer-mounted state-monitoring system of equipment

The engineering equipment is equipped with necessary sensors, which transports operating parameters into the TSC through the WDTS. The mounted computer is designed with sixteen-bit-single-chip microprocessor to compact the volume. The signals are displayed on LCD (Liquid Crystal Display) for the operator. The operator-computer interaction is achieved through the cooperation of LCD and keyboard. First, the operator can transport fault phenomena of equipment into the TSC, then the operator can get technique support from the TSC about the fault phenomena of the equipment. [3]

2.2   GPS satellite receiver device

As a new technique, the GPS has been widely used in all fields. The GPS system consists of twenty-four satellites, which are circulated in six different orbits. The satellites send location signals to the Earth continuously. After receiving the location signals from the satellites, the GPS device on the engineering equipment calculates the location of the equipment automatically [4]. Its location precision can reach at centimetre degree to meet the control requirements of the engineering equipment.

2.3   WDTS

Due to engineering equipment in the mobile state, RTSSEE only through the WDTS can it transport the monitored data into the TSC. The communication schemes are shown in the follow.

2.3.1   Satellite Communication System (SCS)

The main character of the SCS is that it covers widely the whole globe or certain areas. So it is greatly fit to the EETS, but its cost is relatively expensive. To take the maritime affairs communication system as an example, the system's reliability is high and its device's volume is small. Communication frequency adapts L band. Because of its all-directional antenna, the system can communicate well even in terrible working conditions. The system is also capable to work even when engineering equipment is rotating or hoisting. The effective working range of the satellite is at 70°NL—70° SL (south latitude). The data transport process is described in Fig.1. First the working parameters and location information of the equipment are transported to the maritime affairs communication Satellite, and then these data are transmitted to the ground support center. Simultaneously the computers in TSC transport the analyzed results and the equipment position digital map into the Internet. All these data are exchanging in two-channels.

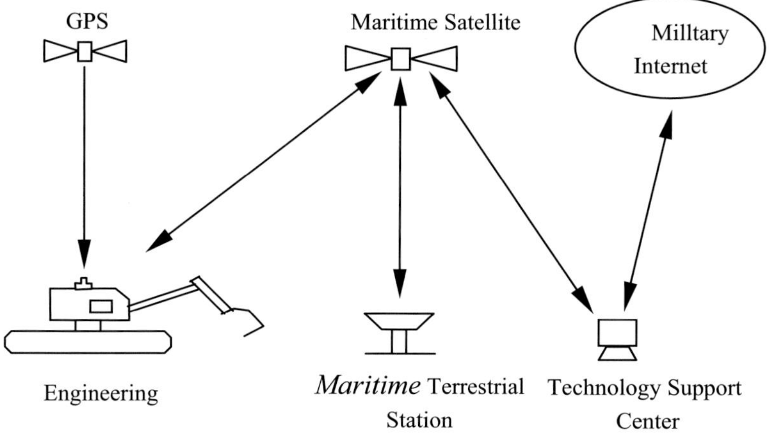

Fig.1 the Engineering Equipment remote technology support system using satellite communication

2.3.2 The wireless communication system based on 802.11b protocol

The 802.11b protocol is the first wireless network transport standard established by IEEE to tackle the wireless network in office and campus, but it is mainly limited to data reading and saving. The transporting speed is only restricted to 2 Mbps. At present, those big companies like 3COM have produced network adapters based on the standard. [5]

For the communication limitations on speed and distance of 802.11 Standard, IEEE group developed the two new protocols, 801.11a and 802.11b. The main difference lies in their MAC layer and physics layer. The physical layer of 802.11b supports two different transport speeds, 5.5Mbps and 11Mbps. 802.11 standard is designed based with eleven-bited modulate chip, but the 802.11b standard is with a new modulation technique.

The 802.11b standard uses dynamic speed excursion to adapt to the changing environments. Its speed can shift automatically with the environment change in the following values: 11Mbps, 5.5Mbps, 2Mbps, and 1Mbps. When its speed is located in 1Mbps or 2 Mbps, the 802.11b protocol can comply with 802.11 protocol.

The wireless communication device based on 802.11b protocol is mounted on engineering equipment. When the working engineering equipment is far apart or separated by barriers from the TSC, the power-amplified device is facilitated between the position of the engineering equipment and support center as described in Fig 2.

When the adaptors are communicated in AD-HOC mode, the system formed is a wireless LAN net. The communication situation is the same as that of a wire LAN net. When two of the network adapters in equipment are changed into Point-Point mode, the two adapters can build a new point-point wireless network. For its high speed data fro and to up to 11Mbps, the wireless network can meet the needs of audio-video conference. For its terrible working conditions, the Video devices are advised to have high-clarity and high-distinguishability such as TANDBERG 500/128, Viewstation 128, and PROXIMA （AV-9350）.

2.3.3 The wireless network communication system based on IEEE1451.2 standard and bluetooth protocol

In most working conditions, sensors are linked through wires, but to engineering equipment for its terrible working conditions such as storm-rain-snow, the sensors designed in wireless mode using IEEE (Institute of Electrical and Electronics Engineers) 1451.2 and bluetooth technologies can be built a wireless network to avoid the limits of wire network system.

AP---access to point, WB---wireless bridge, SA---signal adaptor

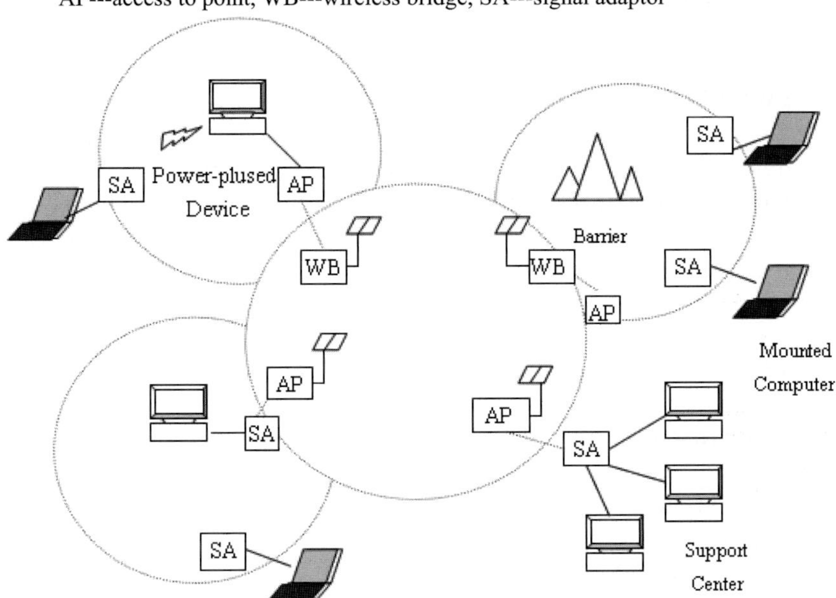

Fig.2 the wireless RTSSEE based on 802.11b standard

The wireless network sensors designed with IEEE1451.2 standard and bluetooth protocol are composed of three parts, the STIM (smart transducer interface module), Bluetooth module, and NCAP (network capable application process). In STIM, IEEE1451.2 is adapted with standard to design the interface of sensors. Analogue signals are changed into standard formatted data. A small memorizer, TEDS（transducer electronic data sheet）, is linked to the specified processor, NCAP（network capable application process）, thus the data can flow by the network protocols.

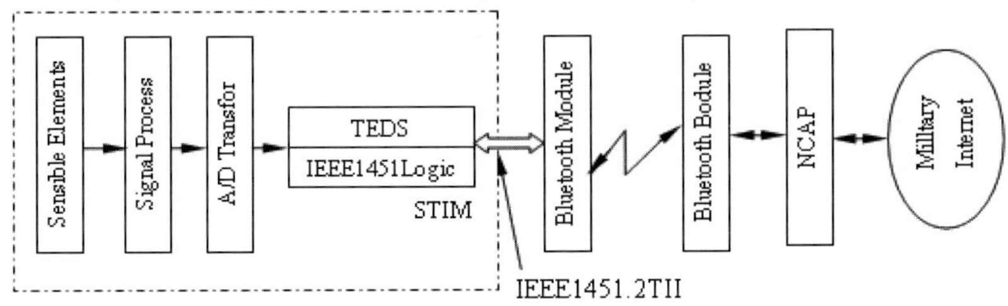

Fig.3 wireless Internet linked sensor system framework

The ten-lined TII (transducer independence interface) of IEEE1451.2 standard is between STIM and buletooth modules. The TII sends the sensor information and accepts the control order from remote servers through the interface linking STIM with NCAP, which is connected to the military network via distributed IP address. The wireless integrated sensor is described in Fig 3.

2.4   The TSC

Not only does the TSC save a lot of technique support information, but also it handles the faults using all kinds of mathematics methods like Wavelet, FFT(fast Fourier transform) etc. The analyzed results instruct the operation of engineering equipment. [6-7]

3.   THE SOFTWARE DESIGN OF RTSSEE

The software part of RTSSEE is constituted of four function modules: the Fault Diagnosis Module, the Video-audio Communication Module, the Online Query Module, and the Data Maintenance Module.

3.1   The Fault Diagnosis Module

The analogue signals including the operation conditions of engineering equipment are transferred to the digital signals, which are transported to the TSC through the wireless network constructed using one of the wireless network building methods described in part one. This module diagnoses the faults of the equipment and gives the correct resolving methods to settle the faults. [8-9]

The fault diagnosis module programmed by Labview CVI has two main functions. The Fig.4 shows the detecting and diagnosing flow of the module. One of its functions is to detect the performances; the other is to diagnose the faults. The fault tree knowledge is memorized in remote service center computer. Firstly, the module realized by fault tree searches for and locates in the fault points of the equipment from root to nodes of the fault tree to find out the fault cell. While diagnosing, the possible correct diagnosis is concluded by comparing the detected signals with the standard signals saved in database, and using the judged strategies saved in database.

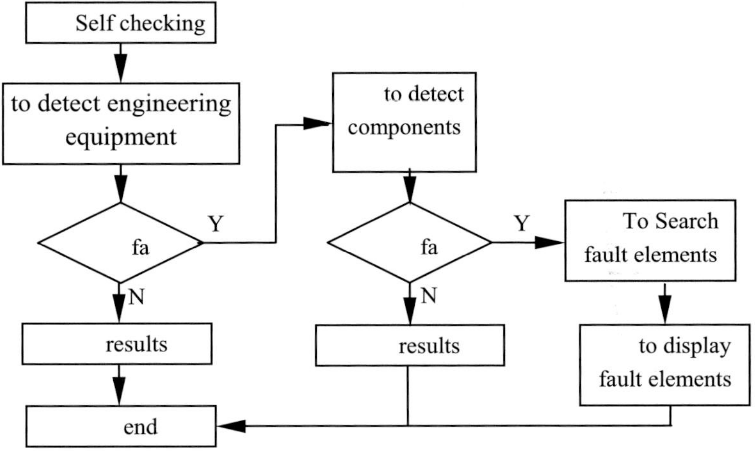

Fig.4 flow chart of fault diagnosis module

3.2   The Video-audio Communication Module

It is required that the realizations of this module need the client computers with the video-audio multimedia hardware devices, such as audio card, earphone, microphone and video camera. It is very convenient for clients and service experts in TSC to communicate video and audio data effectively and timely through the Video-audio Communication Module based on Netmeeting COM components in Windows. The components are developed through the redeveloped functions: online discussing function, delivering message function, sharing program function, white-board program function. [10]

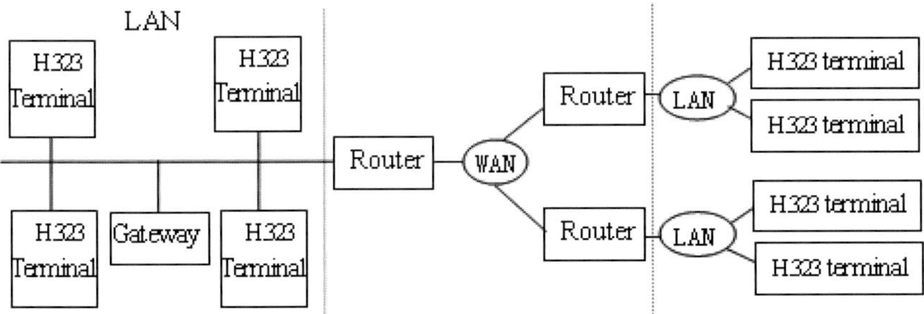

Fig.5 network architecture of video-audio communication system module

The existing construction situation of the LAN net is that it can guarantee the 2Mbps data transport speed between remote users in common conditions. For wireless network communication system is built with Satellite technique, 802.11b protocol, or bluetooth technique, it is possible to reach the minimal required speed(128Kbps) for video and audio data to flow between the clients and center experts in the wireless net constructed by TCP/IP protocol. The module is constructed by T120 data protocol and H.323 communication protocol for the network is united on internet TCP/IP protocol. This module's network structure founded on H.323 video communication protocol is expressed in detail in Fig 5.

3.3   The Online Query Module

The design of this module depends on the realization of the Fault Diagnosis Module. In Fault Diagnosis Database, the history chart of fault diagnosis is built with following fields: Equipment Name, Fault Position, Fault Phenomenon, and Optimum Settlement. When the same or similar fault happens again User Info, Visit Time, users can write fault information with high frequency into the history diagnosis chart to guide diagnosis. [11]

When the same or similar fault happens again, users access to this module, select or input Visit Time, Equipment Name, Fault Position, then users can get directly the history diagnosis records of the fault using the link rules saved in the database in advance. The query optimal algorithm used in this module economizes the visiting time to server greatly, thus the working efficiency of server computer is raised validly. The working mechanism of the module is interpreted in Fig.6.

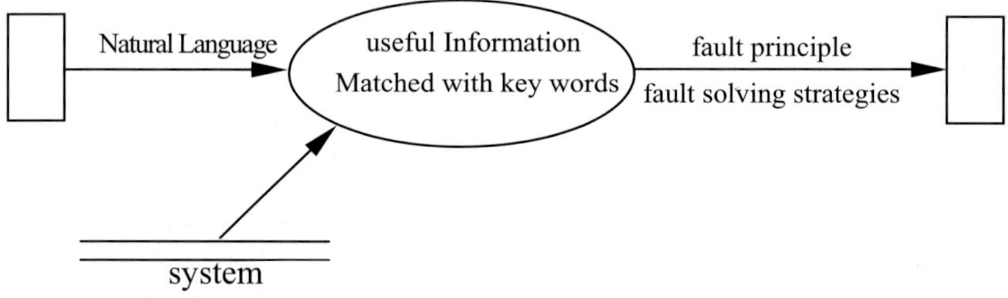

Fig.6 Working Mechanism of the Online Query Module

3.4   The Data Maintenance module

The difficulty of developing this module is increased for visiting users in different levels. For this reason, the module is developed with Client/Server structure by Delphi 5.0. Users visiting this module must install the data maintenance program, and then must pass the identity validation, so the security of this module is guaranteed.

The module's main function is to extend the numbers of equipment in the server database, including the management of users. It provides a developing platform to remote technique support of engineering equipment. The number of remote technique support equipment is increased after the fault tree is built based on the analysis results of faults of the equipment and some characters, including Equipment Introduction, Technique Parameter, Maintenance Guidance, Principle Exposition, Repair Instruction, are offered.

4. CONCLUSIONS

The RTSSEE is an important sub-project in engineering equipment technique support research. The RTSSEE, built based on three kinds of wireless networks, can diagnose the faults on remote equipment, can support the video-audio information exchange and build technique seminar between clients and servers. All these map out a perfect blueprint for engineering equipment remote technique support and highly enforce the engineering equipment support level.

REFERENCES

1. LI Huan-liang ZHANG Qi YANG Xiao-qiang WANG Hai-tao. "State Monitoring System of Construction Machinery Based on Network Platform", *Journal of Tongji University*,Vol.29, pp. 1061-1065, Sep.2001 (In Chinese).
2. YUAN Chu-ming CHEN You-ping ZHOU Zhu-de. "Research on the Architecture of Remote Diagnosis System", *Computer Engineering And Application*, **Vol.18**, pp. 45-46, Sep. 2001. (In Chinese)
3. SONG Bin, HUANG Zhen-guo. "Monitoriog and Measuring through World Wide Web", *Journal Of China Textile University*, **Vol.25**, pp. 104-107, Oct.1999. (In Chinese)
4. HAN Xiao-ming TONG Xiao-fan. "Telemetering Control System of Engineering Machiner", *Journal of Tongji University*,Vol.29, pp. 1077-1081, Sep.2001 (In Chinese).
5. LI Feng-bao LIU Jin GU Tian-xian. "Research on networked sensors technique", *Journal of Transducer Technique*, **Vol.21**, pp. 62-64, July 2002 (In Chinese).
6. CHEN Xiao-hu WANG Han-gong ZHANG Jin-yu ZHANG Ji-ming. "Research on the Romote Fault Diagnosis For Complex Equipment", *Computer Engineering And Application*, pp. 14-15, Oct.2001. (In Chinese)
7. WANG Wen-li DUAN Bao-yan LIU Hong. "Remote Detection and Fault Diagnosis of Machinery Equipment Based on Network", *Manufacturing Industry Automation*, **Vol.21**, pp. 19-21, .June 1999. (In Chinese)
8. KONG Fan-sen DONG Yin-ping. "Fault Diagnosis and Reasoning Based on Fault Tree Knowledge", *Automotive Engineering*, **Vol.23**, pp. 209-213, March 2001. (In Chinese)
9. Ding Ming-ji. "Interconnection between Web Server and DBMS Server", *Computer Application*, **Vol.19**, pp. 18-20, June 1999. (In Chinese)
10. TIAN Dong YAO Zhi-heng etc. "The Development of H.263 and Its Application", *Measure and Control Technique*, **Vol .20**, pp. 5-9, May 2001. (In Chinese)

11. WANG Xiao-ming HOU Ying-wei. "The Design of Expert System Based on Database", *Computer Engineering And Application*, pp. 95-96, Oct.2000. (In Chinese).

METHODS FOR RECONFIGURING MULTI-AGENT AND SELECTING SENSOR IN RECONFIGURING MONITORING SYSTEM BASED ON KNOWLEDGE

Youping Fan, Yunping Chen and Wansheng Sun
Faculty of Electrical Engineering, Wuhan University , Wuhan, China   430072
fyoupingnxinrong@yahoo.com.cn

Abstract: Four problems have been addressed on distribution systems monitoring. First, reconfiguring fault diagnostic agent component group is utilized to realize diagnostic system reconfiguration. Second, an evaluating means and an algorithm for the intelligent multi-agent diagnostic system are presented. Third, method of selecting suitable sensors is analyzed for real-time monitoring of different machining processes and reconfigurable machining systems. Finally, the cubic hierarchy Petri Nets framework structure and a computing algorithm of reconfiguring multi-agent diagnosis system are given. This approach can achieve a self-organizing system that is more robust and flexible in dynamic environment and can be self-updated locally.

Key words: Intelligent manufacturing, multi-agent inspired, immune system, diagnostics.

1.   INTRODUCTION

As the mechanic-electronic equipment system has become more and more complicated, the fault diagnosis based on single artificial intelligent techniques is difficult to represent the knowledge of the new complex system[1]. In modern manufacturing, a reliable monitoring system is essential to reduce downtime, and enable optimum system performance. This happens with reconfigurable machining systems (RMS) where the functionality and capacity change according to the needs of the market. Then another monitoring system has to be developed for the new machining system. This is the issue that will be addressed in this paper.

The layout of the paper is as follows: Section 2 gives the reconfiguring multi-agent system model. Section 3 gives methodology of sensor selection. Reconfiguring theory based on immune regulation will be presented in Section 4. Diagnosing agent-reconfiguring course will be put forward in Section 5. Finally, some conclusions are drawn in Section 6.

2. MULTI-AGENT SYSTEM MODEL

2.1 Basic definition

Definition 1: A multi-agent fault diagnosis system (MAFDS) can be expressed by $MAFDS::=<A, B, F, L>$. In it, A represents an agent group set. B represents possible action set in $MAFDS$. F is a function from B to A. L represents the language to describe the whole $MAFDS$ action in.

Definition 2: $Agent_i::= <ID_i, DK_i, CK_i, MK_i, Bb_i, KA_i, Ps_i$-$AR>$, in which 7-elements group represents agent identification, agent domain knowledge, agent cooperation model, agent mapping model agent information blackboard, other agent knowledge, problem-solver separately. In it, $ID_i ::= <Name, Address, Role>$ is made up of name, address, role, appointing ID of the agent in the whole system. $DK_i ::=<Know$-$Base, Data$-$Base, Deci$-$Base>$ is made up of knowledge-base, data-base, deciding-base. It provides intelligent support for the agent's partial problem solving. $CK_i ::=<PCM, S$-$Model, A$-$Model>$ is made up of planning-cooperation-model, self-model and acquaintance-model.

Where, $<PCM>::=<Task, Evaluation, Initiation>$ can be used in task evaluation, deciding whether it is necessary to cooperate with other agent to complete the task. $<S$-$Model>::=<Action, Object, Condition>$ can be used to define self-action ability of agent, expressed in the form of ability list. Each list item can be expressed as $<ACTION, OBJECT, CONDITION>$. $<A$-$Model>::=<A$-$Name, Task, Inputs, Outputs>$ includes a-name, task, requiring variable and returning results. Each A-$Model$ is expressed as relation list form. Each list item composes an agent cooperation channel. $MK_i::=<Task, ck, dk, mapping>$ is made up of task, cooperation, original layer knowledge, domain layer original knowledge and mapping function. It is a connecting bridge between cooperation layer and domain layer, that is, after cooperation layer plans the task, MK can be used to initiate agent domain layer to complete present solution-acquiring task. $BB_i ::=<M$-$Name, M$-$Information>$ is used to store output information from each model inside Agent. $KA_i ::=<A_j, Q_k, R_j^{\,1}, R_j^{\,2}>$ is made up of 4-element set. In it, A_j is an identifier for agent to solve present tasks. Q_k is sub-questions for A_j to solve. $R_j^{\,1}$ is the action reliability for A_i to make sure that A_j can't solve Q_k while $R_j^{\,2}$ is action reliability for A_j to make sure that A_j can solve Q_k. *Ps-AR*::=*<COMM$_i$, S-IMA, RETR$_i$, CHOS$_i$, RECON$_i$, LEAR$_i$, EVAL$_i$>* Problem solving and agent reconfiguring model is made up of 7 elements set. *COMM$_i$= <Channel, Language, Environment>* *COMM$_i$* in charge of information receiving and sending includes communication channel, language and environment. *RETR$_i$* is in charge of searching for the model or domain field knowledge most similar to the sub-questions through comparability evaluation from *DK$_i$* and *CK$_i$*. *CHOS$_i$* is used to choose proper components from searched model and domain field knowledge. *RECON$_i$* is used to reconfigure a new Agent with chosen components. *LEAR$_i$* is used to adjust *Ck$_i$* and *Dk$_i$* based on sub-questions solving results. *S-IMA =<APC, Mc, Ck, Cp>* Construction model for cell-immunity agent re-engineering controller is made up of antigen presenting chain (*APC*), cell T activation chain (*Mc*), cell T proliferation chain (*Ck*) and differentiation chain (*Cp*).

Definition 3: If *m* is a model system, *p* is an operating set used in *m*, (*M, P*) can be called a model constructing system.

Definition 4: If (*M, P*) is a model constructing system, K={k_1, k_2,..., k_n} is a knowledge set, operating (P) based on *K* has such functions: symbol understanding; symbol choosing; symbol structure building; symbol modifying. (*M, P, K*) is a model intelligent constructing system.

2.2 Multi-agent diagnosis system model

The paper puts forward multi-agent diagnosis management system structure(MAFDMS) (Fig.1)and reconfiguring-multi-agent fault diagnosis system(RMAFDS) (Fan Y.P , 2003).

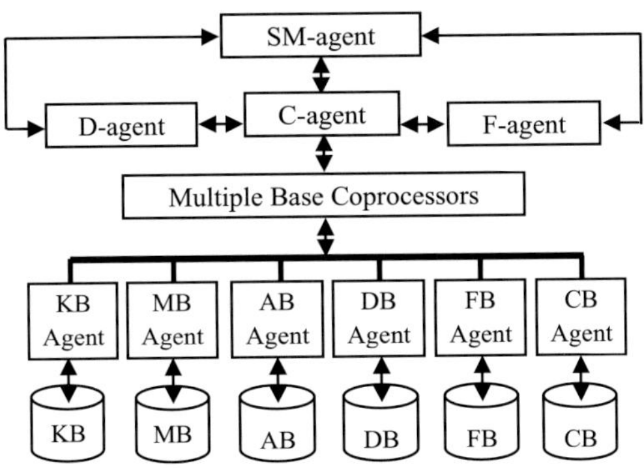

Figure 1. Structure of reconfigurable multi-agent diagnosis system

In a multi-agent diagnosis system, there're 3 different functional system agents: system managing agent(SM-agent), diagnosis agent(D-agent) and function agent(F-agent). SM-agent is the first step and base for any diagnosis in the system's exercising, which mainly takes charge of task distribution, managing and cooperation for the agents in the system, showing the way of thinking and strategy-making in the system diagnosis. D-agent really completes diagnosis for each sub-fault and ascertains the equipment's state. F-agent is consonant with management agent to complete a series of tasks such as user exchanging and signs acquiring. As a special functional agent, communicative agent(C-agent) is presented individually. It is the core of the whole system exercising in cooperation and exchanging among all the agents, managing the system's all-around data zone (knowledge base, model base, approach base, data base, figure base, case base).

3. MEANS OF SENSOR SELECTION

The most effective mathematical tool in explaining these uncertainties and vagueness is fuzzy set theory. A fuzzy set A of universe X is defined by a set or ordered pairs, a binary relation: $A = \{(x, \mu_A(x)) \mid x \in X, \ \mu_A(x) \in [0\ 1]\}$. Where is the membership function that specifies the grade or degree to which any element x in X belongs to the fuzzy set A.

Assume that there are m candidate sensors $(S_1, S_2, \ldots, S_m)$. Furthermore, assume that there are *k* characteristic evaluation criteria(i.e, $C_1, C_2, \ldots, C_m$) for a fault of interest in a given machining process.

Also let W be weighting factors which determine the significance or importance of criteria C_j, where $j = 1,2,\ldots, k$, and R_{ij} be the assigned rating of sensor S_i under criterion C_j, where $I = 1,2,\ldots,m$.

First, the performance ratings of all sensors in the sensor space are evaluated. Under the subjective criteria, the rating scales Rij of sensor Si are assigned in linguistic terms such as worst (W), poor (P), fair(F),good(G), and best(B). This assignment is based on expert knowledge.

Under the objective criteria, the consistence between the desired fault and sensor performance values is used to evaluate a sensor. The similarity degree, SD_{ij}, indicates how well the relevant characteristic of a sensor correlates with the desired parameter. SD_{ij} is given by:

$$SD_{ij} = \left. \int_x \{\min[\mu_{p_i}(x), \mu_{D_J}(x)]\}dx \middle/ \int_x \mu_D(x)dx \right.$$

. Where μ_{Dj} and μ_{pj} are membership functions representing the desired fault and sensor performance, respectively.

Secondly, importance weights of different selection criteria for the given application are evaluated. The rating scale R_{ij} is represented by the trapezoidal fuzzy number $(a_{ij}, l_{ij}, r_{ij}, b_{ij})$, then the central value VT_{ij} of the fuzzy number is defined by the center of gravity (COG), which is given by :

$$VT_{ij} = \frac{b_{ij}^2 + r_{ij}b_{ij} + r_{ij}^2 - l_{ij}^2 - a_{ij}l_{ij} - a_{ij}^2}{3(b_{ij} + r_{ij} - l_{ij} - a_{ij})}$$. Then the normalized outcome of criterion j

with respect to sensor *i*, V_{ij}, can be defined as: $V_{ij} = \dfrac{VT_{ij}}{\sum\limits_{i=1}^{m} VT_{ij}}$, for all *I, j*.

The entropy $E_j = -a\sum\limits_{i=1}^{m} V_{ij} In V_{ij}$ of the set of normalized criteria j. Where

a is a constant defined as $a = \frac{1}{In(m)}$. Which guarantees that the range of E_j is within [0,1]. Entropy is a measure of importance for all evaluation criteria. If the rating scales, R_{ij}, of sensors under a criterion are diversified, the entropy value for that criterion would be low; otherwise, it is high.

The designer's *a priori*, subjective weight $\hat{W}_j$, can be combined with E_j, resulting in the following new importance weight W_j: $W_j = E_j \langle \hat{W}_j$.

The subjective weight $\hat{W}_j$ can be specified using linguistic terms such as very low(VL), low(L), medium(M), high(H), and very high(VH). The final rating FR_i of sensor S_i is obtained by weighting individual ratings R_{ij} under each criterion, i.e.

$FR_i = (1/k)[(R_{i1}\langle W_1)\langle (R_{i2}\langle W_2)\langle ...\langle (R_{ik}\langle W_k)]$, *i=1,2,...,m*. Where *k* is the total number of selected criteria, and *m* the total number of sensors in the sensor space. For simplicity, FR_i can be approximately represented as a trapezoidal fuzzy number, denoted by **FR_i** = (ra_i, rb_i, rc_i, rd_i).

Finally, we can rank the final rating **FR_i** of each sensor to obtain the suitable sensor or sensors for identifying the fault of interest. Many ranking methods have been proposed in the past.

The method in this paper is independent of the type of membership functions and the normality of the function. The integral value of the trapezoidal fuzzy number FR_i can be directly obtained without integration:

$I_T(FR_i) = 1/4 (ra_i + rb_i + rc_i + rd_i)$. The ranking order is determined by the total integral value of the fuzzy number. The larger the value of $I_T(FR_i)$, the higher priority the sensor S_i will have for detecting the given fault.

4. RECONFIGURING THEORY ON AGENT

4.1 Notation and preliminary

According to the area being fault-diagnosed, typical fault symptoms B_i and numbers n can be decided, which can also constitute a fault symptoms set B={B_i}, i=1, 2,..., n. Typical fault symptoms z_j and numbers n will be chosen out, which forms a fault symptoms set $Z=\{Z_j\}$, $j =1,2,\ldots$, m. Evolution producing Agent::= <ID, DK, CK, MK, BB, KA, PS-AR> can be regarded as a non-decisive system. According to Shannon's entropy theory. Diagnosing agent entropy $H_i\{n\}$ can be expressed as:

$$H_i(n) = -\sum_{j=1}^{m} p_{ji} \ln p_{ji}$$

In it, n is the total agent number. P_{ji} is the quotient between symptom numbers while Agent number i diagnosing some fault and total number m in fault symptoms set z. Comparability degree is to describe the similar degree between diagnosing agent *i* and *j*, written as: $A_{ij} = 1/(1+H_{ij}(2))$. In it, $H_{ij}(2) = \frac{1}{m}\sum_{j=1}^{m} H_j(2)$ is mean entropy between diagnosing agent *I* and *j*.

Value choosing limitation for A_{ij} is between [0, 1], The larger number A_{ij} is, the more similar agent i and agent j appear, the closer they are. If $A_{ij}=1$, two agents are completely accordant. If A_i fails to solve present problem T_1, calculation according to T_1 and KA_i leads to starting set $K_1A_1=\{(Tk, A_j, R_j^1, R_j^2)/Tk=T_1\}$, besides, $K_1A_1 \subset KA_i$, So mean corresponding successful ratio is expressed as: $K_1AS_1 = \dfrac{\sum_{j=1}^{|K_1A_1|} R_j^2}{|K_1A_1|}$, $K_1AF_1 = \dfrac{\sum_{j=1}^{|K_1A_1|} (R_j^2-R_j^1)}{|K_1A_1|}$. In it, $|K_1A_1|$ represents element number in set K_1A_1. According to K_1AS_1 and K_1AF_1, K_2A_1 can represent a set for Agent A_j to solve T_1, which is chosen from K_1A_1, what's more, $R_j^2>K_1AS_1$ and $R_j^2-R_j^1>K_1AF_1$. That is totally expressed as:

$K_2A_1=\{(Tk, A_j, R_j^1, R_j^2) \mid Tk=T_1 \wedge R_j^2>AS_1 \wedge R_j^2-R_j^1>AF_1\}$. Besides, $K_2A_1 \subseteq K_1A_1 \subseteq KA_i$. Diagnosing agent density refers to the proportion of Agent i in the same kind of diagnosing agent in the agent colony, written as C_i, so C_i is equal to the total sum/n of diagnosing agents whose similarity degree is larger than λ compared with Agent i. In it, λ is the similarity degree. $\lambda = \dfrac{|K_2A_1|}{|K_1A_1|}$. With every diagnosing agent density calculated, initiation and restriction meditation on agents' reconstructing can be done through choosing mechanism.

Adding probability factor based on density meditation to traditional adaptive proportion choosing mechanism, individual's choosing probability P constitutes of adaptive probability P_f and density restriction probability P_d:

$$p_i = \alpha p_{fi} + (1-\alpha)p_{di}, \quad \text{In it,} \quad p_{fi} = F_{it}(i)\Big/\sum_{j=1}^{n} F_{it}(j), \quad p_{di} = \frac{1}{n}e^{\frac{c_i}{\beta}}.$$

In which α and β are constant meditating factors. $F_{it}(i)$ represents fault number adapted to diagnosis Agent i diagnosing. $\sum\limits_{j=1}^{n} F_{it}(j)$ represents total fault number adapted to the whole agent diagnosing(Fan Y.P, 2003).

4.2 Immune agent algorithm

An individual Immune Agent (ImA) works as follows:

1) System incipience: build local opening database according to past experience and expert knowledge and prestore antibody mode set: $Y=\{y_1, y_2, \ldots, y_n\}$.

2) Receive stimulating vaccine from other ImA and update local knowledge base, antibody mode set after updating is $Y' =\{y_1', y_2', \ldots, y_n'\}$.

3) At time of t=k, antigen sensor collects localized perceiving signals, which is expressed as m number of model signal after filtration and inosculation, written as a set: $X(t)=\{x_1(t), x_2(t), \ldots, x_i(t), \ldots, x_n(t)\}$

4) If $\varpi x_i(k)\notin Y'$, $x_i(k)\in X(K)$, and judging Xi(t) with greater probability of abnormal state, signals for acquiring handling is sent to ImA, preserving waiting state and immune tolerance, and then transferred to 8); Immune repines undertakes if $\forall\ x_i(k)\in Y', x_i(k)\in X(k)$, which demonstrates existence of corresponding settling scheme in local knowledge base.

5) Controller issues action order to executing department and sends executing information to specialized ImA.

6) The mode is used as vaccine issued to pertinent ImA in a way of stimulating signal to direct them for handling similar case.

7) While k=k+1, go to next moment, then to 2).

8) Immune tolerance continues if specialized ImA feedback continues to wait, then transferred to 7); If other ImA sends back reference experience, transferred to 5); Meanwhile, learn and memorize the referred experience, updating local knowledgebase for secondary immune response.

5.   PETRI NETS MODEL FOR RECONFIGURATION

5.1   Cubic hierarchy petri-nets

Distributed real-time multi-agent system can adjust its own function and exchanging way with other agents to adapt to the changing environment and task while the total state has been changed. The whole system's construction realizes dynamic reconstruction, which shows intelligent self-organization and self-adaptation action. The system's dynamic running course can be displayed from 3points, that is, system events state changing, agent function behavior changing and dynamic reconstruction on agents' exchanging way. These three aspects can be reflected in supervisory sub-net CPN, behavior sub-net SPN and running sub-net RPN. Every Petri net takes up some plane of Internet space to constitute a close cubic. Sub-net on the same plane includes different layers, which shows system's partial behavior under different conditions. Next, formalized definition on cubic petri net with multiple layers is given as follows (Lee K and Favrel J, 1985).

Definition 5: Combination and commixing Petri nets includes five elements. *CHPN =(CPN, DSPNS, RPNS, E, S).*

$$CPNS = \bigcup_{i=1} CPN_i, \ DSPNS = \bigcup_{i=1} DSPN_i, \ RPNS = \bigcup_{i=1} RPN_i .$$ In it, *CPNS,*

DSPNS, RPNS is the set of supervisory sub net, diagnosing behavior sub-net and reconfiguring sub net respectively. *E* represents all the events set in the system. *S* represents all the states set in supervisory sub-net. These three sub nets are connected through *E* and *S*.

Definition 6: Supervisory sub net is a colour petri net. *CPN = (P, T, C, F, IN, OUT).* In it, *P, T* represents place set and transition set respectively. $C = E \cup S$ is colour set. $F \subseteq ((P \times T) \cup (T \times P)) \rightarrow C$ is weighed direction arc connecting *P* and *T*. IN = *E* is input events set of *CPN*. Out=*S* is output state set of *CPN*.

Definition 7: Diagnosing behavior subnet is colour petri net. *DSPN=(P, T, C, F, IN, OUT).* In it, *P, T* represents place set and transition set separately. $C = E \cup S$ is colour set, representing every agent behavior way set in the system. $F \subseteq ((P \times T) \cup (T \times P)) \rightarrow C$ is weighed arc connecting *P* and *T*. IN=*S* is input events set in *SPN*. Out=*C* is output state set in *CPN*.

Definition 8: Reconfiguring sub net is a generalized random color petri net, *RPN=(P, T, F, C, OUT$_i$, Tempo).* In it, *P* and *T* represents place set and transition set separately. $T=T_{in} \cup T_{de}$, T_{in} is timely transition set, T_{de} is delaying transition set, $F \subseteq ((P \times T) \cup (T \times P)) \rightarrow C$ is weighed arc connecting *P* and *T*. *C=E* is color set. Out$_i$ is output events set. Tempo is delaying function $\cup OUT_i=E$ for T_{de}.

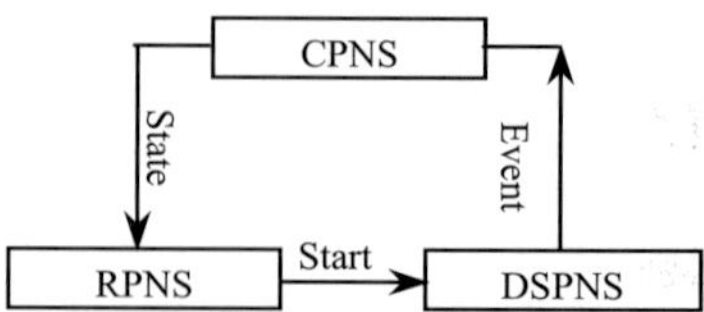

Figure 2. 3 sub-nets relation in cubic hierarchy Petri Nets

Single isolated sub net in the system can't run. Only when 3 kinds of sub net integrated together properly can the system run well. While the system is running ,CPN informs SPN of present system's state. According to the state, agents in SPN choose different behavior function to change system's structure and start up RPN. Every agent reports one event to SPN after finishing sub task. SPN determines next system's running state. Relation among three kinds of sub nets is illustrated in the Figure 2.

5.2 Cubic hierarchy petri-nets algorithm

P_{begin} is source place set, P_{end} is ending-place set, *mark is* stack sequence marking. if mark>0, $CHPN = P_i = P_i^1 \cup P_i^2$, thereinto superscript 1 denotes general place, thereinto superscript 2 denotes macro place(mp_i), T_i is concession transitions set. Algorithm of cubic petri net with multiple layers is given as follows (Moor C J and Whinston B A , 1986/1987):

Step1 Initial setting $M\left(P\middle|P \in P_{begin}\right)=1$, i=1, $P_i = P_{begin}$, $stack_i = \varnothing$, $mark_i = "$.

Step2 Decompose P_i , $P_i = P_i^1 \cup P_i^2$. If $Pi \subseteq P_{end}$, then to Step9; If $stack_i = \varnothing$ and $mark_i = "$, $Pi \subseteq OP_{i-1}$ then $i = i-1$, $P_i^1 = P_i^1 \cup \{mp_{ik}\}$, $P_i^2 = P_i^2 - \{mp_{ik}\}$, $stack_i = \varnothing, mark_i = "$; If $P_i^2 = \varnothing$, then to Step4; If $P_i^2 = \varnothing$, $stack_i = P_i, mark_i =' P'$, $i=i+1$, k=1, then to Step3.

Step3 $P_i = IP_{i-IK}$, then transferred to Step2.

Step4 $T_i = T_i^1 \cup T_i^2 = \{t \in T_i | t \subseteq T_i\} \cup \{mt \in T_i | mt \subseteq P_i\}$, there into superscript 1 denotes general transitions, there into superscript 2 denotes macro transitions, and $P_{is} = \varnothing$. If $T_i = \varnothing$, then to Step9; If $T_i^2 = \varnothing$, then to Step6. If $T_i^2 \neq \varnothing$, $stack_i = T_i, mark_i =' T'$, $i=i+1$, then to Step5.

Step5 Setting $T_i = IT_{it}$, then transferred to Step4.

Step6 Decompose T_i, $T_i^1 = T_i^N \cup T_I^C$, there into subscript N denotes non-collision subset and subscript C denotes collision subset. Confirm $T_i^{N''}$ based on some rule for non-collision subset, and setting: $P_i^N = \cup \{p | p \in t, t \in T_i^N\}$. Select $T_i^{C''}$ based on fact for collision subset, and setting: $P_i^C = \cup \{p | p \in t, t \in T_i^C\}$.

Step7 If $stack_{i-1} \neq \varnothing$ and $mark_{i-1} =' T''$, $T_i \subset OT_{i-1l}$, $i=i+1$, $P_{i-1s} = P_{i-1s} \cup P_i^N \cup P_i^c$, $T_i^2 = T_i^2 - \{mt_{il}\}$, $stack_i = \varnothing$, $mark_i = "$, then to Step4.

Step8 $P_i = P_i^C \cup P_i^N \cup P_{is}$, then transferred to Step2.

Step9 ending.

## 6.	CONCLUSIONS

Through validity definition for reconfiguring multi-agent models in this paper, reconfiguring difficulty was reduced. Meanwhile based on the thought of the structure of diagnosing decision problem and hierarchy in modeling under the reconfiguring system structure of the model, the paper presents cubic hierarchy Petri Nets used as visualized modeling tools to support the nesting modeling, which offers great support for problem definition, problem subdivision, conception test, model generation, model test and model computing algorithm, thereby laying a foundation for distributed intelligent diagnosis of large and complex systems. The model and its modeling method have currently been applied in leakage fault diagnosis system of launch vehicle control system and yield satisfactory results (Fan Youping, 2003). This approach can achieve a self-organizing system that is more robust and flexible in dynamic environment and can be self-updated locally.

ACKNOWLEDGEMENTS

Thank you for Supported by the Doctor Foundation of China (99061116); the Post-doctoral Science Foundation of China.

REFERENCES

1.	Fan Y.P. "Development and Thought of Intelligent Diagnosis Techniques", *Studies in Dialectics of Nature*, **vol.** 17, **no.** 1, pp. 42-46, 2001.
2.	Fan Y.P. "Reconfiguration of fault diagnosis multi-agent based on cell immune response theory", *Journal of system simulation*, **vol.** 15, **no.** 1, pp. 50-55, 2003.
3.	Hunt J, Cooke D. "Learning using an artificial immune system", *Journal of network and computer applications*, **vol.** 19, **no.** 2, pp. 189-212, 1996.
4.	Lee K, Favrel J. "Hierarchical reduction method for analysis and decomposition of Petri nets", *IEEE Transactions on systems, man and cybernetics*, **vol.** 15, **no.** 2, pp. 272-280, 1985.
5.	Moor C J and Whinston B A. "A model of decision-making with sequential information-acquisition-Part I", *Decision Support Systems*, **vol.** 8, pp. 285-307, 1986.

REAL-TIME QUALITY MONITORING TECHNOLOGY FOR POWDER COMPACTING PROCESS

Ge-Wei Chen
Department of Advanced Manufacturing Technology and Engineering, Shenzhen Polytechnic,
Shenzhen 518055, China, Email: cgw2003@sina.com

Abstract: According to the experiments in multiform conditions with different processes, parameters and samples, the corresponding relations in compacting process between vibration signal of tool set system and green density, mass are discovered for the first time. Based on these relations, the theory and method of real-time monitoring for green density and mass during powder compacting process are brought forward in this paper.

Key words: Manufacturing process monitoring, Powder compacting, Time series analysis.

1.   INTRODUCTION

As one of the exactly shaping technologies, powder metallurgy (P/M) technology is the complex technology concerning many subjects and has broad applied foreground. Powder compacting process is the key of the whole P/M production flow, and its monitoring is important and necessary for the automation of P/M production. During powder compacting process, the vibrations of press and tool set caused by pressing force and other forces are representations of compacting process and its characters, and have relationships in a certain extent with parameters of green density and etc. Because of the complexity of compacting system and many inestimable factors in compacting process, there are not absolutely clear causalities between pressing force and green density, pressing force and system vibration, green density and system vibration. At the same time, because mechanical system of compacting system and its movement are very complex, it is very difficult to explain the vibration character of the system in compacting process by matrix movement equation made of mass matrix, damp matrix, stiffness matrix and force matrix.

Time series of vibration signal of powder compacting system represents the system character and its change, thus the compacting process and its change principle can be explained by means of researching, analyzing and processing the time series. According to this clew, this paper samples the acceleration signal of vibration of tool set system in compacting process, makes time series model, and analyses the corresponding relations between model parameters and green density, mass. Based on these relations, the theory and method of real-time monitoring for green density and mass during powder compacting process are brought forward.

There are many researches about powder compacting process in recent years, many of them analyze the mechanical property of green compact based on the compacting parameters such as pressing force, pressing temperature, pressing velocity and etc. This paper gives a new approach to real-time monitor the quality during P/M compacting process.

2.   EXPERIMENT AND ANALYSIS

2.1   Experiment condition

The experimental powder is made of atomised iron power, copper powder, graphite powder and lubricant, in the proportion of 100:2:0.6:0.6. The powder is mixed uniformly.

The sampling system of vibration signal of tool set is made of piezoelectric type acceleration sensor, charge amplifier, data-acquisition board, computer and data processing program.

The basic sizes of compacting specimen are 90mm length, 9mm width and 6mm height. The samples are compacted in oil hydraulic press. The whole compacting process of the specimen should be sampled continuously.

2.2   Density of compact

The experiments are done in 100t press first. The acceleration signal of vibration of die plate is sampled during every compacting process and changed to time series respectively. Altogether 40 compacting processes are sampled and 40 time series are received. The sampling frequency is 400Hz.

Based on the time series, the time series models AR(6) are made respectively. Altogether 40 AR(6) models are made and the densities of corresponding green compacts are measured. Because the density of green compact is generally demanded to be more than 6.70~6.80g/cm^3 in practice, the 6.70~6.80g/cm^3 is selected as the criterion of density. Using the model parameters to make coordinate system, the corresponding densities of green compacts distribute as Fig. 1 and Fig. 2.

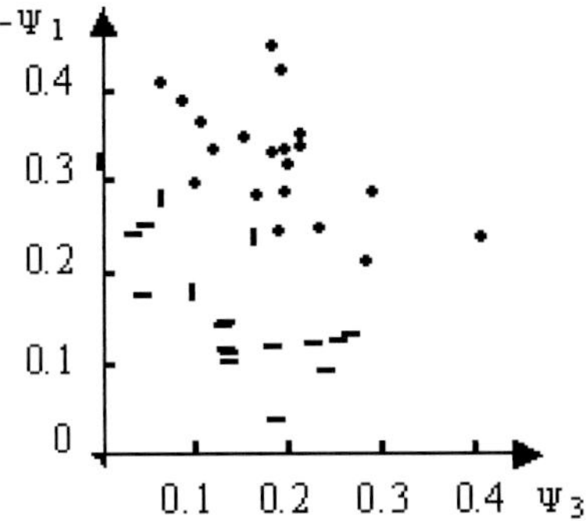

● – – more than 6.80 ; | – – equal to 6.70~6.80 ; — – – less than 6.70

Figure1 .Density distribution of green compact (100t press, sampling frequency 400Hz)

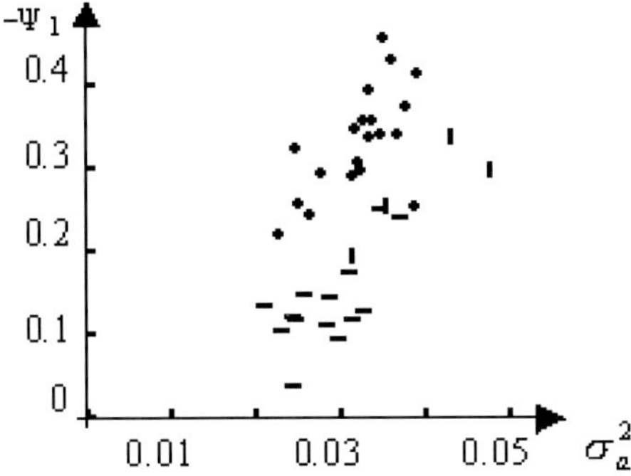

● – – more than 6.80 ; | – – equal to 6.70~6.80 ; — – – less than 6.70

Figure 2. Density distribution of green compact (100t press, sampling frequency 400Hz)

In Fig.1 and Fig. 2, ϕ_1 and ϕ_3 are the first and third parameter of AR(6) model, σ_a is the variance of residual error of AR(6) model. The points which densities more than 6.80g/cm^3 and the points which densities less than 6.70g/cm^3 distribute concentrative respectively, so the density of green compact can be distinguished according to the position of point.

In the same condition, the sampling frequency is changed to 500Hz and the AR(6) models are made. The corresponding densities of green compacts distribute as Fig. 3.

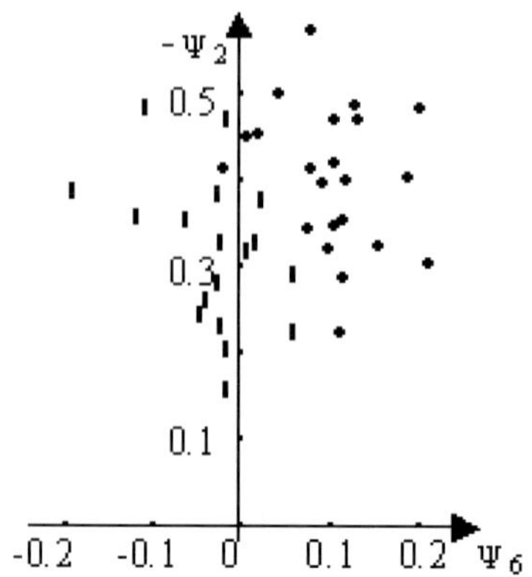

●− − more than 6.80 ; |− − less than 6.80

Figure 3. Density distribution of green compact (100t press, sampling frequency 500Hz)

In Fig. 3, ϕ_2 and ϕ_6 are the second and sixth parameter of AR(6) model. The points which densities more than 6.80g/cm^3 and the points which densities less than 6.80g/cm^3 also distribute concentrative respectively, so the density of green compact also can be distinguished according to the position of point if the sampling frequency is changed.

Using different powder, the density of green compact also can be distinguished.

To do the experiments in 200t press, the sampling frequencies are 400Hz and 500Hz respectively. Making AR(7) models, the corresponding densities of green compacts distribute as Fig. 4 and Fig. 5.

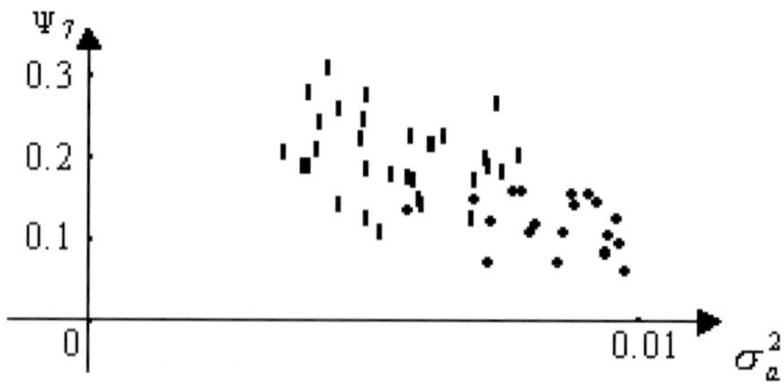

●− − more than 6.80 ; |− − less than 6.80

Figure 4. Density distribution of green compact (200t press, sampling frequency 400Hz)

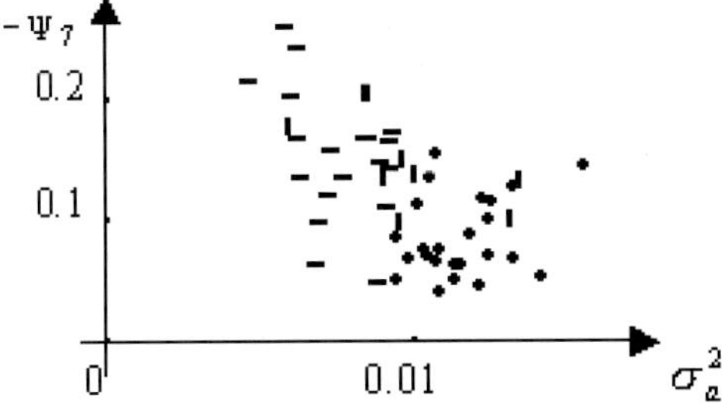

●− − more than 6.70 ; |− − equal to 6.60~6.70 ; −− − − less than 6.60

Figure 5. Density distribution of green compact (200t press, sampling frequency 500Hz)

In Fig. 4 and Fig. 5, ϕ_7 is the seventh parameter of AR(7) model, σ_a^2 is the variance of residual error of AR(7) model. The density of green compact also can be distinguished according to the position of point.

In order to verify the validity of these experiments in practice, the compacting processes of the gears of motorcycle are sampled in 200t press. Altogether 90 compacting processes are sampled. In five of these compacting processes, the densities of green compacts are not eligible. They can be distinguished correctly. In the experiments, the disturbance of environmental vibration has little effect on the result of distinguishing.

2.3　Mass of compact

The mass of green compact also can be distinguished in the experiments.

The compacting processes of 100t press are sampled. The sampling frequency is 400Hz. Making AR(6) models, the corresponding masses of green compacts distribute as Fig. 6 and Fig. 7.

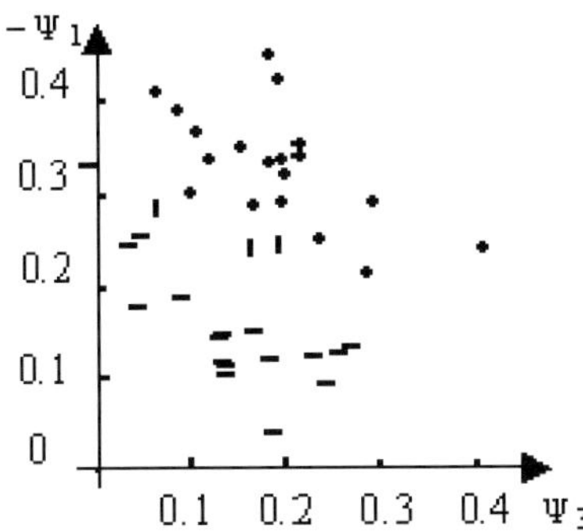

●− − more than 25g ; |− − equal to 23~25g ; −− − − less than 23g

Figure 6. Mass distribution of green compact (100t press, sampling frequency 400Hz)

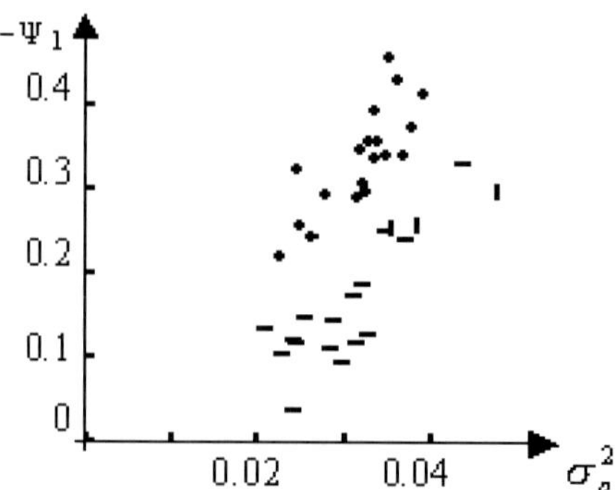

●– – more than 25g ;　|– – equal to 23~25g ;　——– – less than 23g

*Figure 7.*Mass distribution of green compact (100t press, sampling frequency 400Hz)

In Fig. 6 and Fig. 7, ϕ_1 and ϕ_3 are the first and third parameter of AR(6) model, σ_a^2 is the variance of residual error of AR(6) model. The points which masses more than 25g and the points which masses less than 23g distribute concentrative respectively, so the mass of green compact also can be distinguished according to the position of point.

In the same condition, the sampling frequency is changed to 500Hz and the AR(6) models are made. The corresponding masses of green compacts distribute as Fig. 8.

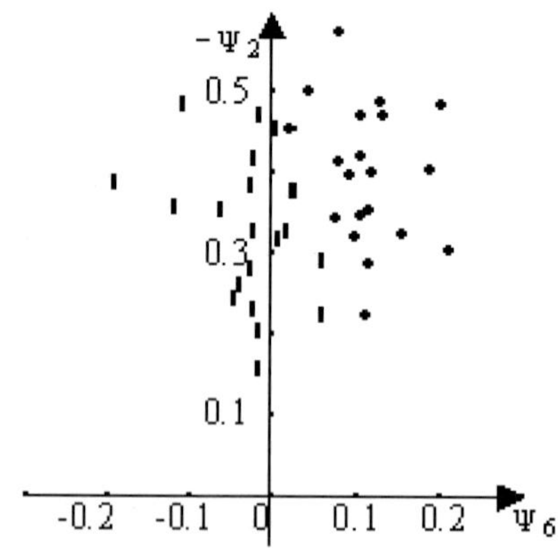

●– – more than 25g ;　|– – less than 25g

*Figure 8.*Mass distribution of green compact (100t press, sampling frequency 500Hz)

In Fig. 8, ϕ_2 and ϕ_6 are the second and sixth parameter of AR(6) model. Same as the density, the mass of green compact also can be distinguished according to the position of point if the sampling frequency is changed.

Using different powder, the mass of green compact also can be distinguished.

To do the experiments in 200t press, the sampling frequencies are 400Hz and 500Hz respectively. Making AR(7) models, the corresponding masses of green compacts distribute as Fig. 9 and Fig. 10.

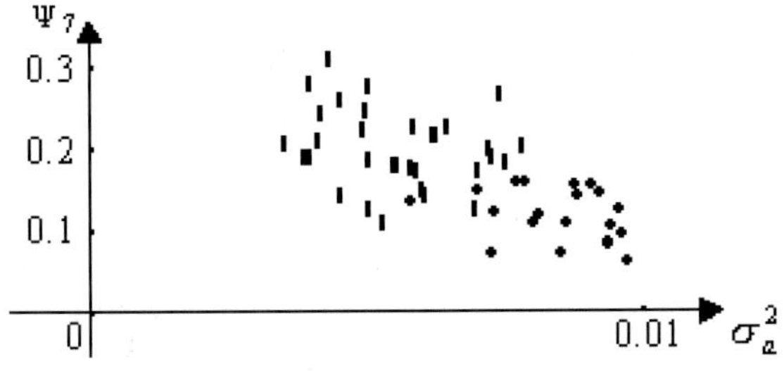

● − − more than 23g ; | − − less than 23g

*Figure 9.*Mass distribution of green compact (200t press, sampling frequency 400Hz)

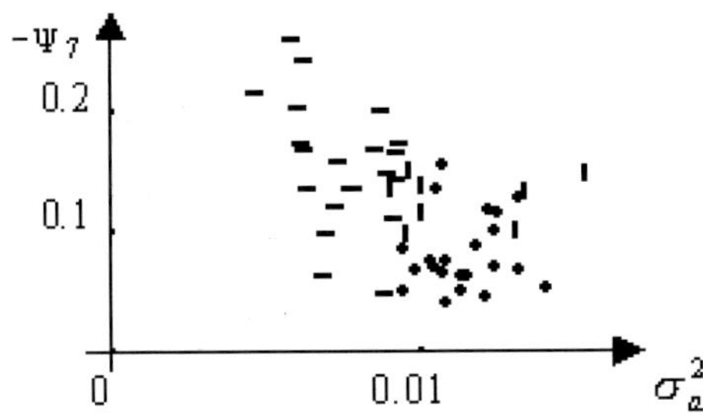

● − − more than 23g ; | − − equal to 21~23g ; — − − − less than 21g

*Figure 10.*Mass distribution of green compact (200t press, sampling frequency 500Hz)

In Fig. 9 and Fig. 10, ϕ_7 is the seventh parameter of AR(7) model, σ_a^2 is the variance of residual error of AR(7) model. The mass of green compact also can be distinguished according to the position of point.

Same to the density, the validity in practice of these experiments is verified by compacting the gears of motorcycle in 200t press.

3.   ACTUALIZING OF MONITORING

According to the experiments in multiform conditions with different presses, parameters and samples, the corresponding relations in compacting process between model parameters of vibration signal of tool set system and green density, mass are discovered. These relations are further verified by fuzzy clustering analysis.

Because not all parameters in one model can be used to distinguish the density and mass of green compact, in order to actualise the real-time monitoring for density and mass in practice, the characteristic quantities of model parameters should be picked up.

There are two purposes of picking up characteristic quantity from model parameters. The one is dimensionality reduction. The other one is to improve the exactness of distinguishing. To pick up characteristic quantity, the principal component method is in common use. For the purpose of distinguishing, picking up two primary characteristic quantities are suitable in general. Applying principal component method to pick up two primary characteristic quantities v_1 and v_2 from the seven parameters of AR(6) model (include the first to the sixth parameter, the variance of residual error) and making two-dimensional coordinate system, the corresponding densities of green compacts distribute as Fig. 11.

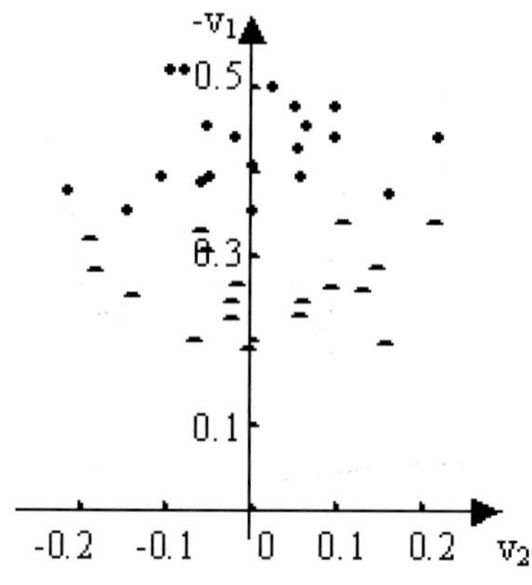

$\bullet - -$ more than 6.80 ; $- - -$ less than 6.80

Figure 11. Density distribution of green compact (100t press, sampling frequency 400Hz)

After picking up characteristic quantities from model parameters, the distinguishing function should be designed. Designing distinguishing function for density need characteristic quantities and the densities of corresponding green compacts. To ensure the exactness of distinguishing, the specimens should be enough. The perception approach could be used to the linear dichotomizing distinguishing. Applying perception approach (1), the linear distinguishing function of characteristic quantities could be design.

$$w(k+1) = w(k) + \tfrac{1}{2}c\{x(k) - x(k)\,\mathrm{sgn}[w^{T}(k)x(k)]\}$$

(1)

In (1), w is weight vector, x is vector of characteristic quantities, k is frequency of iteration and c is constant.

Applying the distinguishing function, the density of green compact could be distinguished based on the characteristic quantities of corresponding model.

The distinguishing function for mass also can be designed based on the characteristic quantities and the masses of corresponding green compacts.

Designing the corresponding software based on these analyses, the real-time monitoring for density and mass of green compact during powder compacting process could be actualized by computer.

4. CONCLUSIONS

According to mathematics analysis and experiments in multiform conditions with different presses, parameters and samples, there are corresponding relations in compacting process between time series model parameters of vibration signal of tool set system and green density, mass. These relations could be used to distinguish the density and mass of green compact.

Based on distinguishing function and characteristic quantities of model parameters, the density and mass of green compact could be distinguished. Further, based on computer, the real-time monitoring for density and mass of green compact during powder compacting process could be actualized.

REFERENCES

1. Yang Shuzi, Wu Ya. *Time Series Analysis in Engineering Application*, Wuhan, Hust Publishing House, 1992.
2. Peter J.Brockwell, Richard A.Davis. *Time Series Theory and Methods*, Beijing, Higher Education Publishing House, 2001.
3. Chen Gewei. *Real-time Monitoring Technology on Quality During Powder Metallurgy Compacting Process and Dynamic Character Analysis*, PhD Thesis, South China University of technology, 2003.

APPLYING GSPN THEORY ON PERFORMANCE ANALYSIS OF THE SERIAL AND PARALLEL PRODUCTION SYSTEMS

Jianhua Gao, Xudong Hu and Ruqing Yang

Department of Mechanical Engineering, Zhejiang Institute of Science and Technology, Hangzhou, P.R. China, 310033, Email: auhnaijoag@sina.com.cn; Department of Mechanical Engineering, Zhejiang Institute of Science and Technology, Hangzhou, P.R. China, 310033, Email: xdhu@zist.edu.cn;   Research Institute of Robotics, Shanghai Jiaotong University, Shanghai, P.R. China, 200030,Email: rqyang@mail1.sjtu.edu.cn

Abstract: A new idea of applying generalized stochastic petri net (GSPN) theory to model the serial and parallel production system is proposed. And one typical discrete event dynamic system (DEDS), turner-unit of palletising system, is taken as a real case to research. Based upon the established GSPN models, the working performances of serial and parallel layout are compared. Furthermore, their differences of working mechanisms including feeding mechanism, coordinating mechanism and monitoring mechanism are discussed. Thus the theoretical basis which is helpful to appraise layout plan and its reasonableness is provided. Meanwhile, the research results show that parallel layout is more advantageous to greatly improve the operational speed of production system than serial one.

Key words: Production system, Layout, Generalized stochastic petri net

1.   INTRODUCTION

The discrete event dynamic production systems including FMS and CIMS have been paid much attention to their analysis, design and optimization in recent years. However, the complexity of systems themselves becomes a barrier to deep research. The serial production systems therefore are generally studied and a great deal of achievements has been obtained [1–6]. The typical modeling methods mainly include queuing theory [1], max-algebra method [2], perturbation analysis method [3], Petri net [4–6], etc. Queuing theory assumes that the information desk (e.g. machining center, feeder) follows a certain distribution. The productivity and mean time of production system can be analyzed by queuing model. But queuing theory is only appropriate for elementary qualitative analyses and it is lack of ability to represent the synchronization in concurrent systems. Max-algebra method abstracts the practical production system to linear algebra model. Although it is suitable for analyzing the sequential line and has a direct relationship to computational aspects of simulation, the strict supposed conditions are usually needed. In addition, it is difficult for max-algebra method to describe some situations such as buffer status, waiting time. Perturbation analysis method has both advantages of simulation and theoretical analysis. It mainly studies the parameter sensitivity of performance index. But for multi-parameter perturbation or large-scale perturbation, it is not easy to simulate operational process. Besides, perturbation analysis method leads to bad approximation when encounter large perturbation. Compare with queuing theory, max-algebra method and perturbation analysis method, owing to the obvious advantages of describing some behaviors such as concurrence, synchronization and conflict, and the explicitness of graphic expression, Petri net has been viewed as an ideal tool of modeling those complex systems with characteristics of concurrence, synchronization and resource share. In spite of that, the application range of traditional Petri net is extremely limited because it doesn't consider time factor [7] and only can express the logic characteristics (e.g. deadlock state). In order to describe the dynamic behaviors of system and obtain some important performance parameters, Time Petri Net (TPN) [8], Stochastic Petri Net (SPN) [9], Generalized Stochastic Petri Net (GSPN) [10] were provided successively. Based upon TPN and SPN, GSPN classifies transitions into two types [11], which are exponential transition and immediate transition. When exponential transition happens, the firing time is a relatively fixed stochastic time determined by exponential distribution. The firing time of immediate transition is zero. Thus GSPN omits the possible transient states, reduces the state numbers used by Markov chain, and simplifies the calculations of time Petri net. Because the sojourn time and switch time can be estimated, some performance indices can be analyzed. In addition, inhibitory arc and random switch are allowed to use in GSPN. These aspects make GSPN more powerful and more flexible than TPN and SPN. GSPN becomes an excellent tool for modeling the complicated systems. GSPN has been used in manufacturing system [12], reliability computation of FMS [13], testability parameters determination [14], fault analysis [**15**], etc. However few literatures can be seen about the GSPN application of modeling and analyzing for the serial and parallel production system. In order to provide the relevant appraisable basis, turner-unit of palletising system, one typical DEDS, is taken as a real case to study the layout effect of serial and parallel production system based on GSPN in this paper.

2. MODELING AND PERFORMANCE ANALYSIS FOR THE SERIAL LAYOUT

2.1 Constitution and the working process

In most of the present palletising lines, which can be viewed as a kind of representative automated production system, the serial-layout turner-unit shown in Fig.1 is generally used. It mainly includes three parts. The first is buffer area which comprises two buffers to guarantee continuous operation of the production line. The second is transferring area which consists of two conveyers to accomplish the task of material feeding, middle location, and output after turning. The third is turning area which contains two turners to rotate bag materials with prescribed angles.

The working process can be simply described as follows: Firstly the bags came from the upstream packing line are feed to the proper position under turners. Then they are turned with the predefined mode to satisfy the requirements. Afterwards those rotated bags are conveyed to pushing unit through the post-buffer. Such operations repeat again.

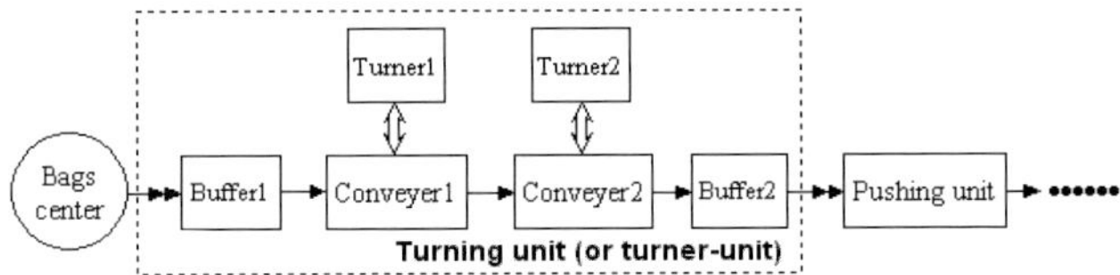

Figure1. One typical production system with the serial layout

2.2 GSPN modeling for the serial layout

Suppose that bags are enough in the production system firstly. Because of the uncertainties existing in the palletising line, those feeding, turning and outputting actions can be viewed as a kind of random serve. So we can assume that feeding time, turning time and outputting time satisfy the exponential distribution. According to GSPN theory [11], the GSPN model of serial-layout system can be established as Fig.2. It comprises 12 places and 8 transitions, and 8 transitions include 4 immediate transitions $\{t_{C1}, t_{C3}, t_{C5}, t_{C7}\}$ and 4 exponential transitions $\{t_{C2}, t_{C4}, t_{C6}, t_{C8}\}$ whose rate are $\{\lambda_{C1}, \lambda_{C2}, \lambda_{C3}, \lambda_{C4}\}$ respectively.

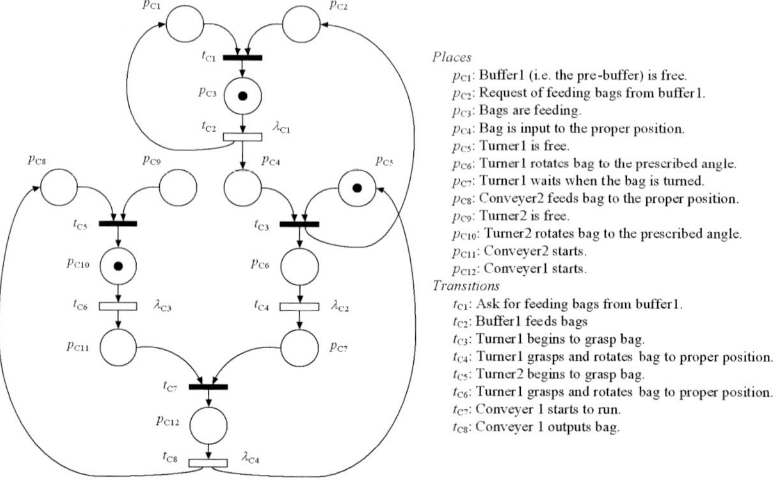

Figure 2. GSPN modeling for the serial layout

2.3　Performance analysis for the serial layout

In the GSPN model of serial-layout system shown in Fig.2, initial marking is $M_{C1}=(001010000100)$, expressed as $M_{C1}=p_{C3}\ p_{C5}\ p_{C10}$. On the condition of $M_{C1}=p_{C3}\ p_{C5}\ p_{C10}$, the exponential transition t_{C2} and t_{C6} are enabled, and the marking M_{C1} is tangible state. Therefore, the fire probability of transition t_{C2} is $\lambda_{C1}/(\lambda_{C1}+\lambda_{C3})$, and that of transition t_{C6} is $\lambda_{C3}/(\lambda_{C1}+\lambda_{C3})$. When t_{C2} fires, one new marking $M_{C11}=p_{C1}\ p_{C4}\ p_{C5}\ p_{C10}$ is generated, which is a vanishing state. Because the immediate transition t_{C3} can be fired, new marking $M_{C12}=p_{C1}\ p_{C2}\ p_{C6}\ p_{C10}$ is also a vanishing state. The next marking $M_{C3}=p_{C3}\ p_{C6}\ p_{C10}$ is a tangible state too, and the exponential transitions t_{C2}, t_{C4} and t_{C6} can be fired, whose fire probabilities are $\lambda_{C1}/(\lambda_{C1}+\lambda_{C2}+\lambda_{C3})$, $\lambda_{C2}/(\lambda_{C1}+\lambda_{C2}+\lambda_{C3})$ and $\lambda_{C3}/(\lambda_{C1}+\lambda_{C2}+\lambda_{C3})$ respectively. When go on applying this method, the whole state-transferring process and their corresponding fire probabilities of the serial-layout GSPN model can be analyzed. As a result, the reachability set can be obtained, and the reachability graph of GSPN, that is the embedded Markov chain (EMC), can be built as Fig.3. In Fig.3, single-ellipse represents vanishing state and double-ellipse expresses tangible state. It can be easily seen that the EMC comprises 10 tangible states $\{M_{C1}, M_{C2}, \ldots, M_{C10}\}$ and 10 vanishing state $\{M_{C11}, M_{C12}, \ldots, M_{C20}\}$. Because vanishing-state transference spends zero time, its steady state probability is zero. From the angle of performance appraise, it is enough to investigate tangible states. So based upon the EMC of GSPN, the reduced embedded Markov chain (REMC) can be obtained when the vanishing states are cancelled, and transition probability matrix can be calculated. When handling the statistical data of 1200 bags of 30 loads, $\lambda_{C1}=35$, $\lambda_{C2}=38$, $\lambda_{C3}=38$ and $\lambda_{C4}=35$ (bags/min) can be acquired. Then the steady probabilities $\{\pi_{C1}, \pi_{C2}, \ldots, \pi_{C10}\}$ can be calculated as Table 1, and each transition probability has already been marked in the REMC. According to the computational equations of GSPN performance indices [11], the analytic results of the serial-layout production system can be calculated as the following.

Utilization ratio of pre-buffer:
$$PROB(p_{C3},1)=\pi_{C1}+\pi_{C2}+\pi_{C3}+\pi_{C5}+\pi_{C6}+\pi_{C9}=57.51\%$$

Utilization ratio of Turner1:
$$PROB(p_{C6},1)+PROB(p_{C12},1)=\pi_{C3}+\pi_{C4}+\pi_{C6}+\pi_{C8}+\pi_{C9}+\pi_{C10}=65.77\%$$

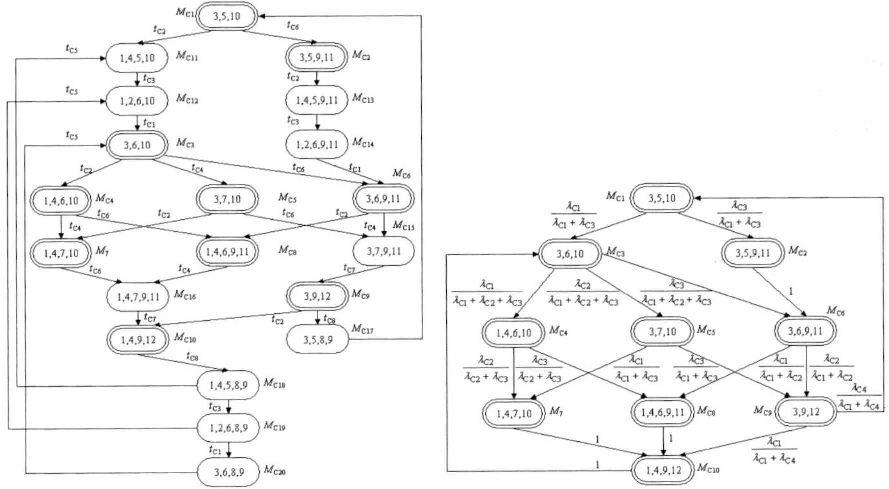

Figure 3. EMC (left) and REMC (right) model of the serial layout

Table 1. Steady-state probabilities of the serial GSPN model

π_{C1}	π_{C2}	π_{C3}	π_{C4}	π_{C5}	π_{C6}	π_{C7}	π_{C8}	π_{C9}	π_{C10}
0.1781	0.1215	0.0134	0.0631	0.0125	0.0521	0.0302	0.1433	0.1975	0.1883

Utilization ratio of Turner2:
$$PROB(p_{C10},1)+PROB(p_{C12},1)=\pi_{C1}+\pi_{C3}+\pi_{C4}+\pi_{C5}+\pi_{C7}+\pi_{C9}+\pi_{C10}=68.31\%$$

Productivity of the serial layout:

$$TR(t_{C8})=(\pi_{C9}+\pi_{C10})\times\lambda_{C4}\times2=27.0060(bags/min)$$

Tact time of the serial layout:

$$T=1/TR(t_{C8})=2.2217(s/bag)$$

According to the analytical result of tact time, it is not difficult to know that the serial-layout system can handle about 1620 bags per hour (BPH). In order to exert the furthest potentiality of layout plan for the production line, a new kind of parallel-layout system is proposed. Some details are explained in the next section.

3. MODELING AND PERFORMANCE ANALYSIS FOR THE PARALLEL LAYOUT

3.1 The Parallel layout and its working process

The proposed parallel-layout is shown as Fig.4. In Fig.4, bags came from feeding unit are stored by twos in pre-buffer, then they are input to turning area at the same time. Two turners handle the bags and complete rotating almost simultaneously. During the whole working process, the restraints between two turners are lessen, and waiting time are reduced. As a result, working efficiency is improved, and higher speed can be realized.

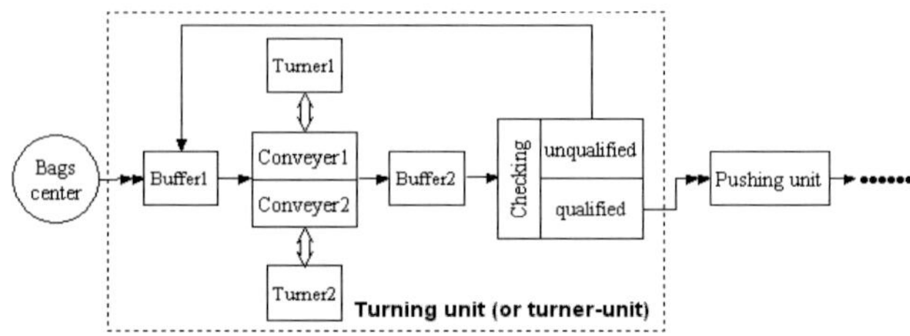

Figure 4. One typical production system with the parallel layout

3.2 GSPN modeling for the parallel layout

Similar to analyzing the serial-layout, the GSPN model of parallel-layout system can be built as Fig.5, which includes 9 places, 4 immediate transitions and 5 exponential transitions.

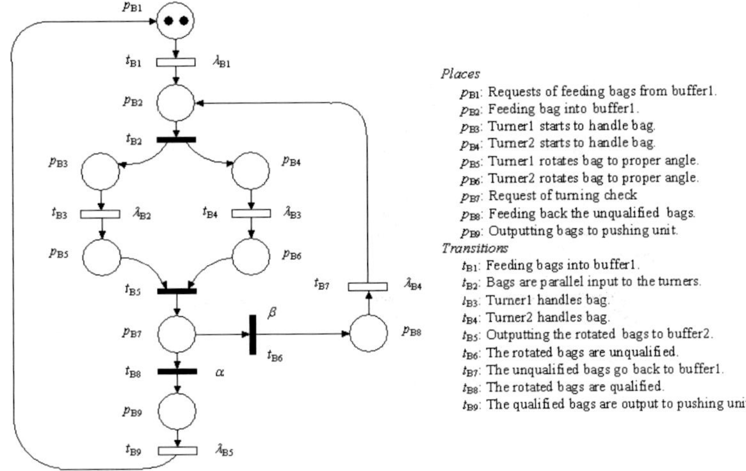

Figure 5. GSPN modeling for the parallel layout

3.3   Performance analysis for the parallel layout

According to Fig.5, the EMC and REMC of parallel-layout system can be

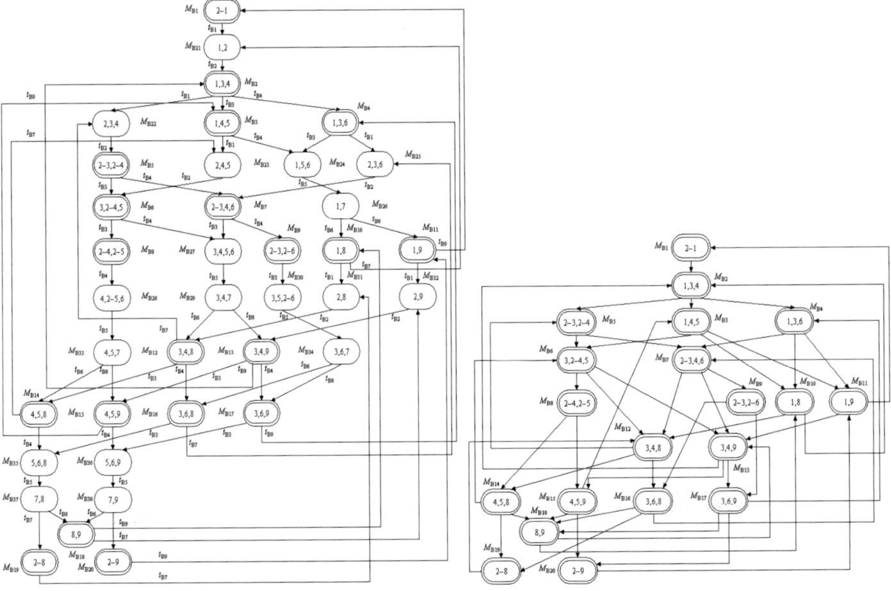

Figure 6. EMC (left) and REMC (right) model of the parallel layout

obtained as Fig.6. The steady probabilities can be calculated as Table 2. The performance analysis results of parallel-layout GSPN model are as follows.

Utilization ratio of Buffer 1: $PROB(p_{B1},1)+PROB(p_{B2},1) = 61.15\%$

Utilization ratio of Turner 1: $PROB(p_{B3},1)+PROB(p_{B5},1) = 77.48\%$

Utilization ratio of Turner 2: $PROB(p_{B4},1)+PROB(p_{B6},1) = 77.48\%$

Productivity of the parallel layout: $TR(t_{B9}) = 34.93(\text{bags/min})$

Tact time of the parallel layout: $T = 1/TR(t_{B9}) = 1.7177$ (s/bag)

Compared with the analytic results of serial layout, no only does the utilization ratio of each device enhance in the parallel layout, but also tact time can reach 1.72 s/bag, which is equivalent to 2090 BPH, 29.01% increased. Obviously, the performance of parallel layout improves greatly.

Table 2. Steady-state probabilities of the parallel GSPN model

π_{B1}	π_{B2}	π_{B3}	π_{B4}	π_{B5}	π_{B6}	π_{B7}	π_{B8}	π_{B9}	π_{B10}
0.1697	0.0384	0.0364	0.0312	0.0647	0.0180	0.0643	0.0118	0.0110	0.0143

π_{B11}	π_{B12}	π_{B13}	π_{B14}	π_{B15}	π_{B16}	π_{B17}	π_{B18}	π_{B19}	π_{B20}
0.1518	0.0107	0.0746	0.0102	0.0867	0.0102	0.1368	0.0376	0.0101	0.0115

4. COMPARISON BETWEEN THE SERIAL AND PARALLEL LAYOUT

4.1 Comparison of the working mechanism

In order to compare the differences which exist between serial and parallel layout, three aspects are selected:

(1) Feeding mechanism. The feeding mechanism of serial layout can be summarize as "one-in one-out", which means that bags are input to turning area one by one and that they are output one by one after being handled. As to the parallel-layout, its feeding mechanism can be described as "two-in two-out", that is to say, two bags are simultaneously sent to turning area and simultaneously output to pushing unit.

(2) Coordinating mechanism. In the serial-layout, the pre-turner is restrained by the post-turner to some degree. Two turners have to work harmoniously with each other. Its coordinating mechanism can be expressed as "cooperative type". But in the parallel-layout, two turners can exert their abilities freely and greatly, and needn't wait. The coordinating mechanism of parallel-layout can be summarized as "isolated type".

(3) Monitoring mechanism. It reflects the ability to cope with potential failures. In the serial-layout, if one of the two turners fails to work, the whole system must be forced to stop running, thus its monitoring mechanism shows "chain type" characteristic. As to the parallel-layout, when one turner fails,

the other turner can still work, system doesn't need to stop running. The parallel-layout monitoring mechanism exhibits "loop type" characteristic. It can also be seen from the GSPN in Fig.5.

4.2   Comparison of the working performance

The left graph of Fig.7 directly shows the performance indices of serial and parallel layout system, in which BUF1, TUR1, TUR2, PRO respectively stands for pre-buffer, turner1, turner2 and productivity. It can be easily seen that each index of the parallel-layout exhibits more excellent than that of the serial-layout. The incremental net value of BUF1 is 3.64%, TUR1 is 11.71%, TUR2 is 9.17%, and PRO increases highly up to 29.34%. The right graph of Fig.7 compares the throughput of serial and parallel layout on condition of 16-hours working system. It is not difficult to see the difference is more and more clear as time goes on. The advantages of parallel-layout are evident.

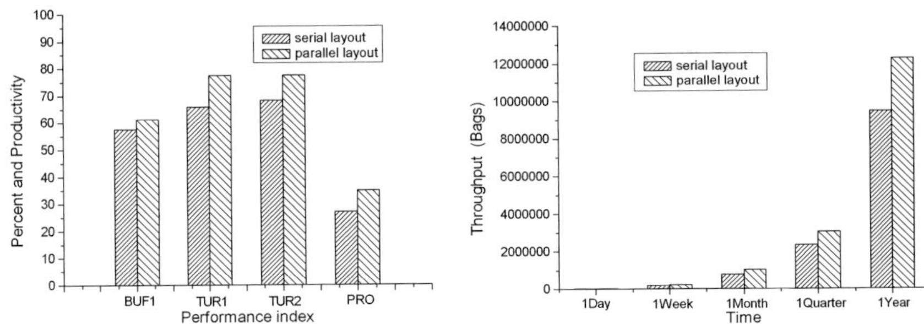

Figure 7. Performance indices and throughput comparison of the serial and parallel layout

5.   CONCLUSIONS

On the basis of applying GSPN theory to model and analyze the typical serial and parallel production system, their differences of working mechanisms including feeding mechanism, coordinating mechanism and monitoring mechanism are discussed. In addition, some working performance indices are compared. The research results show that parallel layout is more advantageous in operational speed than serial layout, and that parallel layout is an available layout to greatly accelerate the production line.

Besides, it must be pointed out that the method applying GSPN to model and analyze the palletising production system can be generalized to other similar industrial production systems.

ACKNOWLEDGEMENTS

We want to give special thanks to the unknown referees for their valuable suggestions on this paper. In addition, the financial support from National 863 Foundation (No. 863-512980106) is gratefully acknowledged.

REFERENCES

1. Vanbracht E. "Performance Analysis of a Serial Production Line with Machine Breakdowns", *Proceedings of the INRIA/IEEE Symposium on Emerging Technologies and Factory Automation*, Paris, France, Oct. 10-13, 1995, pp. 417-424.
2. Zhao QC, Zheng DZ. "Performance Estimations of Production Lines with Stochastic Processing Parameters", *Acta Automatica Sinica*, **vol.23 no.1**, pp.90-93, 1997.
3. Zhang AB, Peng H. "The Performance and Perturbation Analysis in Discrete Event System", *Journal of Xiamen University*, **vol.38 no.1**, pp.21-25, 1999.
4. Kis T, Kiritsis D. "Petri Net Model for Integrated Process and Job Shop Production Planning", *Journal of Intelligent Manufacturing*, **vol.11 no.2**, pp.191-207, 2000.
5. Jiang Z, Zuo MJ, Fung RYK, Tu PY. "Temporized Coloured Petri Nets with Changeable Structure (CPN-CS) for Performance Modelling of Dynamic Production Systems". *International Journal of Production Research*. **vol.38 no.5**, pp.1917-1945, 2000.
6. Furcas R, Giua A, Piccaluga A, Seatzu C. "Modeling Production Systems with Inventory Using Hybrid Petri Nets", *Proceedings of the 2001 IEEE International Conference on Control Applications*, Mexico City, Mexico, Sep. 2001, pp.434-440.
7. Yuan CY. *Theory of Petri Nets*, Beijing, Electronics Industry Press, 1998.
8. Berthomieu B, Diaz M. "Modeling and Verification of Time Dependent Systems Using Time Petri Nets", *IEEE Transactions on Software Engineering*, **vol.17 no.3**, pp.259-273, 1991.
9. Sugasawa Y, Katsumata M, Harada A. "Stochastic Petri Net Model of Distributed Process System with Repair and its Behavior Analysis", *Computers & Industrial Engineering*, **vol.27 no.1**, pp.501-504, 1994.
10. Mizuno O, Hirayama Y. "Application of Generalized Stochastic Petri Net to Quantitative Evaluation of Software Process", *Proceedings of the 1996 IEEE International Conference on Systems, Man and Cybernetics*, Beijing, China, Oct. 1996, pp.3192-3197.
11. Marson MA, Balbo G, Conte G. *Modelling with Generalized Stochastic Petri Nets*, New York, John Wiley & Sons Inc., 1995.
12. Zeng XQ, Wu ZM. "Application of Generalized Stochastic Petri Nets to Manufacturing System", *Acta Automatica Sinica*, **vol.21 no.2**, pp.198-202, 1995.
13. Fan YS, Wu C. "Reliable Index Computation of FMS with Colored GSPN Method", *Computing Technology and Automation*, **vol.16 no.4**, pp.1-5, 1997.
14. Qian YL, Qiu J. "Generalized Stochastic Petri Network for Defining the System-Level Testability Parameters and Figure of Merit", *Systems Engineering And Electronics*, **vol.24 no.5**, pp.4-7, 2002.
15. Mizuno O, Kusumoto S, Kikuno T. "Estimating the Number of Faults Using Simulator Based on Generalized Stochastic Petri Net Model", *Proceedings of the 6th Asian Test Symposium*, Akita, Japan, Nov. 1997, pp.269-274.

OPTIMIZATION DESIGN OF HYDRAULIC SERVO ACTUATORS WITH 3-D.O.F. AND MOTION DECOUPLING

Xinglong Zhu, Jiping Zhou and Jianhua Zhou
(Mechanical Engineering College of Yangzhou University,P.R.China 225009, post box 400180, 86-0514-7978364, xinglongzhu@263.net)

Abstract: Robot actuator is one of very important key parts in robot researching field. Researchers hope an ideal kind of actuators with stronger output torque, lighter weight, smaller volume, motion decoupling and easier control. Once actuators technique is solved, all kinds of robot manipulators can be improved. In this paper, a novel hydraulic servo actuators with three degree of freedom (3-d.o.f.) and motion decoupling is briefly introduced, first. Then a new kind of hybrid chaos optimal algorithm (HCOA) is presented. To minimize the hydraulic servo actuators weight, the results show that HCOA can give satisfactory optimal structural parameters with the constraints of output torque, external diameters, internal diameters and lengths of swing cylinder, diameters of output axis, output angular velocity, correlate dimensions.

Key words: 3-d.o.f. hydraulic servo actuators, chaos optimal algorithm, optimal design, motion decoupling.

1.   INTRODUCTION

Robot manipulators are controlled in order to accomplish the task of assembling, welding and conveying, etc. The conventional control methods, which based on the mathematics modeling or identification modeling, have PID control [1], adaptive control [2], robust control [3], fuzzy control [4], Neural control [5], and hybrid control of above [6-8]. Robot manipulators are consisted of directly drive actuators or hydraulic servo actuators, and most of hydraulic manipulators adopted transferable actuators in heavy duty [4-5,9-10], while rotary actuators scarcely found in design of hydraulic manipulators. Transferable actuators exists some disadvantages as follows: bigger structural dimension, heavier weight and motion coupling. Rotary actuators can implement smaller structural volume, lighter weight and motion decoupling. Since the problem of minimizing the weight of the rotary actuators and raising manipulators implemental efficiency are very significant according to the prior to research [11], a structural optimization algorithm, which bases on chaos theory, is presented.

The paper is organized as follows. The following section introduces briefly the hybrid chaos algorithm. Section 3 validates the algorithm by a typical test functions. Section 4 discusses the structural optimization problem of a novel 3.d.o.f. hydraulic servo actuators to minimize weight. Finally, the algorithm is summarized on the structural optimization problem.

2.   HYBRID CHAOS OPTIMAL ALGORITHM

Chaos is a kind of universal non-linear phenomenon. It seems out-of-order that actually exists regularity. A chaotic variable has three traits. They are randomicity, ergodicity and regularity in chaotic motion [12].

A new hybrid algorithm, which combines the chaos optimization method and SWIFT (Sequential Weight Increasing Factor Technique) approach having an effective convergence property, is proposed in this paper. SWIFT approach that combines the simplex algorithm and the penalty function method transforms the constrained optimization problem into the unconstrained optimization problem by configuring an exact penalty function, and it can search the global/local minimum by the simplex algorithm. Therefore, it can solve nonlinear constraint optimization problem, and the penalty function may be exact non-differentiable. But SWIFT algorithm is prone to fall into the local minimum in searching multi-peak function. On condition that un-reducing the searching space of variable optimized, making use of the ergodicity and randomicity, the hybrid algorithm can help SWIFT approach to skip the local minimum. At the end, it can find the global minimum. The results show that method is simple and easy to implement, and it has high rate of convergence, accuracy and reliability and is effective for the optimization problem. Another feature of the hybrid chaos algorithm is that Hamming distance is acted as the criteria for evaluation. The algorithm is described as follows.

The nonlinear constrained optimization problem is to find x so as to,

$$\min f(x), x = [x_1 \cdots x_n] \in R^n \tag{1}$$

where $x \in F \subseteq S$. The objective function f is defined on the search space $S \subseteq R^n$ and the set $F \subseteq S$ defines the feasible region. Usually the search space S is defined as a n-dimensional rectangle in R^n (domains of variables defined by their lower and upper bounds),

$$l(i) \leq x_i \leq u(i), \quad 1 \leq i \leq n \tag{2}$$

whereas the feasible region $F \subseteq S$ is defined by a set of m additional constraints($m \geq 0$),

$$g_j(x) \leq 0 \text{, for } j=1,\ldots,q, \text{ and } h_j(x) = 0 \text{,for } j=q+1,\ldots,m. \tag{3}$$

SWIFT algorithm will transform the original problem, like in(1-3), into the following unconstrained problem,

$$P(x,r_w) = \min f_i(x) + Q(x,r_w) \tag{4}$$

$$Q(x,r_w) = r_w (\sum_{j=1}^{q} \max[g_j(x),\ 0] + \sum_{j=q+1}^{m} h_j^2(x)) \tag{5}$$

where $\max[g_j(x),0] = \begin{cases} 0 & g_j(x) \leq 0 \\ g_j(x) & g_j(x) > 0 \end{cases}$ $(j=1,\ldots,q)$, r_w is defined as

penalty factor. The flow chart of the algorithm is illuminated as follows (sees figure 1).

3. ONE TEST CASE

An interesting constrained numerical optimization test case emerged recently; the problem [13] is to minimize a function,

$$f(X) = 5.3578547 x_3^2 + 0.8356891 x_1 x_5 + 37.293239 x_1 - 40792.141,$$

subject to three double inequalities,

$$0 \leq 85.334407 + 0.0056858 x_2 x_5 + 0.00026 x_1 x_4 - 0.0022053 x_3 x_5 \leq 92,$$

$$90 \leq 80.51294 + 0.0071317 x_2 x_5 + 0.0029955 x_1 x_2 + 0.0021813 x_3^2 \leq 110,$$

$$20 \leq 9.300961 + 0.0047026 x_2 x_3 + 0.0012547 x_1 x_3 + 0.0019085 x_3 x_4 \leq 25,$$

and bounds,

$$78 \leq x_1 \leq 102 , 33 \leq x_2 \leq 45 , \text{ and } 27 \leq x_i \leq 45 \text{ for } i = 3,4,5.$$

The best solution obtained in 10 runs by Homaifar et al. [14] was,

$$x = [80.49\ 35.07\ 32.05\ 40.33\ 33.34], \text{ and } f(x) = -30005.7.$$

whereas the known global solution is,

$$x = [78.0\ 33.0\ 29.995\ 45.0\ 36.776], \text{with } f(x) = -30665.5.$$

The searched best solution by the hybrid chaos algorithm is,

$$x = [78.000\ 33.025\ 27.088\ 44.938\ 44.942], \text{ with } f(x) = -31022.446.$$

The result shows that the hybrid algorithm has high rate of convergence and accuracy.

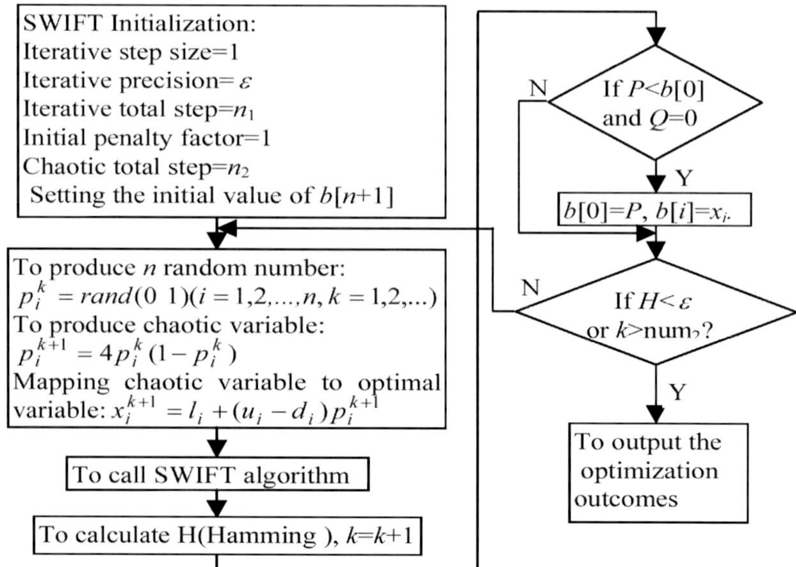

Figure 1.   The flow chart of the algorithm

4. OPTIMIZATION CALCULATION

4.1   Working principal

Fig.2 is the 3-d.o.f. hydraulic servo actuators modeling. Three swing hydraulic cylinders are placed inside the actuators. The actuators can implement the separateness between the position and the pose. If the swing cylinder which rotates x-axis is taken place of the moving cylinder, the axle hole assembling task can be completed. On the basic of the joint, adding a moving cylinder of x-axis, the assembling task of the axle hole, the screw, the key seat hole, the square hole, and the non-regularity hole can be accomplished.

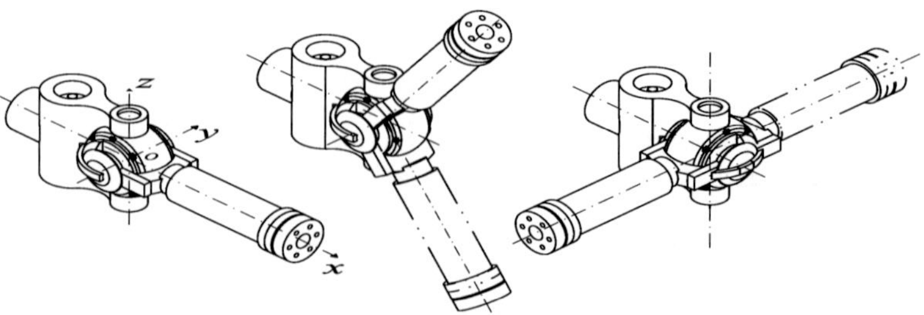

(a) rolling x-axis ±150° (b) pitching y-axis ±60° (c) yawing z-axis ±90°
Figure 2.   The hydraulic servo joint modeling

The structure schematic diagram of the joint shows Fig.3a (*x*-axis motion). Its principal is described as follows. When the pressure oil is pumped from the pump, the pressure oil reaches the servo valve core by the oil lead. Because of the isolation zone between the servo valve core and the following valve pocket, the pressure oil can not entry the swing cylinder, and the servo valve exists in the mesoposition state (sees Fig.3b).

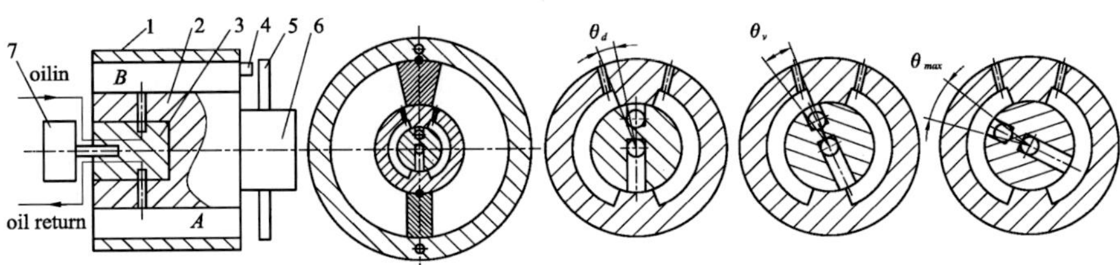

Figure 3. Schematic diagram of the joint (*x*-axis motion), (a) sketch map, 1 swing cylinder body 2 following valve pocket 3 servo valve core 4 magnetic core, 5 magnetic encoder wheel 6 output axis 7 step motor, (b)section, (c) shutting state, (d)opening state, (e)maximum state,

When the step-motor (DC motor) is driven by a rotating signal which the computer sends (positive direction), the motor drives the servo valve core to rotate an angle. The valve port is opened between the servo valve core and the following valve pocket. The pressure oil entries chamber *A* of the swing cylinder, at the same time, the oil of the chamber *B* returns to the oil reservoir by the oil lead. The following valve pocket starts to rotate and drive the output axis with load. With rotating of the following valve pocket, the valve jaw opening between the servo valve core and the following valve pocket is becoming small and shuts at the end. The position tracing is implemented between the servo valve core and the following valve pocket. Otherwise, the motor rotates (negative direction), the position tracing realizes in the same manner. In order to eliminate the position tracing error between the servo valve core and the following valve pocket (for example, the servo valve core rotates 30 angle, because of the dead zone (sees Fig.3c), the motor losses steps, et al., the follow valve pocket rotates 28 angle.), the magnetic encoder wheel of the magnetic encoder is installed on the output axis. The actual rotating angle is measured by the magnetic core of the magnetic encoder. The compensative angle is sent to the motor by the computer. Therefore, the rotating angle of the following valve pocket is ensured to the ideal value. Fig.3d is opening state of the valve jaw opening , and fig.3e is maximum state of the valve jaw opening .

4.2   Structural optimization

The external diameter D_{b3}, D_{m3}, D_{f3}, the internal diameter D_{b2}, D_{m2}, D_{f2}, the output axis diameter D_{b1}, D_{m1}, D_{f1}, the length of the swing cylinder L_b, L_m, L_f are selected as optimal variable (Fig.4), that is,

$$x = [x_1 \; x_2 \; \cdots \; x_{12}] =$$
$$[D_{b3} \; D_{m3} \; D_{f3} \; D_{b2} \; D_{m3} \; D_{f2} \; D_{b1} \; D_{m1} \; D_{f1} \; L_b \; L_m \; L_f],$$

where the unit of all variables is centimeter.

In order to implement the structural optimization of the actuator, to minimize to weight, the objective function is decided to as follows,

$$f(X) = M_x + M_y + M_z \tag{6}$$

where M_x, M_y, M_z are mass of each d.o.f. , respectively. M_x, M_y, M_z are defined as $f_1(x)$, $f_2(x)$, $f_3(x)$,respectively.

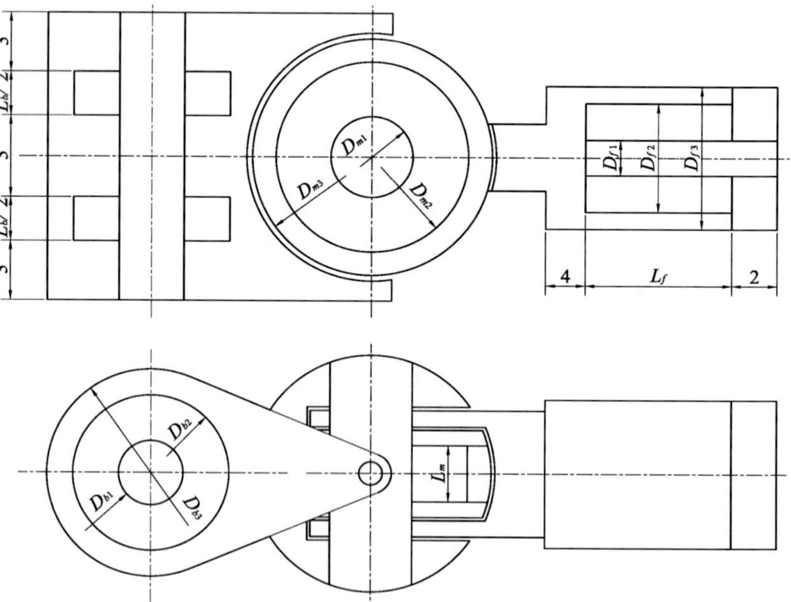

Figure 4. The sketch of 3-d.o.f. hydraulic servo actuators

$$\left.\begin{aligned}
f_1(x) &= M_x = \pi[L_f(\rho_1(D_{f3}^2 - D_{f2}^2) + \rho_2(D_{f2}^2 - D_{f1}^2) \\
&\quad + D_{f1}^2\rho_3) + 6\rho_1 D_{f3}^2]/4 + \Delta M_x \\
f_2(x) &= M_y = \pi\rho_1 D_{m3}^3/6 - \pi[L_m(\rho_3 - \rho_2)(D_{m2}^2 - D_{m1}^2) \\
&\quad - (\rho_3 - \rho_1)D_{f3}D_{m1}^2]/4 + \Delta M_y \\
f_3(x) &= M_z = \pi[(L_b + 9)\rho_1 D_{b3}^2 + (L_b + 9)(\rho_3 - \rho_2)D_{b1}^2 \\
&\quad - L_b(\rho_1 - \rho_2)(D_{b2}^2 - D_{b1}^2)]/4 + \Delta M_z
\end{aligned}\right\} \tag{7}$$

where ΔM_x, ΔM_y, ΔM_z are equivalent mass relational parts respectively. ρ_1, ρ_2, ρ_3 are density of steel, oil and aluminium respectively.

The constraint functions are selected as follows. The output torque constrained is,

$$g_1(x) = 10000 - 0.5(p_0 - \Delta p_x)L_f(D_{f2}^2 - D_{f1}^2)/4 + M_x l_x g ,$$

$$g_2(x) = 11000 - 2 \times 0.5(p_0 - \Delta p_y)L_m(D_{m2}^2 - D_{m1}^2)/4 + (M_x + M_y)l_y g ,$$

and $g_3(x) = 12000 - 0.5 z_2/z_1(p_0 - \Delta p_z)L_b(D_{b2}^2 - D_{b1}^2)/4 + (M_x + M_y + M_z)l_z g$.

where l_x, l_y, l_z are equivalent arm length respectively. Z_1, Z_2 are the number of teeth of the transmission gear respectively. Δp_x, Δp_y, Δp_z are pressure loss on the valve port, respectively. p_0 is the pressure of the hydraulic system. g is gravity acceleration.

The constrained external diameter of the sway cylinder is,

$$g_4(x) = D_{b3} - 12 , \quad g_5(x) = -D_{b3} + 9 , \quad g_6(x) = D_{m3} - 12 ,$$

$$g_7(x) = -D_{m3} + 9 , \quad g_8(x) = D_{f3} - 8 , \text{ and } g_9(x) = -D_{f3} + 5 .$$

The constrained internal diameter of the sway cylinder is,

$$g_{10}(x) = D_{b2} - 8 , \quad g_{11}(x) = -D_{b2} + 5 , \quad g_{12}(x) = D_{m2} - 7 ,$$

$$g_{13}(x) = -D_{m2} + 4 , \quad g_{14}(x) = D_{f2} - 6 , \text{ and } g_{15}(x) = -D_{f2} + 3 .$$

The constrained output axis diameter of each degree of freedom is,

$$g_{16}(x) = D_{b1} - 3 , \quad g_{17}(x) = -D_{b1} + 2 , \quad g_{18}(x) = D_{m1} - 2.5 ,$$

$$g_{19}(x) = -D_{m1} + 2 , \quad g_{20}(x) = D_{f1} - 3 , \text{ and } g_{21}(x) = -D_{f1} + 2 .$$

The constrained length of the sway cylinder is,

$$g_{22}(x) = L_b - 5 , \quad g_{23}(x) = -L_b + 4 , \quad g_{24}(x) = L_m - 3 ,$$

$$g_{25}(x) = -L_m + 2 , \quad g_{26}(x) = L_f - 20 , \text{ and } g_{27}(x) = -L_f + 15 .$$

The constrained correlate dimension is,

$$g_{28}(x) = D_{b3} - D_{b2} - 16 , \quad g_{29}(x) = -(D_{b3} - D_{b2}) + 12 ,$$

$$g_{30}(x) = D_{m3} - D_{m2} - 3 , \quad g_{31}(x) = -(D_{m3} - D_{m2}) + 2.5 ,$$

$$g_{32}(x) = D_{f3} - D_{f2} - 16 , \text{ and } g_{30}(x) = -(D_{f3} - D_{f2}) + 12 .$$

The constrained output angular velocity is,

$$g_{34}(x) = \omega_x = -4q_0 /(L_b (D_{b2}^2 - D_{b1}^2)\pi) + \pi/3 ,$$

$$g_{35}(x) = \omega_y = -4q_0 /(L_m (D_{m2}^2 - D_{m1}^2)\pi) + \pi/3 ,$$

$$\text{and } g_{36}(x) = \omega_z = -4q_0 /(L_f (D_{f2}^2 - D_{f1}^2)\pi) + \pi/3 .$$

where $\omega_x , \omega_y , \omega_z$ are the output angular velocity, respectively. And q_0 is the oil flow through the valve slot.

The structural optimization problem is transformed to be solve the unconstrained optimization problem as follows,

$$P(x,r_w) = \min f(x) + r_w \sum_{j=1}^{36} \max[g_j (x),\ 0] \tag{8}$$

$$\max[g_j (x),0] = \begin{cases} 0 & g_j (x) \le 0 \\ g_j (x) & g_j (x) > 0 \end{cases} \quad (j=1,...,36) \tag{9}$$

Table 1 Comparison of Original design and Optimal design

Optimal variant	D_{b3}/cm	D_{m3}/cm	D_{f3}/cm	D_{b2}/cm	D_{m2}/cm	D_{f2}/cm
Original design	12.00	12.00	8.00	8.00	7.00	6.00
Optimal design	9.000	9.000	5.116	7.467	6.404	3.916
Integer	9.000	9.000	5.100	7.500	6.400	4.000

Optimal variant	D_{b1}/cm	D_{m1}/cm	D_{f1}/cm	L_b/cm	L_m/cm	L_f/cm
Original design	3.00	2.50	3.00	5.00	3.00	20.00
Optimal design	2.670	2.002	2.016	4.001	2.998	16.795
Integer	2.700	2.000	2.000	4.000	3.000	16.800

Substituting the following of known values to Eq.8-9, the hybrid chaos algorithm gives the optimal outcomes (shows table 1). The known values are $\rho_1 = 2.7 \text{g/cm}^3$, $\rho_2 = 0.9 \text{g/cm}^3$, $\rho_3 = 7.8 \text{g/cm}^3$, $p_0 = 5\text{Mpa}$, $q_0 = 160 \text{cm}^3/\text{s}$, $z_2/z_1 = 1.5$, $\Delta M_x = \Delta M_y = \Delta M_z = 0.5\text{kg}$, $\Delta p_x = \Delta p_y = \Delta p_z = 0.7\text{Mpa}$, l_x , l_y , l_z equals 1cm, 8cm,3cm respectively.

From the table 1, we can calculate that total weight of the original actuators and optimal actuators is 11.43kg, 5.67kg respectively. The optimal actuators weight decreases approximately 50.4 ▪ . Substituting the optimal parameters to expressions as follows,

$$T_x = 0.5(p_0 - \Delta p_x)L_f(D_{f2}^2 - D_{f1}^2)/4 = 10836\text{Ncm}$$

$$T_y = 0.5 \times 2(p_0 - \Delta p_y)L_m(D_{m2}^2 - D_{m1}^2)/4 = 11920\text{Nm}$$

$$T_z = 0.5 z_2 / z_1 (p_0 - \Delta p_z)L_b(D_{b2}^2 - D_{b1}^2)/4 = 13452\text{Nm}$$

The calculating results shows that the optimal outcomes satisfy the output torque demand.

5. CONCLUSION

HCOA, which will be capable of searching global solution, is presented based on chaos theory. For the problem of minimizing the actuators weight, HCOA can give optimal structural parameters with the constraints of output-torque and structural dimensions. The optimal actuators can realize strong output torque of 100Nm and motion decoupling, with which the control system can be simplified. The optimal actuators weight decreases approximately 50.4% comparing with the weight of the original actuators.

ACKNOWLEDGEMENTS

This paper is based upon work supported by the educated department of Jiangsu province under Grant 03KJB460161. Many thanks to Dr. Cato Tan of University of Strathclyde for his help. The authors would to thank EASED2004's Europe-Asia symposium and Kluwer Academic Publishers of the United States to give me such opportunity.

REFERENCES

1. Ilse Cervantes, Jose Alvarez Ramirez. "On the PID tracking control of robot manipulators", *Systems & control Letters*, **42.1.**, pp. 37-46, 2001.
2. Pi-Cheng Tung, Sun-Run Wang, Fu-Yee Hong. "Application of MRAC theory for adaptive control of a constrained robot manipulator", *International Journal of Machine Tools & Manufacture*, **40.14.**, pp. 2083-2097, 2000.
3. C Ham,Z Qu,R Johnson. "Robust fuzzy control for robot manipulators", *IEE Proc Control Theory Appl*, **147.2.**, pp. 212-216, 2000.
4. Roya Rahbari, Clarence W de Silva. "Comparison of two inference methods for P-type fuzzy logic control through experimental investigation using a hydraulic manipulator", *Engineering Applications of Artificial Intelligence*, **14.6.**, pp. 763-784, 2001.
5. A Hountras, I antoniadis, A Kanarachos. "Implementation of N-Step-Ahead neurocontrol on a 3-Axes heavy duty hydraulic manipulator", *Mechatronics*, **9.3.**, pp. 235-270, 1999.
6. Y C Chang. "Neural network-based H^∞ tracking control for robotic systems", *IEE Proc Control Theory Appl*, **147.3.**, pp. 303-311, 2000.

7. Krzysztof P Jankowski, Hoda A Elmaraghy. "Robust hybrid position/force control of redundant robots", *Robotics and Autonomous Systems*, **27.3.**, pp. 111-127, 1999.
8. J Y Choi, J S Lee. "Adaptive iterative learning control of uncertain robotics systems", *IEE Proc Control Theory Appl*, **147.2.**, pp. 217-223, 2000.
9. Y ALTINTAS, A J LANE. "Design of an Electro-hydraulic CNC Press Brake", *International Journal of Machine Tools & Manufacture*, **37.1.**, pp. 45-59, 1997.
10. Adolfo Bauchspiess, Sadek C. Absi Alfaro,Leszek A. Dobrzanski. "Predictive sensor guided robotic manipulators in automated welding cells", *Journal of Materials Processing Technology*, **109.1-2.**, pp.13-19, 2001.
11. Zhu Xinglong,Zhou Jiping, Zhou, Yan Jingping. "The Design for a Novel Hydraulic Servo Joint with 3-d.o.f. Vertical Cross and Motion Decoupled", *China Mechanical Engineering*, **13.21.**, pp. 1824-1826, 2002.
12. Fogel D B. "An introduction to simulated evolutionary optimization", *IEEE Trans on Neural Network*, **5.1.**, pp. 3-14, 1994.
13. Michalewicz Z, Schoenauer M. "Evolutionary Algorithms for Constrained Parameters Optimization Problems", *Evolutionary Computation*, **4.1.**. pp.1-32,1996.
14. Homaifar A, S H −Y Lai, X Qi. "Constrained optimization via genetic algorithms", *Simulation*, **62.4.**, pp. 242-254, 1994.

A FRAMEWORK FOR CIGARETTE SUPPLY CHAIN IN THE MONOPOLY ENVIRONMENT
Based on Chinese tobacco industry

Tianfei Li and Linyan Sun
*Management College of Xian Jiaotong University, Xi'an Shaanxi, 710049,029-82673282 ,
LTF@hongta.com*

Abstract:
As a big tax resource, cigarette industry brings about 10% of China's annual income. Similar with that in other industry, the supply chain in cigarette industry also consists of a supplying network, a kernel manufacturer and a distribution network. The supplying network concerns with tobacco suppliers, e.g. tobacco planters, cigarette complement material suppliers and cigarette machine suppliers. The cigarette manufacturers, i.e. a cigarette factory or tobacco group takes the role of the kernel manufacturer. The distribution network involves cigarette companies (wholesalers) and retailers. Due to the local finance is separated from the central finance while the cigarette monopoly management is the responsibility of the local cigarette companies, the distribution network has a strong characteristics of zone-specific and thus is easy to be broken. Meanwhile, as the center of the whole supply chain, the kernel manufacturer is not allowed to build a nation-wide, self-owned distribution network in the market. Therefore, the imperfectness of the monopolistic distribution network could only be made up by the manufacturer's decentralized manufacturing network and distribution-strengthen network, both of which cross multiple supply chain notes.

Key words:
Supply chain management, distribution network, distributed networked manufacture, monopoly management.

1. INTRODUCTION

As a special kind of economic plant, tobacco is double-edged. It is not only an industry but enables a country to reallocate its income as well. On the other hand, because cigarette is deleterious and addicting and for the health of the people, the country should restrain the development of cigarette industry and limit the manufacturing and sale of tobacco products by financial and administrative policies [1]. Almost all the countries in the world impose high taxation on tobacco products and use monopoly management in some phases of the cigarette industry [2]. Nowadays, there are still more than 70 countries take monopoly policies.

As the development of No Smoking Movement in the western developing countries and the law environment the cigarette manufacturers faced becoming more and more grim, cigarette industry shows almost all characteristics of a setting sun. Cigarette monopoly and high taxation enhance the centralization and the globalization of economy makes the global monopolization possible. The process of transferring the tobacco product and market to the countries of the Third World, especially that to the maximum market China, is being accelerated. Philip Morris(P.M.), British American Tobacco Company(B.A.T) and Japan Tobacco Inc.(J.T.I.) have occupied 60% of the market except China. Their annual cigarette output, nearly 42 million boxes is almost 20% more than the total output of the 146 cigarette factories in China. Gallaher, the 10th biggest tobacco company in the world, has cooperated with Shanghai Tobacco Group, China. R.J.Reynolds Tobacco (R.J.R), the 6th biggest tobacco company in the world, has not only built a tobacco chips manufacturing base at Yuxi, China in cooperation with Hongta Group, but also set up Xiamen Huamei Cigarette Company Ltd. to produce cigarettes of "Camel" and "Winston" brands. In the earlier of 1990s, BAT began to produce "Derby" brand cigarettes in cooperation with Wuhu Cigarette Factory which fit for China market's demands [3]. Cigarette industry has form a global manufacturing chain and is further becoming a global supply chain.

The decentralization of manufacturing determines that decentralization and network are the developing trend of the supply chain management. Because of the rapid change of the market and the global economy integration, the survival and development of an enterprise depends not only on the optimization and utilization of the internal resources but on the ability of quick response to the market demands using various resources of other enterprises. The famous British SCM expert Martin Christopher said that to the market there are only supply chains but no enterprise, and the competition in 21st century will occur not among enterprises but among supply chains [4].

2.   EFFECTIVES OF ADMINISTRATIVE MONOPOLY

China has both the biggest cigarette output and consuming market in the world. Its total output is about 1/3 of that of the whole world [5]. In 1998, the cigarette produced all over the world was 119 million boxes and 33.74 million boxes in China, i.e. occupied 28.4%. In 1999, the total output of the world was 110.67 million boxes and that in China occupied 29.7%, i.e.32.87 million boxes. In 2000, cigarette industry brought about more than 100 billion RMB Yuan to China's total financial revenue which takes a proportion of nearly 10% [6]. In 2002 the total profit rose to 140 billion RMB Yuan and the output was 34.449 million boxes. Meanwhile, China is also a weak country in the cigarette industry. For the whole country, cigarette sale is mainly limited within China. For the provinces, due to fiscal decentralization while the cigarette monopoly management is the responsibility of the local cigarette companies, provincial protectionism is very common and thus to some extent the national monopoly policy has become local monopoly policies [7]. Cigarette industry has shown a tendency that the stronger is becoming weaker and weaker. Therefore, from the view of supply chain management, reconstructing tobacco supply chain could increase the utilization rate of the capital, limit the illegal circulation path of cigarette and thus ensure the nation's financial revenue. On the other hand, the effective supply chain competition could help to break the local blockage, drive away coarsely produced cigarette and thus reduce the harmfulness to the health of the smokers and the passive smokers.

The ineffectiveness of government monopoly is the important and fundamental reason why domestic tobacco enterprises have low competition abilities in comparison with their foreign rivals. Essentially, the administrative monopoly of the government is obviously ultra-economy, compulsive and strongly competition-repulsive. It also destroys the order of the market, violates the principle of contract equality and trade liberalization. The contravention to the objective economic discipline results in the role confusion of the government as an enterprise and the long-lasting unreasoning in enterprises' economical activities. The market process is dramatically slowed down and the effect of competition is largely limited. Meanwhile, the monopoly of the government causes various economic or non-economic effects.

3.   EFFECTIVES OF ADMINISTRATIVE MONOPOLY

3.1   Structure of Tobacco Supply Chains

Tobacco supply chains, as shown in Figure 1, consist of supplying networks, kernel manufacturers and distribution networks. Supplying networks mainly involves tobacco leaf suppliers, complement material suppliers and cigarette machine suppliers. As a complementary business process supplier, tobacco science and technology. Kernel manufacturers are cigarette factories or tobacco groups that produce cigarettes.

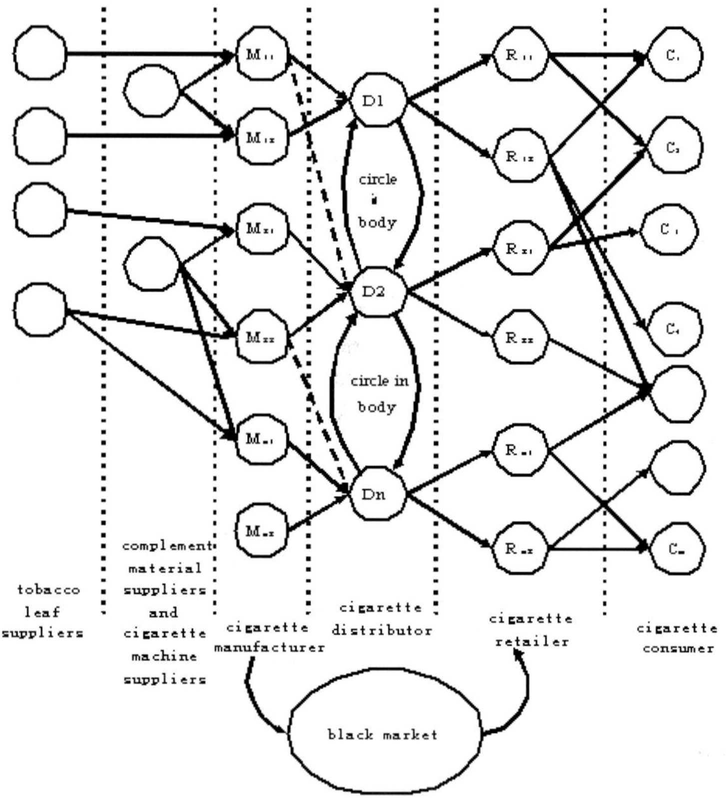

*Figure 1.*Topology of Tobacco Supply Chains

By the end of 2002, there were 123 legal manufacturers in China after some others had been closed, ceased to produce product, merged or changed to produce other products. Distribution networks mainly include tobacco companies in various local zones and retailers. Due to the local finance is separated from the central finance while the cigarette monopoly management is the responsibility of the local cigarette companies, the distribution network has a strong characteristics of zone-specific and thus is easy to be broken. The dashed lines in Figure 1 represent possible break points between tobacco manufacturers and tobacco distributors. As the center of the whole supply chain, a kernel manufacturer is not allowed to build a nation-wide, self owned distribution network in the market. The imperfectness of the monopolistic distribution network could only be made up by the manufacturer's decentralized manufacturing network. As the chained retailers enhance their power in the supply chain, manufacturer could cooperate with retailers and thus build a distribution-strengthen network crossing multiple supply chain notes. This distribution-strengthen network will take place of the cigarette black market exists currently in China.

The supplying network shows obvious characters of the planned economy while the distribution network shows some characters of the market economy. The separation experiment being undertaken of the market management and the market business aims to apart the management functions from the cigarette circulating system. Under the vertical management of the National Tobacco Monopoly Bureau, the local tobacco distribution companies will not manage local cigarette manufacturers any more but only play their roles as wholesalers. Cigarette retailing markets will be the main places where cigarette manufacturers compete for their big market shares.

3.2   Competitive power of Tobacco Supply Chains

3.2.1   Tetrahedron Structure of the Competitive power

There are 4 facets for the competitive power: product power, distribution power, capital power and brand power. Product power is the essential one enabling the kernel enterprise to get competitive advantages and mainly determined by technical and substantial economy activities of the kernel enterprise related to products. Here products is an integrated concept which involves the cigarettes' quantity, catalogue, quality, brands, prices, packages, styles and services. Distribution power (also could be called economy power), is the integrated and intelligent capability of the kernel enterprise to expand its market. Capital power represents the capital operation power of the kernel enterprise. Brand power is the potential immaterial resources of the kernel enterprise and it's determined by the connotation of product brands and services, etc.

The relationships among the 4 facets of competitive power are shown in Figure 2. The product power is the basis of the enterprise development and

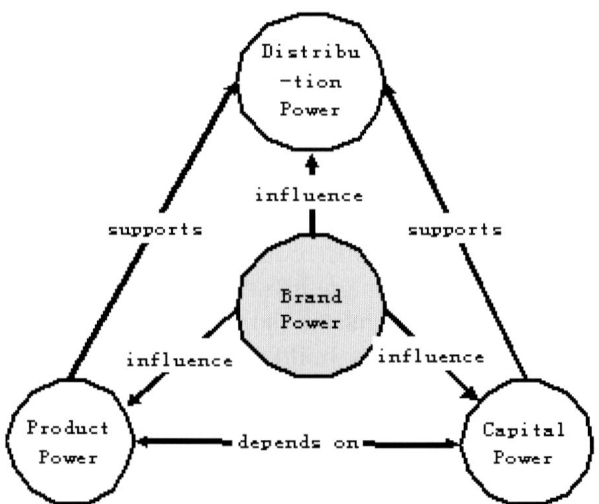

Figure 2. Relationships among 4 facets of competitive power

related to the product manufacturing and management. Distribution power expands the market and thus realizes the value of the products. Capital power ensures that the enterprise has enough resources for its rapid developing. Brand power is the center of the competitive power and determines the results of the other 3 facets.

Firstly, from the view of product power, the product credit standings and features could only be built up by brand reputation. Secondly, from the distribution power, distribution network should take brand reputation as its center. Improved brand reputation will promote the increase of sale and profits meanwhile the expanded market will improve the brand reputation. Thirdly, from the view of capital power, an enterprise with low reputation could not have a large producing scale and thus is not able to withstand various market risks.

3.2.2   Actualities f Tobacco Supply Chain Competitive power in China

Jin presents a systematic measuring and analysis metrics for enterprise competitive power [8]. Large-scale tobacco economy has not been formed in China and weak, scattering, disordering and small are typical characters of not only cigarette and tobacco leaf industry and cigarette distribution system.

Scattering in products means cigarette manufacturers distributes all over China. Every zone in China, whether it has basis and advantages for cigarette industry developing or not, has its local cigarette factory. The cigarette market is not effectively centralized and results in great waste of manufacturing resources. Measured by the world vulgate 4-manufacturer index for tobacco industry, i.e. the sum market share of the 4 biggest cigarette manufacturers in a country, the index in China is 16% while it is as high as 96% in USA. In China, there are only 14 and 30 manufacturers yearly produce more than 500 thousand boxes and 300 thousand boxes respectively. The average yearly output of the 69 main manufacturers in China is only 100 thousand boxes and many small factories only produce 20 or 30 thousand boxes each every year. China's biggest manufacturer, Yuxi Tobacco Factory, produces about 2 million boxes each year. Meanwhile the world's biggest manufacturer, Philip Morris, USA, yearly produces 18.89 million boxes.

Disordering in selling means the cigarette market is not in a good order. Various types of local protection and blockage not only increase the market developing cost of large manufacturers but restrain their developing opportunities as well. The market share of the bigger manufacturers are becoming less and less. On the other hand, fake cigarettes, smuggled cigarettes, cigarettes produced by illegal factories or exceeding quantity quota has existed for a long time and been an important reason for the disordering of China's cigarette market.

Bad capital power means the profit of China's tobacco industry is very low. In 1995 the taxation in USA on its 15-million-box cigarette output was as high as 20 billion US dollars. At the same time, China taxed only 71 billion RMB Yuan (8.55 billion US dollars) on its nearly 35-million-box cigarette quantity. Still in 1995, only 38 of 178 (40%) China's cigarette manufacturers could gain profit more than 1000 RMB Yuan per box (average profit per box in China is 1464 RMB Yuan) and nearly 50% of cigarette producing enterprises are in deficits.

Small in brand means not only scale of China's cigarette manufacturers but also output of single brand is less than that of their foreign rivers. In 2000, the output of Hongmei, the brand with the biggest quantity in China, was 1.005 million boxes and only 1/8 of that of Malboro which is belong to Philip Morris.

4.   RECONSTRUCTION PATTERNS FOR TOBACCO SUPPLY CHAINS

4.1   Business Process Integration in Supply Chains

Process is a series of organized, measurable activities designed for specific quantity of special customers or markets. Business process in a supply chain could cross through the edges of chain members and break the independence of organizations. Supply chain notes are connected by business processes such as customer relationship management, customer service management, customer order executing process, structural network design, inventory management, transportation management, return management, etc[9]. Upward, the kernel manufacturer needs to integrated business processes of tobacco leaf suppliers and take the "manufacturer + tobacco planters" pattern. Downward, the kernel manufacturer should integrate business processes of wholesalers and retailers to form a distribution strengthen network and hence develop a long-term, strategic union with its customers. Internally, the manufacturer should build up an integrated business process suitable for the decentralized manufacturing network.

4.2   Building a Distribution Strengthen Network

There are a lot of methods to be used to improve the distribution power. The most pivotal one is building up a sale network and distribution system in order to better utilize the local network resources of the distributors. Via this network, the kernel enterprise could promote the communications among enterprise, market and customers. When the sale driven is replaced with the marketing driven and the product promotion replaced with relationship building, the distribution power will automatically be improved. A practical way is to skip wholesaler notes in the supply chain and set up a dynamic union with retailers and hence strengthen the chain from the lowest edge backward to the higher layer.

Nowadays, catenation is the tendency of retail developing. Correspondingly, 3 kinds of retailers, i.e. large chained super markets, chained shops and the distribution network notes of wholesalers (local tobacco selling companies), are the main places for the kernel enterprise to strengthen its distribution network.

Among the 3 kinds of retailers mentioned above, large super markets are at a controlling position because cigarettes are not their main profit resources but only sold as a kind of for the customers' convenience. It is great different for chained shops where cigarettes contribute 37% and 27% to the revenue and profit respectively. Therefore, chained shops should be the preferred partners of the cigarette manufacturer. Because the industrial taxation influences the local financial revenue and the relationship between the local tobacco selling company and the local monopoly bureau is very close, the distribution network of the former will prefer to sell local produced products or alien products with much more profits. For instance, Shanghai Tobacco Group has built a nation wide distribution strengthen network by its 'Zhonghua' brand product which will bring the retailers a profit rate higher than 50%. 'Zhonghua' is now almost available at any cigarette retailer all over the China mainland. Meanwhile, having utilized its location advantage, Shanghai Tobacco Group realizes an integrated producing and selling process and has built up a self-owned, formal distribution system in Shanghai and is expanding the system to the rich delta district of Yangtze River.

4.3   Building a Distributed Manufacturing Network

Decentralized manufacturing network is an advanced manufacturing pattern to meet the rapid changed market needs through modification of the business and organization structures of an enterprise [10]. In this network, the manpower, materials, information and process during the product manufacturing could be entirely integrated by using technologies of information, network and computer. The product is quickly transferred from design step to producing step in a reasonable cost so that the ever-varying needs of market and customers could be met[11]. In actual, assisted by information and network technology, a dynamic union entity involving multiple members distributed geographically will be built promptly according to the market needs. This dynamic entity has no walls and is not constrained by space but could be directed as a whole. In a words, decentralized manufacturing network mainly solves two problems, how to respond to the market needs and how to take full advantage of the existing resources, and thus ensures the successive development of the kernel enterprise.

Cigarette manufacturing enterprises could realize producing in remote places through coalition, annexing, uniting, etc. This will break local protection and close the broken distribution network. At present, Hongta Group and two cigarette factories within Yunnan Province, Chuxiong Cigarette Factory and Dali Cigarette Factory, have merged and some joint ventured factories outside Yunnan Province have been founded, e.g. Hongta Changchun Cigarette Factory, Hongta Hainan Cigarette Factory and Hongta Liaoning Tobacco Ltd.

5.   SUMMARY

Cigarettes being regarded as a special kind of product and managed in a peculiar system, its supply chain shows a lot of particularities. The chain is broken because of the benefit collision among the chain members and its competitive power is reduced in essential. In the current monopoly system, the reconstruction of the supply chain conducted by the kernel cigarette manufacturer could amend the broken chain to some extent. Furthermore, the forward expansion of the supply chain and the decentralized manufacturing network is the basis on which the kernel manufacturer could enrich its market share and market centralization degree.

REFERENCES

1.   Kagan, R.A., D. Vogel. "The Politics of Smoking Regulation: Canada, France". the United States. In: Rabin, R.L., Sugarman, S.D (Eds.), *Smoking Policy: law politics and culture*. New York, Oxford University Press,1993.
2.   Sato, Hajime. "Policy and politics of smoking control in Japan". Social Science & Medicine, **vol.49, no.9**, pp.581-600,1999.Wu E, Smith FG. *Title of the Book*, City, Publisher, 1995.
3.   Western manufacturers on China's doorstep. World Tobacco, **Issue 197**,pp.2, 2003.
4.   He, M. K. "Re-understand logistics and supply chain", *China Logistics*, **vol.7, no.3**, pp.8-11, 2000.
5.   World Health Organization. Tobacco or health: a global status report,1997 [on-line]. Available : http://www.cde.gov/needphp/osh/who.whofirst.htm.
6.   Zhao, C. L., Z. L. Li. "Development and countermeasure of Chinese tobacco industry after access WTO", *Statistics and decision*, **vol.5, no.3**, pp.24-26,2002.
7.   Zhou, H. Z.. "Fiscal decentralization and the development of the tobacco industry in China". *China Economic Review* , **vol.11,no.3**, pp.114-133, 2000.
8.   Jin, B. "Theory and methodology of measuring enterprise competitiveness". *China industrial economy*, **vol. 10, no.3**,pp.5-13, 2003.
9.   Lambert, D. M., C. M. Cooper. "Issues in Supply Chain Management". *Industrial Marketing Management*, **vol.29,no.1**, pp.65-83,2000.
10. Montreuil, B, J. M. Frayret, D'Amours, Sophie. "A strategic framework for networked manufacturing". *Computers in Industry*, **vol. 42,no.6**, pp.299-317,2000.
11. Freund B., H. Konig, N. Roth. "Impact of information technologies on manufacturing". *International Journal of Technology Management*, **vol.13,no.3**, pp.215–228,1997.

THE OPTIMIZATION APPROACH OF ASSEMBLY SEQUENCE BASED ON GENETIC ALGORITHM

Chun-Shu Li, Li-Min Dai and Li-Min Wang

Hebei University of Technology, TianJin 300130, E-mail: lics99@eyou.com

Abstract: In this paper, we propose an optimal approach that generates and assesses assembly plans based on genetic algorithm. A randomly initialised population of (possibly non-feasible) assembly sequences may evolve near-optimal assembly plans by the matched crossover and mutation operation. The quality of a feasibly assembly sequence is evaluated based on the theory of set-pair analysis and fuzzy evaluation. And the individual fitness function of assembly plan is introduced. In the end, an example that endorses the soundness of our approach is included.

Key words: Optimization, assembly sequence, genetic algorithm, set-pair analysis

1. INTRODUCTION

Sequence of assembly of a set of parts plays a key role in determining important characteristics of the tasks of assembly and of the finished assembly. In general, there are many different assembly plans for a given product. And the cost and quality of the product can be different through the changing assembly sequences. The choice of assembly sequences to follow in the assembly process is based on a criterion for assessing the quality of each solution. At present, many researches on the assembly sequence planning have been done, and most of them are mainly concentrated on the method of generating assembly sequence, the feasibility and modeling of assembly relationship [1-3]. Chau [4] proposed an assembly system as a tuple of sets. Sequence planning is generated as graph structures, and the generation algorithm of assembly sequences can be implemented as either a forward or backward searching algorithm. The proposed approach will aid the producing engineer to produce the "optimal" assembly plan. Lee and Yi[5] presented an approach to assemblability evaluation based on tolerance propagation, and introduced a concept of clearance, which used to compensate tolerances in the assemblability evaluation. Fujimoto and Ahmed [6] proposed a method for automation of assembly planning. In this method, information entropy of the sequence's factors is considered in the sequence evaluation as follows: directionality, reorientation, tolerance and clearance, support of 3rd part, tooling and flexibility. In this paper, a genetic algorithm for generating optimal assembly plans is introduced based on the theory of set-pair analysis and fuzzy evaluation.

2.   GENETIC ALGORITHM (GAS)

The genetic algorithm is an auto-adaptation probability-searching algorithm, which mimics living beings heredity and evolution and uses population search method to find the optimum [7]. By means of selection, crossover, and mutation, GAs is able to "evolve" optimal solutions for assembly sequence planning. Because of its robustness, GAs shows more unique and better character than other traditional methods used in planning and evaluating. In all essential factors of GAs, the chromosome coding and individual's fitness evaluation are two key factors, which determine whether GAs will scout for the optimal solution or not. Each potential solution to a given problem is coded as a binary or real string that is called chromosome. In this paper, symbol-coding method is adopted to structure the chromosomes. This method accords with a meaningful building black principle. Thus knowledge in solving problems can be used easily in the GAs, and the GAs with other similar algorithms can be utilized favorably.

Generally, GAs starts with a randomly generated initial population of chromosomes. At each step a new population is generated from the current one using two basic operators: crossover and mutation. While crossover combines parent chromosomes to generate offspring chromosomes according to the probability of crossover (P_c), mutation is a local modification of a chromosome according to the probability of mutation (P_m). Chromosomes are selected for crossover and/or mutation based on their fitness value. In this paper, the edge recombination crossover that attaches to orientation restraint is utilized, as well as the insertion mutation and proportional model in GAs of assembly sequences optimization. The GAs flow chart of assembly sequence optimization is shown in Fig.1.

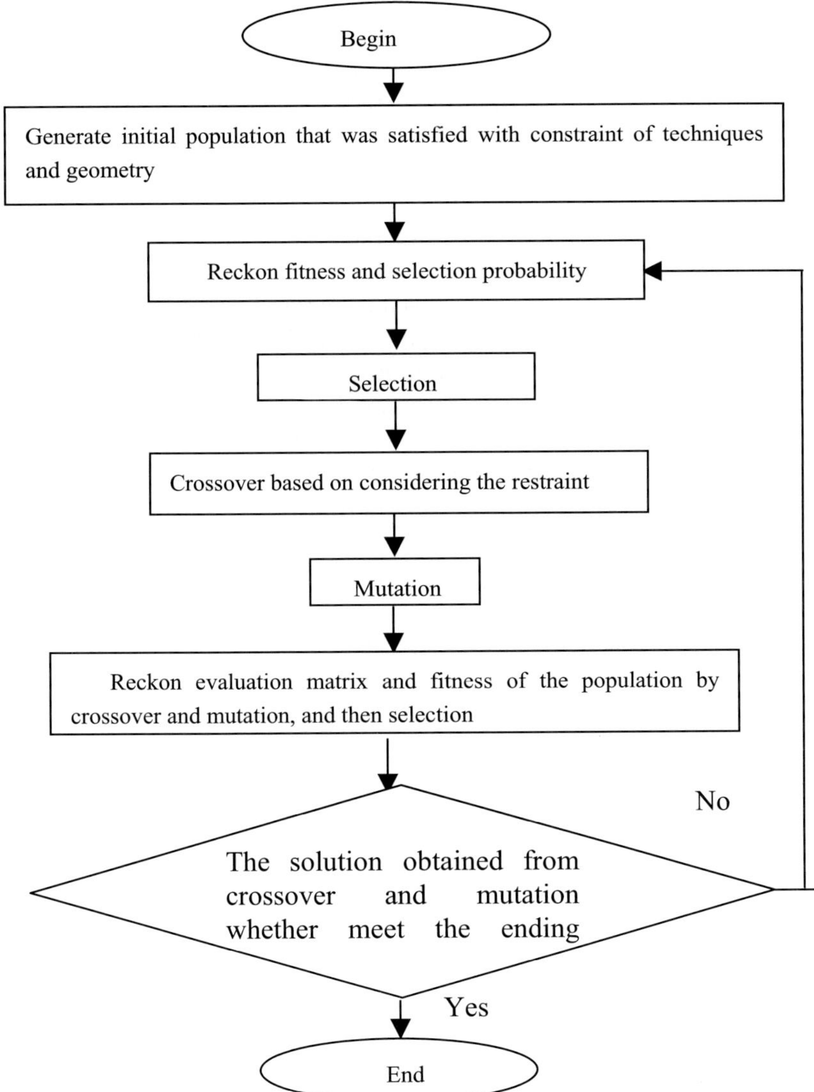

Fig.1   The GAs flow chart of assembly sequence optimization

3.　SET-PAIR ANALYSIS

Individual's fitness evaluations have direct effect on astringency of GAs. As evaluation and decision of assembly sequence is a combinatorial optimization process, therefore a new theory of set-pair analysis has been tried to using in the systematic evaluation of individual's fitness. Set-pair analysis (SPA)[8] is a systematic analysis approach concerning of quantitative analysis of identity-discrepancy-contrary (IDC) of certainty-uncertainty system. The main idea of SPA is to analyze and process objective things' affiliation of certainty and uncertainty as a system of certainty-uncertainty. The affiliation and transformation are analyzed by studying the objective things' situations of identity, discrepancy, and contrary in SPA theory.

Definition 1: Set-Pair is a pair that consists of two sets that have relations with each other. For a set-pair $H=(A, B)$. By analyzing its property under the background of one concrete problem (supposed W), N properties can be obtained, and from which there are S properties possessed both set A and B, and P properties in which A is contrary to B, and F properties in which A is not same as B, nor contrary (where $F=N-S-P$). Based on this analysis we give several definitions as follows:

$$a = \frac{S}{N} \text{ Denote the two sets' identical degree based on } W$$

$$b = \frac{F}{N} \text{ Denote the two sets' discrepancy degree based on } W$$

$$c = \frac{P}{N} \text{ Denote the two sets' contrary degree based on } W$$

Then

$$\mu(W) = \frac{S}{N} + \frac{F}{N}i + \frac{P}{N}j \tag{1}$$

The sign μ denotes the affiliation degree of the two sets, or namely

$$\mu = a + bi + cj \tag{2}$$

Definition 2: set-pair state

In the formula (2), if $c \neq 0$, the ratio of a/c is called set-pair state (*SP* state, $S=a/c$). This state is under the background of an appointed problem, and twelve systematic set-pair state ranking of 3-dimensions can be obtained based on the variety of the value of a, b, c and set-pair state according to reference [8].

4.   THE FITNESS FUNCTION BASED ON SPA

In the paper, the assembly sequence is studied as a certainty-uncertainty system. The set-pair consist of perfect cost level and actual cost level of the assembly sequence being researched. The construction of the affiliation degree μ as follows.

The identical degree a denotes the probability that cost of assembly sequence being studied is consistent with the perfect cost. The contrary degree c denotes the probability that the cost of assembly sequence being studied consistent with passed cost, the discrepancy degree b denotes the probability that the cost of assembly sequence being studied consistent with good cost. Because a and c is a value which can be determined, we give a definition of relative approach degree $\gamma_t = \dfrac{a}{a+c}$ [9].

Because the SP state S and the relative approach degree γ_t are two determined value that represent the certainty degree of the system (in this paper, the certainty degree of the system means the consistent degree that the cost level of the sequence being studied is close to the perfect cost level), the fitness function is constructed by the two value with different weight.

$$f = t_1\gamma_t + t_2 S \qquad\qquad (3)$$

Where, $\gamma_t = a/(a+c)$ denote relative approach degree.

$S = a/c$ denote systematic set-pair state.

f denotes fitness function.

t_1, t_2 denote the weight of the systematic set-pair state and relative approach degree.

a denote identical degree.

c denotes contrary degree.

5.   EXAMPLE

The product being considered is a gear-pump shown in Fig.2. Here, 1- shell-body, 2- peg, 3- back-lid, 4- shaft installing sleeve, 5- driven gear, 6- sealing ring, 7- washer, 8- bolt, 9- front-lid, 10- washer, 11- bolt, 12- sealing ring, 13-shaft installing sleeve, 14-sealing ring, 15- ring gasket, 16- driving gear, 17- peg.

A subassembly $SA1$ is made up of 6 and 3, and another subassembly $SA2$ is made up of 9, 14, 15, and 12.In the Fig.2, 8 and 7 denote the eight bolts and washers that close the body of pump, respectively. 2 and 17 denote the two pegs that connect the lid and the body of pump, respectively. By these analyses, we obtain a product that is made up of 13 components.

The GAs program of optimizing assembly sequences is programmed by MATLAB. Let P_c=0.25,and P_m=0.06. Many constraints are considered in the assembly sequences optimizing, such as the geometrical and technological criteria.

We conducted 20 trials using different, randomly initialised initial populations of 40 chromosomes. The parameters are used in the GAs is P_c=0.25,and P_m=0.06. The GAs has always converged to a feasible sequence. In the trials, the following optimal solution (corresponding to the best feasible sequence) was generated.

13   *SA*2   10   11   17   4   *SA*1   2   7   8

The solution is consistent with the practical sequence, which obtained from long-term experience. And this proves that the GAs possess strong ability of overall scout for optimal solution in the assembly sequences planning. The information obtained by this approach expresses the systematic information comprehensively. The encouraging preliminary results obtained so far confirm the validity of this approach.

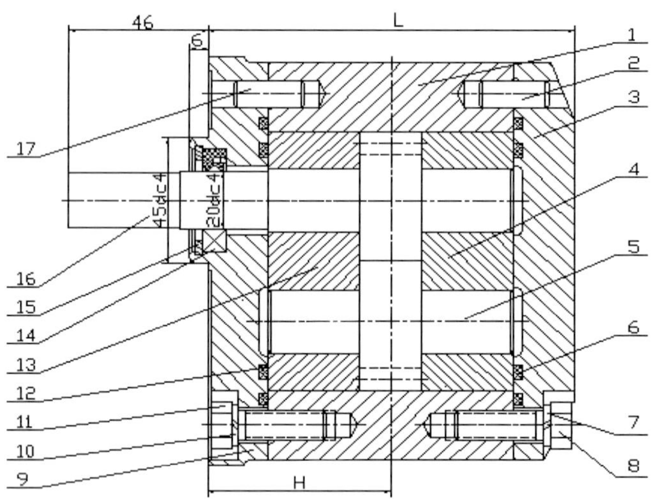

Fig. 2 the assembly graph of gear-pump

6. CONCLUSION

This paper presents an approach for the optimization of assembly sequence based on genetic algorithm. In all essential factors of GAs, the chromosome coding and individual's fitness evaluation are two key factors, which determine whether GAs will scout for the optimal solution or not. In this paper, symbol-coding method is adopted to structure the chromosomes, and individual's fitness function is proposed by the set-pair analysis. Based on the set-pair analysis, an objective and comprehensive solution of assembly sequence planning can be achieved. The validity of this methodology has been proved by an example.

REFERENCES

1. Thomas L. De Fazio and Daniel E. Witney Simplified Generation of All Mechanical Assembly Sequences. *IEEE Journal Of Robotics And Automation*, **Vol RA-3, No.6,** December, 1987,640~658.
2. Luiz S. Homem de Mello, Member IEEE, and Arthur C. Sanderson, Fellow, IEEE.Two Criteria for the Selection of Assembly Plans: Maximizing the Flexibility of Sequencing the Assembly Tasks and Minimizing the Assembly Time Through Parallel Execution of Assembly Tasks. IEEE Transactions On Robotics And Automation, **Vol.7, No. 5**, October 1991.
3. Boothroyd G. Alting L. "Design for Assembly and Disassembly". *Annals of the CIRP.* 1992, **41(2)**. 625~636.
4. H.L Chau, E.J. D errick, H.C Shen, and R.K. Wong "A New Approach for the Specification of Assembly Systems" 0-8186-6995-0/95©1995 IEEE 9~14.
5. Sukhan Lee, Chunsik Yi. "Assemblability Evaluation Based on Tolerance Propagation". *IEEE International Conference on Robotics and Automation* 0-7803-1965-6/95©1995 IEEE. 1593~1598.
6. Hideo Fujimoto and Alauddin Ahmed. "Entropic Evaluation of Assemblability in Concurrent Approach to Assembly Planning". *Proceedings of the 4th IEEE International Symposium on Assembly and Task Planning Soft Research Park*, Fukuoka, Japan. May 28-29, 2001,306~311.
7. Zbigniew Michalewicz. Genetic *Algorithms+Data Structures=Evolution Programs.* Springer-Verlag Berlin Heidelberg 1996.
8. Zhao Ke-qing *Set-pair Analysis and Preliminary Applications.* [M] Zhejiang Science & Technology Publisher. Hangzhou 2000
9. Zhang Bin. *The Fuzzy Set-Pair Analysis Way of Multi-Objective System Decision. Systems Engineering——Theory & Practice* 108~114, July,1997.

THE APPLICATION OF THE FUZZY EVALUATION IN LASER MICRO-MANUFACTURING SYSTEM

Hui-Lai Sun, Shu-Zhong Lin, Jin-Gang Wang, Hong-Lai Li and XM Liu
Mechanical Engineering College, Hebei University of Technology, Tianjin 300130, P.R.China, E-mail: jear@webmail.hebut.edu.cn

Abstract: The article introduces the fuzzy evaluation of optimum project in laser manufacturing system. The application of fuzzy theory in laser manufacturing system was presented. The laser manufacturing system was optical-mechatronics product. And on the system it can realize welding, cutting, marking and so on. By experimenting, several merits of the laser manufacturing system were obtained such as simple structure, easy control and high machining precision.

Key words: Fuzzy evaluation; Laser manufacturing; Optimum design; Laser

1. INTRODUCTION

With development of the science and technology, the design system is more and more complex. More and more variables are required to research and the relation of these variables is complex and fuzzy. The precision in evaluation and reasoning is required. We can get conclusion from facts that in the complex and variable system many factors intervolve each other and it makes the design difficulty [1]. The application of fuzzy theory in the laser manufacturing system makes the designer get one available measures.

The developing of laser technology adds a new field in science and technology. Laser manufacturing technology is new technology that has developed in the 60s-70s; it has many characters such as un-contacting, unpolluting, small hot field, high machining precision. These characters make it be regarded. The laser technology has been applied in mechanism, electronics, chemistry, aviation, medicine, metal material and so on. Laser technology is more economical, advantageous and convenient than other. In order to develop the laser micromachining, in the world many developed country have research the laser technology. In Europe, Japan and USA the government supports the research on laser micromachining and technology concerned. And in the world many enterprises have already researched in the field. In the fields of electronics, semiconductor, communication, chemistry and medicine the laser manufacturing has a stronger competition than conventional technology. In the paper the laser manufacturing system has several advantages, for instance, larger task workspace, simple structure, easy control and so on. The follow introduces the application of fuzzy evaluation in the laser manufacturing system.

2.   DESIGN OF GENERAL PROJECT

2.1   Task and Requirement of design

According to the conditions—the requirement for the laser processing technology, technology lever, the present condition that laser processing machine was used to singular technology processing here and overseas and technology of the same kind overseas. We designed a new type integrated laser processing system, which was used to several functions such as the welding, cutting and laser marking [2]. Concrete requirement were shown in the Table 1.

Table 1. Requirement of the laser manufacturing system

Title of the design topics		Laser processing system	
Requirement		value	importance
necessary conditions	① welding		I
	② cutting		I
	③ laser-marking		I
	④ precision	$\delta \leq 0.01$mm	I
	⑤ depth of marking	$h \leq 0.1$mm	I
lowest conditions	① noisy strength	$\leq$60dB	III
	② operation height	H$\leq$1500mm	IV
	③ manufacture expense	$\leq$200,000RMB	III
	④ overhaul period	$\geq$1,000h	IV

	expectation	① simple operation and maintenance expense ②low maintenance expense

2.2   Functional analysis

According to the laser processing technology, many functions should be included. A series of sub-functions were derived from the general function and solved them separately. And morphologic matrix of laser manufacturing system is formed. It is shown in Table 2.

Table 2. Morphologic matrix of the laser manufacturing system

Sub-function unit		Solution of the sub-function		
		1	2	3
A	Style of welding and cutting	Component fixed	Component move	
B	Welding and cutting system	Parallel system	Common system	Crossing system
C	Marking system	Fixed worktable	Moveable system	
D	Style of laser transmission	Optic line	Optic fiber	
E	Style of control	Control card	Gas dynamic	Liquid dynamic
F	Locating style	chuck	Appropriative tongs	Locating work position
G	Laser	YAG laser	CO_2 laser	Other laser
H	Controller	PC	Industry computer	Appropriative controller

2.3   Combination of all the sub-functions

Total function is decomposed into 8 function units. In order to obtain more design projects, firstly separating, all kinds of the possible solutions which are used to realize function units should be found, Using morphologic matrix to combine the projects, according to the table 2, number of projects formed is [3]:

$$N=2\times3\times2\times2\times3\times3\times3\times3=1944$$

Those projects could be converged to several focuses by removing the incompatible or inter-coordinate ones according to the requirement of design. And the feasible projects would be chosen on the basis of practical experience. Four valuable combination projects then formed on the basis of them.

Principle project I :A_1-B_2-C_1-D_3-E_1-F_1-G_2-H_2

Principle project II :A_2-B_3-C_2-D_2-E_1-F_2-G_1-H_2
Principle projectIII:A_1-B_1-C_1-D_2-E_1-F_1-G_2-H_1
Principle projectIV:A_2-B_3-C_2-D_1-E_1-F_2-G_1-H_1

3. FUZZY EVALUATION OF GENERAL PROJECT

In the designing system of the laser manufacturing system, there are many factors to be considered, which have the relations with each other. And many factors also have some strong fuzziness. Many main factors will be vanished by the minor factors so it will be difficult to compare good and bad sequence of the things of the system and we can't obtain meaningful valuable result if we use the stair fuzzy evaluation model. So the design of this system need the second grade synthetically evaluation in order to select the best project [2, 4].

3.1   Found the set of fuzzy evaluation

The factors are so much that affect the general project, and the affecting degree is incoordination. In theory all factors can be evaluated. The relation among factors is uncertain or fuzzy, so the optimum project needs synthetically evaluation. All the factors affecting project are called factor-set.

Evaluation-set is the result that all judger worked out to the sub-factor-set.

Weight degree is the importance of factor-set to factors in evaluation-set. It is represented by R_{ij}.

The factor of estimation for the mechanical project mainly is the stability. It mainly has four targets and has many influencing factors. The follow is the factors we considered. The first grade judged set and weight degree are gained by the influence degree to the stability by the each single influencing factor of the project. The sub-factor-set, factors, weight-set and weight degree of the mechanical project are shown in Table 3. The weight value was obtained through weighted average that expert kept score to every factor and then weighted averaging these scores.

Table 3. sub-factor-set, factors, weight-set and weight degree of project

sub-factor-set			factors		weight degree (R_{ij})			
title		weight-set(W_i)	title	weight-set (W_{ij})	P I	P II	P III	P IV
U1	functionality	0.30	Style of welding and cutting	0.55	0.15	0.30	0.25	0.30
			Style of control	0.45	0.25	0.20	0.35	0.20
U2	Load-ability	0.15	Welding and cutting system	0.60	0.30	0.10	0.30	0.30
			Marking system	0.40	0.25	0.30	0.15	0.30
U3	utility	0.25	Locating style	0.35	0.20	0.35	0.10	0.35
			Style of laser transmission	0.40	0.24	0.26	0.24	0.26
			controller	0.25	0.20	0.36	0.20	0.24
U4	economy	0.30	laser	1.00	0.18	0.32	0.18	0.32

3.2 Found the first grade judged matrix

In order to use same first grade judged matrix, the factors are stand or fall ranked. The first factor-set and the first factor weight degree can obtain the first fuzzy evaluation set [2]. And this application is basic mathematical fuzzy evaluation knowledge. The matrixes and the introductions were explained detailed in the reference 2. The follow is shown

$$R_1 = W_{1j}R_{1j} = \begin{bmatrix} 0.55 & 0.45 \end{bmatrix} \begin{bmatrix} 0.15 & 0.30 & 0.25 & 0.30 \\ 0.25 & 0.20 & 0.35 & 0.20 \end{bmatrix}$$

$$= \begin{bmatrix} 0.1950 & 0.2550 & 0.2950 & 0.2550 \end{bmatrix}$$

By the same methods, the following formula can be obtained,

$$R_2 = \begin{bmatrix} 0.2800 & 0.1800 & 0.2400 & 0.3000 \end{bmatrix}$$

$$R_3 = \begin{bmatrix} 0.2160 & 0.3165 & 0.1810 & 0.2865 \end{bmatrix}$$

$$R_4 = \begin{bmatrix} 0.1800 & 0.3200 & 0.1800 & 0.3200 \end{bmatrix}$$

3.3 Second grade fuzzy synthetically evaluation

Based on the fuzzy mathematical theory the second grade fuzzy evaluation matrix B_i can be obtained by the first grade judged matrix R_i and weight-set W_i, and the formula explained detailed in the reference 2.

$$B_i = [R_1, R_2, R_3, R_4]^T$$

$$B = W_i B_i = \begin{bmatrix} 0.30 & 0.15 & 0.25 & 0.30 \end{bmatrix}$$

$$\begin{bmatrix} 0.1950 & 0.2550 & 0.2950 & 0.2550 \\ 0.2800 & 0.1800 & 0.2400 & 0.3000 \\ 0.2160 & 0.3165 & 0.1810 & 0.2865 \\ 0.1800 & 0.3200 & 0.1800 & 0.3200 \end{bmatrix}$$

$$= \begin{bmatrix} 0.208500 & 0.278625 & 0.223750 & 0.289125 \end{bmatrix}$$

From the result we can draw a conclusion that the projectIV (0.289125) is the optimum project, because the projectIV value is bigger than others. Thus we ensure the project IV is the adoptive project and the project II is the preparative project. The laser-manufacturing system was obtained according to the evaluating-results. And the manufacturing system was shown in the figure 1.

Figure 1. The laser-manufacturing system according to the evaluating-results

4.   CONCLUSIONS

(1) The research and design were successful application of fuzzy theory in the laser manufacturing system. Scientific and precise evaluating method was the adequate guarantee for which the optimum project was reliable, novelty and advanced.

(2) The optimum project evaluation of laser processing system was a multi-level synthetically evaluation. Evaluating factors should be divided vary

arrangement according to the vary property of product judged, and multi-level synthetically evaluation were adopted.

(3) In the multi-level synthesis evaluating process not only the effect influence of all factors but also the hierarchies and primary-secondary relationships of related factors were considered.

(4) Based on optimum project of evaluating result the laser micro-manufacturing system was obtained. And the system was very satisfied with the design requirement.

ACKNOWLEDGEMENTS

Thanks for the '211' projects of Hebei University of Technology.

REFERENCES

1. Kaufmann A: "Introduce to the Fuzzy Subset". Academic Press, New York 1985
2. HUANG Hong-zhong: *Fuzzy design*, Publishing House of Mechanical Industry 1999
3. LIN Shu-Zhong, SUN Li-Xin, SUN Hui-Lai, et al: "Creative design of an automatic quality evaluating machine for Alkaline Zn/MnO_2 battery". Battery Bimonthly, **Vol.31**, pp. 305-306, November 2001.
4. HUANG JING-YUAN: "Mathematical base". *Optimum Design*, Publishing House of Mechanical Industry 1999.

STUDY ON CONCURRENT DEIGN IN FORMING DIE FOR ITS LIFE

Feng Ni and Jianping Lin
Stamping Center of Shanghai Volkswagen Automotive Co. Ltd, 201805, NiFeng@csvw.com
College of Mechanical Engineering, Tongji University, 200092, jplin1958@sina.com

Abstract: As a research hot point in advanced manufacture techniques, concurrent engineering (CE) has been applied in aeronautics, automobile, mechanical engineering. Forming die is an important technical equipment in auto-body manufacturing. Now, design life of forming die is aimed to make whole die's using life to achieve or exceed design standard. The result is that whole die has to be scraped due to few parts' disabled, but most parts could be used for a long time. It will cause a large amount of waste. In order to solve the problem, concurrent design in forming die for its life is researched in the paper. The general mathematical expression is put forward to define the relation of whole die and parts based on Lebesgue measure and Lebesgue integral in real variable function. Regarding the die used in auto-body manufacturing as a research object, the idea of concurrent design for die's life is discussed in auto-die field for the first time. With the analysis of the design process and information flow of auto-die, the system framework of auto-die concurrent design for its life is built up. As a new method in auto-die design, it will promote the technical development of auto-die and decrease the cost. At the same time, it can be employed in research of mechanical concurrent CAD for life.

Key words: Concurrent design for its life, Lebesgue integral, forming die.

1.   INTRODUCTION

Concurrent Engineering is a systematic method for concurrent design of product and its supporting process [1,2], which is an important research direction in advanced manufacturing engineering and gets fast developed in aeronautics, automobile and mechanical engineering. Die is an important technical equipment, and the auto-body forming dies are the highest technology equipments in the die and mould field. Now, design life of forming die is aimed to make whole die's using life to achieve or exceed design standard. The result is that whole die has to be discarded due to few parts' disabled, but most parts could be used for a long time. It will cause a large amount of waste. For the auto-body die, the price is about several millions or more, so the waste is expensive.

Concurrent Engineering is a systematic method for concurrent design of product and related processes. This method tries to make the engineers consider, in the beginning, all of the factors from concept formation to discard in the lifecycle of product, including quality, cost, schedule and customer's requirement [3]. Concurrent deign in forming die for its life tries to consider the factors of the die life and coordination of each die parts life in one die. According to that design, the life of each part in one die is same or less difference. So in that condition, the die will meets the service life at the lowest cost.

On the research of concurrent engineering in die field, the related papers are issued in 1994 in EI index, which focused on the injection mould [4]. In 1995, research on the concurrent design on CAD of progressive die was developed in Shanghai Jiaotong University, which solved the problem that during the design of traditional progressive die, because of adopting the serial design method, a lot of problems were discovered in the final design or even in the manufacture phase, which caused the design modify, influencing the development cycle and reliability of design [5]. Through EI Index, the research on concurrent deign of auto-body die for its life has no report.

2. GENERAL MATHEMATIC EXPRESSION FOR CONCURRENT DESIGN OF DIE FOR ITS LIFE

Lebesgue measure and Lebesgue integration are important mathematic topics in real variable function. Lebesgue measure is the mensuration of general bounded collection, Lebesgue integration is the sum of Lebesgue measurable function based on Lebesgue measure. Lin Jianping (one of the author of this paper) established the mathematic model of progressive die for its life by applying Lebesgue measure and Lebesgue integration of real variable function [7]. In this paper, the authors have advanced the general mathematic expression on concurrent design of die for its life.

Suppose R is the collection of die parts related to die life, $L(s)$ is a subset life function of die parts on R. If Lebesgue measure $E_s \in R$, then Eq.(1) is the level collection that meets $L_a \leq L(s) \leq L_b$.

$$E_s(L_a \leq L(s) \leq L_b) = \{s \mid (L_a \leq L(s) \leq L_b), \quad s \in R\} \tag{1}$$

Where E_s is Lebesgue measure of die parts subset that meets $L_a \leq L(s) \leq L_b$.

Because E_s is defined on R, $E_s(L_a \leq L(s) \leq L_b)$ is the collection for all of die parts subsets meeting the life of the die. While the structure of a die is the collection of partial die parts and its substructure, and constraint characteristics collection G is another factor infect the die life, which include parts materials, elements quality requirement and guide-set structure request, etc. Therefore, the general mathematic expression of a die on concurrent design for its life that meets G can be described as Eq.(2).

$$E_s = \bigcup_{i=1}^{n} E_i \tag{2}$$

Where E_s is the level collection of n kinds of die parts and its substructures on $[L_a, \ L_b]$, E_i is the measurable subset of the I-kind die parts and its substructures.

Suppose c_i is one of the constraints characteristic on the die, $c_i \in G$ ($0 < i < m$) and $C = \bigcup_{i=1}^{m} c_i$ ($c_i \in G$), then the relationship between the die structure and its parts and its substructure can be described as Eq.(3).

$$\begin{aligned}
P_{die} = \{s \mid \text{take } s_1 \in E_1, & \quad \text{Subject to C};\\
\text{take } s_2 \in E_2, & \quad \text{Subject to C};\\
\text{take } s_3 \in E_3, & \quad \text{Subject to C};\\
\cdots & \quad \cdots \\
\text{take } s_n \in E_n, & \quad \text{Subject to C}\}
\end{aligned} \tag{3}$$

In fact, the general mathematic expression described the relationship between the die and die parts is the conception and abstracting on the concurrent design method of the die for its life. It is also the specific expression of the concurrent design concept for forming die, which give direction and theory for concurrent design of the forming die for its life. Furthermore, the mathematic expression given in this paper is also a new thought to knowledge-oriented die CAD technical development. For auto-body forming die design, it can be thought to be a new design idea and consideration in the forming die designing. By applying this method, the life of forming die parts can get better coordination, which can reduce the maintenance load and cost in the subsequent using and increase the efficiency of auto stamping parts manufacturing.

3.   CONCURRENT DESIGN FRAMEWORK OF AUTO-BODY DIE FOR ITS LIFE

Auto-die is the important technical equipment in auto industry and is the foundation of manufacturing, which belongs to the high value-added products with dense technique. Auto-die technical is one of the technology integration from mechanic, material, electron and auto-control. Die for auto-body contain the highest valuc and technology in the die and mould industry, and auto-die which is the typical, advanced high-technical equipment in manufacture engineering is the most precise and complicated large-scale die.

The auto-die composed by lower die, upper die and blank holder, which include punch, matrix, guide set, gas-kicker set, strengthening rib plate, lifting set and position stop. Concurrent design for die life mainly focuses on the coordinating the life of dies parts and quality of forming surface of punch and matrix. The program of the auto-die design includes the design and calculation of stamping process, the draft of base structure, the design for die structure, drawing check and approve, detail design for die parts, CNC programming etc, which is an integration of design procedure.

Integration design is a complicated procedure, which related to every phase of product design. These phases work independently and cooperate with each other, and much information must be exchanged and shared. The key technology to realize integration design is how to organize varied data and information convenience to the exchange and share in the integration design circumstance. There are many kinds of data and information in the integration design circumstance, such as structured data and non-structured data, relational information and non-relational information. So how to realize the integration of different sort of database and keep them the independence is the key problem to integrate, exchange and share of information among application systems.

According to the main procedure of auto-die design, information flow and the function module, the concurrent design framework for auto-die for its life is advanced in this paper, shown in figure1.

System framework is the skeleton and technique supporting of integration system, which is the basement to realize concurrent design. It is not only make computer aided design, manufacture and quality assurance integrated for sharing information, but also considered the man and computer integrated to realize concurrent CAD with the man as the core. Therefore, the system framework should have a friendly interface, convenient to communication and ability complement between man and machine. In figure1, the system framework includes CAD/CAM software platform, finite element analysis software, knowledge base, life cost evaluation for die parts, design function module of auto-die etc. The partial overlapping between design function modules in the framework indicated concurrent design that the later design have been done before the finish of pre-design. It is also indicated that the design process is synergistic integration working, which is the principal of concurrent CAD for auto-die.

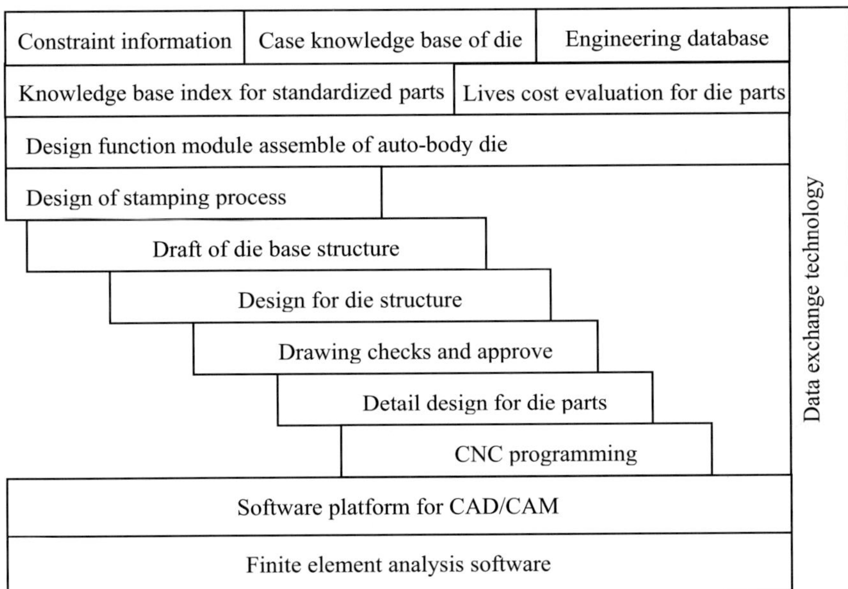

Figure1. Concurrent design framework of auto-body forming die for its life

4.   CONCLUSIONS

Whole life cycle-oriented concurrent engineering is the front edge of advanced manufacture and the important development direction. Concurrent design of auto-die for its life is studied in this paper. The general mathematical expression of the die that considers its parts and structure life are given based on the Lebesgue measure and Lebesgue integration in real variable function. The concurrent design thought based on life coordination is put forward firstly in auto-die field. Based on the analysis of forming die design and information integration, concurrent design framework for product life-oriented auto-die is advanced, which given a new ideal to auto-die design. As a new method in auto-die design, it will promote the technical development of auto-die and decrease the cost. It is significant to the development of auto-die technique and research of mechanical concurrent CAD for its life.

REFERENCES

1. Ahmad K. E., Ganesh M. K., Mansooreh M. "Concurrent engineering deployment: A virtual reality approach". *Integrated Manufacturing Systems*, **vol.4 no.4**, pp.24-28, 1993.
2. Bhamburkar, Suneel S. "Concurrent engineering in manufacture of press tools", *Proceedings of the ASQC 48th Annual Quality Congress*, Las Vegas, USA, 1994, pp.400-406.
3. Pennell J. P., Winner R. I. et al. "Concurrent engineering: An overview for automation", *The system readiness technology conference*, 1989.
4. Lee R.S, Chen YM, Cheng HY and Kuo MD. "A framework of a concurrent process planning system for mould manufacturing", *Computer Integrated Manufacturing systems*, **vol. 11 no.3**, pp.171-190, 1998.
5. Lin Jianping. "Research and Realize on CAD Technologies of Progressive Die Based on Concurrent Engineering", *Shanghai Jiaotong University PH.D Dissertation*, 1998.
6. Qian Peilin *Real Variable Function*, Beijing, Beijing Normal University Press, 1990.
7. Lin J.P., Wang H.X., Ruan X.Y. "Mathematics Model of Concurrent Design for Life of Progressive Die", *Chinese Journal of Mechanical Engineering*, **No. 9**, pp.56-61, 2000.

STUDY ON THE ARCHITECTURE OF RECONFIGURABLE ASSEMBLY SYSTEM[2]

Fanli Meng[1], Xuemei Huang[1,2] , Dalong Tan[1] and Yuechao Wang[1]

1. Shenyang Institute of Automation, Chinese Academy of Sciences, 110015
2. College of Mechanical engineering and Automation, Northeastern University, 110006

Abstract: Reconfigurable Assembly System (RAS) is designed at the outset for rapid change in structure, in order to quickly respond to the needs of market in terms of product variety and quantity. This paper is focus on the architecture of RAS. We propose a hybrid architecture and negotiation protocol for the dynamic scheduling of assembling systems. The architecture is based on multi-agent systems. This architecture combines the advantage of both hierarchical structure and distributed structure. A global management coordinates the whole system and every agent is highly autonomous. Otherwise, we introduce the agent specification. The well-known Contract Net Protocol has been adapted as the coordination model. The purpose of this protocol is to dynamically assign operation to the resources of the assembly system to realize system reconfiguration.

Key words: RAS, architecture, agent specifications, Contract Net Protocol.

1.   INTRODUCTION

In recent years, with the development of the information technology and the tendency to global economy, the manufacturing enterprises have to face the increasing and unforeseen change of market. The demands of the customers have been diversifying; the lifecycle of product has been

[2] Sponsored by National Key Basic Research Plan （973 Plan）(code:2002CB312200)

becoming shorter increasingly; and the manufacturing environments have been getting no more static but dynamic. For those reasons, the manufacturers must adjust their strategy and structure dynamically to respond for the competition and the change of market. Applying distributed autonomous systems to manufacturing has been becoming very important. In the contemporary manufacturing system, the role of the assembly is very important, because the assembly takes up 20%-70% of the whole manufacturing workload and 40%-60% [1] of the whole manufacturing time. Accordingly, to enhance the assembly is a key approach to improve the productivity and adaptability of the manufacturing enterprises.

In recent research, reconfigurable manufacturing system has been widely studied. S.Wayne, and L.Prater, University of Michigan designed a reconfigurable assembly machine for heat exchangers, and they restricted themselves to a modular design that can be readily integrated to the existing system and further revisions are easier to make [2] . Venketesh N Dubey, Bournemouth University and Richard M Crowder, University of Southampton presented a design for a reconfigurable packaging system that could handle cartons of different shape and sizes. The packaging system was based on the principle of reconfigurability, so it was adaptable to all kinds of complex carton geometry[3]. These researches are about reconfigurability of assembly device, and there are also many other ways of studying reconfigurable system. Jin-Lung Chirn and Duncan C.McFarlane, University of Cambridge presented a holonic component-based approach to reconfigurable manufacturing control architecture (HCBA). They developed a reconfigurable manufacturing control architecture, and their study was mainly derived from component-based development (from the software perspective) and holonic manufacturing systems (from the manufacturing perspective)[4]. In this paper we propose a Reconfigurable Assembly System (RAS) based on the concept of multi-agent, and we mainly deal with three problems. In the first part, we discuss the architecture of the system. In the second part, the approach of grouping the devices is given. In the third part, we introduce agent structure and communication specification, then the Contract Net Protocol has been adapted to allocate the proposed

2.   MULTI-AGENT BASED MANUFACTURING SYSTEM

Commonly there are two kinds of multi-agent system. One is the pure distributed architecture. In this kind of architecture, all the agents in the system share the information and knowledge together and they are highly autonomous. This architecture is only applicable for a simple system, because when the system is more complex and the number of agents becomes larger, the commutative information will be excessive. The other is the federated architecture that introduces mediator-based cooperation mechanism. This architecture divides the agents into many different groups according to similarity and by this way the complexity of communication between agents is simplified. However, without global management, it is difficult to attain a global optimal goal. In our research, we propose the hybrid architecture that can ensure the system's flexibility and simplify the system as well. This architecture combines the advantage of both hierarchical structure and distributed structure. A global management coordinates the whole system and every agent is highly autonomous.

2.1　Multi-agent architecture

There are three different kinds of agent in the system as shown in Figure1. Device agent is located at the lower level and represents the assembly device. It controls devices. Cell agent is the upper one to manage the device agent in its own cell and communicates with other cell agent.

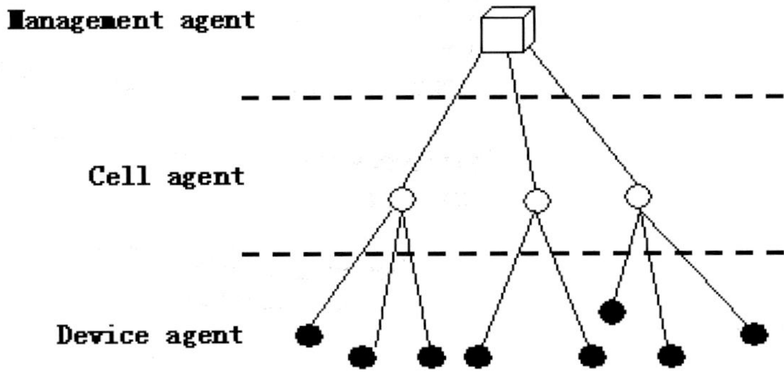

Figure 1 multi-agent architecture

Management agent is the uppermost one which manages whole assembly processes. An upper agent asks lower agent to execute an order and lower agent reply by bids. These three layers form a kind of hierarchy. Every cell agent is an autonomous individual, and it can plan its own process and negotiate with each other to attain a goal without management agent's intervening. We will show the details of the three kinds of agent.

2.2　Management agent

Management agent has mainly three functions:

- Receiving orders: the management agent receives an order from customers. The order is representation of the product that is inputted via computers. Then database of the product is formed.
- Managing the assembly process of a product: the management agent plans the assembly processes according to a product database, that is, describing how to assemble a product with parts. Furthermore, the management agent monitors the accomplishment of tasks.
- Auctioning tasks: when the assembly processes have been formed, the assembly tasks are definite. The management agent will declare these tasks as public bidding.

2.3   Cell agent

Cell agent has five functions:
- Managing the devices: the cell agent has a database about all kinds of information of devices in its own cell, including type of devices, size and interface of reconfigurable modules, capability, current state etc..
- Bidding for tasks from management agent: when the management declares public bidding about assembly tasks, cell agent will bid for tasks according to its capability. Some different tasks can be undertaken by one cell agent By reason of that cell agent represents a group of devices.
- Assigning devices to assembly tasks: according to the undertaken tasks of assembling, the cell agent informs the relevant device agent and makes contracts with them.
- Monitoring accomplishment of tasks: the cell agent is to monitor the status of tasks accomplished by device agent. It must detect whether the device agent finishes its work on time.
- Communicating and coordinating between cell agents: the cells in the reconfigurable assembly line are grouped by logic, but actually disperse in physical location. In some cases, parts reciprocate between two cells in order to assemble a product. So effective communication must be ensured. Furthermore, when a device breaks down, the cell has to resort to the substitute in the other cell.

2.4   Device agent

Device agents exit statically in order to control existing assembly devices such as manipulators, AGVs, parts buffer etc. Note that one device is subject to its own device agent. The device agent has four functions as follows:
- Local planning: a device agent only receives an assembly procedure, but there is no necessary datum to drive the relevant device. So the

    device agent must be able to carry out local planning to decide the driving datum.

- Detecting status: the device agent constantly detects its own status and reports to the cell agent. Detection includes fault detection and accomplishment detection. Fault detection uses the time-out mechanism.
- Communicating with other agents: the device agent will communicate with the cell agent and AGVs. When a assembly procedure is finished, an AGV will be required, the cell agent must communicate with AGVs efficiently. To communicate with the cell agent is to report its status. Note that device agent can't communicate with other device agent directly except AGVs. The cell agent is the medium between device agents Communication.
- Control the physical device: the device agent controls motions of the device according to the result of local planning.

3.   GROUPING ALGORITHM

In this part, we propose an algorithm that is used to form reconfigurable cell by sort-cluster algorithm. The aim is to divide the assembly resources into different groups, so the relevant cell agent can manage every group uniformly. The whole complexity of assembly system is decreased by grouping. Our method is a deduction of King's sort-cluster algorithm, and we change the matrix type according to the concrete purpose. It is unnecessary to calculate the similar coefficient for this method. The only thing is to establish the "assembly resources-product type" matrix, and grouping is accomplished according to this matrix. It is very simple and efficient.

As using the sort-cluster method to form the reconfigurable cell, the "assembly resources-product type" matrix need be established at first. In this kind of matrix, row represents product type, and column represents assembly resources. Supposed the matrix is $M(i \times j)$, the element m_{ij} means whether the i_{th} product needs the j_{th} assembly resource and m_{ij} is a binary digit. If $m_{ij} = 0$, it indicates the i_{th} product doesn't need the j_{th} assembly resource; if $m_{ij} = 1$, it indicates the contrary case. Note that the assembly resources discussed here only include manipulative devices, but do not include the transportation devices (AGVs) and buffer.

The sort-cluster algorithm is described as follows:

Step1 In terms of rows of the "assembly resources-product type" matrix, take out digit 1 and 0 in binary format, and then calculate the value of all the numbers that are made of these binary digits. Every row is reordered in the decreased order. If there are some rows of the same value, they keep their original order;

Step 2 If the whole matrix keeps unchanged, then stop, else go to step 3;

Step 3 In terms of columns of the "assembly resources-product type" matrix, take out digit 1 and 0 in binary format, then calculate the value of all the numbers that are made of these binary digits. Every column is reordered in the decreased order. If there are some columns of the same value, they keep their original order;

Step 4 If the whole matrix keeps unchanged, then stop, else go to step 1;

For example, there are four product types and ten kinds of assembly resources as table 1.

Table 1 assembly resources-product type matrix

	R1	R2	R3	R4	R5	R6	R7	R8	R9	R10
P1	1	1	1	1		1	1	1		1
P2	1	1	1	1	1	1	1	1		1
P3	1	1	1	1		1	1	1	1	1
P4	1	1	1	1	1	1	1	1	1	1

We use the above-mentioned algorithm, and the common devices which are used by all the products (P_1, P_2, P_3, P_4) form a group, so we call this group as the common cell. The rest devices are divided into a few of groups and form several special cells. Because this paper mainly discusses assembly manufacturing, it has its own particular character. A majority of the assembly devices is common device, and a minority is special device. As to this example, the result is that R_1, R_2, R_3, R_4, R_6, R_7, R_8, R_{10} form the common cell. R_5 forms a special cell, because it is only used by P_2, and R_9 that is only used by P_3 forms a special cell. If the common cell includes too many devices, the database will become too large, and we should divide it again into more groups according to some rules such as similarity.

4.   AGENT STRUCTURE

Agent is a piece of software that can act without direct external intervention, and it has some degree of control over its internal state and actions based on its own experience and can communicates with the environment and other agents[5]. In this part, we mainly deal with three problems. The first one is a structure of a single agent, and the second is focuses on agent components and communication specification, and the last is about agent cooperation.

4.1   Single agent structure

- Deliberative agent: a deliberative agent is also called a cognitive agent which adopts the explicit symbol model, including reasoning capability. It keeps the character of the classic artificial intelligence (AI), and it is a knowledge-based system. It is shown as figure2.

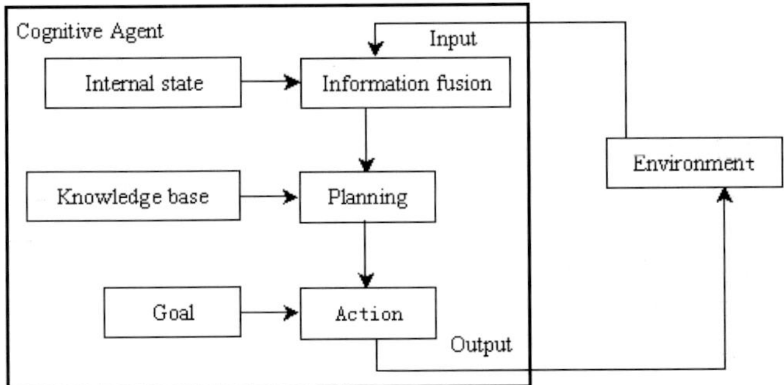

Figure 2 cognitive agent structure

In our research, the management agent and the cell agent is constructed according to this structure. As mentioned in section 2, these two kinds of agents must have planning ability, the management agent must analyze the assembly processes and the cell agent must assign the undertaken tasks. Accordingly, they are knowledge-based agent and have the reasoning capability.

- Reactive agent: a reactive agent doesn't include world model that is represented by symbols. It just uses "if-then" rules to connect the state and the corresponding action. Figure 3 demonstrates this structure.

We introduce this structure to the device agent. Although the device agent is able to execute local planning, this procedure is done by using these "if-then" rules. Because the assembly datum about the product is sent to the device agent, the device agent can produce the driving datum to control the relevant device according these rules. Note that the device agent is very simple.

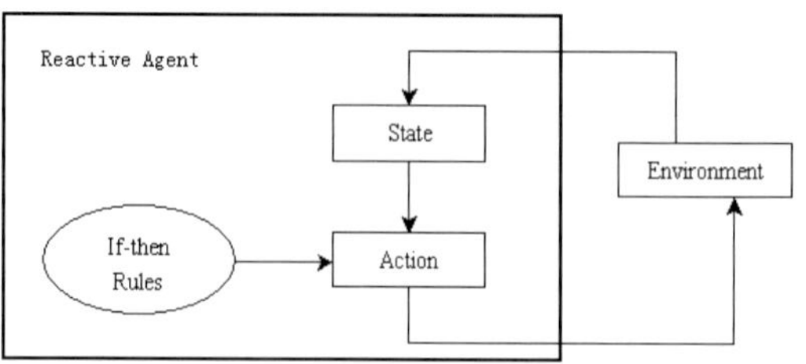

Figure 3 reactive agent structure

4.2   Agent components

Agent is usually composed of several components as shown in figure 4, we explain these modules in detail as follows:

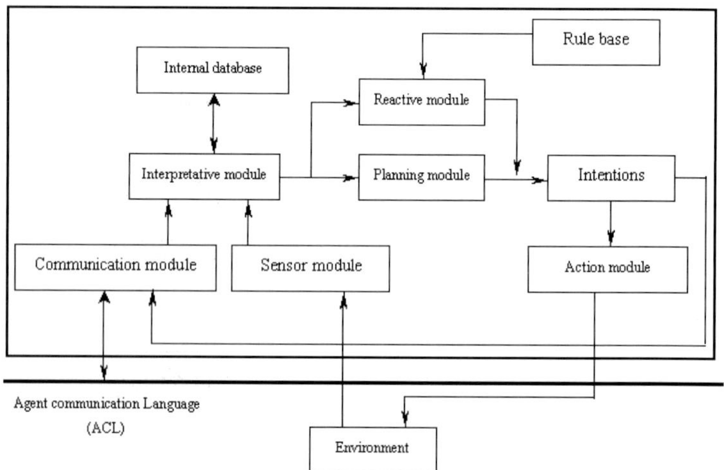

Figure 4 agent components

- Reactive module: this module is to respond to the urgency, it doesn't have reasoning capability; it is a map from sensors to actions.
- Planning module: it is responsible for producing a plan to guide actions. This plan is usually a local plan according its own goal and state or according to its experiences.
- Communication module: this module is important for agent to attain global optimization. The ability of communication is the basic character of agent. It includes language understanding, glossary library, physical communication grammar library etc.
- Interpretive module: this module is to interpret agent communication language in order that the agent can understand the content and the

aim of communication.

- Internal database: a database of agent mainly represents the world model and its own state, including:

    1) Agent's name and address etc;

    2) Its capability name (function module controlled by the agent), and description (keywords, path, run parameter);

    3) The state of the running function module, including running, ready, completing, block;

    4) The current state of agent.

4.3　Agent coopcration

4.3.1　Communication specification (KQML/KIF)

Currently, one of the most important areas for agent is standardization of agent communication. By using these specifications, we effectively codify the basic elements of interaction that can take place between agents. KQML (Knowledge Query and Manipulation language) is a language and protocol for exchanging information and knowledge. It is part of a larger effort, the ARPA Knowledge Sharing Effort which is aimed at developing techniques and methodology for building large-scale knowledge bases which are sharable and reusable[5].KQML is both message format and a message-handling protocol to support run-time knowledge sharing among agents. KQML can be used a language for an application program to interact with an intelligent system or for two or more intelligent systems to share knowledge in support of cooperative problem solving. KIF (Knowledge Interchange Format) provides some grammars to the information content; it is similar to calculus predicate.

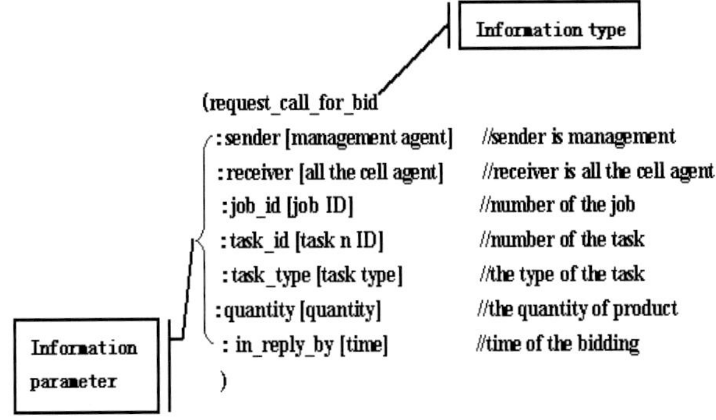

Figure 5 example1

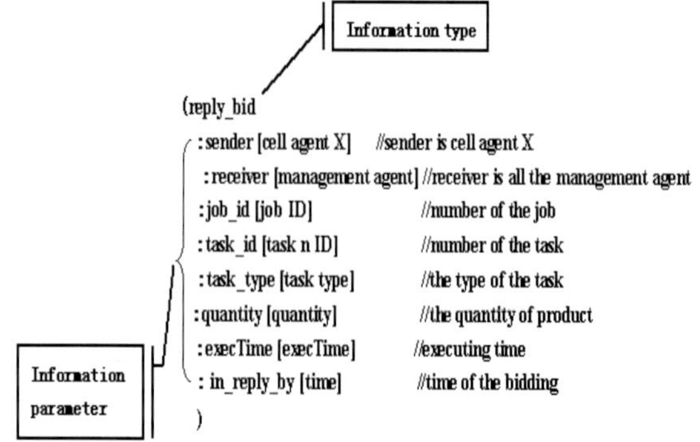

Figure 6 example2

In our study, we comply with the specification, and we use KQML as agent communication language. We define some information types according the Reconfigurable Assembly System. The two examples of the information are given as follows. The first one is that the management sends information about public bidding to all the cell agents as shown in figure5. The second is that cell agent "X" sends information to the management to bid for the task whose ID is "n" as figure6 shows.

4.3.2   Cooperation model

There are many negotiation models for agent cooperation design, such as Contract Net Protocol (CNP)[6], Partial Global Planning (PGP)[7], Constraint-Based Planning (CBP)[8], Case-Based Reasoning (CBR)[9], etc. In our research, we adopt the well-known Contract Net Protocol (CNP) for task allocation in cooperative design. Cooperation capability is considered as an important feature of multi-agent system, and this capability can increase flexibility and reliability. The reconfiguration of system is realized by negotiation between agents. We propose the detailed algorithm according to CNP as follows.

Step1 Management agent receives an order and forms a product database. The order can be input from interface by the users. The product database contains feature parameter, geometrical data, fitting relation between parts, delivery date etc.

Step2 Management agent generates a series of assembly tasks according to the product database.

Step3 Management agent firstly declares public bidding to the common cell agent, and then the common cell agent finds devices capability and type in the local database, and judges if the order can be accepted according to an evaluation model. If the order is not accepted; then go to step4, else go to step5. (Because the common cell agent is crucial to decide what kind of product can be assembled in the system.)

Step 4 Common cell agent communicates with management agent, and informs it to abandon the order.

Step5 Common cell agent bids for the relevant tasks and allocates these tasks to device agents. Then device agent adjusts and replaces the corresponding modules in order to accomplish the undertaken tasks.

Step6 Management agent continues to declare public bidding to special cell agent one by one.

Step7 If one special cell agent accepts an assembly task, it must adjust and replace modules to adapt to the change.

Step8 If all special cell agents bid for tasks and there are some tasks left, the system requires new devices to finish the left assembly processes.

After new devices are added, we only append new data about the devices to the agent structure. We will establish a new device agent, and after registering to the cell agent, it can run in the system. Note that the other parts of the system don't need to be modified, so it is very easy that the whole system realizes reconfiguration.

5. CONCLUSIONS

This paper focuses on modeling of Reconfigurable Assembly Systems (RAS) based on multi-agent systems. A hybrid architecture is proposed as the model of RAS to respond to the needs of market. We introduce the agent specification to ensure that effective interaction can take place between agents. The well-known Contract Net Protocol has been adapted to dynamically assign operation to the relevant assembly devices and realize system reconfiguration.

ACKNOWLEDGEMENTS

The authors would like to thank financial support of National Key Basic Research Plan(973 Plan). The authors would like to thank all member companies for their support and feedback. The authors are grateful to EASED 2004 for communication opportunity.

REFERENCES

1. K Ishil: "Life-Cycle Engineering Design". *J.of Mech. De. ASME Transaction.* 1995
2. N. Bair, T. Kidwai, M. Mehrabi, Y.Koren: "Design of a Reconfigurable Assembly System For Manufacturing Heat Exchangers". University of Michigan, Dept. of Mech. Eng, ERC/RMS. 2000

3. Venketesh N Dubey, Richard M Crowder: "Designing A Dexterous Reconfigurable Packaging System for Flexible Automation". *DETC 2003/DAC-48812*
4. Jin-Lung Chirn, Duncan C. McFrlane: "A Holonic Component-based approach to reconfigurable manufacturing control architecture". Institute for Manufacturing, University of Cambridge. 2000
5. Agent Technology Green Paper. OMG Agent Working Group. 2000, 3
6. FU_SHIUNG HSIEH: "Modeling and Control of Holonic Manufacturing Systems Based On Extended Contract Net Protocol". *Proceedings of the American Control Conference* Anchorage, AK May 8-10, 2002
7. Edmund H. Durfee, Victor R. Lesser: "Partial Global Planning: A Coordination Framework for Distributed Hypothesis Formation". *IEEE Transaction On Systems, Man, And Cybernetics,* **vol 21, no. 5**, September/October 1991
8. Bertrand Allo, Christophe Guettier, Nelly Lécubin: "A Demonstration of Dedicated Constraint-Based Planning within Agent-based Architectures for Autonomous Aircraft". *Proceedings of the 2001 IEEE International Symposium on Intelligent Control,* September 5-7, 2001
9. J. Britanik, M. Marefat: "Distributed Case-Based Planning: Multi-Agent Cooperation for High Autonomy". University of Arizona. 1993

COMBINE AGENT AND HOLON APPROACH FOR MANUFACTURING SCHEDULING IN CAPP/PPC

Shujuan Li and Yan Li
School of mechanical & instrumental engineering, Xi'an University of Technology, Xi'an, P.R.China, 710048, jyxy-ly@xaut.edu.cn

Abstract: Manufacturing scheduling is a difficult problem, particularly when it takes place in an open and dynamic environment. In this paper Agent-based CAPP and PPC integrated architecture under network is proposed. The objective is to overcome the rigidity of conventional hierarchical structures and to introduce new integrated architectures that are able to adapt to a dynamic environment, at same time the agent technology and self-organized method of holonic manufacturing have been used in attempts to resolve scheduling problem, a bidding mechanism based on Contract Net protocol is proposed as a key solution component. Our approach is to combine a bidding mechanism based on Contract Net protocol with a mediation mechanism based on Mediator architecture, for dynamic manufacturing scheduling and rescheduling.

Key words: Agents, CAPP/ PPC, Bidding Mechanism, Contract Net, Holon, Mediator.

1.   INTRODUCTION

Job shop scheduling may be defined as allocating m jobs to n resources satisfying certain objectives. It is a difficult problem because of wide variety of constraints, particularly when it takes place in an open, dynamic environment. In a job shop manufacturing system, rarely do things go as expected. The set of things to do is generally dynamic. The system may be asked to do additional tasks that were not anticipated, and sometimes is allowed to omit certain tasks. The resources available to perform tasks are subject to change. Certain resources can become unavailable, and additional resources introduced. The beginning time and the processing time of a task are also subject to variation. A task can take more time than anticipated or less time than anticipated, and tasks can arrive early or late.

Rising complexity of products, production structures and processing procedures on one side and turbulent market excitations resulting in growing product variety, individualization and shortening time frames on the other are setting new frontiers to the manufacturing business. These factors make the scheduling problem more difficulty. Despite of research efforts and investments in the context of computer integrated manufacturing the existing manufacturing systems are still predominantly. Therefore they cannot adequately conform to these requirements because of their structural rigidity, deterministic approach to decision making in a stochastic or dynamic environment, hierarchical allocation of competencies, and insufficient communication and exploitation of expertise.

In order to face new challenges in manufacturing a shift of the existent manufacturing paradigm from deterministic into a new manufacturing prospect considering natural understanding and concern is needed. Several influencing concepts in this direction have emerged in the last decade: Parunak et al put forward that the manufacturing capabilities (e.g. people, machines, and parts) are encapsulated as autonomous agents. Burke and Prosser proposed that the scheduling problem was decomposed and distributed across a hierarchy of intelligent agents. Maturana, Balasubramanian and Norrie described an integrated planning-and-scheduling approach combining sub tasking and virtual clustering of agents with a modified contract Net protocol. Holonic manufacturing systems, random manufacturing systems, fractal factory and distributed manufacturing system.[1-4]

These concepts influenced research for more effective mastering of complex and dynamic behavior of the system and its environment, especially the holonic manufacturing system concept. In this paper the results of developments of the integrated structure CAPP/PPC (computer aided process planning and Production planning and Control integrated) in holonic manufacturing system are presented.

This paper focus on the dynamic scheduling mechanisms developed in the dynamic manufacturing environment. Section 2 describes the framework of the collaborative control architecture of CAPP/ PPC based on agent in holonic manufacturing system. Section 3 discusses the dynamic scheduling and rescheduling mechanisms applied for job shop in holonic manufacturing system. Section 4 gives concluding remarks. It is show that the framework of the collaborative control architecture is reasonable, can be applied in the integration environment of CAPP/PPC. The scheduling problem can be solved effectively by this way.

2.   THE INTEGRATED ARCHITECTURE OF CAPP/PPC BASED ON AGENT IN HOLONIC MANUFACTURING SYSTEM

2.1   CAPP/PPC integrated architecture based on agent

CAPP and PPC integrated concept is proposed by G. Chryssolouris et al in 1985, it is researched for near 20 years, and many researchers put forward a lots of method and concepts. Such as distributed process planning, non-linear process planning, flexible process planning and dynamic process planning etc. The common observations can be summarized: ① the architecture is the essential issue, ② the role of a human subject is dominant, ③ the information and communication technologies are the enabling technologies, and ④ the transition from highly data-driven to particular information, knowledge and learning driven organization is needed.

Therefore new features of future manufacturing systems have to be developed. The most expected ones are: ① an open multi-level architecture, ② advanced communication capabilities, ③ decentralized decision making, ④ self-organization ability, and ⑤ redefined abilities in terms of autonomy, evolutionary adaptively, re-configurability, cooperativeness, interactivity, task orientation within competence, ability of communication, coordination and cooperation, and learning capability.

In this paper, the CAPP/PPC system architecture was developed based on mediator and Agent structure in holonic manufacturing system. The autonomous agent approach can be used to develop advanced holonic manufacturing systems, its objective was to integrate engineering tools, like CAPP and PPC tools, database systems, or knowledge-based systems, into a truly open system, that is, a system for which users can freely add or remove agents without having to halt or to reinitialise the work in process.

In this section, the general integrated architecture of CAPP and PPC is organized. It consists of CAPP holon and PPC holon. It is a multi-agent architecture for holonic manufacturing developed using the mediator-centric federation architecture. In this particular type of federation organization, agents can link with mediators to find other agents in the environment. Additionally, mediators assume the role of system coordinators be promoting cooperation among agents and learning from the agents' behavior. In other words, mediators are similar to facilitators, with additional functionality such as coordination and learning. In CAPP and PPC integrated system, agents are used to represent manufacturing devices process planning and parts to be fabricated, while mediators are used to coordinate the interactions among agents. In CAPP and production scheduling integrated system, mediator is a distributed decision-making support system for coordinating the activities of a multi-agent system. The CAPP and PPC system architecture is primarily organized through several holons (see figure 1). Each subsystem is connected (integrated) to the system through a special mediator. Each subsystem itself can be an agent-based system (e.g., agent-based time quota determining system, TQDS), or any other type of system like knowledge-based CAPP system, and so on. Agent in a subsystem may also be a autonomous agent at the subsystem level, some of these agents may also be able to communicate directly with other subsystems or the agents in other subsystems.

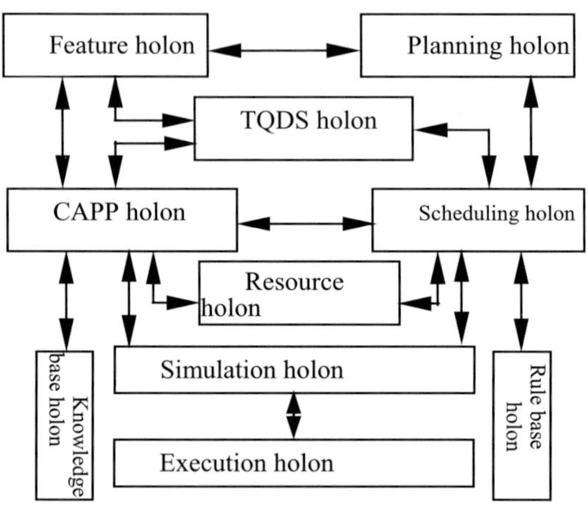

Figure 1. The holon structure of CAPP and PPC integrated system

2.2   The collaborative control archteure of CAPP/ PPC system in holonic manufacturing

The Agents of subsystem are also autonomous Agents, the distributing cooperation control structure among Agents system is shown figure 2. in this model, each mediator can be used to a unattached manufacturing unit, the Agents included mediator compose Agent group, the information processing is accomplished by Agent group, in Agent group, Aa (Adaptation Agent), Oa (Optimization Agent), Ra (Regulation Agent) are set up, these Agents carry coordination control interior of mediator, and the same time each holon has a mediator (Me), it coordinates and solves conflict among mediators, thus the distribution control system is composed. In this model, the bulletin is communication tool among Agents; each holon is connected by database.

In this collaborative control architecture, each holon includes an Agent group, and mediator manages every Agent in Agent group. Each holon is a part of CAPP/PPC system, such as resource holon and TQDS holon. Each mediator of holon accepts information from other mediator of holon in network, it processes information according to mediator own function, the Contract Net protocol is used in this process. And then the results are transmitted by network.

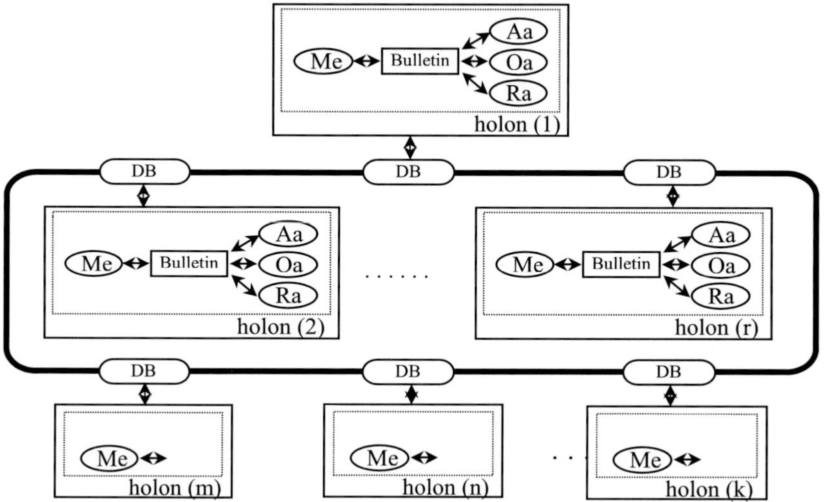

Figure 2. The collaborative control architecture of CAPP and PPC integrated system

As shown in figure 1, CAPP holon and PPC holon can be considered as the main holon of the system, several subsystems such as feature holon; production planning holon may be either agent-based subsystems, these mediators can supply the needed information of CAPP/PPC system by mediators in holon, and insist of agent-based subsystems. Like knowledge base holon can be used to obtain the process planning knowledge, such as machine method agents and machine sequence agents etc. the rule base holon include all the rule of scheduling and evaluation guide line. Such as: Max flow time agents, machine utilization agents etc. each subsystem connects to CAPP/PPC system through a mediator of holon which serves as the coordinator of this subsystem and its only interface to the whole system.

Simulation holon is developed to carry out production scheduling simulation. Each mediator of execution holon is developed to coordinate the execution of process planning and scheduling results. Execution mediator is in general assigned to shop floor.

In current implementation, all scheduling requests come from a feature-based functional design system via feature mediator of feature holon (as shown in figure1). Each product corresponding to a customer order has been decomposed into related manufacturing features. Each manufacturing feature can be realized by a manufacturing task (operation). These manufacturing tasks are organized in a graph data structure representing the sequence to produce the corresponding manufacturing features. Each product or each part is modeled as a part agent containing the information about this product/part including the feature data, due time, its plan, which will be assigned with times, and resources during the scheduling process.

All scheduling/rescheduling processes may take place automatically by implementation, all these scheduling or rescheduling processes are executed manually be human-computer interface. When a rescheduling cannot be completed due to the unavailability of manufacturing resources, some production orders have to be cancelled or delayed.

The main development language is JAVA, and database, knowledge base; rule base is developed using SQL Server 2000. The present implementation is being developed initially as a simulation.

3. SCHEDULING AND RESCHEDULING IN CAPP/PPC INTEGRATED SYSTEM

Machine-centered scheduling and worker-centered scheduling are subjected in scheduling. In this paper, only the machine-centered scheduling mechanism is adapted to CAPP and PPC integrated system. In the system with machine-centered scheduling, first of all, the mediator of resource holon sends a request message to mediator of machine holon at first. It is the machine agents that take the responsibility to negotiate with worker agents and tools agents. In integrated system, parts are also modeled as agents-part agents. In the current implementation, part agents do not communicate directly with resource agents. All manufacturing task requests are sent to the resource holon. And there fore, we do not discuss the communication and negotiation between part agents and resource agents.

3.1 Contract Net Protocol

The Contract Net is a negotiation protocol by Smith (1980) which facilitates distributing subtasks among various agents. The agent wanting to solve the problem broadcasts a call for bids, waits for a reply for some length of time, and then awards a contract to the best offers according to its selection criteria. This protocol has been widely used for multi-agent negotiation. Some researchers have proposed modified versions of contract net protocol for special application, such as the extended contract net protocol (ECNP) proposed by Fischer et al.

The contract net protocol used in CAPP/PPC system has been extended to meet special requirements. For example, in the bidding and mediation process for selecting a machine for a manufacturing task, after receiving the different propositions(bids) from different machine agents, the machine mediator will select a machine agent to perform the task according to its criteria(mainly manufacturing cost criteria) and award a contract to it. Other machines are not selected to perform the task, but they are taken as alternatives that may be contacted (negotiated with) in the future in the case of unforeseen situations such as machine breakdown. This will greatly reduce the rescheduling time when such unforeseen situations take place. The information about the alternative resources (CAPP, machines, workers and tools) will be sent to the part agents with the information about the selected resources. The part agents save these two sorts of information in their knowledge bases. When the selected resources cannot perform the scheduled tasks due to unforeseen situations, the part agent may negotiate directly with alternative resource agents.

3.2   Asynchronous Communication

The mediator-mediator of holon and mediator-agent communications is asynchronous, and the communication mode can he point-to-point (between two agents), broadcast (one to all agents), or multicast (to a selected group of agents). In CAPP and PPC integrated system implementation, we use the multicast mode, which can be extended to point-to point and broadcast. Five types of messages are used: request, inform, announce, bid and notice. They can be grouped into two categories: requests and assertions (request, inform, and notice); call for bids and offers (announce, bid). The messages are formatted in an extended KQML format.

In order to realize the asynchronous communication, each mediator and resource agent has an input message box. All incoming messages are firstly queued in this input message box. These incoming messages are treated according their priority. The message with highest priority is treated first. For the messages with same priority, the first arriving message is treated first.

3.3   Bidding and Mediation Process

As shown in figure 3, when the mediator of machine holon receives a task request from the resource mediator which itself is requested by a part agent, it sends out multicast announce messages to concerned machine agents according to its knowledge about the machine agents registered to it. After receiving announce messages, a machine agent calculates the processing time and cost, verifies its reservation plan, and negotiates with workers agent through the worker mediator and tool agents through tool mediator (described in the following paragraphs). After receiving the propositions (bids) from the worker mediator and the tool mediator, the machine agent sends a bid with the information about the processing time and cost, proposed start time and end time, other free time slots, and the information about the associated worker agent and tool agent to the machine mediator.

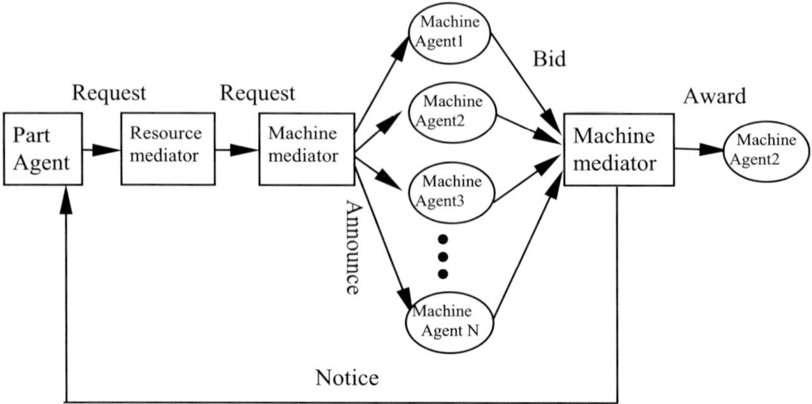

Figure 3. Machine mediator and bidding among machine agents

Waiting for a predefined period after sending out the announce messages to machine agents, the machine mediator analyses the propositions(bids) from different machine agents, and selects one machine agent and its associated worker agent and tool agent(s) to perform the task according its selection criteria based in the cost model. As soon as the decision is made, the machine mediator sends an award message (notice type) to the selected machine agent which then locks and reserves it for the requested production task. This time slot now becomes unavailable to other tasks. At the same time, the machine mediator sends an inform message with the information about the selected resources and the alternative resources directly to the part agent (not through the resource mediator). After receiving the award message, the winning machine agent should also send an award message to the winning worker agent and the winning too agent(s). A more complex mediation mechanism is to be developed to optimize the scheduling for a group of parts or for a given period.

3.4   Scheduling and Rescheduling

As CAPP and PPC integrated system running, the below situation is objected to, the system how quickly respond dynamically to emergent production orders and unforeseen situation.

(1) Rescheduling for new emergent orders

Because the system developed is to be various environments oriented, rescheduling for new emergent orders is quite frequent and very important, when a new emergent order arrives after the normal scheduling is completed, the CAPP mediator production new operation serials based feature mediator TQDS mediator make operation time quickly. The resource mediator will first try to schedule the new task normally by negotiating with resources agents. If it cannot find time slots for scheduling these emergent order, the solution is to release some scheduled orders (release the production plan of the corresponding part agents) for which the scheduled. End time is well before the due time. The scheduled part agents are released one by one until the emergent order is successfully scheduled or no possible part agents can be released. These released part agents are then rescheduled after the scheduling of the emergent order is finished.

(2) Rescheduling after cancelled of orders

For the orders to be cancelled, all that needs doing is only to release the corresponding part agents. In general, the other production tasks may not need to be rescheduled.

(3) Rescheduling for the optimization of the current schedule

If it is found that the schedule dies not effectively utilize the manufacturing resources (e.g. the utilization if the machines is less than a predefined percentage) due to the changing situations, such as cancellation of several orders, earliness of some production talks, and so on, a rescheduling process may be undertaken for all production tasks which have not been started. In certain cases, it may be more economic for the factory (here the job shop) to intentionally and occasionally make several resources (machines, workers etc) non-active before rescheduling.

(4) Rescheduling after tardiness of a task

If the tardiness of a task will affect the next task of the current product, the following tasks of this product have to be rescheduled.

## 4.	CONCLUSION

This paper has presented a simple but efficient dynamic manufacturing scheduling mechanism-combining Agent technology and mediation mechanism in holonic manufacturing system. Agent-based CAPP and PPC integrated architecture under network is put forward. The previous research experience and current simulation results show that such a scheduling mechanism has following advantages.

(1) It is particularly interesting for agent-based manufacturing systems where the environment changes frequently in this type of system, when the environment changes, only the knowledge base of the mediators need to be updated. Learning mechanisms can also be applied only at the mediator level

in holon. Such learning mechanisms also have been developed CAPP and PPC integrated system.

(2) The data-process agents do not need to broadcast request message to all time quota agents but only to send request messages to a selected group of machining method agents through the machining method mediator. Each selected machine method also sends multicast request messages to a selected group of feature agents through the feature mediator. This greatly reduces the communications and therefore reduces the scheduling/rescheduling time.

(3) The bids are sent to the mediators, rather than directly to the agent that sent out the request messages calling for bids, and the mediators execute the selection process. In generally, there are a large number of part agents and resource agents in an agent-based manufacturing, this method simplifies the reasoning mechanisms of the part agents and the resource agents.

ACKNOWLEDGEMENTS

Results proposed in this paper have been developed in research projects supposed by the Natural Science Fund-aided project (Research No: 2000C31) of ShaanXi Province as well as the Science-Tech fund-aided project (Research No: 00JK232) of ShaanXi Provincial Education Department.

REFERENCES

1. Sluga A., Butala P.. "Self-Organization in a Distributed Manufacturing System Bases on Constraint Logic Programming". *Annals of the CIRP*. **Vol.50**, pp.323-326, Jan 2001.
2. Wiendahl H.P., Ahrens V.. "Agent-Based Control of Self-Organized Production Systems". *Annals of the CIRP*. **Vol.46**, pp.365-368, Jan 1997.
3. Rei HINO, Toshimichi MORIWAKI. "Proposal of Holonic Manufacturing System Concept". *The paper collection of Japanese machine institute (C)*. **Vol. 67**, pp.367-373, Jun 2001.
4. Li shujuan, Liyan, et al. "multi-agent based approach for determination of time-quota in integrated environment", *the 10th international manufacturing conference in china*, October 2002, pp.156-157.

AN ALGORITHM OF TOOL PATH GENERATION FOR PARTS WITH REGULAR FEATURES

Luling An, Laishui Zhou, Wei Wei and Rurong Zhou
Nanjing University of Aeronautics and Astronautics, 29th Yudao Street, Nanjing, China, 210016,anllme@nuaa.edu.cn

Abstract: In general, for NC machining of the models with regular features, the relevant tool paths are also arranged in regularity. In this paper, an algorithm is researched of generating tool path attached to the model. That is, tool path is treated as attributes of the model. During modeling a part or a mould, the original feature is formed by Boolean operation with the base material. Then tool path is generated for the original feature. With creation of the model by copying or arraying, the whole tool path is generated accordingly. With the above method, NC tool path generation operation is greatly simplified. Application example is given to show the advantages. Furthermore, the method can be more useful in combination with parametric technology.

Key words: CAD, CAM, tool path, feature, modeling

1.   INTRODUCTION

With the development of CAD/CAM technology, NC milling is applied more and more extensively to the production of mechanical parts, moulds and dies, and aircraft structures, etc. Tool path generation is of vital importance to NC machining.

The common procedures of NC programming are as follows. First, the CAD model, solid model or multi-surface model, is generated with CAD software. Then the tool path of NC milling is planned on the surfaces of the model in terms of the machining parameters with either CAM software or the CAM module in CAD/CAM integrated system, such as CATIA, Pro/E and UG. By post processing, tool path is converted to G codes which control the CNC system to implement machining.

There have been a variety of tool path strategies which deal with the milling of the model surfaces[1][2]. Most of the researches focus on the generation algorithms. These algorithms have been successfully implemented and applied to the production.

For the machining of workpieces with multiple regular features, such as forging dies of straight bevel gears, cylindrical cams and moulds, there are particularities in NC programming. It is not convenient to generate machining tool path with current CAD/CAM software. And if the properties of features are not considered, it is costly to meet the specific precision requirements of the features. Although there have been researches on feature-based NC machining[3][4], they mainly deal with selection of cutter diameters and some machining parameters.

In this paper, aiming at the above problems a new method of machining tool path generation is presented. The tool path is automatically generated at the same time with modeling processes. During modeling a part or a mould, the original feature, e.g. slot or pocket, is formed by Boolean operation with the base material. Then tool path is generated for the original feature. With creation of the model by copying or arraying, the whole tool path is generated accordingly. With the above method, NC tool path generation operations can be greatly simplified and machining precision can be conveniently obtained.

2. FEATURE MODEL AND MANUFACTURING INFORMATION

Feature based product model integrates the geometric and topological information with manufacturing information. It makes the design of product on higher levels and breaks through the barriers of design and manufacturing. To define product information, object oriented technology is adopted. A shape feature is treated as an "object", and the object is described into different levels according to the requirements of design, process planning and manufacturing. Levels constrain and support with each other. The delaminating strategy of feature information is as follows (As shown in Fig. 1):

The first level is used to describe the relationships between features, through which the relationships of each shape feature with the base material (stock) and other shape features are expressed. The functional attributes are attached to subdivide the shape feature classes and supply more manufacturing information to NC programming system.

The second level is used to describe detailed shape feature classes. It is the class pointer of shape features, corresponding to a series of geometric and dimensional constraint rules to describe the features in lower levels.

The third level is feature definition level, which is used to define feature parameters according to the data structures of features provided in the second level. In this level, the design datum and positioning dimension are determined to exactly define the magnitude, position and orientation of features. It is also the parametric benchmark to generate the geometric model description information.

The fourth level is used to define the manufacturing information, which includes machining parameters and tool path. In this level the basic manufacturing information is determined according to the functional attribute information in higher level. Then the machinability of the structure is analyzed through the estimation to its dimensions, precision, material information and process planning in terms of the definition in third level.

The fifth level is used to define the entity geometric model of shape features in terms of the information in the second and third levels. It is the foundation of graphic display through user interaction and the main information of NC programming.

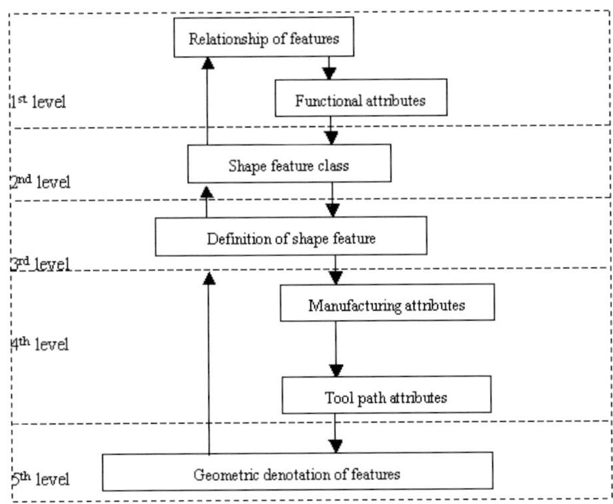

Fig.1 hierarchy of feature information

With the above hierarchy, the design and manufacturing information of product are integrated.

3. FEATURE-BASED MODELING AND TOOL PATH

For parts or moulds with regular cutting features, the final shape is obtained through Boolean operation with the base material (stock), which resembles the machining procedure to some extent. Tool path is taken as the attribute of the feature and is generated with the proceeding of modeling.

Take an example as shown in Fig.2. The base material is a disk, and the original features are two grooves (cutting features). For part designing, the final shape is obtained by circumferential array of the original features. Similarly, the part is produced by machining the grooves one by one.

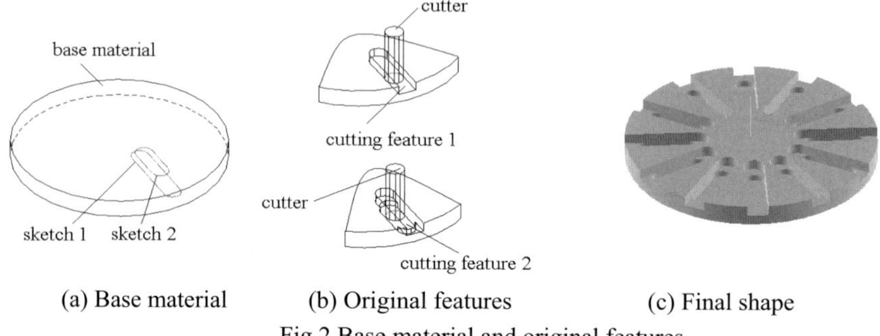

(a) Base material   (b) Original features   (c) Final shape

Fig.2 Base material and original features

The machining tool path of the original feature is generated with solid-model-based strategy. For the original cutting feature volume, the relevant tool path is obtained by common method, e.g. z-level, plane parallel machining, point-to-point, etc., as shown in Fig.3. The tool path is taken as the attribute to the original feature. Then the rest features are generated through copy or array with the related tool path being automatically generated as the attributes of the features.

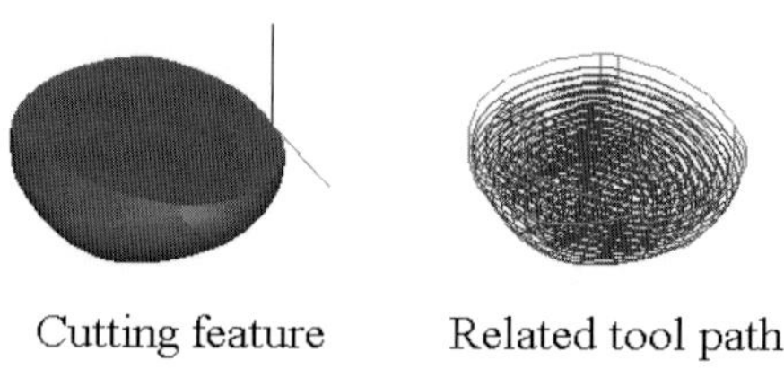

Fig.3 Cutting feature and related tool path

Here we implement the functions on base of ACIS, a powerful geometric engine. ACIS is an object-oriented three-dimensional (3D) geometric modeling engine from Spatial Technology Inc. It is designed for use as the geometry foundation within virtually any end user 3D modeling application. ACIS separately represents the geometry and the topology of objects and provides an open architecture framework for wire frame, surface, and solid modeling from a common, unified data structure [7].

As shown in Fig.4, the attributes of ENTITY in ACIS are used to attach data to entities. As C++ class ATTRIB, which is derived directly from the ENTITY class, provides the data and functionality that all attributes share, for both user-defined attributes and system attributes. User-defined attributes allow the extension of an ACIS geometric model into a true product to attach application-specific data to entities.

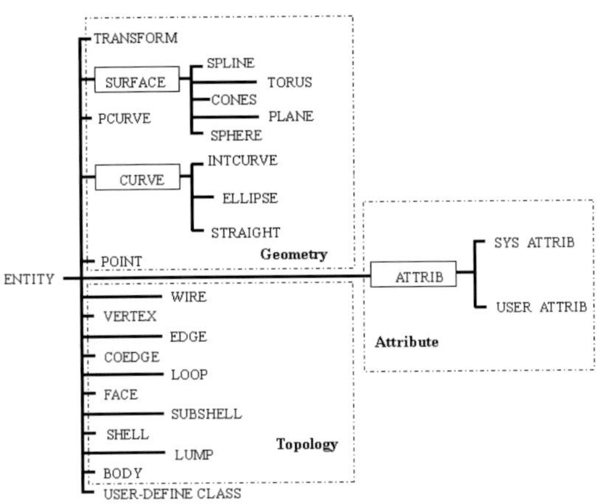

Fig.4 Hierarchy of ACIS Classes

The manufacturing attribute class, Mfg_Attrib is given, which is derived from the ATTRIB class of ACIS. The relation of classes ENTITY, ATTRIB and Mfg_Attrib is shown in Fig.5.The definition of Mfg_Attrib is as follows:

CLASS Mfg_Attrib: public ATTRIB
{
    private:
    BOOL SetToolParam();
          //Set tool parameters
    BOOL SetMfgParam();
      //Set manufacturing parameters
    public:
    BOOL SetMfgTemplate( );
      //Define manufacturing template
    BOOL FeatureToolPath(CLlist&);
    //Generate tool path of original feature
    transf TransfToolPath(CLlist&);
          //Transform tool path
    BOOL LinkToolPath(CLlist&);
      //Link tool path
    BOOL DeleteToolPath(CLlist&);

```
            //Delete tool path
    BOOL CopyToolPath(CLlist&);
            //Copy tool path
        ... ...
}
```

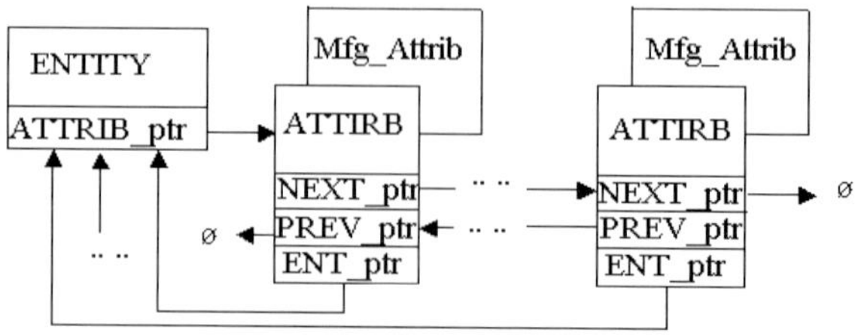

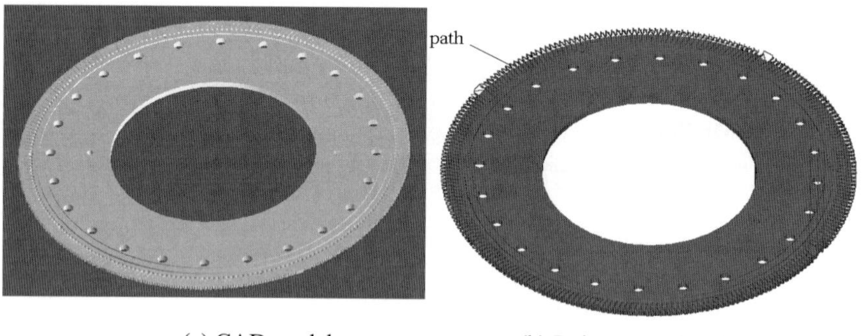

Fig.5 Relation of ENTITY and attribute classes

ACIS entity and features can be related with Mfg_Attrib through double pointers. The functions of class Mfg_Attrib include setting of tools and manufacturing parameters, generation of tool path of the original cutting features, corresponding operations of tool path, such as transformation, copying, deleting and linking, etc.

4. ALGORITHM IMPLEMENTATION AND EXAMPLES

(a) CAD model (b) Relevant tool path

Fig.6 An application example

As shown in Fig.6, it is a mould to form special water pipes used in drip irrigation. Along the circumference of the disk, a number of curvilinear grooves with different depths are distributed, which are to be machined with NC method. Tool path of the original feature is generated with common NC programming or manually editing. Figure 7 shows the tool path for the original feature, which is generated manually. The flow chart of tool path generation is shown in Fig:8.With the above method, a series of mould parts similar to that shown in Fig.6 have been successfully machined and applied to practical usages.

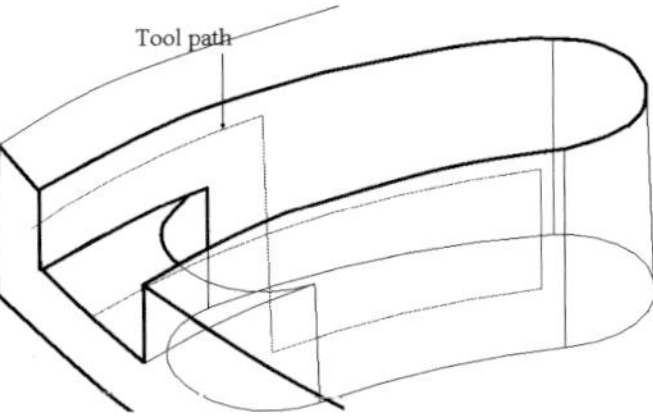

Fig.7 tool path of the original feature

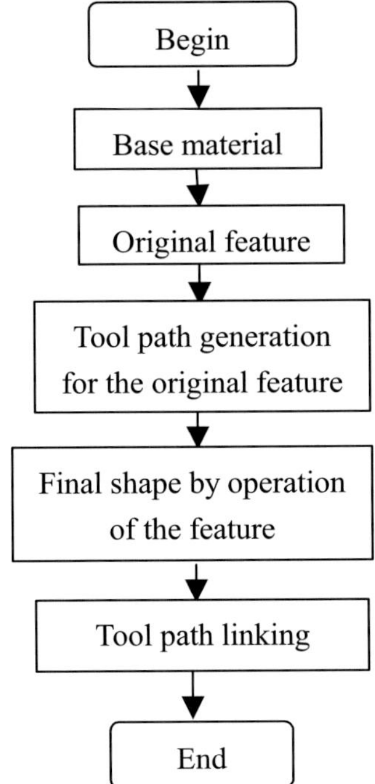

Fig.8 Flow chart of model-attached tool

5.   CONCLUSIONS

For the workpieces with regular machining features, the tool path of the original cutting features can be generated manually or automatically. The final shape of the workpiece is obtained through such operations as arraying and copying to the original features. Accordingly the whole tool path is generated as the attributes of the features. With the above method, NC tool path generation operation is greatly simplified for the workpiece with regular features. The future work is planned to apply the method to parametric CD/CAM system.

References

1. D Dragomatz, S Mann. "Classified bibliography of literature on NC milling path generation". *Computer-Aided Design*. 1997,**29(3)**:239-247
2. Chun-Fong You, Chih-Hsing Chu. "An automatic path generation method of NC rough cut machining from solid models" [J]. *Computers in Industry*, 1995, **26(2)**: 161-173
3. Li Chen, Jian Pu, Xiankui Wang. "A general model for machinable features and its application to machinability evaluation of machine parts" [J]. *Computer Aided Design*, 2002, **34**:239-249.
4. Yuan-Shin Lee, Dhaval Daftari. "Process planning and machining of generic virtual pockets by feature-composition approach". *Computers ind Engng*, 1997,**33(1-2)**:409-412
5. Rosenman M A, Gero JS. Modelling multiple views of design objects in a collaborative CAD environment [J]. Computer- Aided Design, 1996,3(28): 193-205.
6. Fan Yuqing. *Modern Airplane Manufacturing Technology* [M]. Beijing: Beijing University of Aeronautics and Astronautics Press, 2001: 453-471(in Chinese)
7. Spatial Technology. *ACIS Online Documentation* [EB], Spatial Technology, 1998

SIMULATION OF THE MACHINING ERROR OF CNC GEAR SHAPER BASED ON MULTI-BODY SYSTEM THEORY

Shuxin Wang[1], Yanzheng Li[2], and Jintian Yun[3]

[1]*School of Mechanical Engineering, Tianjin University, Tianjin, P.R.C., Email: shuxinw@tju.edu.cn,* [2] *School of Mechanical Engineering, Tianjin University, Tianjin, P.R.C., Email: yz.li@eyou.com,* [3] *School of Mechanical Engineering, Tianjin University, Tianjin, P.R.C., Email: yunjintian@sohu.com.*

Abstract: A general method of error analysis of Multi-body Systems (MBS) is presented, which can be used in error analysis of any project object. Based on this method, this paper analyzed kinematics model of CNC gear shaper and verified by experiment.

Key words: Simulation, multi-body system, CNC gear shaper, machining error

1.   INTRODUCTION

Currently, virtual prototype (VP) technology has become one of the research hotspots in design and manufacture of mechanical system [1-2]. It is a synthetic technology based on special techniques and modeling techniques related to products, information technique, virtual reality technique, AI technique, system technique and management technique, etc. VP can help manufacturers break the bottleneck caused by traditional dependence on hardware prototyping, and gain greater insight earlier in the development cycle, make quantifiable improvements. Some influential VP development tools (software) appeared, such as ADAMS [3] and DADS [4]. They are widely used in many manufacturing companies. But up to now, these kinds of software can only fulfil following functions: interference checking, dynamic demonstration, dynamic characteristics analysis, etc. But they can not simulate the actual working situation of real prototyping, namely, some influencing factors are not considered, such as motion error of machinery, clearance between each component, and so on. Thus, it is essential to study VR by integrating specific engineering object, developing the new kernel technique of VP.

Due to the complexity of configuration and motion of gear cutting machine, it was used as an example in this paper. The goal of this paper is to describe the actual motion situation of each component of machine tools and to simulate the machining error of spur gear generated by CNC gear shaper, mainly focus on the gear tooth profile error. Based on the theory of multi-body systems [5] and the motion of CNC gear shaper, kinematic model of CNC gear shaper was developed. The model included original position errors and motion errors that account for geometric errors owing to manufacture, installation, motion control inaccuracy, etc. Then, the effect of geometric errors on the gear profile is analyzed and simulated. Finally, a series of simulations and experiments results show that this method is feasible.

2.   KINEMATIC MODEL OF MBS CONSIDERING ERROR FACTORS

Multi-body system (MBS) is a complicated mechanical system, which is composed of rigid bodies or flexible bodies by certain means. MBS has strong generality, universality and flexibility, and it is mainly used to solve the dynamics problem of a complex mechanical system. Actually, the error factors that exist in a practical system have still not been considered in this theory. In order to use this method to resolve this problem, the error analysis method of MBS must be developed.

Fig. 1 illustrates the new description method for a pair of adjacent bodies in MBS, in which, ideal position vector q_k^l, position error vector q_k^e, ideal position matrix $[A_{SV}]_P$, position error matrix $[A_{SV}]_{Pe}$, ideal movement vector s_k^l, movement error vector s_k^e, ideal movement matrix $[A_{SV}]_D$, movement error matrix $[A_{SV}]_{De}$ are used to depict the relationship of a pair of adjacent bodies. When the typical body B_K is moving relatively to the adjacent body B_J under the situation with error, the position equation for arbitrary point p of any body B_k can be obtained:

$$\begin{bmatrix} \{p_0\} \\ 1 \end{bmatrix} = \begin{bmatrix} \prod_{t=u}^{0} ([A_{SV}]_P [A_{SV}]_{Pe} [A_{SV}]_D [A_{SV}]_{De}) \end{bmatrix} \begin{bmatrix} \{p_k\} \\ 1 \end{bmatrix} \qquad (1)$$

where,

S —number of adjacent lower body

V —number of typical body

$V = L^t(K), L^u(K) = 1, S = L(V), L^0(K) = K$

L — lower body operator

t, u, K — positive integer number

$\{p_o\}$ —radius vector of point p on the typical body

$\{p_k\}$ —radius vector of point p in the inertial frame

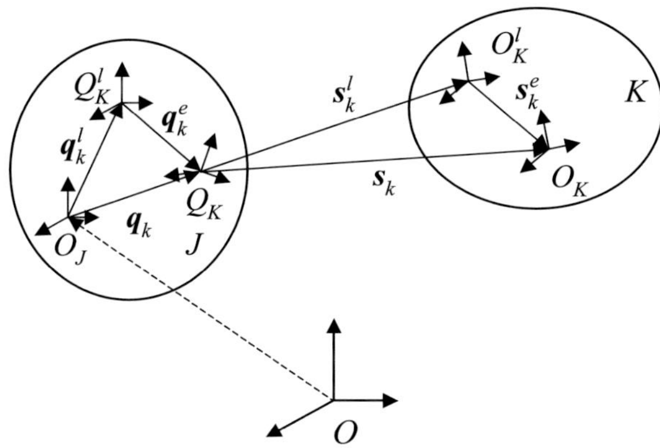

Figure. 1 Geometric relation between adjacent bodies considering errors

Based on the equation (1), it is convenient to deduce the actual position of any point in the inertial frame under the condition of error.

3.   KINEMATICS MODELING METHOD FOR CNC GEAR SHAPER

3.1   Topological structure description for CNC gear shaper

Fig. 2 illustrates the investigated CNC gear shaper. The whole machine is modeled as a   kinematic chain with many bodies connected in series by prismatic and rotational joints. It can be considered a kind of typical multibody system. Because the effect of error, each body has three positional errors $\delta_{VDx}, \delta_{VDy}, \delta_{VDz}$ and three angular errors $\varepsilon_{VDx}, \varepsilon_{VDy}, \varepsilon_{VDz}$ associated with its motion [6]. Fig. 3 and table 1 show the topologic construction of CNC gear shaper and its lower body array, respectively.

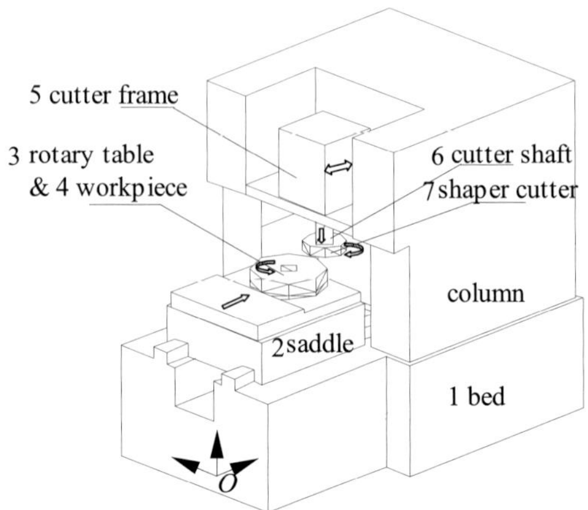

Figure 2. The mechanical structure of CNC gear shaper

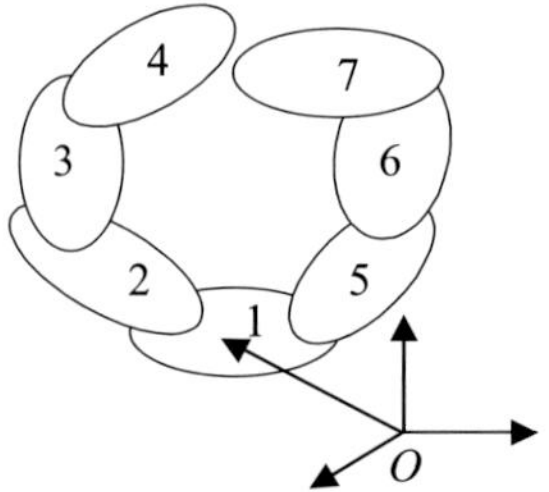

Figure 3. The topologic construction for CNC gear shaper

Table 1. The lower body array for CNC gear shaper

Typical body K	1	2	3	4	5	6	7
$L^0(K)$	1	2	3	4	5	6	7
$L^1(K)$	0	1	2	1	1	5	6
$L^2(K)$	0	0	1	0	0	1	5
$L^3(K)$	0	0	0	0	0	0	1
$L^4(K)$	0	0	0	0	0	0	0

3.2   Kinematics modeling for CNC gear shaper

In Fig. 4, coordinate systems $S(x,y,z)$, $S_p(x_p,y_p,z_p)$, $S_1(x_1,y_1,z_1)$, and $S_2(x_2,y_2,z_2)$ have been set up. Coordinate system S_1 is attached to the shaper cutter. Coordinate S_2 is attached to the generated gear and possesses a center distance a with respect to the coordinate system S_1. Coordinate systems S and S_p are the reference coordinate systems for coordinate systems S_1 and S_2, respectively. In the cutting simulation, the gear shaper cutter rotates about axis z through an angle φ_1 with respect to the reference coordinate system S while the generated gear rotates about axis z_p through an angle φ_2 with respect to the reference coordinate system S_p, meanwhile, the shaper cutter translates a distance l_1 along axis z.

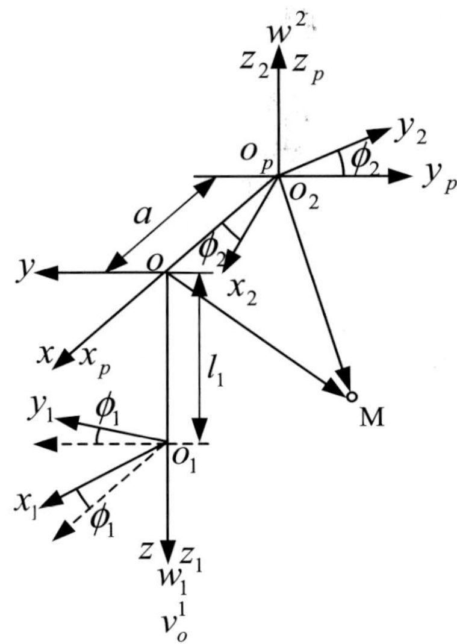

Figure 4. Coordinate system for CNC gear shaper

The transformation matrix between S_1 and S_2 is as follows [7]:

$$M_{21} = M_{2p}M_{p0}M_{01} \qquad (2)$$

When considering geometric error, the reference coordinate system of cutter and workpiece S and S_p can be replaced by the coordinate system of typical body 5 and 2, respectively; the coordinate systems attached to cutter and workpiece are still S_1, S_2. Then,

$$M_{01} = [A_{56}]_D [A_{56}]_{De} [A_{67}]_P \qquad (3)$$

$$M_{p0} = [A_{12}]_{De}^{-1} [A_{12}]_D^{-1} [A_{15}]_P [A_{15}]_{Pe} [A_{15}]_D [A_{15}]_{De} \qquad (4)$$

$$M_{2p} = [A_{34}]_P^{-1} [A_{23}]_{De}^{-1} [A_{23}]_D^{-1} [A_{23}]_P^{-1} \qquad (5)$$

and the surface expression of generated gear is:

$$D = [x_2, y_2, z_2, 1]^T = M_{2p} M_{p0} M_{01} [x_1, y_1, z_1, 1]^T \qquad (6)$$

where, $[x_1, y_1, z_1]^T$ is the coordinate of gear cutter in its own coordinate system.

4.   SIMULATION AND EXPERIMENT

In this section, simulation example is studied to investigate how the error affect the generated gear. Some parameters for the generated spur gear are: teeth number=50; module= 4 mm; pressure angle=20°; suppose all the linear error are 10 μm, the angle errors are 0.01°. In order to verify which error factor affect the machining precision of generated gear prominently, each error factor is substituted into equation(6) each time. As a result, it is concluded that the effect of rotation error relevant to workpiece and cutter is significant. Among these errors, the angular errors around the rotation axis of workpiece and cutter are the main factor that affect the quality of generated gear. Fig. 5 shows the contrast of ideal tooth profile and actual tooth profile (the rotational error are considered).

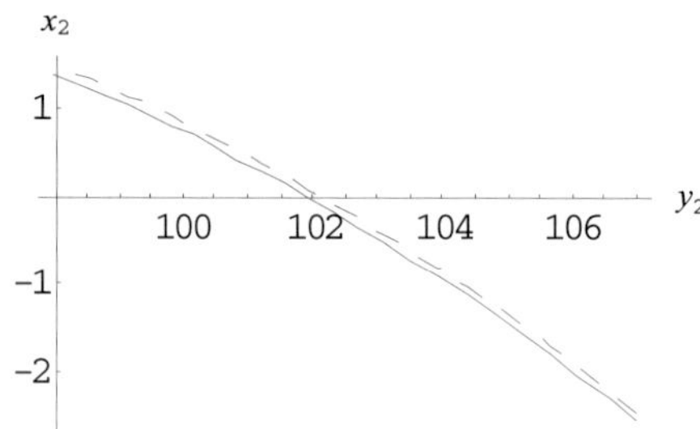

Figure 5. Contrast of ideal and actual tooth profile
 ···· actual tooth profile − ideal tooth profile

Therefor, we devise a experiment to measure the rotational error. By error compensation, the rotational error decrease evidently(see Fig. 6).

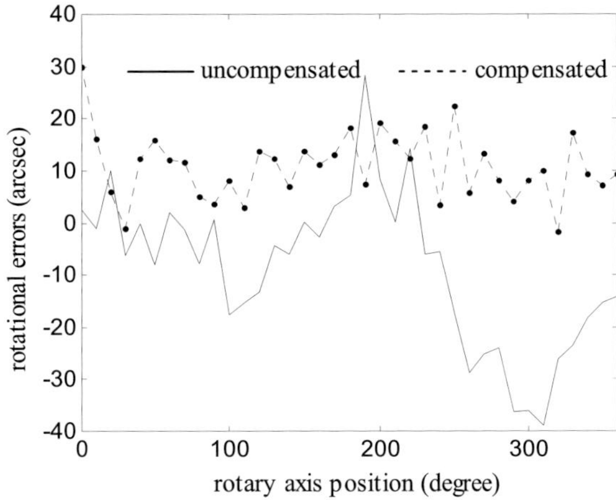

Figure 6. Rotational error of rotary table with and without compensation

The measurement results of generated gear are in Tab. 2.

Table 2. Measurement results of gear (unit: *um*)

	uncompensated	compensated
Tooth profile error	21	13
Accumulated error	30	20
Pitch error	8	6

CONCLUSION

Kinematics model of MBS considering error factors was set up. Based on it, this paper built kinematics model of CNC gear shaper considering error factors that included original position errors and motion errors which account for geometric errors owing to manufacture, installation, motion control inaccuracy, etc. Then, the effect of each error factors on generated gear was analyzed. After compensation, the quality of gear was improved, which proved that this method is effective.

REFERENCES

1. Zorriassatine F., Wykes C., Parkin R. "A Survey of Virtual Prototyping Techniques for Mechanical Product Development". *Proceedings of the Institution of Mechanical Engineers, Part B: Journal of Engineering Manufacture*, **Vol.217**, pp. 513-530, 2003.
2. Zhao W, Zhang L, Li S. "Concurrent Design Methodology Base on the Incremental Virtual Prototyping". *Journal of Computer-Aided Design and Computer Graphics* (in Chinese), **Vol. 15**, pp.475-479, April 2003.
3. *Mechanical dynamics*, INC., *ADAMS Manual*, **V 12**, 2002.
4. DADS Manual Revision 9.0. CADS1, 1999.

5. Huston RL, Liu Y. *Multi-body System Dynamics*, Tianjin University Press, Tianjin, China 1987.
6. Srivastava AK, Veldhuis SC, Elbestawit MA. "Modeling Geometric and Thermal Errors in a Five-axis CNC Machine Tool", *International Journal of Machine Tools & Manufacture*, **Vol. 9**, pp.1321-1327, 1995.
7. Wu X. *Principle of Gear Engagement*, Mechanical Industry Press. Beijing, China, 1982.

MULTI-AGENT SYSTEMS FOR CRACKING SUPPLY CHAIN BULLWHIP EFFECT

Oboulhas Tsahat Conrad Onesime, Xiaofei Xu, Dechen Zhan and Kimutai Kimeli

School of Computer Science and Technology, Harbin Institute of Technology, 13th Flats Room 1303, 92 West Dazhi Street, Nangang District, P.O. Box 773, Harbin 150001, China, E-mail: oboulhas@yahoo.fr

Abstract: In today's highly competitive manufacturing environment, strong emphasis on efficient supply chain management require companies to face product demand that change significantly on a weekly basis, forcing them to hold excessive inventory in order to maintain acceptable customer service levels. In addition, large demand fluctuations make it difficult to manage equipment, transportation, and human resources. The goal of this paper is to use Multi-Agent Systems (MAS), in which agents simulate how companies order, produce and store products in order to improve supply chain (SC) efficiency by reducing amplification of demand variability called "the bullwhip effect" (BE) while keeping low inventories, good customer service and increasing business performance.

Key words: Bullwhip Effect (BE), Supply Chain (SC), Multi-Agent Systems (MAS).

1.   INTRODUCTION

Today's fierce competitive environment and strong emphasis on efficient supply chain management (SCM) require companies to quickly response to market demands. Compounding the challenges of speed, successful manufacturing companies routinely offer a variety of configurations for each product to address diverse market requirements. To manage their nearly constant stream of new or updated products, these companies require complex SC operations. A SC is a network of business units that enables the collection of raw material, its transformation into products and the delivery of these products to consumers through a distribution system. The aim of SCM is to manage these activities so that products go through the business network in the shortest time and at the lowest costs possible [1]. In SCM, it is important to make certain that the information among the SC partners is accurate. Information flow in a SC refer to non-physical flows among SC partners such as demand, order, inventory position, transportation information and finance data. Many manufacturers face product demand that changes significantly on a weekly basis, forcing them to hold excessive inventory in order to maintain acceptable customer service levels. Information distortion or variance amplification from one end of a SC to the other can be called BE. The most effective process for smoothing out the oscillations of the BE will be customers and suppliers understanding what drives demand and supply patterns. Thereafter, both collaboratively work to improve information quality and compress cycle times throughout the entire process. In this paper, we address intelligent agents, which improve SC efficiency by reducing amplification of demand variability while keeping low inventories and good customer service. To achieve this goal, we propose a decentralized coordination mechanism. The remainder of this paper is organized as follows. In Section 2 we introduce the notion of the BE. We also review the existing literature and the proposed methods to counter the effect that have been previously suggested. In Section 3 we present the Multi-Agent System (MAS) and show that agent-based approach potentially offer many advantages for the development of distributed intelligent manufacturing systems. The system architecture and negotiation process among agents are proposed. Section 4 presents the prototype implementation of the multi-agent system, while the conclusion is made in Section 5.

2.   BACKGROUND

2.1   Bullwhip Effect

The "Bullwhip Effect" is a term coined by Lee et al. [2-3] refers to the scenario where the orders to the supplier tend to have larger fluctuations than sales to the buyer and the distortion propagates upstream in an amplified form. Identified as far back as the early 1960's by Jay Forrest in his book Industrial Dynamics [4], it can also in some cases, be called variance amplification or information distortion. The key drive of the bullwhip effect appears to be that the variability of the estimates or the forecasts of customer demand seems to amplify as the orders move up the supply chain from the customer, through retailers and distributors to the manufacturer of the product or service (see figure1). Information distortion from one end of a supply chain to the other can lead to tremendous inefficiencies like:

- Excessive inventory investments throughout the supply chain
- Poor customer services in the form of product unavailability or long backlog
- Lost revenues due to shortages caused by the variations
- Missed production schedules
- Incorrect product forecasts
- Misguided capacity plans
- Inefficient transportation
- High additional costs such as expedited shipment and overtime.

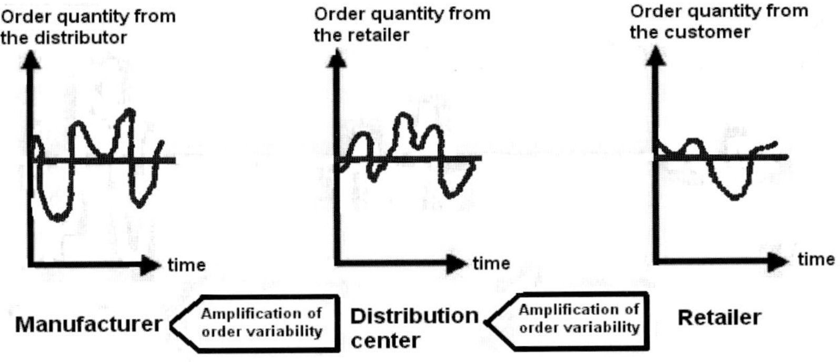

Figure 1. The Bullwhip Effect in supply chain

2.2 Framework for the review

Importantly SCM researchers have demonstrated the existence of the BE, quantified the phenomenon, and proposed methods to counter the effect. Lee et al. [2,3] identified four causes of the BE and proposed methods to counter the effect. Under positively correlated demand model, they derive the optimal order-up-to policy and show that the variance of retail order is strictly larger than that of retail demand. It is well know that variability of production volume may add cost to the production process [5]. Variability reduction methods have been proposed for various problems [6,7,8]. Schmenner [9] provides an historical overview of the problem, including a discussion on how Procter and Gamble have been concerned with bullwhip since 1919. Furthermore, we consider *demand signal processing and non-zero lead times* to be encompassed to the ''*demand amplification*'' or the ''*Forrester Effect*'' after Jay Forrester [4] who encountered the problem in many real-world supply chains and demonstrated it via DYNAMO simulation. The Forrester Effect is noticeable in traditional supply chains as a particular player over-orders in response to genuine changes in demand to account for inventory deviations that result from the production/distribution lead-time. This effect has been used in previous contributions [10,11] to clearly show how order variance amplifies up the supply chain. The Forrester Effect is also encompassed by Sterman's bounded rationality [12], terminology that is common in the field of psychology when used to describe player's sub-optimal but seemingly rational decision-making behaviour. After the identification of the BE, some efforts have been studied to reduce the BE so as to improve the SCM. It is, however, much harder to quantify the impact of bullwhip on the profitability of a company, although Metters [13] has studied the problem with a linear programming approach. He reports that elimination of the Bullwhip Effect can result in a 5% increase in profitability using managerially relevant parameter settings and that this saving can be even higher in a capacity-limited supply chain. Stalk and Hout [14] also report that the production on-costs is proportional to the cube of the deviation about the mean of the production order rate. So variations within the factory are reckoned to be extremely expensive. Some efforts have been studied to reduce the BE so as to improve the SCM. Information sharing can be used to improve the accuracy of demand forecast to as to reduce the process variability along SC. The four revealed key reasons for the occurrence of the bullwhip effect and some possible approaches to counteract are described in Table 1.

Table 1. Main reasons for the bullwhip effect and some possible counteracts approaches

Causes	Description	Counteracts approaches
Demand forecast updating	Demand forecasting updating is a major contributor to the bullwhip effect. The parties of the supply chain build their forecasts on the historical demand patterns of their immediate customers. As the lead times between the re-supply of the items along the supply chain are longer, the fluctuation is more significant.	Avoid multiple demand forecast updates: • Make demand data at a downstream site available to the upstream site • Implement EDI for data sharing • Upstream site use demand data for forecast updates and re-supply for the downstream data • Upstream site's direct access to demand data • Improve operational efficiency to shorten re-supply lead time (JIT)
Order batching	Order batching refers to the practice of placing orders down the supply chain in	Break order batches: • Information sharing (EDI or Internet ordering)

	batches in order to gain economies of scale in set-up activities. The order batching will appear in two different forms: periodic ordering and push ordering.	• Channel Alignment (Discount for truckload assortment, delivery appointments, composite distribution, Third-party logistics) • Operational Efficiency (Reduction in fixed cost of ordering by EDI or EC, CAO, etc.)
Rationing and shortage gaming:	The rationing and shortage gaming occurs when demand exceeds supply. If the producers once have met shortages with a rationing of customer deliveries, the customers will start to exaggerate their real needs when there is a fear that supply will not cover demand.	Eliminate gaming in shortage situations: • When shortage occurs, upstream sites allocate products in proportion to past sales records, in stead of order • Sharing of capacity and inventory and information • Introduction of stringent cancellation policies
Price variations	Price variations refer to practice of the offering products at reduced prices to stimulate demand. The price variations advantage is low price but the disadvantages are: Huge piles of inventory; payment of premium freight rates to transport; etc.	Stabilize prices: Reduce both the frequency and the level of wholesale price discounting. • Channel Alignment (continuous replenishment program (CRP)- plus CAO, Every Day Low Cost (EDLC)) • Operational Efficiency (Every Day Low Price (EDLP), Activity-Based Costing (ABC))

3.   MULTI-AGENT SYSTEM

Intelligent agents are a new paradigm for developing software applications. Franklin and Graesser [15] state that "an autonomous agent is a system situation within and part of an environment that senses that environment and acts on it, over time, in pursuit of its own agenda and so as to effect what it senses in the future". The term "Intelligent Agent" is particularly hard to define because there is no real agreement of exactly what an agent is. The reasons include the fact that the term agent is used in so many fields of computer science and artificial intelligence [16]. However most researchers are in broad agreement with the following agent properties:

Autonomy: The agent's ability to act without direct human intervention. The agent should have control of its own actions and internal state [17].

Flexibility: relates to the agents ability to respond to changes in its environment. Wooldridge and Jennings [18] state that agents, with respect to flexibility, should exhibit certain behavioral types. These behavior types state that an agent should be:

- *Social:* agents interact with other agents (and possibly humans) via some kind of agent communication language [19];

- *Responsive:* They should respond to changes in the environment;

- *Pro-activity:* Agents do not simply act in response to their environment; they are able to exhibit goal directed behaviors by taking the initiative.

An agent-based system may contain any non-zero number of agents. MAS consists of a set of autonomous agents which react according to information received from other agents and the environment. The intelligent agents work collectively to solve problems. Shen et al. [20] discussed the key issues and advantages of applying a MAS architecture for design and manufacturing. Maturana and Norrie [21] pointed out how mediator agent facilitates coordination and negotiation for problem solving in manufacturing environment. Hildum et al., [22] and Sadeh et al., [23] developed an agent-based architecture for supply chain planning. Their case studies demonstrated that MAS is a promising approach to address real world supply chain problems. In this paper, each agent represents an individual firm, which collaborates with other agents to solve a scheduling problem in a supply chain. MAS are successful due to several causes (we only mention those more relevant):

- They are able to solve *big size* problems, especially those where classical systems are not successful.
- They allow different systems to work interconnectively and cooperatively.
- They provide efficient solutions where information is distributed among different places.
- They allow software reusability; therefore there is more flexibility to adopt different agent capabilities to solve problems.

3.1 Multi-agent framework of cracking supply chain bullwhip effect

The architecture of MAS for cracking supply chain bullwhip is shown in Figure 2. This architecture has the following types of agents:

- *Customer agent:* is responsible for agents buying products from their retailers. It makes predictions of future consumption by the corresponding consumer and monitors the actual consumption, and sends information about this to their retailer's agent. Customer agent is very simple, because they only place orders according to a distribution law representing, different kinds of consumption patterns (e.g. constant, constant with little variations, seasonal, increasing, both increasing and seasonal, etc.)

- *Retailer agent*: The goal of the retailer agent is to reduce the turnaround time that the customer experiences, while keeping the inventory costs under control. Commitments of a retailer agent might include service constraints such as 98% of orders fulfilled within a day for top-priority customers. Retailer agent is responsible for acquiring orders from customers; negotiating with customers about prices, due dates and handling customer requests for modifying or cancelling their orders, and the like. When a customer order is changed, that change is communicated to the manufacturer agent. When plans violate constraints imposed by the customer (such as due date violation), the retailer agent negotiates with the customer, the distribution center or and the manufacturer (when there is a cross-dock or the supply chain does not have a distribution center) for a feasible plan. It makes predictions for the cluster and sends these to the manufacturer agent, and monitors the consumption of resources of the consumers in the cluster. If some consumer(s) use more resources than predicted, it redistributes resources within the cluster. If this is not possible, i.e., the total consumption in the cluster is larger than predicted, it will redistribute the resources available within the cluster according to some criteria, such as, fairness or priority, or it may take some other action, depending on the application.

- *Distribution Center agent*: This agent is involved in receiving products from the manufacturing plant and either storing them or sending them right away (cross dock) to the retailer. The main focus here is to reduce the inventory carried and maximize throughput. In a standard distribution center products come in from the manufacturing or supplier plants. They are unloaded and stored in the storage area. When orders come from the retailer, relevant products are removed from the storage area (if the buffer has them or they wait until the products arrive into the buffer) and are sent to the appropriate loading dock where they are loaded and sent to the destination.

- *Manufacturer agent:* receives predictions of consumption and monitors the actual consumption of consumers through the information it receives from the distribution center agent or retailer agent (when there is a cross-dock or the supply chain does not have a distribution center). The main foci here are on optimal procurement of components (particularly common components) and on efficient management of inventory and manufacturing process. If necessary, e.g., if the manufacturer cannot produce the amount of resources demanded by the consumers, the manufacturer agent may notify the consumers about this. This agent is responsible for coordinating the plants and suppliers to achieve the best possible results in terms of the goals of the supply chain, including on-time delivery, cost minimization, and so forth.

- *Supplier agent:* are responsible supply resources (raw materials, parts and so on) to the manufacturing plant. They focus on low turnaround time and inventory. Their operation is characterized by the supplier contracts, which determine the lead-time, flexibility arrangements, cost sharing, and information sharing with customers.

Figure 2. Agent-based Architecture for cracking supply chain bullwhip effect

3.2   Negotiation and communication among agents

Bussmann and Muller [24] define the *negotiation as the communication process of a group of agents in order to reach a mutually accepted agreement on some matter.* For example negotiation with retailer about prices, due dates, and so on. Hence, the cooperative negotiation strategy can be applied to MAS where the individual agents are cooperative and will collaborate in order to achieve a common goal for the best interest of the system as a whole. Agents are connected and communicate using the Internet/Intranet or local networks. To enable effective inter-agent communication, agents that work together have to use an interoperable, platform-independent, and semantically unambiguous communication protocol. The two most widely used agent communication languages (ACL) are the KQML [25, 26] and the FIPA (The Foundation for Intelligent Physical Agents). These languages are highly versatile and therefore can be used in conversations and not only for message passing. The newest language at this moment is the JKQML (Java-based Knowledge Query and Manipulation Language).

4.   CONCLUSION

In this paper we have proposed a multi-agent systems framework for cracking the bullwhip effect that takes place in supply chains. We largely described the causal factors driving the bullwhip effect, which appear complex and have significant economic consequences. Thus, we think that the framework that we have outlined can play an important role of monitoring the information flow and tracking demand variability among supply chain participants by intelligent agents, which are software programs designed to accomplish specific tasks like "real" human agents with specialized skills. A computerized implementation of our results is presently under way. We expect to be able to improve SC efficiency by reducing amplification of demand variability while keeping low inventories and good customer service for the fine Europe-Asia markets.

REFERENCES

1. Lee, H. L., and C. Billington: "Evolution of supply chain management models and practice at Hewlett-Packard Company". *Interfaces*, **Vol. 25**, No. 5, pp. 42-63, September to October 1995.
2. Lee, H.L., Padmanabhan, P., Whang, S.: "Information distortion in a supply chain: The Bullwhip Effect". *Management Science* **43**, pp. 543–558, 1997.
3. Lee, H.L., Padmanabhan, V., Whang, S.: "The bullwhip effect in supply chains". *Sloan Management Review* **38 (3)**, pp. 93–102, 1997.
4. Forrester, J.W.. *Industrial Dynamics*. Published jointly by MIT Press, Cambridge,MA, and J. Wiley & Sons, Inc., NY, pp. 464, 1961.
5. Lee, H. L. and C. Billington. "Material management in decentralized supply chains", *Operations Research*, **41(5)**, pp. 835-847, 1993.
6. Lee, H. and C. Tang. "Variability reduction through operations reversal". *Management Science*, **44 (2)**, pp. 162-172, 1998.
7. Lee, H. and C. Tang. "Modeling the costs and benefits of delayed product differentiation". *Management Science*, **43 (1)**, pp.40-53,1997.
8. Cachon, G. P. "Management supply chain demand variability with scheduled ordering policies", *Management Science*, **45(6)**, pp.843-856, 1999.
9. Schmenner, R.W. "Looking a head by looking back: Swift, even flow in the history of manufacturing". *Production and Operations Management* **10 (1)**, pp.87–96, 2001.
10. Mason-Jones, R., Naim, M.M., Towill, D. R. "The impact of pipeline control on supply chain dynamics", *International Journal of Logistics Management* **8 (2)**, 47–62, 1997.
11. Van Aken, J.E. *On the control of complex industrial organizations*, Doctoral Dissertation, Martinus Nijhoff Social Sciences Division, Leiden, The Netherlands, 1978.
12. Sterman J. D. "Modeling managerial behavior: Misperceptions of feedback in a dynamic decision making experiment", *Management Science*, **35(3)**, pp. 321-339, 1989.
13. Metters, R. "Quantifying the Bullwhip Effect in supply chains". *Journal of Operations Management* **15**, pp. 89 –100,1997.
14. Stalk, G. and Hout, T.M. "Competing Against Time: How Time-Based Competition is Reshaping Global Markets". *The Free Press*, New York, 1990.
15. Franklin, S. and A. Graesser. "Is it an Agent, or Just a Program?: A Taxonomy for Autonomous Agents". *Proceedings of the Third International Workshop on Agent Theories, Architectures, and Languages*, Springer-Verlag, 1996.
16. Jennings N.R. and M. Wooldridge. "Agent Technology: Foundations, Applications, and Markets", ISBN: 3-540-63591-2, Springer-Verlag, pp3-28, 1998.
17. Jennings, N., K. Sycara, K. and M. Wooldridge. "A roadmap of agent research and development". *Journal of Autonomous Agents and Multi-Agent Systems*, **1(1)**, pp.7-38, 1998.
18. Wooldridge, M.J. and N.R. Jennings. "Intelligent Agents: Theory and Practice", *Knowledge Engineering Review*, **10(2)**, pp.115-152, 1995.

19. Genesereth M. R. and Ketchpel S. P. "Software agents", *Communications of the ACM, Special Issue on Intelligent Agents,* **37(7)**, pp.48-53. 1994.
20. Shen, W., F. Matuaran and D. H. Norrie. "MetaMorph II: an agent-based architecture for distributed intelligent design and manufacturing". *Journal of Intelligent Manufacturing,* **Vol.11**, pp.237-251, 2000.
21. Maturana, F. P. and Norrie D.H. "Distributed decision-making using the contract net within a mediator architecture", *Decision Support Systems,* **Vol.20**, pp.53-64, 1997.
22. Hildum, D. W., N.M. Sadeh, T.J. Laliberty, J. McAiNulty, S.F. Smith and D. Kjenstad. "Blackboard Agents for Mixed-Initiative Management of Integrated Process-Planning Production-Scheduling Solutions Across the Supply Chain". *The Ninth Conference on Innovative Applications of Artificial Intelligence,* Providence, RI, 1997.
23. Sadeh, N. M., D.W. Hildum, D. Kjenstad and A. Tsen. "MASCOT: An Agent-Based Architecture for Coordinated Mixed-Initiative Supply Chain Planning and Scheduling", *In Workshop on Agent-Based Decision Support in Managing the Internet-Enabled Supply-Chain, at Agents '99,* pp. 133-138, 1999.
24. Bussmann, S. and J. Muller. "A negotiation framework for cooperating agents". *Proceedings of CKBS-SIG,* Dake Center, University of Keele, pp. 1-17, 1992.
25. Finin T., J. Weber, G. Wiederhold, M. Genesereth, R. Fritzson, J. McGuire, S. Shapiro and C. Beck. "Draft specification of the KQML agent-communication language", *The DARPA knowledge sharing initiative – external interfaces working group,* 1993.
26. Kraus S. "Negotiation and cooperation in multi-agent environments", *Artificial Intelligence,* **94**, pp.79-97, 1997

A 3D DIGITIZING MEASUREMENT SYSTEM BASED ON NOVEL PASSIVE ROBOT

Li-Bin Du, Xiao-Hui Gao, Hao Wang, Jing-Dong Zhao and Hong Liu
(Robotics Research Institute, Harbin Institute of Technology, P.R.China, dulibinhit@yahoo.com.cn)

Abstract: A 3D Digitising Measurement System based on a novel passive robot, together with its measuring strategy, applying field, and components has been introduced at the beginning of this paper. Based on these, the paper presents the design of the passive robot, puts forward a mathematic model of the system, and analyses the robot's working space. This paper realizes chipset design for multi-encoder outputs by utilizing FPGA. In the end, experimental results show the reliability and accuracy of the system .

Key words: reverse engineering; passive robot; digital; reconstruction.

1.   INTRODUCTION

With the development of CAD and CAM, study on reverse engineering technology which process is shown in Figure 1 attracts more attention. As one of the key techniques in reverse engineering which includes digitising, pre-processing of data, triangulation and

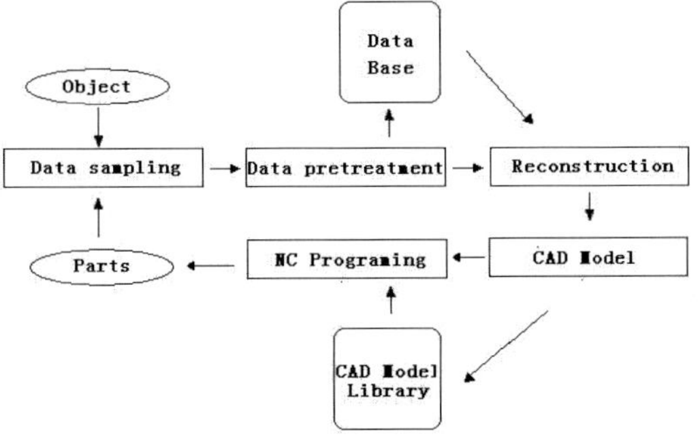

Figure1. Process of reverse engineering

reconstruction of three-dimension solid surface, digitising technology is the first important step in reverse engineering. Three-dimension modeling from reality is the process of creating 3D digital models of real-world objects and environments. Currently there are two kind of digitising methods [1], according to the way the robot reacts with the object. One is contact method, the other is non-contact method. As an example of contact measurement, coordinate measuring machine (CMM) has many advantages such as high precision, high adaptability and so on, but it also has many backwards such as low efficiency, none measurable to soft objects and brittle objects because of the existence of measuring force. Non-contact measurement is divided as optical, ultrasonic, electromagnetic, etc. In this paper, a novel passive robot system for 3D digitising measurement is presented.

2. MEASURING PRINCIPLE

Figure 2 is the schematic diagram of the measuring system. The system is composed of a passive robot, sensor data collecting and processing system. The passive robot carries the sensor and performs the measurement. At the end of the robot there is a laser scanning sensor, based on the double- trigonometry

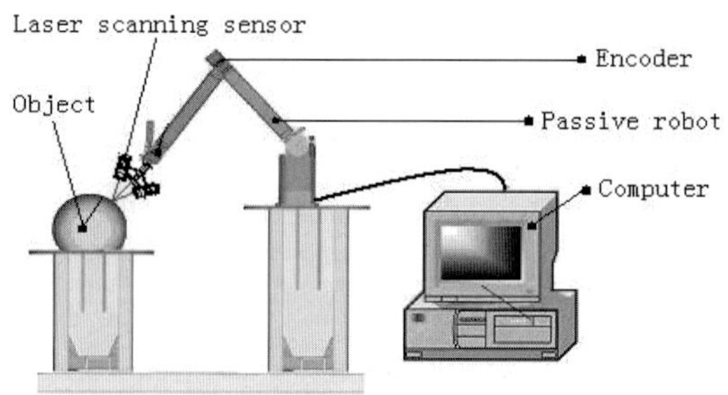

Figure 2. Schematic diagram of the measuring system

theory. The data collecting and processing system consists of two parts. One is for optical encoders which detect the position of robot's joints, and the other is for laser range finder that gives out the distance between the object and the robot. The computer processes the sensor data and reconstructs the object surface.

The measuring process is as follows: When operator handles the end of the robot, 3D scanning sensor will move together with the robot. The position and orientation of the end point of the passive robot will be calculated by forward kinematics through encoders in the robot joints[2- 3]. Thus we can get positions (x, y, z) of the measured object in the passive robot coordinates. And these scattered data can be output in the form of DXF, BIN and TXT.

3.   DESIGN OF THE PASSIVE ROBOT

In fact, the passive robot is a multi-DOF robot without driving actuators. In each of its joints, there assembles an angular sensor for measuring the angle of joints. The end of the robot will be driven by an operator at work.

In order to sample the object completely, the end of the robot must be able to reach more volume and orientation in the space as possible as it can. This means that the robot should possess six DOF at least. Considering more DOFs will increase complexity and decrease the precision of the system, the desirable design is to give the robot six DOF. Every joint is a rotational one with a high-revolution encoder. The characteristics of the passive robot (Fig.3) in this paper are as follows:

(1)Parallel axis structure. It means the axes of link I and link II stagger and parallel. Link I and link II are connected by a rotational joint. The merit of it is that all of the joints can reach the maximal turning angles (joint II is $180°$, others are nearly $360°$ or even more) and enlarge the movement space of the robot greatly [4]. The robot can be folded, which makes it easy to be transported.

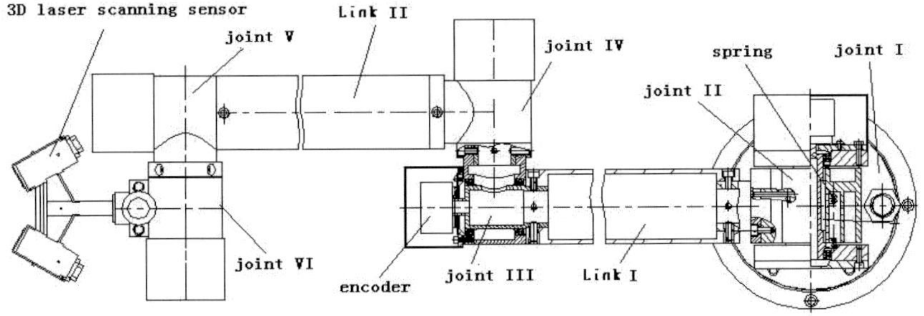

Figure 3. Structure of the passive robot

(2)Modularisation design. The same three pairs of rotational joints are installed, which can be installed or disassembled easily.

(3)Handy operating. The joint of the passive robot is installed with a balancing device to balance the robot's weight itself, which can make the operator handle the robot breezily.

For balancing device, a torsional spring is designed in joint II to balance the weight of link I and link II partly. As the spring is high integrated inside the joint, the whole robot has a good outside appearance.

The parameters [5] of the robot must satisfy the need of movement space and be reasonable, so the design of the arms and the revolution of the position sensors are important parameters which decide the robot's measuring precision. Once the measuring precision is determined, the longer the link is, the higher the revolution of the sensors needed. In this paper, the whole length is 1000mm,which will meet most measurement.

In order to meet the need of precision, two high-precision encoders are installed, which will be used to measure joint I and joint II. The encoder's impulse is 25000 per circle, namely 52″. And it will be 13″ after four-subdivision to the impulse signal. Considering the cost of the passive robot, the last four joints use high capability encorder with hollow axes, which has 2048 impulses per round, namely 10.5′. And it will be advanced to 2.6′ after four-subdivision. Considering the limit condition, the positioning accuracy of the robot can reach 0.094mm.

4.   MODELING OF THE SYSTEM

The passive robot has six degrees of freedom. Figure 4 shows the coordinates of robot based on the robot kinematics. The Denavit-Hartenberg[6] parameters of the robot are shown in Table 1.

Transformation matrix between robot joints can be calculated from Aii-1 (i=1…6) the last transformation matrix A76 represents transformation from coordinate O6 to coordinate O7 (tool coordinate).

Transformation matrix of the robot is:

$$T = A_0^1 A_1^2 A_2^3 \cdots A_6^7 \qquad (1)$$

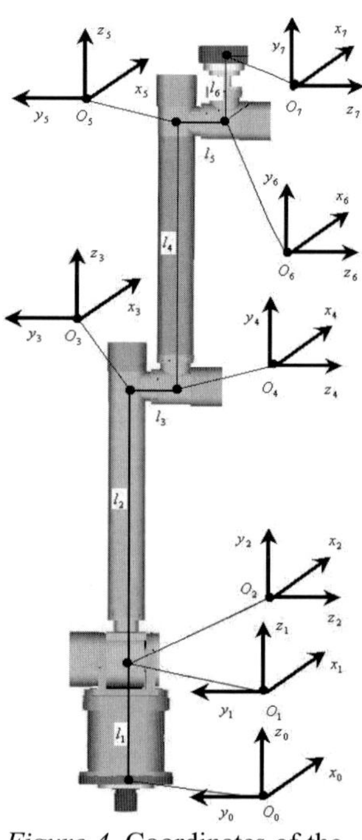

Figure 4. Coordinates of the passive robot

Table 1. D-H parameters of the passive robot

i	α_{i-1}	a_{i-1}	d_i	θ_i
1	0	0	l_1	θ_1
2	90	0	0	θ_2
3	-90	0	l_2	θ_3
4	90	0	l_3	θ_4
5	-90	0	l_4	θ_4
6	90	0	l_5	θ_4

$$A_{i-1}^i = \begin{bmatrix} \cos\theta_i & -\sin\theta_i & 0 & a_{i-1} \\ \sin\theta_i \cos\alpha_{i-1} & \cos\theta_i \cos\alpha_{i-1} & -\sin\alpha_{i-1} & -\sin\alpha_{i-1} d_i \\ \sin\theta_i \sin\alpha_{i-1} & \cos\theta_i \sin\alpha_{i-1} & \cos\alpha_{i-1} & \cos\alpha_{i-1} d_i \\ 0 & 0 & 0 & 1 \end{bmatrix} \qquad A_6^7 = \begin{bmatrix} 1 & 0 & 0 & 0 \\ 0 & 1 & 0 & l_6 \\ 0 & 0 & 1 & 0 \\ 0 & 0 & 0 & 1 \end{bmatrix}$$

5. WORKING SPACE ANALYZING

Working space is an important specification for the passive robot [7], which is decided by the length of robot arms and rotation range of joints. It is necessary to analyze the relation between the maximal inscribing cube in working space and the lengths of robot links and rotation ranges of joints.

The working space of the passive robot is shown in Figure 5. Let a be the length of link I, b be the length of link II, rotation range of joint IV is $\pm\alpha$, thus radii of limit envelope spherical surface r1 and r2 can be calculated:

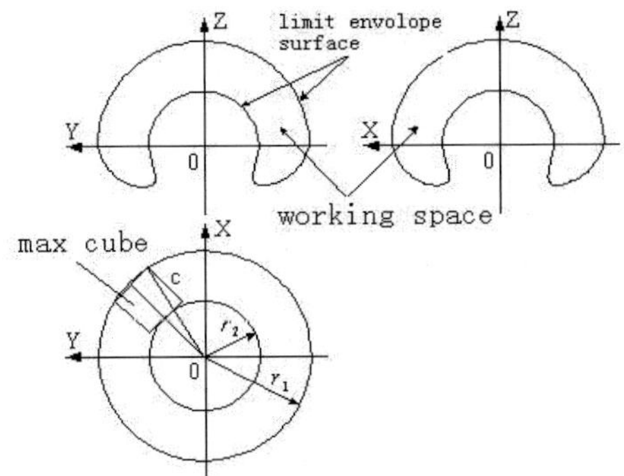

Figure 5. Schematic diagram of test space of the passive robot

$$r_1 = a + b \tag{2}$$

$$r_2 = \sqrt{a^2 + b^2 + 2ab\cos\alpha} \tag{3}$$

in the figure, c is the length of the maximal inscribing cube, then:

$$r_1^2 = \frac{c^2}{4} + (r_2 + c)^2 \tag{4}$$

let r_1 and r_2 substitute into this equation ,we can get c as :

$$c = \frac{4}{5}[\sqrt{(a+b)^2 + \frac{1}{2}(1 - \cos\alpha)ab} - \sqrt{(a+b)^2 - 2(1 - \cos\alpha)ab}\,] \tag{5}$$

When a+b is the fixed length sum, it needs (1-cosα)ab→max to make the cube maximal, namely : (1-cosα)→max and ab→max. The former can reach by increasing rotation range of joint IV. To realize the latter, let a=(a+b)x, then b=(a+b)(1-x), thus :

$$ab = (a+b)^2[\frac{1}{4}-(x-\frac{1}{2})^2] \to \max$$

When x=1/2, ab→max, namely when the length of the first arm is equal to the length of the second arm, the inscribing cube can reach maximum.

6.   SENSOR DATA SAMPLING AND PROCESSING SYSTEM

6.1   Sampling and Processing Unit for Optical Encoders

This unit is used to sample and process encoders of six robot joints and communicate with the computer. This paper realizes chipset design for multi-encoder outputs by utilizing FPGA (Field Programmable Gate Array).

Figure 6 shows the principle diagram of 6 channel encoder processing circuit based on FPGA. The signals of a encoder consist of phase A, B and Z. A and B are square signals for count, and their phase difference is 90 degrees. Z is the pulse signal that signs absolute zero position of encoder. Processing circuit includes subdivision, judgement of phase and reversible counting [8]. As shown in Fig.6, the address line controls multi-channel

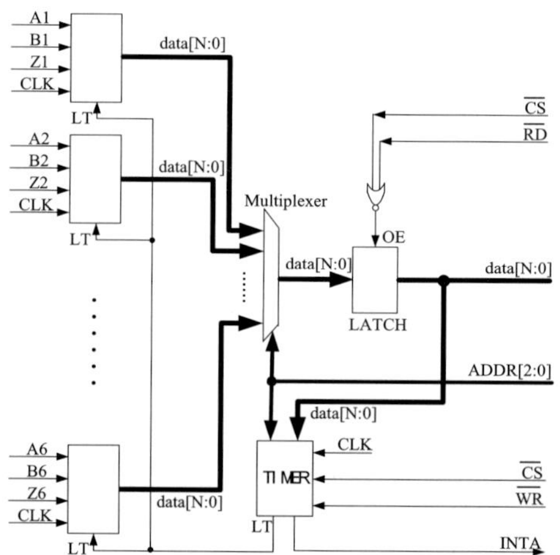

Figure 6. Principle diagram of 6 channel encoder processing circuit based on FPGA

switch to gate each counting signal, which will be exported through tristate buffer controlled by chip selection signal and reading signal. In order to support the computer for reading data by interrupt mode, an interrupt timer is designed. The timer is composed of counters which can load initial value and produce interrupt and control signals.

6.2   Sampling and Processing Unit of CCD Camera

At the end of the passive robot there is a laser range finder, which is composed of a 670nm line-laser generator and two CCD cameras. Double-trigonometry method [9] is used to measure the distance here. Two CCD cameras sample the image synchronously by using a double-channel image collecting card, whose resolution can reach: PAL 768 X 576 X 24BIT.

7.   EXPERIMENT

The robot system is calibrated by using a high-precision coordinate measuring machine (CMM) and the laser range finder is calibrated by means of improved Tsai [10] method. With the measuring system, a plaster head is scanned. Figure 7 shows the reconstructions of plaster head. (a) is the scattered data cloud of 8748 points; (b) is the prototype surface reconstructed with these points by regenerating algorithm.

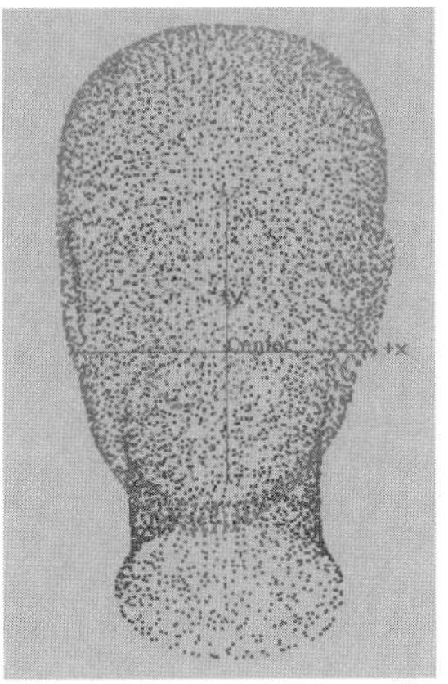
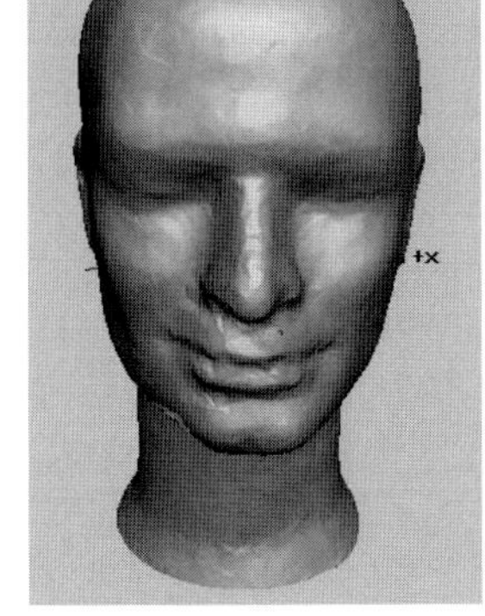

(a) scattered data of the plaster head   (b) reconstructed model of the plaster head

Figure 7. Diagram of head statuary reconstruction

8.   CONCLUSION

The paper presents the design process of a novel passive robot system for 3D digitising measurement. The passive robot is excellent in design, light and has large working space. The model of the passive robot realistically reflects its mechanical structure by mathematic expressions, and satisfies the demands of measuring and calculation. Experimental results prove the effectiveness of the measuring principle and the reliability of this measuring system.

REFERENCES

1. Zhou Ji, Zhou Yan-Hong. *Numerical control machining technology*, National defence industrial publishing company, BeiJing. 2002.1 : 209-249.
2. Cai Zi-xing. *Robotics*, Tsinghua university publishing company, Beijing, China. 2000.
3. A. Nowrowzi et al. "A Overview of Robot Measurement Technology", *The Industrial Robot*, 1988 Dec. : 230-232.
4. Ma Xiang-Feng. *Design of Industrial Robot*, Metallurgy industrial publishing company, 1996:pp.19-89.
5. Li Zhi-jie. *Study on the basical theory of industrial robot space trace moving accuracy*, Harbin, Dissertation of Harbin Institute of Technology, 1993 : pp.24-85.
6. He Xu-Chu. *Basic theory and computing method of general inverse matrix*, Science and technology publishing company of Shanghai ,1985 : pp.47-143.
7. Zhang Chao-Qun. *Research On Kinematic Performance Measurement of Industrial Robots & Theory of Identification of Geometric Parameters*, Harbin, Dissertation of Harbin Institute of Technology, 1998.
8. Han Zhuang-Zhi. "Quadruple analysis of encoder" . *Electron technology applying* ,2002 [**12**]:38~40.
9. Tang Chaos-Wei , Liang Xi-Chang , Zoo Chang-ping . "Laser Two-Triangle Measuring Method of 3D Curved Surface " ,*Metrology transaction*, 1994.4 : pp.99-103
10. R.Y.Tsai. "An efficient and accurate camera calibration technique for 3D machine vision", In Proc. *IEEE Conf. on Computer Vision and Pattern Recognition*, Miami Beach, FL 1986:pp.364~374.e.

RESEARCH OF REMOTE CONTROL SYSTEM BASED ON FIELDBUS AND INTERNET

Tan-Cheng Xie and Ke Hu
Electromechanical Engineering College, Henan University of Science and Technology,Luoyang, Henan 471003,China, xietc@mail.haust.edu.cn

Abstract: Important reasons for connecting Fieldbus systems to IP-based networks (Internet) are the provision of remote access for monitoring and maintenance purposes, but also the inclusion of automation systems into an enterprise-wide management scope. Existing applications of realization model are chiefly based on COM/DOCM/COM+ or CORBA, JAVA, etc. All of them have their own drawbacks. In this paper, we propose a different strategy using the Three Layer Model. This approach is largely Fieldbus-independent and, at the same time, interoperable with existing OPC technology. We discuss the architecture of remote control system and present a modular approach to cope with the variety of available Fieldbus protocols. We further propose a structure for a management information base suitable to represent the server and remote client in Three Layer Model.

Key words: Remote Control, Fieldbus, Internet.

1.   INTRODUCTION

Fieldbus was developed in 1980s, using ad hoc communication protocols, have been developed to interconnect field devices (e.g. sensors, actuators, controllers). It is a network system that is applied in plant fields and can provide serial and digital communication among computerized control devices and control room. It is also an open, digital and multidrop communication network. The technology of fieldbus plants special microprocessors into traditional instruments, and makes them have the capacities of calculating and digitally communicating. Local intelligence and fieldbus together provides new functions such as automatic calibration of field devices, self test, down loading of parameters, etc., with reduced cost of cabling. Using twisted-pair wires as transferring bus, fieldbus connects the field instruments to form a network system. Digitising, interoperating, distributed and fully opening are the characteristics of fieldbus. Nowadays, Profibus, FF, LonWorks, CAN, etc are excellent representatives in fieldbus world. [1]

Along with the economic combination of whole world and universal application of Internet, the information exchange system based on Internet is pervading rapidly from manufacturing realm to business realm, which cover with each aspects in plant including device control, management and so on. It is a great innovation on industrial automation area that makes the possibility of remote control on industry production process. Operator linked produce information on spot into Internet so that he can monitor and maintain producing course and all kinds of parameters without need to attend spot personally. On the other hand, scientist and technician can make advanced process control on remote object using abundant soft &hardware resources on local. Thereupon, lots of manpower and material resources can be saved. As a result of it, how to combine fieldbus technology with Internet consummately will become important research aspect on fieldbus.

2.   REMOTE CONTROL SYSTEM MODEL

2.1   Basic Concepts and Important Technique

In view of the requirement of practical application, remote control system must have several characters including integration, real-time, security, exoteric and irrespective of soft/hard environment; adopt normal standards like OPC, TCP/IP etc; base on the rational practical model of Internet. One of the most convincing reasons for the introduction of field area networks (FANs, fieldbus systems) was the ability to include them into a hierarchical network structure. Still, in the early days of automation networks, this possibility was hardly ever used. Only in the recent past has the interconnection with higher-level networks received more attention, largely driven by the increasing deployment of IP-based LANs and the success of the Internet. The benefits of LAN/FAN connections are evident: direct integration of fieldbus data into comprehensive, well-known data acquisition systems, remote access for monitoring and maintenance purposes, as well as the inclusion of automation systems into an enterprise-wide management scope. Finally, with the advent of the Internet, an interconnection on IP level promises seamless integration into worldwide corporate or public networks with virtually unlimited range and interesting applications in industrial as well as building or home automation. This perspective was also one of the starting points for a current trend to use Internet technology on the field level, thereby substituting the traditional field area networks and flattening the network hierarchy. [2] There is, of course, a severe restriction for the interconnection between fieldbus systems and Internet. Its feasibility is confined to applications where both fixed data rates and predefined response times are not of predominant interest. The reasons are simply the lack of real-time behavior in IP-based networks and the sometimes-unpredictable availability of connections. Hence, real-time as well as safety-relevant applications are better kept local on the FAN level unless appropriate qualities of service can be guaranteed. Another emerging problem with the connection of fieldbus systems to publicly accessible networks is security, which has its own interest but will be left aside here. Recently, several approaches have been put forward to tackle the problem of FAN/Internet connections. We disregard solutions that just connect distant segments of the same fieldbus type by encapsulating their messages and using the LAN or the Internet only as a transport medium. Rather, we focus on hierarchy approaches. All of them have in common that they aim at providing an abstract, if possible unified, interface to different, incompatible fieldbus systems. The details of the fieldbus shall be hidden from the user, who wishes to use standardized tools and readily available commercial off-the-shelf software to access the fieldbus. [3] Apart from this comprehensive endeavour, several other approaches rely on Windows-specific mechanisms like object linking and embedding (OLE), OLE for process control (OPC), or distributed common object model (DCOM) to provide an abstract interface and to distribute objects over a network. Especially for Internet connections, Web interfaces become increasingly popular. The main deficiency of this widely accepted solution for remote access is that the straightforward integration of large networks with several Web access points is difficult. If pieces of information contained in different Web pages (one per system) belong together, they must be correlated by the user, which quickly becomes unwieldy. What is needed instead is a way to represent data from different systems in a form that integrated management solutions understand.

2.2   Our Three Layer model

In order to joint field devices through Internet, we adopt B/S (Browser/Server) structure, make local monitor act as Server. The monitor gathers data on-site, then mapping virtual catalogue so that remote client computer may mode accord interlink present time, has mature convenient to 1 shows our model of remote

access. This with Internet criterion at the furthermore, it technology and realize. Figure realizable control system.

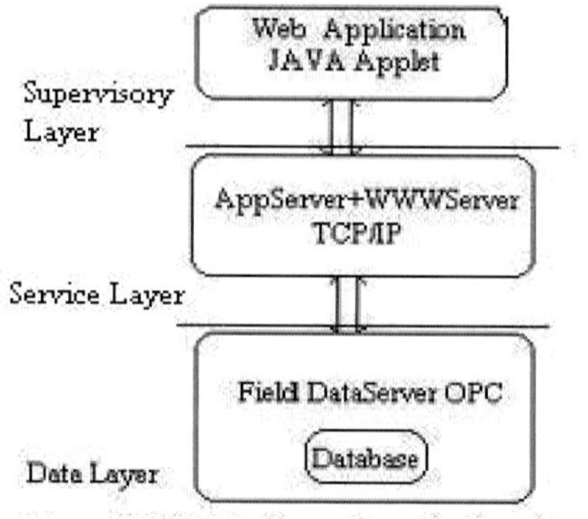

Figure 1. Model of remote control system

2.2.1   Supervisory layer

On the supervisory layer, which contains mainly about Web applications with Java applet embed in HTML pages, user may accomplish many functions aimed at manufactory such as monitor, control and configuration from long distance. All modules are showed as ActiveX control or dynamic mutual pages and are accomplished by Remote Data Service(RDS) client on Internet explorer. RDS client modules were installed along with IE. On the one hand, user does not need to consider how to configure them, on the other hand, RDS break the limit of static data on Web pages and realize dynamic database program on Web completely.

2.2.2   Service layer

On the service layer, which was made up of AppServer and WWW server, user may link up to the supervisory layer through TCP/IP so that control communications, or may link down to the data layer through OPC standard interface. Its modules run on the control of Server. User can access various data source through relevant data access operation.

2.2.3   Data layer

The data layer was made up of various data server on-site which support OPC standard interface. It provided user data save and access in order to select different data source including database, file system, mail system, etc.

3.   SYSTEM DESIGN AND REALIZATION

This system is designed to realize communication between application service program and data server on-site through OPC interface at real-time. Remote client can accomplish data communication with application server and WWW server by means of TCP/IP and HTTP protocol, furthermore, monitor the field process through normal explorer(Browser).

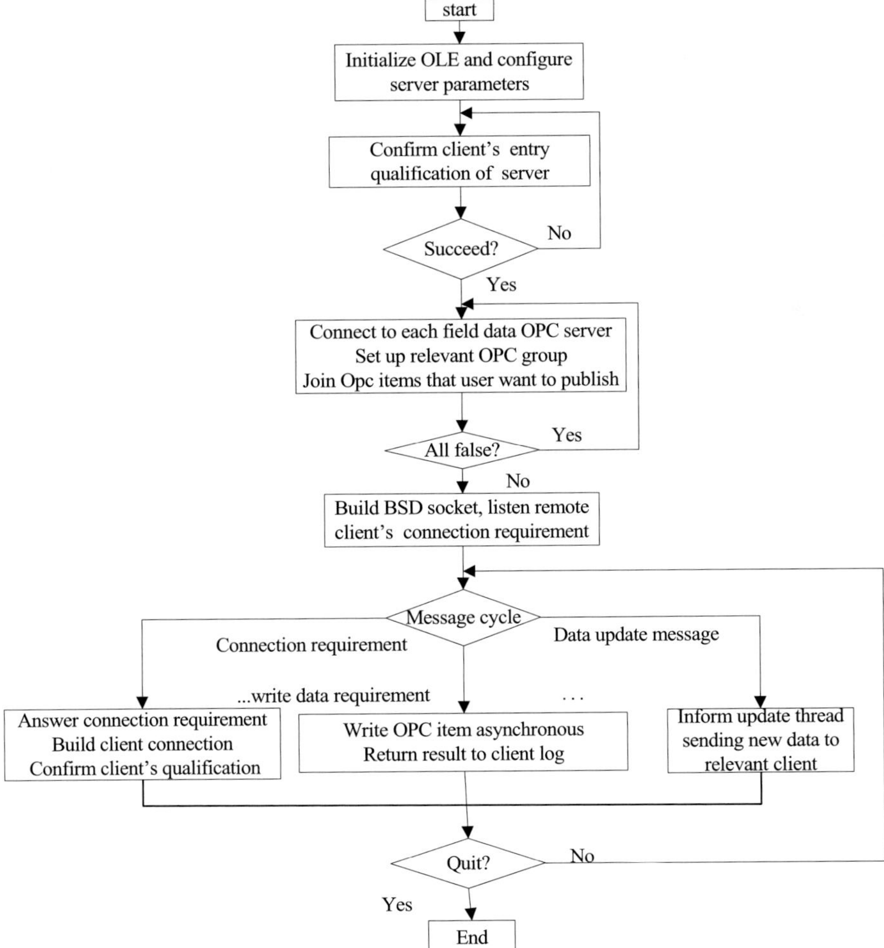

Figure 2 Basic disposal flow chart of Application Server

3.1   Server design

Application server is an independently executable service program which contains several modules design. Figure 2 shows the basic disposal flow chart of server. [4]

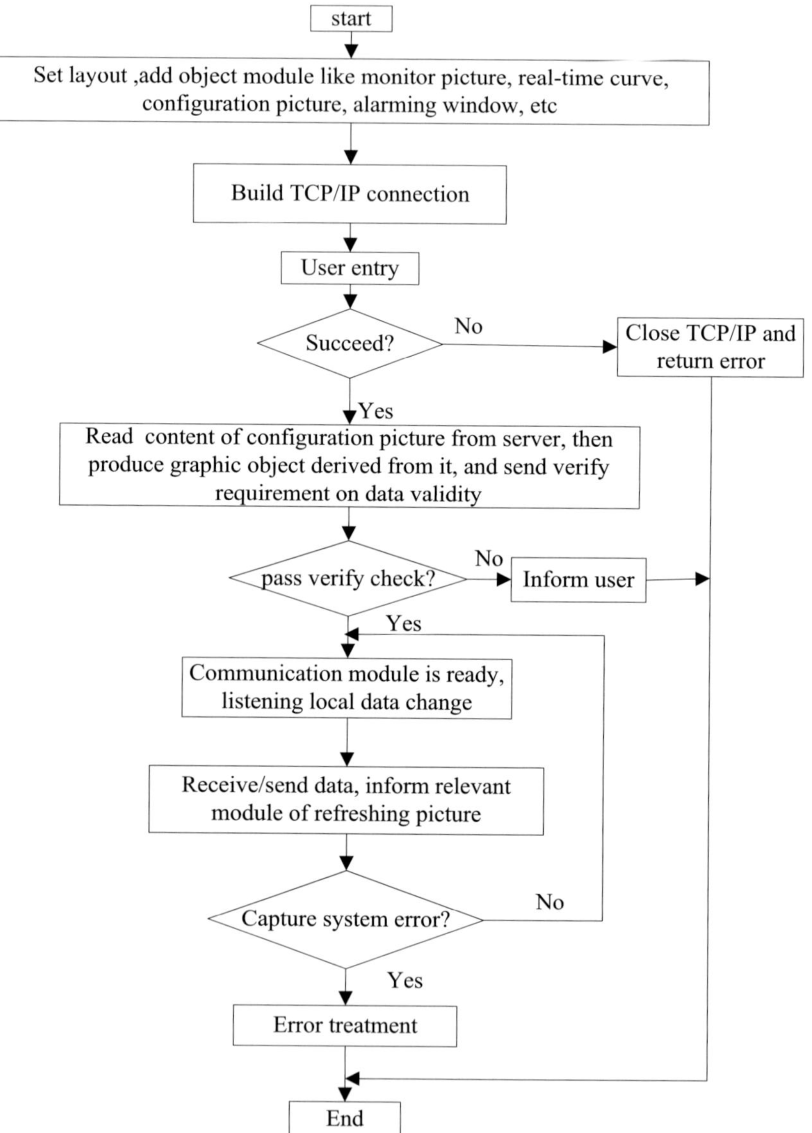

Figure 3 Basic disposal flow chart of Client

3.2   Client design

The client design also contains several aspects just like server. Figure 3 shows the basic disposal flow chart of client. [4]

4.   CONCLUSION AND OUTLOOK

This paper researches realizable model and chief technology of remote control system that based on fieldbus and Internet, proposes a three- layer model and analyses its feasibility and accomplished method.   Fieldbus technology based on Internet makes us achieve remote control of manufactory field. It has opening system structure so that user may extend function easily and link to MIS/CIMS system conveniently .As a result, manufactories will achieve integration of management, control and diagnosis, and enhance competition capabilities greatly.

REFERENCES

1.   Xing JianChun, Yang QiLaing, Wang Ping. "A new Design of DCS Based On PC and Fieldbus", *Proceedings of IEEE TENCON'02*, Beijing, China, Oct. 28-31, 2002, pp. 1416-1419.
2.   Wei Liang: "Research and Application of Profibus Fieldbus in Industry Control System". *Machine Tool Electric Apparatus*, pp.47-51, Feb. 2002.
3.   Chen JiQing, Yang Tao, Jia Bin: "Data Collection of Remote Field Based on Web". *Computer Application*, pp.79-81, June 2001.
4.   Zou YiRen, Ma ZengLiang, Pu Wei. *The Design and Development of Fieldbus Control System*, Beijing ,Defence Industry Publishing House, 2003.

INTEGRATION OF PRODUCT – PROCESS – ORGANISATION FOR ADVANCED ENGINEERING DESIGN

Philippe Girard[1] and Benoît Eynard[2]
[1]LAP – UMR 5131 CNRS, Groupe GRAI
Université Bordeaux 1, 351 cours de la Libération, F.33405 Talence, FRANCE
email: girard@lap.u-bordeaux1.fr, [2]LASMIS – FRE 2719 CNRS
Troyes University of Technology, 12 rue Marie Curie, BP 2060, F.10010 Troyes,
FRANCEemail: benoit.eynard@utt.fr

Abstract: The paper presents a survey on engineering design regarding three main fields: product modelling, design process and design coordination. An advanced approach named IPPOP and aiming at the integration of product-process-organisation is proposed. The IPPOP project aims at supporting design actors in their work by taking into account technological knowledge, dynamic evolution and co-ordination of activities. First, the, modelling of the knowledge used by actors and not only the geometrical product definition is described. Second, the modelling of the dynamic aspect of process and not only the static flow of data is proposed. Finally the co-coordination the collaborative work of design actors involved during the various steps of the product development to increase design performances is presented.

Key words: Product design, design process, organisation, collaborative system.

1.   INTRODUCTION

Since 60's, system theory and more specifically Simon's work on Artificial Science [1],[2] had influence most of research works on engineering design. Many scientific communities are concerned such as engineering, computer science, philosophy, psychology and management in many aspects such as phenomena, nature, cognitive models and computation of design [3], [4]. The key elements of engineering design are: 'humans', 'objects' and 'context' [5]. He shows that these elements interact together: behaviour of objects, object to object interactions, object and context interactions, interactions involving human(s), object(s) and contexts together.

In this paper a survey on engineering design is proposed regarding three main dimensions: product modelling, design process and design organization. Then the integration need is highlighted because these three dimensions have huge interactions during product development process. They should also be handled in the same framework in order to enable a consistent management of the artefacts, the activities and the coordination of engineering design. The proposal of integration of product-process-organisation is detailed in order to argue of an advanced approach of engineering design. The integration is illustrated based on a software kernel of the demonstrator currently developed in the IPPOP project. Finally the approach is summarised and the further works to be carried out are detailed.

2.   ENGINEERING DESIGN SURVEY

The product development process could be viewed as the set of activities aiming to fulfil the design objectives in a specific context. The design objectives concern product specifications. They are constrained by the enterprise organisation [6] and by the design steps and are influenced by technologies or human and physical resources [7].Design is mainly a human activity. Understanding the activities carried out by designers is very complex [8]. So, many design models have been proposed [4]. Perrin [9] classifies those models in five categories:
- a succession of hierarchical steps [10],
- iteration of an elementary design cycle [11], [12],
- the emerging phenomenon of self-organisation [13],
- cognitive processes [14], [15],
- communication and interactive mode [16], [17].

The review on engineering design focuses on product modelling, design process and coordination of design. The need to have a new approach of engineering design is highlighted because of the increasing of collaboration between designers. This paper aims at identifying the characteristics of the integration of product, process and organisation for an advanced approach of engineering design and manufacture.

2.1   Product modelling

Regarding the product modelling many models exist to represent concept, configurations and geometry. Several research works propose product models to support their product knowledge descriptions. The descriptions are generally based on a breakdown into several levels of abstraction. For example, the FBS model (Function-Behaviour-Structure) summarised in [18] use three levels while the domain theory put forward by [19] suggests four levels. The latter introduces a description of the concept of appropriate transformation or operation at process system level while [20] only consider the behaviour of components in order to fulfil product functions. This abstraction level breakdown also applies to the modelling layers suggested by [21]. On the whole, this model develops an analytical representation of product knowledge without any explicit link to the design rationale or the co-ordination of engineering activities. In fact, giving the design results only is insufficient because it means that information on the method used to find the solution is not provided. Some works propose to integrate Design requirements, Product modelling and more or less the design process [22], [23], [24]. But proposed models are generally very conceptual and do not working in the real world during designing.

2.2   Design process

Generic tools for activity or process modelling are available [25]. A basic survey can detail IDEFØ and IDEF3, Petri nets and GRAI nets. IDEFØ provides a model for an efficient and simple use [26]. It also offers a good representation of key elements of activity (input, output, control and mechanism). In [27] an integrated functional representation of concurrent engineering is proposed based on IDEFØ. Concerning IDEF3 [28], it offers description of process flows, precedence and causality relationship of activities and their logical junctions which allow to capture the behaviour and performance of the process. In [29], IDEF3 is used for identifying alternative scenarios that result in the successful completion of product development process. Petri nets [30] propose a structured description of process behaviour and allow performance assessment with associated mathematics tools. In [31], Petri nets are used for modelling data transformation in engineering design. GRAI nets [32] are based on three concepts: state or result, activity and support. In [33] an example of GRAI nets application is proposed.

2.3 Design organisation and coordination

The engineering design process has to be controlled not only in order to respect deadlines and budgets, but also to facilitate co-operation between people [34]. Consequently, the organisational dimension of product engineering design needs to be taken into consideration because the design results depend on efficient collaboration among the people concerned. Today design activities are not performed by single team from a single enterprise, but by several teams from several enterprises. The design teams are set up temporarily until a project is completed, then reassigned to different projects when the initial project is over. The new challenge in product engineering design is therefore to manage virtual design teams [35]. Academic and industrial works on control processes are very active since long time. The last decade some results focus on controlling (i.e. conducting, piloting or coordinating) of design activity [36], [37], [38], [39]. Generally the objective of those works is to control tasks, exchanged and shared data of the design process. One of the original features of GRAI model [37] is its capacity to model decision-making according to product and process considerations. The aim of GRAI model is to organise the engineering design departments according to performance objectives.

2.4 Synthesis

The above survey highlight interested key tool in order to handle and manage the data issued from engineering design. But the key elements of design require to be linked in an integrated approach. Multi-users / multi-abilities environment, market conditions, human and physical resources, management and capitalization of knowledge and know-how should be considered in the same view in order to ensure a consistent understanding of engineering design. The requirements of an integrated approach are:
- to integrate product and process knowledge, in order to improve the co-ordination of design activity,
- to make product and process representing tools evolve, so that the technological knowledge can be capitalised and reused during other design projects,
- to develop and set up an appropriate environment for cooperative design,
- to allow that concurrent design can be managed and coordinated, in order to continuously improve the engineering design performance,
- to deliver a support system.

3.   IPPOP: INTEGRATION OF PRODUCT, PRODUCT AND ORGANISATION

In order to answer to the above requirements a French consortium of research institutes and companies have submitted in 2001 the IPPOP proposal. This proposal has been submitted at the 2nd call of the RNTL program founded by the French government. The IPPOP project relies on research results, market products and partners' know-how in the subject of software development and transfer of technology. The Project aims at federating several research trends that emerge from French research institutes in the subject of product modelling, technological data bases, design process modelling and engineering design co-ordination. The foreseen software components will be made available to manufacturers or teachers, in the form of both synthesis documents and an open–source demonstrator.

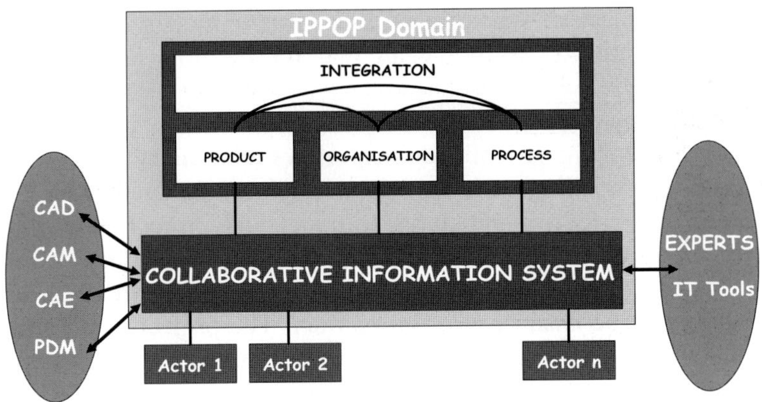

Figure 1: IPPOP demonstrator architecture

The IPPOP project aim is to integrate product, process and organisation dimension (Figure 1). First IPPOP project proposes a product model to formalise the technological knowledge about product (function, structure, behavior, expert view, …). This model should be linked to existing CAD/CAM models, and it can be completed by the designers from its semantic definition level to its geometric definition level. Second IPPOP proposes a design process model to represent the knowledge evolution. This model ensures that the design rationale is traceable and capitalised for further use (both re-use and evolution) through PDM system. Finally, the modelling of coordination decisions could support organisation dimension, for a set of design projects according to external and internal objectives of the enterprise [34]. Those models are computed into software components which can be exploited by the designers. The cooperative and collaborative Information System ensures internal integration (technological knowledge and coordination) and external integration (market software). It will be available on an open-source demonstrator.

3.1   Product knowledge modelling

Design of manufacturing parts involves many expertises with various knowledge and know-how. The Computer Aided Design is a solution to improve the task of each actor. CAD software does not really integrate technological views of the product [41] and much knowledge about the product become implicit and are often lost. Obviously Product Data Management systems propose a way for building a new product model reference, but the material managed by these systems is reduced to closed files. A second point is that CAD systems are mainly configured for mono-actors tasks. They do not allow sharing data with other applications from another skill. They lack of synchronous collaborative modes. Some software editors reacted to this state of the art and try to propose solutions for Collaborative work.

The IPPOP project proposes a new architecture for CAD applications collaboration [42]. Among collaboration views that must be taken into account, product modelling takes an important place. Product knowledge is modelled according to two structures (product and project) while actors collaborate through these structures. The following specifications define the behaviour of this architecture. The product model structure defines the concepts that must be used to model a product. The project data structures are the instantiation of concepts from the product model and thus are linked to the concept they are associated with. Skilled applications reflect the know-how of actors. They manage concept recognized by initiated actors [43]. These applications exchange through messages with the product model structure in order to define concepts they are interested with but also to share concepts from other applications. During design, applications backup its project data in the project data structure and restore them when necessary. Actor's only works with their own knowledge: they exchange with skilled applications through the Graphic User Interface of the application. Except system managers who own applications dedicated to the management of the Product Model other actors are not directly connected to the product model. All is driven by their usual CAD systems.

3.2   Design process modelling

Even if a lot of process modelling approaches has been developed (see section 2.2); there is still a lack of focus on the design process in itself. A previous analysis [44] underlines that there is no explicit understanding of product development process in terms of collaboration, overlapping or dynamic of the various actions led mutually by the design involved actors. There is hence a need for a framework in which dynamics of Concurrent Engineering process can be explicitly analysed [45]. Links with the handled product data are weak or nonexistent. There is no reliable quantification data confidence; no indication about the effective status of the processed data, integration with organisation data (for example available actors and resources) is provided in order to build a complete framework for product designing [46].

The proposed design process model reuse the logical and temporal links from IDEF3, and the semantic of IDEF∅ and GRAI. Of course those concepts are not directly used but are modified in order to fit to our needs. On the other hand, industrial partners involved within the IPPOP project brought their analysis and needs on this topic. They thus lead us to take care of some relevant notion as for example the maturity one. Moreover, the IPPOP context requires some concepts enabling the integration with product and organisation models. The "product data" is the best representative concept allowing the integration of product and process models [47]. Such concepts have been lightened during the requirement specification and analysis phase of the IPPOP project.

Then, the main key elements of the design process model are: activity, constraints, maturity, milestones, product data, objectives, resources, status, transition, and trigger.

3.3   Design organisation modelling

Enhancing performance in engineering relies on the ability to create design environments fitted to each design situation during product development. Our purpose is to model design environments [48] to support the control of an engineering system. Modelling includes tools, models and their procedure of use. A design situation is defined as a state of the system to be controlled. It is characterized by parameters relative to product knowledge, engineering process, human and material resources, designer's competencies... The aim is to set a design environment that is the most adapted to the situation and given objectives for the system. This comes to determine the components that will evolve from an initial state to a final state (Figure 2).

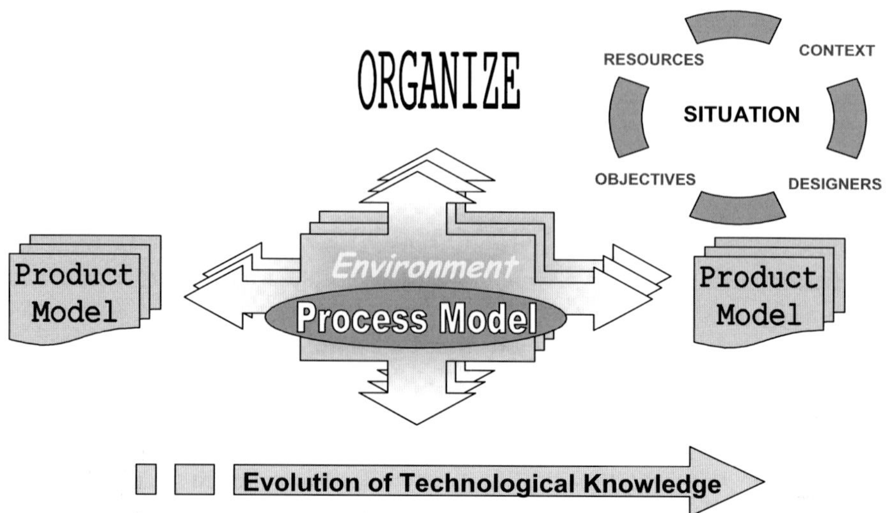

Figure 2: Integration Product-Process-Organization

The model allows defining a design situation, specifying an environment based on situation and objectives and managing the evolution of environments. In that purpose, design environments are described with a static and dynamic point of view. From a static point of view, the environment is the context that allows designers to reach the performance objectives, regarding to both consumers and enterprise expectation. From a dynamic point of view, a procedure to make this environment evolved must be provided according to the real state of the system. A set of performance indicators gives an evaluation for each intermediate situation, in order to act upon the most appropriate parameters.

3.4 Integration of Product-Process-Organisation

The integration of models concerning product, process and organisation has been implemented. Figure 3 details the integration of product, process and organisation data for supporting the coordination of design project.

The graphical user interface in the organisation module shows the running projects and allows the assessment of the available resources for launching a new one. In the product editor shows the attributes change of product data with the link to the CAD file. The object model highlights the link between the project, the input and output product data by the activity, the allocated resources and the successor activity.

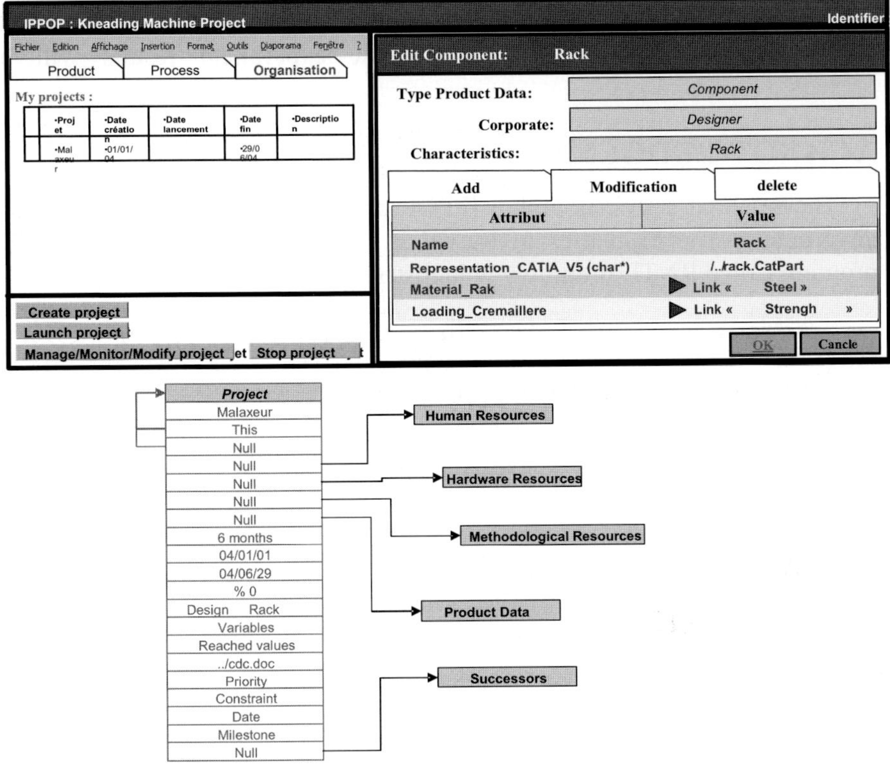

Figure 3: Project coordination

4.   CONCLUSION

This paper presented a survey on engineering design. Three main fields are considered: product modelling, design process and design coordination. This survey underlines that many research works exist on those fields but they do not consider them integrated way. Nowadays this integration is necessary because of increasing of collaboration between designers working in distributed companies. The IPPOP project answers at this requirement. IPPOP consortium has been composed of research institutes from different academic areas to fulfil these integrating needs. The various results of each one have been analysed and integrated models have been proposed. The first developments have been performed and experiments in real context are starting. We have chosen two industrial case studies. First one concerns the joint between engine and flight of an airplane. Its main interest will be on collaborative design, complex and dynamic organization. Second one concerns of an electrical engine. Its main interest will be on conflicts resolution and process coordination. The experiment results will be presented.

ACKNOWLEDGEMENT

The research activities within IPPOP project are founded by the French government. The Ministries of Research and Industries support a specific program called RNTL (http://www.industrie.gouv.fr/rntl/).

FURTHER DETAILS AND POINTS OF CONTACT

For any further information, please contact Dr. Ph. GIRARD, Tel.: +33 5 40 00 24 04, email: girard@lap.u-bordeaux1.fr. Browse also Web Site of the IPPOP project at http://www.opencascade.org:1000/IPPOP/.

REFERENCES

1. Simon HA. *The new science of management decision*, New York, Harper & Row, 1960.
2. Simon HA. *The sciences of the artificial*, MIT Press, 1969.
3. Cross N. "Science and Design methodology: a review", *Research in Engineering Design*, **vol. 5**, pp. 63-69, 1993.
4. Love T. "Philosophy of design: a meta-theoretical structure for design theory", *Design Studies*, **vol. 21**, pp. 293-313, 2000.
5. Love T. "Constructing a coherent cross-disciplinary body of theory about designing and designs: some philosophical issues", *Design Studies*, **vol. 23**, pp. 345-361, 2002.
6. Mintzberg H. *Mintzberg on management: Inside our strange world of organizations*, New York, Free Press, 1989.
7. Wang F, Mills JJ, Devarajan V. "A conceptual approach managing design resource", *Computers in Industry*, **vol. 47**, pp. 169-183, 2002.
8. Gero JS. "An approach to the analysis of design protocols", *Design studies*, **vol. 19**, pp. 21-61, 1998.
9. Perrin J. *Pilotage et évaluation des processus de conception*, Paris, Editions l'Harmattan, 1999.
10. Pahl, G, Beitz, W. *Engineering Design, a systematic approach*, Berlin, Springer-Verlag, 1996..
11. Blessing LTM. *A process-based approach to computer-supported engineering design*, PhD Thesis of University of Twente, Enschede, 1994.
12. Roozenburg NF, Eeckels J. Product Design: *Fundamentals and Methods*, New York, John Wiley & Sons, 1995.
13. Brissaud D, Garro O. "Conception distribuée, émergence", Conception de produits mécaniques, *Méthodes Modèles Outils*, Paris, Hermès, 1998.
14. Ball LJ, Evans J, Dennis I. "Cognitive processes in engineering design: a longitudinal study", *Ergonomics*, **vol. 37**, pp. 1753-1786, 1994.
15. Hacker W. "Improving engineering design-contributions of cognitive ergonomics", *Ergonomics*, **vol. 40**, pp. 1088-1096, 1997.
16. Buccarelli LL. "An ethnographic perspective on engineering design", *Design Studies*, **vol. 9**, pp. 159-168, 1988.
17. Hatchuel A. "Apprentissages collectifs et activités de conception", *Revue Française de Gestion*, pp. 109-119, 1994.

18. Gero JS, Kannengiesser U. "The situated function-behaviour-structure framework", *Design studies*, in press, 2004.
19. Andreasen MM. "The theory of domains", *Proc. of the Workshop on Understanding Function and Function to Form Evolution*, Cambridge University, UK, 1991.
20. Umeda Y, Takeda H, Tomiyama T, Yoshikawa H. *Function, behaviour and structure: Applications of Artificial Intelligent in Engineering*, Berlin, Springer-Verlag, 1990.
21. Grabowski H, Lossack RS, Weis C. "Supporting the design process by an integrated knowledge based design system", *Advances in formal design methods for CAD - Proc. of the IFIP 5.2/5.3 International Workshop*, Mexico City, Mexico, 1995.
22. .Suh NP. *The principles of design, New York*, Oxford University Press, 1990.
23. Zeng Y, Gu P. "A science-based approach to product design theory Part I: formulation and formalization of design process", *Robotics and Computer Integrated Manufacturing*, **vol. 15**, pp.331-339, 1999.
24. Zeng Y., Gu P. "A science-based approach to product design theory Part II: formulation of design requirements and products", *Robotics and Computer Integrated Manufacturing*, **vol. 15**, pp. 341-352, 1999.
25. Vernadat FB. *Enterprise modelling and integration : principles and applications*, London, Chapman & Hall, 1996.
26. Colquhoun GJ, Baines RW, Crossley R. "A state of the art review of IDEFØ", *International Journal of Computer Integrated Manufacturing*, **vol. 6**, pp. 252-264, 1993.
27. Reimann MD, Sarkis J. "An integral functional representation of concurrent engineering" *Production Planning and Control*, **vol. 7**, pp. 452-461, 1996.
28. Mayer RJ, Cullinane TP, DeWitte PS, Knappenberger WB, Perakath B, Wells MS. *IDEF3 process description capture Method, Information Integration for Concurrent Engineering - Compendium Methods Report*; Wright-Patterson Air Force Base, 1992.
29. Larson N, Kusiak A.: "Managing design processes: a risk assessment approach", *IEEE Transactions on Systems, Man and Cybernetics - Part A*, **vol. 26**, pp. 749-759, 1996.
30. Murata T. "Petri nets: properties, analysis and applications", *Proceedings of the IEEE*, **vol. 77**, pp. 541-580, 1989.
31. McMahon CA, Xianyi M, Brown KN, Sims Williams, JH. "A parallel multi-attribute transformation model of design", *Proc. of the ASME Design Engineering Technical Conferences – DETC'95*, Boston, USA, 1995, pp. 341-350.
32. Pun L. *Integrated discrete production control : analysis and synthesis - A view based on GRAI nets*, Amsterdam, Elsevier, 1992.
33. Eynard B, Girard Ph, Doumeingts G. "Control of engineering processes through integration of design activities and product knowledge", *Integration of Process Knowledge into Design Support Systems*, Dordrecht, Kluwer Academic Publishers, 1999.
34. Boujut JF, Laureillard P. "A co-operation framework for product-process integration in engineering design", *Design studies*, **vol. 23**, pp. 497-513, 2003.
35. Pena-Mora F, Hussein K, Vadhavkar S, Benjamin K. "CAIRO: a concurrent engineering meeting environment for virtual design teams", *Artificial Intelligence in Engineering*, **vol. 14**, pp. 203-219, 2002.
36. Klein M. "Integrated Coordination in Cooperative Design", *International Journal Production Economics*, **vol. 38**, pp. 85-102, 1995.
37. Crowston K. "A Coordination Theory Approach to Organizational Process Design", *Organization Science*, **vol. 8**, pp. 157-175, 1997.
38. Conroy G, Soltan H. "ConSERV, a methodology for managing multi-disciplinary engineering design projects", *International Journal of Project Management*, **vol. 15**, pp. 121-132, 1997.
39. Coates G, Whitfield RI, Duffy AHB, Hills B. "Coordination Approaches and Systems – Part II: An Operational Perspective", *Research in Engineering Design*, **vol. 12**, pp. 73-89, 2000.

40. Girard Ph, Doumeingts G. "Modeling of the engineering design system to improve performance", *Computers & Industrial Engineering*, **vol. 46**, pp 43-67, 2004.

41. Dufaure J, Teissandier D. "Geometric tolerancing from conceptual to detail design", *Proc. of the 8th CIRP International Seminar on Computer Aided Tolerancing*, Charlotte, North Carolina, USA, April 28-29, 2003.

42. Sabeur B, Noël F, Tichkiewitch S. "Interface between CAD/CAM Software And an Integrative Design Environment", *Methods and Tools for Co-operative and Integrated Design*, Dordrecht, Kluwer Academic Publishers, 2003.

43. Noël F, Brissaud D. "Dynamic data sharing in a collaborative design environment", *International Journal of Computer Integrated Manufacturing*, **vol. 16**, pp. 546-556, 2003.

44. Nowak P, Rose B, Saint-Marc L, Callot M, Eynard B, Gzara-Yesilbas L, Lombard M. "Towards a design process model enabling the integration of product, process and organization", *Proc of the 5th International Conference on Integrated Design and Manufacturing in Mechanical Engineering - IDMME'2004*, Bath, UK, April 5-7, 2004,.

45. Haque B, Pawar KS, Barson RJ. "The application of business process modelling to organisational analysis of concurrent engineering environments", *Technovation*, **vol. 23**, pp. 147-162, 2003.

46. Rose B, Gzara L, Lombard M. "Towards a formalization of collaboration entities to manage conflicts appearing in cooperative product design", *Methods and Tools for Cooperative and Integrated Design*, Dordrecht, Kluwer Academic Publishers, 2003.

47. Nowak P, Merlo C, Eynard B, Gallet T. "From design process specification towards PDM workflow configuration", *Proc of the 14th International Conference on Engineering Design – ICED'03*, Stockholm, Sweden, August 19-21, 2003.

48. Girard Ph, Robin V, Barandiaran D. "Analysis of collaboration for design coordination", *Proc of the 10th IPSE International Conference on Concurrent Engineering - CE'03*, Madeira, Portugal, July 26-30, 2003.

A UNI-AXIAL TEST RIG DESIGN APPROACH FOR DEFORMATION ANALYSIS OF FLEXIBLE PARTS DUE TO LOCATION TYPE TOLERANCE IN ASSEMBLY

I.A Manarvi and N.P Juster
Department of Design Manufacture and Engineering Management,
University of Strathclyde, Glasgow, UK

Abstract: Previous investigations attempting to establish the relationships between dimensional tolerance and product function have been limited to specific applications such as linkage design and industrial crank rockers. The techniques proposed cannot be extended to other complex parts and assemblies as a generalized tool. In particular little attention has been focused on the effect of tolerance on flexible parts. Present research provides a methodology in which test rigs could be designed and constructed for carrying out experimental investigations on the deformation of flexible part materials prior to manufacturing of the prototype parts. A Uni-Axial test is designed and manufactured to carry out experimental investigations on ABS plastics material used for manufacturing of automotive interior parts. The investigation provides a methodology to observe the influence of location tolerance on the deformation of a rectangular strip.

Key words: Tolerance allocation, Flexible parts, Deformation analysis.

1.   INTRODUCTION

Previous investigations attempting to establish the relationships between dimensional tolerance and product function [1-4] have been limited to specific applications such as linkage design and industrial crank rockers. The techniques proposed cannot be extended to other complex parts and assemblies as a generalized tool. In particular little attention has been focused on the effect of tolerance on deformable parts. Researchers have also contributed significantly to assembly tolerance allocation [5-11]. It included comparative evaluation of composite tolerance specifications in robotic assembly [6], the influence of assembly sequences on tolerance chain detection [9], tolerance charting [10] and influence of design process in assembly [11]. The influence of positioning errors in robotic assembly [6] provided product design aids and robot selection guidelines. Based on these findings design charts were prepared to assist in the selection of robots to increase the chances of successful assembly. The influence of tolerances on assembly sequencing resulted in development of algorithms [7] that could be embedded into a CAD environment for assisting designers in the selection of appropriate tolerances. The tolerance control and propagation scheme (TCP) investigated [8] provided an accurate analysis of tolerances within a product assembly modeller. A framework was proposed [9] for function means modelling and analysis of geometric coupling and detection of potential tolerance chains to avoid tolerance chains. The use of tolerance charting in assembly to verify if assemblies of parts could be made using available resources was investigated [10]. It provided an efficient method to integrate tolerances into product design by using a tolerance-charting methodology at the assembly process planning stage.

Location of hole centres with reference to a datum hole centre is a commonly used dimensioning situation encountered by tolerance design engineers. Inappropriate hole positioning tolerance might result in misalignment of parts and difficulties in assembly in production processes. The research in this paper describes the design and manufacturing of a Uni-Axial test rig, which could be used for experimental investigations by applying tolerance values to observe the deformation of a flexible rectangular strip. The results of experimental investigations were compiled and analysed on the basis of FE simulations results obtained under similar boundary conditions.

2.   UNI-AXIAL TEST RIG CONSTRUCTION

After a number of design iterations, a Uni-axial test rig was designed and manufactured as shown in Figure 1.0 (a) (b) and (c) below. The main parts of the rig included the following:

a) Sheet metal platform base.
b) Two side support baffles
c) Two precision grinded slide rods
d) Lead screw
e) Rotary knob

f) Stationary or fixed platform
g) Movable platform
h) Tolerance pins E and F
i) Dial indicator for tolerance measurement
j) Two dial indicators for vertical support deformation measurement

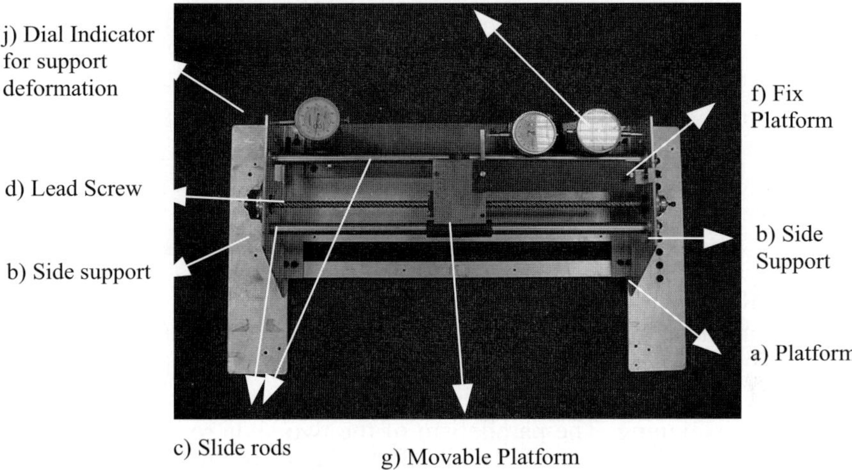

Figure 1.0 (a) Top view Uni-Axial test rig assembly

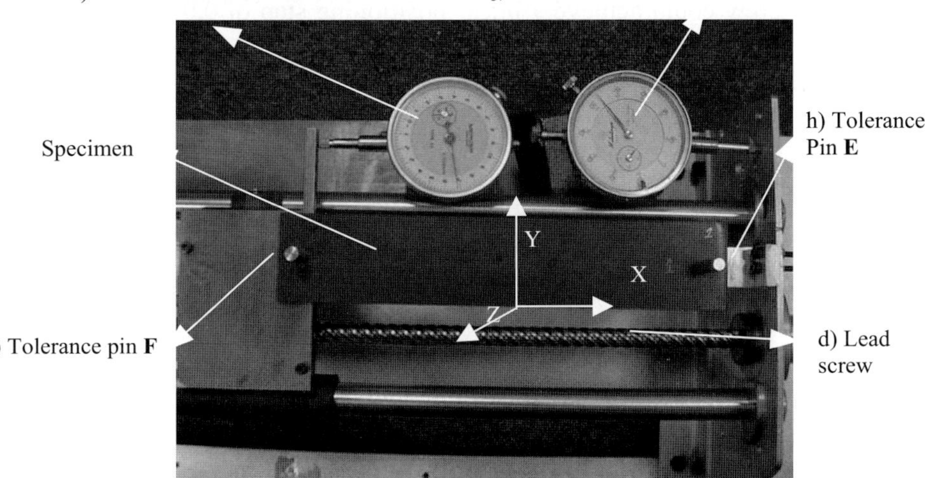

Figure 1.0(b) Close up view Uni-Axial test rig assembly

Figure 1.0 (c) Side view Uni-Axial test rig assembly

The Rig could provide a varying range of positioning tolerances for the specimen under investigation. A fixed and movable platform was used to simulate the datum and related feature location. The movable platform was designed to slide over the lead screw and two precision grinded rods for accurate positioning. The parallelism of the two slide rods was measured to be 0.0038mm over a length of 200mm. Two pins of 1.0+/- 0.1 mm diameter and 30 mm length each were installed on a stationary and movable platform to locate hole centres of a test specimen. The rotary knob at one end of the lead screw could achieve a linear positioning step of 0.02mm. Assembly of the slide rods and positioning platforms was mounted over a light but rigid base with four rubberised pads for resting on the surface plate of a co-ordinate measuring machine. Two dial test indicators were placed adjacent to the mounting walls of precision grinded parallel slide rods and platform assembly to measure any deformation in sidewalls when lead screw was moved. A third dial indicator was positioned against the movable platform to measure the position step tolerance.

3.   PROCEDURE OF EXPERIMENTS

Astryn BR712A material as selected for FE simulations was also used for experimental investigation due to its common use in design and manufacture of automotive interior panels.   Plaques of 200 x 140 x 3 mm size were used for manufacturing of the test specimen. Rectangular strips of 200 x 40 x 1mm of the material were manufactured to determine influence of location positioning tolerance over deformation of specimen. Two holes of 1.0+/- 0.1mm diameter each were drilled along centre axis of specimen at 180mm distance from each other. The specimen was then mounted on the test rig over the pins E and F maintaining 180.0mm distance between hole centres as shown in Figure 1.0 (b). The test rig was then placed on co-ordinate measuring machine surface and datum reference was selected at centre of pin E. X-axis of co-ordinate measuring machine was aligned along the length of specimen and Z-axis was in vertical direction for measurement of deformation. The un-deformed surface of specimen was selected as datum Z. Co-ordinate measuring machine measuring probe pressure was reduced to its minimum value to ensure it did not apply any significant pressure on the specimen surface. A step of 1.0mm was selected in X-axis for measuring Z-axis deformation of specimen. Tolerance values of 0.2, 0.4, 0.6, 0.8, 1.0 and 1.2mm were chosen for experimental investigations. The rotary knob as shown in Figure 1.0 (c) was used to move pin F on the specimen through the required distance thus simulating a manufacturing variation. The dial test indicator attached to movable platform measured linear positioning of pin F centre axis. After applying each tolerance value, the vertical deformation of the specimen in the Z-direction was measured along the centre line of specimen by touch probe. First point was taken closest to pin E and subsequent data was collected while moving towards pin F along X-axis in step of 10.0mm. The results are presented in Figure 2.0 as deformation of specimen in Z-axis at each tolerance value.

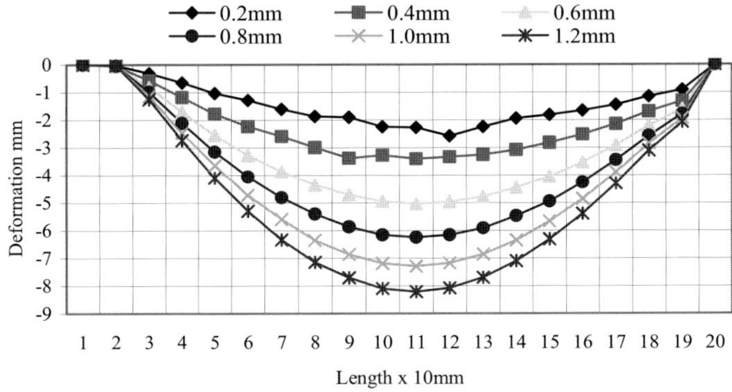

Figure 2.0 Deformation in Z-axis for 1.0mm dia positioning Pins.

4.   INFLUENCE OF POSITIONING PINS DIA ON DEFORMATION

A variety of hole diameters are used in assembly design of mechanical parts. Therefore, at this stage it was considered important to investigate the influence of positioning tolerance over the case study part deformation for different size hole diameters at the same tolerance value and different tolerance values. The pin diameters of 1.0mm, 2.0mm, 3.0mm and 4.0mm were selected. The pins were manufactured to fit the movable platform of Uni-axial test rig and the test specimen holes E and F were altered according to pin sizes for further experimental investigations. The experiments for each pin diameter were conducted according to the outlined procedure and part deformation was measured using Co-ordinate measuring machine.

4.1   Unique tolerance values with different dia pins:

The typical results for 0.2mm, 0.6mm tolerance values and positioning pin diameters of 1.0mm, 2.0mm, 3.0mm and 1.0mm at same tolerance value are shown in Figure 3.0(a) and (b) respectively:

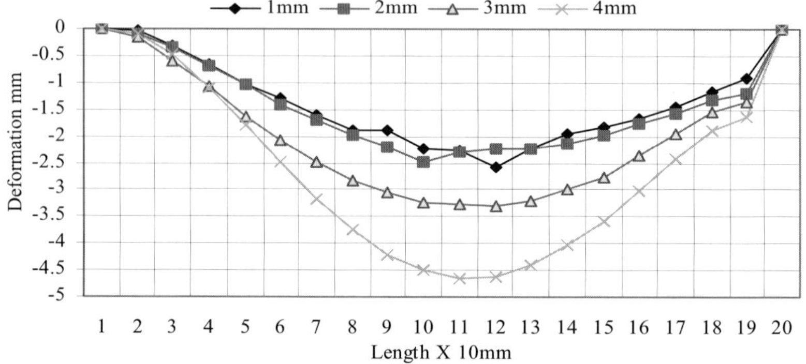

Figure 3.0(a), Deformation of part at 1.0,2.0,3.0 & 4.0mm dia pins at 0.2mm, tolerance

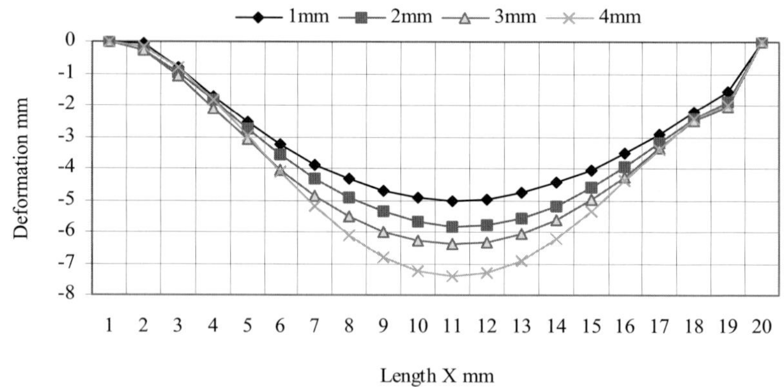

Figure 3.0,(b)) Deformation of part at 1.0,2.0,3.0 & 4.0mm dia pins at 0.6mm tolerance

4.2   Different tolerance values same diameter of pins:

The typical results for tolerance values of 0.2mm, 0.6mm and 1.0mm and positioning pin diameters of 1.0mm, 2.0mm, 3.0mm at different tolerance value are shown in Figure 4.0(a), (b) and (c) below:

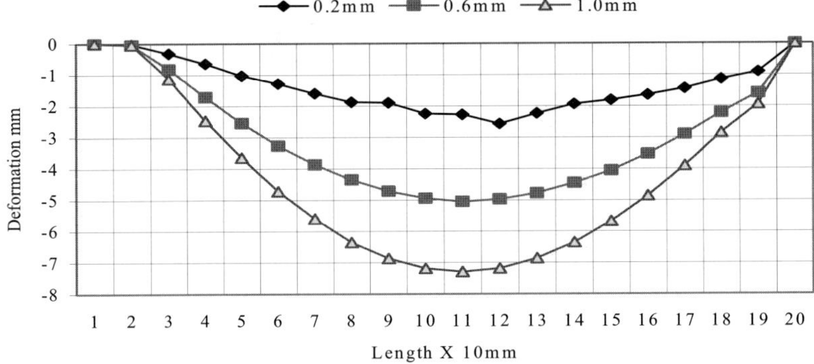

Figure 4.0 (a) Pin dia = 1.0 mm, tolerance = 0.2, 0.6 and 1.0mm.

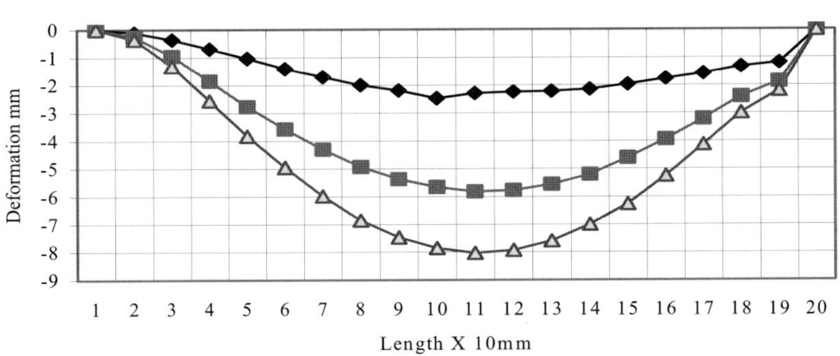

Figure 4.0 (b) Pin dia= 2.0 mm tolerance= 0.2, 0.6 and 1.0mm.

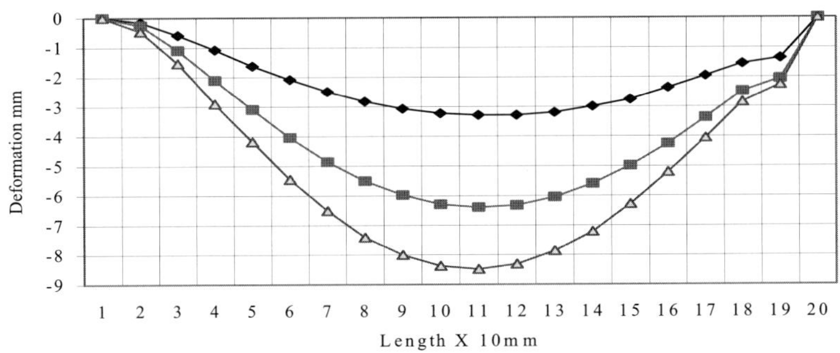

Figure 4.0(c) Pin dia = 3.0 mm tolerance = 0.2, 0.6 and 1.0mm.

4.3   Comparison of FE and Uni-Axial experimental results

It was considered more useful to compare the results for individual tolerance values. These results for tolerance = 0.2 mm, 0.4 mm,0.8mm and 1.0mm for positioning pin diameter = 1.0mm are given in Figure 5.0 (a), (b), (c) and (d) as below:

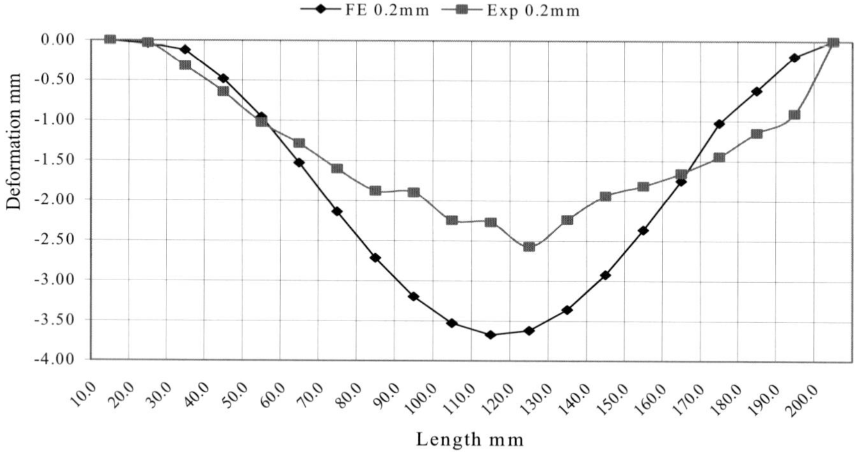

Figure 5.0 (a) Results for tolerance =0.2mm and Pin =1.0mm

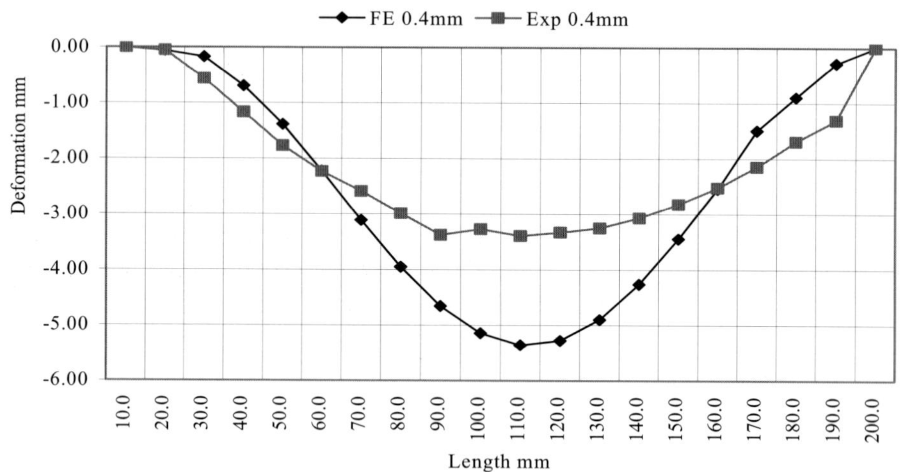

Figure 5.0 (b) Results for tolerance = 0.4mm, Pin = 1.0mm

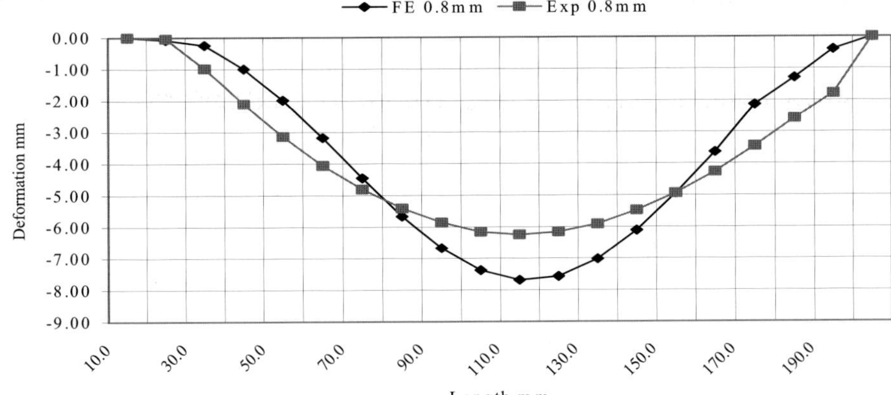

Figure 5.0(c) Results for tolerance = 0.8mm and Pin = 1.0mm

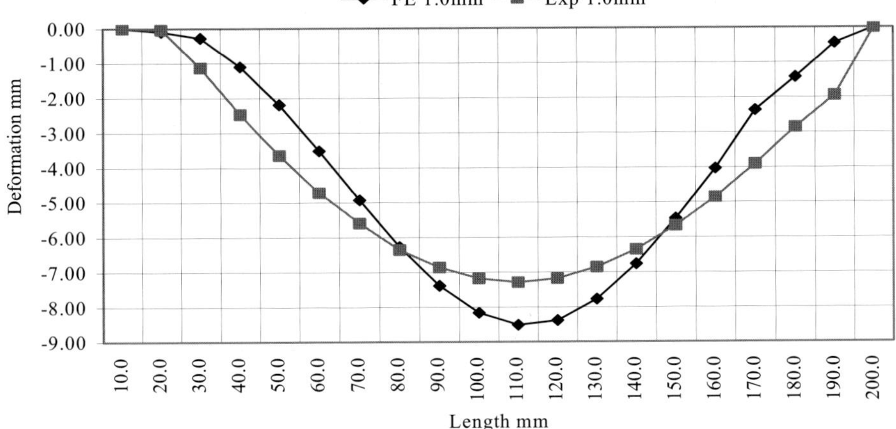

Figure 5.0(d) Results for tolerance =1.0mm and Pin = 1.0mm

The similarity of experimental and FE results showed that design of a test rig approach for observing the deformation behaviour of raw material at early design stage could be used to predict the deformation of flexible parts.

5.   CONCLUSION

The investigation described the design of a Uni-axial test rig approach to observe the deformation behaviour of flexible parts raw materials. It led to following conclusions-

The deformation of the specimen was observed as bell shaped curve with peak value at the centre of specimen for all the selected tolerance values. It gradually reduced when the constrained points were approached on either side in experimental as well as FE results for all tolerance values.

1. The similarity of curves gave evidence of positive relationship between location tolerance and deformation that might create difficulties in assembly function.

2. The maximum deformation value of FE simulations and experimental results differed which was attributable to assuming the material as elastic and Isotropic in FE simulations. However, the test rig approach provided the trends for deformation due to location tolerance.

The approach could also be extended to more complex parts and assemblies.

REFERENCES

1. R. E. Garret, Allen S. Hall, Jr, "Effect of tolerance and clearance in linkage design", *Transactions of the ASME*, February, 1969, pp198-202.
2. Mark M. Benner and Arthur G. Erdman, "Tolerance synthesis method for an industrial crank-rocker application", **DE-Vol. 46**, *Mechanical Design and Synthesis ASME*, pp 463 - 473, 1992
3. Xiancheng Lu and Cheun-Sen Lin, "A proposed method to group the solutions from dimensional synthesis: planar triads for six precision position", **DE-Vol. 46**, *Mechanical Design and Synthesis ASME*, pp 481 - 487, 1992
4. Cheun-Sen Lin, Bao Ping Jia, "Use of resultant in the dimensional synthesis of linkage components: motion generation with prescribed timing", **DE-Vol. 46**, *Mechanical Design and Synthesis ASME*, pp 489 - 496, 1992
5. S. Ranade, E. A. Lehtihet, T. M. Cavalier, "Comparative evaluation of composite position tolerance specifications for patterns of holes", *Proceedings of 33rd International MATADOR conference 1999*, Manchester, pp 533-538
6. Hoda H. ElMaraghy, Waguih H ElMaraghy, "Design specification of parts dimensional tolerance for robotic assembly", *Computers in Industry, 1988*, **Vol 10 No.1** pp 47-59.
7. Jean-Claud Latombe, Randall H Wilson and Frederic Cazals, "Assembly sequencing with tolernaced parts", *Computer Aided Design*, **Vol 29, No.2**, 1997, pp 159-174
8. A. Ashiagbor, H-C. Liu and B. O Nnaji, "Tolerance control and propagation for the product assembler model", *International Journal of Production Research, 1998*, **Vol 36. NO.1** PP 75-93.
9. Rikard Soderberg and Hans Johannesson, "Tolerance chain detection by geometrical constraint based coupling analysis", *Journal of Engineering Design*, **Vol. 10, No. 1**, 1999, pp 1- 24.
10. B.K.A. Ngoi and J.M. Ong, "A complete tolerance charting system in assembly", *International Journal of Production Research, 1999*, **Vol 37. NO.11** pp 2477-2498
11. G.F. Dalgleish, G.E.M. Jared and K.G. Swift, "Design for assembly: influencing the design process", *Journal of Engineering Design, 2000*, **Vol. 11, No.1**, pp17-29

GENERATING MECHANISM AND EXPERIMENTAL STUDY ON IMPACTING WATER JET BY MODULATION OF CHAOS

Xin-Min Zhang[1] and Xiong Duan[2]

1. Department of Mechanical Engineering, Jiaozuo Institute of Technology, Jiaozuo, Henan, 454100, China, Email:Zhangxm@jzit.edu.cn
2. College of Mechanical and Electrical Engineering, China Univ. of Mining and Technology, Xuzhou, Jiangsu, 221008, China

Abstract: Based on the advanced theory of modern scientific chaos, combining with the network theory of unsteady water jet, a network and mathematical model for the chaotic water jet has been presented. This study has been made on the mechanisms and the condition about generating device of impacting water jet using chaotic modulation. The theory and method of generation on impacting water jet through modulation of chaos has been proposed. Laboratory compare experimental investigations were carried out between generating devices of impacting water jet through modulation of chaos and common cavitating jet. The experimental study focused on the erosion effect compare of pure water jet by modulation of chaos and common cavitating jet. In order to investigate the influence of various parameters and conditions on erosion, such as the standoff distance, pump pressure and cutting speed, are considered and the mass loss and cutting depth were measured in the experimental research work. Experiment has shown that impacting water jet through modulation of chaos can efficiently increase ability in erosion and cutting.

Key words: High-pressure water jet, Scientific chaos and engineering chaos, impacting water jet using chaotic modulation, Erosion performance.

1. INTRODUCTION

When high-pressure water jet was first developed, research was focused on mining, because it could increase productivity with lower cost. Since then, the industry has developed along other lines. The use of high-pressure water jets has become much more common in recent years because of high necessity. For their advantage, water-jetting equipments have been developed and used widely. Water jet has become the accepted methods to solve the problems in many industries, such as clearing, breaking, cutting, art, military and medicine, etc.

When time-dependent fluctuations are imposed on steady water jets, unsteady water jets are formed and unsteady effects appear in the erosion mechanism. Researches show that the unsteady water jet is much better than steady water jet in breaking, erosion, cutting and clearing effect[1]. When a continuous water jet impacts on the surface of target perpendicularly, the pressure at the contacting point is called stagnant pressure, but if the water drops impacts on the same target, the transient impacting pressure is called water hammer pressure. Since the water hammer pressure of unsteady water jets is much higher than the stagnation pressure of steady water jets, it is expected that the breaking, erosion and cutting ability of water jets is enhanced. Based on the principle, many generating devices are designed to change steady water jet into water drops, which is called interrupted water jet.

Unsteady water jets have been classified as slug jet or interrupted water jet, oscillating jet, self-oscillating pulsed jet and cavitating jet. Slug jet is a flow composed by a series of slugs and formed by periodical interruption of steady water jets. Lichtarowicz[2], Erdmamn-Jeswiter[3], and Kiyohashi have used a rotating disk with many slits immediately after a nozzle outlet to intermit steady water jets; Peter equipped a nozzle with a reciprocating value to make water jets spout intermittently; Furthermore, Duan Xiong[4] used a two-nozzle type interrupted water jet device to divide a continuous water jet equally into two discontinuous water jets with the same discharge.

A new type of jet, which is called the impacting water jet through modulation of chaos, is developed for the first time[5], and its basic characteristics will be investigated in this paper. This is a new kind of water jet based on modern chaos theory. It is briefly described as that when water jet enter into a tailor-made chaotic oscillating chamber, expansion of the water jet can cause the water jet to attach to a side-wall of the chamber. This phenomenon is called "attaching-wall effect". When the pressure water jet flows in high speed in the chaotic oscillating chamber, a vacuum chaotic pressure is generated, and performance of the chaotic water jet will be improved if the attaching-wall effect is utilized to generate the chaotic pressure oscillations. This can be used to increase the energy and efficiency in breaking, clearing, and cutting as well as the performance.

2.   ESTABLISHMENT OF GENERATING MECHANISM MODEL ON IMPACTING WATER JET BY MODULATION OF CHAOS

Consider a physical model of jet system with double chamber and corresponding network model which are shown in Fig.1. The model will form a three-stage autonomous network system.

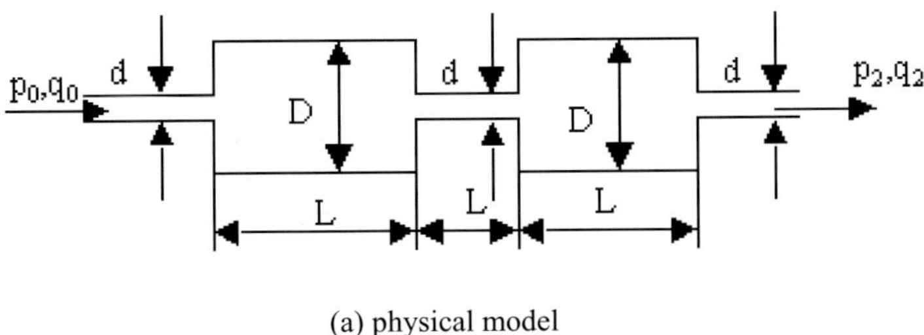

(a) physical model

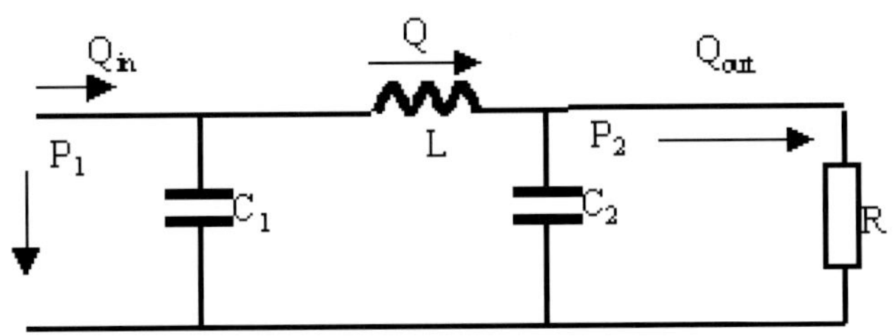

Fig.1 Impacting jet model by modulation of chaos

The state equation can be got from the effect circuit shown in Fig.1 (b):

$$\begin{cases} \dfrac{dp_1}{dt} = \dfrac{1}{C_1}(Q_{in} - Q) \\[2mm] \dfrac{dp_2}{dt} = \dfrac{1}{C_2}\left[Q - R(\dfrac{1}{3}p_2^3 - p_2) \right] \\[2mm] \dfrac{dQ}{dt} = \dfrac{1}{L}(p_1 - p_2) \end{cases} \qquad (1)$$

Where:

P_1, P_2 – – water pressure in the front chamber and back chamber;

C_1, C_2 – – current volume in the front chamber and back chamber ;

Q – – flow in the connecting pipe ;

L – – lagging-flow of the connecting pipe ;

Q_{ib} − − input flow of the device ;

R − − flow-resistance coefficient of the nozzle.

From the above, the model of chaotic jet device built in Fig.1 can be expressed by a three-stage nonlinear calculus equation group, which describes a three-stage nonlinear dynamic system.

3.   ANALYSIS OF DYNAMICS FORM ABOUT IMPACTING JET BY MODULATION OF CHAOS

3.1   Confirmation of equilibrium state in the system

The equilibrium state of the dynamic system which is described by nonlinear dynamic system equation group (1) is a fixed constant state solution of the system. The form of the equilibrium state can be depicted by peculiarity of balance spot or odd spot. The odd spot of nonlinear dynamic system can be got from equation (2). Consider

$$\begin{cases} \dfrac{dp_1}{dt} = \dfrac{1}{C_1}(Q_{in} - Q) = 0 \\[2mm] \dfrac{dp_2}{dt} = \dfrac{1}{C_2}\left[Q - R(\dfrac{1}{3} p_2^3 - p_2) \right] = 0 \\[2mm] \dfrac{dQ}{dt} = \dfrac{1}{L}(p_1 - p_2) = 0 \end{cases} \qquad (2)$$

we have

$$\begin{cases} Q = Q_{in} \\[2mm] Q_{in} - R(\dfrac{1}{3} p_2^3 - p_2) = 0 \\[2mm] p_1 = p_2 \end{cases}$$

that is

$$p_2^3 - 3 p_2 - \frac{3 Q_{in}}{R} = 0 \qquad (3)$$

The solutions of the equation (3) are

$$p_{2,1} = \sqrt[3]{\frac{3 Q_{in}}{2R} + \sqrt{\frac{9 Q_{in}^2}{4R^2} - 1}} + \sqrt[3]{\frac{3 Q_{in}}{2R} - \sqrt{\frac{9 Q_{in}^2}{4R^2} - 1}} \qquad (4)$$

$$P_{2,2} = \omega_1 \sqrt[3]{\frac{3Q_{in}}{2R} + \sqrt{\frac{9Q_{in}^2}{4R^2} - 1}} + \omega_2 \sqrt[3]{\frac{3Q_{in}}{2R} - \sqrt{\frac{9Q_{in}^2}{4R^2} - 1}} \qquad (5)$$

$$P_{2,3} = \omega_2 \sqrt[3]{\frac{3Q_{in}}{2R} + \sqrt{\frac{9Q_{in}^2}{4R^2} - 1}} + \omega_1 \sqrt[3]{\frac{3Q_{in}}{2R} - \sqrt{\frac{9Q_{in}^2}{4R^2} - 1}} \qquad (6)$$

where : $\omega_1 = \dfrac{-1+\sqrt{3}i}{2}$; $\omega_2 = \dfrac{-1-\sqrt{3}i}{2}$.

From above, there are 3 odd spots in the equilibrium state of the non-linear dynamic system(1), which are

$$O_1 : (P_{2,1}, P_{2,1}, Q_{in}) \ ; \ O_2 : (P_{2,2}, P_{2,2}, Q_{in}) \ ; \ O_3 : (P_{2,3}, P_{2,3}, Q_{in})$$

3.2   Relation of stability of equilibrium state in system and diverge

The stability of the equilibrium state in dynamic system depends on the characteristic value λ in the Jacobi matrix $\mathbf{J}$ of the dynamic system, the characteristic equation of Jacobi matrix $\mathbf{J}$ of the system is

$$\begin{vmatrix} \lambda & 0 & \dfrac{1}{C_1} \\ 0 & \lambda - \dfrac{R}{C_2}(p_2^2 - 1) & -\dfrac{1}{C_2} \\ -\dfrac{1}{L} & \dfrac{1}{L} & \lambda \end{vmatrix} = 0$$

That is

$$\lambda^3 - \lambda^2 \frac{R}{C_2}(p_2^2 - 1) + \lambda \frac{C_1 + C_2}{LC_1C_2} - \frac{R}{LC_1C_2}(p_2^2 - 1) = 0 \qquad (7)$$

The solution is

$$\lambda_1 = \sqrt[3]{-\frac{q}{2} + \sqrt{\frac{q^2}{4} + \frac{p^3}{27}}} + \sqrt[3]{-\frac{q}{2} - \sqrt{\frac{q^2}{4} + \frac{p^3}{27}}} - \frac{R}{3C_2}(1 - p_2^2) \qquad (8)$$

$$\lambda_2 = \omega_1 \sqrt[3]{-\frac{q}{2} + \sqrt{\frac{q^2}{4} + \frac{p^3}{27}}} + \omega_2 \sqrt[3]{-\frac{q}{2} - \sqrt{\frac{q^2}{4} + \frac{p^3}{27}}} - \frac{R}{3C_2}(1 - p_2^2) \tag{9}$$

$$\lambda_3 = \omega_2 \sqrt[3]{-\frac{q}{2} + \sqrt{\frac{q^2}{4} + \frac{p^3}{27}}} + \omega_1 \sqrt[3]{-\frac{q}{2} - \sqrt{\frac{q^2}{4} + \frac{p^3}{27}}} - \frac{R}{3C_2}(1 - p_2^2) \tag{10}$$

Where

$$\omega_1 = \frac{-1 + \sqrt{3}i}{2} \quad ; \quad \omega_2 = \frac{-1 - \sqrt{3}i}{2}$$

Consider a kind of simple circumstance without losing the general discussion, that is

$$\frac{q^2}{4} + \frac{p^3}{27} = 0 \tag{11}$$

Now, the equation(7) contains three real roots, among which two are equal, the solution of the equation(7) is

$$\lambda_1 = 2\sqrt[3]{-\frac{q}{2}} - \frac{R}{3C_2}(1 - p_2^2) \tag{12}$$

$$\lambda_2 = \sqrt[3]{-\frac{q}{2}} - \frac{R}{3C_2}(1 - p_2^2) \tag{13}$$

$$\lambda_3 = \sqrt[3]{-\frac{q}{2}} - \frac{R}{3C_2}(1 - p_2^2) \tag{14}$$

Now, we will discuss the odd spots in a simple condition, that is, when

$$3Q_{in} = 2R \tag{15}$$

It's easy to compute that the three odd spots contained in the system are:

$$O_1 : (2, 2, Q_{in}) \ ; \ O_2 : (-1, -1, Q_{in}) \ ; \ O_3 : (-1, -1, \frac{2}{3}R)$$

3.3   Diverge and dynamics quality of the system at the equilibrium state

Now, we only discuss diverge and dynamics quality of the dynamic system(1) at and around the odd spot O_1 : $(2, 2, Q_{in})$.

From the equation (12)~(14), we have

$$\lambda_1 = 2\sqrt[3]{-\frac{q}{2}} + \frac{R}{C_2} \tag{16}$$

$$\lambda_2 = \lambda_3 = \sqrt[3]{-\frac{q}{2}} + \frac{R}{C_2} \tag{17}$$

At the odd spot, we have

$$p = \frac{C_1 + C_2}{LC_1 C_2} - \frac{3R^2}{C_2^{\ 2}} \tag{18}$$

$$q = -\frac{2R^3}{C_2^{\ 3}} + \frac{R(C_1 + C_2)}{LC_1 C_2^{\ 2}} \tag{19}$$

Now, we discuss network parameter's influence within its worth scope on the sign of the Jacobi matrix characteristic value λ of the dynamic system..

Consider $\lambda_1 > 0$, then $2\sqrt[3]{-\frac{q}{2}} + \frac{R}{C_2} > 0$, and the result is : $q < \frac{R^3}{4C_2^{\ 3}}$ .So it's easy to see that when $q < \frac{R^3}{4C_2^{\ 3}}$, we have $\lambda_1 > 0$, and when $q > \frac{R^3}{4C_2^{\ 3}}$, we can easily get $\lambda_1 < 0$.Similarly, consider $\lambda_2 > 0$, we have $\sqrt[3]{-\frac{q}{2}} + \frac{R}{C_2} > 0$, and the result is : $q < \frac{2R^3}{C_2^{\ 3}}$ .So, when $q < \frac{2R^3}{C_2^{\ 3}}$, we have $\lambda_{2,3} > 0$, and when $q > \frac{2R^3}{C_2^{\ 3}}$, we have $\lambda_{2,3} < 0$.

From the analysis and discussion above, we acquire that the matrix of Jacobi characteristic value λ of the dynamic system has three kinds of probable combinations :

① when $q < \dfrac{R^3}{4C_2^{\;3}}$ and $q < \dfrac{2R^3}{C_2^{\;3}}$, we have $\lambda_1 > 0$, $\lambda_{2,3} > 0$, that is,

when $q < \dfrac{R^3}{4C_2^{\;3}}$, we have $\lambda_1 > 0$, $\lambda_{2,3} > 0$. In this condition, the odd spot is

an unsteady saddle spot, and the system generates pitchfork diverge and takes on chaotic quality gradually.

② when $q > \dfrac{R^3}{4C_2^{\;3}}$ and $q < \dfrac{2R^3}{C_2^{\;3}}$, we have $\lambda_1 < 0$, $\lambda_{2,3} > 0$, that is

when $\dfrac{2R^3}{C_2^{\;3}} > q > \dfrac{R^3}{4C_2^{\;3}}$, we have $\lambda_1 < 0$, $\lambda_{2,3} > 0$. In this condition, the odd

spot is also an unsteady saddle spot, and the system generates pitchfork diverge and takes on chaotic quality gradually.

③ when $q > \dfrac{R^3}{4C_2^{\;3}}$ and $q > \dfrac{2R^3}{C_2^{\;3}}$, we have $\lambda_1 < 0$, $\lambda_{2,3} < 0$, that is,

when $q > \dfrac{2R^3}{C_2^{\;3}}$, we have $\lambda_1 < 0$, $\lambda_{2,3} < 0$. At the moment, the odd spot is a

steady node spot.

So we can get the worth scope of network parameters under the above three circumstances about λ in the matrix Jacobi of the dynamic system, as follows:

① when $\dfrac{C_1 + C_2}{LC_1} < \dfrac{9R^2}{4C_2}$, we have $\lambda_1 > 0$, $\lambda_{2,3} > 0$;

② when $\dfrac{4R^2}{C_2} > \dfrac{C_1 + C_2}{LC_1} > \dfrac{9R^2}{4C_2}$, we have $\lambda_1 < 0$, $\lambda_{2,3} > 0$;

③ when $\dfrac{C_1 + C_2}{LC_1} < \dfrac{4R^2}{C_2}$, we have $\lambda_1 < 0$, $\lambda_{2,3} < 0$。

Now, we take network parameters : $C_1 = 1$, $C_2 = 1$, $R = 1.1$, $Q_{in} = 11/15$, $L = 200/537$ to make simulation experiment, and get the emulation result is shown in Fig.2~5. The chart only contains output pressure, flow wave-shape figure and two-dimensions and three-dimension phase-plane figure about impacting jet by modulation of chaos. Using the method of direct observation in which phase-orbits and state variables change with time, we can easily see the chaotic qualities taking on in the system.

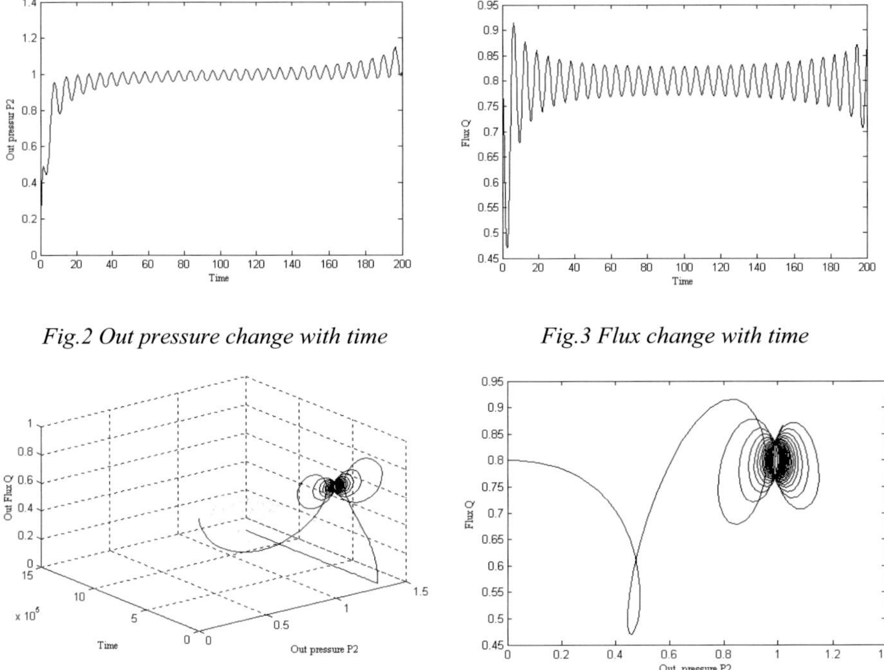

Fig.2 Out pressure change with time   *Fig.3 Flux change with time*

Fig.4   Water jet attractor   *Fig.5 Two-dimensions phase plane chart*

4.   THE EXPERIMENT STUDIES

In order to investigate the quality of generating mechanism on water jet by modulation of chaos, we made a pure water jet erosion effect experiment which is used to compare with that from common cavitating nozzle experiment. Using firebrick as its test piece, the experimental method is "standard test piece on erosion". The volume of the erosion depression made by jet can be indirectly measured with that of Paraffin filled in the depression. The medium in this experiment is water, and we take the reciprocating pump as the pressure resource. In order to eliminate the influence from reciprocating pump which produce impulsive output pressure and output flux impulse, the steady flow device in the pipes between the pump and the nozzle device was added. The result got from this experiment is shown in Fig.6. The two groups of chaotic nozzles used in the experiment is No.1 and No.2 nozzle. No.3 nozzle is a normal structural conic cavitating one with an angle of 120° which is used to make a comparing experiment.

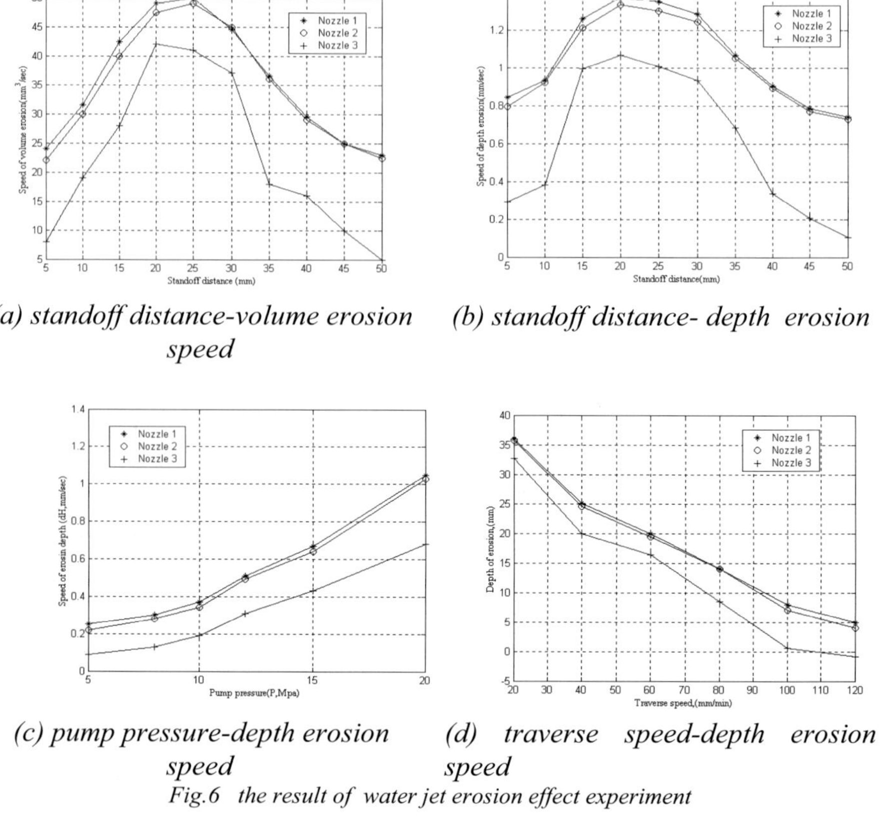

(a) standoff distance-volume erosion speed

(b) standoff distance- depth erosion

(c) pump pressure-depth erosion speed

(d) traverse speed-depth erosion speed

Fig.6   the result of water jet erosion effect experiment

5.   CONCLUSIONS

From the results above discussion, it can be concluded that:

(1) Based on the advanced theory of modern scientific chaos combining with the network theory of unsteady water jet, a network and mathematical model for the chaotic water jet has been presented. We have analyzed the behaviors of chaotic dynamics in detail.

(2) Laboratory compare experimental investigations were carried out between generating devices of impacting water jet through modulation of chaos and common cavitating jet. Whether we observe the experimental results from the effect of standoff distance on volume erosion speed or depth erosion speed, chaotic nozzle exists the "optimal standoff distance". The optimal standoff distance of chaotic nozzle is about 10 to 14 times longer than the output diameter.

(3) If we consider the influence of the pump pressure on the effect of erosion, we observe that the effect of erosion of chaotic nozzle is obvious better than cavitating conic nozzle with an angle of 120°. Its volume erosion speed is about percent of 70 to 110 higher than the latter. The effect of erosion enhances greatly on the exponent principal with the increase of pump pressure at the optimal standoff distance. The effect of erosion of chaotic nozzle is always better than conic cavitating nozzle under various pump pressures in the experiments. However, the advantage is not increased with the pump pressure elevating. It shows that the destruction of material of target body from chaotic nozzle does not linearly rely on the increase of pump resource pressure.

(4) The effect of traverse speed on erosion in the experiment shows that: the effect of erosion of chaotic nozzle is obvious better than cavitating conic nozzle with an angle of 120°, and its maximum depth of erosion is about percentage of 25 to 40 higher than that of the latter. This advantage is lower than that in other situation. From our analysis, we think that it is the increase of traverse speed that shortens the time spent in impacting target body on fixed spot by jet. Meanwhile, because the generation and development of internal and tiny crack of target material under chaotic jet impacting needs a certain time; the molecule of material absorbing energy has a process. Therefore, the chaotic fatigue destructing mechanism on target body has not worked full.

REFERENCES

1. Duan Xiong, "The Net Theory of Unsteady Water jet", China University of Mining and Technology press, Xuzhou, Jiangsu, 221008, China, 1997.
2. Lichtarowicz, A., and Nwachukwu, G., "Erosion by an Interrupted Jet", *Proceedings of 4th International Conference on Jet Cutting Technolog*, BHRA Engineering Cranfield, U.K., **Vol. 1**, Paper B2, 6pp. 1978.
3. Erdmann-Jesnitzer, F., Louis, H., and Schikorr, W., "Cleaning, Drilling and Cutting by Interrupte Jets", *Proceedings of 5th International Conference on Cutting Technology*, BHRA Fluid Engineering, Canfield, U.K., Paper B1, 45pp. 1980.
4. Duan Xiong, Tian Benzhao, Cheng Dazhong and Jin Huikun, "Study on the Erosion Performance of Two-Nozzle Type Interrupted Waterjet", *Proceedings of the International Symposium on New Application on Water jet Technology*, Isinomaki, Japan. October 19~21,1999.
5. Zhang Xin-min "Theory of Generating Device Design and Experimental Study on Impacting Water Jet through Modulation of Chaos", *Doctoral Thesis of China University of Mining and Technology*, Xuzhou, China, April, 2003.

DESIGN ANALYSIS AND MANUFACTUIRNG PLANNING USING ISO 010303 STEP AP224

Rohit Sharma and Dr James X. Gao
Cranfield University, Cranfield, UK

Abstract: Conclusive research studies have statistically proven that a large number of new design projects are actually re-design activities. Most research efforts for providing computer-aided tools for supporting conceptual or early design are concentrated on the design of new products. Manufacturing plans generated when an order was placed last time for manufacturing may quickly become out dated as the manufacturing constraints and environment changes. It is therefore equally important to regenerate and re-evaluate manufacturing plans every time a product is changed, redesigned or reordered. This project is focused on manufacturability analysis and providing re-design/ re-provisioning support for conceptual design of discrete single piece mechanical parts with emphasis on machined processes. A novel hybrid approach using a combination of sequential flowchart logic and an expert system has been developed which uses a user defined feature tree or STEP AP224 file as the input for its analysis. The system allows the designer to estimate manufacturability according to criteria of time and cost and also explore different scenarios of parametric variation in the design. A case study has been presented to validate the systems practicality and usefulness.

Key words: Conceptual design, manufacturing evaluation, computer aided process planning

1.   INTRODUCTION

The manufacturing industry in the twenty-first century is characterized by intense competition with more and more emphasis being placed on improving productivity, product quality and at the same time reducing costs, overheads and time-to-market. These market pressures translate into more optimized designs, manufactured by most optimized manufacturing processes. The most challenging and influential part of the design process is the conceptual design stage. It has been well proven that poor design at the conceptual stage can never be translated into a successful product. It is estimated that 80% of the cost of commercialising the product is determined during the early design stage [1]. The early design stage itself accounts only for a fraction of the total development costs of the product. It is therefore logical to provide the designer with necessary computer-aided tools for designing within cost/ time constraints and facility for cost/ time feedback and deciding on design alternatives.

Given the amount of variables which go into creating a successful design (or re-design), the ability of a designer to arrive at an optimum solution depends upon him indulging in an analytically intense and time consuming iterative cycle. Most of the efforts in providing computer-aided tools for supporting conceptual or early design have concentrated on the design of new products. Conclusive studies have statistically proved that a large number of new design projects are actually redesign activities [2]. Even in green-field design cases where the designer is creating a fresh design the design process may not always be a top-down approach. Further, even when preparing the plans for re-provisioning a particular design (component), it may be beneficial to have a review of the design and the manufacturing processes. This is especially true in view of the dynamic manufacturing technology today. Manufacturing planning at the conceptual or early design stage is the key for designers to evaluate manufacturability and manufacturing criteria and metrics like costs and time. Therefore there is no denying that this field of computer support for redesign has a substantial market. However, there is currently a lack of software tools in conceptual manufacturing planning.

This research proposes a compromise between a conceptual design system and a fully automatic manufacturing planning system for supporting such redesign/ re-provisioning activities. The prototype system presented by the authors builds on their previous successful research efforts to develop a fully automatic manufacturing planning system [3].

1.1   Progressive design

The classic design model described by Phal and Beitz [4] has four stages, i.e., Design Specification, Conceptual Design, Preliminary Design (also called embodiment design or fundamental design in some studies), and Detailed Design. The transition of design from one stage to another is a gradual process and the design evolves from one stage to another rather than an abrupt jump from one stage to the next. Especially in the context of providing computer support for re-design/ re-provisioning process, it is difficult to separate these design stages individually. The authors in their previous work which forms the background for this research [5] had proposed combining the first three design stages together as the Early Design stages. A new term was coined and presented by the authors to describe the evolution of design during the early design stages, i.e. Progressive design. The concept of Progressive Design is especially beneficial when the designer wants to use the same design tools or data across the design life cycle of a product, especially in context of re-design and re-provisioning activities. Progressive design encompasses a much broader scope than the classical definition of Conceptual design. Incorporating manufacturing analysis into progressive design can ensure that the design that reaches the shop floor is manufacturable and within cost/ time limits. The short literature survey presented in this section reviews work done in the field of providing computer support for manufacturability evaluation of the design manufacturability, time and cost evaluation and estimation with special stress on progressive design

2. MANUFACTURABILITY EVALUATION AND COMPUTER-AIDED PROGRESSIVE DESIGN

Manufacturability of a product depends upon the design domain, and a number of other related factors, like the available manufacturing resources and raw material. In general terms, manufacturability can be defined as an indication of the effort required for realizing the product. Given a set of manufacturing resources and product information, the problem of manufacturing evaluation is reduced to determining whether or not the design is manufacturable. If the design is found to be manufacturable, the next step is to determine one of the evaluation metrics. Manufacturability metrics is usually one of the following:

a) Boolean: This is the most basic manufacturability rating. It simply reports if the part "can" or "cannot" be manufactured with the given resources.

b) Cost: This has been an important area to study both, in manufacturing and in marketing communities [6]. Prompt estimation of cost with accuracy is crucial and this is evident by the abundance of research literature on the subject.

c) Time: Every manufacturing process has an associated time. In most cases, it is the manufacturing time that is used to calculate the manufacturing costs.

d) Qualitative measure: Ishii [7] qualified designs using adjective qualifiers: 'excellent', 'very good', 'good', 'poor', 'bad' and 'very bad'. But such measures are hard to interpret and compare, especially if the rating comes from different systems or designers.

e) Abstract: This is similar to the qualitative measures discussed above, but involves each design attribute being assigned a manufacturability index or producability index (PI) instead of a qualifying adjective like qualitative measures. As with the qualitative measures, it can become difficult to interpret or compare designs if the indexes are from different systems.

It has emerged that most manufacturing evaluation systems reviewed are feature-based, directly or indirectly. Manufacturability evaluation is either done through a priori such as DFM guidelines, or through a posteriori such as analysis tools which take the geometric or feature model as the input. The a priori approach is usually rule-based, while a posteriori analysis has often been implemented using a plan-based analysis. The analysis is usually separate from the detailed process plan generation. In most cases the analysis module and process plan generator module are completely separate. The only exception is in some plan-based analysis systems. Some systems integrate design and process planning to allow simultaneous design with process planning [8]. But even in such closely integrated systems, design and process planning are sequential activities and the integration is limited to sharing of a common data model.

Other than evaluating the progressive design on criteria of time, cost and manufacturability discussed above, some systems provided redesign suggestions but these have not been included here as they are more applicable to new design activities and not so useful for redesign activities being considered here.

3.   METHODOLOGY

Based on the literature review and preliminary survey, it was established that there was considerable scope for improving software tools for supporting progressive design/ re-design activities. Therefore a methodology was developed with an aim to provide a deliverable system to support design/ re-design activities.

3.1   International Standard

Another factor which emerged from the literature review was the lack of support for industrial standards (for exchange of information). This was perhaps the main reason why so many of the research systems had limited life outside the research laboratory. This problem of data standards arises from the incompatibilities between different CAD systems.

The most prominent standard emerging is STEP (The Standard for Exchange of Product Data Model) [9]. The standard is divided into domain and application specific Application Protocols (APs). The AP concept is a refinement of the development of subsets which focuses on the end user or application requirements to be satisfied. STEP AP 224 protocol for planning of machined parts using machining features is fairly well developed and commercial application based on this standard are beginning to appear. STEP AP224 was therefore adopted as the standard input format for the system. Pratt [10] provides a good introductory text on STEP.

System Design

It was decided to develop the system as an automatic process planning system. It was summarized from the literature review that most systems for manufacturing analysis were directly or indirectly feature based. The automated manufacturing planning system proposed to be developed, required a detailed feature model as the input. This basic requirement of having a fully validated feature model is precisely the restriction during progressive design.

The AP224 file could be generated from a detailed CAD model. To generate the input for the early or conceptual design stage, it was decided to modify a feature model editor (discussed in detail in next section) to create the feature model for input. The feature model for input could not be generated from 3D solid models (using conventional approaches) as the 3D solid model comes after the progressive design stage. It was therefore decided that the feature model for the input would represent the parametric features with the sole aim of providing information for planning and not for generating the solid model (i.e. visualization).

The system therefore had two separate input methods – a proprietary feature model in the conceptual design sage and a STEP AP224 model in the detailed design stage.

4.   SYSTEM ARCHITECTURE AND IMPLEMENTATION

4.1 Logic Designer

A block diagram of the system is shown in figure 1.

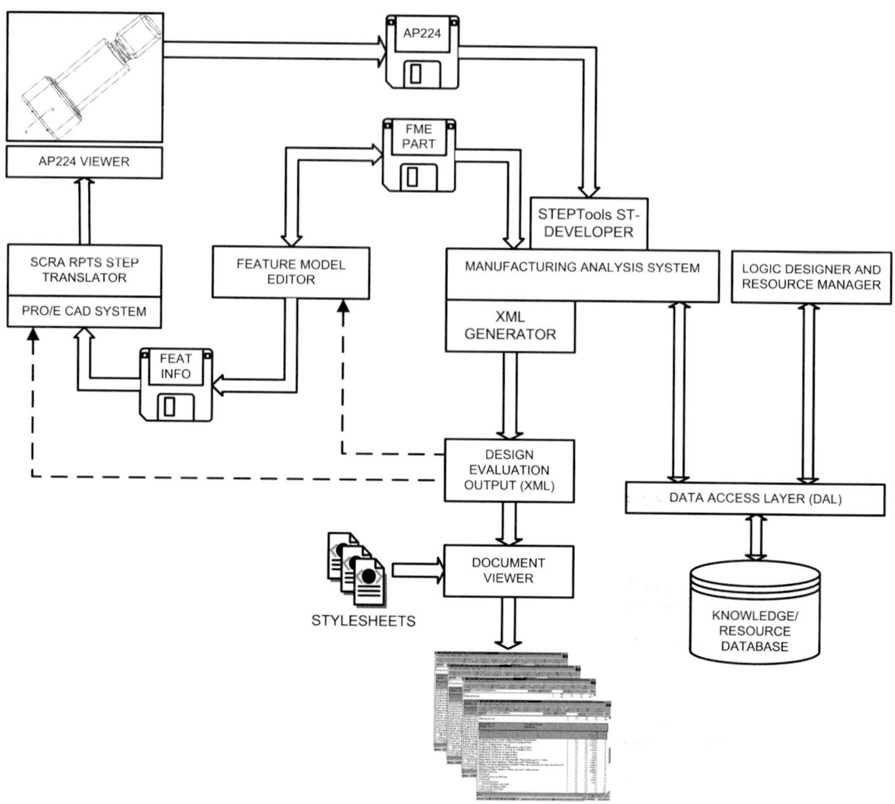

Figure 1 A block diagram of the system

The analytical logic, or more specifically the domain and empirical knowledge is kept in an external database. The captured knowledge is divided into 'Knowledge' and 'data'. Captured knowledge is organized into 'domains'. 'Domains' are totally independent development areas and it is not possible to refer elements of one domain from the other. This logic can be edited and viewed graphically as a flowchart using the LogicDesigner as shown in figure 2.

The user also has the options of having multiple nested logic flowcharts. The point of entry into the logic, or the starting flowchart can be selected/ configured by the user. Further flexibility is offered by allowing nesting of flowcharts.   This approach to storing the logic externally has proved immensely successful not only in making the application flexible and updateable, but also helping in breaking down the domain specific barriers and making the application more generic. The application itself is only a shell to run the logic. Simple development of new analytical logic allows the application to be used in a completely new domain. The logic developed graphically using the LogicDesigner is run (or executed) by the main system.

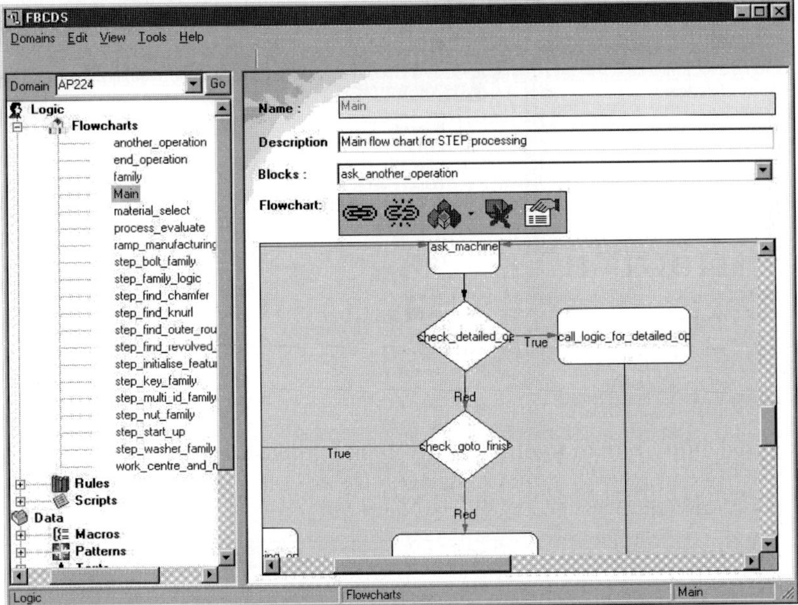

Figure 2 Logic Designer

The logic is captured using a set of standard commands. The system provides an extensive vocabulary of commands for almost every situation. The commands are entered in the Action and Decision Blocks of the flowcharts and used in conjunction with variables. The system also provides commands for integrating with CLIPS expert system. The LogicDesigner is also used to maintain the data which is accessed from the flowcharts. The data is maintained in the form of spreadsheet tables and can be used to hold associated data like machining data, raw material specifications and speed and feed tables.

4.2　Feature Model Editor

The Feature Model Editor (FME) **[11-13]** was developed by the authors as the preferred way of inputting data in a simple planning system for sheet metal products. FME consists of feature libraries, which can be edited by the user. The user can open only one library at a time although any number of part files can be opened simultaneously. The features in the library can be grouped into Families and Sub-Families. This allows for a more logical grouping of the features. The features inherit all the attributes of their parent family or sub-family. This minimizes the number of individual feature attributes as most of the common attributes can be entered at family or sub-family level. Feature instances can be added to the part by simple drag-and-drop operations. The features can also be repositioned within a part using the

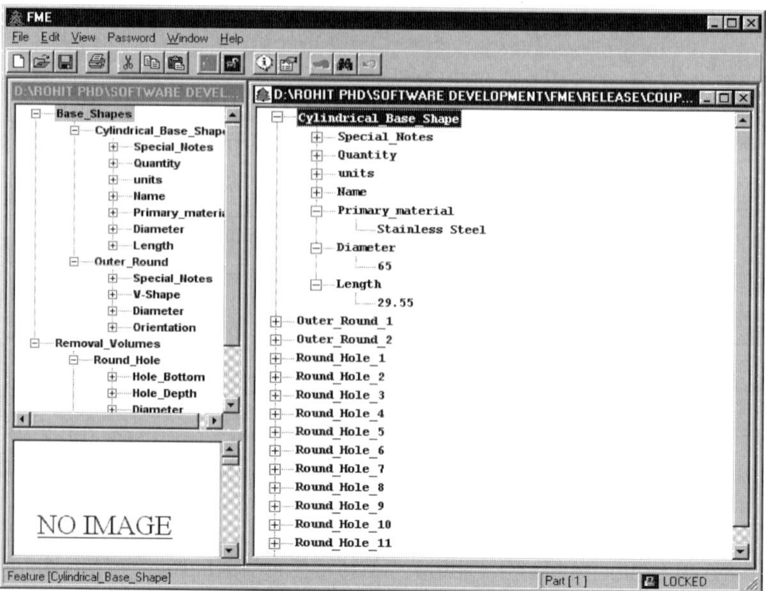

Figure.3. A screenshot of the FME system

drag-and-drop operation. The application also provides facilities for cut-and-paste operations.

The feature tree as well as the library created by the FME is stored as a plan ASCII file. The feature tree has all the constituent features of the part being generated as well the attributes of that feature. It must be noted that the FME does not try to generate a 3D model or a view of the part being designed. The emphasis is only on entering all the information required for planning analysis of the part. Once the template of an existing part family has been generated using the FME, the user can use it as a tool during progressive design to change the part parameters and see the effect it has on manufacturing time and costs by analyzing it. A screenshot of the FME system is shown in figure 3.

Feature-based conceptual design system

A screen shot of the main application screen is shown in figure 4. As shown in the block diagram in figure 1, the main application has two separate software components, which are working in tandem. These components are the Feature-Based Conceptual Design System (FBCDS) and the Document Processor. The FBCDS is the heart of the system and is the shell which runs the logic stored in the external database. The FBCDS is capable of generating the output to the screen, but to make properly formatted documents, the output is passed on to the Document Generator which is an eXtensible Markup Language (XML) engine responsible for generating XML output documents. Only one XML document is generated per planning session, but different views and data can be formatted and presented using XSL stylesheets. XML has been chosen for output documents for easier dissemination of the documents across hardware and software platforms.

The user window is divided into two vertical halves by a splitter bar. The left hand pane displays the tree-view of the FME file read in for analysis while the right hand pane displays the user interaction options and the output. When an FME model is read into the system, each feature node has a checkbox displayed with it. The planner can decided to include or exclude the feature from the planning session by checking/ un-checking the checkbox. This is helpful in quick if-the-else evaluation of the feature models. For example, the user can quickly generate a cost estimate of the complete model, and then regenerate the estimate with a particular feature removed (say a knurl). Comparison of the two versions on basis of cost and machining time would be good guide to decide the trade off. Similarly, editing the parameters of the features and evaluating them could create alternate versions.

5.   ESTIMATING MANUFACTURABILITY CRITERIA

The manufacturing analysis/ evaluation in the prototype system adopts a two-stage hybrid rule-plan approach by combining these two approaches in a sequential manner. In the first stage of analysis, the design is analyzed as per the design rules (developed in the LogicDesigner). This gives a Boolean evaluation of the design i.e. manufacturable or not. In the second stage, a process plan would be generated to determine the time and costs of manufacturing. The system proceeds to the second stage of generating manufacturing metrics only if the design is manufacturable as per the rules and the designer wants a detailed report on the metrics.

For initial tests, five families of commonly used single piece machined components were identified. The five families were 'bolts and fasteners', 'nuts', 'washers and spacers', 'couplings', and 'flanges'. The families were grouped based on their common manufacturing features (and manufacturing processes for realizing these features). The next step was to analyze the most complex members of each family to arrive at a superset of features, which could be used to describe every member of the family. For example, for the bolt family, some of the specimen had a knurled head and some did not. Therefore a knurl feature was included in the set of features used to describe any member of the bolt family. If the knurl feature in the feature tree used to describe the specimen from this family had all zero parameters, it was understood to represent a bolt with a simple head, i.e., no knurling operation was considered. The next step was to develop logic for analyzing the feature tree and deducing the family of the component based on the feature set representing the specimen. Since only these five families were being used as the case study, a more empirical approach based purely on feature count was used. This was done, as efforts put in to develop a generic solution would not have been justified. But since this logic is developed using the LogicDesigner it can very easily be edited to cater for any specific complex algorithm which the user might desire.

A set of DFM type rules were developed. These evaluating rules were also been developed in LogicDesigner and are hence easily editable. They include 'DFM type' rules to see adherence to standard design practices and 'resource' rules which make sure that the given dimensions and tolerances are realizable from the machine data available in the database. Most of these rules require parameters of one or more features of the part in order to evaluate. The net manufacturability evaluation Boolean is the logical AND of all the DFM and resource rules. If all the DFM and resource rules are satisfied then the manufacturability evaluation is reported as TRUE indicating that the component is manufacturable.

5.1   Estimating manufacturing metrics - Cost and time

The logic for estimating the time and manufacturing costs (developed in the LogicDesigner) uses a combination of variant approach and explicit calculations based on the parameters of the features to be machined. This takes into account only the direct manufacturing costs and not the related costs like the overheads and fixed costs of the company. To keep the logic simple, the overhead costs are added as a percentage of the manufacturing costs. The set-up time for the jobs are based on empirical and statistical values which are typical for the kind of job being considered based on the machine being used and the weight of the job which would account for the 'handling' time.

Calculation of the machining time for a particular feature is a straightforward mathematical calculation. For example, in order to calculate the machining time for turning an outer diameter of a cylindrical job like a bolt shaft, the system looks up the material specification v/s feed rate table of the selected machine to calculate the number of cuts required to achieve the final diameter. The resource table for every machine gives the parameters like rpm and feed rate for the given material. It also gives the finishing cut parameters for achieving the desired tolerance and surface finish. Once the staring diameter, finishing diameter and step of each cut is known, the time taken for the machining operation can be calculated. Some of the machine resource tables used in the calculations like the machine setup time, cleaning time, and time between canned cycles are based on empirical valued and hence maybe be specific for each implementation. But since all tables are easily accessible and editable from the LogicDesigner this is not seen as a limitation. The resource manager can update the tables based on the recommended machine values and the feedback from the users of the plans generated by the system. With regular feedback over a period of time, the system can be developed to accurately predict the time/ cost values within limits.

6.   WORKED EXAMPLE AND RESULTS

The case study part used was a reducing flange from the 'flanges' family. The component was selected as it represented a typical component of the family with average complexity. Wire frame representation of the component along with the feature tree as read into the FBCDS system is shown in figure 4.

The feature tree for the component is shown in the left side pane in figure 4. The FME feature library used to generate the FME model was based on

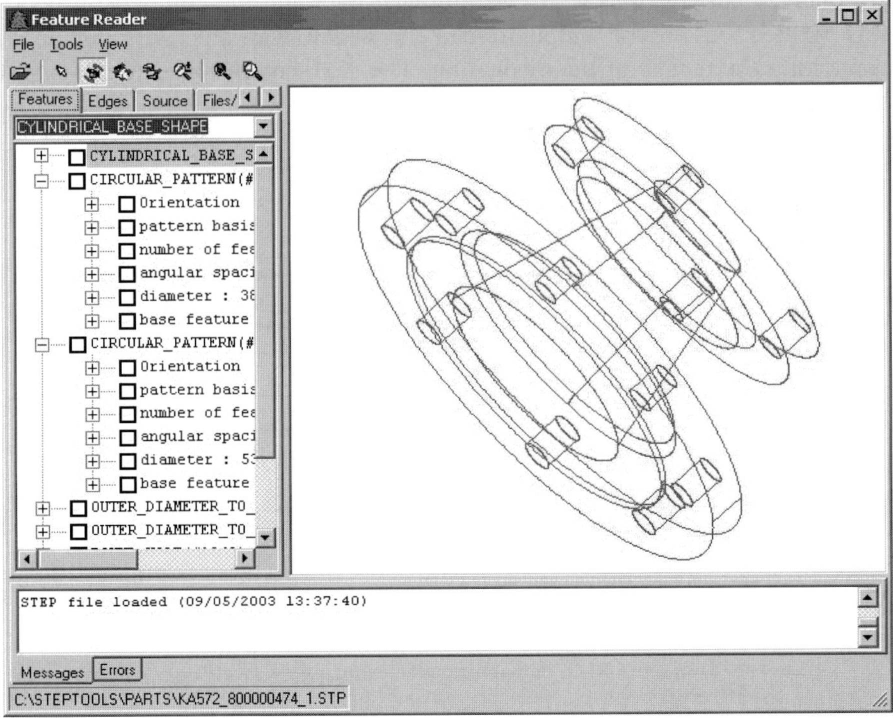

Figure 4 Wire frame representation and feature tree of component

STEP AP224 features. Reading the FME or STEP file yielded exactly the same results except that FME file did not provide any visualization of the product. It may be seen from figure 4, that the features are not arranged in any particular hierarchy. This is similar to the arrangement of AP224 features. However, the features could be arranged in a logical hierarchy purely for the sake of convenience if the user so desires using the feature model editor. But this would have no bearing on the processing logic or the results obtained. However, when arranged in hierarchies, disabling a particular feature by 'un-checking' it would automatically disable all the child features.

The DFM rules for analyzing the component were developed from the data collected from the designers and DFM guidelines and standards. Typical rules included:
– Maximum permissible difference between the major end diameter and the minor end diameter.
– Minimum number of securing bolts (holes) on the flange end depending upon the PID diameter of the flange.

The system derived the parameters required to evaluate these two rules from the features. The standard reference data was available in the resource tables of the LogicDesigner. The results calculated using the feature attributes was compared with the reference data and the comparison result was written out to the output document which was an XML document containing the results of the evaluation. The XML output formatted using a style sheet is shown in figure 5.

Test No. : 10	
Description : **Test for flange length**	
Remarks	**Result**
Maximum permitted flange taper per length as per table. [STEP Feature ID : # 1848] : TAPERED ROUND HOLE	OK

Test No. : 20	
Description : **Test for PID holes**	
Remarks	**Result**
Number of flange securing bolts as per PID table. [STEP Feature ID : # 2099] : PID CIRCULAR PATTERN [STEP Feature ID : # 2179] : PID CIRCULAR PATTERN	OK FAILED

Figure 5 XML output formatted using a style sheet

7. CONCLUSIONS AND FURTHER WORK

The development of integrated design and manufacturing planning systems for supporting conceptual redesign and re-provisioning activities is of critical importance to today's industry. A new system for computer-aided support for early design is presented here. A novel hybrid approach using a combination of sequential flowchart logic combined and an expert system has been developed which uses a user defined feature tree or a STEP AP224 file as the input for its analysis. The system allows the designer to estimate manufacturability metrics like time and cost and also explore different scenarios of dimensional variation in the design. A detailed case study has

been completed to validate the systems practicality and usefulness and has yielded encouraging results. The case studies also concluded that system ought to be capable of meeting the requirements of not only the general re-provisioning activity, but also could be considered for the initial design of new products. Because of the implementation architecture, which allows data and knowledge, capture and updating, the system is generic and is not limited to implementation to one particular manufacturing domain.

It is anticipated that this research will complement further research efforts in the area of supporting intelligent knowledge based conceptual and early design which is already being explored by this research group.

ACKNOWLEDGEMENTS

The authors gratefully acknowledge the support of LSC Group, UK Ltd. during this research, especially Mr Alan Crawford and Mr Ray Muldoon

REFERENCES

1. Boothroyd G., "Product design for manufacture and assembly", *Computer-Aided Design*, **26 (7)**, pp. 505-520, 1994.

2. Salomons O., Slooten F.v., Houten F.J.A.M. and Kals H.J.J., "A computer support tool for re-design", *International Conference on Engineering Design 'ICED 93*, The Hague pp. 1559-1570, 1993.

3. Sharma R. and Gao J.X., "Implementation of STEP AP224 in an automated manufacturing planning system", *Proceedings of the I MECH E Part B Journal of Engineering Manufacture*, **216 (9)**, pp. 1277-1289, 2002.

4. Phal G. and Beitz W., (1997) *Engineering Design: A systematic approach*, Springer, London, UK.

5. Sharma R. and Gao J.X., "A Progressive design and manufacturing evaluation system incorporating AP224", *Computers in Industry*, **47 (2)**, pp. 155-167, 2002.

6. Schreve K., and Basson A.H., "Manufacturing Cost Estimation During Design of Fabricated parts", *Proceedings of the I MECH E Part B Journal of Engineering Manufacture*, **213 (7)**, pp. 731-735(5), 1999

7. Ishii K., "Modeling of Concurrent Engineering Design": *Concurrent Engineering: Automation, Tools and Techniques*, 1993.

8. Feng S.C. and Song E.Y., "Information modeling on conceptual design integrated with process planning, Symposia for Design for Manufacturability" - *The 2000 International mechanical engineering congress and exposition*, Orlando, Florida, USA pp. 160-168, 2000.
9. ISO , Industrial automation systems and integration -- *Product data representation and exchange* -- **Part 224**: Application protocol: Mechanical product definition for process planning using machining features, 2001
10. Pratt, Michael J., "Introduction to ISO 10303 - The STEP Standard for Product Data Exchange." *Journal of Computing and Information Science in Engineering* **Vol 1**, 2001.
11. Sharma R. and Gao J.X., "A Feature Model Editor for Process Planning of Sheet Metal Products", *15th International Conference on Computer Aided Production Engineering CAPE '99*, Durham, UK pp. 151-156, 1999.
12. Gao J.X., Tang Y.S. and Sharma R., "A feature model editor and process planning system for sheet metal products", *Journal of Materials Processing Technology*, **107 (1-3)**, pp. 88-95, 2000.
13. Sharma R. and Gao J.X., "Feature-based progressive design and manufacturing planning in a PDM environment", *16th International Conference on Computer-Aided Production Engineering CAPE 2000*, Edinburgh, UK, 2000.

STUDY ON QUALITY EVALUATION TECHNIQUE FOR RESEARCHING AND DEVELOPING SYSTEM OF MECHANICAL PRODUCTS

Qi Zhang, Cheng-Xian Yang, Huan-Liang Li and Wei Sun
Department of Mechanical Engineering of Engineering Institute of Engineering Corps, PLA. University of Sic. & Tech., Nanjing 210007 China

Abstract:	The Life Cycle Quality (LCQ) of equipment is formed from various processes and stages. The research and development process (RDP) is a very important stage of formalising the LCQ of equipment. To evaluate the quality of the RDP, the defects of design process can be timely detected. Therefore, the technique support for quality control is provided to the RDP of new equipment. The researching and developing system of mechanical products is set up based on the theory and method of operational research and fuzzy mathematics. The quality indexes of the system are described. The relative weight values of the system are calculated. The quality evaluative model of the system is established. The evaluative value of the quality level in the system is worked out in the paper. An example quantified for application is also given.

Keywords:	mechanical products, researching and developing system, operational research, fuzzy set, quality evaluation.

1.   GENERAL DESCRIPTION

During the developing new type construction machinery, the course of the research and development include marketing analysis, feasibility research, scheme verification, engineering design, test, and design approval and qualification phases, etc. The developing process of new product is a systematic engineering. During the development of new product, the whole process from marketing analysis to sale of products is regarded as the developing quality system to be researched. The design process and operational status in the system are directly relative to the quality levels of new product. The quality levels of the new product are the concrete performance of design process management in the researching and developing system. In order to evaluate objectively the quality status of the new product, the quality levels of the designing process and operational status in the system should be evaluated and researched and given a scientific and reasonable evaluating standard. The scientific data is provided for the decisional policy and improvement of the researching and developing system of mechanical products.

2.   SYSTEM DEFINITION

2.1 The Definition of Developing System

The researching and developing system of mechanical products is a very complex one. The overall process from marketing investigation and analysis to example machine test and qualification consists of many subsystems. Developing process is comprised of many influence factors on quality of the system. The system of the research and development can be described as the input action on the developing process of the system through technical flow, information flow, fund flow, material flow, etc. By scientific organization, management, match, and utility of the resources the above mentioned, it can make the system more productive and turn out an objective output conforming to designing requirements and meeting customers' requirements.

2.2 Quality Definition of Developing System

The quality of developing system includes two aspects
(1) The quality status of the objective output of the system, i.e. the quality levels of example machine, a symbol of the degree of entity quality conforming to designing requirements of new product, and a specific performance of superiority or inferiority of objective output.
(2) The designing process quality of the system, it marks whether technical configuration of operational process in the system live up with the documental requirements of quality surveillance in the process of the system is an efficient measure and method to ensure the quality of machine.

3. THE DETERMINATION OF DEVELOPING SYSTEM

3.1 The Constitution of the System

Generally, the researching and developing system is comprised of eleven quality indexes (subsystems), each index is made up of many relative quality factors. The system is divided into Top-layer quality indexes, i.e. A-layer quality indexes, and Sub-top-layer quality indexes, i.e. B-layer quality indexes. The functional flow diagram the system is shown in following fig 1.

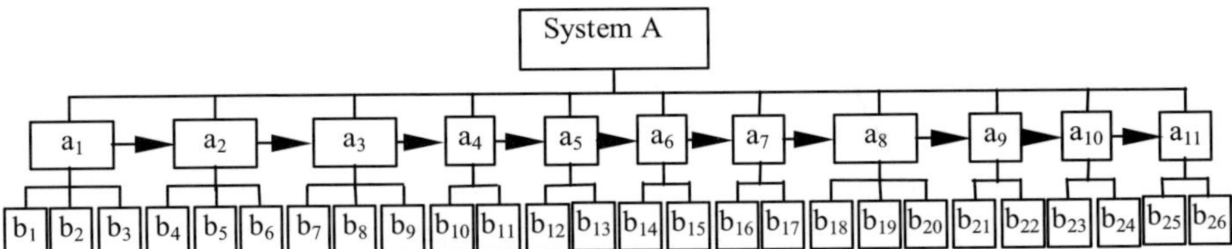

Fig.1 The functional frame diagram of developing system

The quality indexes of the Top-layer (a-layer) in the fig.1 include:
a_1—marketing analysis, a_2—feasibility analysis, a_3—scheme verification, a_4—engineering design, a_5—design review, a_6—technological design, a_7—technological review, a_8—trial production, a_9—test, a_{10}—type approval, a_{11}—batch production.

The quality indexes of sub-top-layer (b-layer) in the fig.1 include:
b_1 —the determination of marketing goal, b_2—the requirements analysis on marketing, b_3—the prediction of economic profit, b_4—the feasibility analysis on key technique, b_5—the investigation of manufacturing capacity, b_6—the supporting analysis on funds, b_7—the determination of functional objective, b_8—the verification of technical and functional indexes, b_9—structural arrangement, b_{10}—the structural design of the system, b_{11}—the design of reliability and maintainability, b_{12}—the drawings review, b_{13}—the standardization audit, b_{14}—technological tool design, b_{15}—the drafting of technological plan and flow, b_{16}— the audit of technological plan, b_{17}— the review of key process and control points, b_{18}—the formulation of production plan, b_{19}—the supply for incoming material before production, b_{20}—the preparation for manufacturing equipment before production, b_{21}— the establishment and commitment of test program, b_{22}—the recording and handling of test data, b_{23}—the drawing up documents of product validation, b_{24}— the analysis on qualification views of experts b_{25}—the requirements audit for batch production, b_{26}—the check of preparation for batch production.

3.2 The Description of quality indexes in the Developing System

3.2.1 Marketing Analysis: the phase includes three aspects:

(1)To specify marketing target, (2) to analyze marketing demands, (3) to analyze marketing profit.

During marketing investigation, marketing at home and abroad should be dynamically analyzed including demands for construction machinery of different type and specifications, statistics analysis on output and sales, at present, and needs for the future marketing. Marketing survey report and predictive analysis report are drawn up.

3.2.2 Feasibility Analysis: The stage also includes three aspects

(1)Feasibility analysis on technique: how to perform and analyze feasibility of the project. (2) Funds supporting analysis, i.e. investment resources and situation of funds in place. (3) Resources supporting analysis, personnel and talent resources supporting, manufacturing capacity and technological supporting of manufacturing equipments, analysis on the quality levels of compatible parts and incoming parts.

3.2.3 Project Verification: This phase also includes three aspects:

(1)Determination of functional objective: To determine the specifications and function a_3 goal, the marketing demands and technical levels of its same type product must be fully taken into account to decide proper indexes, such as, engine power, productivity, walking speed. If these indexes are overestimated, it is hard to reach the indexes. Whereas, these indexes are underestimated, the technical levels of the product developed will lag behind and be lock of competitive on market.

(2)Structure features, overall arrangement, structure is as simple as possible and to reach the functional goal as well. Safety and reliability in working, reasonable overall structure, it should be in accordance with the standards of person-machine engineering.

(3)During qualification of design scheme, concerned experts/ devices should be fully adopted. Design scheme is continuously improved to optimize it as much as possible, and to ensure the functional objective realization of new product.

3.2.4 Engineering Design

After determining design scheme, according to characteristics of designers, the head of project assigns the project to every designer. Advanced technique, new materials, and new workmanship should be intensively applied as much as possible in engineering design. The relationship of technical innovation and heritage has to be exactly tackled in engineering design. Superior Technique and structure in existence can be transplanted into the same type design. The defect in design is improved gradually with technical innovation and development. The advanced design thoughts and method should be fully adopted. CAD on mechanical design privileges to be applied for standardization of design drawings and enhancing working efficient and design quality for designers. Calculation instruction, all drawings, and relative technical documents should be compiled.

3.2.5 Design Review

After engineering design completed, design review should be performed in time by experts on design, technology, and standardization, including calculating instruction examination, i. e. reasonable, scientific, perfect standards of design are audited. Amendment to configuration and usage and management of the drawings should be clearly specified. Design review report should be submitted.

3.2.6 Technological Review

According to design drawings, technological designers should draw up technological scheme and flow. CAAP technique should be applied privileged in technological design to quality levels of design.

3.2.7 Technological Review

After technological design completed, technological review should be preformed timely by concerned expert and important technology are strictly controlled to establish process quality surveillance points and to audit technological supporting capacity, Technological audit report is formed.

3.2.8 Trial Production

After design review and technological review completed, preparation of pre-production should be well prepared in time by production management sectors and to formulate production plan.

3.2.9 Test

After assemble of minor-batch trial production completed, according to the requirements of test program including test condition, test items, test rule, and inspectional standards, etc. test of minor-batch production should be strictly performed, test faults must be scientific divided into different type. The test data must be recorded and figured, The results of test have to be analyzed and evaluated to offer a test report and to a scientific evidence for design approval and qualification.

3.2.10 Design Approval and qualification

According to the design approval and requirements of new product, it should be conducted by concerning expert. Key audit aspects include whether the characteristics indexes of new product meet design requirements or not, functional goal can be realized or not, and possess mass-production capacity or not. Design approval documents including quality guarantee program and quality surveillance program should be mainly qualified to whether to meet requirements or not. Finally conclusion and qualification opinions must be made out.

### 3.1.1	Mass-production

After production approval completed and meeting basic requisites of mass-production, under the analysis of marketing demands and production capacity, Production batch quantities are determined. Before new product is in production, production plan and preparation must be formulated.

## 4.	THE QUALITY EVALUATION MODEL OF DEVELOPING SYSTEM

4.1 The Determination of relative weight of quality indexes

The Determination of relative weight values of quality indexes in sub-top layer (b-layer). The concerned expert mark method is adopted to decide the relative weight values of quality factors. A way-reversed analysis method is applied to calculate the relative weight values of quality indexes in the layer. The method is that absolute marks of single quality index are given by concerned experts, first to compute the relative weight values of b-layer, then to weight values of a-layer. b_j is expressed as the absolute marks of quality index $i(i=1$, 2, ...6). According to principle of one to one, judgmental matrix is established by relative important degree of every quality index to calculate the maximum feature $\lambda_{\max}$ and feature vector of the matrix. i.e., B^T is the weight value vector of every quality factor, while the experts mark, the performance degree of every factor which consists of quality indexes should be fully taken into account to give relative marks. In order to analyze and settle conveniently the problem, factor indexes in b-layer in fig.1 can be simplified to the general formation of frame diagram of quality factors in Fig. 2. There are m concerning experts to mark the n judgmental factor indexes in b-layer.

$b_{ip}^{\ k}=b_i^{\ k}/b_p^{\ k}(i{\neq}p)$ (i,p=1,2,3...n) is expressed as the relative marks of quality indexes.

$b_{ip}^{\ k}$ stands for the marks given by expert k of relative weight value of the factor i to the factor p(while $i=p$,) $b_{ip}^{\ k}=1$, $b_{ip}^{\ k}=1/b_{pi}^{\ k}$, m judgmental matrix can be constructed as follows:

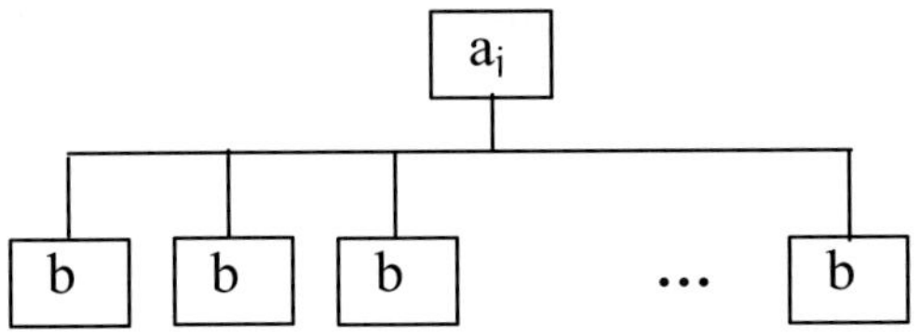

Fig.2 functional frame diagram of quality factors

$$M_b^k = \begin{bmatrix} 1 & b_{12} & b_{13} \cdots & b_{1n} \\ 1/b_{12} & 1 & b_{23} \cdots & b_{2n} \\ \vdots & \vdots & \vdots & \vdots \\ 1/b_{1n} & 1/b_{2n} & 1/b_{3n} \cdots & 1 \end{bmatrix} \quad (k = 1,2,3,\cdots,m) \tag{1}$$

If $B_{ip}(i,p=1,2,3,...,n)$ represents the statistics marks given by expert m of index i to index p, then to calculate mean value and get matrix elements. So General judgmental matrix of b-layer is formed and it can be expressed as follows:

$$M_b = \begin{bmatrix} 1 & B_{12} & B_{13} \cdots & B_{1n} \\ B_{21} & 1 & B_{23} \cdots & B_{2n} \\ \vdots & \vdots & \vdots & \vdots \\ B_{n1} & B_{n2} & B_{n3} \cdots & 1 \end{bmatrix} \tag{2}$$

Corresponding characteristic matrix is $[\lambda_b E\text{-}M_b]$, S.T. $|\lambda_b E\text{-}M_b| = 0$

After calculated, we can obtain maximum characteristic value λ_{bmax} and the value is put into characteristic matrix $[\lambda_b E\text{-}M_b]$ to work out relative weight vector B^T of b-layer, $[\lambda_b E\text{-}M_b] \cdot B^T = 0$, B^T is shown as follows:

$$B^T = \begin{bmatrix} \alpha_{b_1} & \alpha_{b_2} & \alpha_{b_3} \cdots & \alpha_{b_n} \end{bmatrix}^T, \sum_{i=1}^{n} a_{b_i} = 1 \quad 0 < a_{b_i} < 1, \tag{3}$$

According to the method mentioned above, the relative weight values of quality indexes, a_1, a_2, ..., a_{11} are calculated respectively.

In terms of index a1 in fig.1, judgmental matrix is expressed as follows: equation (4):

$$M_b^k = \begin{bmatrix} 1 & b_{12} & b_{13} \\ 1/b_{12} & 1 & b_{23} \\ 1/b_{13} & 1/b_{23} & 1 \end{bmatrix} (k = 1,2,3 \cdots m) \tag{4}$$

After statistic handled of $M_b{}^k$, judgmental matrix is shown in equation (5).

$$M_b = \begin{bmatrix} 1 & B_{12} & B_{13} \\ B_{21} & 1 & B_{23} \\ B_{31} & B_{32} & 1 \end{bmatrix} \tag{5}$$

Characteristic matrix is $[\lambda_{a1}E\text{-}M_b]$, S.T. $\mid \lambda_{a1}E\text{-}M_b \mid =0$. After calculated, we can get $\lambda_{a1max\ and}$ the value is put into $[\lambda_{a1}E\text{-}M_b]\cdot B^T=0$. The relative weight value of quality index a_1 is shown in equation (6) and to apt maximum value, to let$\lambda_{a1}=\lambda a_{1max}$'then

$$B^T = \begin{bmatrix} \alpha_{b_1} & \alpha_{b_2} & \alpha_{b_3} \end{bmatrix}^T \tag{6}$$

In the same way, the relative weight vector and maximum value vector of indexes, a_2, a_3, …, a_{11} can be also calculated. So they are expressed as follows:

$$\lambda^T = \begin{bmatrix} \lambda_{a_1} & \lambda_{a_2} & \lambda_{a_3} & \cdots & \lambda_{a_{11}} \end{bmatrix}^T \tag{7}$$

$$B^T = \begin{bmatrix} \alpha_{b_1} & \alpha_{b_2} & \alpha_{b_3} & \cdots & \alpha_{b_{26}} \end{bmatrix}^T \tag{8}$$

(2)Determination of Top-layer (a-layer) quality indexes vector. In terms of a –layer in Fig.2, S.T.

$$a_j = Int\left[\sum_{k=1}^{m} \alpha_{b_i} b_i^k \right] \qquad (j=1,2,…,11) \tag{9}$$

b_i^k- the marks of expert k for index i in b-layer,
$a_{bi}-$ the relative weight value of index i in b-layer.

For instance, expert relative marks for quality index a_1 in a-layer are calculated in equation (9), then to compute a mean value of index a_1, which comprised of three indexes b_1,b_2,b_3.With the same calculating way above, relative marks of indexes, a_2, a_3, …, a_{11} are also figured out,: The judgmental matrix of quality indexes in a-layer is calculated by the same as equation (1) and equation (2), then to get maximum characteristic value$\lambda_A=\lambda_{Amax}$ and maximum value vector of indexes, as follows.

$$A^T = \begin{bmatrix} \alpha_{a_1} & \alpha_{a_2} & \alpha_{a_3} & \cdots & \alpha_{a_{11}} \end{bmatrix}^T \tag{10}$$

4.2 General Judgmental Matrix of Quality Levels of the System

Marks matrix of quality indexes in a-layer given by m experts is established according to equation (2). So general judgmental matrix of quality levels of the system can be directly constructed by equation (11). As follows:

$$M_Q = \begin{bmatrix} a_{11} & a_{12} & a_{13} & \cdots & a_{1,11} \\ a_{21} & a_{22} & a_{23} & \cdots & a_{2,11} \\ \vdots & \vdots & \vdots & & \vdots \\ a_{m,1} & a_{m,2} & a_{m,3} & \cdots & a_{m,11} \end{bmatrix} \cdot \begin{bmatrix} \alpha_{a_1} \\ \alpha_{a_2} \\ \alpha_{a_3} \\ \vdots \\ \alpha_{a_{11}} \end{bmatrix} \tag{11}$$

The evaluating value of quality levels of the system is some of the every relative mark of quality indexes:

$$A_Q = \alpha_{a_1}\frac{\sum_{k=1}^{m} a_{k1}}{m} + \alpha_{a_2}\frac{\sum_{k=1}^{m} a_{k2}}{m} + \cdots \alpha_{a_{11}}\frac{\sum_{k=1}^{m} a_{k,11}}{m} = \sum_{j=1}^{11}\alpha_{k_j}\left[\frac{\sum_{k=1}^{m} a_{kj}}{m}\right] \tag{12}$$

5.   AN EXAMPLE FOR APPLICATION

While developing a new wheel-type excavator, the quality levels of the system in the process of developing the product is evaluated According to the marks of 26 quality indexes in process of development by concerned experts. The relative weight values vector of the system is calculated as follows:

$$A^T = \begin{bmatrix} 0.10 & 0.10 & 0.15 & 0.20 & 0.05 & 0.15 & 0.05 & 0.05 & 0.05 & 0.05 & 0.05 \end{bmatrix}^T$$

For convenient computation, if the 11 quality indexes are marked by 3 concerned experts, of course, the more expert, the more precise, i.e. m is bigger and better, for the evaluation of quality levels of the system. But Computation is very complex, whereas, it is easy to be calculated by computer. In fact, expert numbers m is impossible too big. Generally, to let $m=5\sim7$, it can meet the requirements of marks by experts. The marks matrix of 3 experts is shown as follows:

$$A_3 = \begin{bmatrix} 0.25 & 0.30 & 0.40 & 0.25 & 0.40 & 0.30 & 0.35 & 0.45 & 0.30 & 0.40 & 0.30 \\ 0.35 & 0.20 & 0.30 & 0.35 & 0.30 & 0.35 & 0.40 & 0.40 & 0.40 & 0.30 & 0.40 \\ 0.30 & 0.25 & 0.50 & 0.40 & 0.40 & 0.45 & 0.45 & 0.30 & 0.40 & 0.40 & 0.30 \end{bmatrix}$$

The evaluating value of the quality levels of the system for a new wheel-type excavator is finally calculated from equation (12) to get $A_Q = 0.347$, Equation(12)is also suitable for the quality evaluation of developing systems of other mechanical products to calculate its evaluating values. Evaluating values are a symbol of superiority or inferiority of quality levels of developing system for a new product.

6.   CONCLUSIONS

This paper provided the technique support for quality control to the RDP of new equipment. The researching and developing system of mechanical products built based on the theory and method of operational research and fuzzy mathematics can make the factory more productive and turn out an objective output conforming to designing requirements and meeting customers' requirements. The system also described the quality indexes, and calculated the relative weight values. An example proved the correctness of all the ways and methods the paper constructed and used.

REFERENCES

1.   GAN Ying-Ai. *Operational Research*, Beijing, Tsinghua University Press, 1992.
2.   *National Military Standard GJB9001A-20001*, Beijing, China Standard Press, 2001.
3.   Moskowitz.H, Kim K. QFD Optimizer. "A Novice Friendly Quality Function Deployment Decision Support System Product Designs", *Computers & Industrial Engineering*, Aug.1997.
4.   YANG Cheng-Xian et al. "Research on the Risk Decision of Quality Inspection for Mechanical Products", *Construction Machinery*, Dec.2002.
5.   CAI Gai-Ping et al. "Fuzzy Judgment of Process Planning for Parts", *Modern Manufacturing Engineering*, Feb.2003.
6.   Don P Closing. "Total Quality Development", *Manufacturing Review*, pp. 108-119, July 1994.
7.   Joseph Srakis et al. "Quality Information Systems in Advanced Manufacturing Environment", *Quality Engineering*, Aug.1996.
8.   Schweyer et al. "Information System Design for Project Management", *Proceedings of International Conference on Industrial Engineering and Production Management*, Morocco.1995.

CONFIGURATION MANAGEMENT USING STANDARD TOOLS
Possibilities, Limits and Development Potential

Marian Mešina and Dieter Roller
Institute of Computer-Aided Product Development Systems, University of Stuttgart
Universitätstr. 38, D-70565 Stuttgart, E-Mail: Dieter.Roller@informatik.uni-stuttgart.de

Abstract: The selection of configuration components in small businesses can be done without any special software, based only on technical and general database knowledge. The most difficult problem in respect to optimal selection of components is the combinational variety. The number of allowed configurations may be very large. The paper presents a solution for selected special tasks of product configuration that allows the reduction of the combinatorial variety. The solutions are demonstrated on an example of a car configuration.

Key words: Rapid prototyping, Engineering on-demand, Combinational variety, Relational databases, Optimisation

1.   OBJECTIVES

Modern products have to be developed quickly and also accomplish many customer-specific requirements. Practical experience from industry shows that approximately 80% of new customer-specific products can be composed as combination of already available components. Product configuration from components is a very successful way to accelerate product development. The importance of these subjects is also mirrored in various publications and research activities in this area (e.g. [1], [2], [3], [5] and [6]). Alongside very powerful and sophisticated approaches [4], which may be used primarily in big companies, partial solutions for the small-business sector are also necessary. **The selection of configuration components in small businesses can be done without any special**

software, based only on technical and general database knowledge. For the practitioners who need computer-supported selection of product components, it is useful to describe, how it is possible to perform this task with generally known standard tools, such as relational databases.

The most difficult problem in respect to optimal selection of components is the combinational variety, which should be taken into account. The number of allowed configurations may be very large.

The goal of this paper is to present such a solution for selected special tasks of product configuration. The solutions will be demonstrated on an example of a car configuration.

1.1   Description of the configuration example

For the construction of a car following components will be used:

Car body (16 models), brakes-system (15 models), gear (15 models), motor (12 models), seat1 (forward left, 13 models), seat2 (forward right, 13 models), settee (back, 12 models), wheels (10 models).

These components are described in eight tables. Each table includes all necessary data for component selection. Assumption: seat1 and seat2 may be selected independently from each other. Therefore it is reasonable to use 2 different seat tables, with the same data.

Without any restriction, with the described components **876,096,000** different cars can be constructed. The table of all combinations cannot be generated and stored explicitly. The necessary memory requirement for such a table is at least 22.8 gigabytes. It is practically not possible to build at first all possible combinations of components and after this to check with help of constraints, which combinations are allowed.

All components have the following attributes: "Price", "Weight" and "ComfortDegree" - as a measure of comfort and quality and moreover component specific geometrical and other technical data.

By the construction of component-subsets, the values of attributes "Price", "Weight" and "ComfortDegree" will be added together.

The goal of this paper is to show, how it is possible to solve the following tasks: treatment of constraints, determination of all allowed combinations and components selection for a car with the optimal ComfortDegree/Price ratio.

2. METHODS AND MEANS

2.1 Design of tables for description of components, selection of necessary attributes

General recommendation:
The number of allowed combinations depends on the constraints and may be very large. The treatment of queries and the solving of configuration tasks can be effective only if all data is processed in RAM of computers. The tables describing the components should be stored in a separate configuration database and should include only data fields, which are necessary for the use of constraints. The names of components should be unique and as short as possible, because short component names reduce memory requirements. Therefore in the component tables very short component names have been used, with 3 or 4 characters. The mapping of short names into full names (e.g. G15 → automatic gear Mod. 15) will be stored in separated tables, if necessary outside of the configuration database.

2.2 Description of configuration-constraints as mathematical or logical formulas

There are various kinds of constraints that should be taken into account:
- Constraints related only to one component, e.g. the customer demand – the motor power should be greater as 60 KW;
- Constrains related to two or more components, e.g. the wheel should be suitable to the car body;
- Constraints related to all components, e.g. customer demand - the weight of the car should be lower as 1200 kg.

Other classification of constraints;
- Physical and technical constraints, e.g. the necessary motor power is depending on the weight of the car – some formula for the description of dependence is necessary;

- Law and security constraints, e.g. the brakes should meet some given demands;

- Marketing and customer-specific constraints – statistical data, which components and combination of components will be with some probability from the customers preferred or rejected.

In the presented example we shall assume, that the constraints may be formulated as mathematical or logical expressions. But the configuration constraints can be given also directly as tuples of allowed or restricted component combinations.

Short description of used constraints.

Two components constraints:

C1   The height of the car body is at least 0.3 m greater than the height of the seat1.

C2   The height of the car body is at least 0.3 m greater than the height of the seat2.

C3   The height of the car body is at least 0.3 m greater than the height of the settee.

C4   The breadth of the car body is at least 0.3 m greater than the breadth of the settee.

C5   The restriction on wheel diameter and wheel broadness: the wheel should   be   suitable   to   the   car   body   wheel   cavity (This constraint uses two formulas)

C6   If the motor power > 70, then wheels comfort degree should be > 1 (a better quality of wheel is recommended)

C7   The motor power should be greater as the flow resistance at the velocity 140 km/h.

C8   The gear should be able to transmit the motor power.

Tree component constraints

C9   The breadth of both seats together should be lower as the breadth of the car body

Constraints coupling attributes of all components

C10 The necessary motor power is depending on the weight of the complete car

C11 The brake force should allow braking the car from velocity 100 km/h to zero on the braking distance 70 m.

The splitting of composed constraints such like e.g. C5, using the "and" operator may be done.

Table 1: Overview about the used constraints

Compo-nent/Con-straints	C1	C2	C3	C4	C5	C6	C7	C8	C9	C10	C11
Brake										X	X
Car_Body	X	X	X	X	X		X		X	X	X
Gear								X		X	X
Motor						X	X	X		X	X
Seat1	X								X	X	X
Seat2		X							X	X	X
Settee			X	X						X	X
Wheel					X	X				X	X
N1	208	208	192	192	160	120	192	180	2704		
N2	143	143	86	76	18	85	144	33	770		

Where

N1: number of combinations without constraints condition

N2: number of combinations with constraints condition

The constraints C3 and C4 are related to the same components, but to different attributes of components.

2.3 Crossover of mathematical or logical constraints into the sets of component-tuples

The constraints will be converted into sets of allowed combinations of components with help of queries. Every constraint can be used in one or more queries. The queries may be constructed based on one or more constraints. The results of the queries should not be very large. This means that the selected data should fit in computers RAM. In the final state of processing we should receive a query including all allowed combinations of all components. If necessary, only the data from this final query may exceed the RAM capacity. It is favourably to determine all allowed combinations of components successively, step by step, through the building of new queries, which are based on already available constraints queries.

The record sets selected with help of queries (for some subsets of components) can be stored as separate tables. These tables are depending on the component tables. After any change of components tables, these deduced tables should be generated again.

Corresponding to constraints C1-C9 the simple queries C1-C9 will be constructed. The constraints may be described with help of SQL.

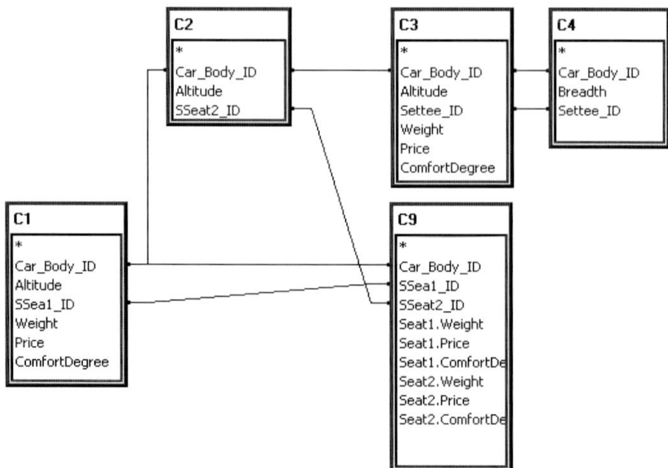

Figure: 1   The Query SUBSYS1, cut out from the QBE-windows

The constraints as selection criteria in the queries will be used. The same component can be used in more than one query. For the selection processes important attributes of some component should be included at least in one query, but if possible, because of memory requirement only in one query.

Based on simple queries C1-C9, 2 following more complex queries, Query SUBSYS1 and SUBSYS2, can be constructed.

This query gives all allowed combinations of subsets of components car body, seat 1, seat 2, and settee. The weight, price and comfort degree of these subsets will be calculated in the new data fields Subs1_weight, Subs1_price and Subs1_ComfortDeg.

In the presented example, the query SUBSYS1 (s. Figure 1) gives 3313 allowed component combinations, from 32,448 all possible combinations. By sorting this query by calculated field Subs1_weight, we can obtain the lower limit for the weight of subsystems 1 (457 kg) and the upper limit for the weight of subsystems 1 (622 kg).

Query SUBSYS2

In the next step the query SUBSYS2 (s. Figure 2) will be constructed. This query gives the sets of allowed combinations of components car body, motor, wheel, gear and brake. The number of allowed combinations is 3765; the number of all possible combinations is 432,000.

The weight, price and comfort degree of these subsets will be calculated in the data fields Subs2_weight, Subs2_price and Subs2_ComfortDegree. The weight, price and comfort degree of a car body are in the calculation from Subs2 fields not taken into account, because these

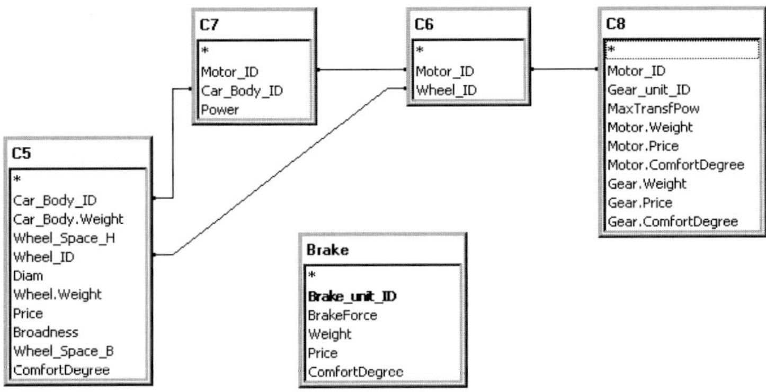

Figure 2: The Query SUBSYS2, cut out from the QBE-windows

data fields have been used already by the calculation weight, price and comfort degree of the subsets Subs1.

We can try to use the queries SUBSYS1 and SUBSYS2 to obtain all allowed component combinations based on constraints C1-C9. Only after that, the constraints C10 and C11 can be applied.

The new query SUBSYS1_and_SUBSYS2 (s. Figure 3) and SQL expression) delivers 340,410 various combinations from all components. We remember that the number of all combination, without using of constraints is 876,096,000.

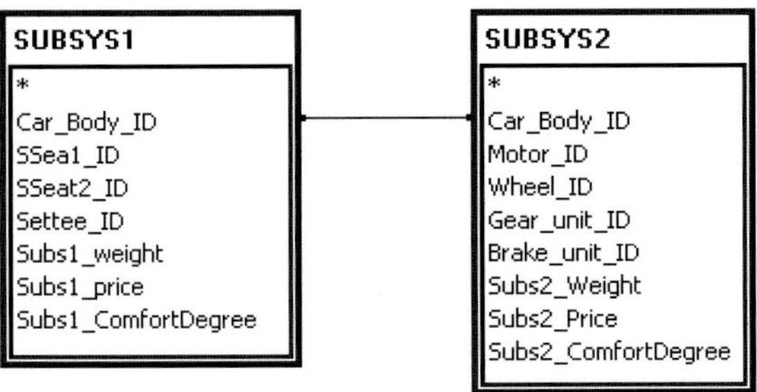

Figure3: The Query SUBSYS1_and_SUBSYS2, cut out from the QBE-windows

The amount of data of query SUBSYS1_and_SUBSYS2 is very large, (only for the names of all components, in all allowed combinations, approximately 6.2 MB are necessary), the further use of this query is possible, but not comfortable.

The constraints C10 and C11 can be used directly only in queries involving all components.

But we can use these constraints in a "weaker" form for subsystem of components, which are present in the query SUBSYS2. The query SUBSYS1 gives the lower limit (457,0) and the upper limit (622) for the subset of components subs1, including car body, seat1, seat2 and settee. Let us denote the weight of some selection of components motor, brake, wheel and gear as G2. The weight of a car having this selection of components is greater or equal to 457+G2 and lower or equal to 622+G2. From the constraints C10 and C11 the two following "weaker" constraints can be derived:

C12 The necessary motor power depends on the **estimated** weight (two values) of the complete car

C13 The brake force should allow to slow the car from velocity 100 km/h to zero on the stopping distance of 70. The force should be calculated with **estimated** weight of car.

The use of new constrains C12 and C13 is an example of how it is possible to derive some constrains concerning only a subset of components from constraints concerning all components.

The constraints C12 and C13 can be used to improve the query SUBSYS2. The modified query SUBSYS2_Mod gives a result with 2767 records, less than the Query SUBSYS2, which has 3765 records. Based on queries SUBSYS1 and SUBSYS2_Mod a new query with the name SUBSYS1_and_SUBSYS2_Mod can be constructed. This query gives 261,024 combinations of all components, instead of 340,410 records without using of constraints C12 and C13.

The results of queries SUBSYS1 and SUBSYS2_Mod can be stored as tables SUBSYS1_Tab and SUBSYS2_Mod_Tab. Instead of query SUBSYS1_and_SUBSYS2_Mod, the SUBSYS1_and_SUBSYS2_Mod_Tab query can be constructed. This query uses instead of queries SUBSYS1 and SUBSYS2_Mod the corresponding tables SUBSYS1_Tab and SUBSYS2_Mod_Tab.

This query gives also 261,024 combinations of all components, but in a very short time, approximately in 8 sec (Athlon PC with clock rate 2.2 GHz) and 17 sec (Intel PC 650Hz and RAMS 256 MB). The use of tables instead of the corresponding queries can accelerate the data selection respectable.

With help of query SUBSYS1_and_SUBSYS2_Mod the table All1, having 261,024 records can be created. On this table the constraints C10 and C11 can be applied. This application will be done with the query Create_All2 (s. Figure 4)

The table All2 includes 31,813 allowed combinations of all components. Such a way, with help of successive using of constraints, the number of combination **has been reduced from 876,096,000 to 31,813.** The table All2 has been generated using only the "technical" constraints. The next more special selection can be done based on customer specific demands, which may be also treated as constrains.

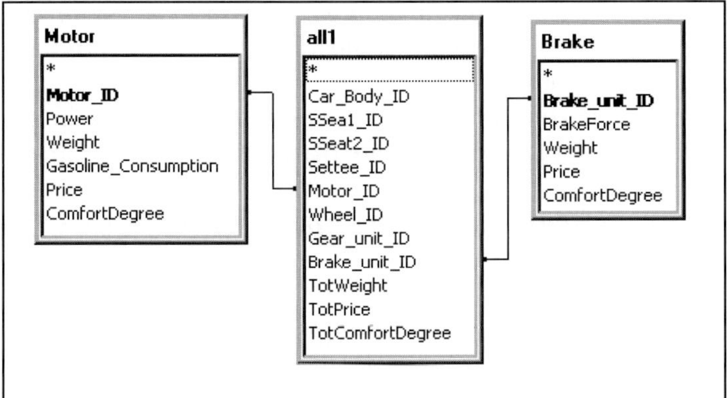

Figure4: The Query Create_All2, cut out from the QBE-windows

2.4 The problem of combinational variety and calculating time versus memory requirements

Considering the car configuring as an example, it has been shown, that the combinational variety of about 876,000,000 combinations can be avoided through successive using of constraints-based queries. In this approach, not allowed combination of components has been neglected. All allowed combinations might be found only in a few minutes. The constraints have allowed reducing the number of combinations considerably. In practise those situations should also be considered, in which the number of combinations is very large even if the constraints have been used. The selected data could exceed the capacity of RAM or even of hard disks. In such a case, the queries should be processed step by step chronologically, in each time step with a selected subset of data.

Example: we should like to find the count of combinations which accomplish only the constraints C10, C11 and to find a combination with the optimal comfort degree / price ratio.

With help of programming, this task can be solved with minimal storage requirement. The component tables will be stored in arrays of VBA programs. In the program all combination of components will be realised through eight nested loops. Every loop corresponds to one component (e.g. brake), the indexes in the loop are corresponding to the names of this component (e.g. B01, B02...). To check all combination and to find the combination with optimal comfort degree / price ratio, the time of approximately 3700 sec (Intel PC with clock rate 650Hz) or 900 sec (Athlon PC with clock rate 2.2 GHz) is necessary. The count of 2,747,968 allowed combinations and the combination with optimal degree / price ratio has been found. In this special case it is not necessary to store all allowed combinations. The criterion "optimal degree / price ratio" can be considered as a special kind of constraint.

Generally the nesting of program loops should be done in such a way, that as many constraints as possible are placed in the external nested loops.

2.5 Solution proposal for the "most difficult case" of configuration

Let us consider the most difficult but also the most interesting case of configuration, in which we have only constraints including attributes from all components. If the number of all combinations is too large, it is not possible to build and afterward to prove all combinations. And it is also not possible to use our above-described approach.

Let us denote (this is an other denotation, than in the former example about car configuration) as C_i the names of components (i=1, 2,..N), where N is the count of components and their corresponding tables. NR_i denotes the number of records in the component table C_i. Every table has a key field denoted $CName_i$. For some special constraints the combinational variety can be reduced. Assumption: The constraints can be defined as equations in the following form:

$$F_i(C_1.\{A_1\}, C_2.\{A_2\},\ldots\ldots C_N.\{A_N\})=0 \tag{1}$$

For i=1, 2,..NC, where NC is the number of constraints of the kind (1), and $C_i.\{A_i\}$ is a set of attributes of the component i involved in the constraints.

Let us assume that the same attributes of the same components are involved in more then one constraint. We can try to calculate these attributes with help of attributes of other components. The number of calculated attributes can be 1, 2 or maximal NC.

For the sake of simplicity we shall denote these components as C_1, $C_2,\ldots C_N$ and these attributes as $C_1.A_1, C_2.A_2,\ldots C_{NC}.A_{NC}$.

$$X(i,k) = Q_{ik} (C_{NC+1}.\{A_{NC+1}\}, C_{NC+2}.\{A_{NC+2}\},\ldots C_N.\{A_N\}) \tag{2}$$

Where i=1, 2,..NC. $X(i,k)$ corresponds to calculated attribute value $C_i.A_i$. In the case of non linear equations (1), there are more then 1 solutions possible. The index k denotes the different solutions X for the same parameters $C_{NC+1}.\{A_{NC+1}\}$, $C_{NC+2}.\{A_{NC+2}\},\ldots C_N.\{A_N\}$. The number of physically reasonable calculated solutions will be denoted r.

Instead of explicit functions Q_{ik}, it is possible to use programs calculating the values $X(i,k)$, corresponding to attributes $C_1.A_1, C_2.A_2,\ldots C_{NC}.A_{NC}$.

Practical use: **A new table SubSyst will be generated. This table has the following structure:**

Table 2: SubSyst

X_1	...	X_NC	$CName\_NC+1$	$CName\_NC+2$	...	$CName\_N$
$X(1,1)$		$X(NC,1)$	ID_{NC+1}	ID_{NC+2}		ID_N
$X(1,2)$		$X(NC,2)$	ID_{NC+1}	ID_{NC+2}		ID_N
..		...	ID_{NC+1}	ID_{NC+2}		ID_N
$X(1,r)$		$X(NC,r)$	ID_{NC+1}	ID_{NC+2}		ID_N
...	...	...	...	...	...	...

The parameters ID_i are key-values identifying the records from component tables. The table SubSyst includes all allowed combinations of components from tables C_{NC+1}, C_{NC+2},...C_N. For every such combination the values $X(i,k)$ will be calculated and stored with all physically reasonable solutions (r-times) in the table Subsyst. If for some combination of components C_{NC+1}, C_{NC+2},...C_N with corresponding attributes the solution X does not exist, such combination should be eliminated

The table Subsyst can be generated with help of programs. If the equations (1) have only 1 solution X, (e.g. the constraints are linear equations), then r=1 and the table Subsyst can be generated with a simple query having calculated fields X_1, X_2,...X_NC

$$X\_1: Q_1(C_{NC+1}.\{A_{NC+1}\}, C_{NC+2}.\{A_{NC+2}\},...C_N.\{A_N\})$$
$$X\_2: Q_2(C_{NC+1}.\{A_{NC+1}\}, C_{NC+2}.\{A_{NC+2}\},...C_N.\{A_N\})$$

...

$$X\_NC: Q_{NC}(C_{NC+1}.\{A_{NC+1}\}, C_{NC+2}.\{A_{NC+2}\},...C_N.\{A_N\})$$

This query can be further used directly or can be used to generate the table Subsyst. The table Subsyst include data fields with calculated values of attributes $C_1.A_1$, $C_2.A_2$,..$C_{NC}.A_{NC}$. To find all allowed combinations of all components it is necessary to compare the calculated values with the attribute values in the tables C_1, C_2,...C_{NC}. This comparison can be done very easy in a new query "AllComb", using the table Subsyst and the tables C_1, C_2,...C_{NC}.

The following SQL-Expression of query "AllComb" describes the example with three calculated fields X1, X2 and X3. The total number of used components is 10.

```
SELECT DISTINCT   C1.Cname_1,   C2.Cname_2,   C3.Cname_3,
Subsyst.Cname_4,          Subsyst.Cname_5,          Subsyst.Cname_6,
Subsyst.Cname_7,          Subsyst.Cname_8,          Subsyst.Cname_9,
Subsyst.Cname_10
FROM Subsyst, C1, C2, C3
```

WHERE　　　$(((\text{Abs}([\text{Subsyst}]![X1]-[C1]![A1]))<[\text{eps}])$　　　AND $((\text{Abs}([\text{Subsyst}]![X2]-[C2]![A2]))<[\text{eps}])$

　　AND $((\text{Abs}([\text{Subsyst}]![X3]-[C3]![A3]))<[\text{eps}]));$

The data fields $X_1,...X_{NC}$ will be calculated numerically. The exact equality to the attribute values $C_1.A_1,...C_{NC}.A_{NC}$ cannot be expected. Instead of equality, the difference lower than some parameter eps will be proven. The parameter eps depends on precision of calculation and should be as low as possible.

There is also other way to use the calculated attributes X: Instead of complete tables $C_1, C_2, ... C_N$, new reduced tables $T_1, T_2, ... T_N$ can be generated. These tables include only such record sets from $C_1, C_2, ... C_N$, whose attribute values correspond to the calculated attribute values X. Generally, the count of record sets NT_i in some reduced table T_i is lower or in the worst case equal to the corresponding count NR_i.

3.　ADVANTAGES OF THE PROPOSED METHOD

- The count of all component combinations, which should be proven to fulfil the constraints equations (1), is the product N_{Tot}.

  $N_{Tot} = NR_1 * NR_2 * ... NR_{NC} * NR_{NC+1} * ... NR_N$

  Where NR_i is the number of records in the table i.
- The number N_{Tot} of necessary combinations will be reduced approximately with the factor R.

  $R = (NT_1 * NT_2 * ... NT_{NC})/(NR_1 * NR_2 * ... NR_{NC})$
- The proposed method guarantees, that **all through the constraints (1) allowed combinations of all components will be found**. The calculating time can be reduced considerable.

4.　THE LIMITS FOR THE USE OF LOW-COST RELATIONAL DATABASES

The size of low cost databases is limited. E.g. the size any MS Access database should not exceed 1 GB. It means that it is not possible to store results of queries greater than 1 GB. But with help of VBA-Programming it is possible to store the result of data selection outside of the database, for example as ASCII operating system files. The use of short component names reduces memory requirements. The low cost systems such as MS Access can be used successfully, if the constraints allow reducing the data volume below these limits.

5.   RESULTS, CONCLUSIONS AND FUTURE RESEARCH

Through the successive use and combination of constraints-based selection queries and if necessary and through the time-successive application of constraints on data subset in programs, the combinational variety with about 10,000,000,000 combinations, can be processed with current PCs in the period from few minutes until few hours. The number and kind of constraints play a important role, how extensive combinational variety can be treated. The analysis of results from constraints queries shows what available components in allowed combinations have been used. This information is a basis for decisions on which components should not be produced or which new compatible components should be developed.

The aim of this paper has been especially to show how it is possible to avoid or minimise the problem of large combinational variety and difficulties conditioned through the limited maximal size of databases, the limited RAM and computation rate. Memory requirements and computing time may be reduced through the successive use of constraints based queries or to these equivalent restriction tables and through the checking of various combinations of components in successive time-steps. This approach is easy for a low number of constraints and kind of components. The aim of further research is to find methods able to generate an optimal set and sequence of queries leading to the table of all allowed component combinations automatically. For the most difficult configuration case in which there are only constraints involving attributes from all components we have proposed a solution for a special case, in which the used constraints allow to calculate one or more attribute value with help of other one (1). This method can be further developed for constraints using instead of equations inequalities. **The advantage of the proposed methods is that actually all allowed combinations of components will be investigated and taken into account.**

REFERENCES

1.   Günter A., Kühn C.: "Knowledge –based configuration – survey and future directions." In: Puppe F., editor*Proceedings of XPS-99*, Wuerzburg, Berlin: Springer 1999. p. 47-66.
2.   Kreuz I.: Die Heuristik, Relevant Knowledge First für wissensbasiertes Konfigurieren in technischen Domänen.Doktorarbeit, Universität Stuttgart 2003
3.   Günter A. (Hrsg.):Wissensbasiertes Konfigurieren.Ergebnisse aus dem Projekt PROKON,Infix-Verlag, Sankt Augustin 1995
4.   Kreuz I., Forchert T., Roller D.: "IRCON Intelligent Configuring SystemIn: Roller D. (Hrsg.)": *Proceedings of the 31th ISATA,* **Volume "Automotive Electronics and New Products"**,Düssedorf Trade Fair, Croydon, England 1998, p. 219-226
5.   Roller, D.; Kreuz I.: "Selecting and parametrising component using knowledge based configuration and heuristic that learns and forgets". *Computer-Aided Design* **35** (2003) 1085-1098

6.   Chan, S,C.; Shek, S.,C; Lee, J.,W.: "Modelling and applying constraint relations in a product family data model." *International Journal of Computer Application in Technology*, **Vol. 12**, No. 1, 1999

THE APPLICATION OF SSE TECHNIQUE IN FAST RECONSTRUCTION

Haipeng Mao, Dinghua Zhang,, Liang Liang, Juan Gu and Qingsheng Wang
Key Lab of Contemporary Design and Integrated Manufacturing Technology, Ministry of Education; PO Box 552, Northwestern Polytechnical University, Xi'an, China 710072; rocmao@263.net

Abstract: Cone beam volume computed tomography (CBVCT) provides a new technique for the reconstruction and realization in reverse engineering, which can be widely used in many fields such as non-destructive test, virtual measurement and so on. After gathering the projection data, we need to reconstruct the slice images. Feldkamp described an approximate reconstruction algorithm for circular cone-beam tomography, which is called the FDK method. For its simplicity, it became the most used algorithm for CBVCT. However , though the frequency of CPU increases very quickly, high resolution reconstruction using the FDK method is still a time consuming process, which will be the bottleneck of real-time imaging. To achieve a high reconstruction speed, some dedicated hardware were implemented. But they are very expensive and lack of scalability. In this paper, we used the SSE technique to tune the core code of reconstruction based on the IA-32 architecture. Thus we could make full and effective use of all the execution units (EU) in the processor, and realize the instruction-level parallelism of the Feldkamp method. This is an easy and cheap upgrade without additional hardware. Moreover, it can be compatible with newer processors. The final simulation result shows that it can save more than 80% reconstruction time compared to the traditional implementation.

Key words: CBVCT, fast reconstruction, SSE, FDK method, reverse engineering

1.   INTRODUCTION

Cone beam volume computed tomography (CBVCT) is a 3D extension of the 2D fan beam tomography, decreasing the scanning time sharply by gathering an entire volume of data at one time, rather than slice-by-slice. It provides a new technique for high-resolution imaging applications, especially metrology and reverse engineering.

Before putting the slice images into use, we need reconstruct them from the gathered 2D projections. There is a great variety of 3D tomographic reconstruction algorithms, which can be separated into analytical and algebraic reconstruction methods [1]. Analytical methods rely on the Fourier Slice Theorem and the Radon Transform, which state that one can reconstruct an object by filtering its X-ray projections by a ramp filter in the frequency domain and then backprojecting the filtered projections into a 3D grid. Algebraic algorithms demand more computing resources than analytic ones. Because a high reconstruction speed is very important in practice, analytic algorithms are always preferred. Feldkamp, Davis, and Kress (1984) described an approximate reconstruction algorithm for circular cone-beam tomography [2], which was called the FDK method. For its simplicity, it has become the most used algorithm for CBVCT [1].

The reconstruction is an expensive process due the huge magnitude of the data. Reimann DA et al. [3] pointed out that reconstructing a data set of projection size N was $O(N^4)$, which will be the bottleneck of real-time imaging.

To achieve a high reconstruction speed, some dedicated hardware were implemented [4 - 5]. A downside for the custom chips is their inflexibility to accommodate the latest algorithmic advances, and besides, they are also quite expensive and therefore inaccessible to researchers and small emerging companies.

In this paper, we propose a feasible, cost-effective and fast cone beam reconstruction method. Our implementation is based on Intel Pentium 4 CPU for its economical price and high performance, which frequency has increased to the GHz level. Moreover, Intel P4 CPU supports Single Instruction Multiple Data (SIMD) technique by MMX/SSE/SSE2. In the following sections we first optimize the traditional FDK method, and then use Intel's SSE technique to parallelism the reconstruction implementation. At last, we prove the correctness and efficiency of the fast reconstruction method.

2. METHOD

2.1 The FDK Algorithm

The analytic FDK algorithm can be summarized as follows [2]:
1. Weight projection data

$$P'_\theta(i,j) = \frac{d_{so}}{\sqrt{d_{so}^2 + u^2 + v^2}} P_\theta(i,j) \tag{1}$$

2. Convolve weighted projection data

$$P^*_\theta(i,j) = P'_\theta(i,j) * h(i) \tag{2}$$

3. Backprojection

$$f(x,y,z) = \int_0^{2\pi} u^2 P^*_\theta(p,q)d\theta$$

$$t = x\cos\theta + y\sin\theta$$

$$s = y\cos\theta - x\sin\theta$$

$$u = \frac{d_{so}t}{d_{so} - s} \tag{3}$$

$$p = ut$$

$$q = uz$$

It is assumed that the projection of the object at angle θ is indexed by detector coordinates i and j. The center of rotation is z axis. The distance from the x-ray focal spot to the rotation axis is d_{so}.

2.2 Overview of Intel Pentium 4 CPU Architecture

The Intel Pentium 4 processor, utilizing the Intel NetBurst micro-architecture, is a complete processor redesign that delivers new technologies and capabilities while advancing many of the innovative features. The Intel processor has multiple execution units that perform operations in parallel. In Pentium 4 CPU, there are two ALU units, one FPU/MMX unit and one SSE/SSE2 unit. The SSE/SSE2 execution unit provides eight independent

128-bit XMM registers and supports six data types: packed double-precision floating-point, packed single-precision floating-point, packed doubleword integers and so on. With SIMD instructions, the P4 processor can process four 32-bit single-precision or two 64-bit double-precision floating-point data in parallel [**6 - 7**]. As a result, using SSE/SSE2 extension can contribute significantly to an overall performance increase.

2.3   Improved Method for Traditional Algorithm

The traditional FDK algorithm implementation is described in Figure 1. When reconstruction process starts, the projection data is loaded one by one. Each projection data is filtered (Line2) and then used to backprojection (Line3~13). After all projection data is finished being backprojected, the reconstruction process ends and the volume data is saved. Suppose a set of M projections is used to reconstruct the N^3 cube, the total reconstruction time can be give as:

$$T_{recon} = M * (T_f + T_b)$$

(4)

Where T_f is the filtering time and T_b is the backprojection time for every projection data. Obviously, in order to reduce the total reconstruction time, we must try to decrease T_f and T_b.

```
1:  for all projections do
2:      filter projection
3:      for all slices do
4:          for all rows do
5:              for each voxel on the row
6:                  calculate the magnified coefficient u
7:                  calculate the weight coefficient w
8:                  calculate the backprojected pixel address p and q
9:                  interpolate the pixel (p, q) value on the filtered projection plane
10:                 accumulate the weighted sum of voxel value
11:             end for
12:         end for
13:     end for
14: end for
```

Figure 1. Traditional FDK algorithm implementation

There are several methods to improve the traditional FDK algorithm. To achieve high reconstruction speed, we apply the following optimization:

✓   Split cone geometry relations from backprojection. It should be noted here that the (u, w, p) occurred at Line 6~9 in Figure 1 only depends on (x, y) while backprojecting. So we can build a parameter lookup table for each projection, that is, the backprojection addresses and weights are calculated in advance. Therefore, the lookup only occurs in the backprojection and avoids repeated calculations in the FDK algorithm.

✓ Implement an incremental algorithm [8]. When calculating the coordinate (t, s) and the backprojection address q, an incremental algorithm is used to reduce the computational time complexity of inner loop, thus we can decrease the backprojection time T_b.

✓ Pre-interpolate the filtered projection before backprojection. After have implemented the above two methods, unfortunately, it is not satisfied with the speedup. This is mainly due to the computational time complexity of the interpolation at Line 9 in Figure 1. Usually bilinear interpolation is used, which is quite sufficient, but consumes the maximum computational time (above 75%). In our implementation, we carry out a pre-interpolation and nearest neighbour interpolation, which is faster than the traditional bilinear interpolation method because of its computational simplicity (which allows a fast MMX/SSE/SIMD implementation).

Several other techniques, such as ROI (Region of Interest) and "links" method [9], can be used to minimize the actual number of operations in the backprojection loop.

Now we take consideration on parallel implementation of improved FDK algorithm. There are mainly two levels of parallelism to general users: program-level parallelism (PLP) and instruction-level parallelism (ILP) [10]. We focus only on the ILP because we hope to get the highest computational efficiency on a single CPU. If we accomplished this, we could break the reconstruction task into constituent part, i.e. according to the projection parallelism, and make different processors compute each part by using multithread technique, etc. Thus we can realize PLP efficiently. To implement fast parallel cone beam reconstruction, we continue the following research and development:

✓ Using SIMD assembler instruction to make SSE/SSE2 unit work to do parallel computation. This method can be used in both filtering and backprojection process, which enhances greatly the algorithm performance.

✓ Memory management optimization. When reading or writing data to memory, use instructions to load/store 16 bytes at a time. Reading or writing 16 bytes to 16-byte aligned memory is the most efficient way to use the memory pipeline.

3.   IMPLEMENTATION

As mentioned before, SSE/SSE2 programming was adopted in our approach. This style is well suited to our problem for the following reasons [11]: 1) In the filtering process, the operations being performed on each column of projection data are the same if we consider the original projections column-wise; 2) In the backprojection stage, the operations being performed on each reconstructed voxel are the same and usually we implement backprojecting calculations row-wise so as to obtain high performance by having quick memory access.

We used Microsoft Visual C++ 6.0 and Intel C++7.1 Compiler as the developing tool, which supports Intel's MMX Technology, SSE and SSE2 extensions through intrinsics or inline assembly. The test system configuration is shown in Table 1.

Table 1. Test System Configuration

Processor	Pentium 4 Processor at 2.0GHz
Memory Size	Kingston 1G DDR
Hard Disk	Seagate 80G
Operating system	Windows XP Professional
SIMD Support	MMX, SSE, SSE2

3.1   Parallel Computation of Filtered Projections

The projection data is represented by a two-dimensional matrix and stored in a one-dimensional single-precision floating-point data array, in which the index increases for x, then y. Since we know the projection size NxN, the memory address of each matrix element can be computed easily. The original filtering algorithm carries out convolving/FFT computations column-wise on each of the matrices and the convolving/FFT operations on one column are the same as, but independent from other columns. For the above reasons and since the projection column size was an integer power of 2, it was decided to parallelism by performing a set of convolving/FFT operations on every four adjacent columns. Before each calculation loop starts, we load four projection points, which are assembled in a packed single-precision floating-point data, from the same row, then push it into the convolving buffer. To operate packed single-precision floating-point data type, we use assembler language to redesign original algorithm code. Thus, we make all execution unit work in parallel and have done the parallel computation of the filtered projections on instruction level.

Since the FFT algorithm works best for vectors which are integer powers of 2, zero padding has been done to keep the vector length of each column as an integer power of 2. Zero padding has the additional benefit of reducing the dishing artifacts inherent in the DFT algorithm. In our implementation, we use 4N zero padding length and achieve satisfied result.

3.2   Parallel Computation of Backprojections

The reconstructed volume can be represented by a three-dimensional matrix and stored in a one-dimensional single-precision floating-point data array, in which the index increases for x, then y and last for z. According to our algorithm optimization proposed earlier, the fast algorithm is as follows:

```
1:  for all projections do
2:      filter projection
3:      pre-interpolate the filtered projection
4:      build a parameter lookup table
5:      for all slices do
6:          for all rows do
7:              for each voxel on the row
8:                  load the magnified coefficient u
9:                  load the weight coefficient w
10:                 load the backprojected pixel address p
11:                 calculate incrementally the backprojected pixel address q
12:                 find the pixel (p, q) value from the interpolated filtered projection
13:                 accumulate the weighted sum of voxel value
14:             end for
15:         end for
16:     end for
17:end for
```

Figure 2. Fast FDK algorithm implementation

Just like parallel implementation in filtering projection, we can perform SIMD operations on pre-interpolating filtered projection, building lookup table, loading parameters, calculating offset address of the sub-pixel and accumulating the weighted sum of voxel data. Every four consecutive voxel data on the same row constitutes a packed single-precision floating-point data, which is processed by the parallel backprojection in one loop.

Backprojection process uses parallel streams of data in the form of magnified coefficient u, weight coefficient w, etc. Arranging these parallel data streams to form a hybrid Structure of Array (SOA) can help SIMD performance by processing 4 parallel data elements at a time. In our implementation, we define this hybrid SOA as follows:

```
struct {float u[4], float w[4], float p[4];} *pHybridLookupTable;
```

Furthermore, the hybrid SOA approach improves the locality of the operands, allowing more efficient use of the memory pipeline.

4.   SIMULATION RESULTS

4.1   Evaluation Methods

To test the reconstruction precision and speed, we must gather projection data along the circular trajectory firstly. Using a CBVCT simulator that we developed, the projection data of simulated phantom, which is modeled by UG/CAD software, can be generated without any noise or distortion.

First we want to compare original data to reconstructed data and determine the reconstruction accuracy of our fast implementation. For this purpose, a CAD model of hollow turbine blade phantom, which is shown in Figure 3, is used. From all projection data, we reconstruct the slice images. We also acquire the actual simulated phantom image that sampled from CAD model directly. Reconstructed and acquired image data are compared by calculating the relative error E. The maximum error of approximating one image X by the corresponding image Y is e_{max} = max(max(X),max(Y))-min(min(X),min(Y)). Now we can calculate the error E between X and Y relative to e_{max}:

Figure 3. Hollow turbine blade phantom

$$E = \frac{1}{N}\frac{1}{e_{max}}\sum_{i=1}^{N}|x_i - y_i|\times 100\% \qquad (5)$$

Where x_i and y_i, i = 1, 2… N, are the pixels of X and Y with not both x_i = 0 and y_i = 0, to exclude the background from the calculations. For we performs a nearest neighbour interpolation instead of a bilinear interpolation in the backprojection stage, a tiny error will be introduced. What we most concerned with is how much precision loss will be caused in our implementation. Denoting E_{fp} as the relative precision error between our fast reconstructed image and the actual phantom image, E_{fo} as the relative error between fast reconstructed image and original reconstructed image, we can calculate E_{fp} and E_{fo} according to Equation 5, and then evaluate the reconstruction accuracy of our fast implementation.

Second we want to evaluate the speedup of our fast reconstruction compared to the traditional method. For this purpose, we have tested the two reconstruction implementations, recorded their running time and then calculated the speedup.

4.2   Simulation Results

We use 300 512^2 projections to reconstruct the 512^3 volume data with a cylinder boundary. Figure 4 presents two images, where part (a) is the simulated phantom image and part (b) is the fast reconstructed image. Table 2 shows the results of relative precision error. It can be seen that E_{fp} is less than 1.2%, while E_{fo} less than 0.07%. This means that our fast reconstruction maintains good precision and is reliable.

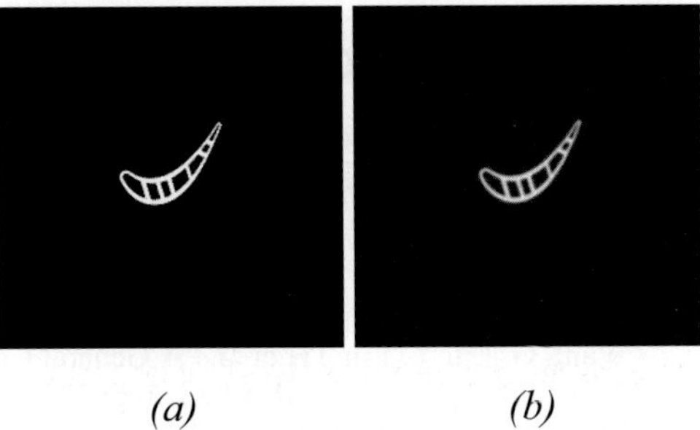

(a)   *(b)*

Figure   4.   Simulated   phantom image and fast reconstructed image

Table 2. Relative precision error at the different Z position

Slice	-200	-100	-50	-25	0	25	50	100	200
E_{fp}(%)	0.98	0.81	0.37	0.30	0.18	0.35	0.58	0.87	1.16
E_{fo}(%)	0.014	0.039	0.041	0.045	0.012	0.065	0.049	0.062	0.022

The compared result of running time for the two reconstruction implementation, including filtering time, is shown in Table 3. It should be noted that we only use SSE technique to make the filtering speedup approach 3.3. SSE is also applied in the acceleration of backprojection and as a whole, the reconstruction speedup reaches 5.797. These results declare that SSE technique contributes much to the speedup of our fast reconstruction implementation

Table 3. Timing result for filtering and reconstruction

Projection Size	300×512^2	Reconstructed Volume Size	512^3
Original Filtering Time(s)	132.934	Original Reconstruction Time(s)	5403
Parallel Filtering Time(s)	40.375	Fast Reconstructed Time(s)	932
Filtering Speedup	3.292	Reconstruction Speedup	5.797

5.   CONCLUSION

We have shown in this paper how SSE technique is used to accelerate cone beam reconstruction. With improved algorithm, fine-tuned assembler code, effective memory management and SSE instructions, we make full and effective use of all the execution units in the Pentium 4 processor, and realize the instruction-level parallelism of the Feldkamp method. A high-speed cone beam reconstruction engine based on the IA-32 architecture is developed and put into practical use. With our reconstruction engine, we can save 70% of filtering time and more than 80% of reconstruction time. Compared to the custom-built hardware, our implementation is very cheap and easy to upgrade with newer processors.

ACKNOWLEDGEMENTS

This project was supported by National Defence Foundation Research Project Grants K1801061008.

REFERENCES

1. Wang G, Liu Y, Lin TH et al. "A General Cone-beam Reconstruction Algorithm", *IEEE Trans. Med. Imaging*, **vol. 12 no. 3**, pp. 486-496, 1993.
2. Feldkamp LA, Davis LC, Kress JW. "Practical Cone-beam Algorithm". *J. Opt. Soc. Am. A*, **vol. 1 no. 6**, pp. 612-619, 1984.
3. Reimann DA, Chaudhary V, Flynn MJ et al. "Parallel Implementation of Cone Beam Tomography", *International Conference on Parallel Processing*, **vol. 2**, pp.170-173, 1996.
4. Luiz M, Felipe MF, Vladimir CA et al. "An FPGA-Based Fan Beam Image Reconstruction Module", *Proc. of the 7th Annual IEEE Symposium on Field-Programmable Custom Computing Machines*, Los Alamitos, CA, USA. Apr. 21-23, 1999, pp. 312-313.
5. Rajan K, Patnaik LM, Ramakrishna J et al. "High-speed Parallel Implementation of a Modified PBR Algorithm on DSP-Based EH Topology". *IEEE Trans. Nucl. Sci.*, **vol. 44 no. 4**, pp. 1658-1672, 1997.
6. *IA-32 Intel® Architecture Software Developer's Manual*, **vol. 1**, Intel Corporation, 2000.
7. *IA-32 Intel® Architecture Optimization Reference Manual*, Intel Corporation, 2003.
8. Cho ZH, Chen CM and Lee SY. "Incremental Algorithms: A New Fast Back Projection Schemes for Parallel Beam Geometries", *IEEE Trans. Med. Imaging*, **vol. 9 no. 2**, pp. 207-217, 1990.
9. Henric T. *Cone-beam Reconstruction Using Filtered Backprojection*, PhD Thesis, Sweden: Linkoping University, 2001.
10. Yu RF, Ning RL and Chen B "High Speed Cone Beam Reconstruction on PC", *Proc. of SPIE*, **vol. 4322**, San Diego, CA, USA, Feb. 19-22, 2001, pp. 964-973.
11. Raman R, Ronald DK et al. "Parallel Implementation of the Filtered Back Projection Algorithm for Tomographic Imaging", *[http://www.sv.vt.edu/xray_ct/parallel/Parallel_7. CT.html]*, Apr. 5, 1995.

CALCULATING ELASTIC MODULUS OF BOTH ENDS FIXED POLYSILICON MICROBEAMS WITH TRAPEZOIDAL CROSS SECTION

Guangping Han[1,2], Kai Liu[1] and Xiuhong Wang[2]

[1]*Xi'an University of Technology, 710048, 5th JING Hua South Road, Xi'an, China.*
E-mail address: kliu@mail.xaut.edu.cn
[2]*Zhengzhou Institute of Aeronautical Industry Management, 450015, Zhengzh, China.*

Abstract: Microelectromechanical systems (MEMS), a rapidly growing multidisciplinary interaction technology, has become a key technology in the 21st century in the first place together with nanometre science and technology (NST), dealing with the design and manufacture of miniaturized machines (MM) or micro mechanical assemblies (MMA) with major dimensions at the scale of tens, to perhaps hundreds, of microns. The difference in scale between micro and macro devices leads to great changes in mechanical properties such as elastic modulus, bending strength, fracture strength, etc. In this paper, a mechanical model of polysilicon microbeams fixed on both ends is investigated, and a formula is deduced calculating elastic modulus of microbeams with trapezoidal cross section by means of size, loads and displacements. Size effects in MEMS upon mechanical properties are studied and elastic modulus function of length, width and thickness is theoretically analyzed, helpful for establishing the theory base of research and development of micro devices.

Keywords: MEMS; Mechanical properties; Elastic modulus; Polysilicon; Size effects

1.   INTRODUCTION

The numbers, information, and computing don't require any particular size, and there is plenty of room at the bottom when the numbers are written and read very small, down to atomic size [1]. In such way, we could store a lot of information in small spaces and make infinitesimal machinery [2]. Based on the thinking described above, MEMS is formed and developed and has been a rapidly growing interdisciplinary technology in recent years, dealing with design and manufacture of miniaturized machines and micro mechanical assemblies with major dimensions at the scale of tens, to perhaps hundreds, of microns [3]. There are three major methods for fabrication of MEMS, the first one appears early in USA, based on silicon and IC (Integrated circuit), the second is LIGA (Lithographie, Galvanoformung, Abformung) technology, developed in Germany, and the last is precision machining, developed in Japan.

MEMS is fast and generally cheap to mass-produce, and the virtues of these tiny machines are many. Scientists have already moved MEMS into various stages of conception and development for making laboratories on chips, data-storage technologies, cell-manipulating gadgets, propulsion systems for micro-satellites and locking mechanisms for nuclear weapons, and many other applications [4]. There is a strong and growing consensus that MEMS will provide a new design technology having an impact on society to rival that of integrated circuits.

The fundamental problems of micro-fabrication and function development in MEMS have been solved so far [5]. But understanding of the mechanical behavior of MEMS materials and structures lags far behind understanding of the electrical behavior and the capability to fabricate new MEMS devices. Thus, Analysis of the mechanical properties is one of the most challenging issues in MEMS [4,5,6].

Polysilicon is one of the most common materials in MEMS because of its excellent mechanical and electronic properties [7]. Elastic modulus takes an important role in design and fabrication of micro-devices. Much more work has been done in recent years to accurately understand elastic modulus and other mechanical properties in MEMS such as microbeam bending test, nanoindentation method, resonant tests and micro-tensile tests [8-14]. It is in urgency to make sure of underlying relations between size and elastic modulus, the experimental value of modulus and the theoretic value as well.

2. SIZE EFFECTS ON MICROSCALE

With the development and applications of nanomaterials, micro devices, microstructures and micro systems, efforts have been made to basic and applied research of size effects in MEMS, promoting the progress of the theory on microscale [15]. The difference in scale between micro and macro devices leads to great changes such as mass, frequency, elastic modulus, bending strength, fracture strength, etc. On microscale, with size decrease down to some extent, the size effects upon physical and mechanical quantities proportional to high power of size, such as L^4, L^3, become weakly comparatively while those upon quantities proportional to lower power of size, such as L^0, L^{-1}, increase relatively. To make it clear, some quantities proportional to size are listed in Table 1. Furthermore, the ratio of surface area to volume gets larger and larger, resulting in great acceleration of heat conduction and chemical reaction and manifest increase in friction force. Thus, redefinition of physical and mechanical quantities should be made in accordance with the shift from macroscale to microscale.

Name	Variable	Expression	Size function	Variable description
Inertia force	f_i	md^2x/dt^2	L^4	m: Mass x: displacement t: time
Mass	m	ρV	L^3	ρ: Density V: volume
Renold's number	R_e	$Lv\rho/\mu$	L^2	v: Velocity μ: viscosity
Heat conduction	Q_c	$\lambda\Delta TA/d$	L	λ: Thermal conduction ΔT: temperature difference A: cross section area d: length
Static electro-force	F_e	$\varepsilon SE^2/2$	L^0	ε: Dielectric constant S: area E: electric field strength
Frequency	ω	$\sqrt{K/m}$	L^{-1}	K: Linear elastic constant

Table1. Size function effects in MEMS

3.   TRAPEZOIDAL CROSS SECTION OF MICROBEAMS

The MEMS technical community, composed of companies, universities, and government laboratories, such as NEXUS (Network of Excellence in Multifunctional Microsystems) in Europe, BSAC (Berkeley Sensor and Actuator Center) in USA, and Tsinghua University in China, has developed many types of test structures to characterize the fabrication process and device performance, among which microbeams, including both ends fixed microbeams and microcantilevers, are usually made and tested to obtain original data of material mechanical properties by means of loads and displacements using proper precision apparatus such as STM (scanning tunneling microscope), AFM (atomic force microscope), and SEM (scanning electronic microscope). Microbeams are fabricated on silicon wafer using LPCVD (lower pressure chemical vapor deposit) and deep silicon etching.

The etching rate of polysilicon is closely relative to its lattice direction and grain orientations. Generally speaking, the etching rate in the <110>

direction is much more than that in the <111> direction, for the theoretic value of elastic modulus is 190 Gpa in the <111> direction while 130 GPa in the <110> direction. Furthermore, feature size of the structure has been found to have important implication for deep etching process. The wider the feature size, the higher the etching rate of silicon is [16]. In order to obtain ideal results, microbeam should be fabricated with high ratios of length to width and length to thickness in bending test of polysilicon microbeams fixed on both ends. Thus the cross section of the microbeams is trapezoid owing to the anisotropic wet etching process and its structure.

4.　CALCULATING ELASTIC MODULUS OF BOTH ENDS FIXED MICROBEAMS

Understanding elastic modulus is very important for the design of electronic devices and MEMS since devices suffer thermal and mechanical stress during service. Measurements of deflection of buckled beams give information on the stress and strain, through which elastic modulus can be calculated.

4.1　Inertia moment of trapezoidal cross section

In material mechanics, the inertial moment is an important quantity, which can be expressed by size [17].

$$I_y = \int_A z^2 \, dA \qquad (1)$$

For trapezoidal cross section (see Figure 1, w_1: upper width, w_2: lower width, e_1: vertical distance between the upper surface and neutral plane, e_2 vertical distance between the lower surface and neutral plane, t: thickness), it is difficult to calculate the inertial moment directly using the formula. Here is a good choice to calculate the moment by parallel axis-moving formula, which can be written as $I_y = I_{y0} + b^2 A$ \qquad\qquad (2)

Where I_{y0} is the inertial moment to shape-center axis, b is the distance between shape-center axis and y axis, that is $b = e_1$ \qquad\qquad (3)

From the definition described above,

$$e_1 = \frac{\int_A z\,dA}{\int_A dA} = \frac{\int_0^t z[(w_2 - w_1)\cdot\frac{y}{t} + w_1]}{\frac{t}{2}(w_1 + w_2)}\,dz = \frac{t(w_1 + 2w_2)}{3(w_1 + w_2)} \quad (4)$$

$$I_y = \int_A z^2\,dA = \int_0^t z^2[(w_2 - w_1)\cdot\frac{y}{t} + w_1]\,dz = \frac{w_1 + 3w_2}{12}t^3 \quad (5)$$

Substituting equations (3), (4), (5) into equation (2), we get

$$I = I_{y0} = I_y - b^2 A$$
$$= \frac{w_1 + 3w_2}{12}t^3 - \frac{w_1^2 + 4w_1 w_2 + 4w_2^2}{18(w_1 + w_2)}t^3 = \frac{w_1^2 + 4w_1 w_2 + w_2^2}{36(w_1 + w_2)}t^3 \quad (6)$$

4.2   Elastic modulus calculated in 3-points bending test

The mechanical model of polysilicon microbeams fixed on both ends in 3-points bending test is shown in Figure 2. The load is applied in the middle point of upper surface along both length and width directions. There are 6 unknown reaction forces on the two ends. The horizontal forces can be ignored under small strains because of no loads being applied in the horizontal direction, thus, there are 4 forces left unknown. In view of symmetry of beam and load, two equations could be obtained

$$R_A = R_B \quad (7) \qquad M_A = M_B \quad (8)$$

Where R_A and R_B are reaction forces on ends in the vertical direction while M_A and M_B are moments.

In addition, subject to constrain conditions of static force balance and elastic deflection, two other equations can be written as follows:

$$R_A + R_B - F = 0 \quad (9) \qquad \theta_A(=\theta_B) = 0 \quad (10)$$

Where θ_A and θ_B are angular deflection on both ends.

From four equations above, 4 reaction forces can be resolved.

$$R_A = R_B = \frac{F}{2} \quad (11), \qquad M_A = M_B = \frac{FL}{8} \quad (12)$$

The deflection at midpoint of both ends fixed microbeam arose from forces and moments on both ends could be given directly using linear elastic deflection addition method as follows:

$$f = \frac{FL^3}{48EI} - \frac{M_A L^2}{16EI} - \frac{M_B L^2}{16EI} = \frac{FL^3}{192EI} \quad (13)$$

Thus elastic modulus can be calculated from load, deflection and size of microbeam (inertial moment can be expressed by length, width and thickness of beam)

$$E = \frac{L^3}{192I} \cdot \frac{F}{f} \quad (14)$$

Where size of microbeam, load and deflection can be measured directly using AFM before and during the 3-points bending test, respectively.

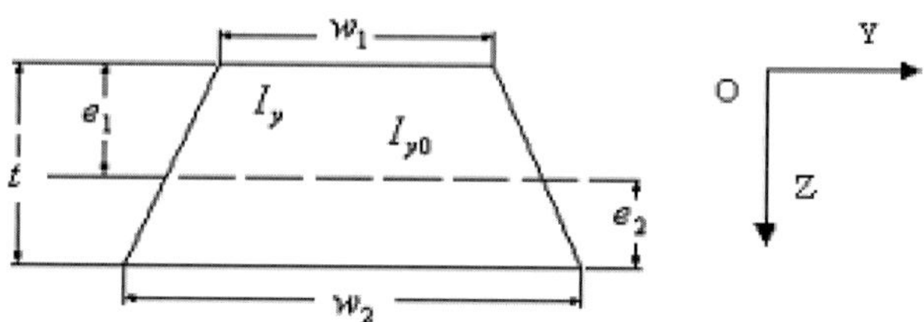

Figure 1. Schematics of trapezoidal cross section

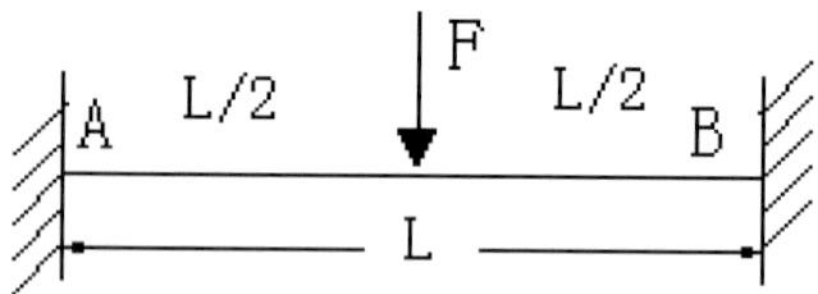

Figure 2. Mechanical model of both ends fixed beam

Using equation (6), (14), the relation expression between elastic modulus and size of both ends fixed polysilicon microbeams with trapezoidal cross section can be deduced in the following:

$$E = \frac{3(w_1 + w_2)L^3}{16(w_1^2 + 4w_1w_2 + w_2^2)t^3} \cdot \frac{F}{f} \qquad (15)$$

Where F/f is the gradient of the force-displacement curve during bending.

4.3   Theoretical value of polysilicon elastic modulus

We can calculate the elastic modulus of an isotropic polycrystalline solid by either the Voigt model for maximum bound or the Reuss model for the minimum bound, and the experimental value should fall into the Voigt-Reuss bounds [18]. The Voigt model is also called the isostrain model because the grains are assumed to have the same strain, following that a polycrystalline solid would behave as a collection of a large number of constituent crystals, each strained an equal amount under varying loads. For polysilicon, the maximum modulus can be calculated as

$$E_{max} = \frac{(C_{11} - C_{12} + 3C_{44})(C_{11} + 2C_{12})}{2C_{11} + 3C_{12} + C_{44}} \qquad (16)$$

Where C_{ij} are the material constants. The Reuss model is an isostress model, assuming that each grain of a polycrystalline solid is under equal stress and is strained a varying amount. For polysilicon, the minimum modulus can be given as

$$E_{min} = \frac{5}{3S_{11} + 2S_{12} + S_{44}} \qquad (17)$$

Where S_{ij} are the material constants. The upper and lower bounds of elastic modulus can be used to evaluate the calculation.

5.   CONCLUSION

Bending tests are often carried out to evaluate elastic modulus of polysilicon used in MEMS. Size effects in MEMS field are analyzed and elastic modulus of both ends fixed polysilicon microbeam with trapezoidal cross section is calculated by size and the gradient of the force-displacement curve based on the assumption that the microbeam follows linear elastic theory of an isotropic material owing to high ratios of length to width and length to thickness. The evaluation of elastic modulus can be applied in 3-points bending test, and we can measure geometry size of polysilicon microbeams, values of force and displacement using precision instruments such as AFM, STM, SEM and TEM, before and during bending test respectively. Furthermore, the upper and lower bounds of elastic modulus are described, to which the calculation value of elastic modulus during bending test could be compared in order to find out whether it is in accordance with the theoretic value.

REFERENCES

1. Richard P. Feynman. "There's plenty of room at the bottom", *Journal of MEMS*, **vol.1, no. 1**, pp. 60-66, 1992.
2. Richard Feynman. "Infinitesimal machine", *Journal of MEMS*, **vol.2, no.1**, pp. 4-14, 1993.
3. J.A. Williams. "Friction and wear of rotating pivots in MEMS and other small scale devices", *WEAR*, **vol.251**, pp. 965-972, 2001.
4. Ivan Amato. " Fomenting a revolution, in miniature", *Science*, **vol.282**, pp. 402-405, 1998.
5. Wen S. Z., Ding J. N. " Study of the fundamental design issues of microelectromechanical systems ", *Chinese J. Mech. Eng.*, **vol.36, no. 7**, pp. 39-42, 2000.
6. Ding J. N., Meng Y. G., Wen S. Z. "Experimental and Theoretical study of Young modulus in micromachined polysilicon films", *Tsinghua science and technology*, **vol.7, no.3**, pp. 270-275, 2002.
7. Yuan W. Z., Ma B. H. *Weijixie yu weixi jiagong jishu*, Xi'an, Northwestern Polytechnical univ. press, 2001.
8. T. Y. Zhang, Y. J. su, M. H. Zhao. "Microbridge testing of silicon nitride thin films deposited on silicon wafers", *Acta materialia*, **vol.48, no.11**, pp. 2843-2857, 2000.
9. Sriram Sundararajan, Bharat Bhushan et al. "Mechanical property measurements of nanoscale structures using an atomic force microscope", *Ultramicroscopy*, **vol.91**, pp. 111-118, 2002.
10. Takahiro Namzu, Yoshitada Isono, Takeshi Tanaka. "Evaluation of size effect on mechanical properties of single crystal silicon by nanoscale bending test using AFM", *Journal of MEMS*, **vol.9, no.4**, pp. 450-459, 2000.
11. Staffan Greek, Fredric Ericson, et al. "In situ tensile strength measurement and Weilbull analysis of thick film and thin film micromachined polysilicon structures", *Thin solid films*, **vol.292**, pp. 247-254, 1997.
12. Toshiyuki Tsuchiya, Osamu Tabata, et al. "Specimen size effect on tensile strength of surface-micromachined polycrystalline silicon thin films", *Journal of MEMS*, **vol.7, no.1**, pp. 106-113, 1998.
13. M.A. Haque, M.T.A.Saif. "Application of MEMS force sensors for in situ mechanical characterization of nano-scale thin films in SEM and TEM", *Sensors and actuators*, **vol. 97-98**, pp. 239-245, 2002.
14. H. Kahn, N. Tayebi et al. "Fracture toughness of polysilicon MEMS devices", *Sensors and actuators*, **vol.82**, pp. 274-280, 2000.
15. Tang Z.A., Wang L.D. "On microscale theory", *Optics and precision engineering*, **vol.9, no.6**, pp. 493-498, 2001.
16. L. Fu, J.M. Miao et al. "Study of deep etching for micro-gyroscope fabrication", *Applied surface science*, **vol.177**, pp. 78-84, 2001.
17. Su Y. F. *Material mechanics*, Beijing, Higher education press, 1984.
18. W. N. Sharpe, Jr. Bin Yuan, and R. L. Edwards. "A new technique for measuring the mechanical properties of thin films", *Journal of MEMS*, **vol.6, no.3**, pp. 193-199, 1997.

A WEB-BASED OPTIMIZATION FOR FUNCTION SYNTHESIS OF PLANAR LINKAGE

Lei Song and Xin Wu
(Electromechanical Engineering College, Henan University of Science and Technology,Luo yang, China, 471039, emailsl@163.com)

Abstract: A new web-based optimization method for function synthesis of planar linkage is developed. The technical features of this method are analyzed and a web-based design mode is confirmed, which combines the characteristics of optimization and web technique. It can accomplish remote design between different locations and simultaneous design with multi-users. The calculate instances show that the method is practicable and effective.

Key words: Planar Linkage, Function Synthesis, Optimization, Web.

1.   INTRODUCTION

Planar linkage has been used in a great variety of machines and devices because it has advantages of simple structure, wide applicability, etc. The design problems for planar linkage on kinematics can be divided into two categories. The first is function generation that the motions of links connected to ground should fit the given function; the other is path generation that the path of a coupler tracer point should fit the given trace. Function synthesis is a basic problem often met in engineering practice. Taking into account a majority of factors affecting mechanism design, the mathematic model of function synthesis of planar linkage is a complex nonlinear constrained system. Optimization is an effective method to solve such problems. At present, function synthesis of planar linkage based on optimization theory has been studied deeply [1-4], but these results are design systems working on single-computer, which are insufficient in applicability and information share, etc.

Web technique is developing rapidly nowadays. Because of having advantages such as huge capacity of information, fast spreading-speed, wide application and huge influence, it attracts more and more attentions now. To carry out the mechanism design system on web can cause the design method and information to bc shared on a same design platform by users who are in different locations. As the system can accomplish remote design between different locations, it can widen the range of application and increase design efficiency. This is a new direction that mechanism design develops. Some researches of web-based mechanism design are discussed [5], but it does not find the researches now to study characteristics and design methods for web-based optimization for function synthesis of planar linkage. Because some new problems will bring out when the system carries out on web, this paper brings forward a new optimization method for function synthesis of planar linkage, based on web.

2.   FEATURES OF WEB-BASED OPTIMIZATION FOR FUNCTION SYNTHESIS OF PLANAR LINKAGE

The basic thinking for function synthesis of planar linkage by optimization theory can be simply expressed as follow: establishing the mathematic model of mechanism optimization with multi-variables and multi-constraints, then approaching optimal result by declining iteration method. The iterative process is also a process of huge amounts of calculation. Calculating time is not same when different mathematic model and iterative process are selected. Usually, design accuracy is considered primarily in a single-machine design system, however, balance between accuracy and time consuming is considered also sometimes [6].

WWW is a multimedia information-inquiring tool on Internet. It works as server/client framework, accomplishes information transmission between server and client, using web server/client communication protocol (Figure 1). Figure 1 indicates that information communication is the essential feature of the web running. Because of the processes of data request, response and communication between server and client if mechanism design system works on web, total time consuming of design for user on client should be the sum of data calculating and communicating. The factor of time consuming to mechanism design emerges unavoidably when considering the influence of system configuration and data transmission between server and client on web, thus the design criterion of web-based optimization for function synthesis of planar linkage can be confirmed as design accuracy and time consuming. Therefore, the methods that are used in single-machine design system are no longer suitable in web-based system. When considering mechanism design system to run on web, the problems such as design objective, constrained conditions, working mode of web server/client must

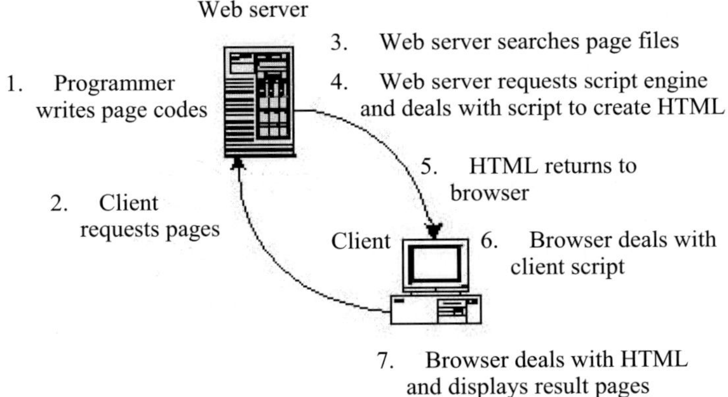

Figure1. Web running mode

be study and define again.

3.   OPTIMIZATION FOR FUNCTION SYNTHESIS OF PLANAR LINKAGE

3.1   Design variables

While function is synthesized, it should be converted to a pattern that planar linkage can identify. Suppose generation function to be $y = f(x)$, $a \le x \le b$ (monotone function, if not, to separate to sections), it is firstly converted to corresponding rotation angle relations of input link and output link.

Suppose initial angle of input and output link is θ_0 and φ_0 respectively, rotation range of them is α and β, then

$$\frac{x-a}{\theta_{0i}} = \frac{b-a}{\alpha} \quad \text{and} \quad \frac{y}{\varphi_{0i}} = \frac{f(b)-f(a)}{\beta}$$

This relationship may also be written as:

$$x = \frac{b-a}{\alpha}\theta_{0i} + a \ \ \text{and} \ \ y = \frac{f(b)-f(a)}{\beta}\varphi_{0i}$$

From equation $y = f(x)$, we can obtain that:

$$\varphi_{0i} = \frac{\beta \times f\left(\dfrac{b-a}{a}\theta_{0i} + a\right)}{f(b)-f(a)} \qquad 0 \le \theta_{0i} \le \alpha$$

θ_{0i} and φ_{0i} count from initial location of input link and output link respectively, $i=1,2,\ldots,n$ (numbers of discrete point).

As shown in Fig.2, geometric position of the mechanism depends on 8 parameters in all, which include lengths of four links L_1, L_2, L_3, L_4, position of joint $A=[a_x, \ a_y]^T$, positional angle β of fixed link AD, rotation angle θ of link AB. In the planar linkage which generates given function $\overline{\varphi}_{0i} = \overline{f}(\theta_{0i})$, where there are relationships as $\overline{\varphi}_{0i} = \overline{\varphi}_i - \overline{\varphi}_0$, $\theta_{0i} = \theta_i - \theta_0$, function synthesis is independent of absolute lengths of links and position of selected coordinate system. Thus $A=[0, \ 0]^T$, $\beta=0$, $L_4=1$, design variables can be decreased to 4 units, that is the follow and shown in Figure 3.

$$X = [x_1, x_2, x_3, x_4] = [L_1, L_2, L_3, \theta_0]$$

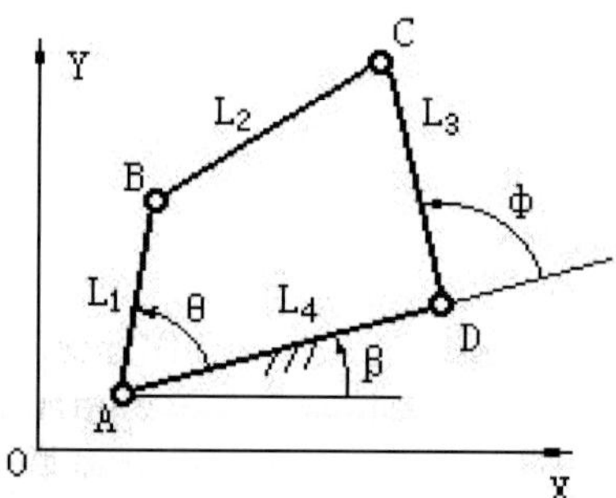

Figure 2. Parameters of dimension and position

3.2 Objective function

Function $\varphi_{0i} = f(L_1, L_2, L_3, \theta_0, \theta_{0i})$ which describes relations of φ_{0i} (rotation angle of output link) and θ_{0i} (that of input link) can be concluded from geometric relations of planar linkage. By comparing with given function $\overline{\varphi}_{0i} = \overline{f}(\theta_{0i})$ to get the minimum of total deviations of φ_{0i} and $\overline{\varphi}_{0i}$ in all positions, the objective function can be given as:

$$F(X) = \sum_{i=1}^{s} \omega_i \left[f(L_1, L_2, L_3, \theta_0, \theta_{0i}) - \overline{f}(\theta_{0i}) \right]^2 \tag{1}$$

When rotation angle of input link AB is θ_{0i}, corresponding angle of output link CD φ_{0i} can be calculated as follow from Fig 3:

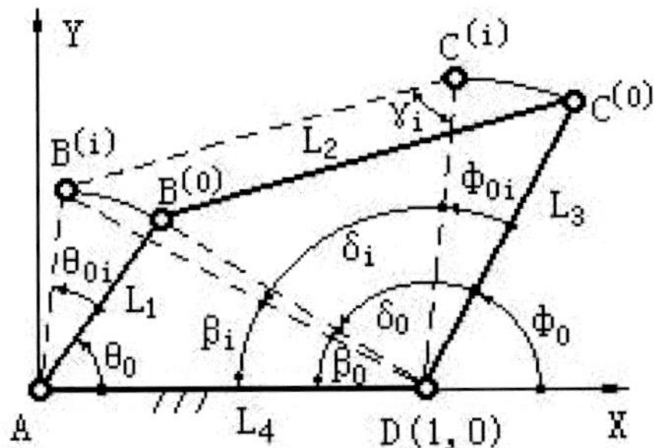

Fig 3. Motion synthesis of function generation

$$\varphi_{0i} = \beta_0 + \delta_0 - \beta_i - \delta_i$$

$$= arctg \frac{L_1 \sin\theta_0}{1 - L_1 \cos\theta_0} + arccos \frac{L_1^2 - L_2^2 + L_3^2 + 1 - 2L_1 \cos\theta_0}{2L_3 \sqrt{L_1^2 + 1 - 2L_1 \cos\theta_0}}$$

$$- arctg \frac{L_1 \sin(\theta_0 + \theta_{0i})}{1 - L_1 \cos(\theta_0 + \theta_{0i})} - arccos \frac{L_1^2 - L_2^2 + L_3^2 + 1 - 2L_1 \cos(\theta_0 + \theta_{0i})}{2L_3 \sqrt{L_1^2 + 1 - 2L_1 \cos(\theta_0 + \theta_{0i})}} \qquad i = 1, \ldots, s \tag{2}$$

Bring equation (2) into (1), objective function can be obtained as

$$F(X) = \sum_{i=1}^{s} \omega_i \left[arctg \frac{x_1 \sin x_4}{1 - x_1 \cos x_4} + arccos \frac{x_1^2 - x_2^2 + x_3^2 + 1 - 2x_1 \cos x_4}{2x_3 \sqrt{x_1^2 + 1 - 2x_1 \cos x_4}} \right.$$

$$\left. - arctg \frac{x_1 \sin(x_4 + \theta_{0i})}{1 - x_1 \cos(x_4 + \theta_{0i})} - arccos \frac{x_1^2 - x_2^2 + x_3^2 + 1 - 2x_1 \cos(x_4 + \theta_{0i})}{2x_3 \sqrt{x_1^2 + 1 - 2x_1 \cos(x_4 + \theta_{0i})}} - \overline{f}(\theta_{0i}) \right]^2$$

ω_i is weighted factor of deviation of each point.

3.3   Constrained conditions

(1) Existent condition of crank-and-rocker mechanism: total lengths of the shortest link and longest link are less than that of other two links, and crank must be the shortest link.

(2) Rotation range limit for transmission angle: transmission angle ranges usually form $45°$ to $135°$ in engineering practice.

(3) Relative size limits: To distribute sizes of links reasonably, it should limit the relative size of each link to the fixed link which is regarded as a reference.

3.4   Optimization method

In mechanical practices, mature techniques need to be applied. As this reason, the optimization method can choose from mature optimization methods that have been validated in practice. Chaos optimization for nonlinear constraint optimization problems is a good choice [6-7]. Its algorithm is simple, practical and high efficiency. Because this optimization method does not have requests for continuity and differentiability of mathematical model, it is easy to realize global optimization. So this method is suitable for the problem.

4.   WEB REALIZATION OF MECHANISM DESIGN

The web-based optimization system for function synthesis of planar linkage that works on server/client framework can accomplish remote design between different locations with data remote transmitting. As the system works based on Internet, it can solve the possibility that more than one user visit the system simultaneity. Evidently, it has more advantages compared with a single-computer design system.

Considering the features of mechanism design, a server/client working mode is used in web-based optimization system. This is a Browser/Server framework based on server, and this can found a flexible system with lower resources. In this mode, tasks of design and maintenance can be performed on server, and legal users on client who have browser can use the system easily. Using ASP (Active Server Pages) technique on this basis, data requests and responses between server and remote client can be realized conveniently. Design parameters are submitted from client, as a request it is sent to server and dealt with on it, then the results are sent back to browser on client. Because it is possible that more than one users visit web server simultaneously and each visit would cause task running, the design task can be distributed to run on client to reduce the server's loads. Thus, when server receives design request from client, it sends design method as a form of client script to client. Then tasks of designing, calculating and results showing accomplish on client.

Usually, web pages created by ASP technique work as explaining execution mode. To increase design efficiency and decrease time consuming, Java Applet technique that can work as compiling execution mode may be used on calculating process of mechanism optimization. Calculating speed of compiling execution is greatly improved than that of explaining execution. As the calculating method is send to client as form of client script, then compiling execution mode ensures secrecy of source files also. Java language is independent of operating system and easy to transplant. It has strong functions to solve kinds of problems of web-based optimization system.

The design results of function synthesis of planar linkage are usually described with parameters of dimension, position, and dynamic motion of the mechanism. All these can be described in web pages created by ASP, Java Applet, and multi-thread technique in detail.

Thus the integral flow chart of web-based optimization system shows in Figure 4.

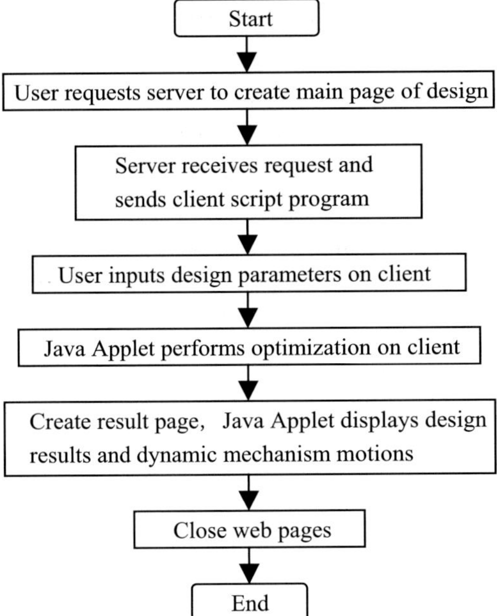

Figure 4. Flow chart of web-based optimization

5.   DESIGN EXAMPLE

To generate a function using the method introduced as above, y=sin (x), $0° \leq x \leq 180°$. Choose the input range to be $180°$, output range $30°$, numbers of discrete point to be 21. Table 1 shows the corresponding relations of rotation angles for input link and output link.

Using objective function and optimization method mentioned above, uploading the design files to a server, designing on a remote client, then optimization errors and results show in Table 1 and Table 2, and details of system running shows in Figure 5.

Total time consuming is less than 5.5s after repetitious tests. Design accuracy of this method can fit the users' requests for common tasks of mechanism design. Considering the integrative features that the mechanism design system demands to work on web, accuracy and time consuming for design are both satisfactory. This method shows characteristics of applicability and practicality.

Table 1. Corresponding rotation angles of input link and output link and errors (Unit: $^\circ$)

Relative rotation angle of crank	Anticipated rotation angle of rocker	Actual rotation angle of rocker	Error of rotation angle of rocker
0.000	0.000	0.000	0.000
9.000	4.693	4.778	0.085
18.000	9.271	9.384	0.113
27.000	13.620	13.734	0.114
36.000	17.634	17.740	0.106
45.000	21.213	21.314	0.101
54.000	24.271	24.368	0.097
63.000	26.730	26.819	0.089
72.000	28.532	28.596	0.064
81.000	29.631	29.651	0.020
90.000	30.000	29.960	-0.039
99.000	29.631	29.526	-0.104
108.000	28.532	28.377	-0.154
117.000	26.730	26.558	-0.171
126.000	24.271	24.125	-0.145
135.000	21.213	21.137	-0.075
144.000	17.634	17.654	0.020
153.000	13.620	13.735	0.115
162.000	9.271	9.438	0.167
171.000	4.693	4.821	0.128
180.000	0.000	-0.049	-0.049

Table 2. Results of optimization

Length of crank	Length of coupler	Length of rocker	Length of fixed link	Starting angle of crank ($^\circ$)
0.100	1.013	0.185	1.000	100.625

6.　CONCLUSION

This paper develops a new optimization method for function synthesis of planar linkage working on web, based on characteristics of optimization and web-running techniques. The new difficulties that appear with design system running on web are studied, in turn, the corresponding problems of objective function, constrained conditions, optimization method, and web-running mode are researched. The calculate instance shows that the method is

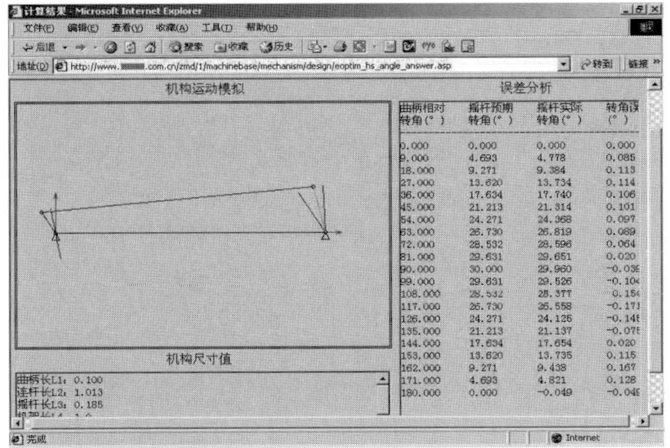

Figure 5. A mechanism design results

practicable and effective. With the web popularisation, the thinking and method may also be applied to other fields of machine design.

REFERENCES

1. I.Prebil, S.Krasna, I.Ciglaric: "Synthesis of Four-bar Mechanism in A Hydraulic Support Using a Global Optimization Algorithm". *Structural and Multidisciplinary Optimization*, **Vol.24**, pp. 246-251, September 2002.
2. Wang zhixing, Chu yue: "The realization method of function synthesis of 4-bar linkage utilizing the method of comparison". *Journal of Machine Design*, **Vol.10**, pp. 7-9, 49-50, October 1997.
3. LI tao, Wang delun: "New approach to function synthesis of planar mechanisms". *Journal of Dalian University of Technology*, **Vol.40**, pp. 721-724, November 2000.
4. Peng zhu: "Synthesis of Planar Four-link Mechanism for Function Generation by Using Chaos Optimization Method". *Journal of Mechanical Transmission*, **Vol.26**, pp. 9-12, January 2002.
5. G.F.Yin, G.Y.Tian, D.Taylor: "A Web-Based Remote Cooperative Design for Spatial Cam Mechanisms". *The International Journal of Advanced Manufacturing Technology*, **Vol.20**, pp. 557-563, October 2002.
6. Zhang Chunkai, Xu liyun, Shao huihe: "Improved Chaos Optimization Algorithm and Its Application in Nonlinear Constraint Optimization Problems". *Journal of Shanghai Jiaotong University*, **Vol.34**, pp. 593-595, 599, May 2000.
7. Luo chenzhong, Shao huihe: "Chaos Search Method for Nonlinear Constraint Optimization". *System Engineering Theory & Practice*, **Vol.8**, pp. 54-57, 90, August 2000.

OPTIMAL DESIGN OF THE FASTENER'S MINIMUM COST

Mingyuan Wu and Wei Yang
Electromechanical Engineering College, Henan University of Science and Technology, Luoyang Henan China, 471039, Email: wmyuan@eyou.com

Abstract: In this paper, according to the idealization mode of the plus-minus term geometric programming with inequality subject and the basic theory of linear programming, the full-range optimal solution is calculated directly in the optimal fastener design at minimum cost. Compared with non-linear programming, the two methods introduced are more feasible and preferable.

Key words: Geometrical Programming, Linear Programming, Fastener, Optimal design.

1.   INTRODUCTION

It always calculates the optimal problem in mechanics by existing software of non-linear programming. Although it is an effective method for complicated optimal problem, non-linear programming has its shortages. The severest shortage is high time complexity. In fact, many optimal problems in mechanics are attributed to geometrical programming; some others can be calculated by linear programming with mathematical treatment. Because the research of geometrical programming and linear programming are fast developed and the programming have lower time complexity, more and more problems can be fast and accurately solved by them[1][2][3].

In this paper, according to the dualization mode of the plus-minus term geometric programming with inequality subject and the basic theory of linear programming with mathematical treatment, the full-range optimal solution is calculated directly in the optimal design of the fastener's minimum cost. Compared with the solution by the mixed penalty function method, the geometrical programming and linear programming is more feasible and preferable.

2.   OPTIMAL DESIGN PROBLEM

Here, fasteners need fix a pressure container, and the fabric and dimension of container are determined. *Cont=(P, A, D)*; *P* is the press of interior gas; *A* is the container's inside diameter; *D* is the central circle's diameter for the fasteners--erection bolts. Hermetic unit is carrier ring. It need ascertain rational bolt's diameter and its number to fasten the container.

Bolt's number is decided by the hermetic requirement, and bolt's diameter need satisfy with the intensity; at the same time, the diameter of container's flange should not too large. So it is an optimal design problem that deciding bolt's dimension and number to satisfy with the multi-requirements and to minimize the fastener's cost.

It can be solved by non-linear programming method, such as the mixed penalty function method. But with mathematical treatment, the geometric programming and the linear programming also can figure out the full-range optimal solution. And they will be more feasible and preferable than non-linear programming method.

3.   MATHEMATICAL TREATMENT

3.1   Objective function

The cost of bolts used in a pressure container (Cn) is decided by one bolt's cost (c) and bolt's number (n). Where $c = M \times d - L$. d is bolt's diameter; M, L can be counted by linear regression method[1][3]. So the optimal design variable is: $X = [d, n]^T = [x_1, x_2]^T$, and the objective function is $Cn = n \times (M \times d - L)$.

3.2 Constraint function

3.2.1 Bolt intensity's constraint

The bolt's allowed load (F) can be counted by regression method, $F = B \times d^u$. Where B, u are undetermined coefficient, decided by the material's allowed load with different diameter in non-previously locking work condition.

To ensure the safe operation in rated pressure, bolts' total allowed load must be greater than or equal to the cover's total press. So the intensity's constraint function is:

$$nF - \alpha \frac{\pi A^2}{4} \times P \geq 0 \qquad \cdots\cdots\cdots\cdots(1)$$

Where α is safe margin coefficient.

3.2.2 Hermetic constraint

To ensure the uniform pressure among bolts, without local leak, the circumference-space should not large than *10d*. So the constraint function is

$$10d - \frac{\pi D}{n} \geq 0 \qquad \cdots\cdots\cdots\cdots(2)$$

3.3 Mathematic models

From the mathematic analysis above all, the mathematic model of optimal design problem of fastener's minimum cost is:

$$\begin{cases} \min F(\vec{X}) = M \times x_1 x_2 - L \times x_2 \\ s \cdot t \quad g_1(\vec{X}) = B \times x_1^u x_2 - \alpha \dfrac{\pi A^2}{4} P \geq 0 \\ g_2(\vec{X}) = 10 x_1 - \dfrac{\pi D}{x_2} \geq 0 \end{cases} \qquad \cdots\cdots\cdots\cdots(3)$$

4. SOLUTION OF THE PROBLEM

Now we put an application illustration into the mathematic model, *Cont=(12.75MPa,120mm,200mm)*. Without the loss of generality, bolt is selected half-bright hexagonal one, which its size is *50mm* decided by the thickness of flange and carrier rings. So Eq.(3) can be rewritten as

$$
\begin{cases}
\min F(\vec{X}) = 0.0202 \times x_1 x_2 - 0.148 \times x_2 \\
s \cdot t \quad g_1(\vec{X}) = 63.92324 \times x_1^{2.13133} x_2 - 158619.335 \geq 0 \quad \cdots \cdots \cdots \cdots \\
g_2(\vec{X}) = 10x_1 - \dfrac{628.3185}{x_2} \geq 0
\end{cases}
$$

(4)

Where $M = 0.0202$, $L = 0.148$, $B = 63.92324$ and $u = 2.13113$, determined by the linear regression method.

Using mathematic analysis and application illustration, the model as Eq.(4) can be computed by geometric programming and linear programming methods[2][3][4][5], stated in the subsections 4.1 and 4.2.

4.1   Geometric programming method

We invert the original model to the dual model of the plus-minus term geometric programming with inequality subject.

4.1.1   The dual programming model

The mathematic model as Eq.(4) is transformed to the geometric programming normalized form:

$$
\begin{cases}
\min g_0(\vec{X}) = 0.0202 \times X_1 X_2 - 0.148 \times X_2 \\
s \cdot t \quad g_1(\vec{X}) = -0.000403 \times X_1^{2.13133} X_2 \leq -1 \quad \cdots \cdots \cdots \cdots (5) \\
g_2(\vec{X}) = 62.83185 X_1^{-1} X_2^{-1} \leq 1
\end{cases}
$$

Then the original problem is changed into solving X_1, X_2. Eq.(5) is a plus-minus term geometric programming with inequality subject. The number of terms about objective function and constraint function $T = 4$ ($T_0 = 2$, $T_1 = 1$, $T_2 = 1$), when variable $n = 2$, constraint $m = 2$, accessible coefficient $d_f = T - (n+1) = 1$, it can be counted that:

$$
\begin{array}{llll}
\sigma_{01} = 1 & C_{01} = 0.0202 & \alpha_{011} = 1 & \alpha_{012} = 1 \\
\sigma_{02} = 1 & C_{02} = 0.148 & \alpha_{021} = 0 & \alpha_{022} = 1 \\
\sigma_{11} = 1 & C_{11} = 0.000403 & \alpha_{111} = 2.13113 & \alpha_{112} = 1 \\
\sigma_{21} = 1 & C_{21} = 62.83185 & \alpha_{211} = -1 & \alpha_{212} = -1
\end{array}
$$

When $\sigma_0 = 1$, the dual programming problem can be formulated as:

Objective function:

$$d(\vec{\delta}) = (\frac{0.0202}{\delta_{01}})^{\delta_{01}} \times (\frac{0.148}{\delta_{02}})^{\delta_{02}} \times (\frac{0.000403\delta_{11}}{\delta_{11}})^{-\delta_{11}} \times (\frac{62.83185\delta_{21}}{\delta_{21}})^{-\delta_{21}}$$

$$\cdots\cdots\cdots\cdots(6)$$

Task:

To find δ_{uj} ($u = 0,1,2; j = 1,2,...,T_u$) so as to maximize $d(\vec{\delta})$.

Constraint function:

$$\begin{cases} s \cdot t \quad \delta_{01} - \delta_{02} = 1 \\ \delta_{01} - 2.13113\delta_{11} - \delta_{21} = 0 \\ \delta_{01} - \delta_{02} - \delta_{11} - \delta_{21} = 0 \\ \delta_{uj} > 0 \quad (u = 0,1,2; j = 1,2,...,T_u) \end{cases} \qquad \cdots\cdots\cdots\cdots(7)$$

There are four unknown dual quantities ($\delta_{01}, \delta_{02}, \delta_{11}, \delta_{21}$) and three equations, so random three quantities can be expressed by the other quantity. Using δ_{02}, the result from Eq.(7) is:

$$\delta_{01} = 1 + \delta_{02}, \quad \delta_{11} = 0.88407\delta_{02}, \quad \delta_{21} = 110.88407\delta_{02} \quad \cdots\cdots\cdots\cdots$$

$$(8)$$

Applying Eq.(8), Eq.(6) transforms into:

$$d(\vec{\delta}) = (\frac{0.0202}{1+\delta_{02}})^{(1+\delta_{02})} \times (\frac{0.148}{\delta_{02}})^{\delta_{02}} \times (0.000403)^{-0.88407\delta_{02}}$$

$$\times (62.83185)^{(1-0.88407\delta_{02})} \qquad \cdots\cdots\cdots\cdots(9)$$

Because Eq.(9) is the function for single variable δ_{02}, the maximal point $d(\delta_{02}{}^{*})$ with δ_{02} can be gained:

$$\max \{ d(\delta_{02}{}^{*}) = 0.90861584 \mid \delta_{02}{}^{*} = 0.3968523347 \}$$

Applying the data δ_{02} into Eq.(8), we can count that:

$$\delta_{01} = 1.396853347, \quad \delta_{11} = 0.350861584, \quad \delta_{21} = 0.649153861$$

Which all of four dual quantities satisfy the non-negative constraint. So $\sigma_0 = 1$ is right.

4.1.2 Relation between original problem and dual programming one

$$\begin{cases} g_0(\vec{X}^{*}) = d(\vec{\delta}^{*}) = 0.90861584 \\ 0.0202x_1x_2 = 1.396853347 \times 0.90861584 \\ 0.148x_1^0 x_2 = 0.396853347 \times 0.90861584 \end{cases} \qquad \cdots\cdots\cdots\cdots(10)$$

The result is x_1=25.7888, x_2=2.4364; that is $\vec{x}^* = [25.7888, 2.4364]^T$, with rounding numbers, $\vec{X}^* = [25, 3]^T$.

It is a local maximal point from the plus-minus term geometric programming. But this is the original problem's optimum solution, the same as the dual programming, because the solution is unique in this problem and solution must exist in practical consideration[2].

4.2   Linear programming method

Firstly, the original model is processed with linearization. The objective and constraint functions are dealt with logarithm.

$$\begin{cases} \min F(\vec{Y}) = y_1 + y_2 - 3.90207 \\ s \cdot t \quad g_1(\vec{Y}) = 2.13113y_1 + y_2 \geq 7.81658 \\ g_2(\vec{Y}) = y_1 + y_2 \geq 4.14046 \\ y_1 \geq 0 \quad y_2 \geq 0 \end{cases} \quad \cdots\cdots\cdots(11)$$

Where y_1=lnx_1, y_2=lnx_2. During logarithm we make a mathematical simplification with ignoring $(-0.148X_2)$ in objective function, which means single bolt's cost will be increased 0.148.

Non-negative slack variable is put in Eq.(11) to turn inequality constraints into equality one, then a normalized form of linear programming is gained.

$$\begin{cases} \min F(\vec{Y}) = y_1 + y_2 - 3.90207 \\ s \cdot t \quad 2.13113y_1 + y_2 - y_3 = 7.81658 \\ y_1 + y_2 - y_2 = 4.14046 \\ y_i \geq 0 \quad (i = 1, 2, 3, 4) \end{cases} \quad \cdots\cdots\cdots(12)$$

The feasible zone of linear programming is a convex set with programming's basic theory[2][5]. Moreover, this set is convex polyhedron, while each peak (or pole) of it matches a basic feasible solution. If there is a unique optimum solution in linear programming, it must be a pole. The number of peaks can't exceed C_n^m (m is the constraint equation; n is the dimension of variables, including slack ones). Because of $m = 2$, $n = 4$ in this problem, the number of basic solution will not exceed $C_4^2 = 6$.

It is clear that every two column vectors of constraint equation (12) is linearly independent:

$$\vec{\alpha_1} = \begin{bmatrix} 2.13113 \\ 1 \end{bmatrix} \quad \vec{\alpha_2} = \begin{bmatrix} 1 \\ 1 \end{bmatrix} \quad \vec{\alpha_3} = \begin{bmatrix} -1 \\ 0 \end{bmatrix} \quad \vec{\alpha_4} = \begin{bmatrix} 0 \\ -1 \end{bmatrix}$$

So they can make up six linearly independent sets. If $\vec{\alpha_3}$, $\vec{\alpha_4}$ are selected, the basic variables are y_3, y_4 and non-basic variables y_1, y_2 should be set zero, which $x_1=x_2=1$ conflicted with practical considerations. So y_1, y_2 as non-basic variables is false.

Now $\vec{\alpha_1}$, $\vec{\alpha_2}$ are selected, the basic variables are y_1, y_2 and non-basic variables y_3, y_4 is set zero, then:

$$\begin{cases} 2.13113y_1 + y_2 = 7.81658 \\ y_1 + y_2 = 0.89051 \end{cases} \qquad \cdots\cdots\cdots\cdots(13)$$

The solution of Eq.(13) is $y_1 = 3.24995$, $y_2 = 0.89051$, then

$$\vec{X}^* = [x_1, x_2]^T = [25.78914, 2.43636]^T$$

5. ALGORITHM COMPARISON AND CONCLUSION

In the paper, we also deal the problem with mixed penalty function method. The solution counted by computer is $\vec{X}^* = [x_1, x_2]^T = [25.671142, 2.46387]^T$, $F(\vec{X})^* = 0.91297$. With round numbers, $\vec{X}^* = [25,3]^T$, $F(\vec{X})^* = 1.071$, the same as the solutions counted above all. But the two programming methods' complexities are all greatly lower than non-linear programming's.

The comparison among three different algorithms proves that the stated two methods are correct and more preferable in the fastener's optimal design. It concludes that, with mathematical treatment and building rational models, many optimal design problems in mechanics can be fast and accurately solved by geometric programming and linear programming methods.

REFERENCES

1. Liu Weixin, *Mechanics Optimal Design*, Beijing, Tsinghua University Press, 1994.
2. Northwestern Polytechnical University. *Mechanical Design*, Higher Education Press, 2001.
3. Chen Lizhuo. *The Method of Mechanics Optimal Design*, Beijing, Metallurgical Industry Press, 1985.
4. Monteiro DC, Adle I. "A polynomial time, primaldual affine scaling algorithm for linear and convex quadratic programming and its power series extension." *Mathematics of Operations Research*, 1990.
5. Ying jinchun. *Modern Design*, Beijing, Mechanical Industry Press, 2000.

STUDY OF HELICAL OIL GROOVED JOURNAL BEARING AND LASER PROCESSING FOR THE GROOVES

Hong-Lai Li, Shu-Zhong Lin, Hui-Lai Sun and Jin-Gang Wang
*School of Mechanical Engineering, Hebei University of Technology, Tianjin, China 300130,
E-mail:jear@webmail.hebut.edu.cn*

Abstract: A new type herringbone grooved journal bearing, which has characteristics of stability, less leakage and considerable load capacity, is proposed in this paper. A special device has been designed to process the helical grooves by laser beam. The rectangular-profile grooves with any helical angle can be processed. There is not distortion on sleeve surface.

Key words: Journal Bearing, Helical Oil Groove, Laser Processing

1.   INTRODUCTION

The bearings used in machinery equipped with high-speed rotating parts, such as high-speed pumps and hard disk drivers, are required to have high-performance. Hydrodynamic bearings are considered a good choice to support rotating shafts due to their constructive simplicity, reliability, efficiency, and low cost. According to the Reynolds equation in three dimensions, if the width of plain journal bearing is at least 4 times the diameter, there is no drop in pressure in the z-coordinate direction. The pressure pattern for a bearing assumes no flow in the z-direction. In fact, the width is less than the ratio of 4 to 1. A modification of derivation must be used to allow for the loss in load-carrying capacity because of side leakage and reduction of pressure. The loss in load-carrying capacity is clearly evident [1]. Beside of this, plain journal bearing is not stable, when its running speed is not much lower than the system's critical speed.

When surfaces slide relative to one another in a fluid medium they produce a viscous drag, which causes the fluid to flow as though it were being pumped. If a shaft has both a right-hand screw on one and a left-hand screw on the other end, it will pump fluid from both ends to the middle. The concept of helical grooves on a rotating shaft has been developed into a pattern known as the herringbone groove journal bearing (HGJB) (Fig.1). Currently, the HGJBs have been applied in high speed and light weight machinery due to their less leakage and improved stability compared to plain cylindrical journal bearings. But, their load capacity is not satisfied. Furthermore, the groove is only a few micrometers deep and cannot be easily obtained by conventional cutting processes. This situation let us in a dilemma.

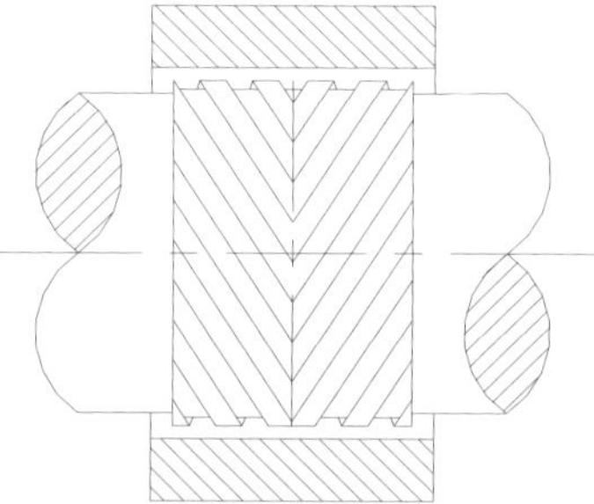

Fig. *1* Herringbone Grooved Journal Bearing

To solve these problems, we have developed a kind of symmetrical pattern journal bearing, which has characteristics of stability, less leakage and considerable load capacity. A special device has been designed to process the helical grooves by laser beam.

2.   EFFECT OF HELICAL OIL GROOVE ON JOURNAL BEARING

The idea of using grooved surfaces in order to produce a pressure distribution in bearing has been shown by Whipple (1949) for more than 50 years. In recent years, HGJBs have been adopted for business machines.

Vohr and Chow (1965) propose a narrow groove theory (NGT) for the infinite number of herringbone-grooved, gas-lubricated, cylindrical journal bearings. They considered the bearing geometry as a series of rectangular shaped steps, so the pressure distribution along the groove-ridge pair was assumed to be linear. The NGT gives fairly accurate characteristics of gas-lubricated HGJB for large number of groove-ridge pairs. At the same time, several studies were accomplished for incompressible fluid films in HGJB. A analysis of aerodynamic HGJBs with small number of grooves was made by D. Bonneau and J. Absi (1994), in which the compressible Reynolds equation was solved by use of Finite Element Method. A comparison of the results for a smooth bearing with previously published results was made and the domain of validity of the NGT was analyzed. They evaluated the load capacity, attitude angle and stiffness coefficients [2]. After that, Kinouchi and Tanaka (1990) used a Finite Element Method for determination of performance characteristics of hydrodynamic HGJBs.

In this paper, we develop a symmetrical pattern journal bearing which is partly grooved on bush surface (Fig.2). The Reynolds equation is solved by use of a Finite Element Method (FEM). We compare the load capacity of this symmetrical pattern journal bearing with HGJBs. Finally, A special device has been designed which drives bush rotating and moving along z-direction at the same. It is convenient to process the helical grooves by laser beam.

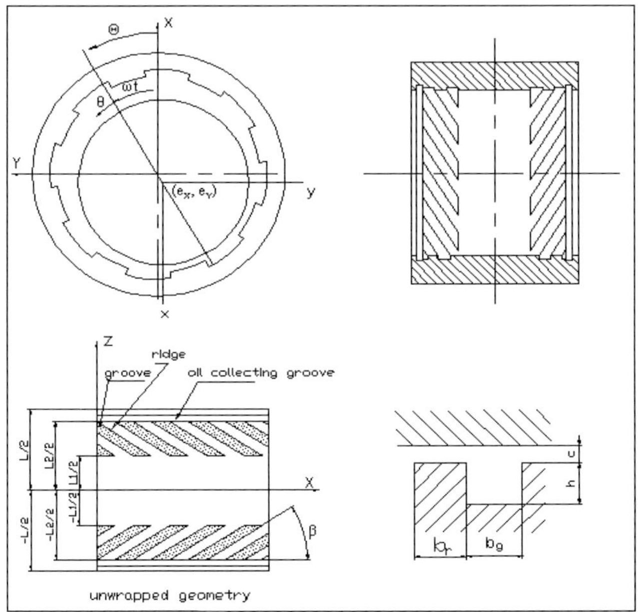

Fig. 2 Symmetrical Pattern Partly Grooved Journal Bearing

The groove pattern is symmetrical with respect to the center of bearing as shown in Fig. 2. It has two groove regions and one smooth region. In each groove region, several rectangular-profile grooves are formed. When the smooth journal rotates, the oil in the gap between the journal and bearing flows along the groove and then accumulates to the center until the viscous and pressure forces are in equilibrium.

Neglecting fluid inertia effects, the governing equation of the pressure field in the lubricating film is the Reynolds equation.

$$\frac{1}{R^2}\frac{\partial}{\partial\theta}\left(\frac{h^3}{12\mu}\frac{\partial P}{\partial\theta}\right)+\frac{\partial}{\partial z}\left(\frac{h^3}{12\mu}\frac{\partial P}{\partial z}\right)=-\frac{\omega}{2}\frac{dh}{d\theta}+\frac{\partial h}{\partial t} \tag{1}$$

where, $x = \theta\cdot R$ and $\Theta = \theta + \omega\cdot t$
film thickness in the ridge

$$h = c + e_x(t)\cos(\Theta) + e_y(t)\sin(\Theta) \tag{2}$$

film thickness in the groove

$$h = c_g + c + e_x(t)\cos(\Theta) + e_y(t)\sin(\Theta) \tag{3}$$

Boundary Conditions.

The pressure field must be continuous in the circumferential direction, i.e.,

$$P(\theta, z, t) = P(\theta + 2\pi, z, t) \tag{4}$$

For a symmetric bearing, the pressure field satisfies specified pressure conditions on the sides, i.e.,

$$P(\theta, L_2/2, t) = P(\theta, -L_2/2, t) = P_a \tag{5}$$
$$P(\theta, L_1/2, t) = P(\theta, -L_1/2, t) \tag{6}$$

where, Pa is the ambient pressure.

Figure 3 shows dimensionless load variation with respect to the eccentricity ratios for various bearing types when L/D =1.0, β=35 deg . Both HGJB and partly HGJB have higher load capacity than the smooth journal bearing for eccentricity ratios less than 0.4. Computations are conducted for the small eccentricity ratios range because the instability occurs easily in this range.

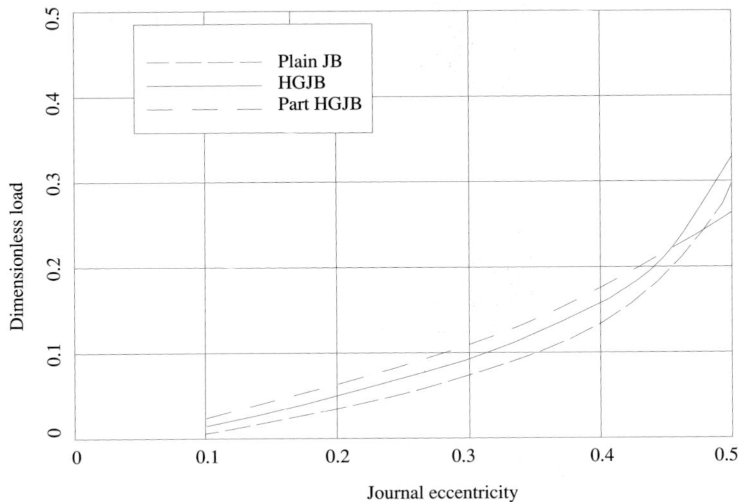

Fig.3 Dimensionless load variation with respect to the eccentricity ratios for various bearing types

_________Nomenclature_________

bg = groove width (m)

br = ridge width (m)

c = nominal film clearance (m)

D = bearing diameter (m)

e_X, e_Y = journal eccentricities (m)

F = actual load capacity

f = $Fc2/\mu\,\omega\,D4$ dimensionless load capacity

h = film thickness (m)

L = bearing length (m)

L1 = smooth part length (m)

L2 = axial extent of grooves (m)

P = film pressure (Pa)

Pa = ambiant pressure (Pa)

R = bearing radius (m)

t = time (s)

(x, y, z) = coordinate system attached to the smooth journal

(X, Y) = inertial coordinate system

β = helix angle (deg)

ε = dimensionless eccentricity

θ = circumferential coordinate rotating with journal

Θ = circumferential coordinate fixed to bearing, ($\theta + \omega\,t$)

μ = fluid viscosity (Pa.s)

ρ = fluid density (kg/m)

ω = journal angular speed (rad/s)

3. THE LASER PROCESSING OF HELICAL OIL GROOVE

The depth of groove is just a few micrometers and cannot be easily obtained by conventional cutting processes. The grooves are made usually by chemical etching process. Because of the difficulty in controllability of the etching process, grooves are not usually generated precisely as designed. The other method to generate grooves is to use small balls rolling inside the sleeve[3]. Circular-profile grooves are generated by using a specially designed tool, which has four small balls retained at the end. The diameter of a circle circumscribing the four balls is a little bit larger than the sleeve diameter. When the tool makes a stroke inside the sleeve, the sleeve is rotated so that the combination of these processes forms circular-profile grooves with desired helix angle. The disadvantages of second method: (1) Only circular-profile grooves can be obtained on the sleeve side. (2) The cutting and clamping forces can cause serious distortions in the finished bearing.

It is convenient to process the helical grooves by laser beam. A special device has been designed, as shown in Fig.4, which drives bush rotating moving along z-direction at the same time. The rectangular-profile grooves with any helical angle can be processed. There is not distortion on sleeve surface. The roughness of the groove surface is much smoother than produced by etching processes.

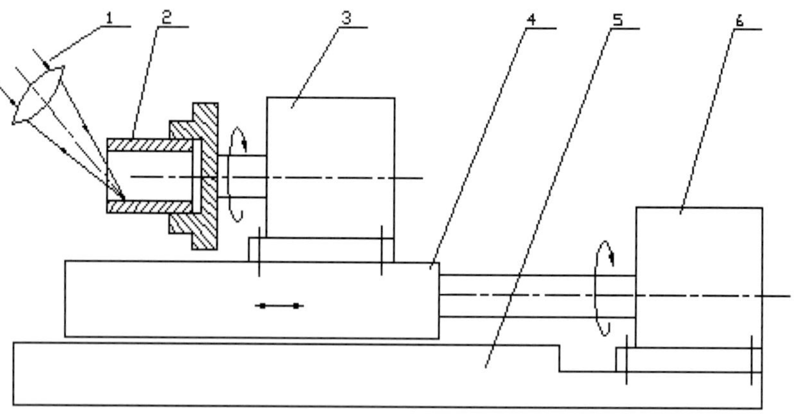

Laser unit. 2 —Workpiece. 3 —Motor. 4 —Roller lead screw. 5 —Base. 6 —Motor.

Fig. 4 Device of helical grooves laser processing

4.   CONCLUSION

Dynamic characteristics of partly HGJB with rectangular-profile grooves have been analyzed. Dimensionless load capacity is compared with HGJB and smooth journal bearing. A special device has been designed to process the helical grooves by laser beam. The following conclusions are drawn.

(1) Both HGJB and partly HGJB have higher load capacity than the smooth journal bearing for eccentricity ratios less than 0.4. There are little differences in load capacity for HGJB and partly HGJB when eccentricity ratios less than 0.35.

(2) The helical oil grooves can be processed by laser beam conveniently.

REFERENCES

1.   Dudley D.Fuller: *Theory and Practice of Lubrication for Engineers* **(Second Edition)**, John Wiley & Sons, Inc. 1984
2.   D. Bonneau, J. Absi: "Analysis of Aerodynamic Journal Bearing with Small Number of Herringbone Grooves by Finite Element Method". *ASME Journal of Tribology*, **Vol.116**, pp 699-704, Apr. 1994.
3.   Kyungphil Kang, Yoonchul Rhim and Kiro Sung: "A Study of Oil-Lubricated Herringbone-Grooved Journal Bearing---Part 1: Numerical Analysis". *ASME Journal of Tribology*, **Vol.181**, pp906-911, Apr. 1996.

AN INTEGRATED PRODUCT AND PROCESS DEVELOPMENT MODEL SUPPORTING ENTIRE LIFE CYCLE

Ying- Kui Gu and Hong-Zhong Huang
School of Mechanical Engineering, Dalian University of Technology, Dalian, Liaoning 116023, P. R. China, Email: guyingkui@163.com, hzhhuang@dlut.edu.cn

Abstract: Product and its development process modeling are very important to product life cycle design. Product's life cycle is a continuous and variational time process, and can be divided into a sequence of life phases. The product status evolves form one to another through these life phases. Some evaluation factors, such as cost, reliability, time and environment impacts can be applied on each life phase to evaluate its corresponding performance. In this paper, in order to improve product quality, minimize the life cycle cost and optimize the design process, the three-dimensional model of life cycle design and the frame of integrated product and process development were developed. Product development process was viewed as an optimization process to find a process set under given evaluation factors such that the set of evaluation indices of entire life cycle is optimized. The state equations of process, energy, material and information of product development were developed. Finally, we provide a brief report of our application of the methodology proposed above to a loader development.

Key words: Life cycle design, process modeling, integrated product and process development, equation of state

1.   INTRODUCTION

Product development is usually considered as the set of activities beginning with the perception of a market opportunity and ending in the production, sale, and delivery of a product [1]. However, because more and more requirements arise from every possible step of a product's life cycle, the concept of product development has been expended such that it includes consideration of the entire life cycle of a product. Therefore, the life cycle design and concurrent engineering have been developed to meet the increasing requirements come from every step of a product's life cycle. Nowadays, quality, cost and development time are the primary factors that can embody the competition capability of a product. The intention of life cycle design is to improve product quality, cut development cost, and shorten development time. Life cycle design inherits the ideology of concurrent engineering, and it is a new design method to take the value of the whole life cycle into account in the initial design stage. By adopting life cycle design, not only the functions and structures of a product, but also the whole process of product life cycle, such as design, manufacturing, assembling, operating, maintaining, recycling, and so on, are to be designed. Therefore, all the factors related to product's life cycle could be programmed and optimized well in design stage.

Nowadays, one has realized that it is the product design stage that plays the most important role in a product's performance during its entire life cycle. Therefore, in design stage, the most important task is to establish integrated product and process development model. Based on the decomposition of product's life cycle, and considering the requirements of customer, producer and environment, many research issues are raised including: design for manufacture (DFM), design for assembly (DFA), design for maintainability (DFMa), design for disassembly (DFD), design for recyclability (DFR), design for quality (DFQ), design for reliability (DFRe), design for environment (DFE), and so on [2-5]. These issues sometimes are referred as Design for "X" (DFX), and are of avail to develop the product in different stages of product's life cycle. However, these methods aim to only one stage of product's life cycle or one performance factor of a product, and the integrity, consistency and interaction of design objectives and constraints are ignored to some extent.

In this paper, the product development process was viewed as an optimization process to find a process set under given evaluation factors such that the set of evaluation indices of entire life cycle is optimized. The three-dimensional model of life cycle design and the frame of integrated product and process development were developed. The state equations of process, energy, material and information of product development were developed.

2.   STRUCTURING THREE-DIMENSIONAL MODEL OF LIFE CYCLE DESIGN

A three-dimensional model of product's life cycle design was developed to assist the integrated product and process development. As shown in Figure 1, the proposed model includes three dimensions: logic (e.g. the constraints and relations in product's life cycle design), factors (e.g. development cost and time, reliability, quality, environment, etc) and stages of life cycle (e.g. design, manufacturing, assembling, maintaining, recycling, etc). There exist some relationships between factor and factor, stage and stage, and factor and stage as follows.

(1) Each stage of life cycle design should embody all the factors in the development process, and each factor should be included in all stages of life cycle design.

(2) There exist some interrelationships in all stages that include the same factor, in all factors of the same stage, and in the different factors of different stages. These relations can be expressed as some constraints in the development process.

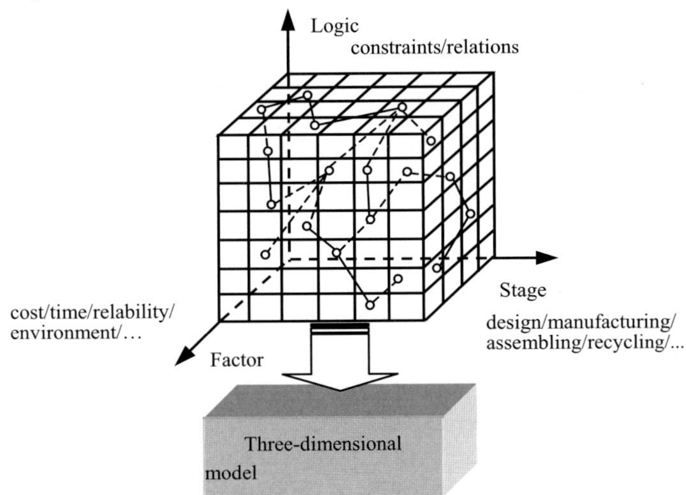

Figure 1. Three-dimensional model of product life cycle design

Therefore, all the factors and stages related to product life cycle should be programmed and optimized well in the design stage. The design shouldn't aim to only one or several stages and factors of life cycle, and the integrity, consistency and interrelations of design objectives and constraints should be taken into account. Points to this, the frame of integrated product and process development supporting entire life cycle was developed in section 3.

3. A FRAME OF INTEGRATED PRODUCT AND PROCESS DEVELOPMENT

Process is a basic unit of activity that is carried out during a product's life cycle. In this sense, the entire development process can be viewed as a set of activities that their physical meanings are varied continuously along with time. Product development is a dynamic, sequential, time-change process P, and each stage of it can be viewed as a process granule P_i. The entire process P can be denoted as

$$P = \{P_i \in P \mid i = 1, 2, \cdots, n\} \tag{1}$$

The granularity of processes depends on the product developer's concern and interest, and the appearance of processes varies along with a product's entering its different life phases [1].

Because process varies along with time, the process granule has time characteristic. Process P is an ordered sequence that extends along with the main thread of product's life cycle.

The process granules in product's life cycle are not isolated. There exist close relations among them. The granule P_i is the evolutional result of the $i-1$ granules before it. At the same time, P_i has important influence on the following $n-i$ processes, and its output is the direct input of the following $i+1$th process. That is to say the activities in one life phase may influence the activities in its posterior life phase. As a result, in order to improve the performance of one life phase, not only should the activities in this life phase be adjusted, the activities in its prior life phases should be adjusted as well. However, in order to simplify the modeling analysis, when the process P_i is analyzed, it must be isolated from the other processes.

To carry out one process of life cycle design has many means, which can produce many design plans. Take the processing of an axis as example, it can be processed by lathing, milling, planning, grinding, an so on. In another example, when we select a sensor, we can adopt series-wound mode or shunt-wound mode based on the type of sensor and the practical measure environment. Moreover, the design of product parameters has very important influence on the performance of product life cycle. Therefore, how to select an optimal means to realize process requirements and how to design product parameters are the key problems of development process programming. However, no matter how to program the development process, or adopting what means to establish the process model, the basic goal is to improve product quality, cut life cycle cost and shorten development time. So, the effective process-programming model should be established to realize the following two intentions, and the different design requirements, designer's preferences should be taken into account too.

(i) Selecting an optimal design plan.

(ii) Optimizing design parameters.

To product's life cycle design, the evaluation factors such as cost, reliability, weight, environmental friendliness, can be selected to evaluate its corresponding performance. Let $C=(c_1,c_2,\cdots,c_j,\cdots,c_m)$ be vector set of evaluation factors of development process.

To a sub-process P_i, its evaluation index vector can be denoted as follows.

$$C_i =\left(c_{i1} \quad \cdots \quad c_{ij} \quad \cdots \quad c_{im}\right)^T =\left(c_1(P_i) \quad \cdots \quad c_j(P_i) \quad \cdots \quad c_m(P_i)\right)^T \tag{2}$$

where $c_j(P_i)$ is the jth index of phase P_i. It could be the degree of membership or degree of satisfaction of certain preference.

We have n evaluation index vectors. The set of evaluation indices of entire life cycle can be expressed as follows.

$$C=\left(C_1 \ C_2 \ \cdots \ C_j \ \cdots \ C_n\right)=\begin{pmatrix} c_1(P_1) & \cdots & c_1(P_i) & \cdots & c_1(P_n) \\ \vdots & & \vdots & & \vdots \\ c_j(P_1) & \cdots & c_j(P_i) & \cdots & c_j(P_n) \\ \vdots & & \vdots & & \vdots \\ c_m(P_1) & \cdots & c_m(P_i) & \cdots & c_m(P_n) \end{pmatrix} \tag{3}$$

Thus, there are various transforms of C that are defined to evaluate the product's overall performance from different perspectives.

To an evaluation index c_j, the integration effect in the entire process can be denoted as follows.

(i) Linear integration.

$$c_j = \sum_{i=1}^{n} c_j(P_i) \tag{4}$$

(ii) Non-linear integration. The non-linear integration effect is complicated. Take the integration effect of product as example, its integration mode can be expressed as

$$c_j = \prod_{i=1}^{n} c_j(P_i) \tag{5}$$

Whether the life cycle design is good or not is not decided only by one evaluation factor. The goal of life cycle design is to make the whole system and all performances optimal by effective process programming.

In the design process, not all sub-processes have the same influence on the whole process. Based on the influence degree of P_i to P, a general weight α_i is proposed and applied to each life phase to evaluate its important degree, and is constrained by

$$\sum_{i=1}^{n} \alpha_i = 1, \ (i = 1,2,\cdots,n) \tag{6}$$

Thus, equation (4), (5) can be substituted by

$$c_j = \sum_{i=1}^{n} \alpha_i c_j(P_i) \tag{7}$$

$$c_j = \prod_{i=1}^{n} \alpha_i c_j(P_i) \tag{8}$$

Moreover, every evaluation factor c_j has different influence on process granule P_i. Based on its effect degree, weight β_{ij} is proposed and applied to each evaluation factor to evaluate its important degree, and is constrained by

$$\sum_{j=1}^{m} \beta_{ij} = 1, \ (j = 1,2,\cdots,m) \tag{9}$$

After the normalization of all evaluation factors, the evaluation vector set C can be transformed into evaluation index C'.

$$C' = \sum_{i=1}^{n} \left[\alpha_i \sum_{j=1}^{m} \left(\beta_{ij} c_{ij} \right) \right] \tag{10}$$

where α_i denotes the influence weight of ith process to the entire process. c_{ij} denotes the influence weight of jth evaluation factor of ith process. β_{ij} denotes the influence weight of jth evaluation factor to process P_i.

Different C (or C') describes the attributes and performances of the product from different aspects. Therefore, the essence of life cycle design is to find a best method to optimize C and realize the function requirements of the process under given evaluation factors. In other words, the modeling of product and its development can be expressed as: "find a process set under given evaluation factors such that C (or C') is optimized".

In another viewpoint, C (or C') can be used to evaluate the development plan. Because of the complexity of development process, and because of the fuzzy and uncertainty of development process, fuzzy comprehensive evaluation can be used to evaluate the design plans. By the evaluation results, the best design plan can be selected.

4.   STATE EQUATION OF LIFE CYCLE

The life cycle process P is a function of time. So, at each moment t, the process has different states. Product's life cycle is a sequential process that realizes the function of each stage under its control mechanism and constraints.

Let x_t, y_t, u_t be the input, output and state of process P_t at the moment t respectively. The process P_t can be defined by the following state equation.

$$\begin{cases} u_{t+1} = f(x_t, u_t) \\ y_t = g(x_t, u_t) \end{cases} \tag{11}$$

Let $U^{t+1}(x_t, u_t)$, X^t and $Y^t(x_t, u_t)$ be the possible value set. The state equation of P_t can be denoted by

$$\begin{cases} U^{t+1} = F(x_t, u_t) \\ Y^t = G(x_t, u_t) \end{cases} \tag{12}$$

where F and G express the mapping of sub-set space from $X \times U$ to U and Y respectively.

The state equation indicates that each process can be viewed as an input-output process under its constraints. That is to say, under the state u_t, the process system can be defined as the integration of time ordered pairs (x_t, y_t), and the necessary condition is that (x_t, y_t) is a segment closure. One has

$$P_t = \{(x_t, y_t)\}, (x_t \in X, y_t \in Y) \tag{13}$$

where (x_t, y_t) is an input-output pair of P_t. Therefore, a complicated process system can be simplified as a simply mathematics model.

In the design process of life cycle, it needs to make the material, energy balance and make the information be accumulated effectively. Moreover, it should embody the environment friendliness in the transform process of material, energy and information. The same as process state equation, the state equation of material, energy and information can be established respectively, and denoted by M^t, E^t and I^t. Under the state u_t, M^t denoted that the use efficiency and the recycle efficiency of material should be higher after the material is transformed. E^t denotes that most of energy is transformed as added value of semi-product (i.e. output). I^t denotes that information transform is an accumulation process, and the accumulation of information will be of avail to the next proceeding of process.

5.   ILLUSTRATIVE CASE

Take the development of an excavator as example to illustrate the design method proposed above. By analyzing the market requirements and functional requirements, the total function of the excavator can be identified. The function decomposition table and the function-physical mapping table can be established too. Thus, we can get many conceptual design solutions.

However, due to the complexity of development and the different development environment, many detail design solutions can be gotten based on the concept solutions. Therefore, the evaluation index C (or C') can be used to evaluate and select design solutions.

Supposed the development process has four sub-processes, i.e. design, manufacturing, assembling and disassembling. The development process can be denoted as follows.

$$P = (p_1, p_2, p_3, p_4) = (\text{design, manufacturing, assembling, disassembling})$$

Each sub-process has different influence on the whole development process. The weights of sub-processes can be given as follows by experts.

$$w = (w_1, w_2, w_3, w_4) = (0.4, 0.3, 0.2, 0.1)$$

The factors, such as excavating force, speed, noise, vibration, convenience of operation, safety, and so on, can be selected as influence factors. The evaluation vector can be denoted as

$$U = (u_1, u_2, u_3, u_4, u_5, u_6)$$
$$= (\text{excavating force, speed, noise, vibration, convenience of operation, safety})$$

Each evaluation factor has different influence on the whole performance. The weights of evaluation factors are gotten as follows by fuzzy comprehensive evaluation.

$$v = (v_1, v_2, v_3, v_4, v_5, v_6) = (0.25, 0.25, 0.1, 0.1, 0.15, 0.15)$$

According to equation (10), the influence of every evaluation factor on the whole process are denoted by

$$e_i = \sum_{j=1}^{4} w_j u_{ij}, \quad i = 1, 2, \cdots, 6$$

where u_{ij} is the influence degree of the ith evaluation factor on the jth process.

We hope that the excavating force is as big as possible, speed as fast as possible, noise as small as possible, convenience of operation as good as possible, and safety as high as possible. That is to say the excavator should have the biggest satisfactory degree. According to equation (10), the evaluation index of the whole process can be given as follows.

$$C' = \sum_{i=1}^{6} v_i e_i$$

The higher the value of C', the better the performance of the excavator.

Supposed that we have two design plans, the satisfactory degree of their evaluation factors are gotten as follows respectively.

$$C_1 = (C_{11}, C_{12}, C_{13}, C_{14}, C_{15}, C_{16}) = \begin{pmatrix} 0.95 & 0.90 & 0.80 & 0.50 \\ 0.88 & 0.80 & 0.80 & 0.20 \\ 0.70 & 0.85 & 0.85 & 0.20 \\ 0.85 & 0.80 & 0.75 & 0.20 \\ 0.85 & 0.80 & 0.70 & 0.70 \\ 0.90 & 0.85 & 0.80 & 0.40 \end{pmatrix}$$

$$C_2 = (C_{21}, C_{22}, C_{23}, C_{24}, C_{25}, C_{26}) = \begin{pmatrix} 0.95 & 0.95 & 0.80 & 0.50 \\ 0.98 & 0.90 & 0.85 & 0.30 \\ 0.70 & 0.85 & 0.80 & 0.20 \\ 0.87 & 0.76 & 0.83 & 0.20 \\ 0.80 & 0.80 & 0.70 & 0.65 \\ 0.88 & 0.83 & 0.80 & 0.70 \end{pmatrix}$$

According to equation (3), the total evaluation index of every evaluation factor in the whole development process can be given as follows respectively.

$$C_1 = (C_{11}, C_{12}, C_{13}, C_{14}, C_{15}, C_{16}) = (0.860, 0.772, 0.725, 0.750, 0.790, 0.815)$$

$$C_2 = (C_{21}, C_{22}, C_{23}, C_{24}, C_{25}, C_{26}) = (0.875, 0.862, 0.715, 0.762, 0.765, 0.831)$$

Thus, the evaluation index (i.e. satisfactory degree) of each design plan was gotten as follows respectively.

$$C_1' = 0.7963, \quad C_2' = 0.8214, \quad C_1' < C_2'$$

So, the design plan 2 is better than plan 1.

6.   CONCLUSIONS

Typical characteristics that can be used to describe development process are: creative and innovative, dynamic, interdisciplinary, strongly interrelated, strongly parallel, interactive, planning intensive, uncertain and risky.

Therefore, many factors related to a product's life cycle should be taken into accounted when the product and its development process models were analyzed and established. In this paper, the three-dimensional model of life cycle design and the frame of integrated product and process development were developed. The state equation of life cycle was established. It not only can provide a systematic approach to accomplish integrated product and process development, but also can provide an integrated decision support mechanism for the product developers. Moreover, it is of avail to improve product quality, cut life cycle cost, shorten development time and use design resource effectively.

ACKNOWLEDGEMENTS

This research was partially supported by the National Natural Science Foundation of China under the contract number 50175010, the Excellent Young Teachers Program of Ministry of Education under the contract number 1766, National Excellent Doctoral Dissertation Special Foundation under the contract number 200232, and Guangxi Liu Gong Machinery Co., LTD, P.R.C.

REFERENCES

1. P. T. Yan, M. C. Zhou and S. Donald. "An integrated Product and process development methodology: concept formulation". *Robotic and Computer Integrated Manufacturing* **15(3)**: pp. 201-210, 1999.
2. J. S. Tang, G. S. Qin and Y. W. Peng, "State-of the art and developing trends of product life cycle integration". *Manufacture Automation* **23(8)**: pp. 1-4, 11., 2001
3. T. C. Kuo, S. H. Huang and H. C Zhang, "Design for manufacture and design for 'X': concepts, applications, and perspectives". *Computers & Industrial Engineering* **41(3)**: pp. 241-260, 2001.
4. W. F. V. D. Vegte, J. P. W. Pulles and J. S. M. Vergeest, "Towards computer-supported inclusion and integration of life cycle processes in product conceptualization based on the processes tree". *Automation in Construction* **10**: pp. 731-740, 2001.
5. China Mechanical Design Canon Committee. China Mechanical Design Canon: **Vol. 1**, *Modern Method of Mechanical Design*. Nanchang, Press of Science and Technology, 2002.

TOOTH CONTACT ANALYSIS OF TI WORM GEARING CONSIDERING BOUNDARY CONDITION

Luqian Duan, Yuehai Sun and Qingzhen Bi

School of Mechanical Engineering, Tianjin University, Tianjin 300072, China ; E-mail: duan_luqian@hotmail.com

Abstract: On the premise of considering the boundary condition of mating zone of the toroidal involute worm gearing (TI worm gearing), judgement condition for meshing limit line is derived based on the theory of gear engagement and the distribution pattern of contact lines on the tooth surface of worm wheel is investigated. The results show that there can be four situations for meshing limit line on the tooth surface of wheel, including inexistence, only one line, two lines and two lines coincidence; if helix angle is equal to or slightly smaller than the angle which makes two meshing limit lines coincide with each other, preferable meshing effect is obtained. Computer simulation confirms validity of the mentioned above.

Key words: TI worm gearing, meshing performance, boundary condition, meshing limit line, contact line

1. INTRODUCTION

The toroidal involute worm gearing (TI worm gearing) is a novel hourglass worm gearing, which is composed of an involute helical gear and an hourglass worm enveloped by conjugate tooth surface, namely involute helical surface. Preliminary analyses show that TI worm gearing is of great advantages for transmission, such as more teeth in mesh simultaneously, lower contact stress of tooth surfaces, easier to form a good lubrication condition and so on [1]. In addition, worm wheel is a simple involute helical gear, and no costly and complicated special hobbing cutter is needed to machine it. If TI worm transmission is applied to a reductor, the new-style TI worm-gear speed reducer will have strong market competition power and cheerful prospect in high-speed and heavy-duty gearing. So many researchers are greatly interested in TI worm gearing. Especially those from Germany, Japan and China made valuable contributions to it. The scholars from Germany [2] firstly proposed TI worm gearing was a practical transmission and did some experiments about it. Japanese scholars [3] confirmed the usefulness of this gearing by the TCA (Tooth Contact Analysis) method and by the experiment. The researchers [1,4] in China preliminarily studied the contact performance of tooth surface based on meshing analysis and drew some valuable conclusions.

Because the distribution pattern of the instantaneous contact lines may disclosure whether meshing performance of a gearing is good or not, there are some researches [1,4] about it. But none of them considers whether contact lines go beyond the boundary of tooth surface or not. So the contact situation and contact zone on the tooth surface of worm wheel can't be given an exact account of.

Therefore, boundary condition is introduced into contact lines analysis in this paper in order to exactly study the meshing performance of tooth surface. For meshing limit line is significant to study distribution pattern of contact lines and select parameters of gearing, judgement of meshing limit line is discussed. And to get ideal meshing effect, distribution situation of contact lines on the tooth surface of worm wheel is also studied. By computer simulation, conclusions will be confirmed to be valuable.

This article is organized as follows. In Section 2, many fundamental equations are deduced according to the theory of gear engagement. Boundary condition is introduced in Section 3. Judgement for meshing limit line is discussed in Section 4. And, on the premise of considering the boundary condition of mating zone, the last Section analyses meshing performance of TI worm gearing.

## 2.	FUNDAMENTAL EQUATIONS

### 2.1	Coordinate systems

Coordinate systems for TI worm gearing are shown in Fig.1. Here, the moveable coordinate system $S_{1(\phi_1)} = [O_1 \; ; \; i_1(\phi_1), \; j_1(\phi_1), \; k_1(\phi_1)]$ and $S_{2(\phi_2)} = [O_2 \; ; \; i_2(\phi_2), \; j_2(\phi_2), \; k_2(\phi_2)]$ are rigidly connected with TI worm and worm wheel, respectively. And the fixed coordinate systems $S=[O\ ;\ \boldsymbol{i},\ \boldsymbol{j},\ \boldsymbol{k}]$ and $S_p=[O_p\ ;\ \boldsymbol{i}_p,\ \boldsymbol{j}_p,\ \boldsymbol{k}_p]$ in space are the original position of $S_{1(\phi_1)}$ and $S_{2(\phi_2)}$, respectively. Vector $\boldsymbol{i}=\boldsymbol{k}_p\times\boldsymbol{k}$; and $\boldsymbol{i}=\boldsymbol{i}_p$; vector $\boldsymbol{j}=\boldsymbol{k}\times\boldsymbol{i}$; vector $\boldsymbol{j}_p=\boldsymbol{k}_p\times\boldsymbol{i}_p$; line OO_p is the center distance of TI worm gearing, that's to say, $OO_p=a$. TI worm and worm wheel rotate about axes $\boldsymbol{k}_1$ and $\boldsymbol{k}_2$ with angular velocities ω_1 and ω_2, respectively. The rotating angles are ϕ_1 and ϕ_2 at some instant, respectively. The directions of rotation correspond to the right-hand worm gearing.

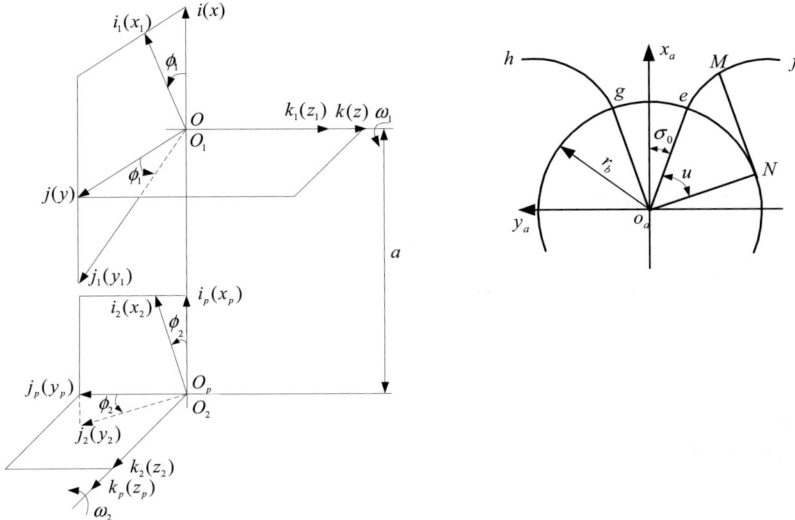

Figure1 Coordinate systems for TI worm gearing

Figure2 End-cross profile of worm wheel

2.2 Worm wheel surface equation

As shown in Fig.2, the worm wheel tooth surface Σ^2 is actually an involute helical surface, and it can be represented in coordinate system $S_{2(\phi_2)}$ as follows [5]:

$$\left.\begin{aligned}
\boldsymbol{r}_L^{(2)} &= x_{2L}\boldsymbol{i}_2 + y_{2L}\boldsymbol{j}_2 + z_{2L}\boldsymbol{k}_2 \\
x_{2L} &= r_b\cos\tau + r_b u\sin\tau \\
y_{2L} &= -r_b\sin\tau + r_b u\cos\tau \\
z_{2L} &= p_2(\sigma_0 + u - \tau)
\end{aligned}\right\} \tag{1}$$

Where u and λ are two variable parameters of surface Σ^2, $\tau = \sigma_0 + u - \lambda$, r_b is the radius of the base circle, p_2 is helix parameter, and it means the lead-per-radian revolution of surface Σ^2, and σ_0 is the half-angle of tooth groove on the base circle.

2.3   Meshing equation

According to the theory of engagement, for conjugate action, two tooth surfaces are in tangency during the meshing process. Therefore, the meshing equation, $\boldsymbol{n} \cdot \boldsymbol{v}_{12} = 0$, should be satisfied. Thus the meshing equation of TI worm gearing may be concluded as follows:

$$u = \frac{p_2^2(\tau - \sigma_0)\sin(\tau - \phi_2) - p_2 r_b i_{21} - r_b^2 \cos(\tau - \phi_2) + ar_b}{(p_2^2 + r_b^2)\sin(\tau - \phi_2)} \tag{2}$$

Where i_{12} is drive ratio, and β_b is helix angle of base circle.

2.4   Contact line equation

Eq.(2) and Eq.(1) yield contact line on the tooth surface of worm wheel as follows:

$$\left.\begin{array}{l} u = \dfrac{p_2^2(\tau - \sigma_0)\sin(\tau - \phi_2) - p_2 r_b i_{21} - r_b^2 \cos(\tau - \phi_2) + ar_b}{(p_2^2 + r_b^2)\sin(\tau - \phi_2)} \\[2ex] x_{2L} = r_b \cos\tau + r_b u \sin\tau \\[1ex] y_{2L} = -r_b \sin\tau + r_b u \cos\tau \\[1ex] z_{2L} = p_2(\sigma_0 + u - \tau) \end{array}\right\} \tag{3}$$

Eq.(3) indicates contact line is a curve corresponding to rotating angle ϕ_2.

2.5   Meshing limit line equation

In terms of Eq.(2), by derivative with respect to time t, the meshing limit line function can be obtained. Then yield meshing limit line equation can be expressed as follows:

$$u = \frac{p_2^2(\tau - \sigma_0) \pm \sqrt{A}}{p_2^2 + r_b^2} \tag{4}$$

Where $A = (p_2 r_b i_{21} - ar_b)^2 - r_b^4$ $\tag{5}$

Eq.(4) and Eq.(1) yield the equation of meshing limit line on surface Σ^2 as follows:

$$
\left.
\begin{aligned}
u &= \frac{p_2^2(\tau - \sigma_0) \pm \sqrt{A}}{p_2^2 + r_b^2} \\
x_{2L} &= r_b \cos\tau + r_b u \sin\tau \\
y_{2L} &= -r_b \sin\tau + r_b u \cos\tau \\
z_{2L} &= -p_2(\tau - \sigma_0 - u)
\end{aligned}
\right\} \qquad (6)
$$

3. BOUNDARY CONDITION

3.1 Span of parameters u, λ, ϕ_2

How to select parameters u, λ, φ_2 is very important to analyze meshing performance of TI worm gearing. The rational span of parameters u, λ, φ_2 determines the trueness of meshing analysis.

(1) Span of parameter u

From Fig.2, we know parameter u is a variable describing involute forming. In the process of actual meshing, involute only exists between tip circle and root circle of tooth surface of involute helical gear. According to the properties of involute, we derive $MN = r_b.u$, so the span of parameter u is as follows:

$$
u \in \left[\frac{\sqrt{r_{f2}^2 - r_b^2}}{r_b}, \frac{\sqrt{r_{a2}^2 - r_b^2}}{r_b} \right]
$$

Where r_{f2} and r_{a2} are root radius and tip radius of gear, respectively.

(2) Span of parameter λ

Parameter λ is a variable describing involute curve performs a screw motion. Distance $p_2\lambda$ along axis z_2 equals face width b_2 of worm whcel. While the coordinate plane $x_2 O_2 y_2$ (shown in Fig.1) is located in the middle section of gear, so the span of parameter λ is as follows:

$$
\lambda \in \left[-\frac{b_2}{p_2}, \frac{b_2}{p_2} \right]
$$

(3) Span of parameter ϕ_2

From Fig.1, we see that contact points only exist some zone between the left and the right of axis x when worm wheel meshs with TI worm. Assuming the working half-angle of single enveloping TI worm gearing is angle ϕ_w, the span of parameter ϕ_2 is as follows:

$$\phi_2 \in \left[-\phi_w\,,\,\phi_w\right]$$

3.2   Boundary of zone of action

The contact lines on the surface Σ^2 can't be beyond the zone determined by geometry of worm wheel. In other words, they only exist in revolving body enclosed by tip cylinder and root cylinder that are cut by two sections whose distance is a face-width of worm wheel. That's to say, points on the contact lines must satisfy the expressions as follows:

$$r_{f2}^2 \leq x_2^2 + y_2^2 \leq r_{a2}^2\,,\text{ and } -\frac{b_2}{2} \leq z_2 \leq \frac{b_2}{2}$$

4.   JUDGEMENT FOR MESHING LIMIT LINE

Position of meshing limit lines varies with variations of parameters of worm gear pair. Sometimes they go through the surface Σ^2, sometimes they don't. Because they are the envelopes of the family of contact lines, studying them is significant to study the distribution pattern of contact lines. Then judgement for meshing limit lines are discussed next.

According to Eq.(4), when $A>0$, there are two meshing limit lines on surface Σ^2; when $A=0$, only one; when $A<0$, none. Because helix angle β is the most active factor affecting position of meshing limit lines, how to select angle β is our emphasis discussed in this paper.

For Eq.(5), let $A=0$. Because $p_2 = r_b/\tan\beta_b$, $\tan\beta_b = \tan\beta \cdot \cos\alpha_t$, and

$$r_b = r_2 \cdot \cos\alpha_t = \frac{m_t \cdot Z_2 \cdot \cos\alpha_t}{2} = \frac{m_n \cdot Z_2 \cdot \cos\alpha_t}{2\cos\beta}\,,\quad\text{as\quad well\quad as}$$

$$\tan\alpha_t = \frac{\tan\alpha_n}{\cos\beta} = \frac{\tan 20^\circ}{\cos\beta}\,,\text{ we can obtain the following equations:}$$

$$\beta_1 = \mathrm{ctg}^{-1}\left\{\frac{1}{i_{21}}\left[\frac{2a}{m_n \cdot Z_2}\cos\beta_1 + \cos\left(\tan^{-1}\frac{\tan 20°}{\cos\beta_1}\right)\right]\right\} \qquad (7)$$

$$\beta_2 = \mathrm{ctg}^{-1}\left\{\frac{1}{i_{21}}\left[\frac{2a}{m_n \cdot Z_2}\cos\beta_2 - \cos\left(\tan^{-1}\frac{\tan 20°}{\cos\beta_2}\right)\right]\right\} \qquad (8)$$

Where r_b and r_2 are base radius and pitch radius of worm wheel, respectively; β_b and β are helix angle of base circle and of pitch circle, respectively; m_t and m_n are transverse module and normal module, respectively; α_t and α_n are transverse pressure angle and normal pressure angle, respectively.

Considering actual boundary condition of gear tooth and by analysis, we can get judgement for meshing limit lines as follows:

(1) When $\beta = \beta_1$ or $\beta = \beta_2$, one meshing limit line on surface Σ^2; in fact, it is a coincidence of two meshing limit lines;

(2) When $\beta_1 < \beta < \beta_2$, no meshing limit line on surface Σ^2;

(3) When $\beta < \beta_1$, two meshing limit lines on surface Σ^2;

(4) When $\beta > \beta_2$, one or two meshing limit lines on surface Σ^2.

5. MESHING PERFORMANCE ANALYSIS

Here, we take an example to analyze meshing performance of TI worm gearing and simulate the analysis results by computer. The parameters of TI worm gearing are given as follows: module $m_n = 5.5$mm; TI worm thread number $Z_1 = 2$; tooth number $Z_2 = 53$; face width $b_2 = 50$mm; center distance $a = 180$mm. According to Eq.(7) and Eq.(8), we can obtain $\beta_1 = 0.9942°$ and $\beta_2 = 7.5264°$.

Five contact lines are calculated and plotted due to five contained teeth of

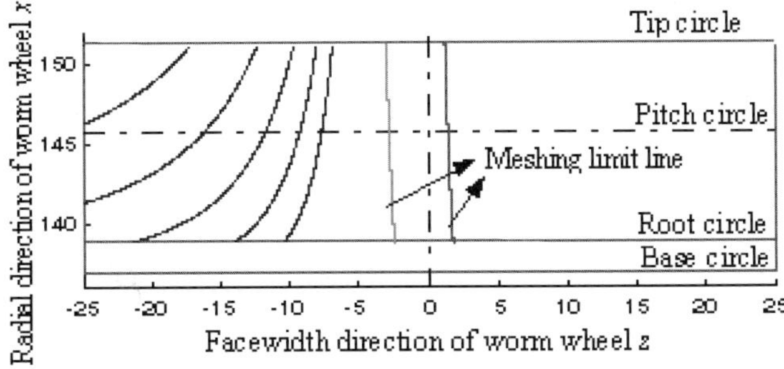

Figure 3 Contact line analysis on gear tooth surface

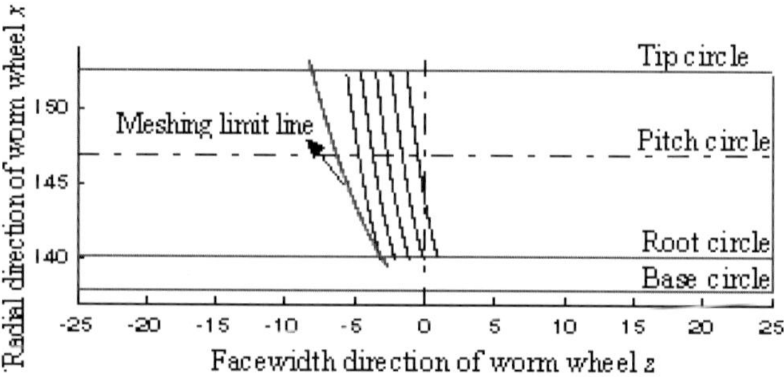

Figure 4 Contact line analysis on gear tooth surface

TI worm gearing. Not only do they show the variations and distribution patterns of contact lines on surface Σ^2 in the process of meshing, but also they display position of contact lines on different tooth surface at the same time when TI worm pair rotates a certain angle. By investigating the distribution patterns of them, we may obtain the meshing performance of worm gear pair.

When $\beta < \beta_1$, there are two meshing limit lines on surface Σ^2, contact lines enveloped by the left meshing limit line distribute on the approach side, while the ones enveloped by the right meshing limit line surpass the geometry of gear tooth surface, shown in Fig.3. Contact lines distribute sparsely, which is beneficial to dissipating heat produced in the process of meshing. But the angles between some contact lines and relative motion velocities are minor, which is adverse to lubrication. In addition, angle β should be as big as possible for TI worm gearing transmitting power, so angle β_1 is not an ideal limit point for angle β.

When $\beta = \beta_2$, contact lines approximate to parallel each other and center near the middle plane of face width of worm wheel. Their distribution is extremely uniform. This is an ideal condition, shown in Fig.4. Such a distribution benefits lubrication, also elimination of heat, and pit corrosion doesn't easily appear on tooth surface.

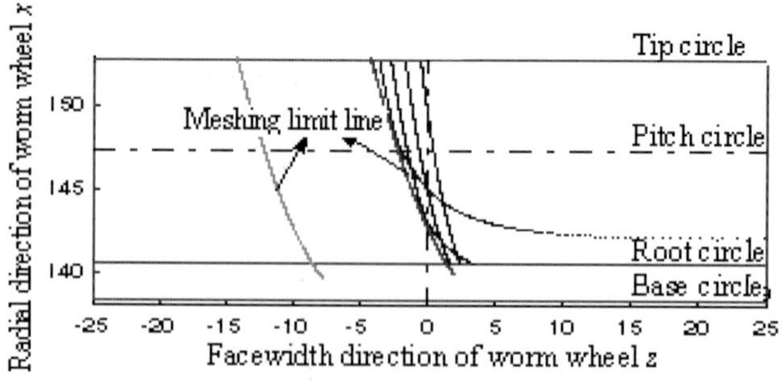

Figure5 Contact line analysis on gear tooth surface

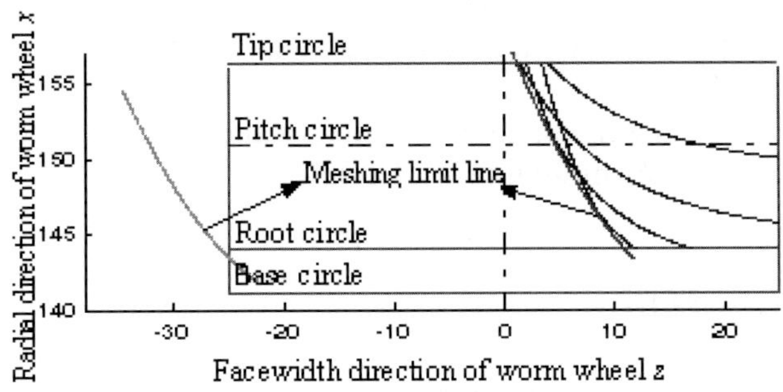

Figure 6 Contact line analysis on gear tooth surface

When $\beta > \beta_2$, there are two cases. For one thing, angle β is slightly bigger than angle β_2, here, let $\beta_2 = 8.5°$, there are two meshing limit lines on surface Σ^2 and there is a empty zone between two lines where no contact lines exist; contact lines enveloped by the right meshing limit line are on surface Σ^2, while the ones enveloped by the left are out of the geometry of gear tooth surface, shown in Fig.5. Compared with the case of angle $\beta = \beta_2$, the position of contact lines on surface Σ^2 slides towards the recess side of tooth surface, and some contact lines intersects each other near the root circle where heat isn't easily dissipated and meshing times increases one time, which will result in accelerating fatigue pitting of tooth surface.

For another, $\beta > \beta_2$, here, let angle $\beta_2 = 15°$. Then only the right meshing limit line appears on surface Σ^2 and the left is out of geometry of worm wheel. Thus the zone on the left side of meshing limit line is non-working area. The intersection of contact lines becomes serious. With meshing limit line sliding towards the recess side, the contact lines locate next to the recess side, shown in Fig.6. It's obvious that such meshing is inferior.

When $\beta_1 < \beta < \beta_2$, there is no meshing limit line on surface Σ^2, and contact lines distribute on the approach side and don't intersect each other.

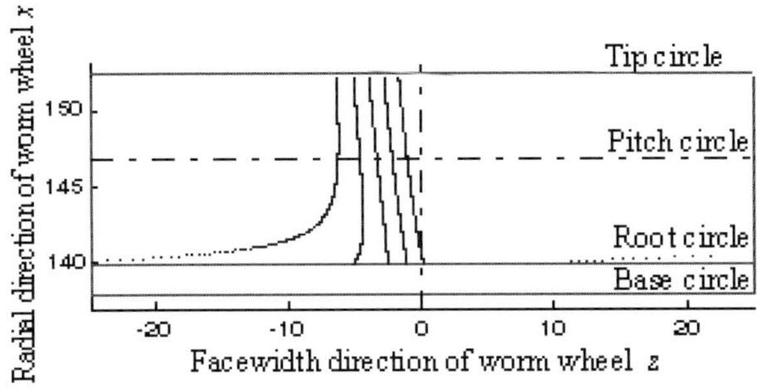

Figure 7 Contact line analysis on gear tooth surface

When angle β is slightly smaller than angle β_2, for most contact lines, their distribution is in an ideal condition, and only the less are warp and present the shape of hyperbola, shown in Fig.7, here, angle $\beta = 7°$. Considering comprehensively, we may let angle β slightly smaller than angle β_2 when it can't equal angle β_2, such meshing isn't too inferior.

6. CONCLUSION

The results of this paper are summarized as follows:

(1) In terms of variation of angle β, there can be four situations for meshing limit line on tooth surface of worm wheel, including inexistence, only one line, two lines and two lines coincidence.

(2) Two angles can make two meshing limit lines coincide with each other. The bigger one is an ideal angle for designing and making a TI worm gearing; when helix angle equals it, the best distribution of contact lines can be obtained on tooth surface; if limited by some conditions, helix angle may slightly smaller than it when they can't equal each other, but better meshing effect is still obtained.

(3) Simulation results show that TI worm gearing has many advantages, such as line contact, many teeth engagement instantaneous, lubricant angle approaching to rectangle and so on. So this gearing will have a cheerful prospect in the near future.

ACKNOWLEDGEMENT

This paper is supported by National Natural Science Foundation of China. (Project No. 50275103)

REFERENCES

1. S. Wang, D. Zhan, et al.. "Tooth contact analysis of toroidal involute worm mating with involute helical gear", *Mechanism and Machine Theory*, **vol. 37**, pp. 685-691, 2002.
2. F. Jarchow, W. Predki. "Tragfähigkeitsoptimierte Schneckentriebe", *VDI-Berichte* **Nr. 488**, pp. 11-22, 1983.
3. Maki Minoru, Okamoto Kazuhiko, Midorikawa. "Study on the hourglass worm gearing whose wheel has the helical teeth(1st report) ", *Transactions of JSME: Part C*, **vol. 61**, pp. 362-366, 1995.
4. Zhang Xikang. "Reasonable selection of helix angle for IT involute worm gearing", Chinese Journal of Mechanical Engineering, **vol. 18**, pp. 86-92, 1982 (in Chinese).
5. Wu Xutang. Theory of gear engagement, Beijing, *China Machine Press*, 1982 (in Chinese).

THE STUDY AND DEVELOPMENT OF THE PACKING SYSTEM OF TWO DIMENSIONAL IRREGULAR POLYGONS

Guolin Duan, Caihong Wang and Jin Cai

Hebei University of Technology,, Tianjin 300130,China E-mail: glduan@hebut.edu.cn

Abstract: Two-dimensional (2D) packing problem of irregular polygons is a very complicated combinatorial optimal one that can only be solved by some heuristics. This paper firstly introduces the architecture of the 2D automated packing system, then describes the procedure and methods solving packing problems using simulated annealing algorithm (SA), genetic algorithm (GA) and critical polygon method, at last, gives the conclusion of reliability and validity of the system through contrasting with a reference. At the same time, another conclusion of different packing problem should use appropriate heuristics to solve is given out.

Key words: Simulated annealing algorithm, genetic algorithm, critical polygon method, packing, irregular polygon

1.   INTRODUCTION

Two-dimensional irregular polygons packing problem is a typical combinatorial optimal one and is very complicated in computing, which arises in a variety of situations including manufacturing, apparel and packing industries, and can't be solved by exact mathematic methods but only by some heuristics. The computer-aided methods can not only save the strength of the worker but also improve efficiency and packing speed of the material. Based on the rectangle material in which the number of the irregular polygons is uncertain and the geometry is arbitrary, we design a packing system of 2D irregular polygons using Visual C++6.0 technology, which can run under the Windows environment. Heuristics are rather problem-specific, but there is no guarantee that a heuristic procedure for finding near-optimal solutions for one NP-complete problem will be effective for another. We use SA, GA and critical polygon method to solve packing problem in the system, the three algorithms can complement each other.

2.   SYSTEM ARCHITECTURE

The 2D irregular polygons packing system is made up of several modules such as man-machine interacting, initial packing data reading, automatic packing and result outputting etc. The architecture of the system is shown in Fig.1.

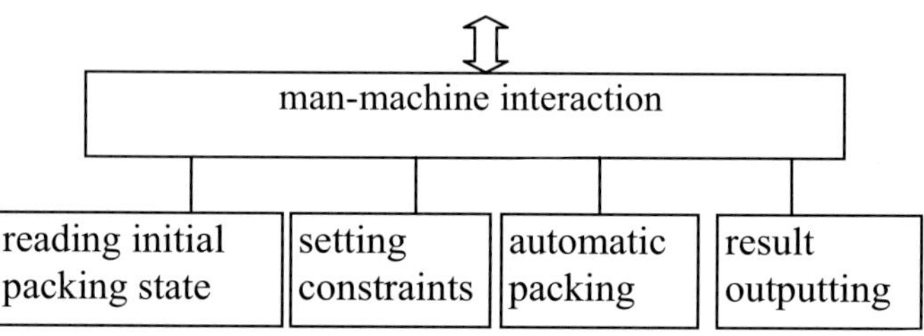

Fig.1 The architecture of the system

(1) Reading initial packing state module.
Firstly initial packing state of the packing components are shown in AutoCAD environment and saved in DXF file style, then the system can get the initial packing state by reading the DXF file.
(2) Setting constraints.
The constraints can be default or values given by user (such as the initial temperature, the termination temperature, the width of the rectangle, rotating the components or not, the rotation angle, the generation number and the initial population scale of GA).
(3) Automatic packing module.
The system can pack the components automatically without human intervention according to the two previous steps. It is impossible to solve all packing problems by one method, and three methods (SA, GA and critical polygon method) are given out in the system in order to satisfy the need of user.
(4) Result outputting module.
The packing result can be output to the screen and also can be saved as DXF file in order to be used in AutoCAD environment.

3.   SIMULATED ANNEALING (SA)

In a typical SA algorithm, the inferior state is accepted with a certain probability (Paccept), so SA can escape from the local optimum, then converge to the global optimum [1].

Some questions should be taken into account when SA is applied to solve optimization problem.

(1) The selection of Initial Temperature

If the initial temperature is too low, the algorithm can't escape from local optimum. If the initial temperature is too high, the searching time is long. So it is necessary to select a proper temperature.

(2) Annealing Schedule

The temperature in a simulated annealing algorithm is controlled by an annealing schedule. The annealing schedule specifies an initial temperature, how often the temperature should be lowered, by how much it should decrease and a final temperature. Simple annealing algorithms have a fixed annealing schedule: the user select an initial temperature, decides a constant number of iterations to be taken at each temperature. In the course of substance annealing, the process of temperature dropping is slow in case it can't form crystal lattice. However, if the temperature value of each step decrease too small, computing time is unbearable, at the same time, if the value of temperature decrease too big, the better result can't be received. So the key to use SA is the selection of proper annealing schedules.

(3) Search Pattern

We can explain search pattern as follows: algorithm changing from one state to another in feasible solution space. The search pattern has a direct effect on algorithm. Generally, in packing problem, geometrical transform is adopted to the components in order to alter the state of them [2].

(4) Multi-Objective Optimization

There are many objectives in application, such as packing density, measures of performance affected by the arrangement of components, assemble consideration and so on, so the formulation of the work presented in this paper incorporates a genetic objective function F, consisting of a weighted sum of the form [1][3]:

$$F = w_1 f_1 + w_2 f_2 + ... + w_i f_i ... + w_n f_n \quad i=1 \sim n \qquad (1)$$

Where fi is the value of the ith objective and wi is the weighted for the ith term. The weights allow the user to specify the relative importance of the multiple objectives.

4.   GENETIC ALGORITHM (GA)

GA is a search algorithm based on the theory of the evolution of species, the solutions of objective function are optimized from one generation to another according to the principles of 'Survival of the fittest' and 'Superior Survival', which are then reorganized to be new parameter combination. In the end the optimal solution can be received. The objective function corresponds to the objective function and the cost function in conventional algorithm, which at the same time can punish the solutions that disobey the constraints [4].

Generally, three operators are used in GA, namely, selection, crossover and mutation [5].

(1) Selection.

The procedure of selection is to select excellent individual from current group, namely natural selection. To a specified problem, each parameter combination (chromosome) corresponds with a function value, namely, adaptability in a specific environment. In the process of evolution, the more adaptive of an individual is, the more probably of generating an individual of next generation. So, a packing solution whose function value is better is superior to the worse one. A replication operator is adopted in selection operation and replication can ensure that an excellent individual of parent appears in the next generation. A replication operator selects an individual from current group and replicated to the group of next generation directly without any change.

(2) Crossover.

The process of crossover is to combine the features of two or more parents to generate one or more offspring, and the solution in GA is often expressed in binary string. The standard crossover operation between individuals of two parents is as follows: firstly, a point is selected from the string. Then the part before the crossover point of the first parent individual is combined with the part after the crossover point of the second parent individual to form a new offspring. Finally, the two individuals of parents are swapped to generate another offspring. Generally, it is random to select a crossover point and the goal of crossover is to produce a new combination. The offspring may have fine properties to adapt themselves to environment better. There is consistency of crossover: Firstly, the procedure of generating an inheritance mask is stochastic, the inheritance mask indicate whether a certain position is inherited information from other parents. Secondly, an individual of parent can inherit information from individual of another parent meeting the inheritance mask when crossover operation is carried out.

(3) Mutation.

A small local change is made in mutation operation, namely, a certain digit (from '0' to 1, or from '1' to '0') in some strings in a binary string changes according to definite probability. The group of species can be ensured to change constantly by importing new genes. The essence of mutation is that mistakes of binary string reproduction appear in the process of replication because of disturbance, thus one or some digits change, individuals with new properties generate. Individuals can escape from local optimum due to mutation. Apparently, a certain digit of mutation as well as several digits is stochastic. To an optimization problem, a binary string is produced at random by coding, and the corresponding function value is then calculated. To use mutation operator can ensure that the probability of a certain specified subspace is not equal to zero while searching solution space. The commonly mutation operator when solving an actual problem is to change the value of one or several digit at random from selected individuals. Then the specified string is continually selected, inherited and mutated until a certain binary string satisfy the need of the optimization problem. In a packing problem, it is the code group not the parameter itself that is operated in GA.

5.　　CRITICAL POLYGON METHOD

To apply AI (Artificial Intelligence) heuristic search technique can translate the packing problem into another problem of searching the optimal route in state space. The efficiency to get better packing result can be enhanced by simulating packing state, gaining packing experience of manpower, adopting heuristic to control the search direction, and restricting the search space.

A raw material area and packing components are given at initial state, a proper component is chosen being arranged on the raw material, thus an initial node is produced and the useable raw material area changes. According to the new state, another component with given constraint is chosen and then a new code is generated. By circulating, a tree is formed when all the components are packed. So, packing problem is translated into finding an optimal route problem by searching the tree [6 –7].

At any time in the process of packing, firstly N packing components are selected according to the constraints, then the availability of raw material after packing the components is calculated, one packing method whose availability of raw material is higher is chosen as a packing style to proceed. Circulating above steps until all the components are packed.

(1) Critical polygon

Parts are expressed as polygons and we can get the polygon that contains all the packing components at any time during the arrangement. So to arrange a new component is to seek the optimal position that two polygons contact but don't overlap each other.

Definition 1

Polygon A whose position is decided is given, polygon B makes a circuit of A without changing its direction and keeps contact with A but not overlaps during the circuit, the received referenced point of B is a polygon, too. Then this polygon is called as a critical polygon B is relative to A.

(2) Composite polygon

The polygon of covering all the packing components need to be sought and the procedure of packing begin from the left bottom according to the constraints.

Definition 2

The polygon formed by raw material boundary and all the arranged components is called critical polygon.

After the position of a new component is decided, the problem of seeking the new critical polygon is changed to another problem of seeking contact point between the new component with the boundary of raw material and the previous critical polygon, then the contact point is handled by operation of pairing, sorting and merging.

(3) Packing availability evaluation function

At a certain time of packing, the critical polygon Ai formed by the arranged components Pi has been known, the accumulation of the material waste is as follows:

$$ADW(A_i) = AREA(A_i) - \sum AREA(P_i) \tag{2}$$

The availability of material at current time is:

$$P(A_i) = 1 - ADW(A_i)/AREA(A_i) \tag{3}$$

Where $AREA(A_i)$ is the area of critical polygon Ai and $\sum AREA(P_i)$ is the area sum of the arranged components.

Using critical polygon method can arrange components in a rectangle material automatically whose number and kind are uncertain by controlling the search depth and search wide according to the different requests on material availability and arrangement speed. This method is flexible and can basically meet the requirements of balance between solution and running time, at the same time the speed is enhanced and material saved.

6.　　EXAMPLES

The system is tested with three algorithms in order to compare efficiency and reliability with each other. Twenty-five 2D irregular polygons of Stefan Jakobs are packed in paper, the width is 17 by GA, and another is 16 by contraction method. Twenty-five 2D irregular components used by Stefan Jakobs[8] are layout in this paper with SA. See Fig. 2.

The initial space usage 0.1868	The final space usage 0.6379
Initial length 51.5000	The final length 15.0823

Fig.2 The result of packing with SA

The parameters is as follows:

(1) The Initial Temperature: An adaptive annealing schedule has strong adaptability, which uses statistical information about the objective function values of the visited states to control the temperature. So we use the following equation to get the initial temperature.

$$T_{start} = -\frac{3\sigma}{\ln p}\,(p = 0.85) \qquad (4)$$

Where σ is the standard deviation of objective function values (obtained by taking a random walk through the state space and accepting every move).

(2) Annealing Schedule: Tnew is controlled by the standard deviation of the function values whose state is already known during the calculating process. As the temperature drops, the distribution of objective function values changes, therefore, n new value of σ is calculated at each temperature.

The temperature is updated using the following equation:

$$T_{new} = T_{current} * e^{\frac{-0.7T_{current}}{\sigma}} \tag{5}$$

(3) Stopping Criteria: We adopt the following equation:

$$T_{end} = \alpha(\alpha = 0.05) \tag{6}$$

Where T_{end} is the final temperature.

The result attained by GA is shown in Fig.3.

The schedule of seeking optimum is as follows:

(1) Selection: The new individual generated by crossover and mutation operations is probably inferior to its parent, so the individual of parent as a part of offspring takes part in the selection procedure in case that degradation occurs, thus the most excellent solution is ensured not to lose. The solution group of offspring is sorted according to solution quality, the one whose required number is ahead of others is regarded as new parent to form new solution group of parent.

(2) Crossover: Multi-point crossover is adopted in this paper, for example, there are two individuals A and B: A ={a1, a2,···, an}; B ={b1, b2,··· ,bn}, the number of crossover is j, j<n; j points are selected as base and i points are selected as base points, if j=3; i=3,6,7; then the new individuals sum after crossover are : A' ={a1, a2, b3, a4, a5, b6, b7, a8, ··· ,an}; B' ={b1, b2, a3, b4, b5, a6, a7, b8, ··· ,bn}.

(3) Mutation: The mutation operation is realized by swapping the two packing irregular polygons at random.

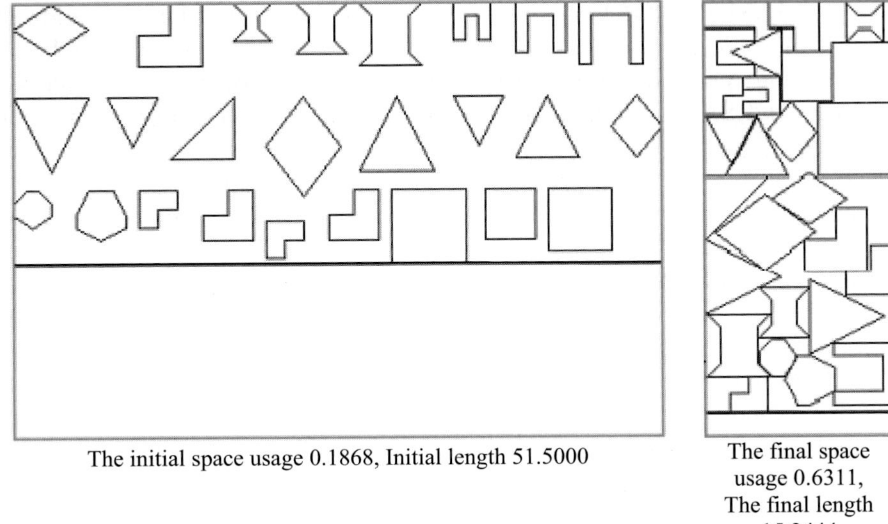

The initial space usage 0.1868, Initial length 51.5000

The final space usage 0.6311, The final length 15.2444

Fig.3 The result of packing with GA

In critical polygon method, the total number of packing components is m=25, every four （N=4） components form a group, then there are 6 groups, namely, n=6(there are five components in last group), each four components are arranged 4 ! =24 times. See Fig.4.

The initial space usage 0.1868, Initial length 51.5000

The final space usage 0.5378, The final length 17.9434,

Fig.4 The result of packing with critical polygon method

7. CONCLUSIONS

Packing problems have been generated extensively in engineering domain and SA is an effective method to solve irregular polygon packing problem. The packing result with SA is the best among above three algorithms.

GA has a wide range of application and can receive more high calculation effectiveness. The selection and crossover operations can improve the individual evolution, while mutation operation keep population a great diversity, the solution space can be searched in a more range at the same time, so the optimal result is excellent.

Critical polygon method has no requirements on the initial packing state of the components because it has no effect on the result, so all the packing components can be overlapped each other completely. The packing result with critical polygon method is inferior to SA and GA because it is combined in a local optimal state.

Because of the complexity of the packing problems, there is no method that is almighty to all of them. So a proper algorithm should be adopted in terms of the specified packing problem. This is why we integrate above three methods with each other in our system.

REFERENCE

1. Szykman S, Cagan J: "A simulated annealing-based approach to three-dimensional component packing". *Transaction of the ASME*, **Vol.117**, pp.308-314,1995.
2. Su Yin, Jonathan Cagan: "An extended pattern search algorithm for three-dimensional component layout". *Transaction of the ASME*, **Vol.122**, pp. 102-1082000.
3. S.zolfghari, M.Liang: " Machine cell/part family formation considering processing times and machine capacities: A simulated annealing approach". *Computer and Engineering*. **1998,34**: 813-823.
4. Ni Weili, Zeng Lin:. "Genetic algorithm and its application for absorbing coating optimization". *Journal of Shanghai University (Natural Science) (in Chinese)*, **Vol.6**, pp. 243-248,1997.
5. Zheneg Haoran, Cao Xianbin, Zhang Jun: "A-multi pattern ecology evolutionary algorithm based on creature's procreation strategy "(in Chinese). *Mini-Micro System*, **Vol.21**, pp.466-468,2000.
6. Luo Wei: "The research and design of computer system for automatic layout of two-dimensional irregular shapes CATLS" (in Chinese). *Computer Engineering*, **Vol.21**, pp. 3-9,1995.
7. Zhang Quanhuo, Fan Huilin: "Design and implementation of a computer-automated layout system (I) ". *Journal of Huaqiao University(Natural Science)* (in Chinese), **Vol.18**, pp.102-106,1997.
8. Stefan Jakobs: "Theory and methodology on genetic algorithms for the packing of polygons". *European Journal of Operational Research*, **Vol.88**, pp. 165-181,1996.

INVENTIVE DESIGN OF CONTINGENT ANTENNA WITH TRIZ

Tuan-Jie Li and Jian-Yuan Jia
School of Electromechanical Engineering, Xidian University, Xi' an 710071, China
E-mail: ltj801@sohu.com

Abstract: TRIZ, the Russian acronym for Theory of Inventive Problem Solving is emerging as a powerful scientific tool that helps decision-makers to make strategic decisions. This paper describes application of TRIZ to design a Contingent antenna. Firstly, the technical contradictions and the principles for resolving technical contradictions were introduced, and the physical contradictions and the classical ways to resolve physical contradictions were discussed. Then a universal process of inventive design based on TRIZ is presented. Finally, as a case study, the inventive design method proposed in this paper was applied to the design of a contingent antenna required to not only have the specified electric capability but also be easy to carry and operate. The contradictions existing in this innovative problem have been found, and all the creative solutions are obtained through solving the problem using the corresponding inventive principles to eliminate these contradictions. The novel projects of contingent antenna are presented, one of which, a telescopic antenna has been manufactured in factory, and the prototype has been tested. The results indicate the new-style contingent antenna can work well and operate easily under the harsh environments, and satisfies the requirements of the electric capability and emergency.

Key words: TRIZ, Inventive design, Creative solutions, Contradiction solution techniques, Inventive principles, Contingent antenna

1.   INTRODUCTION

Solving problems facing 21st century society demands creativity and innovation. Product innovation is the key for enterprise succeeding in the market competition. The product development process may be classified as four stages, i.e. product definition, conceptual design, technical design and detailed design. The core of innovation is to produce novel concepts or work principles with strong market dominance in the stage of conceptual design.

The conventional method of generating new concepts is the trial and error method, similar to the artistic invention depending mainly on afflatus and experiences. The recent researches show TRIZ is a science that allows creative problems in any field of knowledge to be revealed and solved, while developing creative (inventive) thinking skills and a creative personality.

TRIZ, the Russian acronym for Theory of Inventive Problem Solving is emerging as a powerful scientific tool that helps decision-makers to make strategic decisions, TIPS is the corresponding English acronym. TRIZ was developed by Genrich Altshuller and his colleagues in the former USSR starting in 1946 [1], and is now being developed and practiced throughout the world.

TRIZ research began with the hypothesis that there are universal principles of invention that are the basis for creative innovations that advance technology, and that if these principles could be identified and codified, they could be taught to people to make the process of invention more predictable. The research has proceeded in several stages over the last 50 years. Over 2 million patents have been examined, classified by level of inventiveness, and analyzed to look for principles of innovation. The decades of research by Altshuller and his students and colleagues have made it possible people world-wide to become systematic creative thinkers.

TRIZ is a unique knowledge-based technology for accelerated development of design concepts. The power of TRIZ is based on the understanding of the evolution of successful products, generalization of the ways used to solve problems in the most innovative inventions, and ways to overcome psychological barriers. Classical TRIZ is described in books and papers by Genrich Altshuller, the creator of TRIZ, and his followers [2-3].

Three ideals structure TRIZ theory: 1) the ideal design is the desired end result, 2) contradictions assist problem solutions, and 3) innovation can be an organized systematic process. TRIZ moves innovative thought away from being an art and toward a science. In a word, TRIZ is an inventive scientific methodology, which can make the process of thinking and solving problem more scientific, and offers the potential for the most potent solution. TRIZ has proved to be a very strong tool in helping to solve difficult technical problems that require inventive thinking [4-5], that means problems where one or more technical contradictions are involved and which do not have known ways or means of solution. TRIZ has been used for problem solving in business and in other organizations for more than 20 years. Large and small companies are using TRIZ on many levels to solve real, practical everyday problems and to develop strategies for the future of technology.

TRIZ is in use at Ford, Motorola, Procter & Gamble, Eli Lilly, Jet Propulsion Laboratories, 3M, Siemens, Phillips, LG, and hundreds more [6].

2. CONTRADICTIONS AND SOLVING METHODS IN TRIZ

A basic principle of TRIZ is that a technical problem is defined by contradictions. That is, if there are no contradictions, there are no problems. This radical-sounding statement forms the basis for the TRIZ problem solving methods that are fastest and easiest to learn. In traditional engineering thinking, if there is a conflict, it is common to accept a trade-off. TRIZ thinking leads to analysis of the conflict and to solution satisfying both conflicting requirements. TRIZ aims to produce new breakthrough solutions through eliminating or resolving the contradictions, that is we often talk about "design without compromise".

TRIZ defines two kinds of contradictions, "Technical" and "Physical". The benefit of analyzing a particular innovative problem to find the contradictions is that the TRIZ patent-based research directly links the type of contradiction to the most probable principles for solution of that problem.

2.1 Technical contradictions and solving principles

Technical contradictions are the classical engineering "trade-offs". The desired state cannot be reached because something else in the system prevents it. In other words, when something (or parameter A) gets better, something else (or parameter B) gets worse. Classical examples include
- The product gets stronger (good) but the weight increases (bad).
- The bandwidth increases (good) but requires more power (bad).

The technical contradictions can be eliminated or resolved using the contradiction matrix [7]. The matrix tells you which of the 40 inventive principles [2] have been used most frequently to solve a problem that involves a particular contradiction, and is one of the effective and visual tools of TRIZ. The idea of the matrix and its initial development was created and worked out by Genrich Altshuller in the process of research of about 40,000 patents. Altshuller figured out 39 types of engineering parameters on the matrix's axes "Undesired Effect" versus "Feature to Improve" and 40 types of inventive principles. Find the row that most closely matches the feature or parameter you are improving in your "trade-off" and the column that most closely matches the feature or parameter that degrades. The cell at the intersection of that row and column will have several numbers. These are the identifying numbers for the principles of invention that are most likely, based on the TRIZ research, to solve the problem: that is, to lead to a breakthrough solution instead of a trade-off.

2.2   Physical contradictions and solving principles

The concept of physical contradictions is one of the cornerstones of TRIZ. Physical Contradictions are situations where one object has contradictory, opposite requirements. Everyday examples abound:

• When pouring hot filling into chocolate candy shells, the filling should be hot to pour fast, but it should be cold to prevent melting the chocolate.

• Software should be easy to use, but should have many complex features and options.

If the problem is better expressed as a physical contradiction (where one parameter has opposite requirements) rather than a technical contradiction, then the contradiction matrix won't work− it has no entries on the diagonal, so you can't look for "X gets better but X gets worse".

The physical contradictions are resolved with the help of so called "separation principles". TRIZ has 4 classical ways to resolve physical contradictions using the separation principles. These ways traditionally involve use of separation in time, space, upon condition, and between parts and the whole. Each of these may likewise be related in turn to the 40 inventive principles of TRIZ as illustrated in Table 1 [8].

Examination of the 40 principles shows extensive overlap with these 4 methods, since they are based on the same research on the same collection of innovative solutions to a wide variety of problems.

Applying the 4 methods for resolving a physical contradiction will cause us to deal with the cause of the problem and not just with the short-term solution.

Table1. Relationships between the separation principles and the inventive principles

Separation principles	Inventive principles
Separation in space	1, 2, 3, 4, 7, 13, 17, 24, 26, 30
Separation in time	9, 10, 11, 15, 16, 18, 19, 20, 21, 29, 34, 37
Separation upon condition	12, 28, 31, 32, 35, 36, 38, 39, 40
Separation between parts and the whole	1, 7, 25, 27, 5, 22, 23, 33, 6, 8, 14, 25, 35, 13

When using the TRIZ research findings, in general the most comprehensive solutions come from using the physical contradiction formulation, this is why resolving the physical contradiction is regarded as a more general solution than resolving the technical contradiction.

The Separation Principles from the TRIZ methodology can be used to deal with both technical and non-technical problems [9].

3. PROCEDURE OF INVENTIVE DESIGN

The 40 inventive principles and contradiction elimination of TRIZ techniques are good methods for solving engineering innovative design problem with system contradictions. TRIZ does not generate breakthrough solutions by "better brainstorming" or by teaching people to "think creatively". In dealing with contradictions TRIZ generates breakthrough solutions by giving you the tools to find the problem behind the problem, and remove it. The general procedure for inventive design with TRIZ is referred to Figure 1.

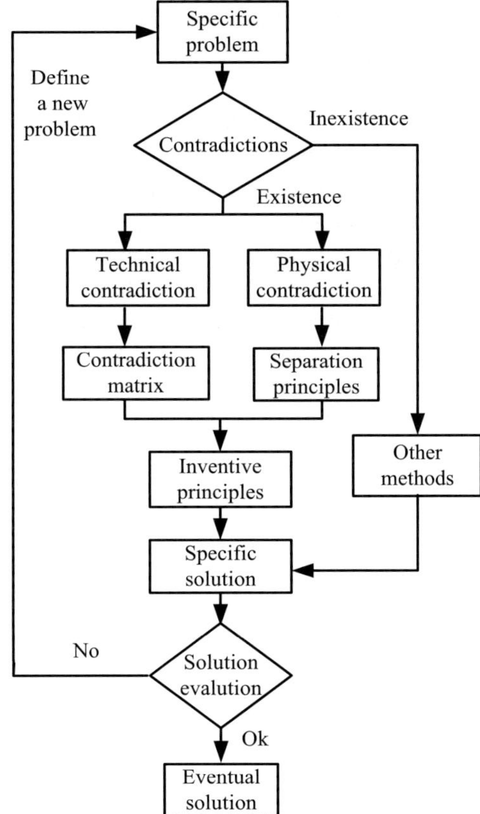

Figure 1. General procedure for inventive design with TRIZ

4.   CASE STUDY: CONTINGENT ANTENNA

The antenna is a metallic apparatus for sending or (and) receiving electromagnetic waves. According to the anti-damage requirements of the electronic warfare, it is very important to insure the information detectability against destroying, and an important means of improving the damage-resistant abilities is to enhance the regenerative abilities.

4.1   Analysis of the problem

An alerting radar can easily be hit the needle because its reflector, antenna base, antenna feeders and so on lie in the immobile and exposed conditions. Once it is destroyed, the information achieving abilities should be resumed as quickly as possible. Its reflector is a paraboloid incised antenna, and inconvenient to be stored, conveyed and installed because of its bigger surface size. So it can not serve as an emergency apparatus.

If the antenna dimension is reduced, it is easy to carry and operate. However, this will lead to the electric capability of the antenna getting worse. For example, the width of the major lobe increases, the antenna directivity gets worse, the antenna gain decreases, the electrical level of the minor lobe increases, and leakage power sharpens, etc. So one dominant physical contradiction for the antenna system is the size of the antenna should be small but it should be big.

4.2   Creative solutions

Since this is a contradiction, the answer will not be a number-that would be the non-TRIZ way of doing a trade-off. Applying the 4 methods for resolving a physical contradiction will cause us to deal with the cause of the problem and not just with the short-term solution.

Careful examination of four separation principles, it is not difficulty to recognize that separation in time and between parts and the whole should be used in this situation.

● Separation in time

With the help of the separation in time we can find the corresponding inventive principles from Table 1 as follows.

(1) Principle 15. Dynamics

A. Allow (or design) the characteristics of an object, external environment, or process to change to be optimal or to find an optimal operating condition.

B. Divide an object into parts capable of movement relative to each other.

C. If an object (or process) is rigid or inflexible, make it movable or adaptive.

D. Increase the degree of free motion.

The creative solutions could be induced from the principle 15. For example, The antenna framework could be changed from the truss structure to a multi-bar telescopic mechanism. The antenna lies in contractive state with smaller dimension when it is stored, and conveyed, and the antenna is expanded with bigger aperture area when it has been installed.

(2) Principle 16. Partial or excessive actions

A. If 100 percent of an object is hard to achieve using a given solution method then, by using "slightly less" or "slightly more" of the same method, the problem may be considerably easier to solve.

The creative solutions could be derived from the principle 16. For example, because the brim of antenna calibre has a smaller effect on the antenna radiation capability, the antenna dimension may be minimised properly. Although the electric capability of the antenna could become worse slightly, the deadweight and the wind resistance to the antenna will be consumedly reduced. The surface accuracy of the antenna is definitely improved.

(3) Principle 20. Continuity of useful action

A. Carry on work continuously; makes all parts of an object work at full load, all the time.

B. Eliminate all idle or intermittent actions or work.

The creative solutions could be obtained by the aid of principle 20. For instance, the section of the antenna radiation pattern is an ellipse, the energies radiated on the four corners of the reflector are very fewer. For the sake of illumination uniformity on the reflector, the reflector shape can be changed into an ellipse through cutting out of the four corners. The results are that the utilization factor of the antenna aperture area increases, the antenna side lobes reduce, and the wind resistance decreases.

(4) Principle 29. Pneumatics and hydraulics

A. Use gas and liquid parts of an object instead of solid parts (e.g. inflatable, filled with liquids, air cushion, hydrostatic, hydro-reactive).

The creative solution could be gained with the help of principle 29, such as the air-antenna with very lighter weight. Its dimension is very smaller after air exhausted and easy to convey, and can be pumped out and puffed out time after time. However, The main disadvantages are the lower precision and worse rigidity.

• Separation between parts and the whole

With the help of the separation between parts and the whole we can find the corresponding inventive principles from Table 1 as follows.

Principle 1. Segmentation

A. Divide an object into independent parts.
B. Make an object easy to disassemble.
C. Increase the degree of fragmentation or segmentation.

The creative solution could be obtained by the aid of principle 1. For instance, the antenna can be divided into some sections that are easy to convey and store. The sections can be knocked together to form a whole reflector.

4.3   Eventual solutions

There are several solutions to this problem, other than those mentioned above. Not all solutions are as good as others. The "best" solution is often some combination of several other solutions. The resulting solutions to this problem are as follows.

Figure 2. The expanding-contracting antenna

One is the expanding-contracting antenna using a multi-bar extending mechanism as the antenna framework. It has been manufactured in factory, the prototype has been developed, as shown in Figure 2, and the confirmation test has been conducted. The results indicate this type of contingent antenna can work well and be operated easily under the execrable circumstances.

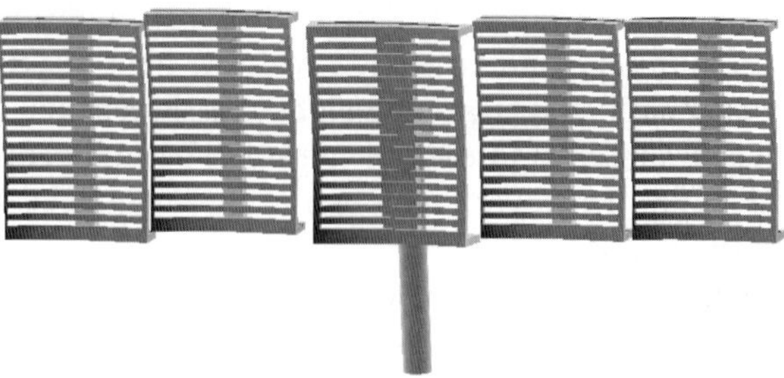

Figure 3. The mosaic antenna

The other is the mosaic antenna, as illustrated in Figure 3, in which the whole reflector is constructed by some sections knocked together.

5.   CONCLUSIONS

This paper describes application of TRIZ to design a Contingent antenna. The novel projects of contingent antenna are presented, one of them has been manufactured in factory, the prototype has been developed, and the confirmation test has been conducted. The test results closely matched with the expected purpose. The case study shows how powerful TRIZ is. We can conclude from the work done in this paper as follows.

(1) TRIZ is a valid method for solving real world product design problems. New breakthrough concepts that eliminate contradictions can be found routinely by applying TRIZ technology. If we could have used TRIZ at the beginning of our studies to solve the product design problems, then we would have greatly decreased the time and energy spent through trial and error method.

(2) Resolving the physical contradiction is a more general solution than resolving the technical contradiction.

(3) In dealing with contradictions TRIZ generates breakthrough solutions by giving the tools to find the problem behind the problem, and remove it.

(4) We should look for the integration of different design tools and methodologies to increase product design effectiveness and productivity in the future.

REFERENCES

1.   G. S. Altshuller: *Creativity as an Exact Science*, Gordon &Breach, New York 1988.
2.   E. Domb and K. Tate: "40 Inventive Principles with Examples". *TRIZ Journal*, July 1997.
3.   R. Gennady: "40 Inventive Principles in Quality Management". *TRIZ Journal*, March 2003.
4.   L.R. Noel: "A Proposal to Integrate TRIZ into the Product Design Process". *TRIZ Journal*, November 2002.
5.   R. Graham: "TRIZ and Design". *TRIZ Journal*, December 2002.
6.   C. Denis and L. Philippe: "Intuitive Design Method (IDM), a New Approach on Design Methods Integration". *TRIZ Journal*, October 2000.
7.   D.S. Semyon: "A few Words about the Altshuller's Contradiction Matrix". *TRIZ Journal*, July 1997.
8.   D. Mann and R. Stratton: "Physical Contradictions and Evaporating Clouds". *TRIZ Journal*, April 2000.
9.   J. Hipple: "The Use of TRIZ Separation Principles to Resolve the Contradictions of Innovation Practices in Organizations". *TRIZ Journal*, August 1999.

ANALYSIS OF COMPACTION PROCESS BETWEEN IMPACT COMPACTING MACHINE AND ROADBED

Qian Liu, Jin-Gang Wang,Gen-Qun Cui Yi-Min Wu and Zhi-Ge Zhou
School of Mechanical Engineering, Hebei University of Technology, Tianjin 300130,China,
liuqiann@sina.com

Abstract: The non-linear soil mechanics model based on Mohr-Coulomb yield criteria is deducted, and impaction effect of non-column wheels of compacting machine is calculated using MSC/NASTRAN. The comparison between calculation and experimentation indicates that the simulating calculation satisfied the engineering precision and it can replace experimentation to measure the compaction effect. In the meantime, the impaction effect of two kinds of wheels is simulated and several factors are discussed.

Key words: Compacting machine, roller, FEM, Non-linear.

1.   INTRODUCTION

With the rapid development of the large-scale highways construction, the engineering machine—road roller used for impacting the roadbed receives more and more attention. The traditional kinds of rollers have low efficiency in compaction and with them the roadbed must be dig many times then compacted layer and layer. In order to improve the work efficiency, we developed a new type roller—impact compacting machine (ICM). The roller of the compacting machine is steered by means of a tractor and will punch the roadbed soil. The roller has two non-column petallike wheels. When the wheels are rolling along with the tractor, their centers of gravity will move cycloidally. Because the difference between the maximal and the minimal radius of the wheel's curvature is very large, its center vertical translations will change greatly. With the wheel's huge inertia the roller performs the impaction to roadbed soil. The impaction effect is related to the wheel's velocity and the outline shape. In order to receive the most ideal impaction effect, we must study the difference on compaction between different curves, and then we can choose the best. If we compare them through experiment, many time and money are needed. In this paper, we simulated the impacting process between the wheel and the roadbed soil with FEM method.

The simulation must make certain the behavior of soil under forces from outside. Being a building material, soil has the longest history. But till today, the character of soil has not been drastically mastered yet because of its complexity. This paper developed the soil model of mechanics based on Mohr-Coulomb Yield Criteria, and calculates the strains and stresses in everywhere of the roadbed soil with nonlinear FEM by MSC/NASTRAN; the results are rather satisfying. The example testified that the computer simulation could replace practical experiments efficiently, so the money and time cost on experiments will be drop down wonderfully.

2.   THE SOIL NON-LINEAR FEM MODEL

2.1   The soil mechanics model

The stress-strain function of the ideal materials based on Mohr-Coulomb yield criteria is shown [1, 2] as followed:

$$\frac{1}{3}I_1 \sin\varphi + \sqrt{J_2}\,\sin\!\left(\theta + \frac{\pi}{3}\right) + \frac{\sqrt{J_2}}{\sqrt{3}}\cos\!\left(\theta + \frac{\pi}{3}\right)\sin\varphi - C\cos\varphi = 0 \quad (1)$$

Where: $\varphi -$ internal frictional angle, $C -$ cohesion, $I_1 -$ the first invariant of the stress tensor, J_2、$J_3 -$ the second and third invariants of the deviator, $\theta -$ can be deduced from $\cos 3\theta = \sqrt{2}\,J_3 \big/ \tau_8^3$, $\tau_8 -$ octahedral shear stress.

The soil elasto-plastic model could be defined based on soil mechanic properties. When the soil strain is in the elastic range or unloaded, the stress-strain relation can be expressed as followed:

$$\Delta\theta_{ij} = \frac{\Delta I_1}{9K}\delta_{ij} + \frac{1}{2G}\Delta S_{ij} \quad (2)$$

or:

$$\Delta\sigma_{ij} = K\Delta\theta_8\delta_{ij} + 2G\Delta\varepsilon_{ij} \quad (3)$$

Where: K − bulk modulus, G − shear modulus, $\Delta\theta_{ij}$ − strain increment, $\Delta\sigma_{ij}$ − stress increment, $\Delta\varepsilon_{ij}$ − increment of strain deviator invariant, ΔS_{ij} − increment of stress deviator invariant, $\Delta\theta_8$ − octahedral strain increment, δ_{ij} − Kronecker[3] delta symbol.

The above soil model is used in MSC/NASTRAN calculation.

2.2   The non-linear FEM equation

The FEM method could deal with non-linear problems. The translation, stress, strain all abides by the mechanic relations of compatibility, equilibrium and physics. The non-linear FEM equation in total strain style[4] is shown as followed:

$$ku = f \tag{4}$$

or in component strain style:

$$k_t \Delta u = \Delta f \tag{5}$$

Where: k − secant stiffness matrix, u − displacement vector, f − load vector, k_t − tangent stiffness matrix, Δu − the step size in the calculation.

k, k_t and f are functions of u, ε and σ .Through a series of linear processes, the approximate values will be got. The relationship of k and u is used in function (4), the equation becomes the followed style:

$$[k(u)]u = f \tag{6}$$

or:

$$[\Delta k(u)]\Delta u = \Delta f \tag{7}$$

Where: Δk − increment stiffness matrix.

The u in the above equation is unknown. In order to solve the nonlinear equation, we can select an approximate value to receive k and Δk , then return to the above equation. In this way, the non-linear equation will be changed to linear equation, and after n times calculations, u_n generally is not the precision. So we define:

$$e_r = f - [k(u_n)] \qquad u_n \neq 0 \tag{8}$$

Where: e_r is the error that decides the solution exact degree. The iterative destination is to decrease the error and find an ideal approximation.

3.   FEM ANALYSIS OF COMPACTION PROCESS

3.1   Analysis and simplify

The structure of the compacting roller is shown in figure 1, and the wheel's whole working cycle is shown in figure 2. Figure 2a shows the wheel's initial statement, in which the wheel's center of gravity is in the low

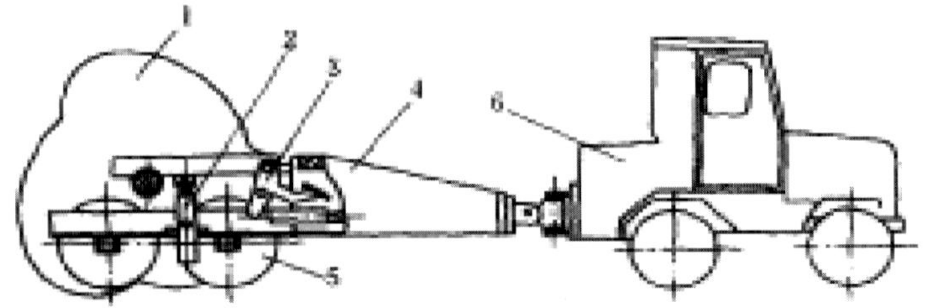

Figure 1. The structure of the compaction machine

1. compaction wheel   2.raising frame   3.cousion frame   4.machine frame   5.drive wheel   6.tractor

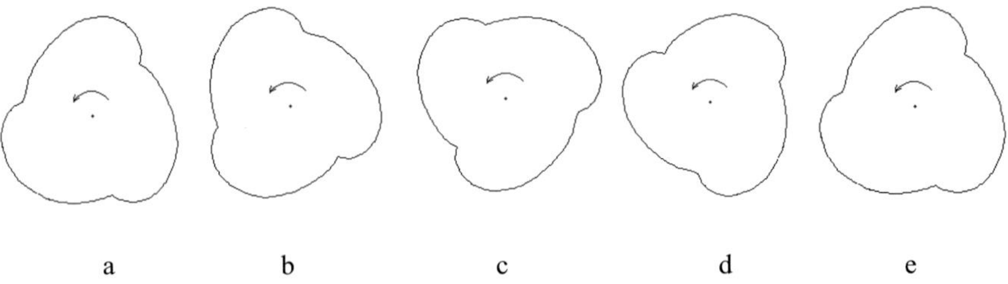

Figure 2. The compaction wheel's working cycle

position; figure 2b shows the wheel begins to rolling and the center position changes from the low to the high, this is a process for accumulating energy; figure2c shows the center has reached the high position and the wheel has the most potential energy; figure 2d shows the wheel drops down from the high to the low to perform compacting process; figure 2e shows the wheel return the initial statement and the whole working cycle is over. The interaction between the wheel and the soil is very complicated. Firstly, when the center of gravity of the wheel drops down to impact the roadbed, hundreds of tons force will be produced to make the roadbed much distortion, which belongs to nonlinear problem; secondly, the soil of the roadbed is a particular kind of material, in which the stress and strain relationship is differ from other materials; thirdly, the reciprocity between the wheel and the soil belongs to contacting problem, in the beginning, the wheel compacts the soil with one line, then with area, and the area is larger and larger. Based on above all, the FEM software- MSC/PATRAN, MSC/NASTRAN with powerful pre and post treatment performance is used to analyze this problem.

3.2   Soil and wheel FEM model

The soil is a kind of solid that extends infinitely except upwards. We take a hexahedron, which is 2 meters thick, 3 meters wide, and 4 meters long, for instead. Five surfaces of the hexahedron except the top are fixed to simulate the roadbed with soil in this direction. The whole elements of the soil are defined octahedron that meets non-linear requirements. Because the properties of soil taken from different depth are different, we must define the different characters of every layer of soil.

The wheel is made up of several boards acting as shells; their FEM model is quite simple. When meshing, we use quadrangular and triangular elements. Figure 3 shows the FEM model.

3.3   Particular transition element

It is a very complicated process in which the soil is compacted by the wheel, so we used the element GAP (clearance element) supported by MSC/NASTRAN to describe the interaction model between soils and wheel.

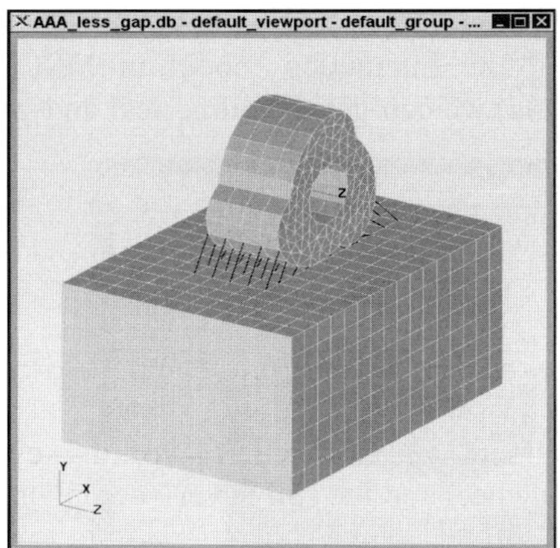

1.   *Figure 3*. The whole FEM model of soil and wheel

There are two kinds of internal force in GAP: one is axial stress; the other is transverse friction. When two points come into conjunction, a large value is used to describe the close stiffness, and the shearing strength is not bigger than static friction –product of axial stress and static friction factor. When two points are separating from each other, a little value is used to describe the expanding stiffness, and the shearing strength is disappearing.

3.4   The interaction FEM model of soil and wheel

The interaction between soil and wheel belongs to contacting problem. From the wheel begins to reaching the soil to the soil gets the most sedimentation, the direction and values of the interface force are varied, so the initial conditions are difficult to define. But when the wheel begins to impact the soil, the rotating rate of the wheel can be deduced by the tractor's running velocity. By this way, the soil sedimentation caused by the wheel's inertia force can be calculated.

The FEM model of the wheel, soil and the interaction of them are defined respectively with the initial conditions, the whole FEM model including soil and wheel is received as shown in figure 3.

4.   4. THE COMPARISON BETWEEN CALCULATIONS AND EXPERIMENTS

4.1 The comparison and analysis between simulations and experiments

After calculation of the interaction model in MSC/NASTRAN with modified Newton method, we can get the numerical and graphical results of

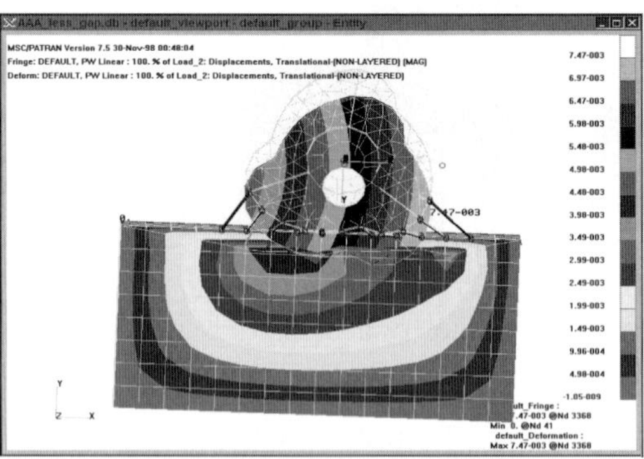

Figure 4. The soil vertical distortion

soil distortions shown in figure 4. Figure 5 showed the comparison between calculation and experiment of the soil's sedimentation. The soil's displacement curve get from the experiment is smooth. The simulation corresponds to the experiment curve mostly except that some positions have deviation such as A. there are two reasons: one is that divided elements are not so small that the continuity of the calculation is not good; the other

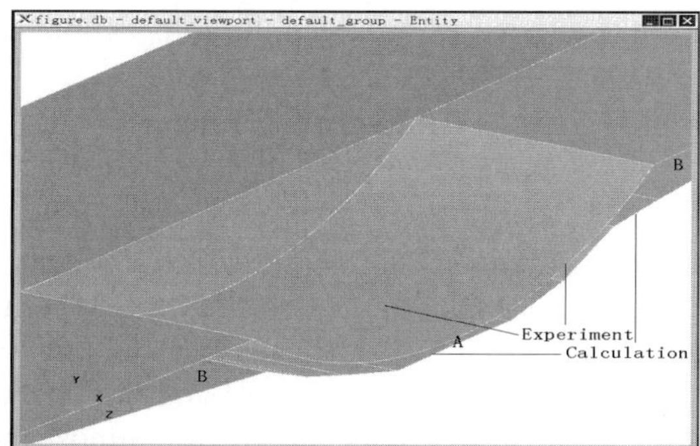

Figure 5. Comparison between calculation and experiment

reason is the effect of the GAP. The nodes connected to the GAP have remarkable distortion, but the other nodes have only little distortion because they didn't connect to the GAP. Because of the correlation between nodes, the curve in B district has produced remarkable distortion.

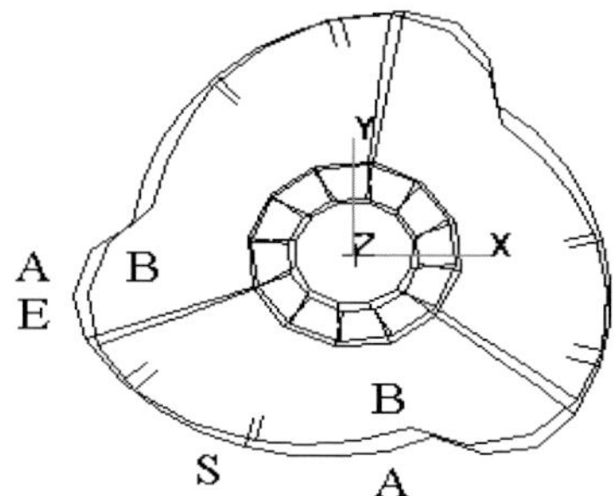

Figure 6. Two wheels with different outline shape

In order to estimate the differential from the calculation and the experiment, we used the ratio of the two distortion volumes of the soil to define the error. The error is R :

$$R = \left|1 - \frac{V_1}{V_2}\right| = \left|1 - \frac{0.2659}{0.2441}\right| = 10.893\% \tag{9}$$

where: $R-$ error, V_1- distortion volumes calculation of soil, $\quad V_2 \quad -$ distortion volumes experiment of soil.

It is high in absolute value, but for the soil with so complicated properties, the error still meets the engineering requirements. So the non-linear FEM method is efficiency in simulating the compaction effect.

4.2 The comparison of two different wheels with different outline shape

There are two kinds of wheels with different outline curves shown in figure 6. A is made in South Africa; B is designed by Hebei university of technology and trial-manufactured by Tangshan Special Automobile Factory China. From figure 6, we know that the maximal and minimal radius of the two wheels is almost the same, and their weighs are almost the same. But curve of wheel A is smoother than B. for wheel B, the transition from the maximal radius to the minimal radius or from minimal to maximal is very rapid. Under the same speed and initial conditions, the compaction calculations of the two wheels have been done by the FEM method, and the comparison of compaction effect between two different wheels is shown in figure 7. From this figure, we can know that the compaction effect of the wheel with A is better than that of B. this indicates that if the maximal radius moves to minimal one rapidly, the curvature is large and the soil first connected to the wheel will produce larger distortion while the soil later

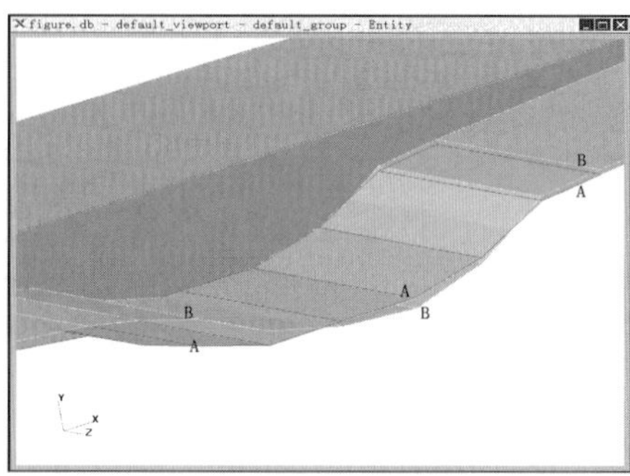

Figure 7. Comparison of compaction effect of two different wheels

connected to the wheel will produce little distortion relatively. The little the distortion is, the poor the compaction effect is. So when we design the new compaction rollers, we should make the front range of the wheel curve smoother to get more compaction areas and make the rear increase the curvature to get more compaction depth.

5.   CONCLUSIONS

In this paper, the impaction effect of non-column wheels of compacting machine is calculated using MSC/NASTRAN. The comparison between calculation and experimentation indicates that the simulating calculation satisfied the engineering precision. Therefore, the computer simulation can replace practical experiments efficiently, so the money and time cost will be drop down wonderfully. By simulating the compaction effect, we also found the wheel's optimum outline shape. In conclusion, being a new way to handle soil mechanics problems, FEM is feasible and efficient.

REFERENCES

1.   Parry, R.H.G. " Stress-Strain Behavior of Soils", *Proc. Roscoe Memorial Symposium,*
2.   Cambridge, May.14-18,1971,pp.110-115.
3.   Oda, J. "Application of the Point Matching Method to the Three-Dimensional Elastic
4.   Contact Problems", *Bull. J.S.M.E.*, **vol.17.** , pp. 1129-1134, 1974.
5.   Lysmer J. "A Finite Element Method for Seismology,   Method in Computational
6.   Physics", *Academic Press*, **vol.11.** , pp. 181-216, 1972.
7.   The MacNeal-Schwendler Corporation, Robust Design. The CAE Challenges, '97
8.   MSC China User's Conference Proceedings, Beijing, China, Oct.10-15,1997, pp.1-8.

RESEARCH ON MODELING FOR RECYCLING STRATEGY OF WASTE PRODUCTS

Haihong Huang, Zhifeng Liu, Shuwang Wang, Guangfu Liu and Weixiang Guo

P.O. Box 2, School of Mechanical and Automobile Engineering, Hefei University of Technology, 193 Tunxi Road, Hefei, Anhui 230009, China

Abstract: To describe the recycling process scientifically, IDEF3 method is adopted to establish the process model. A conclusion that recycling strategy decision-making is the most crucial step of recycling analysis can be made from the model. The decision-making model is presented based on the metrics of reuse, service, remanufacturing, and disassembly. The information model is established by analyzing the informational requirement of the decision-making model. Finally, a case study of recycling the waste air-conditioners is given.

Key words: recycling strategy; decision-making model; life cycle; End-of-life

1.   INTRODUCTION

Manufacturing, as the stanchion of contemporary civilization of industry, have extremely accelerated the development and advancement of human society and enriched both material and spiritual civilization. But the quantity of waste products is increasing steadily under the current manufacturing mode without recycling the castoff, along with the improvement of the level of consumer and the high speed of technical renovation. The fact that resources and energy have been excessively depleted has become an austere actuality mankind have to face [1]. According to the related research results, there are about 4 millions refrigerators, 5 millions washing machines and 5 millions television sets will be out of use now only in China [2]. If the waste products composed by the diversiform materials are directly sent to landfill or incineration without proper disposal, those will pollute the atmosphere, soil, and water resources and waste the resources in the castoffs at the same time. To alleviate the burden of environment and resource, many nations distributes the laws early or late to control the manufacturing industry to reduce the depletion of resources and resolve the environmental problem. On one hand business enterprises have to enhance the usage efficiency of resources and energy and decrease or eliminate the harm to the environment during the process of production and usage of products; on the other hand the manufacturer must take the responsibility of recycling their products [2].

Recycling the waste products can not only alleviate the pressure on the environment, but also save limited energy and resources through improving the efficiency of usage of materials and products. Nowadays, there are two main approaches; one is the design for recycling, which takes the recycling strategies of products at the stage of recycling into consideration at the stage of design [3], and the other is the research on the techniques of disassembly and recycling, which aims to lower the cost, heighten the profit and lessen the environmental impact of recycling by adopting the optimal and reasonable recycling technologies. The former makes the product recyclable aboriginally; and the latter is a kind of remediation of current manufacturing to some extent, which is discussed mainly in this paper.

Currently, Recycling technology of waste products has attracted steadily increasing attention in the world. Researches have mainly focused on the planning of disassembly, recycling processes assessment of specific products, recycling technologies analysis and recycling devices development etc. To acquire maximum economical profit and engender minimum environmental pollution in the recycling process, many unit models of recycling analysis and relative tools have been developed [4 - 5]. While the decision-making of recycling strategies has been the bottleneck of recycling industrialization, this paper researches on the recycling strategy model to make the recycling economical feasible and environmental friendly and presents the information requirement of the decision-making model.

2. RECYCLING PROCESS ANALYSIS

2.1 Recycling strategies

The waste products to be recycled must be inspected firstly to find out the basic usage information. Recycling decision-making is essential after the inspection, which points out the depth of disassembly and the parts to be disassembled and decides the recycling strategies of products [6]:

Reuse: Reuse is the second hand usage of the parts or components as originally designed. At this scenario, it has no environmental impact brought by manufacturing and disposal; and it economizes the resources and decreases the cost of manufacturing at the same time.

Service: Parts or components or products can be reused through servicing simply or substitution of parts. It is another effective way to extend the life of the products by repairing or rebuilding the components or parts.

Remanufacturing: Remanufacturing is a process in which reasonably large quantities of similar products are brought into a central facility and disassembled. Parts from the products need be remachined, which are assembled on the assembly line of the same or other products.

Material recycling: Material recycling reclaims materials with or without disassembly. Recycling with disassembly mostly adopts manual methods to separate the components. Disassembly into material fractions increases the value of the materials recycled by removing material contaminants, hazardous materials or high value components. Recycling without disassembly uses mechanical shredding technologies reduce the size of compound materials of products or components which then are separated using separation methods based on magnetic, density or other properties of the compounds of materials.

Disposal: This end-of-life strategy is to landfill or incinerate the products or components or parts with or without energy recovery.

2.2   Recycling process

Pre-treatment such as cleaning the products and eliminating the liquid or noxious substances before the recycling is necessary to wipe off the external contamination and appurtenance influencing the recycling. For instance, refrigeration mixture in the refrigerator such as CFCs must be removed at the stage of pre-treatment to avoid the second pollution.

Disassembly, as one of the most important end-of-life technologies, aims to obtain higher benefit, since economic profit becomes the most crucial factor to drive the recycling industry. Firstly, the parts or components reused, remanufactured or serviced are separated from the product by manual or destructive disassembly. Planning of disassembly path has been discussed much including full and aimed disassembly [2]. Secondly, the parts or components containing noxious materials are disassembled to avoid farther pollution and assure the physical healthiness of worker. Thirdly, parts comprising single material are separated to obtain the pure recycled material that has higher value.

Material recycling, as an effective way for material regeneration, is adopted when farther disassembly operation is not economical. Firstly, the products totally or the components are shredded with the mechanisms to the size requested by the following separation technologies that are used to separate the compound comprising several different materials. The usual shredding technologies include milling, cut, clash, and low temperature freezing shredding etc. The usual separation technologies include wind power separation, griddle separation, whirling electricity separation, magnetic force separation and floatation separation etc. The recycling process planning and analysis have been discussed [7]. Then the separated single material or compatible compound materials are regenerated, and the remains without value of regeneration are disposed.

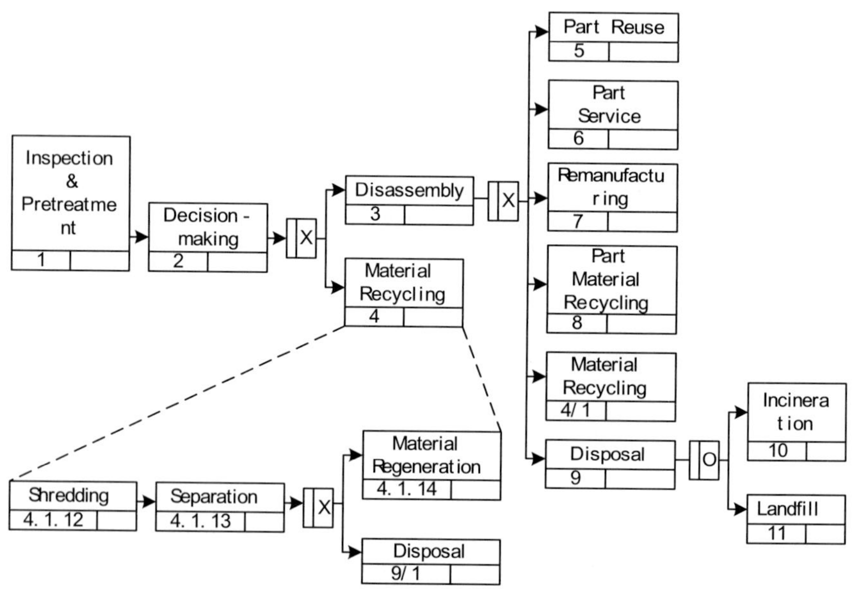

Figure 1. Recycling process flow chart (IDEF3)

To illustrate the process of recycling of waste products, IDEF3 method is adopted to establish the recycling process flow chart shown as in figure 1. The OSTN chart and the specific explanation of all the units of behavior are omitted as limited to the length of this article.

3.   RECYCLING STRATEGIES DECISION-MAKING

Decision-making, as the most crucial step in recycling as demonstrated in Figure 1, determines the recycling strategies of the products.

3.1   Several metrics

The *reuse value removal rate* (R.R) denotes the reuse capability of parts and components defined as in Equation (1). Reuse is economically viable when the dismantler is able to remove the components or parts with high reuse value quickly. If the reuse value is higher and the time of disassembly is shorter, the R.R is higher, which is used to be the metric to decide whether the specific parts or components are reused.

$$R.R = \frac{V_{reuse}}{T_{disassembly}} \tag{1}$$

V_{reuse} is the reuse value of the parts or components, and $T_{disassembly}$ that is same as the following is the time spending on dismantling the parts or components to be reused from the products.

The *service value removal rate* (R.S) denotes the service capability of parts and components defined as in Equation (2). Service is economically viable only when the reuse value is high, the service cost is low, and the components or parts are able to be dismantled quickly. The reuse value is higher and the service cost is lower and the time of disassembly is shorter, the R.S is higher, which is used to be the metric to decide whether the specific parts or components are serviced.

$$R.S = \frac{V_{reuse} - C_{service}}{T_{disassembly}} \tag{2}$$

$C_{service}$ is the service cost of the parts or components to be serviced.

The *remanufacturing value removal rate* (R.M) is adopted to denote the remanufacturing capability of the components or parts, which is defined as in Equation (3) to be used as the metric to decide whether the parts and components to be remanufactured.

$$R.M = \frac{V_{use} - C_{remanufacturing}}{T_{disassembly}} \tag{3}$$

V_{use} is the value of the remanufactured parts or components; $C_{disassembly}$ is the cost of remanufacturing. The R.M is controlled by the three factors: the value, the cost, and time of disassembly. Remanufacturing is economically viable only when the use value is high, the cost is low, and the components or parts are able to be dismantled quickly.

The *disassembly value removal rate* (R.D) is used to denote the disassembility of the parts and components shown as in Equation (4).

$$R.D = \frac{M_{material}V_{material}}{T_{disassembly}} \tag{4}$$

$M_{material}$ is the weight of material, and $V_{material}$ is the value of the unit material. The time of disassembly, the material weight and the value of the materials determine the cost and economic viability of disassembly. The time is shorter, the weight is lighter, and the value is higher, then the R.D is higher. The R.D is used to be one of the metrics to decide whether the parts or components to be dismantled. Following two aspects must be taken into consideration together with the R.D metric:

One is the material compatibility and currently known separation technologies. It is an effective recycling strategy to dismantle the part manually and recycle and regenerate the material solely if the material of the part is incompatible with other materials of the product and the material compound is hardly separated with current separation technologies when the product is shredded to apropos size. Although metal is incompatible with plastic, known technologies such as magnetic force separation and gravity separation can separate the two materials easily. While for the incompatible plastic compound comprising PP and ABS, there are no effective technologies nowadays to separate the two, and then dismantles the part or component with PP or ABS is a reasonable way.

The other is toxicity of materials and harmful substances of the products. The parts or components containing the toxic materials or noxious substances have to be dismantled manually to farther disposal, as that do harm to humans and contaminate the environment and other components.

3.2 Decision-making model

Recycling strategies analysis aims to obtain maximum economic profits as well as the environmental profits, which is useful not only at the stage of recycling to determine the recycling strategies but also at the stage of design to analyze the recycling capability. Based on the above discussion, the decision-making model is illustrated in Figure 2, which is mainly from the viewpoint of recycling using the several metrics mentioned above.

4. INFORMATION MODEL

An information model has to be established to provide the recycling strategies decision-making model the essential data, as that requires information involving all the life cycle of products.

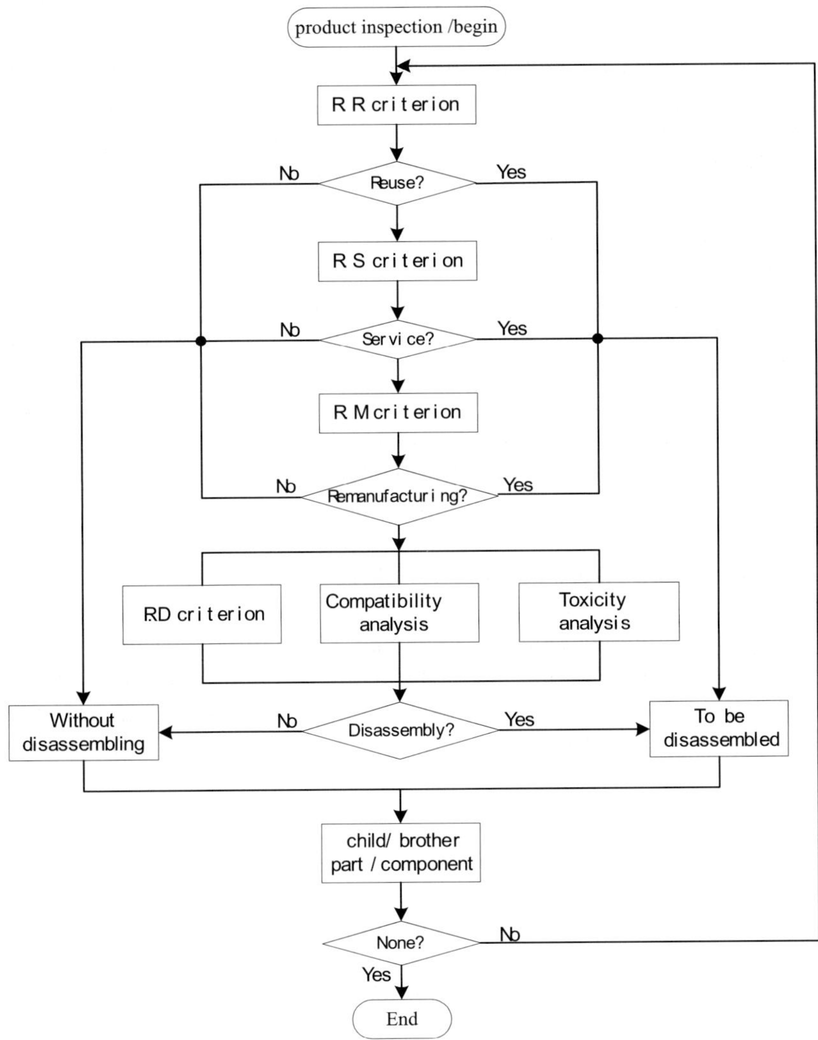

Figure 2. Recycling strategies decision-making model

4.1　Information requirement

The decision-making model requires the data including the following:
✓　Design data: Design data mainly include the design life cycle of the products and components, variety of materials, configuration, size, shape, and weight of products etc. These data, as the basic information, is vital to the recycling strategies analysis.

✓ Structure data: Structure data include hierarchy of assembly of products, assembly relation and constraint relation between the parts, and variety of fastener etc.

✓ Parts data: Parts data include the basic information of parts such as type, shape, weight, position, and material. These data is pivotal to support the plan of disassembly, toxicity analysis, compatibility analysis, and plan of separation technologies.

✓ Usage data and service data: Recycling capability may be changed in the stage of usage, as the uncertain factors such as work environment and cycle and users always influence the data of products. Especially, substitution of parts, components addition or subtraction, and structure change at the stage of service must be recorded. These data is useful to the strategies decision-making of reuse, service and remanufacturing.

✓ Assistant data: Assistant data include the data required by the calculation of reuse value, service cost, and remanufacturing cost and benefit, as well as the information of disassembly time and value of materials etc.

4.2   Establishing information model

To acquire the above data, Bras linked qualitative measures of re-manufacturability to engineering information embodied in CAD systems and accordingly developed a Computer Aided Design tool [8]. Klausner have investigated embedded sensors to monitor the conditions of possible reused component [9].

With the development of CAD software, it is an effective way to embody the data in CAD systems. The information model of recycling strategies decision-making is presented in Figure 3. The visible model called as "hierarchy-net" based on our former researches expresses the hierarchy of assembly with tree structure as a whole and the relation of assembly with the net graph in detail at each level of the tree. The hierarchy-net is the abstract reflection of the assembly that presents both connection relations between different parts at the same hierarchies and subordinate relation between the different hierarchies [2].

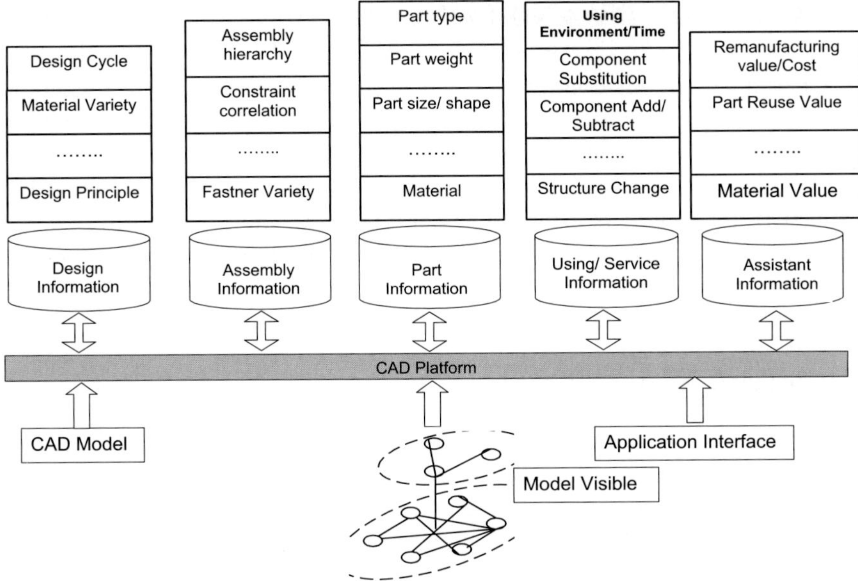

Figure 3. Information model of recycling strategies decision-making

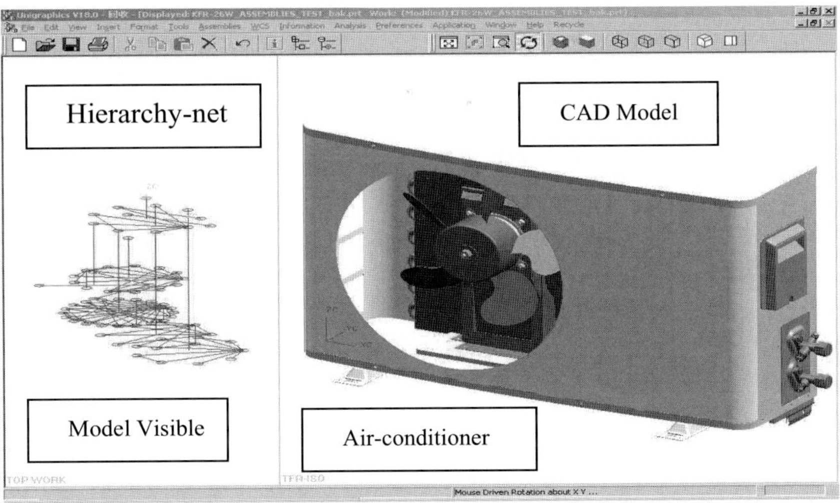

Figure 4. Information model of air-conditioner recycling strategies decision-making

5. CASE STUDY AND CONCLUSION

A certain waste air-conditioner was used in our case study. The detailed analysis processes are omitted, which shows that the decision-making model works well. The information model based on UGII is shown as in Figure 4.

The recycling strategies of the air-conditioner go: the compressor is serviced to reuse that is aimed disassembled, and the refrigerant CFCs are disposed to be harmless. The rest is shredded totally, and some separation technologies such as gravity separation, magnetic force separation are used to separate the compound comprising copper, steel and plastic.

ACKNOWLEDGEMENTS

The authors would like to thank all the collaborators and colleagues from the Institute of GD&M Engineering in HFUT, and thank the sponsor of this research, the National Natural Science Foundation, China (Granted No. 50375044), the Excellent Youth Foundation of Anhui Province and the Natural Science Foundation of Anhui Province (Granted No. 01044105).

REFERENCES

1. Kosuke Ishii. "Design for Environment and Recycling: Overview of Research in the United States", *http://mml.stanford.edu/Research/Papers/1998/1998.CIRP.LCES.ishii*
2. Pan Xiaoyong. "Analysis and Research on Key Technology of Product's Disassembly under Three–dimension Environment", *PhD Thesis*, HFUT, pp. 1-30, 2003.
3. A. Kriwet etc. "Systematic Integration of Design-for-Recycling into Product Design", *International journal of production economics*, **Vol. 38**, PP. 15-22, 1995.
4. Stewart Coulter etc. "Designing for Material Separation: Lessons from Automotive Recycling", *ASME Journal of Mechanical Design*, **Vol. 120**, pp. 501-509, 1998.
5. Catherine M. Rose, and Kosuke Ishii etc. "Product End-of-Life Strategy Categorization Design Tool", ***Journal of Electronics Manufacturing***, **Vol. 9.1**, pp. 41-51, 1999.
6. Catherine M. Rose, and Ab Stevels. "Metrics for End-of-Life Strategies", *2001 IEEE International Symposium for Electronics and the Environment Conference*, Denver, CO, May 2001. http://mml.stanford.edu/Research/Papers/2001/papers01.html
7. Wang Shuwang etc. "Study on Materials Recycling Process Planning Based on Agent", *Computer Integrated Manufacturing System*, **Vol. 9**, pp. 1001-1005, 2003.
8. B. Bras. "Integrated Product and De- and Remanufacture Process Design", *Proc. NSF Design and Manufacturing Grantees Conference*, Monterrey, Mexico, 1998, pp. 67-68.
9. M. Klausner, and W. M. Grimm etc. "Sensor-Based Data Recording of Use Conditions for Product Takeback", *Proc. IEEE International Symposium for Electronics and the Environment*, Oak Brook, Illinois, 1998, pp. 138-143.

DEVELOPMENT OF COMPUTER AUTO COLOR MATCHING DESIGN SYSTEM FOR TOOTH REPARATION

Dan Jiang[1], Gaoting Wu[2] and Lancheng Wang[3]

School of Mechanical and Power Engineering, Shanghai JiaoTong University,No. 1954, Hua Shan Rd., jiangdan@sjtu.edu.cn [1,2]; The Department of Information Management, Nanjing Political College, No. 2053, Huangxing Rd, wanglancheng@163.com [3]

Abstract: In the process of tooth reparation, the color reparation has become a very important part. So it is necessary to create a computer auto color matching design system, which will help the teeth sufferers to get natural and brilliant teeth efficiently. In this research Tristimulus Values Color Matching Method (TVCMM) is adopted for the particularity of tooth reparation and color material. The emendation algorithm of controlling Same Color with Different Spectrum is used to improve precision of color matching.

Key words: Computer color matching, tristimulus values, tooth reparation

1.   INTRODUCTION

Along with the raising of modern material life and cultural level, the aesthetic level of people also rises continuously. For modern people, tooth reparation is not only for the demand of its function, but also for beauty. The color of tooth reparation has become one of the most important judge standards in oral cavity reparation. Natural, lifelike, overall collocation and harmonious is a consistent goal that patients and the doctors pursue together.

Porcelain powder has been a kind of the material of tooth reparation in common use [1]. It is one of crucial factor whether the color of artificial tooth is nature, lifelike and concordant match with neighbouring tooth. But the disadvantages of traditional colorimetry by eyes can often not make the results satisfied.

Therefore it has been the direction of tooth reparation technology to induct the basic color concept into oral cavity reparation field, research and control various influences on color match between reparation body and natural tooth by computer. This paper synthesized the relative factors influencing tooth color matching, studied the color matching algorithm and algorithm optimization in computer auto color matching design system.

2. THE BASIC COLOR MATCHING METHOD OF TOOTH REPARATION

2.1 The condition of model establishment

In the tooth color database established in advance, series A in VITA colorimetric board and near colors are added because of their pertinence at Chinese people [2]. The simplified model can reduce the difficulty in research and make for constructing system frame.

2.2 The choice of color matching methods

In general, the method of computer matching colors can be divided into tristimulus values and full spectrum [3], [4].

Tristimulus values computer color matching method is established by Allen based on the foundation of Park and Steams. Its basic principle is to mix some kinds of pigments to match standard color, make the difference of Tristimulus values between match color and standard color minimum. Like in (1).

$$(\Delta X, \Delta Y, \Delta Z) \to (0,0,0) \tag{1}$$

Tristimulus values method has a lot of obvious advantages. The primary one is that the aberration of match result is 0 under the invariance light source. And color matching software development difficulty is less in Tristimulus values method. But its obvious shortcoming is the aberration value greater under other light sources. So in this paper controlling same color different spectrum is used in computer color matching system.

Full spectrum color matching method is to directly fit spectrum curve to match color, make the difference of the spectrum curves between match color and standard color minimum. Like in (2). Here, $\omega\,(\lambda_j)$ is a weight factor. $\Delta R(\lambda_j)$ is the difference of spectrum reflectivity between goal color and match color on wavelength λ_j.

$$\sum \omega^2(\lambda_j)[\Delta R(\lambda_j)]^2 \to \min \tag{2}$$

Full spectrum color matching needs to compare the reflection curve of whole spectrum, so matching precision relative higher. It is suitable for 3 or more color materials mixed to match color. But entire system is more complex, consume longer for its algorithm.

Tristimulus values method is suitable for 2 or 3 kinds of color materials to match colors. Entire system is simpler than full spectrum. The research in this paper is based on Tristimulus values method according to the requirement of porcelain powder mixing method and system feasibility.

2.3   The major algorithm of color matching

In clinical oral cavity reparation, 2 or 3 kinds of color porcelain powders are usually used to mix to reach requirement of patient. This system mixes 3 kinds of porcelain powder according to certain proportion based on the mathematic model with Kubelka-Munk principle. Porcelain powder density matrix is given with Tristimulus values color matching method [2], [5]. Like in (3).

$$c = (TED\phi)^{-1} TED\left[f^{(a)} - f^{(t)}\right] \tag{3}$$

The visible lights are defined in the scope of 380-760nm, spacing 10nm. Matrix and parameters are defined as follows:

$$T = \begin{bmatrix} \overline{x}_{380} & \overline{x}_{390} & \cdots & \overline{x}_{760} \\ \overline{y}_{380} & \overline{y}_{390} & \cdots & \overline{y}_{760} \\ \overline{z}_{380} & \overline{z}_{390} & \cdots & \overline{z}_{760} \end{bmatrix} \qquad \overline{x}, \overline{y}, \overline{z} \text{ are the match functions}$$

$$E = \begin{bmatrix} E_{380} & 0 & \cdots & 0 \\ 0 & E_{390} & \cdots & 0 \\ \vdots & \vdots & & \vdots \\ 0 & 0 & \cdots & E_{760} \end{bmatrix} \qquad Ei \text{ is spectrum power distribution}$$

$$
f^{(a)} = \begin{bmatrix} f(R)^{(a)}_{380} \\ f(R)^{(a)}_{390} \\ \vdots \\ f(R)^{(a)}_{760} \end{bmatrix}
\qquad
f^{(t)} = \begin{bmatrix} f(R)^{(t)}_{380} \\ f(R)^{(t)}_{390} \\ \vdots \\ f(R)^{(t)}_{760} \end{bmatrix}
$$

R is the spectral reflectance of porcelain powder. The superscript an expresses tooth color sample, and superscript t expresses matching color result, in which, in which, $f(R) = (1-R)^2 / (2R)$.

$$
D = \begin{bmatrix} d_{380} & 0 & \cdots & 0 \\ 0 & d_{390} & \cdots & 0 \\ \vdots & \vdots & & \vdots \\ 0 & 0 & \cdots & d_{760} \end{bmatrix}
\qquad \text{where } d_i = \left[dR/df(R) \right]_i
$$

$$
\phi = \begin{bmatrix} \phi^{(1)}_{380} & \phi^{(2)}_{380} & \phi^{(3)}_{380} \\ \phi^{(1)}_{390} & \phi^{(2)}_{390} & \phi^{(3)}_{390} \\ \vdots & \vdots & \vdots \\ \phi^{(1)}_{760} & \phi^{(2)}_{760} & \phi^{(3)}_{760} \end{bmatrix},
\qquad
c = \begin{bmatrix} c_1 \\ c_2 \\ c_3 \end{bmatrix}
$$

Φ is the unit value of k/s of porcelain powder. The subscript is wavelength, and another mark is the order number of porcelain powder. c_1, c_2, c_3 represents the density of porcelain powder.

After the first matching is completed by formula (3), iterative method is used to improve to make calculation result more approaching matched result till in tolerance. Iteration revises formula likes in (4).

$$
\Delta c = (TED\phi)^{-1} \Delta t \tag{4}
$$

where Δc and Δt are defined as follows:

$$
\Delta c = \begin{bmatrix} \Delta c_1 \\ \Delta c_2 \\ \Delta c_3 \end{bmatrix}
\qquad
\Delta t = \begin{bmatrix} \Delta X \\ \Delta Y \\ \Delta Z \end{bmatrix}
$$

$\Delta c_1, \Delta c_2, \Delta c_3$ are the iteration correction of density. $\Delta X, \Delta Y, \Delta Z$ are the differences of Tristimulus values between matching color and matched color.

3. THE ALGORITHM OF IMPROVING PRECISION

3.1 The color matching emendation

In the most conditions, the differences are always existed between the sample colors and matching colors under specific light sources. For improving the precision of color matching, it is necessary to control the difference of Tristimulus values under reference illumination body between colors as possible. The emendation calculation is made first for calculating the index of same color different spectrum under different illumination body.

The difference of Tristimulus values between the sample color and the result color mixed with porcelain powders is supposed to 0 in calculating index of same color different spectrum under different illumination body by formula (3). So multiply emendation for matching results is adopted before ΔE is calculating.

$$f_X = X_1/X_2 \qquad f_Y = Y_1/Y_2 \qquad f_Z = Z_1/Z_2$$

Tristimulus values of matching result 2 are X_2', Y_2', Z_2' under tested illumination body. The formula of Tristimulus values after emendation likes in (5).

$$X_2'' = f_X X_2' \qquad Y_2'' = f_Y Y_2' \qquad Z_2'' = f_Z Z_2' \qquad (5)$$

ΔE is calculated by the results above for measuring same color different spectrum of these two kinds of results.

3.2 The control of same color different spectrum

The index of same color different spectrum MI is an important parameter used to evaluate the effect [6]. It means that the color difference ΔE_1 is 0 under reference D65 of light source for a pair of ideal match samples. ΔE_2 and ΔE_3 under test light sources CWF or A are used to define the indexes of same color different spectrum MI_2 and MI_3. In actual application, it is difficulty to make $\Delta E_1 = 0$. So the index of same color different spectrum of matching sample under 3 kinds of light source can suppose as following:

$$\text{D65:} \qquad MI_1 = \Delta E_1 \tag{6}$$

$$\text{CWF:} \qquad MI_2 = \{\Delta E_1^2 + [\Delta E_2 - \Delta E_1]^2\}^{1/2} \tag{7}$$

$$\text{A:} \qquad MI_3 = \{\Delta E_1^2 + [\Delta E_3 - \Delta E_1]^2\}^{1/2} \tag{8}$$

The total index of same color different spectrum is used to evaluate the matching result. Suppose:

$$MI^2 = MI_1^2 + MI_2^2 + MI_3^2 \tag{9}$$

Combining (6), (7), (8) into (9), yields the formula (10).

$$MI^2 = 5\Delta E_1^2 + \Delta E_2^2 + \Delta E_3^2 - 2\Delta E_1 \times \Delta E_1 - \Delta E_1 \times \Delta E_1 \tag{10}$$

After coordinate, yields the formula (11).

$$MI = \left[\Delta E_1^2 + \Delta E_2^2 + \Delta E_3^2\right]^{1/2} \tag{11}$$

MI is very important for color matching precision. The less MI value is, the lower the grade of same color different spectrum is, the better the effect of matching color. Therefore the system realizes the grade control of same color different spectrum in color matching program and improve the precision of color matching.

4.   CONCLUSION

This paper researches the computer color matching application on tooth reparation. The conclusion is as follows:

(1) Tristimulus values color matching method is validated and realized to be suitable for computer color matching in tooth reparation.

(2) Precision controlling method is studied to solve the problem of same color different spectrum in Tristimulus values color matching.

ACKNOWLEDGEMENTS

Acknowledgements go to Dr. Zhu, the Ninth People Hospital, for collaboration in the field of medical application.

REFERENCES

1. JW. Xu: *Oral Cavity reparation*, People Sanitation Press, Beijing 1999.
2. SQ. Tang, *Chromatics*, Beijing Science and Technology University Press, Beijing 1990.
3. XC. Wang, FK. Zhou: "The Compare of Tristimulus values and Full Spectrum Color Matching" . *Optics Precision Engineering*, **Vol.7, No.2**, pp 13-16, 1999.
4. E. Allen: "Basic Equations used in Computer Color Matching [J]". *Opt. Soc. Am*, **Vol.56**, pp. 1256-1259, 1996.
5. R. Kuehni: *Computer Color Formulation, lexington Book[M]*, M. A. Lexington 1975.
6. A. Jungman, J. Conno: "Linear and Quadratic Optimization Algorithms for Computer Color Matching". *Color Res App*, pp. 40-42, 1988.

STRUCTURE DESIGN FOR AUTOMATIC CONVEYING

Yinghou Lou[1], Yanan Jiang[2] and JinLiu[3]

[1,2]*Department of Mechanical , University of Ningbo, Ningbo, P.R.C. 315211*
louyinghou@email.nbu.edu.cn
[3] *Department of Mechanical , University of Shanghai, Shanghai, P.R.C 20072*

Abstract: Automation is inevitable for manufacturing. With the perfection of machining equipment, automatic conveying becomes the key factor of manufacturing system automatization. There are two ways for realizing automatic conveying. One way is to increase the capabilities of transport devices, and the other is to improve the performances of transported objects. The former has been making great efforts and therefore much progress has been made. But now machine intelligent is still very poor, which results in complex equipment and vast investment. The later could be achieved by changing product structure partly, proving product conveyance performance, reducing demand for transport facility, and simplifying its structure. Studies in this aspect is overlooked, however it is very efficient for little input and more output. Based upon above, an idea is presented, realizing automatic conveying by changing product structure properly under the precondition of no influencing product function, improving product conveying performance, and reducing demands for facility. A method for product structure design is also proposed based on the view of automated transportation by investigating relationship between product structure and links of automated transportation. Applications are illustrated.

Key words: Automatic conveying, product design, DFX

1. INTRODUCTION

With the rapid developments of technologies of computer and automation, automation becomes inevitable for manufacturing industrial. It is not difficult to realize automation for process itself with the increasing functional and sophisticated facilities, so automatization of auxiliary jobs, e.

g. loading/unloading, transporting, orientating, identifying and controlling etc., is the key factor for the manufacturing system automation, simultaneously processing time is greatly reduced, time and cost apart from processing become more and more important. Give an example of mechanical manufacturing, in work shop, only 5%~10% of the time is used for processing and inspecting, and the rest 90%~95% is in the state of storing, loading/unloading, transporting and waiting, material handling activities generally account for 30% to 40% of production costs [1], therefore, material handling comes to be the important source of future profit to some extent.

Automatic conveying is primary for material handling, and is the hardware basis of forming manufacturing system with each isolated automatic islands (e.g. processing equipment, machining center, workstation, storing/retrieving, etc.) and performing material exchange, therefore it is also the premise for composing any automatic manufacturing system. There is an experiential statement in automatic assembly field, "If a part can be conveyed automatically, it can generally be assembled automatically." In a word, automatic conveying directly affects system on manufacturing level, response ability and overall benefit.

With the increasing of competition, product design has turned from design for pure function to design for life cycle, and automatic conveying is of vital importance to time to market and production cost, so it is quite necessary to carry out researches on this way.

2.   HISTORICAL BCKGROUND

2.1   Methods for structure design

Being lack of global considerations and information exchanges among the perspectives from different fields, traditional sequential product design results in design corrections and takes long time and huge resources. It is mainly based on product functions and its performances, and purposes of related theories is for best usage characteristics, such as optimum design, reliability design, finite element analysis, etc..

With the increasing of competition, quality is no longer the exclusively factor, T (Time), C (Cost), Q (Quality), S (Service) are equivalently important. Production is not terminated by selling out, but coming through the whole life cycle. For improving market response and competition, product design should be considered every stages and links in life cycle. CE (Concurrent Engineering) puts forward a philosophy of integration and concurrent product and process design, DFX (Design For X) is the key technology in it. DFX refers to computer-aided tools for a certain application field, which can help designer considering the effects of subsequent jobs by the early design decisions, it is one of the most effective method for CE. X represents various factors in product life cycle, such as designing, analyzing, processing, manufacturing, assembling, disassembling, testing, maintaining and supporting [2-4].

Many progresses related to DFX have been made by lots of researches [5], concerning with manufacture process, mainly there are DFM (Design For Manufacture), DFA (Design For Assembly) and DFC (Design For Cost). DFM is used for stage of detailed design, considering product process ability, resources restriction, manufacturability, manufacturing time and cost, etc. DFA is used for laying down assembly process plan, considering disassembly feasibility, optimizing assembly route, examining assembly interference by simulating. DFC refers to reducing cost as much as possible on the precondition of satisfying demands of user.

Researches above are concentrated on process realization, i.e. process ability, assembly feasibility. Auxiliary jobs necessary to process are little considered, which play an important role in system realizing, efficiency and reliability.

2.2　Automatic conveying

On account of backward controlling technology, automatic conveying is mainly used for rigid transfer line before 1960s, content of design is automatic mechanism and production line, the typical applications are auto assembly line and engine box producing line. Subsequently, corresponding with the development of technologies of computer, communication, numerical control, and studies of flexible manufacturing and computer integrated manufacturing, research center has turned to improving conveying equipment functions and its controlling and scheduling, like using AGV (Automated Guided Vehicle), robot, linear motor as conveying device, studying machine intelligence, optimum response route of AGVs [6-9]. These researches greatly enhance functions and performances of conveying equipment, but inevitably leading to structure complication and cost increasing.

3.   IDEA OF DESIGN FOR AUTOMATIC CONVEYING

From analysis in section 2,we know that automatic conveying can be achieved by improving conveying equipment. In contemporary flexible manufacturing system, AGV and robot are widely used for material conveying. Nevertheless, the level of machine intelligence presently is still very low, and it results in complex structure of conveying equipment and corresponding control system, difficult maintenance, vast investment and confined application. Based on the philosophy of CE, structure design should consider all factors in product life cycle, naturally includes material conveying. Product (part) is the object of conveying, so investigating automatic conveying certainly cannot apart form object of conveying. Sound structure could greatly improves characteristics of conveying, thereby reduces the demands for conveying equipment.

Therefore a new idea for realizing automatic conveying is presented, that is based on links of automatic conveying, investigating relationship between product structure and conveying characteristics, properly changing product structure with the precondition of not influencing product function, and reducing difficulty for conveying, simplifying conveying equipment, reducing facility invest.

4.   METHODS AND APPLICATIONS OF DESIGN FOR AUTOMATIC CONVEYING

4.1   Methods of design for automatic conveying

Effects of product structure on automatic conveying are determined by characters of product and its parts, which not only influence the efficiency of automatic conveying, but also its rationality and feasibility. Structure characters comprise shape, dimension, part numbers, physical character, and structure complexity. Conveying also relates with state characters, which comprise stability, security and location ability in static and directional stability in move.

Product structure influences on automatic conveying in many ways, the most direct and efficient effect reflects on the overall design. When design is being carried out, types and numbers of parts should be cut as many as possible, because types determine the flexibility of conveyance and numbers directly affect the quantity of conveying. Therefore this is the first of all problems for product design. Furthermore, effects of product structure reflect on loading/unloading, conveying manner, efficiency, reliability, identify and control, etc. Fundamental principles below must be followed in design procedure.

Loading/unloading principle: Structure should be convenient for material feeding, contour outline should be simple and regular, avoiding acute prominence, round shape is recommended, thereby parts can be separated easily, avoiding enwinding and hitching; Structure should benefits to simplifying material chute, automatic orientating and feeding; Structure should be facilitated for manipulator clamping and holding up, using universal clamp, and identifying and control.

Conveying manner principle: structure should satisfy with demands for different conveying manner. Small parts should be ensured reliable sliding or rolling for chute conveying, and be capable of correct locating for tray conveying; complex or large parts should be provided with credible base for directly conveying, if collateral fixture used, it should be convenient for locating and fixing.

Efficiency principal: apart from advancing conveying efficiency by choosing appropriate manner, structure should be capable of improving feeding efficiency when hopper adopted, fixing efficiency when collateral fixture used, allowing high conveying speed and increasing contemporary conveying numbers.

Reliability principal: partial structure design should be benefit to eliminate phenomena of arch forming and difficult separating in hopper, prevent jamming, hanging, hitching, enwinding, stacking, etc., ensure stability in move state, avoid relatively position changing, and reliability of auto orientating, locating and controlling too.

Identify and control principal: identify is used for obtaining product information, which works according to the features of product, thus features determine the complexity of the recognizing system. For instance, identifying posture of product in conveying, structure features determine the numbers and location of light point when light point identifying method is used. If this method is not satisfied for the complexity of product, complicated mode matching image-identifying method had to be adopted.

4.2   Applications of design for automatic conveying

Illustrations of parts improving conveying characteristics through partial modification as follow [2]:

Avoiding hitching, enwinding and stacking. Parts (fig.1 a, b, c) tend to enwind when automatic feeding, and are difficult to be separated. It may be changed to closed structure. The means (fig.1 c) is generally used for avoiding hitching. Parts with thin edge are ease to stack; it may be solved by adding flange (fig.2).

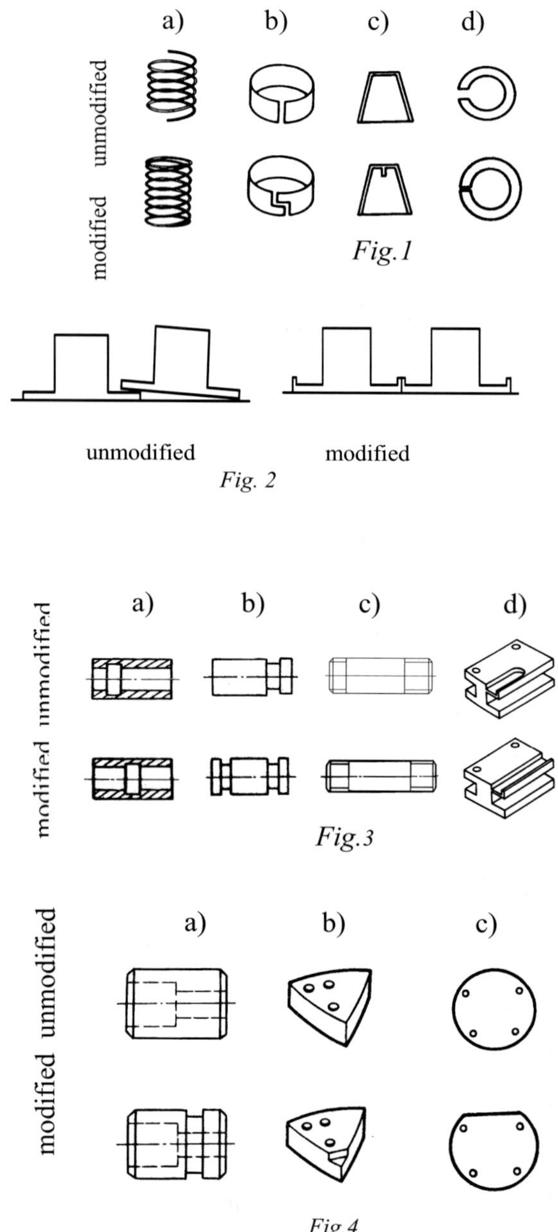

Fig.1

Fig. 2

Fig.3

Fig.4

Improving characteristics of automatic orienting. The most effective method is taking symmetry configuration, if not, magnifying asymmetry fraction purposely (fig.3, fig.4 c).

Improving identifying characteristics. Part identifying is achieved by its feature, and it would become easy by identifying simple feature instead of complex feature. Part (fig.4 a) external feature instead of interior feature, parts (fig.4 b c) identifying by easy recognizable notch or gap.

5.   EVALUATION OF DESIGN FOR AUTOMATIC CONVEYING

Evaluation of structure on automatic conveying is concrete and all round. Results would be changed by many factors (e.g. shapes and dimensions, conveying manner, etc.), even great difference may take place for partial alteration. So it should make a concrete analysis of concrete structure. However, in the final analysis, structure can be evaluated on automatic conveying characteristics, mainly conveying cost, efficiency and reliability.

Conveying costs mostly depend on equipment investment and operation cost, which relate to the equipment's complexity and universality resulted from product structure. Conveying efficiency is determined by the manner of conveying and other accessorial works. Reliability is mostly relied on orienting, locating, conveying stability and control system, etc.

Evaluation can be carried out with expert system or ANN, or refer to that of DFA, DFM [**10**].

REFERENCES

1.   Matson, JO, et al: " EXCITE: Expert Consultant for In-plant Transportation Equipment". *Int.J.Prod.Res.*, **Vol.30**, pp.1969-1983, 1992
2.   Boothroyd, G, et al.: *Product design for manufacture and assembly*. Marce41 Dekker, Inc. , pp.178-198, 1994
3.   Zha XF, Lim SYE and Fok SC: "Integrated Design and Assembly Planning: A Survey". *Int. J. Adv. Manuf. Technol.*, pp.664-685, 1998
4.   XIONG GL, et al.: "Research and Application of Concurrent Engineering in China". *CIMS*, **Vol.6** ,pp.1~6,45, 2000
5.   Tsai,CK, et al.: " Design for Manufacture and Design for "X": Concepts, Applications, and Perspectives". *Computers &industrial engineering*, pp.241_260,2001
6.   Beamon, BM: "System reliability and congestion in a material system". *Computer & Industrial Engineering*, pp.673-684, 1999
7.   Dutta, FT and Tanchoco JMA: "Comparison of dynamic routing techniques for automated guided vehicle system". *Int. .J. Prod. Res.*, **Vol.33**, pp.2653-2669, 1995
8.   Hao, G and Lai, KK: " Solving the AGV Problem via a Self-organizing Neural Network" *. Journal of the Operational Research Society*, **Vol.47**, pp.1477-1493, 1996
9.   Huang, C: "Design of Material Transportation System for Tandem Automated Guided Vehicle Systems". *Int. J. Prod. Res*, **Vol.35**, pp.943-953,1997

10. Shehab, E M and Abdalla, HS : "Manufacturing Cost Modeling for Concurrent Product Development". *Robotics and computer Integrated Manufacturing,***Vol.17** pp. 341-353, 2001

AUTHORS INDEX

780